ADVANCES IN CHEMICAL PHYSICS

VOLUME XCIII

Advances in
CHEMICAL PHYSICS

New Methods in
Computational Quantum Mechanics

Edited by

I. PRIGOGINE

University of Brussels
Brussels, Belgium
and
University of Texas
Austin, Texas

and

STUART A. RICE

Department of Chemistry
and
The James Franck Institute
The University of Chicago
Chicago, Illinois

VOLUME XCIII

AN INTERSCIENCE® PUBLICATION
JOHN WILEY & SONS, INC.
NEW YORK • CHICHESTER • BRISBANE • TORONTO • SINGAPORE

CONTRIBUTORS TO VOLUME XCIII

KERSTIN ANDERSSON, Department of Theoretical Chemistry, Chemical Center, Lund, Sweden

DAVID M. CEPERLEY, National Center for Supercomputing Applications and Department of Physics, University of Illinois at Urbana, Champaign, Illinois

MICHAEL A. COLLINS, Research School of Chemistry, Australian National University, Canberra, Australia

REINHOLD EGGER, Fakultät für Physik, Universität Freiburg, Freiburg, Germany

ANTHONY K. FELTS, Department of Chemistry, Columbia University, New York, New York

RICHARD A. FRIESNER, Department of Chemistry, Columbia University, New York, New York

MARKUS P. FÜLSCHER, Department of Theoretical Chemistry, Chemical Center, Lund, Sweden

K. M. HO, Ames Laboratory and Department of Physics, Iowa State University, Ames, Iowa

C. H. MAK, Department of Chemistry, University of Southern California, Los Angeles, California

PER-ÅKE MALMQVIST, Department of Theoretical Chemistry, Chemical Center, Lund, Sweden

MANUELA MERCHÁN, Departamento de Química Física, Universitat de Valéncia, Valéncia, Spain

LUBOS MITAS, National Center for Supercomputing Applications and Materials Research Laboratory, University of Illinois at Urbana, Champaign, Illinois

STEFANO OSS, Dipartimento di Fisica, Università di Trento and Istituto Nazionale di Fisica della Materia, Unità di Trento, Italy

KRISTINE PIERLOOT, Department of Chemistry, University of Leuven, Heverlee-Leuven, Belgium

W. THOMAS POLLARD, Department of Chemistry, Columbia University, New York, New York

BJÖRN O. ROOS, Department of Theoretical Chemistry, Chemical Center, Lund, Sweden

LUIS SERRANO-ANDRÉS, Department of Theoretical Chemistry, Chemical Center, Lund, Sweden

PER E. M. SIEGBAHN, Department of Physics, University of Stockholm, Stockholm, Sweden

WALTER THIEL, Institut für Organische Chemie, Universität Zürich, Zurich, Switzerland

GREGORY A. VOTH, Department of Chemistry, University of Pennsylvania, Philadelphia, Pennsylvania

C. Z. WANG, Ames Laboratory and Department of Physics, Iowa State University, Ames, Iowa

INTRODUCTION

Few of us can any longer keep up with the flood of scientific literature, even in specialized subfields. Any attempt to do more and be broadly educated with respect to a large domain of science has the appearance of tilting at windmills. Yet the synthesis of ideas drawn from different subjects into new, powerful, general concepts is as valuable as ever, and the desire to remain educated persists in all scientists. This series, *Advances in Chemical Physics*, is devoted to helping the reader obtain general information about a wide variety of topics in chemical physics, a field that we interpret very broadly. Our intent is to have experts present comprehensive analyses of subjects of interest and to encourage the expression of individual points of view. We hope that this approach to the presentation of an overview of a subject will both stimulate new research and serve as a personalized learning text for beginners in a field.

I. Prigogine
Stuart A. Rice

CONTENTS

ix

ADVANCES IN CHEMICAL PHYSICS

VOLUME XCIII

QUANTUM MONTE CARLO METHODS IN CHEMISTRY

DAVID M. CEPERLEY

National Center for Supercomputing Applications and Department of Physics, University of Illinois at Urbana–Champaign, Illinois

LUBOS MITAS

National Center for Supercomputing Applications and Materials Research Laboratory, University of Illinois at Urbana–Champaign, Illinois

CONTENTS

Advances in Chemical Physics, *Volume XCIII*, Edited by I. Prigogine and Stuart A. Rice.
ISBN 0-471-14321-9 © 1996 John Wiley & Sons, Inc.

ABSTRACT

We report on recent progress in the development of quantum Monte Carlo methods, including variational, diffusion, and path-integral Monte Carlo. The basics of these methods are outlined together with descriptions of trial functions, treatment of atomic cores, and remaining problems, such as fixed-node errors. The recent results for atoms, molecules, clusters, and extended systems are presented. The advantages, achievements, and perspectives demonstrate that quantum Monte Carlo is a very promising approach for calculating properties of many-body quantum systems.

I. INTRODUCTION

This review is a brief update of the recent progress in the attempt to calculate properties of atoms and molecules by stochastic methods which go under the general name of quantum Monte Carlo (QMC). Below we distinguish between basic variants of QMC: variational Monte Carlo (VMC), diffusion Monte Carlo (DMC), Green's function Monte Carlo (GFMC), and path-integral Monte Carlo (PIMC).

The motivation for using these methods to calculate electronic structure, as opposed to methods of expanding the wavefunction in a basis, arises from considerations of the computational complexity of solving the Schrödinger equation for systems of many electrons. By the complexity we simply mean the systematic answer to the question: How long does it take to compute some property of a system to a specified absolute error? So the complexity is the study of the function $T(\varepsilon, \ldots)$, where T is the needed computer time and ε is the error. The absolute magnitude of T depends, of course, on such features as the type of computer, the compiler, and the skill of the programmer, which are difficult to specify systematically. But the basic scaling of T with the required error, with the number of electrons, and with the type of molecule should be independent of such details.

The error ε in this expression must be the true error (i.e., all the systematic and statistical errors). Uncontrolled approximations cannot be allowed; otherwise, the complexity problem is not well posed. Chemistry is unique in that, first, there is a well-tested, virtually exact theory (the Schrödinger equation), and second, the mean-field estimates of chemical energies are often surprisingly accurate. Unfortunately, *very* accurate estimates are required to provide input to real-world chemistry, since much of the interesting chemistry takes place at room temperature. Currently, the level of chemical accuracy is considered to be ≈ 1 kcal/

mol. However, in many cases, higher accuracy is necessary; for example to calculate energy differences (say, between two isomers, energy levels, or a binding energy) to better than room temperature requires an error of $\varepsilon \leq 100\,\mathrm{K} \approx 0.01\,\mathrm{eV} \approx 0.35\,\mathrm{mH} \approx 0.1\,\mathrm{kcal/mol}$. Of course, there are many phenomena for which even higher accuracy is required (e.g., superconductivity).

To obtain errors of 1 kcal/mol or better, it is essential to treat many-body effects accurately and, we believe, directly. Although commonly used methods such as the density functional theory within the local density approximation (LDA) or the generalized gradient approximation (GGA) may get some properties correctly, it seems unlikely that they, in general, will ever have the needed precision and robustness on a wide variety of molecules. On the other hand, methods that rely on a complete representation of the many-body wavefunction will require a computer time that is exponential in the number of electrons. A typical example of such an approach is the configuration interaction (CI) method, which expands the wavefunction in Slater determinants of one-body orbitals. Each time an atom is added to the system, an additional number of molecular orbitals must be considered, and the total number of determinants to reach chemical accuracy is then multiplied by this factor. Hence an exponential dependence of the computer time on the number of atoms in the system results.

Simulation methods construct the wavefunction (or at positive temperature the N-body density matrix) by sampling it and therefore do not need its value everywhere. The complexity then usually has a power-law dependence on the number of particles, $T \propto N^{\delta}$, where the exponent typically ranges from $1 \leq \delta \leq 4$, depending on the algorithm and the property. The price to be paid is that there is a statistical error which decays only as the square root of the computer time, so that $T \propto \varepsilon^{-2}$.

Recently, very accurate QMC calculations have been reported on few-electron systems with H, He, and Li atoms and on many-electron systems in the jellium model. QMC results are rapidly approaching chemical accuracy on much more complicated systems, such as clusters of carbon and silicon, so the method is quickly becoming of practical importance. This progress is coming about through improvements in methods (e.g., the use of pseudopotentials and fermion path integral methods), programming advances (interfacing to standard chemistry packages for building high-quality trial wavefunctions), and advances in computer hardware (parallel computation).

We do not mean to imply that QMC has rigorously been shown to have a more favorable complexity; this is the crux of the infamous fermion sign problem of QMC that we discuss later. Rather, we argue that QMC has a number of desirable features which, even if the fermion sign problem is not solved, indicate that the method will still be useful:

1. QMC has a favorable scaling with the size of the system, with computational demands growing as $\approx N^3$.

2. One can introduce thermal effects naturally, both for electrons and ions and zero-point motion for ions.

3. Besides energies, QMC can compute properties such as the optical and electric response, geometries, and so on.

4. QMC has general applicability for both isolated systems such as molecules and extended systems such as solids.

5. QMC has been shown to reach chemical accuracy or beyond for small systems.

Therefore, QMC has many of the ingredients of a method that can really "solve" the computational quantum many-body problem. We know of no other general methods with these characteristics. In addition, QMC and PIMC methods offer new ways of understanding chemical concepts and translating that understanding into a computationally efficient approach.

The idea of using a statistical approach for quantum many-body problems was mentioned rather early by both Wigner [1] and Fermi. Serious application began with McMillan's [2] calculation of liquid helium by the VMC method. Simultaneously, Kalos et al. [3, 4] had developed the GFMC methods, which go beyond the variational approximation. Ceperley [5] generalized the VMC method to treat fermions in 1978 and generalized the importance–sampled GFMC methods to fermion systems in 1980 [6]. Anderson introduced the fixed-node approximation to avoid the fermion sign problem in 1975 and did the first simple molecular applications [7]. The first major applications to electronic systems were performed by Ceperley and Alder on the electron gas model [6] and solid hydrogen [8]. These authors also introduced the release-node method to go beyond the fixed-node approximation for small and medium-sized systems [9] and applied it to systems of up to 54 electrons.

We will not exhaustively review previous applications and methods, as there is a recent book on the subject [10] as well as reviews [11, 12] with details of methods and overviews of many applications. There are also very recent reviews by Anderson on rigorous QMC calculations for small systems [13] and on fixed-node applications [14]. The focus here is on examining to what extent QMC could perform calculations of chemical accuracy for larger chemical systems, to assess recent developments relevant to this quest, and to point out the remaining fundamental problems. We summarize only a few of the computational results that have been obtained, point to changes since previous review articles were written, and present our point of view with regard to future applications. We do not discuss the application of QMC methods to study vibration energies in atoms, quantum effects of nuclear motion, quantum Monte Carlo for real-time dynamics (see the chapter by Mak and Egger), or path integral calculations of single electrons in classical liquids [15].

The review is organized as follows. In the next section we introduce the three main methods: VMC, DMC, and PIMC. In the following section we describe the forms and optimization of trial wavefunctions. Then we discuss the treatment of atomic cores. Next, we outline selected applications to atoms, molecules, clusters, and a few results for extended systems. We conclude with prospects for future progress.

II. QUANTUM MONTE CARLO METHODS

Here we summarize the various quantum Monte Carlo methods that have been used for calculations of electronic structure.

A. Variational Monte Carlo

In variational Monte Carlo (VMC), one samples, using the Metropolis rejection method, the square of an assumed trial wavefunction, $|\psi_T(R)|^2 / \int dR |\psi_T(R)|^2$, where $R = \{\mathbf{r}_i\}$ are the coordinates of all the particles (possibly including their spin coordinates.) Using the sampled coordinates, one can calculate any simple matrix element with respect to the trial wavefunction. In the most common example, the estimate of the variational energy is taken as an average over the sampled points:

$$E_V = \lim_{M \to \infty} \frac{1}{M} \sum_{i=1}^{M} \psi_T(R_i)^{-1} \mathcal{H} \psi_T(R_i) , \qquad (1)$$

where $\{R_i\}$ with $1 \leq i \leq M$ are points sampled from the distribution $|\psi_T(R)|^2 / \int dR |\psi_T(R)|^2$. The variational energy is obtained as the average of the local energy $E_L(R) = \psi_T^{-1}(R) \mathcal{H} \psi_T(R)$. The zero variance principle applies: As the trial function becomes more accurate, the fluctuations in the local energy are reduced. The trial wavefunction is then chosen either (1) to minimize the variational energy, (2) to minimize the dispersion of the local energy (the variance), or (3) to maximize the overlap with the exact ground state. Any of these criteria could lead to a good trial function, but there are important differences in using them in practice. The reweighting method [11, 16] is used to carry out this optimization efficiently.

VMC is a wavefunction-based QMC method and hence is the most closely related to standard basis-set approaches. In methods such as configuration interaction, based on expanding the wavefunction in Slater determinants, correlation appears indirectly through sums of products of one-body orbitals. However, in VMC, correlation can be put into a trial wavefunction directly, once the problem of doing the expectation value integrals is solved. Using the pair product (Jastrow) trial function, the correlation is included directly. On the other hand, one pays the price of having a statistical error from Monte Carlo integration, which implies scaling of the computer time as ε^{-2}. Fortunately, the prefactor can be reduced with good trial functions. The complexity of VMC versus the number of electrons is quite favorable, as it scales as N^3, which is similar to scaling of mean-field approaches such as Hartree–Fock (HF). This enabled Ceperley [5] in the first VMC calculations to deal with 162 electron systems. The dominant piece as $N \to \infty$ is evaluating the determinants during the random sampling.

The difficulty with VMC is exactly identical in spirit to all the problems of traditional methods: the basis-set problem. Although the wavefunction is vastly improved in VMC, it is difficult to know when the wavefunction form is sufficiently flexible, and therefore it is always necessary to show that the basis-set limit of a given class of trial function has been reached. Moreover, the accuracy of energy in no way implies accuracy of other properties. One can assume that many of the variational errors cancel out in going from one system to another, but it is not very hard to find counterexamples. With the current class of wavefunctions it seems that we are far from getting chemical accuracies from VMC when applied to systems more complex than the electron gas or a single atom. In addition, in VMC one can waste a lot of time trying new forms rather than have the computer do the work. This problem is solved in a different way in the next two methods we discuss.

B.   Diffusion Monte Carlo and Green's Function Monte Carlo

DMC goes beyond VMC in that the wavefunction is sampled automatically during the Monte Carlo process but without an analytic form being generated. The mathematical basis of DMC is that the operator, $\exp(-\tau \mathscr{H})$, acting on any initial function, will filter out the lowest-energy eigenfunction of $\mathscr{H}$ from any initial state with given symmetry. Hence the following procedure is iterated until convergence is reached:

$$\phi(R, t + \tau) = \int dR' \langle R| \exp[-\tau(\mathcal{H} - E_T)]|R'\rangle \phi(R', t) , \qquad (2)$$

where E_T is an adjustable trial energy. If we interpret the initial state as a probability distribution, this process can easily be seen to diffuse the points and cause them to branch (split or disappear). The branching originates in renormalization of the kernel in Eq. (2), which comes from the potential energy term. We already see the difficulty with the DMC method: The wavefunction cannot be interpreted as a probability distribution because it has both positive and negative regions for more than two electrons.

The simplest way around this problem is the fixed-node (FN) approximation introduced by Anderson [7]. Using the nodes of some good trial function, we put an infinite potential barrier at those nodes. Then we can use the projection technique in one nodal region at a time (in fact, there are typically only two of them [11]) to solve for the energy and wavefunction. This additional potential has no effect if the nodes happen to be in the right location; otherwise, it can be shown always to increase the ground-state energy [8]:

$$E_0 \leq E_{FN} \leq E_V . \qquad (3)$$

Hence we find the best wavefunction consistent with an assumed set of nodes. The nodes are not exactly known except for the simplest systems. However, we can also go beyond the fixed-node approximation, as will be mentioned later.

In 1974, Kalos [4] introduced the idea of importance sampling by asking the following question: What is the expected number of walkers resulting from a walker at position R_0? The answer is seen to be $\phi_0(R_0)$, the ground-state wavefunction. With importance sampling we try to reduce the branching by putting in the best estimate of $\phi_0(R)$. Therefore, we work with the distribution $f(R, t) = \psi_T(R)\phi(R, t)$ and the new Green's function: $\psi_T(R)\psi_T^{-1}(R')\langle R'| \exp(-\tau\mathcal{H}|R\rangle$. With importance sampling, branching is greatly reduced. A way of seeing this in detail is by taking the continuous time limit of the iteration process and writing down the evolution equation for the importance sampled distribution $f(R, t)$. Ceperley and Alder [6] showed that

$$\frac{\partial f(R, t)}{\partial t} = \lambda \left[\sum_{i=1}^{N} \nabla_i^2 f(R, t) - \nabla_i(f(R, t)\nabla \ln|\psi_T|^2) \right]$$
$$- (E_L(R) - E_T)f(R, t) , \qquad (4)$$

where $\lambda = \hbar^2/2m$. This equation is the basis for the diffusion Monte Carlo approach. The three terms on the right-hand side correspond to diffusion, drift, and branching, respectively. The branching now is with respect to the local energy and thus is under control. The details of the algorithm and its application to molecules can be found elsewhere [19]. Umrigar et al. [20] have recently studied very carefully issues concerning the diffusion Monte Carlo method and its speedup.

GFMC is a very similar algorithm (but developed earlier [4]) which has no time step error as it samples not only the wavefunction but also the Green's function itself. It is to be preferred when highly accurate results are needed and computer time requirements are not overwhelming.

We show in the next section that using the simplest nodes (a single Hartree–Fock determinant) gives more than 90% of the correlation energy for first-row atoms and dimers (Fig. 1) [21] and essentially 100% of the binding energy for dimers. One can do better by using multiconfiguration nodes or even nodes from the natural orbital determinant.

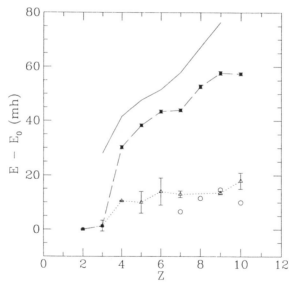

Figure 1. Error in the nonrelativistic total energy (in millihartrees) for the first-row atoms of atomic charge Z. The upper line is from coupled cluster calculations [27]. The dashed lines with symbols and error bars denote VMC calculations [30], with a single Slater determinant and the correlation factor with 17 variational parameters. The dotted lines with symbols are results of fixed-node DMC calculations with a single Slater determinant. The circles are CI calculations [83, 84].

However, although using many Slater determinants does allow very good upper bounds, as more and more atoms are added to a system the computer time will grow exponentially quickly (to keep the same accuracy), and hence it is really a solution only for small systems.

C. Path-Integral Monte Carlo

PIMC is mathematically similar to DMC and shares many of its advantages [23]. In fact, it goes further, since a trial function is not specified and the method generates the quantum distribution directly from the Hamiltonian. We define PIMC to be a QMC method that is formulated at a positive temperature. Instead of attempting to calculate the properties of a single quantum state, we sum over all possible states, occupying them according to the Boltzmann distribution. This might sound hopeless, but Feynman's [22] imaginary-time-path integrals makes it almost as easy as DMC. The imaginary-time paths, instead of being open-ended as they are in DMC, close after an imaginary time $\beta = (k_B T)^{-1}$, where T is the temperature. Because of the absence of boundaries in imaginary time, it is not necessary to have a trial wave function. In DMC, the trial wavefunction is needed to start the paths and to predict the future outcome of a path, but if the whole world line of a path is there, it is not needed. A very important advantage of PIMC is that all observables are obtained exactly, while in DMC only energies are gotten correct. The rest have a bias caused by the importance function, and removal of the bias causes additional uncertainties.

Instead of having imaginary time evolution as in DMC, one keeps the entire path in memory and moves it around. PIMC uses a sophisticated Metropolis Monte Carlo method to move the paths. One trades off the complexity of the trial function for more complex ways to move the paths [23]. One gains in this trade-off because the former changes the answer while the latter changes only the computational cost.

Particle statistics come in rather differently in PIMC. A permutation operation is used to project Bose and Fermi symmetry. (Remember that in DMC the fixed-node method with an antisymmetric trial function was used.) The permutations lead to a beautiful and computationally efficient way of understanding superfluidity for bosons, but for fermions, since one has to attach a minus sign to all odd permutations, as the temperature approaches the fermion energy a disastrous loss of computational efficiency occurs. There have been many applications of PIMC in chemistry, but almost all of them have been to problems where quantum statistics (the Pauli principle) were not important, and we do not discuss those here. The review article by Berne and Thirumalai [10] gives an overview of these applications.

Recently, there has been some progress in generalizing the path-integral method to treat fermion systems, which is called restricted PIMC (RPIMC) [24]. One can also apply the fixed-node method to the density matrix. The fermion density matrix is given by

$$\rho(R, R_0; t) \propto \sum_P \int_{PR_0}^R dR \exp\left\{-\int_0^\beta dt \left[\frac{1}{4\lambda} \left(\frac{dR}{dt}\right)^2 + V(R(t))\right]\right\}, \quad (5)$$

where the integral (dR) is over all continuous paths starting at PR_0 and ending at R, with the restriction that

$$\rho(R(t), R_0; t) \neq 0, \qquad 0 \leq t \leq \beta. \tag{6}$$

P is a permutation of atoms with the same spin and necessarily must be even because of the restriction [$V(R)$ is the potential energy] [23]. The exact density matrix will then appear both on the left-hand side of Eq. (5) and implicitly in the restriction on the right-hand side of Eq. (6). This implies that there exists a restriction which does not have a fermion sign difficulty. In the fixed-node approximation, a trial density matrix is used for the restriction on the right-hand side of Eq. (6).

In the most relevant calculation to date using this method, Pierleoni et al. [25] placed 64 electrons and 64 protons in a periodic box and cooled it down to temperatures as low as $5000\,K \approx 0.5\,eV$. At higher temperatures and pressures the expected behavior for a correlated hydrogenic plasma was recovered. But at the lowest temperature, evidence for a first-order phase transition, where the electrons spontaneously went from an ionized state to a molecular H_2 state, was seen. This transition had been conjectured, but the simulations have been the first strong evidence.

Although the computer time requirements were large (several months on a workstation), the human input to the calculation was much less than in very similar studies with VMC and DMC (at zero temperature) [26]. RPIMC appears to be a very promising direction for constructing a black-box program for many-electron systems where correlation may be important. Also, PIMC seems to lead more easily to a physical interpretation of the results of a simulation, although very little work has yet been done on understanding the restricted paths of fermion systems. As an example, PIMC could lead to a more direct understanding of bond formation, electron pairing, and localization.

There are several technical difficulties with RPIMC. The first is that the time step is smaller than $1\,H^{-1}$, on the order of 1 million degrees. Hence to work down to a temperature of $1000\,K$ takes on the order of

1000 points on the path, which is rather slow. A more serious difficulty is to come up with automatic ways of generating accurate restrictions.

III. TRIAL WAVEFUNCTIONS

One of the main advantages of the Monte Carlo method of integration is that one can use any computable trial function, including those going beyond the traditional sum of one-body orbital products (i.e., linear combination of Slater determinants). Even the exponential ansatz of the coupled cluster (CC) method [27, 28], which includes an infinite number of terms, is not very efficient because its convergence in the basis set remains very slow. In this section we review recent progress in construction and optimization of the trial wavefunctions.

The trial function is very important for both the VMC and DMC methods. That one needs a good trial function in VMC is obvious. There are at least two reasons for having a good trial function in DMC. First, as mentioned earlier, the error in the nodes of the trial wavefunction gives rise to the fixed-node error. Second, it is utilized for the importance sampling, which increases the efficiency of the DMC simulations substantially (typically, by more than two orders of magnitude) by decreasing the energy fluctuations resulting from sampling the local energy instead of potential energy. One can show [29] that the DMC error bar depends on the trial function as

$$\varepsilon \approx \left[\frac{2(E_V - E_0)}{\tau P} \right]^{1/2}, \tag{7}$$

where τ is the time step and P is the number of steps on the walk. Thus it is advantageous to improve the variational energy as long as that is not too costly in computer time per step.

Currently, the ubiquitous choice for the trial function is of the *Slater–Jastrow* or *pair-product form*. It is a linear combination of spin-up and spin-down determinants of one-body orbitals multiplied by a correlation factor represented by an exponential of one-body, two-body, and so on, terms [16, 30]:

$$\Psi_T(R) = \sum_n d_n \, \text{Det}_n^\uparrow [\{\phi_\alpha(\mathbf{r}_i)\}] \, \text{Det}_n^\downarrow [\{\phi_\beta(\mathbf{r}_j)\}]$$

$$\times \exp(U_1 + U_2 + U_3 + \cdots). \tag{8}$$

The sum of determinants in Eq. (8) can accommodate multiconfiguration wavefunctions, which are especially important in systems with a near-degeneracy features. Perhaps the simplest manifestation of this is in the

Be atom [16,20] where inclusion of an additional configuration has a very significant impact on the final energies.

The terms in the correlation factor can be expressed formally as

$$U_1 = \sum_i u_0(\mathbf{r}_i) , \tag{9}$$

$$U_2 = \sum_{i<j} u_{ee}(\mathbf{r}_i, \mathbf{r}_j) + \sum_{i,I} u_{eI}(\mathbf{r}_i, \mathbf{r}_I) , \tag{10}$$

$$U_3 = \sum_{i<j<k} u_{eee}(\mathbf{r}_i, \mathbf{r}_j, \mathbf{r}_k) + \sum_{i<j,I} u_{eeI}(\mathbf{r}_i, \mathbf{r}_j, \mathbf{r}_I) + \sum_{i,I<J} u_{eII}(\mathbf{r}_i, \mathbf{r}_J, \mathbf{r}_I) , \tag{11}$$

and so on (lowercase index letters denote electrons and uppercase letters correspond to ions).

One of the important features of the Slater–Jastrow form is that it can describe the electron–electron cusp in the wavefunction directly and efficiently. The cusp is an important nonanalytic feature of the true wavefunction whenever any two charged particles approach each other. The electron–electron cusp region gives rise to the *dynamical correlation*. Because electrons repel each other, by introducing the cusp term, the electronic density is spread from high-density to low-density regions [9, 31]. To get the density back to the optimal value, which is usually rather close to the mean-field density, it is necessary either to reoptimize the orbitals (which can be rather difficult) or to optimize the one-body term simultaneously with the electron–electron and higher-order terms. Therefore, the one-body term U_1 is retained in the correlation factor, although formally it can be absorbed into the orbitals.

One can include the leading terms of three-particle nonanalytic points with logarithmic terms from the Fock expansion, but their impact on the variational energy is small [20]. Several forms for the correlation terms have been proposed and tested. For example, Umrigar et al. [16] used a Padé form of polynomials in a linear combination of electron–electron (r_{ij}) and electron–ion (r_i) distances,

$$U_2 + U_3 = \sum_{i,j,I} \frac{P(r_{iI}, r_{jI}, r_{ij})}{1 + Q(r_{iI}, r_{jI}, r_{ij})} , \tag{12}$$

which was then further refined [20]. Schmidt and Moskowitz [30] also provided an interpretation of the correlation part in terms of an *average backflow* and used products of powers of transformed distances $a(r) =$

$r/(1 + br)$:

$$U_2 + U_3 = \sum_{i,j,I} \sum_{klm} c_{klm} a(r_{iI})^k a(r_{jI})^l a(r_{ij})^m , \tag{13}$$

with b and $\{c_{klm}\}$ as variational parameters. Mitas [32] used a similar form, but instead of higher powers he introduced a tempering scheme for the Padé constant which controls the steepness of the transformed distance, and for the nonanalytic part employed a separated exponential term [34]

$$U_2 + U_3 = -\sum_{i<j} \frac{c}{\gamma} e^{-\gamma r_{ij}} + \sum_{i,j,I} \sum_{k,l,m} c_{klm} a_k(r_{iI}) a_l(r_{jI}) b_m(r_{ij}) , \tag{14}$$

where

$$a_k(r) = \left(\frac{\alpha_k r}{1 + \alpha_k r} \right)^2 , \quad \alpha_k = \alpha_0/2^{k-1} , \quad k > 0 , \tag{15}$$

$$b_m(r) = \left(\frac{\beta_m r}{1 + \beta_m r} \right)^2 , \quad \beta_m = \beta_0/2^{m-1} , \quad m > 0 , \tag{16}$$

with $a_0(r) = b_0(r) = 1$, while $\{c_{klm}\}$, α_0, and β_0 are variational parameters. Very recently, a systematic expansion of the correlation factor (8) in polynomial invariants has been proposed by Mushinskii and Nightingale [33].

These forms are good at capturing the dynamical part of the electron–electron correlation, as shown in Fig. 1. Typically, one obtains about 85% of the correlation energy. To obtain the same amount of the correlation with an expansion in determinants, one would need a large basis set that generates an enormous number of determinants even for a rather small number of correlated electrons.

An important step for getting high-quality trial function is the optimization process. One usually takes a set of configurations (Monte Carlo samples of electron positions) from previous runs and minimizes the variational energy or the fluctuations of the local energy [16]:

$$\sigma^2 = \frac{\int \Psi_T^2(R)[\Psi_T^{-1} \mathscr{H} \Psi_T - E_V]^2 \, dR}{\int \Psi_T^2(R) \, dR} = \frac{1}{M} \sum_{i=1}^M [\Psi_T^{-1} \mathscr{H} \Psi_T - E_V]^2 , \tag{17}$$

where M is the number of configurations (for simplicity we omitted the reweighting factors). With the new trial function, new configurations are generated and the optimization process is repeated until the improvement in σ^2 becomes small.

It is also important to consider the fact that orbitals, which are generated by a one-body approach, are not necessarily optimal when the correlation is included [20, 31]. In general, the reoptimization of orbitals in the presence of correlation is an unsolved task. For small systems such as atoms or small molecules, one can use an expansion in a suitable basis set and reoptimize the expansion coefficients [20]. For larger systems the number of expansion coefficients grows rapidly. In addition, for larger systems the number of sampling points [M in Eq. (17)] used for optimization must grow in order to find a stable minimum. Therefore, the computational demands grow rapidly and currently make the orbital optimization very slow.

Grossman and Mitas [55] tried another approach for improving the orbitals. For small silicon molecules they replaced the Hartree–Fock orbitals by the natural orbitals which diagonalize the one-body density matrix. The correlated one-body density matrix was calculated within the multiconfiguration Hartree–Fock using standard quantum chemistry approaches. Natural orbitals improved the agreement with experimental binding energies by about a factor of 2, with resulting discrepancies of 1–2% (i.e., 0.05 eV/atom).

In uniform systems there has been more progress on forms of the trial function, going beyond that of Eq. (8) [56]. Translation symmetry greatly reduces the possible forms of wavefunctions. Recently, Kwon et al. [99] carried out calculations of the two-dimensional electron gas with wavefunctions, including a backflow effect derived by a current conservation argument [35]. Backflow affects the nodes of the trial wavefunction, so by optimizing it, one can lower the fixed-node energy. The backflow trial function is given by

$$\Psi_T(R) = \sum_n d_n \, \mathrm{Det}_n^{\uparrow} \left[\{\phi_\alpha(\mathbf{x}_i)\}\right] \mathrm{Det}_n^{\downarrow} \left[\{\phi_\beta(\mathbf{x}_j)\}\right] \exp(U_1 + U_2 + U_3 + \cdots),$$

(18)

where the quasi-particle "coordinates" $\mathbf{x}_i$ and $\mathbf{x}_j$ in the Slater determinants are given as

$$\mathbf{x}_i = \mathbf{r}_i + \sum_{k \neq i} f(r_{ik})(\mathbf{r}_i - \mathbf{r}_k),$$

(19)

where $f(r)$ is a variational function. The argument of the one-body orbital is a "dressed" position of the electron: It is a sum of its actual position plus a correction that depends on the positions of remaining electrons. The results have shown that a significant amount of the fixed-node error (up to 90%) was recovered by using backflow nodes in the two- and three-dimensional homogenous electron gas.

In addition to a speedup in efficiency and higher accuracy, calculations with the trial functions (8) have also brought an important insight into the nature of the electron–electron correlation: Once the nodes of the wavefunction are sufficiently close to the exact ones (e.g., by using a few configurations and/or optimized orbitals), about 85% (or more) of the correlation can be described by rather simple analytical forms [Eqs. (12)–(16)] of order 20–30 variational parameters. This is observed for all systems studied: atoms, molecules, medium-sized clusters, surfaces, and solids—some of these included more than 200 valence electrons. In this way QMC has helped our understanding of electron–electron correlation and has demonstrated a significant gain in efficiency for describing these many-body effects.

IV. TREATMENT OF ATOMIC CORES IN QMC

The core electrons pose a problem for QMC methods because the core energy is much higher than chemical energies and the relevant distance scale of core states is much smaller. It has been shown [29] that the scaling of computer time grows $\approx Z^6$ with the atomic number, Z. Obviously, all-electron calculations quickly become unfeasible (at least to reach a fixed accuracy on the energy) as Z increases.

Shown in Fig. 2 [21,36] is the Monte Carlo efficiency as a function of Z using GFMC and DMC. In the GFMC algorithm the statistical efficiency is seen to scale at $Z^{-9.5}$, while in DMC it scales as $Z^{-5.2}$. The GFMC method has particularly unfavorable scaling with Z compared with DMC, presumably because very small steps are taken. While all-electron calculations of Li and Be are within our stated chemical accuracy of 0.01 eV, clearly the all-electron algorithm cannot be used for heavier atoms and reach the needed accuracy. For the QMC methods to become practical for heavy-atom systems, one has to deal with core degrees of freedom in a different way.

The core electrons create two basic problems. The first is that the very small size of the core region requires a different strategy for sampling the core region; otherwise, the time step that controls the movement of electrons will scale as Z^{-2}. Although this might be difficult technically, it is not the primary obstacle. One can modify the propagator [20] so that it reflects the strong localization of the core charge and thus to a large extent avoid substantial slowing down of the simulations.

Far more severe are the local energy fluctuations caused by the strong potentials and large kinetic energies in the core. Because of a rapidly

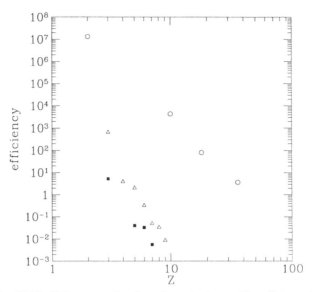

Figure 2. QMC efficiency as a function of atomic charge. The efficiency is defined as $1/(\text{CPU time} \cdot \sigma^2)$. The filled squares are for GFMC calculations of the dimers (LiH, BeH, . . . , FH) and the triangles are for GFMC calculations of the homonuclear diatomics ($Li_2, B_2, . . . , O_2$) both on the CRAY YMP [21]. It is seen that the efficiency of GFMC calculations scale as $Z^{-9.5}$. The circles are DMC calculations of Hammond [36]. The efficiency of those results scale as $Z^{-5.2}$.

changing density, it is very difficult (although, perhaps, not impossible) to design a trial function that can decrease these fluctuations. Even though correlation is relatively less important in the core, on the absolute scale it is still very large. The core, because of the high density, large potentials, and large kinetic energy, is always the strongest fluctuating term of the local energy.

First, we mention briefly two methods in which the core electrons are kept but their deleterious effects are reduced somewhat. In the damped core approach introduced by Hammond et al. [37], the efficiency was improved by dampening the core energy fluctuations. In this method the wavefunction is written as a product of the core and valence determinants, and electrons in the core region are treated variationally while the valence space is treated by the DMC method. There is a smooth transition between the valence and core region by suppressing the DMC branching with a smooth cutoff function. Although the resulting energy depends to a certain extent on the particular choice of the cutoff, successful calculations have been carried out for the atoms C, Si and Ge [37]. Carlson et al. [38] tried to smooth out the core energy fluctuations by decreasing the strength of the electron–electron interaction inside the core region. The total energy was corrected for the average value of the missing interaction. Tests on Li and Li_2 led to an increase in efficiency by a factor of ≈ 6. There has been no further development of this method. Both approaches lack a systematic way of understanding the transferability of the algorithm from one chemical environment to another. The *core-retained* methods will be much slower than the *pseudopotential* methods because the computational effort of sampling energy fluctuations with Z^2 scaling remains. Fortunately, for most valence properties the core remains practically inert and has a negligible impact on the valence properties. This fact can be used to eliminate the core electrons from the calculations and replace them with effective core Hamiltonians.

A. Nonlocal Pseudopotentials

In LDA calculations, pseudopotentials (or effective core potentials) are almost always used to increase the efficiency of calculations, even for calculations involving hydrogen. This allows smoother wavefunctions, which in turn reduces the number of basis functions. It has been found that transferability (the ability of a pseudoatom to mimic a full-core atom) is governed by norm conservation [39], and pseudopotentials are constructed so that the pseudo-orbitals match the full-core orbitals outside the core.

Almost all pseudopotentials [40], including those which were used in QMC calculations, were generated by mean-field approaches (a notable exception is the work of Dolg et al. [41]). It is not obvious that pseudopotentials constructed in LDA or HF are appropriate for more exact approaches. Acioli and Ceperley [43] showed that the transferability for correlated wavefunctions is achieved if the sequence of one-body, two-body, and so on, density matrices for the pseudoatom and the full-core atom match outside the core region. The most important effects are contained in the one-body density matrix, which can be expressed as a diagonal sum of its natural orbitals. Therefore, the pseudoatom has to be constructed so that its natural orbitals (both occupied and unoccupied) match those of the full-core atom in spatial regions where atoms can overlap. Also, the relevant part of the energy spectrum of the full-core atom should match that of the pseudoatom. When this idea was tested on several first-row atoms [44], it was found that the natural orbitals generated with DMC were very similar to those obtained from CI calculations. Having determined the full-core natural orbitals, the next task is to find a valence-only Hamiltonian that yields the same natural orbitals and the experimentally measured atomic spectrum.

The error introduced by the pseudopotentials for valence properties depends on the size of valence space. One can also increase the accuracy by taking into account the most important effect omitted in the rigid-ion pseudopotential, namely, the polarizability of the core. Polarizability, which is important for improving accuracy beyond 0.1 eV, is straight-forward to introduce in DMC [42, 44].

B. Local Pseudo-Hamiltonians

Bachelet et al. [47], in the *pseudo-Hamiltonian* approach, proposed to replace the action of the core on the valence states by an effective single-electron Hamiltonian. The most general one-electron Hamiltonian, which is local, spherically symmetric, and Hermitian, has a local effective ionic potential and a spatially varying radial and tangential mass. Outside the atomic cores the potential becomes coulombic and the mass becomes the usual scalar constant mass. The freedom in the effective ionic potential, and the tangential and radial mass, can be used to tune the pseudo-Hamiltonian to mimic the action of the core electrons on the valence electrons. The approach has the great advantage that the resulting valence Hamiltonian is local and all virtues of the DMC method apply immediately. For example, the fixed-node approximation gives an upper bound, and release-node calculations can then converge to the exact answer.

The disadvantage of the pseudo-Hamiltonian is that one does not have very much flexibility in matching the core response to valence electrons with different angular momentum because the restrictions on the mass tensor are too severe, especially for first-row and transition metal atoms (i.e., for the cases with strong nonlocalities). In particular, for transition metals it is not possible to use an Ar core because the first electron must always go into an s state [48]. In fact, this is of secondary importance since for accurate calculations, which are the aim of QMC, one has to include $3s$ and $3p$ states in the valence space for the $3d$ transition elements.

However, for the second row, which exhibits relatively small nonlocality effects, a good accuracy pseudo-Hamiltonians can be constructed. Reference 47 gives results for several atoms and dimers. New pseudo-Hamiltonian parameterizations for several elements from the first two rows were calculated very recently [49]. Li et al. [50] used a pseudo-Hamiltonian to carry out a DMC calculations on solid silicon, which resulted in excellent agreement with experiment for the cohesive energy. This demonstrated for the first time the feasibility of the DMC calculations on solids other than hydrogen.

C. Nonlocal Pseudopotentials and DMC

The usual form of a valence-only Hamiltonian is

$$\mathcal{H}_{val} = \mathcal{H}_{loc} + W , \tag{20}$$

with the local part given by

$$\mathcal{H}_{loc} = \sum_i \left[-\frac{1}{2} \nabla_i^2 + \sum_I v_{loc}(r_{iI}) + \frac{1}{2} \sum_{j \neq i} \frac{1}{r_{ij}} \right]. \tag{21}$$

The nonlocal pseudopotential operator W includes pseudopotentials $v_l(r)$ for a small number of the lowest symmetry channels, labeled by l (usually spd):

$$\langle R|W|R' \rangle = \sum_{I,i} \sum_l \frac{2l+1}{4\pi} v_l(r_{iI}) \frac{\delta(r_{iI} - r_{iI}')}{r_{iI} r_{iI}'} P_l(\hat{\mathbf{r}}_{iI} \cdot \hat{\mathbf{r}}_{iI}') , \tag{22}$$

where P_l is the Legendre polynomial. Therefore, the valence states of different symmetry experience different potentials in the core region. The variational Monte Carlo can accommodate such Hamiltonians without major problems, and Fahy et al. [31,51] used nonlocal pseudopotentials for the first VMC simulations of solids.

The nonlocality, however, is a problem for the DMC simulations because the matrix element for the evolution of the imaginary-time diffusion is not necessarily positive. For realistic pseudopotentials the matrix elements are indeed negative and thus create a sign problem (even for one electron), with consequences similar to those of the fermion sign problem (see, e.g., work of Bosin et al. [49]).

To circumvent this problem it was proposed by Hurley and Christiansen [53] and by Hammond et al. [54] to define a new transformed effective core potential by a projection onto a trial function:

$$V_{eff}(R) = \Psi_T^{-1}(R) \int dR' \langle R|W|R' \rangle \Psi_T(R) . \tag{23}$$

The new effective potential is explicitly many-body but local and depends on the trial function. We were able to show [52] that the energy $\mathcal{H}_{val}^{eff}$ converges quadratically to the exact energy of $\mathcal{H}_{val}$ as the trial function converges to the exact eigenstate. However, the DMC energy with V_{eff} will not necessarily be above the true eigenvalue of the original $\mathcal{H}_{val}$ and will depend on the quality of $\Psi_T(R)$. In addition, we have also pointed out [52] that the meaningful solutions of this Hamiltonian are those where the wave function vanishes at the nodes of Ψ_T since the effective Hamiltonian will diverge at the nodes of Ψ_T. Hence we need to write the fixed-node Hamiltonian as

$$\mathcal{H}_{\mathrm{val}}^{\mathrm{eff}} = \mathcal{H}_{\mathrm{loc}} + V_{\mathrm{eff}}(R) + V_{\infty}[\Psi_T(R) = 0] , \qquad (24)$$

where the last term, which is infinite on the subspace for which $\Psi_T(R) = 0$, assures that the nodes of the solution will coincide with the nodes of $\Psi_T(R)$. A study of the projection and fixed-node errors for B, Al, Ga, and In atoms has been carried out by Flad et al. [45].

The speedup resulting form the use of pseudopotentials can be demonstrated on the example of iron atom [32, 57]. Table I gives total energies, typical values for the dispersion of the local energy, and decorrelation time κ for obtaining an independent sample of energy normalized to the all-electron case; and finally, the efficiency is proportional to $1/\kappa\sigma^2$. It is evident that with increasing core size, the efficiency but also the systematic errors introduced by the pseudopotentials increase. For the given case of iron, the best compromise (if we accept the accuracy level 0.1 eV) is the Ne core. This comparison gives a qualitative picture between various choices of valence space and should not be taken as definitive: To some extent one can always change some of these factors through improvement of the trial function, more efficient sampling, and so on.

A number of VMC and DMC calculations of atomic, molecular, and solid systems have been carried out by this approach. This includes *sp* and transition element atoms [32, 46, 52] silicon and carbon clusters [55, 58], nitrogen solids [59], and diamond [60]. Our experience indicates that with a sufficient number of valence electrons one can achieve a high *final* accuracy. This, however, requires using 3s and 3p in the valence space for the 3d elements and, possibly, 2s and 2p states for elements such as Na. Once the core is sufficiently small, the systematic error of the fixed-node

TABLE I

Comparison of the Fe Atom Calculations with All-Electron, Ne-Core, and Ar-Core Pseudopotentials

	All-Electron	Ne-Core	Ar-Core
E_{HF}	-1262.444	-123.114	-21.387
E_{VMC}	$-1263.20(2)$	$-123.708(2)$	$-21.660(1)$
σ^2	≈ 50	1.54	0.16
$\kappa/\kappa_{\mathrm{all}}$	1	≈ 0.3	≈ 0.05
Efficiency	0.02	2.1	125
Valence errors	0	$\approx 0.1\,\mathrm{eV}$	$\approx 0.5\,\mathrm{eV}$

approximation is larger than the systematic error from pseudopotentials and their subsequent projection in the DMC algorithm. Of course, developments of better trial functions or better pseudopotentials could change these errors. The accuracy of pseudopotentials is one of the important factors which should be thoroughly tested. Rudin et al. [61] have recently showed that commonly used pseudopotentials reproduce all-electron results for the N_2 dimer with excellent accuracy.

Interestingly, ten Haaf, van Bemmel, and co-workers [62, 63] have shown that for a lattice model, it is possible to modify the effective Hamiltonian in such a way that the resulting energy is an upper bound. One can write the nonlocal operator as a sum of two pieces:

$$\langle R'|W|R\rangle = \langle R'|W_+|R\rangle + \langle R'|W_-|R\rangle \,, \tag{25}$$

where $\langle R'|W_-|R\rangle$ are these matrix elements for which $\langle R'|W|R\rangle \Psi_T(R)\Psi_T(R') > 0$ and vice versa for W_+. Then it is possible to construct the following Hamiltonian:

$$\mathcal{H}_{\text{val}}^{\text{eff}*} = \mathcal{H}_{\text{loc}} + \int dR' \langle R'|W_+|R\rangle + V_{\text{eff}-}(R) + V_\infty[\Psi_T(R) = 0] \,, \tag{26}$$

where

$$V_{\text{eff}-}(R) = \Psi_T^{-1}(R) \int \langle R'|W_-|R\rangle \Psi_T(R') \, dR' \,. \tag{27}$$

We can repeat the proof of the original paper [62] for electrons in continuous space and show that the energy of $\mathcal{H}_{\text{val}}^{\text{eff}*}$ will be an upper bound to the eigenvalue of $\mathcal{H}_{\text{val}}$. However, the straightforward application of this will have some new features. In particular, the variance of the energy used for the DMC propagation will not go to zero even in the limit of the exact trial function since W_+ is sampled directly. One can understand this from the simple example of one p electron in the field of an ion with repulsive s pseudopotential and attractive potential for all higher angular momenta [e.g., the C^{3+} ion in the ^2P (p^1) state]. The s pseudopotential has a zero contribution to the energy of the p state since the negative and positive contributions from the projection integral cancel exactly. However, by evaluating W_- exactly while sampling W_+ in the actual Monte Carlo, we obtain zero only after averaging over many Monte Carlo samples. This means that the walkers might experience large fluctuations of energy, especially in a region close to the ion where pseudopotential is large. Until such calculations are done, it is not clear whether these complications will be minor for many-electron systems.

This claim can be, in fact, generalized: Whenever there is a nonlocal term in the Hamiltonian, its exact sampling will produce an estimator with nonzero variance, even in the limit of the exact trial function. On the other hand, projection of the nonlocal part onto the trial function has the zero-variance property; however, for a nonexact trial function the upper bound property is not guaranteed.

V. EXCITED STATES

The calculation of excited-state energies has been attempted only occasionally with QMC methods. The simplest situation is to determine the excitation energy from the state of one symmetry to a state of different symmetry (e.g., the 1s-to-2p excitation in hydrogen). Since both states are ground states within their symmetries, one can do fixed-node calculations for each state individually and get individual upper bounds to their energies.

There are several problems with this approach. In the two separate calculations the statistical error is on the whole system, while the desired energy difference (say, of the gap) is a single-particle excitation. Thus a method that calculates the excitation energy directly, rather than as a difference of two independent calculations, would be preferable. In addition, the difference in energies is not bounded. If the nodes of the two states are of roughly comparable accuracy, one hopes that the difference will be accurate as well, but a substantial systematic error can occur, particularly since excited-state trial functions are known less precisely. The final problem with this method is more serious: One would like to calculate the energy difference between states with the same symmetries (e.g., 1s and 2s states of the hydrogen atom). One can perform the fixed-node calculation with a 2s trial function, but the result may be above or below the correct answer and the state can collapse into the 1s state. One needs to maintain orthogonality with lower states.

Ceperley and Bernu [64] introduced a method that addresses these problems. It is a generalization of the standard variational method applied to the basis set: $\exp(-t\mathcal{H})\,\Psi_\alpha$, where Ψ_α is a basis of trial functions $1 \le \alpha \le m$. One performs a single-diffusion Monte Carlo calculation with a guiding function that allows the diffusion to access all desired states, generating a "trajectory" $R(t)$, where t is imaginary time. With this trajectory one determines matrix elements between basis functions: $N_{\alpha,\beta}(t) = \langle \Psi_\alpha(t_1)|\Psi_\beta(t_1 + t)\rangle$ and their time derivatives. Using these matrix elements one can determine a sequence of upper bounds to the first m excited states. The bounds decrease exponentially fast and monotonically to the exact energies. Since the same MC data are used for all the states, some correlation of energies coming from the various states is built in. Since the bounds converge to the exact energies one has a systematic way of getting more-and-more-precise energy differences by increasing t. The statistical error will also increase because of the "sign problem," so in practice one may not be able to converge. The states are kept orthogonal, just as they are with the usual HF method.

Bernu et al. [65] used this method to calculate some excited states of molecular vibrations. Kwon et al. [66] used it to determine the Fermi liquid parameters in the electron gas. Correlation of walks reduced the errors in that calculation by two orders of magnitude. The method is not very stable and more work needs to be done on how to choose the guiding function and analyze the data, but it is a method that, in principle, can calculate a desired part of the spectrum from a single Monte Carlo run.

VI. EXACT FERMION METHODS

Quantum Monte Carlo techniques do not yet solve rigorously the many-fermion problem because of the sign problem. To map the quantum system onto a purely probablistic process for fermions seems to require some knowledge of the wavefunction before we start. Solving this problem would be a major advance in computational quantum mechanics. There are a variety of claims in the literature concerning what the fermion problem is. A coherent formulation of the "fermion problem" is stated in the introduction in terms of complexity: How much computer time T will it take to compute a given property to a specified error ε? The error includes the effects of all the systematic and statistical errors and must be of chemical interest. The fermion simulation problem is to find a method to calculate the properties of a many-fermion system that converges as $T \propto N^{\delta}\varepsilon^{-2}$, where δ is some small power (say, $\delta \leq 4$) and N is the number of atoms or electrons. Although there has been some work on this problem in the last few years, the solution is not in sight.

The fixed-node method does not qualify as a solution since the systematic errors are not under control, so they cannot be made arbitrarily small. There have been some recent attempts to parameterize the nodal surface and then determine the parameters dynamically. This will reduce the fixed-node errors, probably by an order of magnitude. However, it does not solve the problem since the error, even if it is smaller, is still uncontrolled. For a satisfactory solution one would have to parameterize, in a completely arbitrary fashion, the nodal surface.

In practice, there are several ways of generalizing the Slater determinant nodes. With backflow one maps the coordinates into new quasicoordinates, which then go into the orbitals of the determinant. This slows the calculation down by a factor of the number of electrons. In the second approach, one takes a sum of determinants, say, those coming from a CI calculation. The difficulty with this strategy is that as molecules get bigger, the number of possible determinants grows exponentially in the number of atoms, so this is not a feasible solution to the complexity problem. What has been learned to date is that backflow works well for homogeneous systems and multiconfiguration wavefunctions work well for nearly degenerate systems. A systematic way of putting these together has not yet been attempted.

Two related QMC methods without systematic errors are the transient estimate and release node methods. Both of them advance the wavefunction with the exact (antisymmetric) fermion Green's function. They differ in that a transient estimate starts from the trial function, whereas the release node starts from the fixed-node solution. Neither has the fixed-node restriction, so they introduce a minus-sign weight on the random walk whenever two electrons exchange or, more correctly, whenever the walks cross the nodal surface of the trial function an odd number of times. This gives rise to an exponentially growing signal-to-noise ratio. The error is under control, but the needed computer time is not. Suppose that one chooses the number of random walks and the projection time optimally to achieve a given error. It has been shown [64] that the total error will then be related to the computer time as

$$T \propto N^\delta \varepsilon^{-\alpha} , \tag{28}$$

where $\alpha = 2 + 2(E_F - E_B)/E_{gap}$. Here $E_F - E_B$ is the fermion ground-state energy relative to the boson ground-state energy, and E_{gap} is the energy gap between the fermion ground state and the next-highest fermion state that has some component in the trial wavefunction. For a boson calculation we see that $\alpha = 2$; this is the usual Monte Carlo convergence. But once the fermion sign is introduced, for large N the ground-state energy is extensive in the number of particles, $E = \mu N$, so that $T \propto N^\delta e^{-\ln(\varepsilon)\mu N}$. It is no better than an explicit basis-set method such as CI.

Not to leave the reader with the impression that transient estimate and release node calculations are not useful, let us mention briefly some results obtained with these methods. In 1980 a good convergence was obtained on the electron gas with up to 54 electrons [6]. Some of these calculations have recently been redone by Kwon [99] using better wavefunctions and DMC algorithms. Ceperley and Alder [9] also studied some small molecules (LiH, H_3). Recently, Caffarel and Ceperley [67] used maximum entropy methods to analyze the transient estimate energy more efficiently and rigorously, achieving a result for LiH that agrees with experiment within 0.2 mH for the ground-state energy, far better than chemical accuracy. There is no very strong reason to think that maximum entropy will change the complexity, but clearly there is room for improving the accuracy of these methods. These are valuable methods to see how good the fixed-node results are, and for local potentials they can be implemented with very little additional programming, but without future developments they will always be too slow for sufficiently large systems.

Any exact scheme where the walkers propagate independently will have the above-mentioned sign problem and unfavorable complexity. To do better, one somehow has to couple the positive and negative walkers. We call a positive walker one that starts out in the positive region of Ψ_T and a negative walker one that starts out in the negative region of Ψ_T. If a positive and a negative walker approach each other, the future contribution of the pair is nearly zero, so they can be canceled out. The correct nodal surface should be established dynamically by the annihilation of pairs of walkers. This idea was originally tried out by Arnow et al. [68]. Anderson [13] reviewed recent developments using cancellation methods, including some of his own very accurate calculations on few-electron systems based on the walker cancellation strategy. Bianchi et al. have also studied the cancellation algorithms for few-fermion systems [69].

Essentially the simple cancellation works only for few-electron systems because the number of walkers needed to establish the nodal surface dynamically rise exponentially with the number of electrons. For stability one needs many close approaches between positive and negative walkers; one has to fill up the relevant part of phase space. Unfortunately, the size of phase space grows exponentially with the number of electrons. Some schemes reduce the size of phase space by using various symmetries, such as permutation, translation, rotation, or reflection symmetry [13]. Again, this helps only for small systems. Anderson estimates that the computer time needed for a stable algorithm increases by a factor of 10 for each additional electron [13]. Hence the cancellation schemes so far invented do not solve the fermion problem as we have defined it in terms of complexity. Zhang and Kalos [70] and Liu et al. [71] have recently devised a scheme that forces pairs close together, thus increasing the chance for annihilation and stabilizing the cancellation. If this can be done sufficiently well, it could change the complexity and allow for calculations on larger systems.

There has been some recent work on applying auxiliary-field techniques to continuum systems. In these techniques the pair interaction between electrons is replaced, using the Stratonovitch–Hubbard transformation, with an interaction between electrons and a random potential, thus reducing the many-body problem to a mean-field problem. The auxiliary-field technique is used extensively for lattice models (e.g., for the Hubbard model) but until recently has not been used in the continuum. One advantage of this approach is that for spin-symmetric ground states (such as the Hubbard model at half-filling), there is no fermion sign problem. However, the repulsive electron–electron interaction brings in a new sign, with difficulties similar to the ones with an ordinary fermion sign [72]. Fahy and Hamann [73] introduced the fixed-node-like method, keeping only determinants with a positive projection on a Slater determinant. Zhang et al. [74] have recently shown how to make this much more efficient with a DMC branching algorithm. Silvestrelli et al. [75] have applied a similar technique (without fixed-node approximation) to an electron gas and to the H_2 molecule. They have found that it is not as efficient as DMC except in the limit of a very high density. Wilson and Gyorffy [76] recently generalized this approach to the relativistic electron gas. Until more accurate auxiliary-field techniques are developed, it is difficult to assess the prospects of these approaches for reaching the chemical accuracy.

VII. APPLICATIONS

A. Atoms and Small Molecules

There were many QMC calculations of the energy of atoms and small molecules. Most of these have been reviewed previously by Lester et al. [77] and by Anderson [13, 14]. We have chosen only few of these to highlight the recent achievements.

Among the most accurate calculations by any method were QMC simulations of the $H_2 + H$, HeH, and HeHe systems by Anderson and collaborators [13, 78, 79]. A high-accuracy-correlated wavefunction with the exact Green's function algorithm, which did not rely on the fixed-node approximation and had a zero time step error, was used. Because the system is small, one can deal with the fermion problem using a direct brute-force method: The low-dimensional configuration space can be filled with walkers. Using other algorithm improvements, one can achieve final results of unprecedented precision. For example, the exact ground state of the H–H–H system was estimated with an error bar of 0.0004 eV. In addition, a part of the energy surface for the $H_2 + H \rightarrow H + H_2$ reaction was evaluated together with the barrier height 9.61(1) kcal/mol. This illustrates one of the advantages of the QMC method: One can internally estimate real error bars. An earlier calculation [9] using GFMC with a release node obtained a barrier height of 9.65 ± 0.08 kcal/mol. The difference in error bars is due to the improved trial functions and algorithms and the much greater computational resources available today. High accuracy was also obtained on other systems, such as estimation of LiH total energy with an 0.05-mH error bar [80].

A systematic study of the first-row atoms and ions by variational Monte Carlo has been carried out by Schmidt and Moskowitz [30, 81]. These papers were important for several reasons. First, they provided a rather simple form of the correlation factor [see Eq. (13)], based on the work of Boys and Handy [82], essentially validating their early insight. The interpretation of this correlation term as an average backflow also helped to explain the physical roots of the success of this form. Second, they showed that this rather simple form with a few variational parameters can recover about 70% or more of the correlation systematically through the entire first row of periodic table, as shown in Fig. 1. Third, the results also demonstrated the impact of the near-degeneracy effects that were largest for the Be atom (getting $\approx 68\%$ of the correlation), while for the Ne atom the resulting correlation was about 85% of the exact value. Further study included evaluation of the first-row ionization potentials and electron affinities, with very good agreement with experiment [81].

Other high-accuracy all-electron calculations have been carried out by Umrigar et al. [16, 20]. Calculations of the Be atom with a two-configuration trial function and very accurate correlation factor with 109 variational parameters produced the best variational energy, with a statistical error bar of 0.03 mH and an energy higher than the estimated exact energy by only ≈ 0.2 mH. Calculations for the Ne atom obtained energy above the estimated ground state by ≈ 15 mH, which competes with the best variational result (≈ 10 mH above the estimated exact energy). The latter results has been obtained by Rizzo et al. [83] by a CI calculation that included *spdfghi* basis functions.

Other interesting calculations of Be^{2+}, Ne^{8+}, Be, and Ne atom have been carried out by Kenny et al. [85] in which they evaluated perturbationally the relativistic corrections to the total energies. In particular, they found that the Breit correction is systematically larger in the Dirac–Fock approximation and calculated the most accurate values of relativistic corrections for the Ne atom to date. These results demonstrate another useful capability of the correlated wavefunction produced by QMC: to estimate relativistic effects. Similar study within the VMC method has been done on examples of Li and LiH [86]. We expect that an important future application will be to carry out similar calculations of transition metal atoms where relativistic effects have a significant impact on the valence energy differences (e.g., 0.5 eV for the $s \rightarrow d$ transfer energy for the Ni atom) [87]. Only very few estimations of relativistic corrections for correlated wavefunctions are available for these elements [88, 89].

Other QMC capabilities have been demonstrated on the calculation of Li $2^2S \rightarrow 2^2P$ oscillator strength with significantly better agreement with experiment when compared with previous calculations [90]. Various quantities for a few electron atoms and molecules, including the electric response constants, were evaluated by Alexander et al. [91]. Study of vibrational properties of molecules has been advanced by Vrbik and Rothstein for the LiH molecule using DMC estimators for the derivatives of energy with respect to the ion positions [92].

B. Transition Metal Atoms

The $3d$ transition metal atoms are rather difficult systems for traditional quantum chemistry methods and for the LDA approach. There are several complications: a very compact high-density $3d$ shell, the near-degeneracy of the $3d$, $4s$, and $4p$ levels, and as we have mentioned, the relativistic effects. In addition, the $3s$ and $3p$ electrons occupy the same region of space as the $3d$ electrons, and several previous calculations have shown that these states must be included in the valence space if accurate results are desired. The first QMC calculation of the correlation energy and ionization potentials for transition metals atoms Sc and Y were carried out by Christiansen [93].

Table II shows the QMC calculated energies of the iron atom [57] compared with LDA [94] and coupled cluster [89] calculations. These were systematic calculations of the $3d$ atom with Ne-core scalar relativistic pseudopotentials derived within the multiconfiguration Hartree–Fock,

TABLE II
VMC and DMC First Ionization Potential, Electron Affinity, and Excitation Energies (eV) of the Fe Atom As Compared with Experiment and Other Calculations

	First Ionization Potential	$^5D \rightarrow {}^3F$	$^5D \rightarrow {}^5F$	Electron Affinity
HF	6.35	7.94	2.06	−2.36
LSDA	7.93	3.04	0.10	—
CCSD(T)[a]	7.79	—	1.07	−0.16
VMC	7.61(6)	4.73(6)	0.84(4)	−0.72(6)
DMC	7.67(6)	4.24(9)	0.84(6)	−0.03(9)
Experiment	7.87	4.07	0.87	0.15

[a] Coupled cluster calculations.

and gave results very competitive or better than the CI or CC [89] calculations, with an average discrepancy from experiment of ≈ 0.15 eV.

The electron affinity, which is very small for the Fe atom (0.15 eV), has so far not been reliably calculated. However, even the essentially zero affinity obtained is a tremendous improvement from the uncorrelated value of -2.36 eV. One of the reasons for the small remaining errors is that only simple trial functions were used. In particular, the determinants were constructed from Hartree–Fock orbitals. It is known that the Hartree–Fock wavefunction is usually more accurate for the neutral atom than for negative ion, and we conjecture that the unequal quality of the nodes could have created a bias on the order of the electron affinity, especially when the valence correlation energy is more than 20 eV. One can expect more accurate calculations with improved trial functions, algorithms, and pseudopotentials.

C. Clusters

Very recently, we were able to carry out simulations of much larger systems using nonlocal pseudopotentials. The rapid scale-up shows the power of QMC to calculate properties of much large systems. Silicon clusters provided interesting examples for testing the performance of QMC and for the study of correlation energy as a function of the size of the cluster. There are experimental data available for clusters of up to seven atoms which allowed for a direct comparison. There was also a controversy between the results of LDA calculations of Röthlisberger and co-workers [95] and theoretical arguments of Phillips [96] concerning the proper treatment of correlation and structural stability of Si_{13}. Phillips argued that correlation should stabilize the icosahedral structure against a lower C_{3v} symmetry trigonal capped antiprism (Fig. 3). We have carried out a systematic study of Si clusters with sizes between 2 and 20 atoms [55] to observe the structural trends. The comparison of binding energies with LDA, HF, and experimental results are shown on Fig. 4. The QMC results are within a few percent (0.2 eV/atom) of experimental data, decreasing the error of LDA by a factor of almost 5. These calculations also give insight into the impact of the correlation for various isomers (Fig. 3): the icosahedron has indeed a larger correlation energy, but the C_{3v} ground-state structure is still lower by almost 4 eV. Another remarkable fact was the observation that the correlation energy of the 20-atom cluster was very close to that found in silicon bulk crystal.

Even more interesting are carbon clusters. Raghavachari and his co-workers [97] discovered that mean-field methods have led to dramatically different results for the low-energy C_{20} isomers. The structures of these isomers are very different: a ring (D_{10h}) symmetry that is essentially

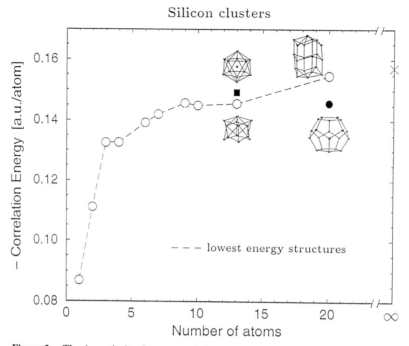

Figure 3. The (negative) valence correlation energy for silicon clusters as a function of the number of atoms in the cluster. The dashed line connects the values that belong to the lowest-energy structures. The filled square and circle correspond to the icosahedron and dodecahedron structures, respectively. The cross corresponds to the estimated correlation of the silicon crystal in diamond structure.

a one-dimensional system, a bowl (C_{5v}) where hexagons and a pentagon lie in a slightly curved plane, and a cage (C_2) which is a distorted dodecahedron. All these structures have closed shells with gaps ≈ 1 eV. LDA calculations predicted that the cage was the most stable structure, with the bowl and ring above by 1.6 and 3.8 eV, respectively. Surprisingly, GGA (Becke–Lee–Yang–Parr functional), which is supposed to be a small correction to LDA, completely reversed this ordering. Other GGAs with different exchange-correlation functionals do not provide much more useful information: the results varied in a nonsystematic way. An ambitious attempt by the CCSD(T) method with $\approx 10^7$ single and double excitations self-consistently and $\approx 10^{10}$ triples perturbatively was done by Taylor and co-workers [98]. However, a rather restricted basis set [58], which recovered about 75% of the valence correlation energy, required extrapolations that did not allow for a clear-cut prediction. The results indicated both the bowl and cage as possible lowest-energy candidates.

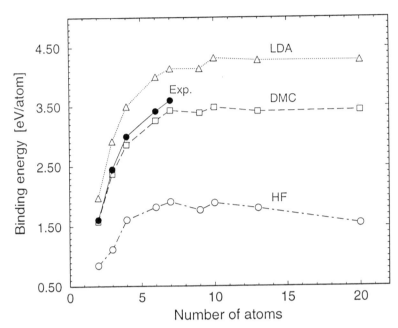

Figure 4. Binding energy of silicon clusters from HF, LDA, and DMC methods compared with experiment. (From Ref. 55.)

Recent QMC calculations [58] show that the correlation energy of the bowl and cage are very similar and thus the bowl is favored because of its lower electrostatic energy. The ring has a smaller correlation energy than the cage by about 3.8 eV, but with the most favorable electrostatic contribution it is placed between the bowl and cage in the overall energy ordering. This was perhaps the first time that the QMC method was used as a predictive tool for large molecular systems. The obtained correlation energies from the QMC [58] and CCSD(T) [98] methods are compared in Table III. The computer time for the QMC calculation of a C_{20} system is about 2 h for VMC and about 40 h for DMC on the Cray C90 (error bar 0.01 eV/atom).

D. Extended Systems

We mention here briefly calculations on extended systems that have obvious relevance to chemistry and the development of QMC. Recent calculations of the two- and three-dimensional electron gas by Kwon et

TABLE III
Comparison of Valence Correlation Energies (a.u.) for C_{20} in the Ring,
Bowl, and Cage Geometries

	Ring	Bowl	Cage
CCSD(T)	−2.63	−2.70	−2.77
VMC	−2.735(3)	−2.888(4)	−2.888(4)
DMC	−3.24(1)	−3.34(1)	−3.36(1)

al. [99] shed some light on the nodes of the many-body wavefunction by using the backflow wavefunction. The use of backflow modified trial functions has led to significantly lower variational energy and lower fixed-node energy, indicating that backflow modifications are important for the uniform electron gas. The improvement in the nodes was especially evident for the high-density gas, while many-body terms in the correlation factor were more important for the low-density case. Fahy et al. [31,51] first performed variational calculations of solid silicon and carbon. Li et al. performed the DMC calculations of solid silicon using a pseudo-Hamiltonian [50]. Fixed-node DMC calculations with local pseudopotential for solid Ge have been carried out by Rajagopal et al. [100]. These authors have also introduced a technique to decrease the finite-size effects by using a more balanced sampling of the Brillouin zone.

There has also been some progress in expanding the solid-state calculations to transition metal oxides. A VMC attempt to evaluate the cohesive energy of the Ni-O solid was carried out by Tanaka [101]. Although the agreement with experiment was very good, it was mainly because of a cancellation of errors, as the absolute accuracy was limited by a rather simple form of trial functions. The first DMC calculations of copper oxide ($CaCuO_2$) were tested by Mitas [102] with 164 electrons in the simulation cell and Ne-core pseudopotentials on both Ca and Cu. For a nonmagnetic state, the resulting cohesive energy per $CaCuO_2$ unit was 12(1) eV. The error bar is rather large because of a large total energy, but the attempt demonstrates the feasibility of such calculations and illustrates the need for new computational strategies (parallel computing) and further method development to increase the performance. Both of these calculations are paving the way toward simulations of transition metal compounds where the accurate treatment of exchange and correlation is essential for a proper understanding of these systems.

There has been some interesting progress on the excited-state calculations in insulating solid systems. Mitas and Martin [59] calculated an excited state for a solid by evaluating an exciton in a compressed nitrogen insulating solid. A similar calculation was also carried out for diamond

[60]. The comparison with LDA results and experiment shows that calculations of excitons provide valuable and accurate information about the band gaps, which can be estimated from exciton energies. This technique opens new possibilities for band structure studies of insulating solids, where for excited states a proper treatment of correlation is even more important than for ground states.

An alternative method for calculating the gap has been tested by Engel et al. [103] on a model of the two-dimensional electron gas in a periodic potential. He estimated the gap by adding an electron to the system and evaluating the difference. The estimations were in excellent agreement with an alternative many-body method.

VIII. CONCLUSIONS

In this short review we have pointed out only very few of the basic issues involving the simulation of chemical systems with Quantum Monte Carlo. What has been achieved in the last few years is remarkable: very precise calculations of small molecules, the most accurate calculations of the electron gas, silicon and carbon clusters, solids, and simulations of hydrogen at temperatures when bonds are forming. New methods have been developed as well: high-accuracy trial wavefunctions for atoms, molecules, and solids, treatment of atomic cores, and the generalization of path-integral Monte Carlo to treat many-electron systems at positive temperatures.

The reader may wonder why we have focused much of this review on the computational complexity. We have chosen complexity because that is the key to making these theoretical calculations an indispensable tool to scientists and engineers. It is often far from the minds of the computational scientists that real-world applications are much more complex than isolated molecules with a few electrons. A prototypical application is a chemical reaction of a large molecule in solution. Here, Car–Parrinello-like [104] methods have a very large impact on condensed matter chemistry and physics. They have this impact because almost any real application involves thousands if not millions of electrons. Any method that can treat the real complexity without reductionism (since that requires a highly trained theoretician) has an inherent advantage and utility. However, not only must the method be able to treat large systems, it must do so with high accuracy and reliability and must be able to calculate a whole spectrum of physical properties. For the moment that can only be done at the mean-field (LDA) level, which though often surprisingly accurate, is equally often inadequate.

The algorithms of the future will have at least three important components: accuracy, efficiency, and generality. One can see that current mainstream methods fulfill at most two of these requirements. Accurate and efficient methods are not general (e.g., analytical methods for low-dimensional or simplified models). Those that are general and efficient are not accurate (mean-field methods). Finally, general and accurate approaches are not efficient for large systems (such as the CI method).

It seems that QMC has all the requisites to become the method of choice in the future, as, we believe, it can fulfill all three of these requirements. Clearly, QMC's ability to scale up and treat the many-body effects directly is invaluable. But QMC also has many other attributes. For example, it is straightforward to include thermal, zero-point, or classical nuclear effects in PIMC. Certainly, until the fermion sign problem is solved, there is always a question mark hanging over the field: Is the method a fundamental advance, or is it merely a candidate for the most accurate approximate scheme currently known?

There are, of course, many opportunities and challenges for improvement of the basic methodology aside from the sign problem. From a practical point of view, more research is needed to determine better pseudopotentials since it appears that they will determine the quality of results for systems with atoms heavier than Be. What we have not discussed in depth is how important it is to compute a variety of physical relevant quantities. QMC has, for too long, been focused on calculating the ground-state energy. Path integrals are in this respect better than DMC because the trial function is eliminated and one can compute estimators directly. For that reason, they have given more insight into qualitative features such as long-range order and linear response properties. But PIMC has yet to be applied to chemically relevant systems involving many correlated electrons. Although there has been some method development, large standardized packages are not yet in existence for QMC, and as a consequence, applications have been much more limited. There has been some development in the theory of how to compute spectra, forces, optimized wavefunctions, and geometries, but calculations of these are not yet routine. With the coming of more powerful computers and algorithms, that day is not far off.

ACKNOWLEDGMENTS

We thank all of our colleagues that sent us preprints and reprints of their work and we apologize that we have not had enough space to discuss them in more detail. We are grateful to Jeffrey C. Grossman for reading the manuscript. We are supported by the National Science Foundation through the National Center for Supercomputing Applications and through grant DMR-91-17822 by the Department of Physics at the University of Illinois, and L.M. is partially supported by Department of Energy grant DEFG02-91-ER45439.

REFERENCES

1. E. P. Wigner, *Phys. Rev.* **40**, 749 (1932).

2. W. L. McMillan, *Phys. Rev.* **138**, 442 (1965).

3. M. H. Kalos, *Phys. Rev.* **128**, 560 (1962); *J. Comput. Phys.* **2**, 257 (1967).

4. M. H. Kalos, D. Levesque, and L. Verlet, *Phys. Rev. A* **9**, 2178 (1974).

5. D. M. Ceperley, *Phys. Rev. B* **18**, 3126 (1978).

6. D. M. Ceperley and B. J. Alder, *Phys. Rev. Lett.* **45**, 566 (1980); B. J. Alder, *Phys. Rev. Lett.* **45**, 566 (1980).

7. J. B. Anderson, *J. Chem. Phys.* **63**, 1499 (1975).

8. D. M. Ceperley, *Physica B* **108**, 875 (1981).

9. D. M. Ceperley and B. J. Alder, *J. Chem. Phys.* **81**, 5833 (1984).

10. B. L. Hammond, W. A. Lester, Jr., and P. J. Reynolds, *Monte Carlo Methods in Ab Initio Quantum Chemistry*, World Scientific, Singapore, 1994.

11. D. M. Ceperley and M. H. Kalos, in *Monte Carlo Methods in Statistical Physics*, K. Binder, Ed., Springer-Verlag, Berlin, 1979.

12. K. E. Schmidt and M. H. Kalos, in *Monte Carlo Methods in Statistical Physics II*, K. Binder, Ed., Springer-Verlag, Berlin, 1984; K. E. Schmidt and D. M. Ceperley, in *Monte Carlo Methods in Statistical Physics III*, K. Binder, Ed., Springer-Verlag, Berlin, 1992.

13. J. B. Anderson, in *Understanding Chemical Reactivity*, S. R. Langhoff, Ed., Kluwer, Dordrecht, The Netherlands, 1995.

14. J. B. Anderson, *Intern. Rev. Phys. Chem.*, **14**, 85 (1995).

15. B. J. Berne and D. Thirumalai, *Ann. Rev. Phys. Chem.* **37**, 401 (1986).

16. C. J. Umrigar, K. G. Wilson, and J. W. Wilkins, *Phys. Rev. Lett.* **60**, 1719 (1988).

17. D. M. Ceperley, *J. Stat. Phys.* **63**, 1237 (1991).

18. J. W. Moskowitz, K.E. Schmidt, M. A. Lee, and M. H. Kalos, *J. Chem. Phys.* **78**7, 349 (1982).

19. P. J. Reynolds, D. M. Ceperley, B. J. Alder, and W. A. Lester, *J. Chem. Phys.* **77**, 5593 (1982).

20. C. J. Umrigar, M. P. Nightingale, and K. J. Runge, *J. Chem. Phys.* **99**, 2865 (1993).

21. R. P. Subramaniam, M. A. Lee, K. E. Schmidt, and J. W. Moskowitz, *J. Chem. Phys.* **97**, 2600 (1992).

22. R. P. Feynman, *Statistical Mechanics*, Addison-Wesley, Redwood City, Calif., 1972.

23. D. M. Ceperley, *Rev. Mod. Phys.*, **67**, 279 (1995).

24. D. M. Ceperley, *Phys. Rev. Lett.* **69**, 331 (1992).

25. C. Pierleoni, D. M. Ceperley, B. Bernu, and W. R. Magro, *Phys. Rev. Lett.* **73**, 2145 (1994).

26. D. M. Ceperley and B. J. Alder, *Phys. Rev. B* **36**, 2092 (1987).

27. M. Urban, R. J. Bartlett, and S. A. Alexander, *Intern. J. Quantum Chem. Symp.* **26**, 271 (1992).

28. J. Cizek, *J. Chem. Phys.* **45**, 4256 (1966); *Adv. Chem. Phys.* **14**, 35 (1969).

29. D. M. Ceperley, *J. Stat. Phys.* **43**, 815 (1986).

30. K. E. Schmidt and J. W. Moskowitz, *J. Chem. Phys.* **93**, 4172 (1990).

31. S. Fahy, X. W. Wang, and S. G. Louie, *Phys. Rev. B* **42**, 3503 (1990).

32. L. Mitas, in *Computer Simulation Studies in Condensed-Matter Physics V*, D. P. Landau, K. K. Mon, and H. B. Schuttler, Eds., Springer-Verlag, Berlin, 1993.

33. A. Mushinskii and M. P. Nightingale, *J. Chem. Phys.* **101**, 8831 (1994).

34. Z. Sun, P. J. Reynolds, R. K. Owen, and W. A. Lester, Jr., *Theoret. Chim. Acta* **75**, 353 (1989).

35. R. P. Feynman and M. Cohen, *Phys. Rev.* **102**, 1189 (1956).

36. B. L. Hammond, to be published.

37. B. L. Hammond, P. J. Reynolds, and W. A. Lester, Jr., *Phys. Rev. Lett.* **61**, 2312 (1988).

38. J. Carlson, J. W. Moskowitz, and K. E. Schmidt, *J. Chem. Phys.* **90**, 1003 (1989).

39. D. R. Hamann, M. Schlüter, and C. Chiang, *Phys. Rev. Lett.* **43**, 1494 (1979).

40. W. E. Pickett, *Comput. Phys. Rep.* **9**, 115 (1989); M. Krauss and W. J. Stevens, *Ann. Rev. Phys. Chem.* **35**, 357 (1984).

41. M. Dolg, U. Wedig, H. Stoll, and H. Preuss, *J. Chem. Phys.* **86**, 866 (1987).

42. E. L. Shirley, L. Mitas, and R. M. Martin, *Phys. Rev. B* **44**, 3395 (1991).

43. P. Acioli and D. M. Ceperley, *J. Chem. Phys.* **100**, 8169 (1994).

44. P. A. Christiansen, *J. Phys. Chem.* **94**, 7865 (1990).

45. H. J. Flad, A. Savin, M. Schultheiss, A. Nicklass, and H. Preuss, *Chem. Phys. Lett.* **222**, 274 (1994).

46. P. A. Christiansen, *J. Chem. Phys.* **95**, 361 (1991).

47. G. B. Bachelet, D. M. Ceperley, and M. G. B. Chiocchetti, *Phys. Rev. Lett.* **62**, 2088 (1989).

48. W. M. C. Foulkes and M. Schluter, *Phys. Rev. B* **42**, 11505 (1990).

49. A. Bosin, V. Fiorentini, A. Lastri, and G. B. Bachelet, *Phys. Rev. A* **52**, 236 (1995).

50. X.-P. Li, D. M. Ceperley, and R. M. Martin, *Phys. Rev. B* **44**, 10929 (1991).

51. S. Fahy, X. W. Wang, and S. G. Louie, *Phys. Rev. Lett.* **61**, 1631 (1988).

52. L. Mitas, E. L. Shirley, and D. M. Ceperley, *J. Chem. Phys.* **95**, 3467 (1991).

53. M. M. Hurley and P. A. Christiansen, *J. Chem. Phys.* **86**, 1069 (1987).

54. B. L. Hammond, P. J. Reynolds, and W. A. Lester, Jr., *J. Chem. Phys.* **87**, 1130 (1987).

55. J. C. Grossman and L. Mitas, *Phys. Rev. Lett.* **74**, 1323 (1995).

56. R. M. Panoff and J. Carlson, *Phys. Rev. Lett.* **62**, 1130 (1989).

57. L. Mitas, *Phys. Rev. A* **49**, 4411 (1994).

58. J. C. Grossman, L. Mitas, and K. Raghavachari, *Phys. Rev. Lett.* **75**, 3870 (1995).

59. L. Mitas and R. M. Martin, *Phys. Rev. Lett.* **72**, 2438 (1994).

60. L. Mitas, in *Electronic Properties of Solids Using Cluster Methods*, T. A. Kaplan and S. D. Mahanti, Eds., Plenum Press, New York, 1995.

61. S. P. Rudin, M. M. Steiner, and J. W. Wilkins, *Bull. Am. Phys. Soc. Ser. II.* **40**, 711 (1995), and to be published.

62. D. F. B. ten Haaf, H. J. M. van Bemmel, J. M. J. van Leeuwen, W. van Saarloos, and D. M. Ceperley, *Phys. Rev. B* **51**, 13039 (1995).

63. H. J. M. van Bemmel, D. F. B. ten Haaf, W. van Saarloos, J. M. J. van Leeuwen, and G. An, *Phys. Rev. Lett.* **72**, 2442 (1994).

64. D. M. Ceperley and B. Bernu, *J. Chem. Phys.* **89**, 6316 (1988).

65. B. Bernu, D. M. Ceperley, and W. A. Lester, Jr., *J. Chem. Phys.* **93**, 552 (1990).

66. Y. Kwon, D. M. Ceperley, and R. M. Martin, *Phys. Rev. B* **50**, 1684 (1994).

67. M. Caffarel, F. X. Gaeda, and D. M. Ceperley, *Europhys. Lett.* **16**, 249 (1991); M. Caffarel and D. M. Ceperley, *J. Chem. Phys.* **97**, 8415 (1992).

68. D. M. Arnow, M. H. Kalos, M. A. Lee, and K. E. Schmidt, *J. Chem. Phys.* **77**, 5562 (1982).

69. R. Bianchi, D. Bressanini, P. Cremaschi, and G. Morosi, *Comput. Phys. Commun.* **74**, 153 (1993).

70. S. Zhang and M. H. Kalos, *Phys. Rev. Lett.* **67**, 3074 (1991).

71. Z. Liu, S. Zhang, and M. H. Kalos, *Phys. Rev. E* **50**, 3220 (1994).

72. M. Boninsegni and E. Manousakis, Phys. Rev. B **46**, 560 (1992).

73. S. B. Fahy and D. R. Hamann, *Phys. Rev. Lett.* **65**, 3437 (1990).

74. S. Zhang, J. Carlson, and J. Gubernatis, *Phys. Rev. Lett.* **74**, 3652 (1995).

75. P. L. Silvestrelli, S. Baroni, and R. Car, *Phys. Rev. Lett.* **71**, 1148 (1993).

76. M. T. Wilson and B. L. Gyorffy, *J. Phys. Condensed Matter* **7**, 1565 (1995).

77. W. A. Lester, Jr., and B. L. Hammond, *Ann. Rev. Phys. Chem.* **41**, 283 (1990).

78. D. L. Diedrich and J. B. Anderson, *Science* **258**, 786 (1992).

79. A. Bhattacharya and J. B. Anderson, *J. Chem. Phys.* **49**, 2441 (1994).

80. B. Chen and J. B. Anderson, *J. Chem. Phys.* **102**, 4491 (1995).

81. K. E. Schmidt and J. W. Moskowitz, *J. Chem. Phys.* **97**, 3382 (1992).

82. S. F. Boys and N. C. Handy, *Proc. R. Soc. London Ser. A* **311**, 309 (1969).

83. A. Rizzo, E. Clementi, and M. Sekiya, *Chem. Phys. Lett.* **177**, 477 (1991).

84. D. Feller and E. R. Davidson, *J. Chem. Phys.* **90**, 1024 (1990).

85. S. D. Kenny, G. Rajagopal, and R. J. Needs, *Phys. Rev. A* **51**, 1898 (1995).

86. H. Bueckert, S. M. Rothstein, and J. Vrbik, *Chem. Phys. Lett.* **190**, 413 (1992).

87. K. Raghavachari and G. W. Trucks, *J. Chem. Phys.* **91**, 1062 (1989).

88. C. W. Bauschlischer, Jr., *J. Chem. Phys.* **86**, 5591 (1987).

89. M. Urban, J. D. Watts, and R. J. Bartlett, *Int. J. Quantum Chem.* **52**, 211 (1994).

90. R. N. Barnett, R. J. Reynolds, and W. A. Lester, Jr., *Int. J. Quantum Chem.* **42**, 837 (1992); R. N. Barnett, E. M. Johnson, and W. A. Lester, Jr., *Phys. Rev. A* **51**, 2049 (1995).

91. S. A. Alexander, R. L. Coldwell, G. Aissing, and A. J. Thakkar, *Int. J. Quantum Chem. Symp.* **26**, 213 (1992).

92. J. Vrbik and S. M. Rothstein, *J. Chem. Phys.* **96**, 2071 (1992); J. Vrbik, D. A. Legare, and S. M. Rothstein, *J. Chem. Phys.* **92**, 1221 (1990).

93. P. A. Christiansen, *J. Chem. Phys.* **95**, 361 (1991).

94. T. V. Russo, R. L. Martin, and P. J. Hay, *J. Chem. Phys.* **101**, 7729 (1994).

95. U. Röthlisberger, W. Andreoni, and P. Giannozzi, *J. Chem. Phys.* **92**, 1248 (1992).

96. J. C. Phillips, *Phys. Rev. B* **47**, 14132 (1993).

97. K. Raghavachari, D. L. Strout, G. K. Odom, G. E. Scuseria, J. A. Pople, B. G. Johnson, and P. M. W. Gill, *Chem. Phys. Lett.* **214**, 357 (1994).

98. P. R. Taylor, E. Bylaska, J. H. Weare, and R. Kawai, *Chem. Phys. Lett.* **235**, 558 (1995).

99. Y. Kwon, D. M. Ceperley, and R. M. Martin, *Phys. Rev. B* **48**, 12037 (1993).

100. G. Rajagopal, R. J. Needs, A. James, S. D. Kenny, and W. M. C. Foulkes, *Phys. Rev. B* **51**, 10591 (1995); *Phys. Rev. Lett.* **73**, 1959 (1994).

101. S. Tanaka, *J. Phys. Soc. Japan* **62**, 2112 (1993).

102. L. Mitas, *Bull. Am. Phys. Soc.* **39**, 87 (1994), and to be published.

103. G. E. Engel, Y. Kwon, and R. M. Martin, *Phys. Rev. B.* **51**, 13538 (1995).

104. R. Car and M. Parrinello, *Phys. Rev. Lett.* **55**, 2471 (1985).

MONTE CARLO METHODS FOR REAL-TIME PATH INTEGRATION

C. H. MAK

Department of Chemistry, University of Southern California, Los Angeles, California

REINHOLD EGGER

Fakultät für Physik, Universität Freiburg, Freiburg, Germany

CONTENTS

I. INTRODUCTION

The wide availability of fast and cheap computing power in the past decade has promoted intense activities in computational chemical physics. A great deal of effort has been devoted to developing computational methods for condensed-phase reactions in chemistry, physics, and biolo-

Advances in Chemical Physics, Volume XCIII, Edited by I. Prigogine and Stuart A. Rice.
ISBN 0-471-14321-9 © 1996 John Wiley & Sons, Inc.

gy. Because quantum mechanical effects are essential for the understanding of many processes in these disciplines, the development of quantum mechanical simulation methods in particular has become the focus of much recent interest. In this review we deal with one particular class of quantum mechanical simulation methods commonly known as quantum Monte Carlo (QMC) techniques.

Generally, QMC methods are based on a path-integral description of quantum mechanics [1] and can thus be divided into two broad categories. Imaginary-time QMC methods are designed to compute equilibrium properties of quantum systems [2, 3]. This branch of QMC is by now well developed for almost all types of quantum many-body systems, with the exception of fermions and certain quantum spin systems. On the other hand, real-time QMC methods are aimed at computing dynamical properties of quantum systems. Compared to imaginary-time QMC methods, real-time QMC techniques are still quite primitive. The lag in the development of real-time QMC methods is due to the notorious dynamical sign problem. In this article, we summarize some of the recent progress in real-time QMC simulations, mainly from our perspective in this field. Reviews of earlier work on the dynamical sign problem can be found in Refs. 4 and 5.

The origin of the sign problem in real-time QMC simulations is easy to understand. In a QMC simulation, the path integration is implemented by stochastic sampling over all possible paths $x(t)$ the system may take to go from some initial state to some final state [1]. This stochastic sampling is guided by a weight $\exp\{S[x(t)]\}$, where $S[x(t)]$ is the action of the path. In the computation of dynamical quantities, the action S is necessarily complex valued, and because of this one has to deal with interference effects. The intense phase cancellations between different paths are then responsible for the dynamical sign problem. The manifestation of the sign problem is a small signal-to-noise ratio that vanishes exponentially with increasingly long paths. In other words, any stochastic real-time QMC simulation will eventually become unstable at long enough time. Although the dynamical sign problem is cumbersome, interference effects are at the heart of quantum mechanics and they cannot be eliminated from the problem entirely. The real question is then whether one is able to delay the sign problem and shift the instability out to sufficiently long times to do meaningful calculations.

There are two main approaches toward extracting real-time information from QMC simulations. The first is a direct real-time computation as described in the rest of this article. The second approach, which we describe only briefly here, starts out with a conventional imaginary-time QMC simulation. The numerical data derived from the imaginary-time calculation are then analytically continued to extract real-time information [6, 7]. This approach seems to yield reliable results for spectral functions (i.e., frequency-dependent properties) in a number of cases, but appears to be impractical for the direct computation of time-dependent dynamical quantities. Various methods have been suggested for performing the required analytic continuation of the numerical data, which is an ill-posed problem in the strict mathematical sense. Recent efforts have focused on the use of maximum entropy methods which were originally developed for image recovery problems. This approach combines prior knowledge (e.g., sum rules or positivity of spectral functions) with the QMC simulation data in a information-theoretic framework based on Bayes' theorem. A concise account of this method together with application to the Anderson model can be found in Ref. 6. However, there are certainly cases where even maximum entropy fails, and all approaches based on imaginary-time simulation followed by analytic continuation are plagued by the problem of ground-state dominance: The low-temperature equilibrium properties of the system are dominated by the ground state, whereas the dynamics requires information about low-lying excitations. Furthermore, if one is directly interested in real-time information instead of spectral functions, it is generally much more reliable to start out directly with real-time QMC. In the following, we restrict our attention to real-time QMC.

Clearly, if one is able to control the dynamical sign problem, there are numerous interesting and important applications of real-time path integration methods. Focusing on kinetic processes and reactions in condensed phase, we draw attention to two general classes of problems. The first concerns the motion of some quantum particle on a potential surface around a single local minimum (i.e., tunneling processes connecting different minima are not important). For this class of problems, a variety of techniques, such as coherent-state path integrals [8], wave-packet propagation methods [9], and many others [4, 5, 10, 11] are sufficient. The dynamical sign problem appears to be well under control for these types of processes. However, if tunneling is important, most of these methods are inapplicable, and the numerical evaluation of the real-time path integral is much more challenging. Many researchers have therefore abandoned the use of stochastic QMC methods for tunneling problems in favor of exact diagonalization techniques combined with optimized reference systems (so that longer Trotter time slices can be used) [12, 13]. Although these methods avoid the sign problem completely, they nonetheless impose an intrinsic upper time limit on the simulations due to the growth of computational time with the size of the basis set. In the cases discussed below, we show that contrary to prevailing skepticism, the direct real-time QMC method is able to go to time scales that are of chemical or physical relevance for a number of interesting tunneling problems.

Many of the modern real-time path integration methods are outgrowth of the work of Vonogradov and Filinov [14]. The original approach is similar to the more widely known stationary-phase Monte Carlo (SPMC) method later put forward by Doll and co-workers [15, 16]. A conceptually similar variant has also been proposed by Makri and Miller [17]. The basic idea in the SPMC method is to synthesize part of the phase interference analytically by a stationary-phase-like approximation instead of putting all the load on the stochastic sampling. The result of this is a filtering function that biases the Metropolis trajectory [18] toward regions where the action is nearly stationary [16]. This idea has been applied successfully to several quantum mechanical problems [15–17, 19]. We point out that even in cases when the SPMC filtering techniques are applicable, the sign problem still becomes exponentially uncontrollable with increasing system size [20], but the relevant time limit can be made long enough for many practically interesting calculations.

To treat tunneling problems (which is our primary goal here), one has to go beyond ordinary SPMC techniques. The reason is that tunneling can turn the problem originally defined in a continuum position space into a quasidiscrete problem because the fundamental events in the deep-tunneling limit are quantal motions between localized states through classically forbidden regions [21]. Since a discrete problem does not possess any stationary paths (in real space at least), SPMC methods are not useful in that limit.

In the following we present a general concept for performing real-time QMC simulations that can potentially eliminate the sign problem. This general strategy is able to treat both continuum and discrete problems, and it contains the SPMC method as a special case. The practical problem that remains is the actual algorithm by which this strategy is carried out. In the past we have realized this concept using a simple algorithm to study condensed-phase reactions in the deep tunneling regime. The models we examine are of the spin-boson type [22], where a system characterized by discrete ("spin") states is coupled to a dissipative Gaussian ("boson") bath representing the condensed-phase environment. We will show that because of the specific features of these models, the general strategy can be implemented in a natural and straightforward manner. We should emphasize, however, that the application of this general strategy to other problems remains a challenging task.

Our very first attempts on the QMC simulations of the dynamics of the dissipative two-state spin-boson model employed an exact mapping from the discrete representation back to a continuous-spin representation. This transformation allowed us to carry out a conventional SPMC simulation (SPMC methods generally require a continuous reaction coordinate). The tunneling problem was partially alleviated by contour distortion techniques [23, 24]. The general strategy outlined in Section II, however, allows one to work directly in the numerically more convenient discrete basis, and the superiority of this method over SPMC has been demonstrated unambiguously for tunneling problems [25, 26]. Subsequently [27–29], we adopted an optimized version of the same algorithm. In Section III we give a detailed account of this optimized method for dissipative tight-binding systems. All of these approaches have been critically evaluated in Ref. 27, and it was shown that the optimized method is much more powerful than all earlier approaches, allowing us to go out to much longer real times than before.

Applications are then presented in Section IV. These examples should served as a guide as to what kinds of problems can be studied with these techniques and the limitations and possibilities for these methods. We present three examples; (1) a dynamical test of the "centroid" quantum transition-state theory for electron transfer (ET) reactions in the crossover regime between adiabatic and nonadiabatic electron transfer, (2) the primary electron transfer reaction in bacterial photosynthesis, and (3) the diffusion kinetics of a Brownian particle in a periodic potential. Finally, Section V offers an outlook and a perspective of the current status of the field from our vantage point.

II. GENERAL STRATEGY

We shall be concerned with the computation of real-time quantities (e.g., correlation functions or time-dependent occupation probabilities). The standard approach in a QMC simulation of such dynamical quantities consists of constructing a suitably discretized path integral expression for the quantum mechanical propagator of the system

$$K(x, x'; t) = \int \mathcal{D}\{x_i\} \exp(iS[\{x_i\}]) , \qquad (2.1)$$

where x_i is the system coordinate on the ith Trotter time slice (x may be either a continuous or a discrete variable). Then the time-dependent property $A(t)$ of interest is expressed in terms of a ratio,

$$\langle A(t) \rangle = \frac{\int dx \int dx' \int dx'' \rho(x, x'') K(x, x'; t) A(x', t) K(x', x''; -t)}{\int dx \int dx' \int dx'' \rho(x, x'') K(x, x'; t) K(x', x''; -t)} , \qquad (2.2)$$

where ρ is a density matrix that specifies how the system was prepared initially. Putting Eq. (2.1) into (2.2), a stochastic estimate for $\langle A(t) \rangle$ can be expressed symbolically as

$$\langle A(t) \rangle = \frac{\Sigma_{x_i} P[x_i]\Phi[x_i]A[x_i]}{\Sigma_{x_i} P[x_i]\Phi[x_i]} , \qquad (2.3)$$

where the weight function consists of a positive-definite part P and a phase factor Φ. Again, x_i stands for the system configuration at Trotter slice i. The phase factor is an oscillatory function of the coordinates $\{x_i\}$ and is responsible for the dynamical sign problem. Assuming that there are no further exclusivity problems in the numerator so that $A[x]$ is well behaved, we can analyze the sign problem in terms of the variance of the denominator,

$$\sigma^2 \approx \frac{1}{N}(\langle \Phi^2 \rangle_P - \langle \Phi \rangle_P^2) , \qquad (2.4)$$

where N is the number of MC samples taken and averages are taken only with P as the weight function.

For the sake of simplicity, let us consider the case where $\Phi = \pm 1$. (This would be directly applicable to the fermion sign problem [30], which arises in imaginary-time QMC sampling of many fermionic systems. Extension to general real-time QMC simulations where Φ is a complex factor is straightforward.) In this case, $\langle \Phi^2 \rangle = 1$ always and the variance of the signal is controlled entirely by the size $|\langle \Phi \rangle|$. Under normal circumstances, $|\langle \Phi \rangle|$ can be very small, for the reasons that we have described earlier, and the simulation becomes unstable at long time.

Remarkably, one can achieve considerable progress by *blocking states together*. By this we mean instead of sampling single states along the MC trajectory, we can consider sampling sets of states that we will call blocks. Under such a blocking operation, the stochastic estimate for $\langle A(t) \rangle$ takes the form

$$\langle A(t) \rangle = \frac{\Sigma_{b_j} (\Sigma_{x_i \in b_j} P[x_i]\Phi[x_i]A[x_i])}{\Sigma_{b_j} (\Sigma_{x_i \in b_j} P[x_i]\Phi[x_i]} , \qquad (2.5)$$

where one first sums over the configurations belonging to a block b_j in an exact (nonstochastic) way and then invokes the MC method to perform the summation over the blocks. Of course, there is considerable freedom as to how to choose this blocking and we exploit this freedom to reduce the sign problem as much as possible.

Now we analyze the variance σ'^2 of the denominator of Eq. (2.5). We first define new sampling functions in terms of the blocks, which are then sampled stochastically:

$$P'[b_j] = \left| \sum_{x_i \in b_j} P[x_i]\Phi[x_i] \right| , \qquad (2.6)$$

$$\Phi'[b_j] = \text{sgn}\left(\sum_{x_i \in b_j} P[x_i]\Phi[x_i] \right) . \qquad (2.7)$$

It is then easy to show that

$$\frac{|\langle\Phi'\rangle|}{|\langle\Phi\rangle|} = \frac{\Sigma_{x_i} P[x_i]}{\Sigma_{b_j} P'[b_j]} ,$$

and with the inequality

$$\sum_{b_j} P'[b_j] = \sum_{b_j} \left| \sum_{x_i \in b_j} P[x_i]\Phi[x_i] \right|$$

$$\leq \sum_{b_j} \left| \sum_{x_i \in b_j} P[x_i] \right| = \sum_{x_i} P[x_i] ,$$

we see that for any kind of blocking $|\langle\Phi'\rangle| \geq |\langle\Phi\rangle|$. Furthermore, since $\langle\Phi'^2\rangle = \langle\Phi^2\rangle = 1$, we conclude that

$$\sigma'^2 \leq \sigma^2 \qquad (2.8)$$

(i.e., *the signal-to-noise ratio is always improved upon blocking configurations together*). Based on similar arguments, it is then straightforward to show that it is most advantageous to form blocks that mostly have the same sign. On the other hand, the worst blocking one could possibly choose would be to group the configurations into two blocks, one with positive sign and the other with negative sign. In this case, blocking yields no improvement whatsoever [i.e., "$\leq$" becomes "$=$" in Eq. (2.8)]. The same arguments carry over to the real-time case where Φ is an arbitrary complex phase factor [25, 31]. Along a similar line of reasoning, a partial path summation scheme for numerical real-time simulations was suggested by Winterstetter and Domcke [32].

For historical reasons, let us relate these ideas to earlier approaches to the dynamical sign problem. For the sake of concreteness, consider the Green's function for a system described by the generalized (continuous or discrete) coordinate x:

$$G(x_f, t_f; x_i, t_i) = \int_{x(t_i)=x_i}^{x(t_f)=x_f} \mathcal{D}x(t)e^{iS[x(t)]/\hbar} , \qquad (2.9)$$

where S is the usual action. Let us now introduce some reference path $\bar{x}(t)$ subject to the boundary conditions $\bar{x}(t_i) = x_i$ and $\bar{x}(t_f) = x_f$. In terms of $\bar{x}$ and fluctuations y, which obey the boundary conditions $y(t_i) = y(t_f) = 0$,

$$G(x_f, t_f; x_i, t_i) = e^{iS[\bar{x}(t)]/\hbar} \int_0^0 \mathcal{D}y(t)e^{i\delta S[\bar{x}(t);y(t)]/\hbar} , \qquad (2.10)$$

with

$$\delta S[\bar{x}(t); y(t)] \equiv S[\bar{x}(t) + y(t)] - S[\bar{x}(t)] . \qquad (2.11)$$

We can do this for any reference path $\bar{x}(t)$ fulfilling the correct boundary conditions. Summing over all possible reference paths $\bar{x}$ (which are then renamed to x), one finds that [25]

$$G(x_f, t_f; x_i, t_i) = \int \mathcal{D}x(t)e^{iS[x(t)]/\hbar}D[x(t)] . \qquad (2.12)$$

Here the *filtering function* $D[x]$ is obtained from summing over all fluctuations from the reference path $x(t)$:

$$D[x(t)] = \int_0^0 \mathcal{D}y(t)e^{i\delta S[x(t);y(t)]} . \qquad (2.13)$$

Although Eq. (2.12) differs from the original formula (2.9) only by the filtering function $D[x]$, it is important to point out that $D[x]$ is *not* necessarily identically unity for every $x(t)$. The practical usefulness of this formulation lies in the availability of certain analytical approximants for D.

For a continuous system coordinate, one can now derive from Eq. (2.13) the stationary-phase Monte Carlo (SPMC) method put forward originally by Doll and co-workers [15, 16]. Restricting fluctuations around $x(t)$ to small values by inserting a Gaussian envelope into Eq. (2.13), we have the lowest-order filtering function

$$
\begin{aligned}
D_0[x] &= \int \mathcal{D}y \exp\left[-\frac{1}{2\varepsilon_0^2} \int dt y^2(t) + i \int dt y(t) \frac{\delta S[x]}{\delta x(t)} \right] \\
&= \exp\left[-\frac{\varepsilon_0^2}{2} \int dt \left(\frac{\delta S[x]}{\delta x(t)} \right)^2 \right],
\end{aligned}
\tag{2.14}
$$

where we have taken a time-independent variance for the Gaussian envelope and Taylor-expanded the difference δS defined in Eq. (2.11). In the second equality, the Gaussian functional integration has been performed, and the resulting expression clearly shows that only regions where the phase is approximately stationary are important to the integral in Eq. (2.12). Therefore, the filtering function D_0 forms the basis for a biased Metropolis walk [15]. In this scheme the SPMC method is seen to be a special case of the general blocking strategy. Blocking corresponds to summing over paths under a Gaussian envelope in the local vicinity of a given configuration, rendering information from this configuration somewhat nonlocal.

A similar blocking idea is easily extended to deep-tunneling systems, where x is now a quasidiscrete coordinate. In the Trotter-discretized version, a given path $x(t)$ corresponds to a "spin" configuration. A possible blocking scheme could then take into account the local fluctuations around a given spin configuration by summing all configurations with, say, one spin flipped as compared to the reference configuration. This would result in the first-order fluctuation scheme of Ref. 25. More generally, one could define configurations with k spins flipped as the kth-order fluctuations and then sum up all fluctuations analytically up to a certain order to form the filtering function. The filtering function would then read

$$
D_k[x] = \sum_{j=0}^{k} \varepsilon^j \sum_{y_i} e^{i\delta S[x; y_i]},
\tag{2.15}
$$

where ε is a suitable expansion parameter. This filter has to be computed for every trial move during the Metropolis walk, and to be practical, one is restricted to considering some small value for k (typically, $k = 1$ or 2). There is a notable difference between this discrete filtering method and the conventional SPMC technique. While SPMC methods generally introduce some systematic error when $\varepsilon_0 > 0$, the discrete filtering becomes exact for $\varepsilon = 1$. In this limit, the filtering function merely produces an oversampling of configuration space, which does relieve the sign problem to some extent.

At this point we should note that neither the discrete nor the conventional version of the SPMC filter was used in the work that we review in Section IV. In fact, for the problems discussed below, it is possible to sum up a much larger subset of configurations analytically by exploiting certain symmetry properties. The resulting blocking scheme based on this exact analytical summation is what we referred to earlier as the *optimized* filter. This technique is discussed in great detail in the following section and in Refs. 27 and 29.

III. REAL-TIME QMC FOR SPIN-BOSON DYNAMICS

In this section we present a specific application of the general blocking strategy outlined above. We confine ourselves to the tunneling dynamics of a quantum particle in a condensed-phase environment, in the framework of spin-boson models. Since the seminal work of Caldeira and Leggett [33], system-plus-reservoir models have generated a lot of interest and can be considered the paradigm for the proper inclusion of dissipation into quantum mechanics. These models have been applied to a large variety of processes in chemistry, physics, and biology [22]. We focus on the case of a discrete system coordinate, as is appropriate in the low-temperature deep-tunneling regime. Detailed studies have been carried out for the cases of $N = 2$, 3, and ∞ available states for the tunneling particle. Popular applications for such models are found in the fields of macroscopic quantum coherence phenomena [34], tunneling of interstitials in solids [35], and more generally, interaction of a tunneling degree of freedom with the relevant elementary excitations in the system, such as phonons, excitons, or conduction electrons. Specific applications of these models to chemical physics and biophysics are given in Section IV.

In the rest of this section we give an in-depth description of how the optimized blocking is done for spin-boson systems. To make the algorithm more accessible to those readers who wish to implement similar methods, we have provided all the necessary details in this section. For those readers who do not wish to go through the mathematics in detail, we give a very concise overview of the essential ideas now.

In spin-boson models, the tunneling particle is described by a tight-binding picture and the solvent by a Gaussian fluctuating bath. Our QMC simulation starts by expressing a dynamical quantity of interest in the form of Eq. (2.2). The forward- and reverse-time propagators are then discretized in the usual way. The first crucial point to make is that because the bath is harmonic (although infinite in dimensionality), it can be traced out analytically. This constitutes the first stage of the blocking strategy. After the bath has been traced out the problem reduces to an isomorphic spin system dressed with complex interactions that are nonlocal in time. This first stage of blocking already provides immense savings over the explicit Monte Carlo sampling of all the bath modes. Unfortunately, taken alone it proves to be insufficient in almost all cases in its ability to circumvent the sign problem, except for very short times. Further blocking is therefore necessary to achieve a reasonable signal-to-noise level for practical calculations.

The next stage of blocking eliminates half of the remaining degrees of freedom from the problem. This stage proves to be crucial to the sign problem and allows one to go out to substantially longer real times. It constitutes the optimized blocking strategy referred to previously. The basis for this second stage of blocking is again related to the harmonic nature of the bath. The blocking starts from forming linear combinations of the forward- and reverse-time paths into a symmetric (quasiclassical) component and an antisymmetric (quantum fluctuation) component. One observes then that the quasiclassical path is linearly coupled to the quantum fluctuations but not to itself. In other words, for every quantum fluctuation path, a trace can be performed analytically over all possible quasiclassical paths. This immediately reduces the number of variables in the system immensely, and the QMC sampling is now performed over the remaining quantum fluctuations alone. As we will demonstrate by explicit examples in Section IV, the optimized blocking strategy enables many QMC calculations to be performed up to time scales that are of interest to chemically relevant problems.

A. Spin-Boson Models

Next we present the dynamical simulation techniques for the kinetics of a dissipative particle with a finite number of accessible discrete states. For simplicity, the discussion is restricted to the case of two states, $N = 2$. The generalization to any finite N is completely straightforward, and for the case $N = \infty$ one may find the necessary modifications in Ref. 29. For $N = 2$, we have the often-studied spin-boson Hamiltonian [34]

$$
\begin{aligned}
H = {} & H_0 + H_B + H_I \\
= {} & -\frac{\hbar\Delta}{2}\sigma_x + \frac{\hbar\varepsilon}{2}\sigma_z + \sum_\alpha \left[\frac{p_\alpha^2}{2m_\alpha} + \frac{1}{2}m_\alpha\omega_\alpha^2 \left(x_\alpha - \frac{C_\alpha}{m_\alpha\omega_\alpha^2}\sigma_z \right)^2 \right].
\end{aligned} \tag{3.1}
$$

Here H_0 describes the bare two-state system having an intrinsic tunneling matrix element Δ [in electron transfer (ET) theory, this is twice the electronic coupling between the two sites] and an external bias ε, which introduces an asymmetry between the two localized energy levels (in ET theory, this is the difference in the potentials of two redox states), and σ_x and σ_z are Pauli matrices. The Gaussian environment is described by harmonic oscillators $\{x_\alpha\}$ which are bilinearly coupled to the spin operator σ_z. These oscillators need not correspond to real physical modes. They might be effective harmonic degrees of freedom representing the Gaussian statistics of some collective macroscopic bath polarization. Such is the case for many electron transfer reactions, where individual microscopic modes might be highly nonlinear, but the overall solvent influence can still be accurately modeled by a harmonic oscillator bath [36].

Within this model, the bath is completely specified in terms of the spectral density,

$$J(\omega) = \frac{\pi}{2} \sum_\alpha \frac{C_\alpha^2}{m_\alpha \omega_\alpha} \delta(\omega - \omega_\alpha) , \qquad (3.2)$$

which has a quasicontinuous form in most real-life situations. The spectral density then determines the bath correlation function [36]:

$$L(z) = \frac{a^2}{\pi \hbar} \int_0^\infty d\omega \, J(\omega) \frac{\cosh[\omega(\hbar\beta/2 - iz)]}{\sinh[\omega\hbar\beta/2]} , \qquad (3.3)$$

defined here for complex-valued times $z = t - i\tau$. The distance between the localized states is given by a length scale a, and $\beta = 1/k_B T$. Of great practical importance is the ohmic spectral density,

$$J(\omega) = (2\pi\hbar K/a^2)\omega e^{-\omega/\omega_c} . \qquad (3.4)$$

This spectral density has a characteristic low-frequency behavior $J(\omega) \sim \eta\omega$, where η is the usual ohmic viscosity. The system–bath coupling strength can then be measured in terms of the dimensionless Kondo parameter K, and time scale of bath motions is described by a cutoff frequency ω_c. For many problems in low-temperature physics, this cutoff frequency is taken to be the largest frequency scale in the problem. In the case of electron transfer, the same spectral density with some intermediate value for ω_c is most appropriate for a realistic description of many polar solvents [37]. Our method is, however, by no means restricted to ohmic damping. Numerical examples for colored noise can be found in Ref. 24.

Two dynamical quantities of interest for this model are the symmetrized time correlation function

$$C(t) = \mathrm{Re}\langle\sigma_z(0)\sigma_z(t)\rangle$$
$$= Z^{-1}\,\mathrm{Re}\,\mathrm{Tr}(e^{-\beta H}\sigma_z e^{iHt/\hbar}\sigma_z e^{-iHt/\hbar})\,,$$

(3.5)

with $Z = \mathrm{Tr}\,e^{-\beta H}$, and the time-dependent occupation probabilities for the two sites $P_+(t)$ and $P_-(t)$, which can be expressed in terms of a single function

$$P(t) = P_+(t) - P_-(t) = \langle e^{iHt/\hbar}\sigma_z e^{-iHt/\hbar}\rangle$$

(3.6)

with the initial condition $P(0) = 1$. From a comparison of Eqs. (3.5) and (3.6) we observe that the two quantities differ only by the way the system is prepared initially. For $P(t)$, the system is held fixed in the $+$ state until $t = 0$, with the bath being unobserved (factorized initial condition). On the other hand, the more realistic situation for many experiments is represented by equilibrium states of the total system at $t = 0$ as described by $C(t)$.

B. Discretization of the Path Integral

To compute the two-state dynamics numerically, we employ a discretized path-integral representation of the dynamical quantities [23, 24, 38]. The correlation function $\langle\sigma_z(0)\sigma_z(t)\rangle$ can be regarded as the probability amplitude for a sequence of steps in the complex-time plane. In particular, one propagates along the Kadanoff–Baym contour γ defined in Fig. 1: $z = 0 \to t \to 0 \to -i\hbar\beta$ and measures σ_z at $z = 0$ and $z = t$. Of

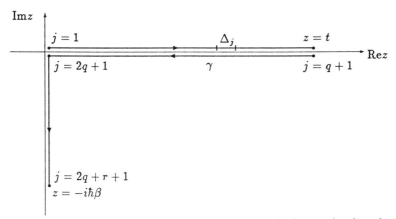

Figure 1.   Discretization of the Kadanoff–Baym contour γ in the complex-time plane.

course, there are several other possible choices of this contour due to the cyclic structure of the trace. For the quantity $P(t)$, one simply omits the imaginary-time branch (the resulting contour is the Feynman–Vernon contour). Clearly, if one is interested in the zero-temperature limit for $C(t)$, the method would become impracticable. In that limit one should use the Keldysh contour instead of the Kadanoff–Baym, contour. For the Keldysh contour the imaginary time branch is replaced by two real-time paths extending to $t_0 \to -\infty$. However, this approach has not been tested in numerical implementations so far.

To parametrize the paths, we use q uniformly spaced discrete points for each of the two real-time paths and r points for the imaginary-time path (see Fig. 1). Hence there are $M = 2q + r$ points in total. The time discretizations are

$$
\Delta_j = \begin{cases}
t/q , & 1 \le j \le q , \\
-t/q , & q + 1 \le j \le 2q , \\
-i\hbar\beta/r , & 2q + 1 \le j \le 2q + r \equiv M ,
\end{cases}
\tag{3.7}
$$

and the complex time after i steps is $z_i = \sum_{j=1}^{i-1} \Delta_j$. The construction of the discretized path integral then proceeds by inserting complete sets

$$
\begin{aligned}
1_i &= \sum_{\sigma_i = \pm 1} \int \prod_\alpha dx_{\alpha,i} |\sigma_i, \{x_{\alpha,i}\}\rangle\langle\sigma_i, \{x_{\alpha,i}\}| \\
&= \int d\mathbf{r}_i |\mathbf{r}_i\rangle\langle\mathbf{r}_i|
\end{aligned}
\tag{3.8}
$$

at each discretization point z_i ($i = 2, \ldots, M$). The vector $\mathbf{r}_i$ represents the state $\sigma_i = \pm 1$ of the two-level system as well as the environmental degrees of freedom $\{x_{\alpha,i}\}$.

To disentangle the short-time propagator, we use a (symmetrized) Trotter formula,

$$
\exp(-iH\Delta_j/\hbar) = \exp(-iH'\Delta_j/2\hbar) \exp(-iH_0\Delta_j/\hbar)
$$

$$
\times \exp(-iH'\Delta_j/2\hbar) + \mathcal{O}(\Delta_j^3[H_0, H']) ,
$$

where $H' = H - H_0$ is diagonal in the representation (3.8). The error in this split-up vanishes in certain limits, as discussed in Ref. 12. The free path H_0 leads to the short-time propagator

$$
K(\sigma_j, \sigma_{j+1}) = \langle\sigma_{j+1}| \exp(-i\Delta_j H_0/\hbar)|\sigma_j\rangle ,
\tag{3.9}
$$

which can be evaluated exactly. Thus we arrive at

$$\langle \sigma_z(0)\sigma_z(t')\rangle = \frac{\int d^N \mathbf{r} e^{S[\mathbf{r}_1, \ldots, \mathbf{r}_M]}\sigma'\sigma''}{\int d^N \mathbf{r} e^{S[\mathbf{r}_1, \ldots, \mathbf{r}_M]}}, \qquad (3.10)$$

which converges to the required continuum limit as the number of discretizations $M \to \infty$. The discretized action $S[\mathbf{r}_1, \ldots, \mathbf{r}_M]$ is a complex-valued sum of the actions picked up in separate parts of the contour γ. To compute the correlation function in Eq. (3.10), one can average over all pairs of spins $\{\sigma', \sigma''\}$ separated by a time t' along the contour γ. This allows us to compute the dynamical quantities for all times $t_k = kt/q$ (where $k = 0, \ldots, q$) from a single Metropolis trajectory. Because of the cyclic structure of the trace in Eq. (3.5), $\mathbf{r}_{M+1} \equiv \mathbf{r}_1$.

Since the bath is made up of harmonic oscillators, it can be eliminated analytically (provided that one is only interested in the reduced density matrix of the reaction coordinate). After performing this elimination, the bath-plus-coupling part H' of the Hamiltonian leads to an *influence functional* $\Phi[\sigma]$ in terms of the spins $\{\sigma_i\}$ alone [38, 39], and the correlation function takes the form

$$\langle \sigma_z(0)\sigma_z(t')\rangle = \frac{1}{Z}\sum_{\{\sigma\}}\exp\left(-\Phi[\sigma] + \sum_i \ln K(\sigma_i, \sigma_{i+1})\right)\sigma'\sigma'', \qquad (3.11)$$

where $Z = \Sigma_{\{\sigma\}} \exp(\cdots)$ (the exponent will be referred to as "the action" henceforth). In the continuum limit $M \to \infty$, the nonlocal influence functional is given in terms of the bath correlation function $L(z)$ introduced in Eq. (3.3), and one finds with $\sigma_i \to \sigma(z)$ [40]

$$\Phi[\sigma] = \int_\gamma dz \int_{z' < z} dz'\, \sigma(z)L(z - z')\sigma(z')/4 . \qquad (3.12)$$

The integrations in the complex-time plane are ordered along the Kadanoff–Baym contour γ. Note that $L(z)$ fulfills the important symmetry relation

$$L(z - i\hbar\beta) = L(-z), \qquad (3.13)$$

which implies certain symmetry properties of the influence functional. This symmetry relation is also responsible for the detailed balance relation between forward and backward rates in such a system.

In discretized form the influence functional is then given by

$$\Phi[\sigma] = \frac{1}{2} \sum_{j,k=1}^{N} \sigma_j L_{jk} \sigma_k / 4 , \tag{3.14}$$

with the complex-valued influence matrix L_{jk}. We observe from Eqs. (3.11) and (3.14) that the (discretized) spin-boson problem is isomorphic to a classical one-dimensional Ising model with long-range complex-valued interactions [23, 38]. The critical behavior at zero temperature found by Chakravarty [41] is a consequence of the algebraic nature of the long-range spin interactions at $T = 0$. At finite temperature, however, there is no phase transition or ordered phase (localization).

For the numerical approach, the influence matrix is taken as the average value of the interaction $L(z)$ between two points z_j and z_k:

$$L_{jk} = L_{kj} = \int_{C_j} dz_j' \int_{C_k} dz_k' \, L(z_j' - z_k') \qquad \text{(for } j > k\text{)} , \tag{3.15}$$

where C_i is one discretization on the contour centered at point z_i,

$$\{z \in C_i | z_i - \Delta_{i-1}/2 < z < z_i + \Delta_i/2\} .$$

The remaining time integrations in Eq. (3.15) can be carried out easily, and one finds with $\Delta_{jk} = z_j - z_k$,

$$L_{jk} = Q(\Delta_{jk} + (\Delta_j + \Delta_{k-1})/2) + Q(\Delta_{jk} + (-\Delta_{j-1} - \Delta_k)/2)$$
$$- Q(\Delta_{jk} + (-\Delta_{j-1} + \Delta_{k-1})/2) - Q(\Delta_{jk} + (\Delta_j - \Delta_k)/2) ,$$

where $Q(z)$ is the twice-integrated bath correlation function with $Q(0) = 0$ [i.e., $d^2 Q(z)/dz^2 = L(z)$]. Of course, this function exhibits the same symmetry property (3.13). Finally, the diagonal elements are given by

$$L_{jj} = 2Q((\Delta_{j-1} + \Delta_j)/2) ,$$

which is only of relevance for $N > 2$ (since $\sigma_i^2 = 1$ for the two-state problem).

C. Elimination of Quasiclassical Degrees of Freedom

Since the action for each spin path is complex valued, a stochastic Monte Carlo evaluation of the resulting isomorphic Ising chain has to deal with the inevitable dynamical sign problem. Relying on the strategy outlined in Section II, we wish to block configurations together in an optimal way.

This amounts to taking care, as much as possible, of the sign problem in a nonstochastic way. This is achieved here upon switching to a new spin representation (which is also of much use in analytical calculations [42,43]). Similar to previous numerical treatments [44, 45], We introduce the sum and difference coordinates of the forward (σ_j) and backward (σ'_j) real-time spin paths and rename the imaginary-time spins ($\bar{\sigma}_m$),

$$\eta_j = (\sigma_j + \sigma'_j)/2 \,,$$

$$\xi_j = (\sigma_j - \sigma'_j)/2 \,, \qquad (3.16)$$

$$\bar{\sigma}_m = \sigma_{2q+m} \,,$$

where $\sigma'_j = \sigma_{2q+2-j}$ in the old notation ($j = 1, \ldots, q+1$). Thus we first relabel the spins on the three pieces of the Kadanoff–Baym contour and then build the said linear combinations. A simple physical meaning can be attached to these new spins: $\{\eta_j\}$ describe the propagation along the diagonal of the reduced density matrix and can thus be identified with *quasiclassical paths*, while $\{\xi_j\}$ describe the off-diagonaliticity of the reduced density matrix and can be identified with *quantum fluctuations* [43]. According to the definitions (3.16), the new spins can take on three possible values $\xi, \eta = -1, 0, 1$ (for general N, there are $2N + 1$ possible values), but they are not independent variables because either ξ_j or η_j has to be 0 for the same j.

Written in terms of the new spins, after some algebra the influence functional takes the form [27]

$$\Phi[\xi, \eta, \bar{\sigma}] = \frac{1}{2} \sum_{m,n=2}^{r} \bar{\sigma}_m Y_{mn} \bar{\sigma}_n + \frac{1}{2} \sum_{j,k=1}^{q} \xi_j \Lambda_{jk} \xi_k$$

$$+ i \sum_{j>k=1}^{q} \xi_j X_{jk} \eta_k + \sum_{j=1}^{q} \sum_{m=2}^{r} \xi_j Z_{jm} \bar{\sigma}_m \qquad (3.17)$$

$$+ \eta_1 \left(\sum_{m=2}^{r} \bar{\sigma}_j [L_{2q+m,1} + L_{2q+m,2q+1}] \right).$$

The elements of the matrices appearing in Eq. (3.17) are

$$Y_{mn} = L_{2q+m,2q+n}/4$$

$$\Lambda_{jk} = \text{Re } L_{jk}$$

$$X_{jk} = \text{Im } L_{jk}$$

$$Z_{jm} = L_{j,2q+m}/2 \, .$$

It is worth mentioning that these matrices, with the exception of Z_{jm}, are all real valued. The meaning of the five terms in Eq. (3.17) is as follows. The first term describes a self-interaction within the imaginary-time segment. Due to the second term, the dissipative bath will damp out the quantum fluctuations, and the system is likely to be found in a diagonal state characterized by $\xi = 0$. This point will later be important with regard to the choice of a suitable Monte Carlo weight. The third term is a bilinear interaction between quasiclassical paths and quantum fluctuations, and the fourth term describes a similar interaction between the imaginary-time spins and the quantum fluctuations. The last term is a preparation term, coupling η_1 to the imaginary-time spins. Note that only the second and third terms show up in a computation of $P(t)$. Remarkably, there is no self-interaction in the quasiclassical paths, and they are coupled only linearly to the other degrees of freedom.

This observation is absolutely crucial for our computational procedure, since it allows for an exact treatment of the quasiclassical paths. *For any given quantum fluctuation path, the path summation over all allowed quasiclassical paths can be carried out in an exact manner*; hence the blocking idea has been realized in an efficient way. To elucidate this, we first examine the free action due to H_0. The imaginary-time contribution can be put into the matrix elements Y_{mn} simply by adding $\frac{1}{2} \ln \tanh(\hbar\beta\Delta/2r)$ to $Y_{m,m+1}$ and $Y_{m+1,m}$; in case an external bias ε is present, the action has to include an additional term $(\hbar\beta\varepsilon/2r) \Sigma_m \bar{\sigma}_m$ [24]. Regarding the real-time paths, we proceed in a different way. For the isolated two-state system, we have (in terms of the original spins) a product of the form [cf. Eq. (3.9)]

$$\prod_{j=1}^{q} K(\sigma_j, \sigma_{j+1}) K^*(\sigma_j', \sigma_{j+1}') \, . \tag{3.18}$$

If we now switch to the $\{\xi, \eta\}$ representation and perform the summation over all η spins (while keeping the ξ configuration frozen), we obtain a *matrix product* for Eq. (3.18). Of course, one has to account for the η-dependent terms in the influence functional (3.17) during this procedure. In the end, the complex-valued contribution of all these terms for a given ξ configuration takes the form

$$\mathscr{I}[\xi] = \sum_{\eta_1 = 0, \pm 1} \sum_{\eta_{q+1} = 0, \pm 1} \langle \eta_1 | \mathbf{V}^{(1)} \cdots \mathbf{V}^{(q)} | \eta_{q+1} \rangle \, , \tag{3.19}$$

where the (3×3) matrices $\mathbf{V}^{(j)}[\xi]$ are defined by [in the general case, these are $(2N - 1) \times (2N - 1)$ matrices]

$$\langle \eta_j | \mathbf{V}^{(j)} | \eta_{j+1} \rangle = [K \times K^*](\eta_j, \eta_{j+1}, \xi_j, \xi_{j+1}) \exp\left(-i\eta_j \sum_{k>j}^{q} X_{kj}\xi_k\right).$$

(3.20)

There is a slight modification of $\mathbf{V}^{(1)}$ due to the fifth term in (3.17), as described in Ref. 27. Each of the matrices $\mathbf{V}^{(j)}$ depends on all ξ_k spins with $k \geq j$; however, the "free" part $K \times K^*$ is determined by ξ_j and ξ_{j+1} alone. Clearly, $\mathscr{I}[\xi]$ can be evaluated with a simple matrix multiplication routine, leading to a numerically exact and efficient treatment of the quasiclassical paths. Note that the remaining part of the influence functional is real valued—with the exception of the fourth term in Eq. (3.17), which is generally very small—indicating that much of the dynamical sign problem has been relieved by treating the numerically problematic quasiclassical paths in an exact manner.

In effect, the factor $\mathscr{I}[\xi]$ contains all contributions from H_0 and, in addition, the third and fifth terms of the influence functional (3.17). The correlation function can thus be written as

$$\langle \sigma_z(0)\sigma_z(t') \rangle = \frac{1}{Z} \sum_{\{\xi\}} \sum_{\{\bar{\sigma}\}=\pm 1} \mathscr{I}[\xi] \exp\left(\frac{1}{2} \sum_{m,n=2}^{r} \bar{\sigma}_m Y_{mn} \bar{\sigma}_n\right.$$
$$\left. -\frac{1}{2} \sum_{j,k=1}^{q} \xi_j \Lambda_{jk} \xi_k - \sum_{j=1}^{q} \sum_{m=2}^{r} \xi_j Z_{jm} \bar{\sigma}_m\right) \sigma' \sigma''.$$

Therefore, we are left with the task of summing over the imaginary-time spins $\{\bar{\sigma}_m\}$ and the quantum fluctuations $\{\xi_j\}$, which is conveniently done via a Monte Carlo (MC) sampling.

D. Monte Carlo Sampling

The suitable MC weight for the imaginary-time spins is straightforward,

$$\mathscr{P}_{\text{imag}}[\bar{\sigma}] \sim \exp\left(-\frac{1}{2} \sum_{m,n} \bar{\sigma}_m Y_{mn} \bar{\sigma}_n - \text{Re} \sum_{j,m} \xi_j Z_{jm} \bar{\sigma}_m\right).$$

Since the influence functional forces the quantum fluctuations to stay near the diagonal of the density matrix, we first try to use the following MC weight for the quantum fluctuations:

$$\mathscr{P}_{\text{real}}[\xi] \sim \exp\left(-\frac{1}{2} \sum_{j,k} \xi_j \Lambda_{jk} \xi_k - \text{Re} \sum_{j,m} \xi_j Z_{jm} \bar{\sigma}_m\right).$$

The problem with this weight, however, arises for small system–bath couplings where the damping of the quantum fluctuations becomes weak. In this case, the sampling would become very inefficient for small system–bath coupling. Hence, in a next step, we try the product

$$W[\xi] \sim \mathscr{P}_{\text{real}}[\xi] \times |\mathscr{I}[\xi]| ,$$

where $\mathscr{I}[\xi]$ has been defined in Eq. (3.19). This weight function considers both the damping of the quantum fluctuations due to the influence functional *and* the integrated-out quasiclassical paths.

Unfortunately, there is another problem with this weight [29]. This problem arises since for correlation functions such as $\langle \sigma_z(0)\sigma_z(t) \rangle$ one has to compute the ratio of two quantities. It turns out that certain spin configurations only contribute to the denominator but not to the numerator (and vice versa). Due to this exclusivity problem, one will not be able to access all relevant spin configurations $\{\xi\}$ contributing to the numerator when using $W[\xi]$ alone as the Monte Carlo weight. We can circumvent this problem using standard ideas of importance sampling [46]. First, we observe that for the numerator one has to compute such terms as [where α_1, $\alpha_2 = \pm$]

$$\mathscr{I}^{(k)}_{\alpha_1,\alpha_2}[\xi] = \sum_{\eta_1=0,\pm 1} \sum_{\eta_{q+1}=0,\pm 1} \langle \eta_1 | H_{\alpha_1}(\xi_1)\mathbf{V}^{(1)} \cdots \mathbf{V}^{(k-1)}$$

$$\times H_{\alpha_2}(\xi_k)\mathbf{V}^{(k)} \cdots \mathbf{V}^{(q)} | \eta_{q+1} \rangle . \tag{3.21}$$

The projection operators $H_+ = (1 + \sigma_z)/2$ and $H_- = (1 - \sigma_z)/2$ onto the two spin values have the η-representation (for a given ξ)

$$H_+(\xi = 0) = \begin{pmatrix} 1 & 0 & 0 \\ 0 & 0 & 0 \\ 0 & 0 & 0 \end{pmatrix} , \qquad H_-(\xi = 0) = \begin{pmatrix} 0 & 0 & 0 \\ 0 & 0 & 0 \\ 0 & 0 & 1 \end{pmatrix} ,$$

$$H_\pm(\xi = \pm 1) = \begin{pmatrix} 0 & 0 & 0 \\ 0 & \frac{1}{2} & 0 \\ 0 & 0 & 0 \end{pmatrix} .$$

Finally, an appropriate positive-definite Monte Carlo weight function can be constructed:

$$\tilde{W}[\xi] \sim \mathscr{P}_{\text{real}}[\xi] \left(\sum_{k=1}^{q} \sum_{\alpha_1,\alpha_2=\pm} |\mathscr{I}^{(k)}_{\alpha_1,\alpha_2}| \right) . \tag{3.22}$$

Using $\tilde{W}[\xi]$ as the weight allows us to carry out an efficient Monte Carlo sampling of ξ spins.

Our QMC algorithm employs single-particle Metropolis moves as well as moves that allow kinks to translate along the spin chain [26]. Single-particle moves attempt to change one spin ξ_k to a new value, whereas kink moves attempt to change two adjacent spins simultaneously. The imaginary-time spins are samples from $\mathscr{P}_{\text{imag}}[\bar{\sigma}]$ using single-particle moves (i.e., one tries to flip a single spin $\bar{\sigma}_m$). During one MC pass, the single-particle moves are attempted once for every spin, and the kink moves are attempted for every pair of spins with $\xi_k \neq \xi_{k+1}$. Typical acceptance ratios for these types of moves are $\approx 15\%$, and we usually take samples separated by five MC passes. This ensures that the MC samples are sufficiently uncorrelated, since roughly half of the spins have been assigned new values between two subsequent samples. Numerical results are then obtained from several 10,000 to 100,000 samples, so that statistical errors are well below 5%. To ensure that the Trotter error is sufficiently small, one has to keep the discretization numbers q and r large enough. This is checked by systematically increasing these numbers until convergence is reached. For all results presented here, the Trotter error is small compared to the statistical errors due to the stochastic MC sampling.

IV. APPLICATIONS

To demonstrate the usefulness (and also the current limitations and challenges) of these real-time QMC simulations, we now give several examples. For more detailed information on these applications, we refer the interested reader to the original papers.

A. Dynamical Test of the Centroid Quantum Transition-State Theory for Electron Transfer Reactions

Electron transfer (ET) reactions are ubiquitous in many chemical and biological processes [47]. All redox reactions occur by the tunneling of an electron from the electronic state localized on the donor to the one localized on the acceptor. However, the ET rate is not determined by the electronic coupling (tunneling matrix element) alone. Due to the often strong interaction between the electron and the solvent, ET requires a specific large-scale reorganization of the environment, usually achieved through equilibrium fluctuations in the solvent degrees of freedom. The conventional and very successful Marcus ET theory [47] incorporates the environmental fluctuations using linear response theory to describe the solvent polarization. In fact, the Marcus model is completely equivalent to the spin-boson Hamiltonian (3.1).

Within this model, the free-energy surfaces are parabolic in the coordinates that describe the motion of the environment, and as a consequence of the Franck–Condon principle, ET can occur only at the intersection of the reactant and product free-energy surfaces. The reaction rate is first determined by the free-energy change necessary for the environmental rearrangement to get to the intersection region and then by the electronic coupling Δ. For large electronic coupling the reaction is adiabatic and the ET rate can be computed classically, whereas for small electronic coupling the process is nonadiabatic and Fermi's golden rule holds [48]. Whether a process is adiabatic or nonadiabatic depends on the relative magnitude of the electronic coupling and the typical frequency of the solvent motions [i.e., the ratio Δ/ω_c for the case of spectral density (3.4)].

Most practical schemes for computing *classical* reaction rates make use of a factorization of the reaction rate into an equilibrium factor and a dynamical factor [49, 50]. A similar factorization of *quantum* reaction rates into an equilibrium centroid density and a dynamic factor has been proposed by Voth and co-workers (VCM) [51] based on ideas suggested earlier by Gillan [52]. The basic assumption of their quantum transition-state theory (QTST) is that the quantum rate is governed by the quantum flux evaluated for the transition state for which the centroid of the imaginary-time path is at its highest free-energy position. Using the centroid of the electron path, this concept seems to work at least in two limits: (1) in the nonadiabatic limit the QTST approach proves to be equivalent to the golden rule formula [26]; and (2) in the context of an equilibrium theory, QTST allows one to extract seemingly correct results in both the nonadiabatic and adiabatic limits for the case of a sluggish bath [53].

From an analysis of a double-well system, Gillan has observed that the equilibrium density of the centroid of the imaginary-time path for the reaction coordinate $q(\tau)$,

$$q_0 = (\hbar\beta)^{-1} \int_0^{\hbar\beta} d\tau \, q(\tau) \,,$$

could be used to obtain a good approximation for the quantum rate [52]. VCM have extended this idea in their general factorization scheme of the quantum rate expression [51]

$$x_A k(\bar{t}) = \int dq \, \rho(q)\nu(q;\bar{t}) \,, \qquad (4.1)$$

where $x_A = Q_A/Q$, $Q = \mathrm{Tr}\, e^{-\beta H}$, and $Q_A = \mathrm{Tr}(h_A e^{-\beta H})$ with a reactant projection operator h_A. Furthermore, $\rho(q)$ is the centroid density and $\nu(q; \bar{t})$ is a dynamical factor, with $\bar{t}$ denoting the plateau time [36, 49]. In Eq. (4.1), the factorization emphasizes the role of the equilibrium density for finding the centroid at position q,

$$\rho(q) = Q^{-1} \int \mathcal{D}q(\tau)\delta(q - q_0) \exp(-S_E[q(\tau)]) , \qquad (4.2)$$

where S_E is the action of the imaginary-time path alone. The remaining real-time path integrals are lumped into a dynamical factor $\nu(q; \bar{t})$. Based on an analysis of the inverted parabolic barrier, VCM argued that at the transition state q^*, defined to be the position where $\rho(q)$ has a minimum, the dynamical factor $\nu(q; \bar{t})$ would have a sharp maximum. Therefore, the rate would be dominated by $\rho(q^*) \times \nu(q^*; \bar{t})$ at the transition state if $\nu(q; \bar{t})$ decays away from q^* faster than $\rho(q)$ increases.

In the spin-boson model, the (electronic) reaction coordinate is represented by a discrete variable $q(\tau) = \sigma(\tau) = \pm 1$ and the centroid is restricted to the range $-1 \le q_0 \le +1$. Surprisingly, the electron flux, which is the product $\rho(q) \times \nu(q; \bar{t})$, is generally not sharply peaked around the transition state for this model. Indeed, in the nonadiabatic limit it can be shown rigorously [26] that this product is *constant for all values of q* instead of being sharply peaked at q^*. If the quantum flux proves to be invariant for all q even as one goes away from the nonadiabatic limit, we are allowed to use the convenient QTST prescription for calculating the electron transfer rate again,

$$x_A k(\bar{t}) = \rho(q^*)\nu(q^*; \bar{t}) \int_{-1}^{+1} dq = 2\rho(q^*)\nu(q^*; t) . \qquad (4.3)$$

We refer to this as the *QTST estimate for the quantum electron transfer rate*. Dynamical simulations allow us to test whether this formula applies to ET reactions in general. Here we have employed the electronic coordinate $\sigma(\tau)$ to define the reaction coordinate and the centroid. An approach employing a nuclear reaction coordinate and centroid has been put forward in Ref. 37, leading to conclusions similar to those for the case of an electronic centroid.

We have studied the usefulness of this factorization in the crossover region between the nonadiabatic and adiabatic limits [26]. The starting point for these rate calculations is the exact quantum expression for the forward rate [54],

$$k(\bar{t}) = \frac{2}{\hbar\beta Q_A} \text{Im Tr}(h_A e^{-\beta H} h_B e^{-i\bar{t}H/\hbar} h_B e^{-i\bar{t}H/\hbar}), \qquad (4.4)$$

where h_A and $h_B = 1 - h_A$ are the population operators for the reactant and product states. The Hamiltonian H has been specified in Eq. (3.1), and Eq. (4.4) can be evaluated numerically following the method in Section III. We will see that the full quantum rate computed by QMC simulations is actually *larger* than the QTST estimate discussed above. This enhancement can be attributed to dynamical quantum effects which are missed by the centroid QTST formulation. The appropriate enhancement factor $\mathcal{L}_e$ is defined as the true rate (as observed in the simulations) divided by the QTST estimate. In general, this enhancement factor depends on the specific choice of the reaction coordinate, and since we are using the *electron* path $\sigma(\tau)$ as reaction coordinate, a subscript e has been added.

Numerical results are shown in Fig. 2. Since ET reactions usually involve strong system–bath coupling, we have taken an ohmic spectral density (3.4) with $K = 5$ and $\Delta/\omega_c = \frac{1}{10}$. In the classical limit, this would correspond to a system with a classical reorganization energy $\lambda = 100\Delta$. Taking $k_B T = \Delta$ and a symmetric system $\varepsilon = 0$, this corresponds to typical parameter values near the nonadiabatic limit. Hence the quantum flux is expected to be approximately independent of the electronic centroid. The free energy $F(q) = -\ln \rho(q)/\beta$ is shown in Fig. 2a, as computed from imaginary-time MC simulations employing umbrella sampling. The real-time simulation results for the dynamical factor are shown in Fig. 2b, and one finds a plateau after a short initial transient. In Fig. 2c, the plateau values are plotted as functions of the electronic centroid. This shows that the dynamical factor is indeed sharply peaked around the transition state $q^* = 0$. However, in Fig. 3 when we plot the product $\rho(q) \times \nu(q; \bar{t})$, which is proportional to the quantum flux, we observe that the flux is *not* peaked around $q^* = 0$ as conjectured by QTST. Also quite remarkably, even for such a small Δ/ω_c, the nonadiabatic expectation of a centroid-independent flux is not reproduced either. The flux increases dramatically when going away from the transition state, and two well-developed peaks at $q \approx \pm 0.75$ can be observed. These peaks are also predicted by a higher-order real-time perturbation theory [26], and similar behavior can be observed for an asymmetric system as well.

The appearance of these peaks is a purely dynamical effect not captured by QTST, and a simple estimate for the enhancement factor gives (for the simulation parameters considered here) $\mathcal{L}_e \approx 4$ from the

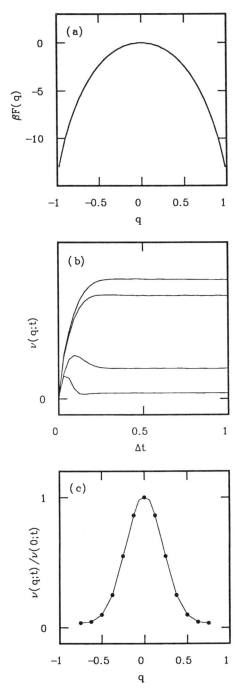

Figure 2. Monte Carlo results for a symmetric electron transfer system: (a) centroid free energy using 100 discretization points for the imaginary-time path; (b) dynamical factor for several values of the centroid q (from top to bottom, $q =$ 0, 0.125, 0.375, and 0.625); (c) plateau value as a function of q.

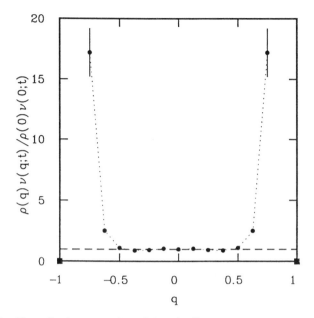

Figure 3. Normalized quantum flux $\rho(q) \times \nu(q;\bar{t})$. Vertical bars indicate one standard deviation. The dashed line denotes the QTST estimate; the dotted curve is a guide to the eye only. The squares indicate that the flux goes to zero at $q = \pm1$.

area under the peaks. Therefore, the true quantum rate is significantly enhanced over the QTST estimate, even though we have considered a parameter regime where QTST is expected to give excellent predictions. This shows that important quantum corrections to the rate are not captured by QTST, due to the neglect of certain dynamical tunneling effects in the bath. In addition, a direct application of QTST to the inverted tunneling regime is not possible when using the electronic centroid since $-1 \leq q_0 \leq 1$. A possible cure for this problem has been suggested by Stuchebrukhov and Song [55], who analytically continued the electronic centroid into the complex plane, similar to the treatment of the zero-mode integral in macroscopic quantum decay theories [50]. We wish to stress that using a proper nuclear centroid in the original QTST formulation does not seem to improve the situation either [37], and we are led to the conclusion that QTST is not always able to make accurate rate predictions for ET reactions. In a sense, the QTST centroid theory appears to be of limited use for electron transfer reactions in the low-temperature region, where it has to be regarded as an uncontrolled approximation.

B. Primary Charge Separation in Bacterial Photosynthesis

The problem of bacterial photosynthesis has attracted a lot of recent interest since the structures of the photosynthetic reaction center (RC) in the purple bacteria *Rhodopseudomonas viridis* and *Rhodobacterias sphaeroides* have been determined [56]. Much research effort is now focused on understanding the relationship between the function of the RC and its structure. One fundamental theoretical question concerns the actual mechanism of the primary ET process in the RC, and two possible mechanisms have emerged out of the recent work [28, 57–59]. The first is an incoherent two-step mechanism where the charge separation involves a sequential transfer from the excited special pair (P*) via an intermediate bacteriochlorophyll monomer (B) to the bacteriopheophytin (H). The other is a coherent one-step superexchange mechanism, with P^+B^- acting only as a virtual intermediate. The interplay of these two mechanisms can be studied in the framework of a general dissipative three-state model ($N = 3$).

From experimental studies to date, one cannot easily rule out either possibility. In fact, conflicting interpretations have been derived from different experiments. The detection of transient population in the P^+B^- state has been considered by many to be the key experimental evidence that can differentiate between the two mechanisms. Many transient absorption spectroscopic experiments [60, 61] could not detect bleaching in the B^- band, lending support to the coherent superexchange mechanism. On the other hand, some other experiments [62] point toward a two-step process since a fast second rate constant corresponding to the $P^+B^-H \rightarrow P^+BH^-$ transition was detected. The picture emerging from a comparison of the molecular dynamics (MD) simulations for the RC carried out so far [63, 64] is equally murky. Different groups arrive at conflicting conclusions.

However, MD simulation performed to data share one common point—they have all independently validated a three-state spin-boson model description of the RC. In effect, the protein environment can be described in terms of a harmonic oscillator bath with an approximately ohmic spectral density (3.4) where $\omega_c = 165\ \mathrm{cm}^{-1}$. The bare three-state Hamiltonian is

$$H_0 = \begin{pmatrix} E_1 & -K_{12} & -K_{13} \\ -K_{12} & E_2 & -K_{23} \\ -K_{13} & -K_{23} & E_3 \end{pmatrix}, \qquad (4.5)$$

where the electronic coupling (tunneling matrix element) between states i

and j is denoted by K_{ij}. In our simulations we generally take $K_{13} = 0$, $K_{23} = 4K_{12}$, and $E_3 = -2000 \, \text{cm}^{-1}$, with $E_1 = 0$ by convention. This parametrization, as well as the limitations for this model, are explained in great detail in our original papers [28]. The spectral density has a simple relationship with the familiar reorganization energy λ_{13} between states 1 and 3 in the classical limit, and one can characterize the bath completely by the two parameters ω_c and λ_{13}.

In the recent past we have carried out detailed simulations of the primary charge separation kinetics [28]. The time dependence of the occupation probabilities on the three relevant chromophores has been computed as a function of the P^+B^- energy for many combinations of the electronic coupling and the reorganization energy. The primary objective of these studies was to determine the parameter values that are required for the dynamics to be consistent with characteristic experimental observations. We consider the following to be key experimental characteristics: (1) the P^+B^- population (P_2) is small throughout ($\lesssim 20\%$); (2) the charge separation rate is roughly $(3 \, \text{ps})^{-1}$ at room temperature; and (3) the rate increases about twofold at cryogenic temperatures.

Taking our parametrization at face value (noting that there is still substantial disagreement in the community regarding the appropriate parameters for the wild-type RC), the simulations provide a map of the various possible parameter regions in which the RC could actually operate. Of course, due to the very limited knowledge one currently has about the relevant parameters, our conclusions might have to be modified once more detailed electronic structure calculations of the binding energies and electronic couplings become available. However, we believe that our findings are indeed of a generic nature and of importance to the real RC. The simulations reported below have been carried out up to a few picoseconds, the same time scale as the experimentally determined ET rate. The sign problem makes simulations at much longer time scales unstable.

We have found two distinct regions that yield charge separation dynamics with the correct experimental characteristics. The first is characterized by a P^+B^- state energy E_2 lower than about $-600 \, \text{cm}^{-1}$ and down to about $-1300 \, \text{cm}^{-1}$. Within this region, the dynamics largely agrees with predictions from conventional nonadiabatic theory for the sequential mechanism. Simulation results for the transient populations on the three electronic states at several temperatures are shown in Fig. 4 for $E_2 = -666 \, \text{cm}^{-1}$. We indeed observe dynamic characteristics largely in accord with the experiments. In this region the results can be rationalized in terms of two sequential activationless transfers [28]. The second region is characterized by a P^+B^- state energy E_2 higher than $+666 \, \text{cm}^{-1}$ (we

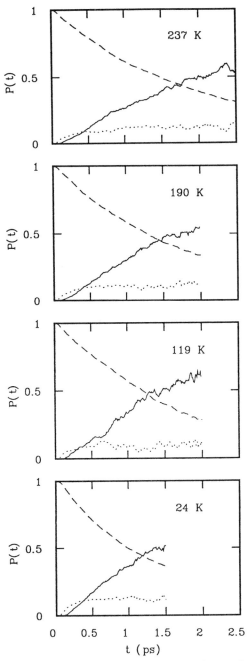

Figure 4. Simulation results for the transient occupation probabilities for $E_2 = 666\ \text{cm}^{-1}$, $K_{12} = 28\ \text{cm}^{-1}$, and $\lambda_{13} = 4000\ \text{cm}^{-1}$ at four temperatures after photoexcitation of the special pair at time zero: P* (dashed lines), P^+B^- (dotted lines), and P^+H^- (solid lines).

67

have performed simulations for E_2 up to $+3000\,\mathrm{cm}^{-1}$). Within this region, the dynamics agrees qualitatively with the superexchange mechanism. In Fig. 5 we show some results for $E_2 = +666\,\mathrm{cm}^{-1}$ and rather large electronic couplings, $K_{12} = 140\,\mathrm{cm}^{-1}$ with $K_{23} = 4K_{12}$. The noise due to the dynamical sign problem is now more severe since one has highly oscillatory bias factors caused by large energy differences. Nevertheless, one can observe the correct experimental characteristics as well.

In addition, there seem to be signatures of biexponential behavior in our data, as observed in recent experiments [65]. In the past, such biexponential dynamics has been explained in terms of inhomogeneity in the sample [65]. We emphasize that biexponential behavior has a purely dynamical origin in our simulations, and the underlying mechanism has been identified recently [66]. Interestingly for the data shown in Fig. 5, the rates turn out to be a factor of about 7 slower than expected from usual nonadiabatic superexchange rate estimates. This is caused by adiabatic corrections due to the huge electronic couplings used in the simulations. The data in Fig. 5 are consistent with an activationless superexchange mechanism, but the actual rates are not well described by nonadiabatic theory for low-lying intermediate state energy E_2. It remains a challenge for theory to explain completely this factor of 7 observed in the simulations in this region. But as one moves the intermediate state energy up further, the process expectedly becomes increasingly nonadiabatic. For $E_2 \gtrsim 2000\,\mathrm{cm}^{-1}$, our simulation data suggest that nonadiabatic rate theory becomes essentially exact. These two regions are depicted in Fig. 6 in the form of a schematic phase diagram.

Finally, it is important to stress that the RC does *not* seem to operate via a combined sequential/superexchange mechanism as proposed in Refs. 67 and 68. Our simulation indicate clearly that it is either a pure sequential or a pure superexchange transfer (But again, this assertion is obviously subject to our choice of the model parameters.) We emphasize that this does not imply that for $E_2 \approx E_1$ it is impossible to achieve a fast ET rate. On the contrary, it is possible to achieve a 3-ps transfer at room temperature with only minor population accumulation on the P^+B^- state. However, we have not found any parameter set in this region that can satisfy all the experimentally observed characteristics simultaneously at *both* high and low temperatures.

C. Diffusion of Brownian Particle in a Periodic Potential

As the third application, we consider the motion of a Brownian particle in a periodic potential. Schmid [69] was the first to study this problem, and many others have since then applied a variety of techniques to this model [70–72]. The importance of the model stems from its widespread

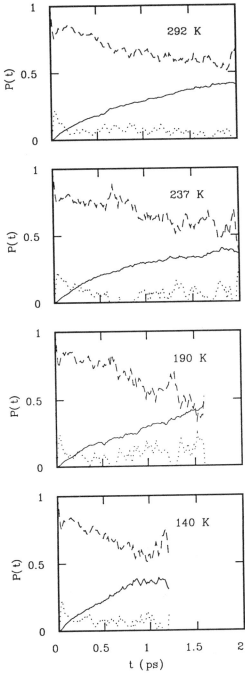

Figure 5. Simulation results for the transient occupation probabilities for $E_2 = +$ 666 cm^{-1}, $K_{12} = 140$ cm^{-1}, and $\lambda_{13} = 2000$ cm^{-1}.

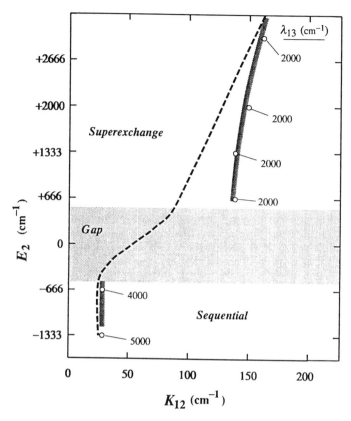

Figure 6. Phase diagram of the RC compiled from QMC simulation data, correlating K_{12} with E_2 (thick lines). The proper reorganization energies are indicated on the phase diagram for those parameter sets that yield dynamics consistent with experiments (open circles). Nonadiabatic predictions for the K_{12} yielding a room-temperature rate of $(3\,ps)^{-1}$ are also shown for comparison (dashed curve).

applications in chemistry and physics. For instance, it has been used to describe the dynamics of a Josephson junction [73], the transport properties of an interacting one-dimensional many-fermion system (Luttinger liquid) [74], electron transfer in a charge-transport chain [75], and (in an idealized fashion) diffusion on the surface of a crystal or among the interstitials in the bulk [76, 77]. An exact solution is still not available, and a numerical study of this model is clearly of interest. While early numerical studies [45] were performed directly in position space, we found it numerically more convenient to exploit an exact duality transformation [69, 71] and work with the tight-binding limit instead [29]. This duality links the extended weak corrugation problem to the opposite instanton limit of very high barriers, where the particle coordinate becomes quasidiscrete. One can express all transport quantities of the Brownian particle in the weak corrugation problem in terms of the mobility and the diffusion coefficient of the dual tight-binding (TB) particle, and we therefore discuss only the latter model in what follows.

In effect, we are dealing with a Hamiltonian like Eq. (3.1) with infinitely many states, $N = \infty$. The nearest-neighbor hopping matrix elements are Δ, and one can apply an external bias ε, which induces a drift on the TB particle. In our simulations we have taken an ohmic spectral density (3.4) with a finite cutoff frequency ω_c. The two transport quantities of interest are the *nonlinear mobility*

$$\mu(\varepsilon) = \lim_{t \to \infty} \langle q(t) \rangle / \varepsilon t \qquad (4.6)$$

and the *diffusion coefficient* (at zero external bias)

$$D = \lim_{t \to \infty} \langle q^2(t) \rangle / 2t . \qquad (4.7)$$

Clearly, these two quantities can be obtained from a direct computation of the first and second moments of the TB position operator, and we demonstrate below that one can indeed reach the long-time limit required to extract these transport quantities numerically from Eqs. (4.6) and (4.7). Here we confine ourselves to the linear case $\varepsilon = 0$, where the Einstein relation provides a simple check for our data. The Einstein relation, which has recently been proved for the general quantum case [72], connects the diffusion coefficient with the linear mobility.

Our QMC simulation technique [29] is very similar to the one outlined in Section III. One starts out with a path integral expression for the first or second moment, integrates out the bath oscillators as well as the quasiclassical degrees of freedom, and then samples only the quantum fluctuations. We have taken a factorized initial condition, which is numerically simpler to handle (the initial correlations are of subleading importance [22]). Focusing on the diffusion coefficient, we have computed the mean-squared displacement directly in real time. Typical results are shown in Fig. 7. One observes a linear regime after a short initial transient, and from the slope one can then extract the diffusion coefficient. The Kondo parameter K measures the system–bath coupling strength. There is an interesting zero-temperature localization quantum phase transition at $K = 1$ [69]. Due to the logarithmic behavior of the bath correlations at zero temperature, our present algorithm allows us to work with finite temperatures only. But signs of this localization transition are already apparent from the simulations.

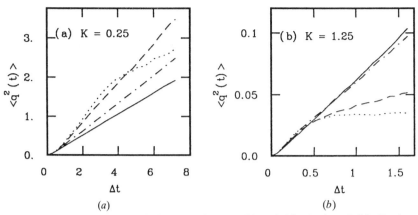

Figure 7. Mean-squared displacement for $\omega_c = 5\Delta$ and (a) $K = \frac{1}{4}$ and (b) $K = \frac{5}{4}$ at several temperatures: $k_B T/\hbar\Delta = 0.156$ (dotted line), 0.625 (dashed line), 2.5 (dash-dotted line), and 5 (solid line).

A summary of the results for the temperature dependence of D are shown in Fig. 8. Here the solid curves denote the predictions of lowest-order perturbation theory in Δ, which gives a T^{2K-1} temperature dependence as $T \to 0$ [22]. This simple analytical theory reproduces the simulation data for $K > 1$ and for high temperatures at $K < 1$. However, for $K < 1$ and low temperatures (which is the most interesting regime in real applications), perturbation theory breaks down. The diffusion coefficient always goes to zero as $T \to 0$ and exhibits a maximum at a finite temperature. The quality of these simulation data does not permit extraction of the asymptotic low-temperature behavior of D, but mobility calculations reveal a complicated temperature-dependence at intermediate Δ/ω_c values [78, 79]. If Δ/ω_c is either very large or very small, one finds anomalous exponents instead, namely a $T^{2/K-2}$ temperature dependence, reminiscent of the critical behavior present in this model.

V. CONCLUSIONS

The field of real-time QMC simulations has seen a rapid evolution during the past few years, and it seems to offer more exciting and interesting opportunities than ever. The idea of sampling blocks of states rather than single states can circumvent the inevitable sign problems to a large extent, and we have presented several nontrivial applications from the realm of electron transfer reactions, bacterial photosynthesis, and dissipative transport in linear chains. There are also interesting applications in mesoscopic physics or field theory, on which we will report shortly.

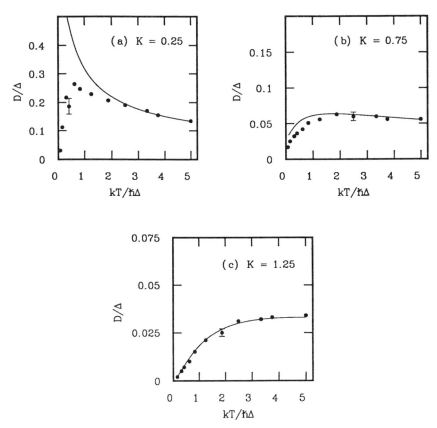

Figure 8. Diffusion coefficient D/Δ (in units of the squared lattice spacing) extracted from the long-time portion of the mean-squared displacement for $\omega_c = 5\Delta$ and (a) $K = \frac{1}{4}$, (b) $K = \frac{3}{4}$, and (c) $K = \frac{5}{4}$ as a function of temperature T. Solid lines indicate results of lowest-order perturbation theory. Note the change in scale as K is increased.

Another fascinating application of these ideas can be found in the fermionic sign problem, which plagues even imaginary-time simulations of interacting many-fermion models. Of great interest for an understanding of high-T_c superconductors, for example, is the complete determination of the phase diagram of the two-dimensional Hubbard model or related strongly correlated systems. Our preliminary studies for these systems indicate that the fermionic sign problem is substantially reduced, and indeed reliable information can be extracted from QMC simulations even at very low temperatures.

ACKNOWLEDGMENTS

Work described in this review has been supported by the National Science Foundation under grant CHE-9216221 and the Young Investigator Awards Program (CHE-9257094), by the Camille and Henry Dreyfus Foundation under the New Faculty Awards Program and the Camille Dreyfus Teacher-Scholar Award, by the Sloan Foundation, and the Petroleum Research Fund administered by the Americal Chemical Society. Computational resources furnished by the IBM Corporation are gratefully acknowledged.

REFERENCES

1. R. P. Feynman, *Rev. Mod. Phys.* **20**, 367 (1948); R. P. Feynman and A. R. Hibbs, *Quantum Mechanics and Path Integrals*, McGraw-Hill, New York, 1965.

2. See, for example, J. E. Gubernatis, Ed., *Proceedings of the Conference on Frontiers of Quantum Monte Carlo*, *J. Stat. Phys.* **43**, 729 (1986).

3. K. Binder, Ed., *Monte Carlo Methods in Statistical Physics*, Topics of Current Physics, Vol. 7, Springer-Verlag, Berlin, 1986.

4. D. Thirumalai and B. J. Berne, *Ann. Rev. Phys. Chem.* **37**, 401 (1986).

5. N. Makri, in *Time-Dependent Quantum Molecular Dynamics*, J. Broeckhove and L. Lathouwers, Eds., Plenum Press, New York, 1992.

6. J. E. Gubernatis, M. Jarrell, R. N. Silver, and D. S. Sivia, *Phys. Rev. B* **44**, 6011 (1991).

7. E. Gallicchio and B. J. Berne, *J. Chem. Phys.* **101**, 9909 (1994).

8. T. L. Marchioro and T. L. Beck, *J. Chem. Phys.* **96**, 2966 (1992).

9. D. K. Hoffman and D. J. Kouri, *J. Phys. Chem.* **96**, 1179 (1992).

10. D. Thirumalai and B. J. Berne, *J. Chem. Phys.* **79**, 5029 (1984); **81**, 2512 (1984).

11. E. C. Behrman, G. A. Jongeward, and P. G. Wolynes, *J. Chem. Phys.* **79**, 6277 (1983).

12. N. Makri, *Chem. Phys. Lett.* **193**, 435 (1991); M. Topaler and N. Makri, *Chem. Phys. Lett.* **210**, 285 (1993); N. Makri, *J. Phys. Chem.* **97**, 2417 (1993).

13. M. Takasu, in *Quantum Monte Carlo Methods in Condensed Matter Physics*, M. Suzuki, Ed., World Scientific, Singapore, 1993.

14. A. P. Vinogradov and V. S. Filinov, *Sov. Phys. Dokl.* **26**, 1044 (1981); V. S. Filinov, *Nucl. Phys. B* **271**, 717 (1986).

15. J. D. Doll, D. L. Freeman, and M. J. Gillan, *Chem. Phys. Lett.* **143**, 277 (1988); J. D. Doll, T. L. Beck, and D. L. Freeman, *J. Chem. Phys.* **89**, 5753 (1988).

16. J. D. Doll and D. L. Freeman, in *Lasers, Molecules and Methods*, Advances in Chemistry and Physics, Vol. 73, J. O. Hirschfelder, R. E. Wyatt, and R. D. Coalson, Eds., Wiley, New York, 1989.

17. N. Makri and W. H. Miller, *J. Chem. Phys.* **89**, 2170 (1988).

18. N. Metropolis, A. W. Rosenbluth, M. N. Rosenbluth, A. H. Teller, and E. Teller, *J. Chem. Phys.* **21**, 1087 (1953).

19. A. M. Amini and M. F. Herman, *J. Chem. Phys.* **96**, 5999 (1992).

20. A. M. Amini and M. F. Herman, *J. Chem. Phys.* **99**, 5087 (1993).

21. U. Weiss, H. Grabert, P. Hänggi, and P. Riseborough, *Phys. Rev. B* **35**, 9535 (1987).

22. U. Weiss, *Quantum Dissipative Systems*, Series in Modern Condensed Matter Physics, Vol. 2, World Scientific, Singapore, 1993, and references therein.

23. C. H. Mak and D. Chandler, *Phys. Rev. A* **41**, 5709 (1990); **44**, 2352 (1991).

24. R. Egger and U. Weiss, *Z. Phys. B* **89**, 97 (1992).

25. C. H. Mak, *Phys. Rev. Lett.* **68**, 899 (1992).

26. C. H. Mak and J. N. Gehlen, *Chem. Phys. Lett.* **206**, 130 (1993); R. Egger and C. H. Mak, *J. Chem. Phys.* **99**, 2541 (1993).

27. R. Egger and C. H. Mak, *Phys. Rev. B* **50**, 15210 (1994).

28. R. Egger and C. H. Mak, *J. Phys. Chem.* **98**, 9903 (1994); C. H. Mak and R. Egger, *Chem. Phys. Lett.*, in press.

29. C. H. Mak and R. Egger, *Phys. Rev. E* **49**, 1997 (1994).

30. E. Y. Loh, J. E. Gubernatis, R. T. Scalettar, S. R. White, D. J. Scalapino, and R. L. Sugar, *Phys. Rev. B* **41**, 9301 (1990).

31. T. D. Kieu and C. J. Griffin, *Phys. Rev. E* **49**, 3855 (1994).

32. M. Winterstetter and W. Domcke, *Phys. Rev. A* **47**, 2838 (1993).

33. A. O. Caldeira and A. J. Leggett, *Phys. Rev. Lett.* **46**, 211 (1981); *Ann. Phys. (NY)* **149**, 374 (1983); **153**, 445(E) (1983).

34. A. J. Leggett, S. Chakravarty, A. T. Dorsey, M. P. A. Fisher, A. Garg, and W. Zwerger, *Rev. Mod. Phys.* **59**, 1 (1987).

35. H. Wipf, D. Steinbinder, K. Neumaier, P. Gutsmiedl, A. Magerl, and A. J. Dianoux, *Europhys. Lett.* **4**, 1379 (1987).

36. D. Chandler, in *Liquids, Freezing and the Glass Transition*, Les Houches 51, Part 1, D. Levesque, J. P. Hansen, and J. Zinn-Justin, Eds., Elsevier/North-Holland, Amsterdam, 1991.

37. R. Egger, C. H. Mak, and U. Weiss, *J. Chem. Phys.* **100.**, 2651 (1994).

38. D. Chandler and P. G. Wolynes, *J. Chem. Phys.* **74**, 4078 (1981).

39. R. P. Feynman and F. L. Vernon, *Ann. Phys. (NY)* **24**, 18 (1963).

40. H. Grabert, P. Schramm, and G.-L. Ingold, *Phys. Rep.* **168**, 115 (1988).

41. S. Chakravarty, *Phys. Rev. Lett.* **49**, 681 (1982).

42. R. Egger, C. H. Mak, and U. Weiss, *Phys. Rev. E* **50**, R655 (1994).

43. A. Schmid, *J. Low Temp. Phys.* **49**, 609 (1982).

44. R. E. Cline, Jr., and P. G. Wolynes, *J. Chem. Phys.* **88**, 4334 (1987).

45. B. A. Mason and K. Hess, *Phys. Rev. B* **39**, 5051 (1989).

46. J. P. Valleau and S. G. Whittington, in *Statistical Mechanics, Part A: Equilibrium Techniques*, Vol. 5 of *Modern Theoretical Chemistry*, B. J. Berne, Ed., Plenum Press, New York, 1977.

47. R. A. Marcus and N. Sutin, *Biochim. Biophys. Acta* **811**, 265 (1985).

48. J. Ulstrup, *Charge Transfer Processes in Condensed Media*, Springer-Verlag, New York, 1979.

49. D. Chandler, *Introduction to Modern Statistical Mechanics*, Oxford University Press, New York, 1987; D. Chandler, *J. Chem. Phys.* **68**, 2959 (1978); J. A. Montgomery, Jr., D. Chandler, and B. J. Berne, *J. Chem. Phys.* **70**, 4056 (1979).

50. P. Hänggi, P. Talkner, and M. Borkovec, *Rev. Mod. Phys.* **62**, 251 (1990).

51. G. A. Voth, D. Chandler, and W. H. Miller, *J. Chem. Phys.* **91**, 7749 (1989).

52. M. J. Gillan, *J. Phys. C* **20**, 3621 (1987).

53. J. N. Gehlen and D. Chandler, *J. Chem. Phys.* **97**, 4958 (1992); J. N. Gehlen, D. Chandler, H. J. Kim, and J. T. Hynes, *J. Phys. Chem.* **96**, 1748 (1992).

54. G. A. Voth, D. Chandler, and W. H. Miller, *J. Phys. Chem.* **93**, 7009 (1989).

55. A. A. Stuchebrukhov and X. Song, *J. Chem. Phys.* **101**, 9354 (1994).

56. J. Deisenhofer, O. Epp, K. Miki, R. Huber, and H. Michel, *J. Mol. Biol.* **180**, 385 (1984).

57. R. A. Marcus, *Chem. Phys. Lett.* **146**, 13 (1988).

58. M. Bixon, J. Jortner, M. E. Michel-Beyerle, A. Ogrodnik, and W. Lersch, *Chem. Phys. Lett.* **140**, 626 (1987).

59. Y. Hu and S. Mukamel, *J. Chem. Phys.* **91**, 6973 (1989).

60. J. L. Martin, J. Breton, A. J. Hoff, A. Migus, and A. Antonetti, *Proc. Natl. Acad. Sci. USA* **83**, 957 (1986).

61. C. Kirmaier and D. Holten, *Proc. Natl. Acad. Sci. USA* **87**, 3552 (1990).

62. W. Holzapfel, U. Finkele, W. Kaiser, D. Oesterhelt, H. Scheer, H. U. Stilz, and W. Zinth, *Chem. Phys. Lett.* **160**, 1 (1989).

63. M. Marchi, J. N. Gehlen, D. Chandler, and M. Newton, *J. Am. Chem. Soc.* **115**, 4178 (1993).

64. A. Warshel and W. W. Parson, *Ann. Rev. Phys. Chem.* **42**, 279 (1991).

65. Y. Jai, T. J. DiMagno, C. K. Chan, Z. Wang, M. Du, D. K. Hanson, M. Schiffer, J. R. Norris, G. R. Fleming, and M. S. Popov, *J. Phys. Chem.* **97**, 13180 (1993).

66. J. N. Gehlen, M. Marchi, and D. Chandler, *Science* **263**, 499 (1994).

67. M. Bixon, J. Jortner, and M. E. Michel-Beyerle, *Biochim. Biophys. Acta* **1056**, 301 (1991).

68. C. K. Chan, T. J. DiMagno, L. X. Q. Chen, J. R. Norris, and G. R. Fleming, *Proc. Natl. Acad. Sci. USA* **88**, 11202 (1991).

69. A. Schmid, *Phys. Rev. Lett.* **51**, 1506 (1983).

70. F. Guinea, V. Hakim, and A. Muramatsu, *Phys. Rev. Lett.* **54**, 263 (1985).

71. M. P. A. Fisher and W. Zwerger, *Phys. Rev. B* **32**, 6190 (1985).

72. U. Weiss and M. Wollensak, *Phys. Rev. B* **37**, 2729 (1988); U. Weiss, M. Sassetti, T. Negele, and M. Wollensak, *Z. Phys. B* **84**, 471 (1991).

73. V. Ambegaokar, U. Eckern, and G. Schön, *Phys. Rev. Lett.* **48**, 1745 (1982).

74. C. L. Kane and M. P. A. Fisher, *Phys. Rev. B* **46**, 15233 (1992).

75. D. N. Beratan, J. N. Betts, and J. N. Onuchic, *Science* **252**, 1285 (1991).

76. C. P. Flynn and A. M. Stoneham, *Phys. Rev. B* **1**, 3966 (1970).

77. F. Capasso, Ed., *Physics of Quantum Electron Devices*, Springer-Verlag, New York, 1990.

78. K. Leung, R. Egger, and C. H. Mak, *Phys. Rev. Lett.* (in press).

79. P. Fendley, A. W. W. Ludwig, and H. Saleur, *Phys. Rev. Lett.* **74**, 3005 (1995).

THE REDFIELD EQUATION IN CONDENSED-PHASE QUANTUM DYNAMICS

W. THOMAS POLLARD, ANTHONY K. FELTS, AND
RICHARD A. FRIESNER

Department of Chemistry, Columbia University, New York

CONTENTS

Advances in Chemical Physics, Volume XCIII, Edited by I. Prigogine and Stuart A. Rice.
ISBN 0-471-14321-9 © 1996 John Wiley & Sons, Inc.

I. INTRODUCTION

Calculation of the quantum dynamics of condensed-phase systems is a central goal of quantum statistical mechanics. For low-dimensional problems, one can solve the Schrödinger equation for the time-dependent wavefunction of the complete system directly, by expanding in a basis set or on a numerical grid [1, 2]. However, because they retain the quantum correlations between all the system coordinates, wavefunction-based methods tend to scale exponentially with the number of degrees of freedom and hence rapidly become intractable even for medium-sized gas-phase molecules. Consequently, other approaches, most of which are in some sense approximate, must be developed.

The most straightforward approximate approach to dealing with large dynamical systems is to treat them classically (i.e., to solve Newton's equations for the positions and momenta of the atoms of the system). This method (molecular dynamics) has been highly successful in many problems, not only for heavy particles, but also for lighter particles at high temperature, where quantum interference effects are in some sense coarse grained in a coupled multidimensional problem. While the performance of classical trajectory calculations has yet to be rigorously quantified for complex, anharmonic, multidimensional problems, there has been remarkable success in obtaining reasonable answers for a wide variety of chemical problems using these techniques. However, there are clearly problems where classical mechanics will not do—the motion of electrons is an obvious example—as well as processes, such as vibrational relaxation, for which the quality of the classical approximation is still unclear.

If only a small part of the system needs to be treated quantum mechanically, one attractive approach is the semiclassical surface-hopping methodology, developed by Tully and co-workers in the early 1970s [3–6]. Here the quantum subsystem is treated with basis-set methods of some sort while the remainder of the system is described classically, evolving under Newton's equations in the field of the subsystem. The tricky aspect of this method is the coupling between the classical and quantum parts of the system. This can be accomplished either via a formalism worked out by Pechukas [5, 7–9] or by methods developed by

Tully [3]. Such methods have been highly successful, for example, in predicting the quantum dynamics of the solvated electron [5]. However, they are fundamentally incapable of capturing the effects of quantum mechanical tunneling by the bath particles, often quite important, and do not treat the development and evolution of quantum coherences with complete accuracy [9].

A rather different starting point for dynamical modeling is based on the ability to analytically solve quantum-mechanical problems for harmonic oscillators. This fact is used in deriving the many variants of quantum transition-state theory, in which corrections to the classical expressions are determined by modeling the force field in the transition-state region as a harmonic bath coupled to a small-dimensional reactive surface [10–12]. Such theories can be reasonably accurate if one has a good physical picture of the transition state, as, for example, in the Marcus theory of electron transfer. However, there are many problems (the solvated electron being a good example) where the physical picture is much less clear, and construction of the transition state would be exceedingly difficult.

Over the past five years there has been a great deal of progress in utilizing Feynman's path-integral formulation of quantum mechanics to compute quantum dynamical observables [13–20]. This project is much more difficult than the evaluation of equilibrium quantities, where the path integral is formulated in imaginary time and the integrand is always real and positive [21]. As a result, the latter problem essentially amounts to quadrature of a smoothly varying, highly multidimensional function, and straightforward Monte Carlo methods have been applied successfully here. Although the method is still computationally expensive, it is not wildly more so than a classical simulation on the same system would be, and can provide benchmark quantum mechanical results for a given potential.

In contrast, the calculation of dynamical quantities via real-time path-integral methods involves similarly multidimensional integrals, but with highly oscillatory integrands [15]. The earliest work in this area employed brute force Monte Carlo methods, typically for one-dimensional systems; not surprisingly, these proved difficult to extend to long times, many dimensions, or unfavorable parameter regimes. Improvements utilizing clever sampling and filtering schemes have followed, and progress has clearly been made in this direction, although the computational cost remains very substantial [13]. Perhaps the most attractive methods that have been developed to date involve the use of either classical or adiabatic reference systems in the path integral, in order to minimize integration over the oscillatory regions of the integrand [15–20, 22].

These methods appear quite promising, although tests in a wider variety of parameter regimes will be necessary before their power can be fully evaluated.

In the present article we discuss a complementary approach that has its roots in yet another fundamental field, statistical mechanical perturbation theory, originating in the work of Zwanzig and Mori in the early 1950s [23, 24]. The basic object here is the reduced density matrix, (i.e., the density matrix of a quantum subsystem obtained by integrating the complete density matrix over the bath degrees of freedom) [25, 26]. As in the surface-hopping approach, one chooses to treat just a small part of the system with accurate quantum methods, thus allowing arbitrarily complicated small-dimensional problems to be modeled accurately. Unlike the surface-hopping methods, however, here the bath can also be treated quantum mechanically, albeit approximately via low-order perturbation theory. Perhaps most important, the method can be made highly computationally efficient, so that rather complex problems can be solved quickly. This is of particular importance when one is trying to model real physical phenomena, as opposed to carrying out proof-of-concept benchmark calculations to demonstrate a new methodology. It is not obvious how effectively one would be able, for example, to optimize potential functions via real-time path-integral calculations. At the very least, rapid reduced-density-matrix calculations would allow one to explore the parameter space of the problem and construct a good zeroth-order model before proceeding to more accurate calculations; at best, density matrix methods can themselves be made quantitative, so that no additional computational methods are necessary.

In fact, reduced-density-matrix modeling has been used in this fashion for more than 30 years in the field of NMR spectroscopy [27]. There the reduced-density-matrix equation of motion, known as the Bloch–Redfield relaxation equation [28–30], remains the standard approach to predicting NMR observables from a physical model. Typically, one models the quantum subsystem as a small number of coupled spins, so that solution of the Redfield equation itself is essentially trivial. This methodology has been highly successful from the point of view of confronting theory and experiment. Although the reduced-density-matrix picture has long been important in the theory of optical spectroscopy as well [31–34], the Redfield equation has found somewhat more limited use [35], and applications to chemical dynamical systems have only begun to appear fairly recently [36–43].

However, as it is applied to this wider range of systems, the validity of Redfield theory becomes more problematic than in NMR. A canonical example is the spin-boson Hamiltonian—two electronic levels linearly coupled to a harmonic bath—which is, for example, a good model for electron transfer processes [44]. A formulation of the spin-boson dynamics as a two-state Redfield problem was developed by Silbey and co-workers many years ago [45–47]. For some regions of parameter space, as we shall see below, this is an excellent approximation. However, in other regimes, a two-state Redfield model fails qualitatively. For more complicated Hamiltonians—for example, a three-state spin-boson Hamiltonian—the parameter regimes in which Redfield theory is valid are not known.

In view of the computationally trivial nature of a two-state Redfield model, one could ask whether a less drastic reduction of the problem, (e.g., inclusion of a reaction coordinate as well as the two electronic states) would (1) lead to better answers and (2) be computationally tractable. The goal of the present paper is to explore these two questions in some detail. Our answers to both will be in the affirmative; still, new algorithms and mathematical formalisms will be necessary to realize the simultaneous goals of computational tractability and reasonable accuracy for difficult parameter regimes. Although these methods are not yet completely worked out, the initial results are quite promising, as will be shown below.

The paper is organized as follows. In Section II we briefly summarize reduced-density-matrix theory and derive the Redfield equation. In Section III.C we present our methods for efficiently solving the Redfield equation for large quantum subsystems, including the approximations necessary to obtain the matrix elements of the Redfield tensor and the time propagation algorithms used to advance the differential equation. In Section IV one application of the method (a model of electron transfer through DNA) is discussed in some detail. In Section V we consider the use of canonical transformations to render a general class of system–bath Hamiltonians to a weak-coupling form, so that the Redfield equation can be applied to obtain the exact dynamics. We conclude by outlining the future directions of this research program.

II. REDUCED-DENSITY-MATRIX THEORY

In condensed-phase dynamics, we are typically interested in the detailed behavior of just a small part of a large system, so we partition the total system into the subsystem of interest, s, and a surrounding bath, b. The total Hamiltonian is then written

$$\mathbf{H} = \mathbf{H}_s + \mathbf{H}_b + \mathbf{V} \, , \tag{1}$$

where $\mathbf{H}_s$ and $\mathbf{H}_b$ are the Hamiltonians of the isolated subsystem and the bath, which interact through the coupling operator, $\mathbf{V}$. Since this complete system is closed, its state may be completely described by a wavefunction, $|\Psi\rangle$, which evolves according to the time-dependent Schrödinger equation,

$$\dot{\Psi}(t) = -i\mathbf{H}\Psi(t) \ . \tag{2}$$

Equivalently, the system is described by its density matrix, $\rho = |\Psi\rangle\langle\Psi|$, which evolves under the quantum mechanical Liouville equation,

$$\dot{\rho}(t) = -i[\mathbf{H}, \rho(t)] = \mathscr{L}\rho(t) \ . \tag{3}$$

Here $\mathscr{L} = [\mathbf{H}, \ldots] = \mathscr{L}_s + \mathscr{L}_b + \mathscr{L}_v$ is the Liouville "superoperator" [23, 24].

When the subsystem is considered separately from the bath, however, its state can no longer be written as a wavefunction; instead, it is completely described only by its reduced density matrix, σ, obtained from the full density matrix by integrating over the bath variables, that is, [25–27],

$$\sigma = \mathrm{Tr}_b\{\rho\} \ . \tag{4}$$

Obtained in this way, σ will share with ρ the necessary properties required of a physically meaningful density matrix, namely, hermiticity ($\sigma = \sigma^\dagger$), normalization [$\mathrm{Tr}(\sigma) = 1$], and positivity. Positivity is the property that the eigenvalues of σ are nonnegative; this is necessary since the diagonal matrix elements of σ, in any basis, represent physical probabilities.

A. The Redfield Equation

In principle, the reduced density matrix σ can be obtained at any point in time by evolving the complete density matrix ρ under the Liouville equation and then integrating out the bath, as in Eq. (4). However, this gains nothing; defining the reduced density matrix is useful only if there is an equation of motion for σ directly.

The standard methods of obtaining equations of motion for reduced systems are based on the projection-operator techniques developed by Zwanzig and Mori in the late 1950s [23, 24]. In this approach one defines an operator, $\mathscr{P}$, that acts on the full density matrix ρ to project out a direct product of σ and the thermalized equilibrium density matrix of the bath, ρ_b^{eq}, that is,

$$\mathscr{P}\rho = \rho_b^{\text{eq}} \otimes \text{Tr}_b\{\rho\} = \rho_b^{\text{eq}} \otimes \sigma , \tag{5}$$

$$\rho_b^{\text{eq}} = e^{-\beta \mathbf{H}_b}/\text{Tr}_b\{e^{-\beta \mathbf{H}_b}\} . \tag{6}$$

Using this and the complementary projection operator $\mathscr{Q} = (1 - \mathscr{P})$, one formally obtains an exact equation of motion for σ, written [48]

$$\dot{\sigma}(t) = \mathscr{L}_s\sigma(t) + \int_0^t \mathscr{G}(\tau)\sigma(t - \tau) \tag{7}$$

and known as the *generalized master equation*. Here the complete dynamics of the complete bath are still included in the generalized "friction" kernel,

$$\mathscr{G}(\tau) = \text{Tr}_b\mathscr{L}_v e^{\tau \mathscr{Q}\mathscr{L}}\mathscr{Q}\mathscr{L}_v\rho_b^{\text{eq}} , \tag{8}$$

where $\mathscr{L}_v = [\mathbf{V}, \ldots]$.

At this point, one has to make approximations to proceed further. The standard approximations leading to the Redfield equation have been discussed by numerous authors [26–30, 48–53]. First, one assumes that the bath correlation time [i.e., the time scale on which $\mathscr{G}(\tau)$ goes to zero] is short enough that the upper limit of integration in Eq. (7) may be effectively extended to infinity. Second, the system–bath coupling is neglected in the exponent of the friction kernel [Eq (8)], so that only terms to second order in the system–bath coupling, $\mathbf{V}$, are retained in Eq. (7). Finally, one adopts the Markov approximation, replacing $\sigma(t - \tau)$ by $\sigma(t)$ on the assumption that σ evolves slowly compared to the decay of $\mathscr{G}(\tau)$. One can now evaluate the integral in Eq. (7) independently of the state of the subsystem, leaving

$$\dot{\sigma}(t) = \mathscr{L}_s\sigma(t) + \mathscr{R}\sigma(t) = \mathscr{L}_R\sigma(t) . \tag{9}$$

This result is the Redfield–Liouville–von Neumann equation of motion or, simply, the Redfield equation [29, 30, 49–53]. Here the influence of the bath is contained entirely in the Redfield relaxation tensor, $\mathscr{R}$, which is added to the Liouville operator for the isolated subsystem to give $\mathscr{L}_R$, the dissipative Redfield–Liouville superoperator (tensor) that propagates σ. Expanded in the eigenstates of the subsystem Hamiltonian, $\mathbf{H}_s$, Eq. (9) yields a set of coupled linear differential equations for the matrix elements of σ,

$$\dot{\sigma}_{i,j}(t) = -i\omega_{i,j}\sigma_{i,j}(t) + \sum_{k,l} R_{i,j,k,l}\sigma_{k,l}(t) .\tag{10}$$

B.   The Redfield Relaxation Tensor

The Redfield tensor $\mathcal{R}$ is defined in terms of stationary correlation functions of the system–bath coupling operator, $\mathbf{V}$, evolving under the bath Hamiltonian, $\mathbf{H}_b$. Thus the dynamics of the bath are retained in Eq. (9), the only assumptions being that the bath is in thermal equilibrium and that its dynamics are independent of the state of the system beyond some correlation time, τ_c, short compared to the rate of change of σ. The tensor element $R_{i,j,k,l}$ can be written [26, 42]

$$R_{i,j,k,l} = \Gamma^+_{l,j,i,k} + \Gamma^-_{l,j,i,k} - \delta_{l,j}\sum_r \Gamma^+_{i,r,r,k} - \delta_{i,k}\sum_r \Gamma^-_{l,r,r,j} ,\tag{11}$$

where

$$\Gamma^+_{l,j,i,k} = \frac{1}{\hbar^2}\int_0^\infty d\tau\, e^{-i\omega_{i,k}\tau}\langle \mathbf{V}_{l,j}(\tau)\mathbf{V}_{i,k}(0)\rangle_b ,\tag{12}$$

$$\Gamma^-_{l,j,i,k} = \frac{1}{\hbar^2}\int_0^\infty d\tau\, e^{-i\omega_{l,j}\tau}\langle \mathbf{V}_{l,j}(0)\mathbf{V}_{i,k}(\tau)\rangle_b\tag{13}$$

are Fourier–Laplace transforms of correlation functions of matrix elements of the system–bath coupling operator and the brackets represent a trace over the thermalized bath; $\mathbf{V}_{i,j}$ is a matrix element of $\mathbf{V}$ between system eigenstates $|i\rangle$ and $|j\rangle$, and $\mathbf{V}_{i,j}(\tau) = e^{i\mathbf{H}_b\tau}\mathbf{V}_{i,j}e^{-i\mathbf{H}_b\tau}$.

The physical interpretation of the Redfield tensor elements in the eigenstate representation is straightforward: Each $R_{i,j,k,l}$ element in Eq. (10) is essentially a rate constant that relates the rate of change of density matrix element $\sigma_{i,j}$ to the current value of each other matrix element $\sigma_{k,l}$. Thus $R_{i,i,i,i}$ is the relaxation rate for the population of eigenstate $|i\rangle$ and the $R_{i,i,j,j}$ represent the rate of population transfer from each state $|j\rangle$ to state $|i\rangle$. The elements $R_{i,j,i,j}$ represent the dephasing rates of the off-diagonal density matrix elements, $\sigma_{i,j}$. A common assumption, known as the *secular approximation*, asserts that population and coherence transfer processes are uncoupled, that is, that the diagonal and off-diagonal density matrix elements evolve independently.

An important property of the Redfield tensor is that when the bath is treated quantum mechanically, the physically required detailed-balance conditions are naturally satisfied, so that the subsystem relaxes to thermal equilibrium with the bath. This becomes evident from the general property of finite-temperature quantum correlation functions that [24]

$$\int_{-\infty}^{\infty} dt \, e^{-i\omega_{i,j}t} \langle F_a(t)F_b(0) \rangle_b = e^{-\beta\hbar\omega_{i,j}} \int_{-\infty}^{\infty} dt \, e^{-i\omega_{i,j}t} \langle F_b(0)F_a(t) \rangle_b \, . \quad (14)$$

This property ensures that the ratio of forward and reverse rate constants between the populations of states $|i\rangle$ and $|j\rangle$ is always

$$\frac{R_{i,i,j,j}}{R_{j,j,i,i}} = \frac{\Gamma_{j,i,i,j}^{+} + \Gamma_{j,i,i,j}^{-}}{\Gamma_{i,j,j,i}^{+} + \Gamma_{i,j,j,i}^{-}} = e^{-\beta\hbar\omega_{i,j}} \, . \quad (15)$$

1. Separation of the System–Bath Coupling

To evaluate the correlation functions in Eqs. (12) and (13), it is usual to complete the separation of the system and bath by decomposing the system–bath coupling into a sum of products of pure system and bath operators. This allows the correlation functions of the system–bath coupling to be replaced, without loss of generality, by correlation functions of bath operators alone, evolving under the uncoupled bath Hamiltonian. Moreover, as we have previously pointed out [39, 40], this decomposition of the system–bath coupling make it possible to write the Redfield equation in a highly compact form, without explicit reference to the Redfield tensor at all.

We write, then,

$$\mathbf{V} = \sum_a \mathbf{G}_a \mathbf{F}_a \, , \quad (16)$$

where the operators $\mathbf{G}_a$ and $\mathbf{F}_a$ act only on functions of the system and bath variables, respectively. Inserting this form into Eqs. (12) and (13) gives

$$\Gamma_{l,j,i,k}^{+} = \frac{1}{\hbar^2} \sum_{a,b} (G_a)_{l,j}(G_b)_{i,k}(\Theta_{a,b}^{+})_{i,k} \, ,$$

$$\Gamma_{l,j,i,k}^{-} = \frac{1}{\hbar^2} \sum_{a,b} (G_a)_{l,j}(G_b)_{i,k}(\Theta_{a,b}^{-})_{l,j} \, , \quad (17)$$

where

$$(\Theta_{a,b}^{+})_{i,j} = \int_{0}^{\infty} d\tau \, e^{-i\omega_{i,j}\tau} \langle \mathbf{F}_{a}(\tau)\mathbf{F}_{b}(0) \rangle_{b} \, ,$$

$$(\Theta_{a,b}^{-})_{i,j} = \int_{0}^{\infty} d\tau \, e^{-i\omega_{i,j}\tau} \langle \mathbf{F}_{a}(0)\mathbf{F}_{b}(\tau) \rangle_{b} \, . \tag{18}$$

Note that when the bath operators $\{\mathbf{F}_{a}\}$ are Hermitian, $\Theta_{a,b}^{+}$ and $\Theta_{b,a}^{-}$ are Hermitian conjugates, that is,

$$(\Theta_{a,b}^{+})_{i,j} = (\Theta_{b,a}^{-})_{j,i}^{*} \, . \tag{19}$$

The correlation functions in the definitions of $\Theta_{a,b}^{+}$ and $\Theta_{a,b}^{-}$ involve only bath operators, evolving under the bath Hamiltonian, and therefore represent properties of the bath that are independent of its interaction with any particular system.

2. Factorization of the Redfield Tensor

In most applications of the theory to date, the solution of the Redfield equation has required first the explicit calculation of the Redfield tensor elements [Eq. (11)]; given these, Eq. (10) could be solved as an ordinary set of linear differential equations with constant coefficients, either by explicit time stepping [41, 42] or by diagonalization of the Redfield tensor [37, 38]. Since there are N^4 such tensor elements for an N-state subsystem, the number of these quantities can become quite large. Because of this, until recently most applications of Redfield theory have been limited to small systems of two to four states, or else assumptions, such as the secular approximation, have been used to neglect large classes of tensor elements.

We recently demonstrated that when the system–bath coupling is written in the sum-of-products form [Eq. (16)], a substantial amount of structure is imparted to the Redfield tensor, $\mathscr{R}$, allowing the product $R\sigma$ to be rewritten without reference to the individual tensor elements at all [39]. Instead, $\mathscr{R}\sigma$ is replaced by commutators of σ and the system operators defined in Eqs. (16) and (18). The result is

$$\dot{\sigma} = -i[\mathbf{H}_{s}, \sigma] + \sum_{a} \{[\mathbf{G}_{a}^{+}\sigma, \mathbf{G}_{a}] + [\mathbf{G}_{a}, \sigma\mathbf{G}_{a}^{-}]\} \, , \tag{20}$$

where

$$(G_a^+)_{i,j} = \sum_b (G_b)_{i,j} (\Theta_{a,b}^+)_{i,j} , \qquad (21)$$

$$(G_a^-)_{i,j} = \sum_b (G_b)_{i,j} (\Theta_{b,a}^-)_{i,j} . \qquad (22)$$

When all of the operators $\mathbf{G}_a$ and $\mathbf{F}_a$ are Hermitian, this result simplifies to

$$\dot{\sigma} = -i[\mathbf{H}_s, \sigma] + \sum_a \{[\mathbf{G}_a^+ \sigma, \mathbf{G}_a] + \text{h.c.}\} , \qquad (23)$$

where h.c. denotes the Hermitian conjugate and the superscript plus and minus label the newly defined system operators, $\Theta_{a,b}^+$, $\Theta_{a,b}^-$, $\mathbf{G}_a^+$, and $\mathbf{G}_a^-$, by their correspondence to the Liouville space operators, Γ^+ and Γ^-.

This result demonstrates that the Redfield tensor can be applied to any operator *without* the explicit construction of the full tensor. The $\mathbf{G}_a$ and $\mathbf{G}_a^+$ are ordinary operators, so numerical evaluation of the right side of Eq. (20) in an N-dimensional basis involves only the storage and multiplication of $N \times N$ matrices. The full Redfield tensor, in contrast, is an $N^2 \times N^2$ operator in Liouville space, whose application to an $N \times N$ density matrix requires $O(N^4)$ scalar multiplications, as opposed to the $O(N^3)$ operations required to multiply $N \times N$ matrices. The ability to apply the Redfield tensor to the density matrix using Eqs. (20) or (23) therefore allows a significant savings in both computer time and memory, particularly as N becomes large.

C. Dynamical Semigroup Approach

The final form of the Redfield equation [Eq. (20)] is superficially similar to the equation of motion that arises in the axiomatic semigroup theory of Lindblad, Gorini et al. [48, 54–57]. They showed that the most general equation of motion that preserves the positivity of the density matrix must have the general form

$$\dot{\sigma} = -i[\mathbf{H}_s, \sigma] + \frac{1}{2} \sum_a \{[\mathbf{W}_a \sigma, \mathbf{W}_a^\dagger] + [\mathbf{W}_a, \sigma \mathbf{W}_a^\dagger]\} . \qquad (24)$$

Here the $\{\mathbf{W}_a\}$ are operators of the subsystem and the superscript dagger denote the Hermitian conjugate. The Redfield equation can be written in this form only when an additional symmetrization of the bath correlation functions is performed [48]. Note that this alternative equation also expresses the dissipative evolution of the density matrix in terms of $N \times N$

operators, and therefore has the same favorable computational properties as our result [58, 59].

Despite their similarity, the two results are quite distinct; in particular, the axiomatic semigroup equation guarantees, by design, that an arbitrary initial density matrix will remain positive definite, whereas the Redfield equation does not. The consequences of this have been debated, but the issue is still open. Although it is clearly desirable that the density matrix remain positive definite (since its eigenvalues represent physical probabilities), it is not clear that it is appropriate to expect the equations of motion to guarantee this property for any initial density matrix; to do so requires making additional approximations beyond those of Redfield theory. Suarez has pointed out that although the initial conditions for reduced-density-matrix propagation assume a completely factorized system–bath density matrix, the equations of motion implicitly allow transient correlations between the system and bath to arise, and that the Redfield equation also preserves positivity if the initial σ is consistent with these correlations [60]. Thus all reduced density matrices are not necessarily physically reasonable when correlations with the bath exist (as they must if the system–bath coupling is nonzero). Indeed, Pechukas has pointed out that for general correlated initial conditions [53, 61], the evolution of the reduced density matrix need not remain positive at all [62]. The issue of how to choose an initial σ that remains positive under the Redfield equation is an open question, but the work of Suarez suggests that typical deviations from positivity are quite limited and scale with the strength of the system–bath coupling.

III. NUMERICAL SOLUTION OF THE REDFIELD EQUATION

A. Overview

There are three important issues to consider in the numerical solution of the Redfield equation. The first is the evaluation of the Redfield tensor matrix elements $R_{i,j,k,l}$. To obtain these matrix elements, it is necessary to have a representation of the system–bath coupling operator and of the bath Hamiltonian. Two fundamental types of models are used. First, the system–bath coupling can be described using stochastic fluctuation operators, without reference to a microscopic model. In this case, the correlation functions appearing in the formulas for $R_{i,j,k,l}$ are parameterized functions of time whose functional forms are typically chosen so that the integrals can be done analytically. Alternatively, one can calculate the $R_{i,j,k,l}$ from a microscopic model (e.g., a bath of harmonic oscillators linearly coupled to the system). In this case, the correlation functions may often be more complicated, necessitating the numerical evaluation of matrix elements (although analytical approximations of various types could be used as well). These two alternatives are discussed further in Section III.B.

The second issue is the method of solution of Eq. (10). One approach is to diagonalize the Redfield–Liouville tensor, $\mathscr{L}_R$, numerically, for example via the QR algorithm for complex matrices [38]. If there are N quantum levels in the system, there will be N^2 density matrix elements, and hence the Redfield matrix will be an $N^2 \times N^2$ matrix. As the computational effort for QR diagonalization scales as the third power of the dimension of the matrix [63], the work involved in this approach will grow as N^6, while computer memory requirements will scale as N^4. This precludes calculations involving a large number of system levels unless one uses a parallel supercomputer (and even in this case, the QR algorithm is notoriously difficult to parallelize efficiently). The alternative approach is to solve the equations by iterative methods, involving either time stepping or the iterative determination of a few relevant eigenvalues and eigenvectors of $\mathscr{L}_R$. We shall discuss these methods, which allow considerable reduction in computational effort as compared to the QR approach, in more detail below.

The third issue is the overall structure of $R_{i,j,k,l}$ and the impact of this structure on the development of an efficient algorithm. At first glance, the use of iterative solution methods would appear to be of limited utility in solving the Redfield equation, since $\mathscr{L}_R$ is a dense matrix, so that each matrix-vector multiply (i.e., $\mathscr{L}_R \sigma$) will be expensive, involving on the order of $\sim N^4$ multiplications. However, as pointed out in Section II.B.2, for the most commonly assumed forms of the system–bath coupling, a factorization of the tensor operator is possible such that an efficient matrix-vector multiply can be devised [Eq. (20)]. The cost of the product is thereby reduced to order $\sim N^3$ operations, a very large improvement if one needs to consider a large number of quantum states (e.g., $N \sim 100$–1000). At the same time, the memory requirements are reduced to $\sim N^2$, a dramatic reduction compared to the N^4 size of the full Redfield tensor.

B. Bath Models

1. Harmonic Oscillator Bath

An important feature of the reduced-density-matrix approach is that it allows the bath to be treated at different levels of approximation. In the Redfield equation, the bath enters only through the correlation functions of the coupled bath variables in Eq. (18). This means that a substantial part of the complexity of a realistic condensed-phase environment is irrelevant to the relaxation of the embedded subsystem, at least when the assumptions underlying Redfield theory are met. As a consequence of this, one can always identify a simpler effective bath that has the same relaxation properties as any actual environment.

A general and convenient choice is to model the bath as a set of independent harmonic oscillators linearly coupled to the system. The Hamiltonian for a collection of oscillators of unit mass and frequencies $\{\omega_j\}$ is

$$\mathbf{H}_b = \frac{1}{2} \sum_j (\mathbf{p}_j^2 + \omega_j^2 \mathbf{q}_j^2), \tag{25}$$

where $\mathbf{q}_j$ and $\mathbf{p}_j$ are the position and momentum of the jth oscillator. One then assumes that displacements of these oscillators are coupled to a system operator, $\mathbf{G}$, according to

$$\mathbf{V} = \mathbf{G} \sum_j g_j \mathbf{q}_j, \tag{26}$$

where g_j is a density-weighted linear coupling for all oscillators of frequency ω_j. The bath operator that enters Eq. (18) is thus $\mathbf{F} = \Sigma_j \, g_j \mathbf{q}_j$ and its autocorrelation function is

$$\langle \mathbf{F}(t)\mathbf{F}(0)\rangle_b = \frac{1}{2} \sum_j \frac{g_j^2}{\omega_j} Q_j (e^{-\omega_j t} + e^{-\beta\hbar\omega_j} e^{i\omega_j t}) \tag{27}$$

$$= \frac{1}{2} \sum_j J(\omega_j) \left[\coth\frac{\beta\hbar\omega_j}{2} \cos(\omega_j t) - i \sin(\omega_j t) \right]. \tag{28}$$

Here $Q_j = 1/(1 - e^{-\beta\hbar\omega_j})$ is the partition function for oscillator j and

$$J(\omega) = \pi \sum_j \frac{g_j^2}{\omega} \delta(\omega - \omega_j) \tag{29}$$

is the spectral density of the bath.

2. Classical Bath

Except for formal models, one usually does not have the complete quantum mechanical Hamiltonian of a large multidimensional system. On the other hand, there are many reasonable classical potential functions for condensed-phase systems, and solving the classical equations of motion for very large systems is quite practical. This makes it attractive to consider replacing the quantum correlation functions in Eq. (18) with classical correlation functions obtained from molecular dynamics simulations [42, 43]. For instance, if the coupling is linear in some system coordinate Q (i.e., $\mathbf{V} = \mathbf{Q}\mathbf{F}$), the unspecified bath variable F is just the force exerted by the bath on coordinate Q. In the MD simulation, one would therefore measure the autocorrelation function of this force to obtain $\langle \mathbf{F}(0)\mathbf{F}(t) \rangle_b$ for use in Eq. (18) [42, 43]. Similarly, if the bath is coupled to the energy of some system state $|i\rangle$, so that $\mathbf{V} = |i\rangle \mathbf{F} \langle i|$, the autocorrelation function of the fluctuations of this energy about its mean are measured [64, 65].

One issue that arises when classical correlation functions are used is that they do not satisfy the detailed-balance relation Eq. (14) (because they are even with respect to $t = 0$) and hence cannot produce a Redfield tensor that lets the subsystem come to thermal equilibrium. Therefore, before being introduced into Eq. (18), the classical results must be modified to satisfy detailed balance. Unfortunately, there is no unique way to accomplish this, and a handful of different approaches are found in the literature.

The traditional approach is to make contact between classical correlation functions and the symmetrized quantum correlation functions, $\langle [\mathbf{F}(t)\mathbf{F}(0)]_+ \rangle_b$ [66]. Using Eq. (14), one sees that the Fourier transforms of a quantum correlation function and its symmetrized counterpart satisfy

$$\int_{-\infty}^{\infty} dt\, e^{-i\omega_{i,j}t} \langle \mathbf{F}_b(0)\mathbf{F}_a(t) \rangle_b = \frac{1}{1 + e^{-\beta\hbar\omega_{i,j}}} \int_{-\infty}^{\infty} dt\, e^{-i\omega_{i,j}t} \langle [\mathbf{F}_a(0)\mathbf{F}_b(t)]_+ \rangle_b .$$

$$(30)$$

Because of this, it is standard practice to correct the Fourier transforms of classical correlation functions with the factor $1/(1 + e^{-\beta\hbar\omega_{i,j}})$. This also assures that the relaxation rates for quantum and classical correlation functions converge in the limit of high temperature, as they should. In terms of the notation used above, this correction can be applied using the definitions

$$(\Theta_{a,b}^{+})_{i,j} = \frac{1}{1 + e^{\beta\hbar\omega_{i,j}}} \int_{-\infty}^{\infty} dt\, e^{-i\omega_{i,j}t} \langle \mathbf{F}_a(t)\mathbf{F}_b(0) \rangle_{cl} , \qquad (31)$$

$$(\Theta_{a,b}^{-})_{i,j} = \frac{1}{1 + e^{-\beta\hbar\omega_{i,j}}} \int_{-\infty}^{\infty} dt\, e^{-i\omega_{i,j}t} \langle \mathbf{F}_a(t)\mathbf{F}_b(0) \rangle_{cl} , \qquad (32)$$

where $\langle\ \rangle_{cl}$ indicates the classical correlation functions.

The most important property possessed by the operators $\Theta_{a,b}^{+}$ and $\Theta_{a,b}^{-}$

Eq. (31) is that

$$
\begin{aligned}
(\Theta_{a,b}^{+})_{i,j} &= (\Theta_{a,b}^{+})_{j,i} e^{\beta\hbar\omega_{j,i}} \\
&= (\Theta_{a,b}^{-})_{j,i} \, .
\end{aligned}
\tag{33}
$$

This relation is, itself, sufficient to satisfy the detailed-balance-condition and define a Redfield tensor that allows the system to come to thermal equilibrium. It has thus been used directly by some as an ad hoc correction for classical or stochastic correlation functions [38].

The third alternative is to use the classical correlation functions to define an equivalent quantum mechanical harmonic bath. This approach was pioneered by Warshel as the *dispersed polaron method* [67, 68]. More recently, this idea has been used in studies of electron transfer systems in solution [64] and in the photosynthetic reaction center [65, 69] (see also Ref. 70). This approach is based on the realization that the spectral density describing a linearly coupled harmonic bath [Eq. (29)] can be obtained by cosine transformation of the classical time-correlation function of the bath operator [Eq. (28)]. Comparing the classical correlation function for the linearly coupled harmonic bath [Eqs. (25) and (26)],

$$
\langle F(t)F(0)\rangle_{\mathrm{cl}} = \sum_{j} \frac{J(\omega_{j})}{\beta\hbar\omega_{j}} \cos(\omega_{j}t) \, ,
\tag{34}
$$

with the symmetrized quantum correlation function of the same system [Eq. (28)],

$$
\langle [\mathbf{F}(t)\mathbf{F}(0)]_{+} \rangle_{b} = \frac{1}{2} \sum_{j} J(\omega_{j}) \coth\frac{\beta\hbar\omega_{j}}{2} \cos(\omega_{j}t) \, ,
\tag{35}
$$

we see that

$$
\begin{aligned}
&\int_{0}^{\infty} dt \cos(\omega t)\langle [\mathbf{F}_{a}(0)\mathbf{F}_{b}(t)]_{+} \rangle_{b} \\
&= \frac{1}{2}\beta\hbar\omega \coth\frac{\beta\hbar\omega}{2} \int_{0}^{\infty} dt \cos(\omega t)\langle F_{a}(0)F_{b}(t)\rangle_{\mathrm{cl}} \, ,
\end{aligned}
\tag{36}
$$

which can be used with Eq. 30 to obtain the unsymmetrized correlation functions required to define $\Theta_{a,b}^{+}$ and $\Theta_{a,b}^{-}$ (Eq. 18). Alternatively, the bath coupling constants can be obtained directly from the classical correlation function through

$$g(\omega) = 2\beta\hbar\omega^2 \int_0^\infty dt \, \cos(\omega t) \langle F_a(0)F_b(t) \rangle_{cl} \,, \tag{37}$$

and used in Eq. (27). Often, the dynamics of a weakly coupled bath can reasonably be expected to be essentially harmonic, and in those cases this procedure should be somewhat more meaningful than the first two approaches described above. For instance, each of these corrections introduces some temperature dependence to the resulting Redfield tensor; however, because the classical correlation functions are themselves a function of temperature, the temperature dependence of the corrected functions cannot generally be expected to be meaningful. Mapping onto an effective quantum harmonic bath, on the other hand, will capture the correct temperature dependence to the extent that the harmonic-bath assumption is valid.

A more interesting point was made by Bader and Berne, who noted that the vibrational relaxation rate of a classical oscillator in a classical harmonic bath is identical to that of a quantum oscillator in a quantum harmonic bath [71]. On the other hand, when the relaxation of the quantum system is calculated using the corrected correlation function of the classical bath [Eq. (31)], the predicted rate is slower by a factor of $\frac{1}{2}\beta\hbar\omega \coth(\beta\hbar\omega/2)$, which can be quite substantial for high-frequency solutes. The conclusions of a number of recent studies were shown to be strongly affected by this inconsistency [42, 43, 72]. Quantizing the solvent by mapping the classical correlation functions onto a quantum harmonic bath corrects the discrepancy.

3. Stochastic Bath

When detailed information about the nature of the bath is lacking or unimportant for the problem at hand, it can be convenient to treat the bath stochastically, that is, as a source of random fluctuations of the subsystem Hamiltonian [30, 35, 38, 48]. In a stochastic treatment the bath operators in **V** are replaced with classical random variables.

The system–bath coupling is then a random quantum operator and may be written generally as

$$\mathbf{V}(t) = \sum_a f_a \mathbf{G}_a F_a(t) \,. \tag{38}$$

The random variables, $\{F_a(t)\}$, are normalized fluctuations about some mean interaction, that is,

$$\langle F_a(t) \rangle = 0 \quad \text{and} \quad \langle F_a(0)F_a(0) \rangle = 1 \, .$$

The strength of the bath coupling to each system variable is described by the coupling constants $\{f_a\}$ and, because they enter at second order, the rate constant for the dissipation process arising from each term in Eq. (38) will be proportional to f_a^2. The only important properties of the $\{F_a(t)\}$ are their autocorrelation and cross-correlation functions, $\langle F_a(0)F_a(t) \rangle$ and $\langle F_a(0)F_b(t) \rangle$, which enter the definition of the Redfield tensor in Eq. (18). These, like the classical correlation functions discussed earlier, do not satisfy the detailed-balance relation and must be corrected in the same way. It is convenient, but not necessary, that the variables be chosen to be independent, so that the cross-correlation functions vanish.

Often, the most that is known about a bath is a relevant relaxation time. In that case, the use of a simple exponential correlation function might be appropriate. If one makes the further assumption that the autocorrelation functions of the random variables decay rapidly compared to all other important time scales in the system, Θ_a^+ simplifies to [38, 40]

$$(\Theta_a^+)_{i,j} = \tau_c^{(a)}/(1 + e^{\beta \hbar \omega_{i,j}}) \, , \tag{39}$$

where $\tau_c^{(a)}$ is the characteristic decay time of the autocorrelation function of $F_a(t)$ and the standard correction for non-quantum mechanical correlation functions [Eq. (31)] has been applied. In practice, the form [38, 40].

$$(\Theta_a^+)_{i,j} = (\Theta^+)_{i,j} = 1/(1 + e^{\beta \hbar \omega_{i,j}}) \tag{40}$$

can then be used, so that the $\{\tau_c^{(a)}\}$ are included implicitly in the phenomenological fluctuation parameters, $\{f_a\}$. This is the fast-bath approximation adopted for the cases considered in Section IV.

C. Time Propagation

The evolution of the density matrix in time requires the solution of the equation of motion, Eq. (10) or (20); in a basis set of N states, this represents a system of N^2 coupled linear differential equations for the individual density matrix elements. It is most natural to consider the equation in Liouville space, where ordinary operators ($N \times N$ matrices) are treated as vectors (of length N^2) and superoperators such as $\mathcal{R}$ and $\mathcal{L}_R$, which act on operators to create new operators, become simple matrices (of size $N^2 \times N^2$). In the Liouville space notation, Eq. (9) would be written

$$\dot{\sigma}_J = \sum_K L_{J,K}\sigma_K \,, \tag{41}$$

where $L_{J,K}$ is a matrix element of $\mathscr{L}_R$,

$$L_{J,K} = i\Omega_J\delta_{J,K} + R_{J,K} \tag{42}$$

and the indices J and K each run over the N^2 matrix elements of σ; the $\{\Omega_J\}$ are the N^2 transition frequencies $\{\omega_{i,j}\}$ among the eigenvalues of $\mathbf{H}_s$ [compare with Eq. (10) or (20)]. The formal solution of Eq. (41) would be

$$\sigma(t) = e^{\mathscr{L}_R t}\sigma(0) \,. \tag{43}$$

The expansion of the $\mathscr{L}_R$ in a Liouville space basis set yields a complex asymmetric matrix, $\mathbf{L}$ [see, e.g., Eq. (54)]. If this matrix can be diagonalized, the exponential operator in Eq. (43) can be applied in the form

$$\sigma(t) = \mathbf{M}e^{\Lambda t}\mathbf{M}^{-1}\sigma(0) \,, \tag{44}$$

where Λ is the diagonal matrix of the eigenvalues of $\mathbf{L}$ and $\mathbf{M}$ is the solution of the eigenvalue equation, $\mathbf{LM} = \mathbf{M}\Lambda$. In fact, in the limit of zero temperature $\mathbf{L}$ is not diagonalizable at all (i.e., a similarity transformation $\mathbf{MLM}^{-1}$ can at most reduce $\mathbf{L}$ to block-diagonal Jordan form). Even at higher temperatures, calculation of the time-evolved density matrix by Eq. (44) is often numerically unstable.

We have recently described a technique, the short-iterative Arnoldi propagator, for evolving the density matrix under Eq. (41) by solving the equation of motion explicitly in time [39]. This method, a generalization on the efficient short-iterative-Lanczos propagator introduced by Park and Light to solve the time-dependent Schrödinger equation [2, 73], allows fairly large time steps to be taken with high accuracy. The central idea is to form an approximate exponential propagator at each time step, defined in the n-dimensional Krylov space spanned by $n-1$ applications of the Redfield–Liouville tensor to the current density matrix. This approach is especially effective for "stiff" systems of dynamical equations, which contain a wide range of intrinsic time scales; the approximate exponential propagator handles this well because much of the rapid short-time dynamics can often be completely contained within the Krylov subspace. The long-range electron transfer system discussed below is a good example of such a system; in the most extreme case, processes with

rates spanning six orders of magnitude are treated in the same calculation. In fact, a more elaborate implementation of the same idea has also
been demonstrated to be useful for more general systems of stiff
differential equations [74].

1. Short-Iterative-Arnoldi Propagator

Given a symmetric matrix $\mathbf{L}$ and an initial vector $\mathbf{v}_0$, the Lanczos
algorithm can be used to construct an orthonormal basis set $\{\mathbf{v}_j\}$ that
spans the n-dimensional space defined by $n-1$ applications of $\mathbf{L}$ to $\mathbf{v}_0$
(the Krylov space) [75]; the special property of this basis is that it gives $\mathbf{L}$
a tridiagonal representation, simplifying its subsequent diagonalization.
However, since the Liouville space representation of the Redfield–
Liouville tensor $\mathscr{L}_R$ is complex asymmetric, we use instead Arnoldi's
generalization of the Lanczos algorithm to define a basis that gives $\mathbf{L}$ an
upper Hessenberg representation, $\mathbf{h}$ [74, 76, 77]. The method is defined
by the recurrence relation

$$h_{j+1,j}\mathbf{v}_{j+1} = \mathbf{L}\mathbf{v}_j - \sum_{i=0}^{j} h_{i,j}\mathbf{v}_i \,, \tag{45}$$

where $\mathbf{v}_i$ is the ith orthonormal basis vector and $h_{i,j}$ is a matrix element of
$\mathbf{L}$ in this basis. This transformation of $\mathbf{L}$ may be written, $\mathbf{L} \approx \mathbf{V}\mathbf{h}\mathbf{V}^T$, where
$\mathbf{V}$ is the orthogonal matrix whose jth column is $\mathbf{v}_j$. The $\{\mathbf{v}_j\}$ are
constructed iteratively; starting from vector $\mathbf{v}_j$, the next basis vector $\mathbf{v}_{j+1}$
is obtained by the following sequence of steps:

$$
\begin{aligned}
&1. \quad \mathbf{u}_{j+1} = \mathbf{L}\mathbf{v}_j \\[4pt]
&2. \quad h_{i,j} = \mathbf{v}_i^T \mathbf{u}_{j+1} \qquad \text{for } i = 0, \dots, j \\[4pt]
&3. \quad \mathbf{u}_{j+1}' = \mathbf{u}_{j+1} - \sum_{i=0}^{j} h_{i,j}\mathbf{v}_i \\[4pt]
&4. \quad h_{j+1,j} = (\mathbf{u}_{j+1}'^T \mathbf{u}_{j+1}')^{1/2} \\[4pt]
&5. \quad \mathbf{v}_{j+1} = \mathbf{u}_{j+1}'/h_{j+1,j} \,.
\end{aligned}
\tag{46}
$$

The matrix elements $h_{i,j}$ of the upper Hessenberg representation of $\mathbf{L}$ are
thus automatically generated during the construction of the vectors $\{\mathbf{v}_j\}$.
The essence of the short-iterative-Arnoldi propagator is to form an
explicit representation of the exponential operator $e^{\mathscr{L}_R t}$ in the n-dimen-
sional Krylov space based on the initial density matrix, $\sigma(t)$.

Identifying $\mathbf{v}_0 = \sigma(t)$, the basis $\{\mathbf{v}_j\}$ and approximate Redfield–Liouville matrix $\mathbf{h}$ are constructed as described above. After diagonalizing $\mathbf{h}$, the application of the approximate propagator to $\sigma(t)$ is written

$$\sigma(t + \Delta t) = e^{\mathscr{L}_R \Delta t} \sigma(t) \approx \mathbf{V} \mathbf{W} e^{\Lambda \Delta t} \mathbf{W}^{-1} \mathbf{V}^T \mathbf{v}_0 , \qquad (47)$$

where

$$\mathbf{h} \mathbf{W} = \mathbf{W} \Lambda ; \qquad (48)$$

that is, Λ is the diagonal matrix of the n eigenvalues of $\mathbf{h}$ and $\mathbf{W}$ is the corresponding $n \times n$ eigenvector matrix; note also that since $\mathbf{h}$ is not Hermitian, $\mathbf{W}$ is not unitary.

Practical Issues. Experience with this method has shown that a Krylov space size, n, on the order of 10 to 20 is practical [39, 40].; larger spaces can permit larger time steps to be taken, but this is largely offset by the increased numerical overhead incurred. Note that n is independent of the dimensionality of Liouville space, N^2, which may be as large as 10,000 or more. The maximum time interval for which the approximate propagator is used is chosen so that the propagated density matrix has an arbitrarily small projection outside the Krylov space. It is useful to consider the vector of coefficients for the Krylov space basis vectors after propagation for a given time t,

$$\mathbf{c}(\Delta t) = \mathbf{w} e^{\Lambda \Delta t} \mathbf{w}^{-1} \mathbf{V}^T \mathbf{v}_0 , \qquad (49)$$

such that

$$\sigma(t + \Delta t) = \mathbf{V} \mathbf{c}(\Delta t) . \qquad (50)$$

For a sufficiently short propagation time, perturbation theory predicts that the coefficients will grow in as

$$c_k(\Delta t) = \frac{1}{k!} (\Delta t)^k \prod_{j=0}^{k-1} h_{j+1,j} \qquad (51)$$

for $k \geq 1$; this is actually the same result as that obtained for a symmetric L [73]. Assuming that the coefficients decrease monotonically with increasing index k, the constraint that $c_{n-1}(\Delta t)$ be less than some cutoff c_{max} assures that the projection of the final density matrix onto the Krylov space basis vectors of order higher than n is also less than the cutoff.

In the course of relaxation, the number of system eigenstates with significant population typically decreases as the simulation progresses. Since the amount of effort required in each propagation step is proportional to N^3, a great deal of additional efficiency can be gained by truncating the size of the system periodically. A practical algorithm, followed in Refs. 39 and 40, is to truncate the density matrix to N' states when all matrix elements involving states $N' + 1$ to N are below a specified cutoff.

2. Other Methods

Berman and co-workers have recently shown how to use a generalized high-order polynomial expansion to represent a nonunitary exponential propagator [59]. In particular, they have applied it in the context of the axiomatic semigroup theory, discussed above [54–57, 60], to propagate a density matrix on a phase-space grid [59, 78]. This method represents a generalization of the authors' previous highly effective Chebysheff polynomial propagator for the time-dependent Schrödinger equation [1, 2, 79]. Like that method, it can be applied to very high order and thus take very large time steps; in fact, the entire propagation can often be performed as a single time step. The technique is directly applicable to the solution of the Redfield equation as well, since it requires only that the generator of the propagator (e.g., the Redfield–Liouville tensor) can be applied efficiently to the density matrix. While it is likely that each will ultimately be found better suited for certain classes of problems, the trade-offs in convenience and efficiency between this technique and the short-iterative-Arnoldi propagator remain to be explored.

IV. APPLICATIONS OF THE STOCHASTIC MODEL

A. Two-Level System in a Fast Stochastic Bath

Before going on to consider more complicated systems, we review here some of the basic behavior of a two-state quantum system in the presence of a fast stochastic bath. This highly simplified bath model is useful because it allows qualitatively meaningful results to be obtained from a density matrix calculation when bath correlation functions are not available; in fact, the bath coupling to any given system operator is reduced to a scalar. In the case of the two-level system, analytic results for the density matrix dynamics are easily obtained, and these provide an important reference point for discussing more complicated systems, both because it is often possible to isolate important parts of more complicated systems as effective two-level systems and because many aspects of the dynamics of multilevel systems appear already at this level. An earlier discussion of the two-level system can be found in Ref. 80. The more challenging problem of a two-level system in a general harmonic bath (the spin-boson system) is addressed in Section V.

Let the system be described by the Hamiltonian $\mathbf{H} = \mathbf{H}_0 + \mathbf{V}(t)$, with

$$\mathbf{H}_0 = \Delta(|1\rangle\langle 1| - |2\rangle\langle 2|) + J(|1\rangle\langle 2| + |2\rangle\langle 1|) \tag{52}$$

and

$$\mathbf{V}(t) = |1\rangle F_1(t)\langle 1| + |2\rangle F_2(t)\langle 2|) . \tag{53}$$

Here $F_1(t)$ and $F_2(t)$ are independent normalized random variables that satisfy $\langle F_a(t)\rangle = 0$ and $\langle F_a(t)F_b(0)\rangle = f^2\delta_{a,b}$. Using Eqs. (10), (11), (17), and (40), the Redfield–Liouville tensor $\mathscr{L}_R$ is expressed in the Liouville space basis $\{|1\rangle\langle 1|, |2\rangle\langle 2|, |1\rangle\langle 2|, |2\rangle\langle 1|\}$ as the matrix

$$\mathbf{L} = \begin{pmatrix} -KA & A & \sqrt{AB} & \sqrt{AB} \\ KA & -A & -\sqrt{AB} & -\sqrt{AB} \\ K\sqrt{AB} & -\sqrt{AB} & -(AM + 2B) + 2iE & AM \\ K\sqrt{AB} & -\sqrt{AB} & AM & -(AM + 2B) - 2iE \end{pmatrix}, \tag{54}$$

where

$$A = f^2 J^2 / 2E^2 ,$$

$$B = f^2 \Delta^2 / 2E^2 ,$$

$$M = \tfrac{1}{2}(1 + K) ,$$

$$K = e^{-2\beta E} ,$$

$$E = \sqrt{J^2 + \Delta^2} ;$$

that is, $\pm E$ are the eigenvalues of $\mathbf{H}_0$. The eigenvalues of $\mathbf{L}$ are the relaxation rates of the system, and we can identify that value with the smallest real part as the rate of equilibration of the level populations. The nonzero eigenvalues of $\mathbf{L}$ are the roots of the cubic equation,

$$\lambda^3 + 4(B + AM)\lambda^2 + 4[(B + AM)^2 + E^2]\lambda + 8AE^2M = 0 . \tag{55}$$

Although the general roots of Eq. (55) are easily obtained, there are two limiting cases in which particularly simple and useful results are found. The first is for the symmetric case, $\Delta = 0$. In this limit, the population and coherence equations decouple and the population relaxation rate in the site representation is precisely the *coherence* decay rate

$(1/T_2)$ in the eigenstate representation, which is

$$k = \tfrac{1}{2}(f^2 M - \sqrt{f^4 M^2 - 16 J^2}) \,, \tag{56}$$

where the thermal factor, M, varies from $\tfrac{1}{2}$ for $T \ll J$ to unity when $T \gg J$. In the weak (underdamped) and strong (overdamped) fluctuation regimes, respectively,

$$k = \begin{cases} \tfrac{1}{2} f^2 M & (f^2 < 2J/M) \,, & (57) \\ 4J^2/f^2 M & (f^2 \gg 2J/M) \,. & (58) \end{cases}$$

In the underdamped regime the rate is thus largely independent of J, except through the thermal factor M, and is proportional to the fluctuation strength.

The overdamped limit gives the golden rule rate, with k proportional to the square of the coupling, J^2, and inversely proportional to the damping strength, f^2. Note that the *population* relaxation rate in the eigenstate representation $(1/T_1)$ is simply $f^2 M$ in both regimes. Thus, while the expected relation $1/T_2 \geq 1/2T_1$ is satisfied in the weak-fluctuation regime [27], the unphysical result $1/T_2 < 1/2T_1$ is found in the overdamped case. This is manifested in the development of negative eigenvalues for the density matrix, reflecting physically the fact that the system–bath coupling is too great to be treated at the level of the second-order perturbation theory underlying the Redfield equation [60].

The other case is that of strongly asymmetric systems, in which $\Delta \gg J$. A simple solution for the roots of Eq. (55) is then obtained by treating the coupling, J, perturbatively. The site-population relaxation rate is found to be

$$k = 4 \frac{J^2 f^2 M}{f^4 + 4\Delta^2} \,, \tag{59}$$

which simplifies in the weak and strong fluctuation limits to

$$k = \begin{cases} \dfrac{J^2 f^2 M}{\Delta^2} & (f^2 \ll 2\Delta) \,, & (60) \\[3mm] \dfrac{J^2 M}{f^2} & (f^2 \gg 2\Delta) \,. & (61) \end{cases}$$

Here, the thermal factor, M, goes from $\tfrac{1}{2}$ for $T \ll \Delta$ to unity when $T \gg \Delta$. For the strongly asymmetric system, the rate is always proportional to the square of the coupling but depends strongly on the energy gap, Δ, only in the underdamped case.

B. Long-Range Electron Transfer in DNA/Metal Complexes

One initial application of the techniques described above has been to construct a model for long-range electron transfer that includes both tunneling and conduction-band-like transport in a consistent treatment [40]. This work was stimulated by the controversial experiments of Barton, Turro, and co-workers, who attracted attention some years ago [81, 82] by reporting experiments that appeared to demonstrate extraordinarily efficient long-range photoelectron transfer between metal–ligand complexes intercalated into a DNA double helix. In more recent work, for example, the Barton group obtained evidence of nanosecond-scale electron transfer over a distance of 40 Å between covalently linked DNA/metal-complex adducts [83].

Because of the indirect nature of these experiments, there has been some debate about whether transfer over these distances is at all reasonable on such a short time scale. A substantial body of theoretical and experimental work has shown that long-range electron transfer in both biological and other systems occurs readily by quantum-mechanical tunneling [84–87]. Standard theories for such transfer (such as McConnell's superexchange theory [88, 89]) all predict that the rate should decrease exponentially with increasing donor–acceptor separation R [i.e., $k_{ET} \propto \exp(-\beta R)$, where β is a falloff rate that increases with the height of the barrier] [84, 90]. In fact, Dutton and co-workers have asserted that the distance dependence of all electron transfers in proteins are well described by a *single* value for β in the neighborhood of $1.4 \, \text{Å}^{-1}$ when the various experiments are carefully corrected for temperature dependence and other factors [87]; reported values of β in the literature actually range between 0.6 and 2.0 [86]. Extrapolating from these data, a 40-Å electron transfer would be expected to have a maximum rate slower than once per second. Similarly, Barton and co-workers, using measurements of the driving force, transfer distance and a lower bound for the rate, arrived at an estimate for β of $\leq 0.2 \, \text{Å}^{-1}$ in their system. Attempts to force the standard perturbative superexchange model on this unusual system would require assumptions that are inconsistent with the theory itself: for instance, that the donor state be nearly degenerate with the orbitals of the bridging molecules.

A density matrix approach makes it possible to examine more general models for systems like this [38,39]. For instance, unlike superexchange theory, this approach admits models that include strong couplings between the closely associated species in the intercalated DNA/donor–acceptor complex. More important, the density matrix description allows a proper treatment of quantum coherences, which is critical in bridging the tunneling and free-conduction regimes of transport. On the one hand, systems are easily constructed in which electron transfer is well described by the semiclassical golden rule with the effective donor–acceptor coupling defined by superexchange theory. However, as the temperature increases or the donor energy approaches the energy of the bridge orbitals, the electron can also be thermally promoted into the bridge, whereupon bath fluctuations cause rapid dephasing into the spatially delocalized, wavelike states of the bridge, ultimately followed by the slower uptake of the electron into the acceptor. In this case a much longer range, potentially distance-independent transfer becomes possible. However, the competition between this and the nonadiabatic (tunneling) mechanism has never been explored systematically.

Regarding the simulation of the thermally activated process, it is important to note that the mere presence of spatially delocalized states in the bridge does not, in itself, predict efficient long-range transfer; the action of dephasing processes is crucial. The initial electron wavepacket in the bridge is a coherent linear combination of the wavelike bridge eigenstates that project onto the donor; although each component has, in principle, some projection onto the acceptor state as well, destructive interference among these competing channels effectively localizes the wavepacket in the bridge until it is dephased into an incoherent sum of its components. To capture these competing processes in a unified way, a density matrix formulation of the problem is required.

1. Hamiltonian

The model is based on the standard tight-binding Hamiltonian consisting of a donor, a number of bridge sites, and an acceptor, all coupled to form a linear chain. In addition, a single linearly coupled oscillator is included, representing a high-frequency vibrational coordinate coupled to the electron transfer. The lack of detailed information about this system makes it appropriate to treat the bath stochastically. Thus

$$\mathbf{H} = \mathbf{H}_s + \mathbf{V}(t) , \tag{62}$$

where

$$\mathbf{H}_s = |D\rangle(\mathbf{h}_D(\mathbf{Q}) + \varepsilon_D)\langle D| + |A\rangle(\mathbf{h}_A(\mathbf{Q}) + \varepsilon_A)\langle A| + \sum_{j=1}^{N_B} |j\rangle(\mathbf{h}_j(\mathbf{Q}) + \varepsilon_j)$$

$$\times \langle j| + |D\rangle J_D\langle 1| + |N_B\rangle J_A\langle A| + \sum_{j=2}^{N_B} |j-1\rangle J_B\langle j| \tag{63}$$

represents the isolated electron transfer system and $\mathbf{V}(t)$ is a time-dependent potential arising from interactions with a thermal bath. $|D\rangle$ and $|A\rangle$ are the states in which the electron is localized on the donor and acceptor sites, respectively, while in states $\{|1\rangle \cdots |N_B\rangle\}$ the electron is on one of the N_B intervening bridge sites. Each site has a minimum energy ε_j and is coupled to its nearest neighbors in the fashion of a Hückel Hamiltonian. J_D is the coupling of the donor to the first bridge site, J_A the coupling of the last bridge site to the acceptor, and J_B the coupling of adjacent bridge sites. The single associated oscillator, of unit mass and frequency ω, is described by

$$\mathbf{h}_j(\mathbf{Q}) = \frac{1}{2}\mathbf{P}^2 + \frac{1}{2}\omega^2\left(\mathbf{Q} - \frac{g_j}{\omega^2}\right)^2 . \tag{64}$$

The vibronic coupling g_j causes a displacement of g_j/ω^2 in the position of the oscillator minimum in state $|j\rangle$.

The system–bath coupling is described by the time-dependent term

$$\mathbf{V}(t) = \mathbf{G}_Q F_Q(t) + \mathbf{G}_D F_D(t) + \mathbf{G}_A F_A(t) + \sum_{j=1}^{N_B} \mathbf{G}_j F_j(t) , \tag{65}$$

where the system operators are

$$\begin{aligned}
\mathbf{G}_Q &= (f_{od}\mathbf{Q} + f_d\mathbf{Q}^2) , \\
\mathbf{G}_D &= |D\rangle f_D \langle D| , \\
\mathbf{G}_A &= |A\rangle f_A \langle A| , \\
\mathbf{G}_j &= |j\rangle f_B \langle j| , \qquad j = 1 \cdots N_B ,
\end{aligned} \tag{66}$$

and $F_Q(t)$, $F_D(t)$, $F_A(t)$, and $F_{j=1,N_B}(t)$, are independent normalized random variables representing the influence of the bath. In the calculations of Ref. 40, the bath was treated in the fast-bath limit (see Section III.B.3).

Three distinct types of relaxation processes are found in this model [Eq. (65)]. Bath-induced fluctuations of the site energies are controlled by the parameters f_D, f_A, and f_B; physically, these lead to the dephasing of site–site coherences or, equivalently, to population relaxation among the delocalized eigenstates of the coupled electronic system. Similarly, f_{od} and f_d control the strength of fluctuations arising through the operators $\mathbf{Q}$ and $\mathbf{Q}^2$. Since $\mathbf{Q}$ couples harmonic oscillator eigenstates which differ by a single quantum number, fluctuations of $\mathbf{Q}$ lead to vibrational population decay (T_1 relaxation) and are referred to as "off-diagonal" fluctuations.

Similarly, the $\mathbf{Q}^2$ term produces fluctuations of the oscillator frequency, resulting in pure vibrational dephasing (T_2' relaxation); because $\mathbf{Q}^2$ has diagonal matrix elements in the harmonic oscillator basis, it is said to contribute "diagonal" fluctuations. In each case, the rate of the associated relaxation process is proportional to f_a^2.

Parameters were varied to highlight the competition between adiabatic and nonadiabatic transfer. In the reference system, the donor and acceptor energies were kept around $2000 \, \text{cm}^{-1}$ below the bridge; for comparison, thermal energies ($k_B T$) ranged up to about $200 \, \text{cm}^{-1}$ (around room temperature). Site–site couplings were in the range 200–$300 \, \text{cm}^{-1}$, which is reasonable for closely associated chemical units [65, 91, 92]. When included, the oscillator was given a frequency of $1000 \, \text{cm}^{-1}$.

The total number of basis states required to represent the state of the system depended strongly on the vibrational–electronic coupling; here, a vibrational basis of 5–10 states typically sufficed, giving an overall basis size of up to 80 states; with cutoffs to eliminate negligible density matrix elements, however, only up to about 40 states were typically retained in an actual simulation. For systems without the oscillator, construction and diagonalization of the Redfield tensor was practical, in which case the lowest eigenvalue could be identified as the electron transfer rate. Otherwise, time-stepping simulations were done. The numerical methodology described in Section III made it possible to simulate systems directly with transfer times from the picosecond to the millisecond regime, despite the presence of subpicosecond time scales in the problem; it should be pointed out that this would have been very challenging, if not impossible, to treat by even the most efficient real-time path-integral methods [13, 16, 20].

2. Results

Time-Dependent Results. To explore the nature of the long-range electron transfer dynamics via the thermally activated (adiabatic) and tunneling (nonadiabatic) channels available in this model, systems containing between two and six bridge sites were considered, with a variety of parameter values for the electronic couplings, energy gaps, and amplitudes of the fluctuation operators. The detailed behavior of the transport kinetics as a function of the various parameters is described in detail in Ref. 40; here we summarize just the main conclusions of this work.

Figure 1 shows the time-dependent populations in a typical system of six bridge sites. When the coupling between the donor and bridge was turned on at $t = 0$, there was a sudden jump of population into the bridge.

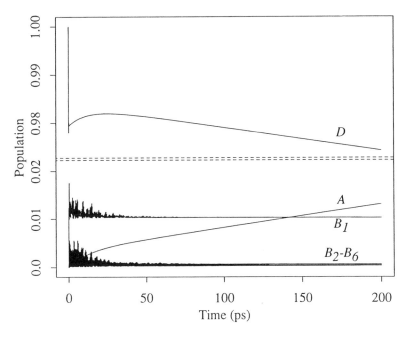

Figure 1. Time-dependent populations for each site of a typical eight-site system. Plotted are the populations of the donor site (D), the acceptor (A), and each of the bridge sites (B_1–B_6). The site energies were $\varepsilon_D = -2000\ \mathrm{cm}^{-1}$, $\varepsilon_B = 0\ \mathrm{cm}^{-1}$, and $\varepsilon_A = -2300\ \mathrm{cm}^{-1}$. Site–site coupling energies were $J_D = J_A = 200\ \mathrm{cm}^{-1}$ and $J_B = 300\ \mathrm{cm}^{-1}$. A single 1000-$\mathrm{cm}^{-1}$ oscillator was included with vibronic couplings $g_D = -g_A = -10\ \mathrm{cm}^{-1}$ and $g_B = 0\ \mathrm{cm}^{-1}$. For this figure, stochastic fluctuations of only the donor and acceptor site energies were included, with $f_D^2 = f_A^2 = 49$ ($f_B = f_{od} = 0$).

Because the dynamics in this particular bridge is underdamped, the population that enters the bridge oscillates rapidly among all the bridge sites. The oscillations are gradually dampened to a steady level over the first hundred picoseconds, due both to the dephasing of the quantum coherences by bath fluctuations and to destructive interference with additional population amplitude coming from the donor.

An interesting observation to make here is that the dephasing of the bridge is fairly rapid despite the fact that there are no explicit fluctuations of the bridge site energies in this system. Instead, dephasing in the bridge arises indirectly, through the weak coupling of the (fluctuating) donor and acceptor sites to the bridge. As this demonstrates, the connection between the parameters of the model Hamiltonian, defined in the diabatic site representation, and specific processes seen in the simulations is often indirect. For instance, although some explicit dephasing must be

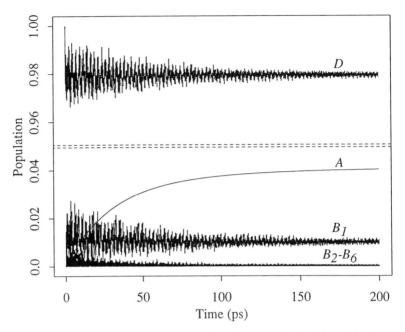

Figure 2. Time-dependent site populations for the system of Fig. 1, but with site-energy fluctuations *only* in the acceptor ($f_A^2 = 49$). Plotted are the populations of the donor site (D), the acceptor (A), and each of the bridge sites (B_1-B_6).

included for there to be a definable nonadiabatic rate constant, there is no way to introduce the necessary dephasing of the donor–acceptor transition without also introducing fluctuations into the bridge. In Fig. 2, time-dependent populations are plotted for a system identical to that of Fig. 1, except that the sole source of dephasing is in fluctuations of the acceptor site energy. Here the oscillations of the bridge site populations are much longer lived; in fact, the dephasing of the bridge and donor is inefficient enough that tunneling continues to dominate even at temperatures where the dynamics of the first system is overwhelmingly adiabatic.

Temperature and Distance Dependence. The most important feature of this model is the distance dependence predicted for long-range transfer through the activated channel. Figure 3 is an Arrhenius plot of the temperature-dependent transfer rates for systems of increasing bridge lengths (number of bridge sites). As the temperature was decreased (as $\beta = 1/k_B T$ increases) the rate in each system reached an asymptotic value that represented direct tunneling from the donor to the acceptor. At

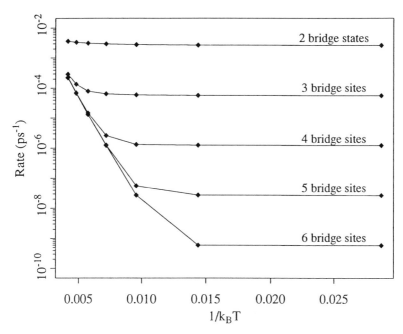

Figure 3. Arrhenius plot of the temperature-dependent transfer rates for systems with an increasing number of bridge sites. k_B is Boltzmann's constant. Site energies were $\varepsilon_D = -2000\,\mathrm{cm}^{-1}$, $\varepsilon_B = 0\,\mathrm{cm}^{-1}$, and $\varepsilon_A = -2300\,\mathrm{cm}^{-1}$, while site–site couplings were $J_D = J_A = 200\,\mathrm{cm}^{-1}$ and $J_B = 300\,\mathrm{cm}^{-1}$. A single 1000-cm^{-1} oscillator was coupled to each site, with $g_D = -g_A = -10\,\mathrm{cm}^{-1}$ and $g_B = 0\,\mathrm{cm}^{-1}$. Site-energy fluctuations and off-diagonal vibrational fluctuations were also included, with $f_D^2 = f_A^2 = f_B^2 = 49$, $f_{od}^2 = 25$, and $f_d^2 = 0$.

higher temperatures, the classic Arrhenius behavior of an activated process was found. Moreover, the rates for systems with more than three bridge sites exhibited an identical temperature dependence in this regime. To see more clearly what happened, these results are replotted in Fig. 4 as a function of bridge length at selected temperatures. Here it is clear that at any temperature chosen, the transfer rate becomes *distance independent* for a sufficiently long bridge.

The temperature-dependent rates, k, at each bridge length were analyzed using the modified Arrhenius equation

$$k = k_{NA} + \nu e^{-\beta \Delta E^\ddagger}, \tag{67}$$

where k_{NA} is the limiting nonadiabatic (tunneling) rate at low temperature, ν the Arrhenius preexponential factor, and $\Delta E^\ddagger$ an effective activation energy. Table I shows the results of this analysis for the data of

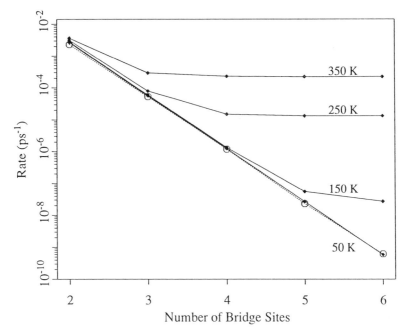

Figure 4. The data of Fig. 3, replotted as a function of bridge length at selected temperatures. In addition, rates calculated for a 300-K system with site-energy fluctuations only in the acceptor ($f_D = f_B = f_d = f_{od} = 0$, and $f_A^2 = 49$) are plotted with a dashed line and open circles.

TABLE I

The Arrhenius Analysis [Eq. (67)] of the Temperature-Dependent Rates Shown in Fig. 3[a]

Number of Bridge Sites	k_{NA} (ps^{-1})	ν (ps^{-1})	$\Delta E^{\ddagger}$ (cm^{-1})
2	2.656×10^{-3}	0.0047	391
3	5.888×10^{-5}	0.1158	1515
4	1.238×10^{-6}	0.2151	1674
5	2.662×10^{-8}	0.2158	1681
6	5.899×10^{-10}	0.2197	1684

[a] Note that the analysis is somewhat ill defined in the two- and three-bridge-site systems because of the dominance of tunneling, even at the highest temperatures considered.

Figs. 3 and 4. The tunneling rate decreases exponentially with distance, as it should. In the weak-coupling (golden rule) limit, the rate goes as the square of the effective coupling of the donor and acceptor, J_{NA}, proportional to $J_D J_A J_B^{N_B-1}/\varepsilon_{B,D}^{N_B}$ in this system. Therefore, k_{NA} falls off like $\exp(-\beta N_B)$, with $\beta = -2\ln(J_B/\varepsilon_{B,D})$.

The Arrhenius equation decomposes the activated rate into the preexponential factor ν and the activation energy, $\Delta E^{\ddagger}$. The latter just reflects the energy gap between the donor and the bridge, $\varepsilon_{B,D}$, or more precisely, the energy difference between the ground state of the donor and the lowest eigenstate associated with the bridge. Thus $\Delta E^{\ddagger}$ is always a bit lower than $\varepsilon_{B,D}$ because the electronic couplings, J_B, cause the degenerate zero-order bridge states to split into a band of delocalized eigenstates, whose width approaches $4J_B$ as the length of the bridge increases.

The more interesting part of the adiabatic rate is captured in the preexponential factor, ν, which represents the high-temperature limit of the activated rate. This depends in a nontrivial way on a number of factors, including the donor-bridge and acceptor-bridge energy gaps, the couplings between adjacent sites, and most critically, the dephasing of the site–site electronic coherences by the stochastic bath. For the purposes of discussion, however, it is useful to decompose it (very crudely) into a sequential kinetic scheme: promotion of electron amplitude to the bridge, dephasing of the electron in the bridge, and uptake by the acceptor. Then the effects of modifying the system can be analyzed in terms of the impact on one or more of these various steps.

Vibrational and Electronic Fluctuations. The critical role of relaxation processes in opening the adiabatic channel is shown dramatically in Fig. 4, in which the high-temperature (300 K) transfer rates are plotted for a system identical to that of Fig. 2, except that *only* the acceptor is directly coupled to the bath. Despite sufficient thermal energy to populate the bridge, it is clear that the activated channel was essentially shut down, leaving only long-range tunneling as an effective transfer mechanism. Although the coupling of the acceptor to the bridge contributes some fluctuations to the rest of the system, the resulting value of ν was apparently so low that either $\varepsilon_{B,D}$ would have to be smaller or the distance between D and A much greater in order for the activated rate to become competitive with tunneling.

Of the various relaxation mechanisms at work in this system, fluctuations of the site energies ar the most important. The rates of donor-bridge and acceptor-bridge transfer and the dynamics within the bridge are all strongly dependent on these fluctuations, and this dependence can be understood by appealing to the simple two-level results presented in Section IV.A. The donor-bridge and acceptor-bridge dynamics fall in the underdamped asymmetric limit ($\Delta \gg J$), for which the population transfer rate is proportional to $J^2 f^2 / \Delta^2$ as long as the fluctuation strength f^2 is less than 2Δ. The bridge, on the other hand, is a multisite generalization of the symmetric two-level system ($\Delta = 0$). For the two-level system, the dynamics are underdamped for $f^2 < 4J/(1 + e^{-2\beta J})$, with relaxation dynamics proportional to $f^2(1 + e^{-2\beta J})$; in the overdamped limit, relaxation goes as J^2/f^2. The activated dynamics of the full system were qualitatively well described by these relationships, that is, rate increased with $J^2_{\{D,A\}}$ and $f^2_{\{D,B,A\}}$, because of the effects on the donor-bridge and acceptor-bridge kinetics. Also, the rate *dropped* weakly with increasing J_B as predicted for the underdamped regime; note that this is unlike the long-range tunneling rate, which increases with J_B [40]. Finally, the expected turnover from underdamped dynamics (completely distance-independent rate) to overdamped dynamics, (weak decrease in the rate with bridge length) was observed as either J_B was decreased or f_B increased.

In contrast, the rate was relatively insensitive to either of the vibrational relaxation mechanisms in the regimes studied. Because of the high frequency of the included oscillator ($1000 \, \text{cm}^{-1}$), excited vibrational states were significant here only in the donor and acceptor sites, where they provided a more efficient gateway into and out of the bridge. Thus vibrational coherences were not particularly important and vibrational dephasing (through f_d) had little effect on the rate. Increasing the vibrational population relaxation rate (proportional to f^2_{od}) increased the activated rate only up to point, indicating that electronic process were rate limiting here. Presumably, however, strongly coupled lower-frequency modes could be capable of trapping electrons in individual sites, and in this type of system vibrational relaxation could become crucial.

3. Discussion

The electron transfer model presented here recalls the process of charge transport in semiconductors; that is, a "conduction band" is populated by a thermalized electron, which then moves freely through the semiconductor via wavelike **k** states. While the possibility of semiconductor-like electron transfer in biological systems was first raised many years ago by DeVault and Chance [93], it has never been found experimentally; in fact, there was reasonable skepticism that nature would choose such a mechanism in natural biological systems [84]. The density matrix method allows one to construct a model in which the conditions for such a process can be clarified and investigated in a detailed way.

The first such condition, naturally, is that some sort of conduction band of delocalized wavelike states exists in the bridge. Such a band should be found whenever the site–site couplings among the bridge states are not significantly smaller than the spread of site energies in the bridge. However, as shown in Ref. 40, even for a bridge of irregularly spaced sites (in energy) and weak couplings, an only weakly distance-dependent process can remain and dominate over long-range tunneling. A second requirement is that the donor-bridge energy gap be small enough to allow the bridge to be populated. This work showed that the actual population of the bridge need not be substantial and that energy gaps of a few thousand wavenumbers should be sufficient to see the process at room temperature.

Finally, the model underscores the critical importance of the system–bath interactions that lead to thermalization and the dephasing of electronic coherence in the bridge. It would be surprising if this were a bottleneck for a large molecule in a condensed phase; it seems more likely that bath fluctuations are stronger than considered here. As our analysis emphasized, however, it is the relative magnitudes of the site–site couplings and the electronic dephasing processes in the bridge that determine whether the electron can move in a wavelike manner or propagate diffusively. These conditions are likely to be quite different in biological systems than in systems previously considered (such as crystalline semiconductors).

With respect to the experiments of the Barton group, it is not unreasonable to expect that the conditions for band formation exist in a DNA duplex, in which the bridge nominally consists of nearly identical stacked molecular units: namely, the purine (adenine and guanine) and pyrimidine (thymine and cytosine) base pairs. Relatively strong coupling between stacked bases of either type is expected [65, 91, 92]; weaker purine–purine or pyrimidine–pyrimidine coupling between bases on complementary strands might also be significant, but relatively little work has been done to provide good estimates of the energies and couplings in such a system. In the DNA system studied by Barton [83], the relevant energy is that between the photoexcited donor state, nominally the lowest excited state of the large conjugated dipyridophenazone ligand of the Ru donor complex, and a reduced nucleotide base in the intercalation site. While the energy of the former can be estimated from the absorption spectrum of the donor, the latter is unknown.

4. Connection to Other Treatments

In Marcus theory [91, 94], the bath is represented by an overdamped harmonic coordinate, coupled linearly to the donor and acceptor states and representing the collective polarization of the environment around the donor–acceptor pair. The system–bath coupling is characterized by the bath reorganization energy. In the work described here [40] the donor and acceptor are treated essentially as single quantum levels and the bath appears only indirectly through its effects (stochastic fluctuations) on the system Hamiltonian. Thus no reorganization energy appears, nor is there an activation energy for the bath reorganization that typically precedes electron transfer [91, 94]. In the pure tunneling limit, a reduction of the reactants to single levels may be justified because the process is fast enough that the bath is essentially frozen during the transfer (see also Ref. 95). Our model, however, is supposed to represent a more general case.

In fact, canonical transformation methods allow one to define an equivalent, albeit temperature-dependent weak-coupling system from a general Marcus type of Hamiltonian, where the solvent activation barrier and reorganization energy are incorporated into renormalized values of the donor–acceptor energies and coupling [46, 96]. It is in this sense that parameters in our picture are best understood. To use this model to predict absolute reaction rates given experimentally or theoretically derived parameters would probably require that such a transformation be made. Alternatively, an explicit solvent coordinate can always be incorporated in the same way as the high-frequency mode already included; in this case the connection to the Marcus picture once again becomes transparent.

V. CANONICAL TRANSFORMATIONS: REDUCING THE STRENGTH OF THE SYSTEM–BATH COUPLING

A. General Strategy

The success of the reduced-density-matrix methodology hinges largely on the ability to find an appropriate separation of the system into subsystem and bath. If the subsystem dynamics are to be obtained using the Redfield equation, the primary consideration is that the resulting system–bath coupling be weak enough to treat with the second-order perturbation theory that underlies this approach. This can be evaluated rigorously if the starting point is a microscopic system–bath Hamiltonian. More generally, the observation that loss of positivity is semiquantitatively associated with the strength of the system–bath coupling (see Ref. 60) provides a heuristic for judging when this coupling becomes too large to treat properly without additional measures.

In this section we suggest a general twofold transformation strategy for reducing strong-coupling systems to weak-coupling form and demonstrate its application to the spin-boson Hamiltonian (i.e., a two-level quantum system coupled linearly to a bath of harmonic oscillators). The methods presented below are, in fact, directly generalizable to a much broader range of systems; in particular, general N-state systems or systems with polynomial potential surfaces are immediately admissible. Also, while the coupling of the system to the harmonic bath has to be linear in the bath displacements, it can be nonlinear in the system coordinates.

First, the new numerical technology available for density matrix propagation allows substantially larger systems to be treated than before. As a result, it becomes a feasible strategy to incorporate one or two of the most strongly coupled bath degrees of freedom directly into the system. In some cases the choice of an appropriate coordinate is obvious; in proton transfer systems, for instance, the donor–acceptor atom separation is very strongly coupled to the proton transfer, so it makes more sense to incorporate this into the system, if at all possible, than to treat it as a bath coordinate [97, 98]. It is probably more typical that the bath consists of a large number of weakly coupled coordinates, each equally important; in this case it may be possible to define a suitable collective coordinate artificially. For a bath of linearly coupled oscillators, in particular, one can always define a collective coordinate in which all of the coupling to the system is localized [99]. The Marcus theory of electron transfer, in which the bath is represented by a single collective "polarization" coordinate, is derived from the spin-boson Hamiltonian in this way [44]. Pollak recently used a similar idea in the context of classical activated rate theory [100].

Second, canonical transformation methods may be employed to diagonalize the system–bath Hamiltonian partially by a transformation to new ("dressed") coordinates. Such methods have been in wide use in solid-state physics for some time, and a large repertoire of transformations for different situations has been developed [101]. In the case of a linearly coupled harmonic bath, the natural transformation is to adopt coordinates in which the oscillators are displaced adiabatically as a function of the system coordinates. This approach, known in solid-state physics as the small-polaron transformation [102], has been used widely and successfully in many contexts. In particular, Harris and Silbey demonstrated that many important features of the spin-boson system can be captured analytically using a variationally optimized small-polaron transformation [45–47]. As we show below, the effectiveness of this technique can be broadened considerably when a collective bath coordinate is first included in the system directly.

B. Spin-Boson Hamiltonian

We address ourselves here to the case of an asymmetric two-level system (TLS) coupled linearly to a bath of harmonic oscillators, the spin-boson system. This system has been studied extensively as the prototype of a quantum system in a dissipative environment [13, 14, 20, 45–47, 103–

110]. Despite its apparent simplicity, its behavior is rich and the dynamics of the two-level system can be obtained analytically only in particular limiting cases (see Ref. 107 for a comprehensive review). At high temperature, transitions from one quantum state to the other proceed by activated crossing of the barrier that arises from the coupling to the bath; below a certain crossover temperature, transitions occur instead by tunneling through the multidimensional barrier. The dynamics of the low-temperature tunneling are sensitively dependent on the strength and nature of the bath coupling, with the transition rate typically going exponentially to zero as the coupling strength is increased.

The Hamiltonian of the spin-boson system is

$$\mathbf{H} = 2\Delta\sigma_z + 2J\sigma_x + \frac{1}{2}\sum_j^N (\mathbf{p}_j^2 + \omega_j^2 \mathbf{q}_j^2) + 2\sigma_z \sum_j^N g_j \mathbf{q}_j ,\tag{68}$$

where σ_z and σ_x are Pauli spin-$\frac{1}{2}$ operators of the TLS, and $\mathbf{q}_j$ and $\mathbf{p}_j$ are the position and momentum operators of the jth bath oscillator. As discussed earlier, the linearly coupled bath is usually described in terms of its spectral density function,

$$J(\omega) = \pi \sum_j \frac{g_j^2}{\omega} \delta(\omega - \omega_j) .\tag{69}$$

For a continuous distribution of oscillators, this is often assumed to be of the general form

$$J(\omega) = \eta\omega^n, \qquad \omega < \omega_c \tag{70}$$

$$J(\omega) = \eta\omega^n e^{-\omega/\omega_c} ,\tag{71}$$

where η is a friction constant and ω_c a cutoff frequency. In particular, $n = 1$ describes the ohmic bath [104, 111], which has a number of interesting properties, among which are that it can result in complete suppression of tunneling, even at zero temperature, for η beyond a critical value [103, 105]. A convenient measure of the overall strength of the TLS–bath coupling is the classical binding energy

$$\begin{aligned} S &= \frac{1}{2}\sum_j \frac{g_j^2}{\omega_j^2} \\ &= \frac{1}{2}\int_0^\infty \frac{J(\omega)}{\omega} d\omega . \end{aligned}\tag{72}$$

For a symmetric system ($\Delta = 0$), S can also be thought of as the classical activation barrier between the two quantum states.

When the overall coupling is weak, the Redfield equation can be applied directly to this Hamiltonian. Then the only part of the bath that affects the dynamics are oscillators of frequency $\omega_j = 2\sqrt{J^2 + \Delta^2}$, resonant with the splitting of the bare two-level system. In fact, if we let $f^2 = g_j^2 / [2\omega_j(1 - e^{-\beta\hbar\omega_j})]$, the dynamics are identical to that of the simple two-level system examined in Section IV.A and it is clear that effects such as complete suppression of tunneling cannot be obtained. What is missing is that, at stronger couplings, the nonresonant bath oscillators become important as well. This is clear from the results of Laird and Skinner, tho derived a Redfield-like equation of motion for the two-level system that was accurate to fourth order in the system–bath coupling and depended on the three-time correlation functions of the bath coordinates [112, 113]. Our approach, in contrast, will be to use canonical transformations to obtain a "renormalized" system–bath Hamiltonian in which these interactions are resolved in a static way in the Hamiltonian itself.

C. Variational Canonical Transformations

Canonical transformations are used to reformulate the Hamiltonian of a system in new coordinates that render it more amenable to analysis [101]. The transformation of $\mathbf{H}$ is defined by a unitary operator, $\mathbf{U}$, that acts to produce a new Hamiltonian, $\tilde{\mathbf{H}} = \mathbf{U}^\dagger \mathbf{H} \mathbf{U}$. Because $\mathbf{U}$ is unitary, it is guaranteed to preserve the eigenvectors and eigenvalues of the original Hamiltonian. When $\mathbf{H}$ is written in terms of some set of variables (operators) $\{\mathbf{A}_j\}$, then $\tilde{\mathbf{H}}$ will have the same form when written in terms of the transformed variables $\{\tilde{\mathbf{A}}_j = \mathbf{U}^\dagger \mathbf{A}_j \mathbf{U}\}$, each of which is in general a function of all the untransformed variables [i.e., $\tilde{\mathbf{A}}_j = f(\{\mathbf{A}_j\})$]. To make progress, then, we use the latter relationship to write the transformed Hamiltonian $\tilde{\mathbf{H}}$ in terms of the untransformed variables, $\{\mathbf{A}_j\}$ (see Ref. 101).

A unitary operator can always be defined in terms of a Hermitian generator, $\mathbf{S}$, such that

$$\mathbf{U} = e^{-\mathbf{S}}. \tag{73}$$

This form is convenient because the transformed operators can then be obtained using the commutator expansion [32, 101]

$$\begin{aligned} \tilde{\mathbf{A}} &= \mathbf{U}^\dagger \mathbf{A} \mathbf{U} \\ &= \mathbf{A} + [\mathbf{A}, \mathbf{S}] + \frac{1}{2!}[[\mathbf{A}, \mathbf{S}], \mathbf{S}] + \frac{1}{3!}[[[\mathbf{A}, \mathbf{S}], \mathbf{S}], \mathbf{S}] + \cdots . \end{aligned} \tag{74}$$

Here, and below, transformed operators and the corresponding renormalized parameters are marked by a tilde. The superscript dagger indicates the Hermitian conjugate.

Variational Optimization. The power of the canonical transformation strategy is greatly enhanced when the transformation is defined in terms of adjustable parameters that can be variationally optimized. The present goal is to minimize system–bath coupling, hopefully to the point that it can be treated effectively with perturbation theory; however, this statement needs to be made more specific before the optimization problem is well defined. A standard approach is to appeal to the Gibbs–Bogoliubov variational principle, which states that A, the Helmholtz free energy of some Hamiltonian, $\mathbf{H} = \mathbf{H}_0 + \mathbf{V}$, is a lower bound on the energy $A_0 + \langle \mathbf{V} \rangle_0$, the free energy of uncoupled Hamiltonian, $\mathbf{H}_0$, plus the expectation of the coupling term, $\mathbf{V}$ [101]. That is,

$$A = -\frac{1}{\beta} \mathrm{Tr}(e^{-\beta \mathbf{H}}) \le -\frac{1}{\beta} \mathrm{Tr}(e^{-\beta \mathbf{H}_0}) + \mathrm{Tr}(e^{-\beta \mathbf{H}_0} \mathbf{V}) / \mathrm{Tr}(e^{-\beta \mathbf{H}_0})$$
$$= A_0 + \langle \mathbf{V} \rangle_0 .$$
(75)

Typically, we formally remove the term $\langle \mathbf{V} \rangle_0$ from this relation by incorporating it into $\mathbf{H}_0$; the Gibbs–Bogoliubov inequality [Eq. (75)] is then simply $A \le A_0$. Also, in this way the final system–bath coupling is defined to affect the system only through fluctuations, that is, at second order.

Thus a transformation that minimizes $\tilde{A}_0$, the free energy of the uncoupled transformed system, $\tilde{\mathbf{H}}_0 = \tilde{\mathbf{H}}_s + \tilde{\mathbf{H}}_b$, hopefully minimizes the effects of the system–bath coupling term. If the transformation depends on some set of parameters $\{c_j\}$, this minimum is defined by the solution of the simultaneous equations

$$\frac{\partial \tilde{A}_0}{\partial c_j} = 0 , \qquad j = 1, 2, \ldots .$$
(76)

One limitation on the use of the variational approach in analytic theory is the necessity of having analytic expressions for $\tilde{A}_0$ and $\langle \tilde{\mathbf{V}} \rangle_0$, so that Eqs. (76) can be formulated. In cases where $\tilde{A}_0$ cannot be obtained at general temperatures, progress can often still be made by minimizing the lowest eigenstate of $\tilde{\mathbf{H}}_0$, which corresponds to minimizing the zero-temperature free energy [101, 114].

Since there is not generally an analytic form for either the eigenstates of $\tilde{\mathbf{H}}_0$ or for the Helmholtz free energy, we will treat the problem numerically, which requires that the Hamiltonian be represented in some basis set, $\{|\phi_j\rangle\}$. A natural basis set to use is the direct product of the two states of the TLS and the set of harmonic oscillator wavefunctions for each bath oscillator. In terms of the eigenstates of $\tilde{\mathbf{H}}_0$,

$$\tilde{\mathbf{H}}_0 |\psi_k\rangle = \varepsilon_k |\psi_k\rangle , \tag{77}$$

the free energy is

$$A_0 = -\frac{1}{\beta} \sum_k e^{-\beta \varepsilon_k} . \tag{78}$$

Its derivatives with respect to the variational parameters $\{c_j\}$ are

$$\frac{\partial A_0}{\partial c_j} = -\beta \sum_k e^{-\beta \varepsilon_k} \frac{\partial \varepsilon_k}{\partial c_j} , \tag{79}$$

and the derivatives of the eigenvalues themselves are expressed in terms the variation of the matrix elements of $\tilde{\mathbf{H}}_0$ with respect to the parameters,

$$\frac{\partial \varepsilon_k}{\partial c_j} = \sum_{m, n} \psi_{m,k} \psi_{n,k} \frac{\partial (\tilde{\mathbf{H}}_0)_{m,n}}{\partial c_j} , \tag{80}$$

where m and n run over the basis functions and $\psi_{m,k}$ is the coefficient of basis function m in eigenstate k.

D. Small-Polaron Transformation

When the system–bath coupling is linear in the bath coordinates, as in the spin-boson Hamiltonian, the physical interpretation is that the minimum position of each bath oscillator is shifted proportionately to the value of the system variable to which it is coupled. The small-polaron transformation redefines the Hamiltonian in terms of oscillators shifted adiabatically as a function of the system coordinate; here the system coordinate is σ_z, so that the oscillators will be implicitly displaced equally but in opposite directions for each quantum state. Note that in the limit that the TLS coupling J vanishes, this transformation completely separates the system and bath. This makes it an effective transformation for cases of small coupling, and it has in fact been long and widely used in many types of physical problems, although typically in a nonvariational form [102]. Harris and Silbey showed that while simple enough to handle analytically, a variational small-polaron transformation contained the flexibility to treat the spin-boson problem effectively in most parameter regimes (see below) [45–47].

The transformation is defined by the generator

$$\mathbf{S} = i \sum_{j=1}^{N} \mathbf{p}_j \sigma_z \frac{2f_j}{\omega_j^2} . \tag{81}$$

Using Eq. (74), one obtains the transformed variables.

$$\tilde{\sigma}_x = \sigma_x \cosh \alpha - i\sigma_y \sinh \alpha$$

$$= \tfrac{1}{2}\sigma_+ e^{-\alpha} + \tfrac{1}{2}\sigma_- e^{\alpha} ,$$

$$\tilde{\sigma}_z = \sigma_z , \tag{82}$$

$$\tilde{\mathbf{q}}_j = \mathbf{q}_j - \sigma_z \frac{2f_j}{\omega_j^2} ,$$

$$\tilde{\mathbf{p}}_j = \mathbf{p}_j ,$$

where

$$\alpha = i \sum_{j=1}^{N} \mathbf{p}_j \frac{2f_j}{\omega_j^2} . \tag{83}$$

The transformed Hamiltonian is

$$\tilde{\mathbf{H}} = \tilde{\mathbf{H}}_s + \tilde{\mathbf{H}}_b + \tilde{\mathbf{V}} , \tag{84}$$

where the new system Hamiltonian,

$$\tilde{\mathbf{H}}_s = 2\Delta\sigma_z + 2\tilde{J}\sigma_x + E_0 , \tag{85}$$

contains the renormalized TLS coupling

$$\tilde{J} = J\langle e^{\pm\alpha}\rangle_b = J \exp\left(-\sum_{j=1}^{N} \frac{f_j^2}{\omega_j^3} \coth\frac{\beta\hbar\omega_j}{2}\right) \tag{86}$$

and the energy shift

$$E_0 = \frac{1}{2}\sum_{j=1}^{N} \frac{f_j^2 - 2f_j g_j}{\omega_j^2} . \tag{87}$$

The transformed bath Hamiltonian still consists of the original uncoupled oscillators,

$$\tilde{\mathbf{H}}_b = \frac{1}{2} \sum_{j=1}^{N} (\mathbf{p}_j^2 + \omega_j^2 \mathbf{q}_j^2) . \tag{88}$$

The system–bath coupling, however, is now

$$\tilde{\mathbf{V}} = \mathbf{F}_+ \sigma_+ + \mathbf{F}_- \sigma_- + 2\mathbf{F}_z \sigma_z , \tag{89}$$

with the bath operators

$$\mathbf{F}_+ = J e^{-\alpha} - \tilde{J} ,$$

$$\mathbf{F}_- = J e^{\alpha} - \tilde{J} , \tag{90}$$

$$\mathbf{F}_z = \sum_{j=1}^{N} (g_j - f_j) \mathbf{q}_j .$$

It is clear that the polaron shifts, $\{f_j\}$, directly reduce the linear TLS–bath couplings, so that $\mathbf{F}_z$ would vanish entirely for $f_j = g_j$; this is the traditional (nonvariational) small-polaron transformation used in solid-state physics [102]. This choice is not generally optimal, however, because it leads to larger $\mathbf{F}_+$ and $\mathbf{F}_-$ couplings. In calculations presented below, the $\{f_j\}$ are always variationally optimized.

Results. We can evaluate the effectiveness of the small-polaron transformation by considering a system of just two modes, for which exact free energies can also be obtained using traditional basis-set methods. The results are shown in Fig. 5 as a function of the bath frequency and the TLS coupling, J. Here the $\{f_j\}$ in Eqs. (83)–(87) were varied to minimize the free energy $\tilde{A}_0$ of the zero-order Hamiltonian, $\tilde{\mathbf{H}}_s + \tilde{\mathbf{H}}_b$, after the transformation. For ease of comparison, the normalized solvation free energy [109] of the TLS is plotted, $\Delta\mu = (\tilde{A}_0 - A_0)/S$. The results are reasonably good for small ($J < S$) and large ($J > S$) TLS coupling, as well as the case of a high-frequency bath ($\omega_c \gg S$). However, as pointed out previously [109], the transformation becomes completely ineffective when the bath frequency is low, the so-called *adiabatic bath limit*. Here the variational polaron treatment essentially predicts that the system and bath decouple completely as the bath cutoff frequency ω_c goes to zero, which

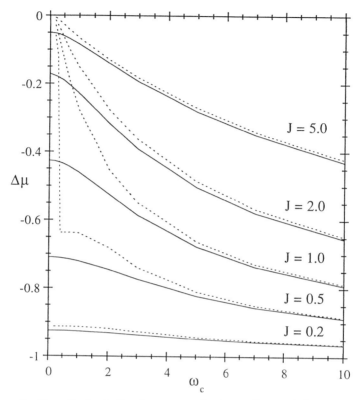

Figure 5. Normalized solvation free energy of a two-oscillator spin-boson system as a function of the bath cutoff frequency, ω_c, and the two-level system coupling, J. The system–bath coupling is fixed at $S = 1$ and temperature at $k_B T = 0.2$. The bath oscillators in each case have frequencies ω_c and $\omega_c/2$, and coupling constants $g = (\omega_c S)^{1/2}$ and $g/2$, respectively. Minimized free energies for the zero-order system following a variational polaron transformation are plotted with dashed lines; the exact free energies obtained by basis-set methods are plotted with solid lines.

is unphysical for this bath. A general solution of the problem therefore requires a different approach.

E. Effective Adiabatic Approaches

The partition function of the multidimensional spin-boson Hamiltonian becomes analytically solvable in the limit that the maximum frequency of the bath oscillators goes to zero (more precisely [115], as $\beta \hbar \omega_c \ll 1$); this is the adiabatic bath of Chandler and Wolynes [116, 117]. Here the sluggishness of the oscillators allows the use of a simple configurational average in the calculation of the partition function; for the linearly coupled harmonic bath, this reduces to a one-dimensional integral over a collective bath coordinate, multiplied by the partition function of the harmonic bath.

Carmeli and Chandler proposed treating the general spin-boson Hamiltonian by variationally identifying an *effective* adiabatic system to approximate the original system [109, 110]. The effective system was obtained by allowing both the total system–bath coupling, S, and the TLS coupling, J, to vary freely in order to minimize the free energy $A_b + A_{EA} + \langle V_{EA} \rangle$. In particular, by allowing S to vary, they were able to get the exact partition function in both the adiabatic and high-frequency bath limits. They [109] and others [115] demonstrated that this method was quite accurate for calculating thermodynamic partition functions of both single and multimode systems in both weak- and strong-coupling regimes.

Despite this success, the effective adiabatic approach has limitations. First, the method's accuracy for thermodynamic properties does not extend uniformly to the calculation of the real-time dynamics of the spin system. Although qualitatively accurate results are obtained when the effective adiabatic system is used as the reference system for a Redfield treatment of the two-level system, significant deviations from the Monte Carlo simulation results remain [110]. Moreover, this treatment of the dynamics could not be applied at all in some parameter regions because of formal difficulties. A second, a more fundamental limitation, is that the method does not generalize straightforwardly to quantum systems of more than two states. This is because the path-integral formulation of the thermodynamic partition function, on which the method is based, include contributions with negative signs in the case of multilevel quantum systems, and thus the Gibbs–Bogoliubov variational principle does not apply [115].

Coalson has proposed an *extended adiabatic method*, which is similar in spirit but somewhat different in implementation: in particular, it is naturally extended to multilevel quantum systems [115]. Rather than varying parameters to find an optimal adiabatic system, nonadiabatic corrections are defined for an adiabatic reference system; each correction term adds another effective harmonic degree of freedom to the configuration-space integral for the partition function. This was compared with the effective adiabatic method and found to give comparably excellent results for thermodynamic properties; however, it has not been used to calculate dynamical properties.

F. Inclusion of an Effective Bath Coordinate

The success of the effective adiabatic approach suggests that including a collective bath coordinate in the system may be a practical way to make sure that the low-frequency bath case is handled correctly. To do this, we redefine the bath coordinates $\{\mathbf{q}_j\}$ to create a single special coordinate, $\mathbf{Q}$, that will be incorporated into the system; we will call this the *effective bath coordinate* (EBC). A general N-dimensional orthogonal rotation of the bath coordinates can be defined by specifying $\frac{1}{2}N(N-1)$ independent planar (two-dimensional) rotations. By convention, the EBC will be constructed as the first coordinate, $\tilde{\mathbf{q}}_1$, in the transformed space.

An individual orthogonal rotation of angle θ, in the space of two coordinates, $\mathbf{q}_a$ and $\mathbf{q}_b$, is defined by the generator $\mathbf{S}_{a,b}(\theta) = -i\theta(\mathbf{q}_a\mathbf{p}_b - \mathbf{q}_b\mathbf{p}_a)$. The resulting transformation is

$$\tilde{\mathbf{q}}_a = \mathbf{q}_a \cos \theta + \mathbf{q}_b \sin \theta \;,$$

$$\tilde{\mathbf{q}}_b = -\mathbf{q}_a \sin \theta + \mathbf{q}_b \cos \theta \;,$$

$$\tilde{\mathbf{p}}_a = \mathbf{p}_a \cos \theta - \mathbf{p}_b \sin \theta \;,$$

$$\tilde{\mathbf{p}}_b = \mathbf{p}_a \sin \theta + \mathbf{p}_b \cos \theta \;. \tag{91}$$

The EBC is thus defined by specifying $N-1$ rotation angles, $\{\theta_2, \theta_3, \ldots, \theta_N\}$, describing successive planar rotations in the space of $\mathbf{q}_1$ and each of the other bath coordinates. Writing $\mathbf{U}_a(\theta) = \exp[-\mathbf{S}_{1,a}(\theta)]$, the overall transformation operator is

$$\mathbf{U}_{\mathrm{EBC}} = \mathbf{U}_2(\theta_2)\mathbf{U}_3(\theta_3)\cdots\mathbf{U}_N(\theta_N) \;. \tag{92}$$

Although $N-1$ planar rotations are sufficient to define an arbitrary EBC, performing these rotations alone will also mix the other bath coordinates and introduce explicit couplings among them, where initially the bath oscillators were uncoupled. Since the remaining $\frac{1}{2}(N-1)^2$ unspecified parameters in the N-dimensional orthogonal rotation can be chosen arbitrarily without affecting the EBC, we always use them to reorthogonalize the transformed bath.

The EBC is a linear combination of the original coordinates,

$$\mathbf{Q} = \tilde{\mathbf{q}}_1 = \sum_{j=1}^{N} a_j\mathbf{q}_j \;, \tag{93}$$

with expansion coefficients

$$a_1 = \cos \theta_2 \cos \theta_3 \cdots \cos \theta_N \;,$$

$$a_2 = \sin \theta_2 \;,$$

$$a_3 = \cos \theta_2 \sin \theta_3 \;, \tag{94}$$

$$a_j = \cos \theta_2 \cos \theta_3 \cdots \cos \theta_{j-1} \sin \theta_j , \qquad 2 < j \leq N .$$

If a particular set of coefficients $\{a_j\}$ is desired, Eq. (94) is easily inverted to obtain the corresponding set of rotation angles.

With the definition of the effective bath coordinate, the spin-boson Hamiltonian can be rewritten as

$$\tilde{\mathbf{H}} = \tilde{\mathbf{H}}_s + \tilde{\mathbf{H}}_b + \tilde{\mathbf{V}} , \tag{95}$$

where the system Hamiltonian,

$$\tilde{\mathbf{H}}_s = 2\Delta\sigma_z + 2J\sigma_x + \tfrac{1}{2}(\mathbf{P}^2 + \Omega^2\mathbf{Q}^2) + 2\sigma_z G\mathbf{Q} \tag{96}$$

now contains an oscillator, the EBC, of frequency Ω and linear coupling G. The transformed $(N-1)$-dimensional bath,

$$\tilde{\mathbf{H}}_b = \frac{1}{2} \sum_{j=2}^{N} (\mathbf{p}_j^2 + \tilde{\omega}_j^2\mathbf{q}_j^2) , \tag{97}$$

is coupled linearly to *both* the TLS and the EBC, through

$$\tilde{\mathbf{V}} = 2\sigma_z \sum_{j=2}^{N} \tilde{g}_j\mathbf{q}_j + \mathbf{Q} \sum_{j=2}^{N} c_j\mathbf{q}_j . \tag{98}$$

Thus the original coupling of the TLS and bath, in the $\{g_j\}$, has been distributed into the new TLS couplings $\{\tilde{g}_j\}$, the EBC–bath couplings $\{c_j\}$ and TLS–EBC coupling G. The TLS–bath coupling here is similar to the original coupling, although hopefully much smaller; the EBC–bath coupling is *off-diagonal* in the language introduced in Section IV.B.1, giving rise to vibrational relaxation in the newly defined coordinate. Equations (95)–(98) will be referred to as the *EBC Hamiltonian*.

1. Polaron Transformation of the EBC Hamiltonian

Since the remaining TLS–bath coupling, in the $\{\tilde{g}_j\}$, can be made much smaller than the original coupling, a small-polaron transformation of Eq. (95) should be substantially more effective than before. Moreover, since the new EBC–bath couplings, the $\{c_j\}$, are also linear in the bath coordinates, they are also treated naturally by a small-polaron transformation [114]. To do this, the original generator [Eq. (81)], is generalized to

$$\mathbf{S} = i \sum_{j=2}^{N} \frac{\mathbf{p}_j}{\tilde{\omega}_j^2} (2f_j\sigma_z + b_j\mathbf{Q}) , \tag{99}$$

so that the oscillators are now shifted implicitly as a function of two system coordinates, σ_z and $\mathbf{Q}$.

The TLS variables transform as before, while the vibrational variables now go as

$$\tilde{\mathbf{Q}} = \mathbf{Q} \,,$$

$$\tilde{\mathbf{P}} = \mathbf{P} + \sum_j \mathbf{p}_j \frac{b_j}{\tilde{\omega}_j^2} \,,$$

$$\tilde{\mathbf{q}}_j = \mathbf{q}_j - \frac{1}{\tilde{\omega}_j^2} (2f_j \sigma_z + b_j \mathbf{Q}) \,, \tag{100}$$

$$\tilde{\mathbf{p}}_j = \mathbf{p}_j \,.$$

Applying this transformation to Eq. (95), the final transformed EBC Hamiltonian is found to be

$$\tilde{\tilde{\mathbf{H}}} = \tilde{\tilde{\mathbf{H}}}_s + \tilde{\tilde{\mathbf{H}}}_b + \tilde{\tilde{\mathbf{V}}} \,, \tag{101}$$

where the transformed system Hamiltonian

$$\tilde{\tilde{\mathbf{H}}}_s = 2\Delta\sigma_z + 2\tilde{J}\sigma_x + \tfrac{1}{2}(\mathbf{P}^2 + \tilde{\Omega}^2\mathbf{Q}^2) + 2\sigma_z\tilde{G}\mathbf{Q} + E_0 \,, \tag{102}$$

now contains the renormalized TLS coupling

$$\tilde{J} = J\langle e^{\pm\alpha}\rangle_b = J\exp\left(-\sum_{j=2}^{N} \frac{f_j^2}{\tilde{\omega}_j^3}\coth\frac{\beta\hbar\tilde{\omega}_j}{2}\right) \,, \tag{103}$$

the renormalized EBC frequency

$$\tilde{\Omega}^2 = \Omega^2 + \sum_{j=2}^{N} \frac{b_j^2 - 2b_jc_j}{\tilde{\omega}_j^2} \,, \tag{104}$$

the renormalized TLS–EBC coupling

$$\tilde{G} = G + \sum_{j=2}^{N} \frac{b_jf_j - \tilde{g}_jb_j - f_jc_j}{\tilde{\omega}_j^2} \,, \tag{105}$$

and the energy shift

$$E_0 = \frac{1}{2} \sum_{j=2}^{N} \frac{f_j^2 - 2f_j \tilde{g}_j}{\tilde{\omega}_j^2}.$$ (106)

Here α is defined as before [Eq. (83)], except that the bath frequencies of the EBC Hamiltonian appear. The final transformed bath Hamiltonian, in raw form, is

$$\tilde{\tilde{\mathbf{H}}}_b = \frac{1}{2} \sum_{j=2}^{N} (\mathbf{p}_j^2 + \tilde{\omega}_j^2 \mathbf{q}_j^2) + \sum_{j=2}^{N} \sum_{k=2}^{N} \frac{b_j b_k}{\tilde{\omega}_j^2 \tilde{\omega}_k^2} \mathbf{p}_j \mathbf{p}_k.$$ (107)

Here kinetic couplings created among the bath oscillators by this transformation still need to be resolved by finding the new normal modes of the bath; this will affect the values of the bath coupling constants in $\tilde{\mathbf{H}}_s$, but not the form of the final expressions.

The most important result, of course, is the final system–bath coupling; this takes the form

$$\tilde{\tilde{\mathbf{V}}} = \mathbf{F}_+ \sigma_+ + \mathbf{F}_- \sigma_- + 2\mathbf{F}_z \sigma_z + \mathbf{F}_Q \mathbf{Q} + \mathbf{F}_P \mathbf{P},$$ (108)

written in terms of the bath operators

$$\mathbf{F}_+ = (Je^{-\alpha} - \tilde{J}),$$

$$\mathbf{F}_- = (Je^{\alpha} - \tilde{J}),$$

$$\mathbf{F}_z = \sum_{j=2}^{N} (\tilde{g}_j - f_j)\mathbf{q}_j,$$ (109)

$$\mathbf{F}_Q = \sum_{j=2}^{N} (c_j - b_j)\mathbf{q}_j,$$

$$\mathbf{F}_P = \sum_{j=2}^{N} \frac{b_j}{\tilde{\omega}^2} \mathbf{p}_j.$$

To use this final result in the Redfield equation, all of the autocorrelation and crosscorrelation functions of these five bath operators will be required to define the $\{\Theta_{a,b}^{\pm}\}$ matrices in Eq. (18); these can all be obtained by standard techniques [32]. For reference, the required correlation functions are summarized in the Appendix. Note that the operators $\mathbf{F}_+$ and $\mathbf{F}_-$ are not Hermitian, so that the general form of the Redfield equation [Eq. (20)] must be used.

2. Final Results

We can now reexamine the two-mode system of Fig. 5, defining an effective bath coordinate as described above; the final bath therefore consists of just a single oscillator. Two different treatments are possible at this level, depending on how the polaron transformation is done. In the simpler case the two variational parameters are $\theta_{1,2}$, which defines the effective bath coordinate, and f_2, which treats the bath–TLS coupling; b_2, which treats the EBC–bath coupling, is held at zero. The results of this treatment are shown as the upper curves in Fig. 6. It is clear that by

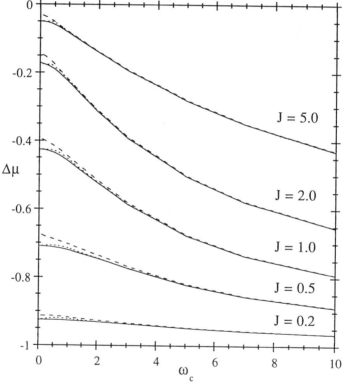

Figure 6. Normalized solvation free energy for the system of Fig. 5 as a function of the bath cutoff frequency, ω_c, and the two-level system coupling, J. Here an effective bath coordinate (EBC) was first included in the system and a variational polaron transformation applied to the resulting TLS–bath and EBC–bath couplings. The dashed lines indicate the minimum free energies obtained when only the TLS–bath coupling was treated by the second transformation; the dotted line shows results when both were treated. The exact free energies are plotted with solid lines.

including a collective bath coordinate in the system, one can indeed remove the gross errors of the simple polaron treatment. In fact, the calculated free energies are now qualitatively accurate throughout and essentially quantitative in most regions.

The second set of results in Fig. 6 was obtained by optimizing all three available variational parameters, $\theta_{1,2}$, f_2, and b_2. The results show that this allows the adiabatic limit to be treated exactly. As a result, accurate energies are now obtained in essentially all regions. The most noticeable remaining errors are found when the bath cutoff frequency, ω_c, and the TLS coupling, J, are both close to the total binding energy of the bath, S; this is the regime where any approximation method would be expected to face the most difficulty.

Perhaps more interesting than the free energy of the transformed system is the magnitude of the final system–bath coupling. It is encouraging that in most cases examined the total binding energy of the system and bath was reduced by at least a factor of 10 when the full transformation was used; even in the worst case ($\omega_c = J = S$) a reduction of more than a factor of 3 was obtained. Although these are only initial results for a relatively limited set of systems (in particular, the high-temperature regime was not considered), the results strongly validate the ideas proposed here and the prospects for obtaining accurate spin-boson dynamics with the Redfield equation are quite strong.

VI. CONCLUSIONS

We have shown in this paper how the Redfield equations can be solved efficiently when a large number of quantum levels is included in the system; the CPU and memory scaling of our algorithms with system size are such that, using massively parallel supercomputers, calculations with thousands of quantum levels will become a real possibility. If a sufficient number of levels are included explicitly, it should be possible to obtain accurate results for a wide variety of complex systems by determining an appropriate set of canonical transformations. Of course, this is highly nontrivial task in practice and the work presented in this paper represents only an initial attempt along these lines. However, the bottleneck to general applicability is probably more in conceptual understanding of what formulation of the system will yield reliable results as opposed to necessary computational effort.

There are likely to be fundamental limitations to the Redfield approach that will render certain types of problems intractable. For example, if the bath is highly anharmonic and strongly coupled to the system, the use of a harmonic representation of the bath may be unacceptable; in this case it may not be possible to obtain the correlation functions needed to evaluate the Redfield tensor (indeed, it is not obvious that any of the methods in current use will be feasible for this type of problem). This issue can be explored by carrying out numerically accurate basis set calculations on moderate-dimensional (three- to six-degree-of-freedom) systems and comparing the results with a harmonic approximation for the bath variables. Recent developments in our laboratory have made such basis set calculations quite tractable [118, 119].

From the point of view of accuracy, the recent progress of Makri and co-workers using path-integral methods (in which many of the themes emphasized here, such as an intelligent splitting of the system and bath, are also important) suggests that these methods are likely to yield the most reliable results given access to unlimited CPU time [15–20]. However, the Redfield methods are capable of substantially greater computational efficiency if a useful system–bath partitioning, where the various approximations inherent in the Redfield treatment are satisfied, can be devised. Over the next several years, it will be interesting to compare the Redfield approach with optimized canonical transformations to the path integral results to determine quantitatively what the level of accuracy and CPU advantage actually are.

A wide variety of problems are amenable to the Redfield methodology in addition to those discussed here. Some of the most important, in our view, are as follows. First, problems involving the interaction of strong laser fields with a condensed-phase system are often difficult to solve because the construction of a small, physically intuitive zeroth-order quantum subsystem Hamiltonian is difficult; the numerical methods described above will make it possible to expand the size of the quantum subsystem and allow the problem to be attacked much more easily. A second class of problems involves relaxation of complex systems (e.g., vibronic or vibrational relaxation of a molecule in a liquid) [42, 43, 72]. A third class of problems would be concerned with chemical dynamics in which the system could not be described easily by a single reaction coordinate, for example, general proton transfer reactions [98] or the isomerization of retinal in bacteriorhodopsin [120]. A low-dimensional system probably is adequate for these cases, but a nontrivial number of quantum levels will still be required.

To apply Redfield theory to these real-world condensed-phase problems, it will clearly be essential to construct appropriate reduced models from classical simulations, as in the dispersed polaron method of Warshel [67, 68]. Carrying this process out for a general multilevel, as opposed to a two-level, system will necessitate considerable study as to the most efficient numerical implementation. Work along these lines has recently been reported by Levy and co-workers for a vibrational relaxation problem [42] and by both Chandler and for Schulten the initial electron transfer in the photosynthetic reaction center [65, 69]. In principle, there is no reason why this type of modeling cannot be developed successfully into a general and practical formulation.

In summary, we believe that the multilevel Redfield methods described here will have broad applicability to a variety of important condensed-phase problems and will be the method of choice for a subset of these problems due to their computational efficiency as compared to more expensive path-integral approaches while making a minimal sacrifice of accuracy. However, a number of goals must be met before the most challenging and important set of these problems becomes amenable to such treatment, including further development of the canonical transformation methods described briefly here and construction of a robust methodology for passing from atomic-level simulations on complex systems to a tractable reduced-density-matrix model. Work along these lines is currently in progress in our research group.

ACKNOWLEDGMENTS

W.T.P. acknowledges the support of a National Science Foundation Postdoctoral Research Fellowship in Chemistry. This work was supported in part by a grant from the Department of Energy (DE-FG02-90ER14162).

APPENDIX: BATH CORRELATION FUNCTIONS

The system–bath coupling operator [Eq. (108)] in the final effective-bath Hamiltonian [Eq. (101)] has five terms with five distinct bath operators; each is either a sum of the bath positions, of the bath momenta, or an exponentiated sum of the bath momenta. For reference, we summarize here the autocorrelation and crosscorrelation functions of these three classes of operator; from these, the various functions needed to define the Redfield tensor [Eq. (18)] are easily obtained. The Heisenberg operators here all evolve under the bath Hamiltonian,

$$\mathbf{H}_b = \frac{1}{2} \sum_{j=1} (\mathbf{p}_j^2 + \omega_j^2 \mathbf{q}_j^2) \, ;$$

that is, $\mathbf{A}(t) = e^{i\mathbf{H}_b t} \mathbf{A} e^{-i\mathbf{H}_b t}$. Using standard methods [32], the following results are obtained:

$$\mathbf{A} = \sum_j g_j \mathbf{q}_j, \qquad \mathbf{B} = \sum_j f_j \mathbf{q}_j,$$

$$\langle \mathbf{A}(t)\mathbf{B}(0)\rangle_b = \frac{1}{2}\sum_j \frac{g_j f_j}{\omega_j} Q_j (e^{-i\omega_j t} + e^{-\beta\hbar\omega_j} e^{i\omega_j t}), \tag{110}$$

$$\mathbf{A} = \sum_j g_j \mathbf{p}_j, \qquad \mathbf{B} = \sum_j f_j \mathbf{p}_j,$$

$$\langle \mathbf{A}(t)\mathbf{B}(0)\rangle_b = \frac{1}{2}\sum_j g_j f_j \omega_j Q_j (e^{-i\omega_j t} + e^{-\beta\hbar\omega_j} e^{i\omega_j t}), \tag{111}$$

$$\mathbf{A} = \sum_j g_j \mathbf{q}_j, \qquad \mathbf{B} = \sum_j f_j \mathbf{p}_j,$$

$$\langle \mathbf{A}(t)\mathbf{B}(0)\rangle_b = \frac{i}{2}\sum_j g_j f_j Q_j (e^{-i\omega_j t} - e^{-\beta\hbar\omega_j} e^{i\omega_j t}), \tag{112}$$

$$\mathbf{A} = \exp\left(i\sum_j \gamma_j \mathbf{p}_j\right), \qquad \mathbf{B} = \exp\left(i\sum_j \eta_j \mathbf{p}_j\right),$$

$$\langle \mathbf{A}(t)\mathbf{B}(0)\rangle_b = \langle \mathbf{A}\rangle_b \langle \mathbf{B}\rangle_b \exp\left[-\frac{1}{2}\sum_j \gamma_j \eta_j \omega_j Q_j (e^{-i\omega_j t} + e^{-\beta\hbar\omega_j} e^{i\omega_j t})\right],$$

$$\tag{113}$$

$$\mathbf{A} = \sum_j g_j \mathbf{q}_j \qquad \mathbf{B} = \exp\left(i\sum_j \gamma_j \mathbf{p}_j\right),$$

$$\langle \mathbf{A}(t)\mathbf{B}(0)\rangle_b = -\frac{1}{2}\langle \mathbf{B}\rangle_b \sum_j \gamma_j g_j Q_j (e^{-i\omega_j t} - e^{-\beta\hbar\omega_j} e^{i\omega_j t}), \tag{114}$$

$$\mathbf{A} = \sum_j g_j \mathbf{p}_j \qquad \mathbf{B} = \exp\left(i\sum_j \gamma_j \mathbf{p}_j\right),$$

$$\langle \mathbf{A}(t)\mathbf{B}(0)\rangle_b = \frac{i}{2}\langle \mathbf{B}\rangle_b \sum_j \gamma_j g_j \omega_j Q_j (e^{-i\omega_j t} + e^{-\beta\hbar\omega_j} e^{i\omega_j t}), \tag{115}$$

where

$$\left\langle \exp\left(i\sum_j \gamma_j \mathbf{p}_j\right)\right\rangle_b = \exp\left(-\frac{1}{4}\sum_j \omega_j \gamma_j^2 \coth \beta\omega_j/2\right) \tag{116}$$

and $Q_j = 1/(1 - e^{-\beta\hbar\omega_j})$ is the partition function of bath oscillator j. The other required correlation functions are derived from these in a straightforward way through the properties that

$$\langle \mathbf{A}(t)\mathbf{B}(0) \rangle_b = \langle \mathbf{A}(0)\mathbf{B}(-t) \rangle_b \tag{117}$$

$$= \langle \mathbf{B}^\dagger(0)\mathbf{A}^\dagger(t) \rangle_b^*, \tag{118}$$

whereby

$$\langle \mathbf{A}(0)\mathbf{B}(t) \rangle_b = \langle \mathbf{B}^\dagger(t)\mathbf{A}^\dagger(0) \rangle_b^*,$$

$$\langle \mathbf{B}(t)\mathbf{A}(0) \rangle_b = \langle \mathbf{A}^\dagger(-t)\mathbf{B}^\dagger(0) \rangle_b^*, \tag{119}$$

$$\langle \mathbf{B}(0)\mathbf{A}(t) \rangle_b = \langle \mathbf{A}^\dagger(t)\mathbf{B}^\dagger(0) \rangle_b^*.$$

REFERENCES

1. R. Kosloff, *J. Chem. Phys.* **92**, 2087 (1988).
2. C. Le Forestier, R. H. Bisseling, C. Cerjan, M. D. Feit, R. Friesner, A. Guldberg, A. Hammerich, G. Jolicard, W. Karrlein, H.-D. Meyer, N. Lipkin, O. Roncero, and R. Kosloff, *J. Comput. Phys.* **94**, 59 (1991).
3. J. C. Tully, *J. Chem. Phys.* **93**, 1061 (1990).
4. P. J. Kuntz, *J. Chem. Phys.* **95**, 141 (1991).
5. F. Webster, P. J. Rossky, and R. A. Friesner, *Comput. Phys. Commun.* **63**, 494 (1991).
6. D. F. Coker and L. Xiao, *J. Chem. Phys.* **102**, 496 (1995).
7. P. Pechukas, *Phys. Rev.* **181**, 166 (1969).
8. P. Pechukas, *Phys. Rev.* **181**, 174 (1969).
9. F. Webster, E. T. Wang, P. J. Rossky, and R. A. Friesner, *J. Chem. Phys.* **100**, 4835 (1994).
10. W. H. Miller, *J. Chem. Phys.* **62**, 1899 (1975).
11. P. Wolynes, *Phys. Rev. Lett.* **47**, 968 (1981).
12. P. Hänggi, P. Talkner, and M. Borkovec, *Rev. Mod. Phys.* **62**, 251 (1990).
13. C. H. Mak and D. Chandler, *Phys. Rev. A* **44**, 2352 (1991).
14. R. Egger, C. H. Mak, and U. Weiss, *J. Chem. Phys.* **100**, 2651 (1994).
15. N. Makri, *Comput. Phys. Commun.* **63**, 389 (1991).
16. M. Topaler and N. Makri, *J. Chem. Phys.* **97**, 9001 (1992).
17. N. Makri, *J. Chem. Phys.* **95**, 10413 (1991).
18. M. Topaler and N. Makri, *Chem. Phys. Lett.* **210**, 285 (1993).
19. M. Topaler and N. Makri, *Chem. Phys. Lett.* **210**, 448 (1993).
20. M. Topaler and N. Makri, *J. Chem. Phys.* **101**, 7500 (1994).
21. B. J. Berne and D. Thirumalai, *Ann. Rev. Phys. Chem.* **37**, 401 (1986).
22. R. A. Friesner and R. M. Levy, *J. Chem. Phys.* **80**, 4488 (1984).
23. R. Zwanzig, *Statistical Mechanics of Irreversibility*, Vol. 3 of *Lectures in Theoretical Physics*, Interscience, New York, 1961, p. 106.

24. B. J. Berne, *Time-Dependent Properties of Condensed Media*, Vol. 8B of *Physical Chemistry: An Advanced Treatise*, Academic Press, New York, 1971. p. 539.

25. U. Fano, *Rev. Mod. Phys.* **29**, 74 (1957).

26. K. Blum, *Density Matrix Theory and Applications*, Plenum Press, New York, 1981.

27. C. P. Slichter, *Principles of Magnetic Resonance*, Springer-Verlag, New York, 1990.

28. R. K. Wangsness and F. Bloch, *Phys. Rev.* **89**, 728 (1953).

29. A. G. Redfield, *IBM J. Res. Develop.* **1**, 19 (1957).

30. A. G. Redfield, *Adv. Magn. Reson.* **1**, 1 (1965).

31. L. Allen and J. H. Eberly, *Optical Resonance and Two-Level Systems*, Dover, New York, 1987.

32. W. H. Louisell, *Quantum Statistical Properties of Radiation*, Wiley, New York, 1973.

33. M. Schubert and B. Wilhelmi, *Nonlinear Optics and Quantum Electronics*, Wiley, New York, 1986.

34. D. Lee and A. C. Albrecht, *Adv. Infrared Raman Spectrosc.* **12**, 179 (1985).

35. D. A. Wiersma, *Adv. Chem. Phys.* **47**, 421 (1981).

36. R. Friesner and R. Wertheimer, *Proc. Natl. Acad. Sci. USA* **79**, 2138 (1982).

37. J. M. Jean, G. R. Fleming, and R. A. Friesner, *Ber. Bunsenges. Phys. Chem.* **95**, 253 (1991).

38. J. M. Jean, R. A. Friesner, and G. R. Fleming, *J. Chem. Phys.* **96**, 5827 (1992).

39. W. T. Pollard and R. A. Friesner, *J. Chem. Phys.* **100**, 5054 (1994).

40. A. K. Felts, W. T. Pollard, and R. A. Friesner, *J. Phys. Chem.* **99**, 2929 (1995).

41. A. M. Walsh and R. D. Coalson, *Chem. Phys. Lett.* **198**, 293 (1992).

42. F. E. Figueirido and R. M. Levy, *J. Chem. Phys.* **97**, 703 (1992).

43. H. Gai and G. A. Voth, *J. Chem. Phys.* **99**, 740 (1993).

44. J. N. Onuchic, D. N. Beratan, and J. J. Hopfield, *J. Phys. Chem.* **90**, 3707 (1986).

45. R. Silbey and R. A. Harris, *J. Chem. Phys.* **80**, 2615 (1985).

46. R. A. Harris and R. Silbey, *J. Chem. Phys.* **83**, 1069 (1985).

47. R. Silbey and R. A. Harris, *J. Phys. Chem.* **93**, 7062 (1989).

48. N. G. van Kampen, *Stochastic Processes in Physics and Chemistry*. North-Holland, Amsterdam, 1992.

49. P. N. Argyres and P. L. Kelley, *J. Chem. Phys.* **134**, 98 (1964).

50. J. Albers and J. M. Deutch, *J. Chem. Phys.* **55**, 2613 (1971).

51. B. Yoon, J. M. Deutch, and J. H. Freed, *J. Chem. Phys.* **62**, 4687 (1975).

52. B. Fain, *Theory of Rate Processes in Condensed Media*, Springer-Verlag, New York, 1980.

53. V. Romero-Rochin, A. Orsky, and I. Oppenheim, *Physica A* **156**, 244 (1989).

54. V. Gorini, A. Kossakowski, and E. C. G. Sudarshan, *J. Math. Phys.* **17**, 821 (1976).

55. G. Lindblad, *Commun. Math. Phys.* **48**, 119 (1976).

56. E. B. Davies, *Quantum Open Systems*, Academic Press, New York, 1976.

57. H. Spohn, *Rev. Mod. Phys.* **52**, 569 (1980).

58. M. Berman and R. Kosloff, *Comput. Phys. Commun.* **63**, 1, (1991).

59. M. Berman, H. Tal-Ezer, and R. Kosloff, *J. Phys. A* **25**, 1283 (1992).

60. A. Suarez, R. Silbey, and I. Oppenheim, *J. Chem. Phys.* **97**, 5101 (1992).

61. V. Romero-Rochin and I. Oppenheim, *Physica A* **155**, 52 (1989).

62. P. Pechukas, *Phys. Rev. Lett.* **73**, 1060 (1994).

63. G. Strang, *Linear Algebra and Its Applications*, Academic Press, New York, 1980.

64. J. S. Bader, R. A. Kuharski, and D. Chandler, *J. Chem. Phys.* **93**, 230 (1990).

65. M. Marchi, J. N. Gehlen, D. Chandler, and M. Newton, *J. Am. Chem. Soc.* **115**, 4178 (1993).

66. D. W. Oxtoby, *Adv. Chem. Phys.* **47**, 487 (1981).

67. A. Warshel, Z. T. Chu, and W. W. Parson, *Science* **246**, 112 (1989).

68. A. Warshel and W. W. Parson, *Ann. Rev. Phys. Chem.* **42**, 279 (1991).

69. D. Xu and K. Schulten, *Chem. Phys.* **182**, 91 (1994).

70. C. Zheng, J. A. McCammon, and P. G. Wolynes, *Chem. Phys.* **158**, 261 (1991).

71. J. S. Bader and B. J. Berne, *J. Chem. Phys.* **100**, 8359 (1994).

72. M. Bruehl and J. T. Hynes, *Chem. Phys.* **175**, 205 (1993).

73. T. J. Park and J. C. Light, *J. Chem. Phys.* **85**, 5870 (1986).

74. R. A. Friesner, L. S. Tuckerman, B. C. Dornblaser, and T. V. Russo, *J. Sci. Comput.* **4**, 327 (1989).

75. J. K. Cullum and R. A. Willoughby, *Lanczos Algorithms for Large Symmetric Eigenvalue Computations*, Vol. 1, *Theory*, Birkhauser, Boston, 1985.

76. W. E. Arnoldi, *Quart. Appl. Math.* **9**, 17 (1951).

77. Y. Saad, *Linear Algebra Appl.* **34**, 269 (1980).

78. A. Bartana and R. Kosloff, *J. Chem. Phys.* **99**, 196 (1993).

79. H. Tal-Ezer and R. Kosloff, *J. Chem. Phys.* **81**, 3967 (1984).

80. R. Wertheimer and R. Silbey, *Chem. Phys. Lett.* **75**, 243 (1980).

81. M. D. Purugganan, C. V. Kumar, N. J. Turro, and J. K. Barton, *Science* **241**, 1645 (1988).

82. G. Orellana, A. K. Mesmaeker, J. K. Barton, and N. J. Turro, *Photochem. Photobiol.* **54**, 499 (1991).

83. C. J. Murphy, M. R. Arkin, N. D. Ghatlia, S. Bossman, N. J. Turro, and J. K. Barton, *Science* **262**, 1025 (1993).

84. J. J. Hopfield, *Proc. Natl. Acad. Sci.* **71**, 3640 (1974).

85. G. L. Closs and J. R. Miller, *Science* **240**, 440 (1988).

86. J. R. Winkler and H. B. Gray, *Chem. Rev.* **92**, 369 (1992).

87. C. C. Moser, J. M. Keske, K. Warnche, R. S. Farid, and P. L. Dutton, *Nature* **355**, 796 (1992).

88. H. M. McConnell, *J. Chem. Phys.* **35**, 508 (1961).

89. P. W. Anderson, *Phys. Rev.* **79**, 3507 (1950).

90. D. N. Beratan and J. J. Hopfield, *J. Am. Chem. Soc.* **106**, 1584 (1984).

91. M. D. Newton and N. Sutin, *Ann. Rev. Phys. Chem.* **35**, 437 (1984).

92. J. Jortner, *J. Am. Chem. Soc.* **102**, 6676 (1980).

93. D. DeVault and B. Chance, *Biophys. J.* **6**, 825 (1966).

94. R. A. Marcus and N. Sutin, *Biochim. Biophys. Acta* **811**, 265 (1984).

95. S. S. Skourtis and J. N. Onuchic, *Chem. Phys. Lett.* **209**, 171 (1993).

96. W. T. Pollard and R. A. Friesner, *J. Chem. Phys.*, manuscript in preparation.

97. A. Suarez and R. Silbey, *J. Chem. Phys.* **94**, 4809 (1991).

98. D. Li and G. A. Voth, *J. Phys. Chem.* **95**, 10425 (1991).

99. M. C. M. O'Brien, *J. Phys. C.* **5**, 2045 (1972).

100. E. Pollak, *J. Phys. Chem.* **95**, 10235 (1991).

101. M. Wagner, *Unitary Transformations in Solid State Physics*, North-Holland, New York, 1986.

102. G. D. Mahan, *Many-Particle Physics*, Plenum Press, New York, 1990.

103. A. J. Bray and M A. Moore, *Phys. Rev. Lett.* **49**, 1545 (1982).

104. A. O. Caldeira and A. J. Leggett, *Phys. Rev. Lett.* **149**, 374 (1983).

105. S. Chakravarty and A. J. Leggett, *Phys. Rev. Lett.* **52**, 5 (1984).

106. A. J. Leggett, *Phys. Rev. B.* **30**, 1208 (1984).

107. A. J. Leggett, S. Chakravarty, A. T. Dorsey, M. P. A. Fisher, A. Garg, and W. Zwerger, *Rev. Mod. Phys.* **59**, 1 (1987).

108. E. C. Behrman and P. G. Wolynes, *J. Chem. Phys.* **83**, 5863 (1985).

109. B. Carmeli and D. Chandler, *J. Chem. Phys.* **82**, 3400 (1985).

110. B. Carmeli and D. Chandler, *J. Chem. Phys.* **89**, 452 (1988).

111. A. O. Caldeira and A. J. Leggett, *Physica* **121A**, 587 (1983).

112. B. B. Laird and J. L. Skinner, *J. Chem. Phys.* **94**, 4391 (1991).

113. B. B. Laird and J. L. Skinner, *J. Chem. Phys.* **94**, 4405 (1991).

114. P. E. Parris and R. Silbey, *J. Chem. Phys.* **86**, 6381 (1987).

115. R. D. Coalson, *J. Chem. Phys.* **92**, 4993 (1990).

116. D. Chandler and P. G. Wolynes, *J. Chem. Phys.* **74**, 4078 (1981).

117. K. S. Schweitzer, R. M. Stratt, D. Chandler, and P. G. Wolynes, *J. Chem. Phys.* **75**, 1347 (1981).

118. R. A. Friesner, J. A. Bentley, M. Menou and C. Le Forestier, *J. Chem. Phys.* **99**, 324 (1993).

119 J. Antikainen, C. LeForestier, and R. A. Friesner, *J. Chem. Phys.* **102**, 1270 (1995).

120. R. A. Mathies, S. W. Lin, J. B. Ames, and W. T. Pollard, *Ann. Rev. Biophys. Biophys. Chem.* **20**, 491 (1991).

PATH-INTEGRAL CENTROID METHODS IN QUANTUM STATISTICAL MECHANICS AND DYNAMICS

GREGORY A. VOTH

Department of Chemistry, University of Pennsylvania, Philadelphia, Pennsylvania

CONTENTS

Advances in Chemical Physics, *Volume XCIII*, Edited by I. Prigogine and Stuart A. Rice.
ISBN 0-471-14321-9 © 1996 John Wiley & Sons, Inc.

I. INTRODUCTION

In this article we review a formulation of quantum statistical mechanics based on the path centroid variable in Feynman path integration [1, 2]. The primary theoretical and conceptual basis for the discussion is a series of five papers by Cao and Voth [3–7] and an early Communication [8]. The five papers are referred to throughout as Papers I–V. This article is not, however, about Feynman path integrals in general. There exist a number of excellent books (see, e.g., Refs. 1, 2, and 9–12) and review articles (see, e.g., Refs. 13–20) on the subject of path integrals. Only the elements of path integral theory necessary for the discussion of centroid theory are described here. It should also be noted that there is a significant body of literature on the *effective potential method*, which is related to the first part of this review on equilibrium properties, going back to the seminal work of Feynman and Hibbs [1, pp. 279–286], Giachetti and Tognetti [21, 22], and Feynman and Kleinert [10, 23–25]. The effective potential theory will be derived as a particular limit of a general centroid-based formulation of quantum equilibrium statistical mechanics. While the seminal papers on the effective potential are cited in their proper context throughout this review, the numerous applications of the method (see below) are not presented in detail.

Before beginning the discussion of the path centroid perspective, some general background material on path integrals is in order. The Feynman path-integral representation of quantum mechanics [1], and in particular of quantum statistical mechanics [2], is a powerful and natural way of describing physical processes. Indeed, the path-integral method has proven to be advantageous in the theoretical and computational analysis of many problems in condensed matter chemical physics (for reviews of numerical and analytical path-integral methods in chemical physics, see Refs. 17–19) and solid-state physics (see, e.g., Ref. 20). In this approach, the canonical partition function Z is expressed as functional integral over all possible paths $q(\tau)$ such that [2]

$$Z = \int \cdots \int \mathcal{D}q(\tau) \exp\{-S[q(\tau)]/\hbar\}, \qquad (1.1)$$

where the exponential weighting of the paths is determined by the imaginary time action functional [2]

$$S[q(\tau)] = \int_0^{\hbar\beta} d\tau \left\{ \frac{m}{2} \dot{q}(\tau)^2 + V[q(\tau)] \right\}. \tag{1.2}$$

For simplicity, the equations above have been written for a single quantum particle in one dimension. A generalization to multiple particles and/or dimensions is straightforward.

Path integrals are particularly useful for describing the quantum mechanics of an equilibrium system because the canonical distribution for a single quantum particle in the path integral picture becomes isomorphic with that for a classical ring polymer of quasiparticles [17–19, 26] (cf. Fig. 1). In the discretized path-integral representation, the partition function for a quantum particle is given by the expression

$$Z = \lim_{P \to \infty} \left(\frac{mP}{2\pi\hbar^2\beta} \right)^{P/2} \int dq_1 \cdots \int dq_P \exp[-\beta V_P(\mathbf{q})], \tag{1.3}$$

where m is the particle mass, β equals $1/k_B T$, and the isomorphic quasiclassical polymer potential V_P is given by

$$V_P(\mathbf{q}) = \sum_{i=1}^{P} \left[\frac{mP}{2\hbar^2\beta^2} (q_i - q_{i+1})^2 + \frac{V(q_i)}{P} \right]. \tag{1.4}$$

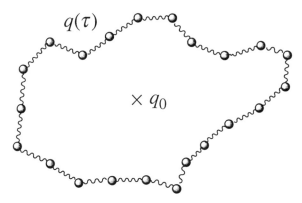

Figure 1. Discretized Feynman path $q(\tau)$ on the imaginary-time interval $0 \leq \tau \leq \hbar\beta$. The classical quasiparticles are shown by the dark circles, which form an isomorphic ring polymer. Each quasiparticle interacts with its two nearest neighbors through effective harmonic forces and with the external potential through the term $V(q_i)/P$ [cf. Eq. (1.4)]. The centroid variable q_0 defined in Eqs. (1.5) and (1.6) is also shown.

In this discrete representation, the coordinates $\{q_i\}$ describe the positions of the classical quasiparticles and have the cyclic property such that $q_{i+P} = q_i$ (cf. Fig 1). Each quasiparticle is harmonically bound to its two nearest neighbors, and it "feels" the interaction potential through the term $V(q_i)/P$. In numerical applications, a finite value of the discretization parameter P is used which is large enough that a suitable numerical convergence is obtained [17–19]. In the discretized representation, the path integral formalism has allowed for the numerical simulation of highly nontrivial quantum systems using path-integral Monte Carlo (PIMC) or molecular dynamics (PIMD) techniques [17–19].

The focus of this article, however, is on a more specialized topic in path integration—the path centroid perspective. One of the many interesting ideas suggested by Feynman in his formulation and application of path integrals was the notion of the path centroid variable [1], denoted here by the symbol q_0. The centroid is the imaginary time average of a particular closed Feynman path $q(\tau)$, which, in turn, is simply the zero-frequency Fourier mode of that path, that is,

$$q_0 = \frac{1}{\hbar\beta} \int_0^{\hbar\beta} d\tau\, q(\tau) \,. \tag{1.5}$$

In the discretized path-integral picture (i.e., for finite P), the path centroid variable is equivalent to the center of mass of the isomorphic polymer of classical quasiparticles (cf. Fig. 1) such that

$$q_0 = \frac{1}{P} \sum_{i=1}^{P} q_i \,. \tag{1.6}$$

Feynman noted that a quantum mechanical *centroid density* $\rho_c(q_c)$ can be defined for the path centroid variable, which is the path sum over all paths with their centroids located at some point in space denoted by q_c. Specifically, the formal expression for the centroid density is given by

$$\rho_c(q_c) = \int \cdots \int \mathcal{D}q(\tau)\delta(q_c - q_0) \exp\{-S[q(\tau)]/\hbar\} \,. \tag{1.7}$$

The centroid density formally defines a classical-like effective potential; that is [1, 3, 21–23]

$$V_c(q_c) = -k_B T \ln [\rho_c(q_c)/(m/2\pi\hbar^2\beta)^{1/2}] \,. \tag{1.8}$$

The quantum partition function in Eq. (1.1) is then obtained by the integration of the centroid density over all possible configurations of the centroid such that

$$Z = \int dq_c \, \rho_c(q_c) = \left(\frac{m}{2\pi\hbar^2\beta}\right) \int dq_c \, \exp[-\beta V_c(q_c)] \,. \qquad (1.9)$$

It should be noted that the centroid density is distinctly different from the coordinate (or particle) density $\rho(q) = \langle q|\exp(-\beta H)|q\rangle$. The particle density function is the diagonal element of the equilibrium density matrix in the coordinate representation, while the centroid density does not have a similar physical interpretation. However, the integration over either density yields the quantum partition function.

Feynman used the definition of the centroid density along with a simple approximation for the action functional in Eq. (1.2) to derive an expression for a quasiclassical partition function [1]. The latter function is expressed as an integration over an effective Boltzmann factor, which, in turn, depends on a variational effective potential determined at each value of the centroid variable using the Gibbs–Bogoliubov variational principle [1, pp. 303–307; 2, pp. 86–96]. In subsequent work, several authors [21–23] improved upon the Feynman's original approach by using a more physically accurate variational harmonic reference system to describe the imaginary time path fluctuations around the centroid variable. The effective harmonic frequency and effective potential are again determined at each value of the centroid variable, resulting in an approximate expression for the centroid density in Eq. (1.7). An approximate variational partition function can then be determined in such a theory by virtue of Eq. (1.9). The latter approach, which defines the effective potential theory referred to previously, has proven to be very fruitful in a number of areas of condensed matter theory. Some applications are for quantum solids, [27–30], nonstandard Hamiltonians (i.e., spin lattices) [31, 32], nonlinear quantum chains [33], static [34] and dynamic [35–39] correlations, quantum phase-space [40], path-integral simulation methods [30, 41], and quantum-activated rate theory [42–44].

As stated earlier, the focus of the present article is not on the effective potential theory but on a general centroid-based methodology for studying quantum equilibrium and, even more significantly, dynamical properties of condensed matter systems [3–8]. In this methodology, the path centroid variable is cast as the *central* statistical and dynamical variable in equilibrium systems. Accordingly, the equilibrium density function associated with the centroid variable defines the statistical distribution with which to average equilibrium and dynamical quantities. In this spirit, an infinite-order renormalized diagrammatic perturbation theory can be developed for the centroid density and related quantities

[3], and the quantum mechanical rules for computing equilibrium averages and imaginary time correlations can be appropriately reformulated [3, 34, 45]. This formalism has significant mathematical and conceptual advantages and can be generalized to a phase-space perspective [5]. The calculation of quantum dynamical time correlation functions can also be formulated within the context of the path centroid variable [4–6, 8]. In this approach, the centroid perspective is introduced into the calculation of real-time position and velocity correlation functions and the classical-like dynamical equations for the centroid variable are identified. This theory, called *centroid molecular dynamics* (CMD), is formulated in terms of general interaction potentials. Moreover, the computational effort in CMD scales with system size in the same way as a classical molecular dynamics simulation [4–6]. Various numerical algorithms have been developed for efficient CMD simulations [6] and numerical studies on both prototype models and for more realistic many-body systems have been carried out [4, 5, 46–48]. A more mature path centroid perspective [42–44, 49] for quantum-activated dynamics and quantum transition-state theory supports and complements the CMD approach to dynamics, allowing for the study of very complex quantum activated rate phenomena. The basic formulation and two recent applications [46, 50] of the centroid approach to activated dynamics will also be summarized in the main text.

Before proceeding to the main body of the article, however, it is necessary first to address the important question of whether any of this matters in the grand scheme of things. Indeed, any theoretical formulation or methodology should be capable of demonstrating a certain degree of success in the actual *calculation* of physical properties. Otherwise, it runs the risk of being empty formalism, no matter how interesting or formally pleasing it may be. The centroid-based formalism discussed in the present article does not suffer from this deficiancy. For example, it has led to considerable progress on at least two significant and long-standing problems in chemical physics: (1) the computation of quantum dynamical properties in many-body systems and (2) an intuitive and computationally useful quantum mechanical formulation of transition-state theory. One strength of the centroid-based methods is that they are applicable to general systems described by complicated many-body potentials. It must be stated, however, that the methodology gives *approximate*, not exact, results for such systems.

The broad topics of this article are organized into sections as follows: In Section II the centroid density–based formulation for calculating equilibrium properties in quantum mechanical systems is described. Then in Section III the centroid approaches for computing dynamical properties of equilibrium systems are discussed, while in Section IV the centroid picture of quantum-activated dynamics and quantum transition-state theory is summarized. Concluding remarks are given in Section V. Within each section are found a number of subsections covering a variety of subtopics, some of which are described in detail.

II. EQUILIBRIUM PROPERTIES

One conclusion that can be reached from the early work on effective potentials [1, 21–23], the work of Cao and Voth [3–8], as well as the centroid density–based formulation of quantum transition-state theory [42–44, 49] is that the path centroid is a particularly useful variable in statistical mechanics about which to develop approximate, but quite accurate, quantum mechanical expressions and to probe the quantum–classical correspondence principle. It is in this spirit that a general centroid density–based formulation of quantum Boltzmann statistical mechanics is presented in the present section. This topic is the subject of Paper I, and the emphasis in this section is on *analytic* theory as opposed to computational approaches (cf. Sections III and IV).

The centroid-based description of equilibrium (and dynamical) properties is based in part on the key notion that the centroid density should occupy the same role as the Boltzmann distribution in classical statistical mechanics. In fact, by concentrating on the centroid density as the central statistical distribution, a formally exact diagrammatic expansion [3, 51] for the centroid density can be developed which turns out to be simplified from the point of view of the relevant diagram topologies and is also particularly amenable to powerful renormalization techniques. The diagrammatic theory draws the formal connection between various variational effective potential expressions for the centroid density which have been derived by others [21–23] and specific diagram resummation and renormalization strategies. As a result, a systematic approach to improve on the effective potential result of the variational centroid density theory emerges. A considerable amount of time will be devoted to the diagrammatic methods in the present article, due to the central practical and formal importance of the centroid density in the overall formalism.

Before proceeding, it is necessary to define a centroid-constrained imaginary-time propagator [3] (i.e., the correlation function of quantum path fluctuations with respect to the position of the centroid variable

$q_0 = q_c$). This correlation function is defined as

$$C_c(\tau, q_c)$$
$$= \frac{\int \cdots \int \mathscr{D}q(\tau)\delta(q_c - q_0)(q(\tau) - q_0)(q(0) - q_0) \exp\{-S[q(\tau)]/\hbar\}}{\int \cdots \int \mathscr{D}q(\tau)\delta(q_c - q_0) \exp\{-S[q(\tau)]/\hbar\}}.$$

(2.1)

Since the centroids of the paths $q(\tau)$ in this correlation function are constrained to be at q_c, the paths can be rewritten as $q(\tau) = q_c + \tilde{q}(\tau)$, where $\tilde{q}(\tau)$ is the quantum path fluctuation variable with respect to the centroid. A Fourier decomposition of the paths $\tilde{q}(\tau)$ can now be introduced such that

$$\tilde{q}(\tau) = \sum_{n \neq 0} \hat{q}_n e^{-i\Omega_n \tau},$$

(2.2)

where the summation is over all nonzero integers and Ω_n is the Matsubara frequency, defined by $\Omega_n = 2\pi n/\hbar\beta$. Throughout this article, a fully multidimensional notation has been avoided in the interest of notational simplicity, except where necessary. The reader is referred to the original papers [3–6] for such generalizations, which are, for the most part, straightforward.

The correlation function in Eq. (2.1) differs from the usual Euclidean time position correlation function $C(\tau)$ because only paths with centroids at q_c contribute to the centroid-constrained propagator $C_c(\tau, q_c)$. However, one can obtain $C(\tau)$ by averaging the centroid-constrained propagator over the normalized centroid density $\rho_c(q_c)/Z$; that is,

$$C(\tau) = \langle q(\tau)q(0)\rangle = \langle C_c(\tau, q_c) + q_c^2\rangle_{\rho_c},$$

(2.3)

or, equivalently,

$$\tfrac{1}{2}\langle |q(\tau) - q(0)|^2\rangle = C(0) - C(\tau) = \langle C_c(0, q_c)\rangle_{\rho_c} - \langle C_c(\tau, q_c)\rangle_{\rho_c}.$$

(2.4)

It should be noted that the imaginary time correlation function in Eq. (2.4) provides a measure of the localization of quantum particles in condensed media [17–19, 52]. From this point on, the notation $\langle \cdots \rangle_{\rho_c}$ denotes an averaging by integrating some centroid-dependent function over the centroid position q_c weighted by the normalized centroid density $\rho_c(q_c)/Z$. An alternative method for defining the correlation function

$C_c(\tau, q_c)$ is through functional differentiation of a generating functional based on the centroid density [3].

A. Diagrammatic Representation of the Centroid Density

Because of its central importance in the centroid-based theory, a systematic study is presented in this section of the perturbation expansion, or equivalently, the cumulant expansion of the centroid density. This expansion will be shown to have a one-to-one correspondence with a diagrammatic representation. It will be demonstrated that diagrammatic classifications and topological reductions result in the renormalization of the vertices and lines and thus lead to a set of self-consistent equations. The previous effective potential results based on an optimized harmonic reference potential [21–23] can readily be derived from the renormalization scheme and thus systematically improved through higher-order terms. It is important to note that the small parameter in the diagrammatic expansion for the centroid density is $\hbar^2$, so the underlying classical limit of the theory is the *exact* classical Boltzmann density function for any given system.

In general, it can be specified that the Euclidean time action functional for a reference system has a quadratic form in the path fluctuations variable $\tilde{q}(\tau)$ such that

$$S_{\text{ref}}[\tilde{q}(\tau)] = \int_0^{\hbar\beta} d\tau \left\{ \frac{m}{2} \dot{\tilde{q}}(\tau)^2 + V_{\text{ref}}[\tilde{q}(\tau)] \right\}$$
$$= \sum_{n=-\infty}^{\infty} \frac{|\hat{q}_n|^2}{2\alpha_n}, \tag{2.5}$$

where α_n defines the reference centroid-constrained propagator [cf. Eq. (2.1)] such that

$$\alpha(\tau) = \sum_{n \neq 0} \alpha_n e^{-i\Omega_n \tau} \tag{2.6}$$

and $\alpha_0 = 0$ due to the centroid constraint. Two well-known quadratic models are the free-particle reference system, where $\alpha_n^{-1} = m\beta\Omega_n^2$, and the linear harmonic oscillator (LHO) reference system, where $\alpha_n^{-1} = m\beta(\Omega_n^2 + \omega^2)$, with ω being the intrinsic LHO frequency.

With the quadratic reference system in hand, the centroid density can be expressed as

$$\rho_c(q_c) = \rho_{c,\text{ref}}(q_c) \langle \exp(-\beta \overline{\Delta V}) \rangle_{c,\text{ref}}, \tag{2.7}$$

where $\overline{\Delta V}$ is the imaginary-time average

$$\overline{\Delta V} = \frac{1}{\hbar\beta} \int_0^{\hbar\beta} d\tau \, \Delta V[q_c + \tilde{q}(\tau)] \tag{2.8}$$

and $\Delta V = V - V_{\text{ref}}$ is the deviation of the real potential V from the reference potential V_{ref}. The symbol $\langle \cdots \rangle_{c,\text{ref}}$ in Eq. (2.7) denotes a centroid-constrained path-integral average in the reference system. Since the centroid-constrained propagator of the reference potential $\alpha(\tau)$ uniquely defines an infinite set of Gaussian averages over the Fourier modes $\{\tilde{q}_n\}$, one can equivalently denote the centroid-constrained average in the reference system with the symbol $\langle \cdots \rangle_\alpha$.

The first step in the formulation of the complete diagrammatic theory for the centroid density is to Taylor expand the average in Eq. (2.7), that is,

$$
\begin{aligned}
\langle \exp(-\beta \, \overline{\Delta V}) \rangle_\alpha &= \sum_0^\infty \frac{1}{n!} \left\langle \left[-\frac{1}{\hbar} \int_0^{\hbar\beta} d\tau \, \Delta V(q_c + \tilde{q}(\tau)) \right]^n \right\rangle_\alpha \\
&= \sum_{n=0}^\infty \frac{1}{n!} \left\langle \left[-\frac{1}{\hbar} \int_0^{\hbar\beta} d\tau \int \frac{dk}{2\pi} \Delta \hat{V}(k, q_c) e^{ik\tilde{q}(\tau)} \right]^n \right\rangle_\alpha ,
\end{aligned}
\tag{2.9}
$$

where $\Delta\hat{V}(k, q_c)$ is the spatial Fourier transform of the difference potential $\Delta V(q_c + \tilde{q})$ with respect to the variable $\tilde{q}$. The subscript n in Eq. (2.9) differs from that used in the Fourier expansion in Eq. (2.6). Since $\tilde{q}(\tau_i)$ in the reference system is a Gaussian variable with zero mean for any imaginary time τ_i, the cumulant expansion of a linear combination of those variables truncates at second order, giving

$$\left\langle \prod_{i=1}^n e^{ik_i\tilde{q}(\tau_i)} \right\rangle_\alpha = \exp\left\{ -\left[\frac{1}{2} \sum_{i=1}^n k_i^2 \alpha(0) + \frac{1}{2} \sum_{i \neq j}^n k_i k_j \alpha(\tau_i - \tau_j) \right] \right\}, \tag{2.10}$$

in which $\alpha(\tau)$ is defined by Eq. (2.6). There is no linear term in Eq. (2.10) because $\langle \tilde{q} \rangle = 0$ according to the definition of the variable $\tilde{q}(\tau)$ as the path fluctuation with respect to the centroid. The latter property considerably simplifies the cumulant expansion and resulting diagrammatic analysis. By substituting the Taylor expansion of Eq. (2.10) into Eq. (2.9), and transforming back into coordinate space, one arrives at the result [3]

$$\langle \exp(-\beta \overline{\Delta V}) \rangle^c_{\text{ref}} = \sum_{n=0}^{\infty} \sum_{m=0}^{\infty} \frac{(-1)^n}{n! \, m!} \prod_{i'=1}^{n} \int_0^{\hbar\beta} \frac{d\tau_{i'}}{\hbar}$$

$$\times \left[\frac{1}{2} \sum_{i=1}^{n} \alpha(0) \, \partial_i^2 + \frac{1}{2} \sum_{i \neq j}^{n} \alpha(\tau_i - \tau_j) \, \partial_i \, \partial_j \right]^m \Delta V, \tag{2.11a}$$

where

$$\Delta V \equiv \prod_{l=1}^{n} \Delta V[q_c + \tilde{q}(\tau_l)]|_{\tilde{q}=0} \tag{2.11b}$$

and the partial derivative symbol is defined to be $\partial_i \equiv \partial / \partial \tilde{q}(\tau_i)$ applied at imaginary time τ_i. The imaginary-time integrals in Eq. (2.11a) are understood to be integrations over all imaginary times that appear in the expansion. After the derivatives are performed in the expression, the potential difference terms are evaluated at the position of the centroid q_c.

At this point, a diagrammatic representation of Eq. (2.11) can be introduced to analyze the perturbation expression and to classify the various terms [3]. An examination of the expansion in Eq. (2.11) reveals the basic composition of the terms. Specifically, the diagrams consist of two basic elements, vertices and lines. Each vertex is associated with a Euclidean time τ_i which is to be integrated from $\tau_i = 0$ to $\hbar\beta$, and the potential, or its derivatives, are evaluated at the position of the centroid. Each line connecting two vertices at times τ_i and τ_j denotes a centroid-constrained propagator $\alpha(\tau_i - \tau_j)$ in the quadratic reference system. Whenever a line connects to a vertex, a spatial derivative is applied to the potential with the order of the derivative being equal to the number of lines that connect to the vertex. A negative sign is assigned to each vertex. The value of a diagram is the product of all the composing elements that multiply a symmetry coefficient as determined by the topological structure of the diagram.

With the definitions above in hand, one can establish a one-to-one correspondence between each distinct perturbation term and each diagram [3]. The expansion series of Eq. (2.11) is thus the collection of all topologically different diagrams and all possible combinations. A well-known graph theorem [53, 54] states that an infinite series of all possible topologically different diagrams and their combinations is equal to the exponential of all possible topologically different connected diagrams. For a connected diagram, any two vertices are linked to each other by at least one line or one path of lines. Therefore, one can formally express the centroid density as

$$\rho_c(q_c) = \rho_{c,\text{ref}}(q_c) \exp(\mathscr{F}) , \qquad (2.12)$$

where $\mathscr{F}$ is given by

$$\mathscr{F} = \circ \; + \; \bigcirc \; + \; \bigcirc\!\!\!\!\times\!\!\!\!\bigcirc \; + \; \triangleleft\!\!\!\!\bigcirc \; + \cdots . \qquad (2.13)$$

The underlying diagrammatic techniques employed in this theory have appeared many times in the literature (e.g., the Mayer cluster expansion [55] and the Feynman diagrams [56]). In the case of the centroid density, the diagrammatic approach is not restricted by the functional form of the potential and proves to be particularly advantageous in analytical studies. The topological reduction performed in the present case is equivalent to a diagrammatic representation of the cumulant expansion. Without a diagrammatic representation, the cumulant expansion becomes complicated at higher orders and there are a large number of cancellations in the diagrammatic approach not explicitly apparent in the cumulant relations. Moreover, all the diagrams of $\mathscr{F}$ are closed due to the fact that $\alpha_0 = 0$ (i.e., the centroid constraint). This feature of Eq. (2.13) simplifies the analysis considerably and suggests that the centroid perspective is a good starting point for the analytical calculation of equilibrium quantities in quantum statistical mechanics.

B. Diagram Renormalization

The diagrammatic representation of the centroid density enhances one's ability to approximately evaluate the full perturbation series [3]. For example, one can focus on a class of diagrams with the same topological characteristics. The sum of such a class results in a compact analytical expression that includes infinite terms in the summation. A very useful technique in such cases is the renormalization of diagrams [57, 58]. This procedure can be applied to the vertices to define the effective potential theory diagrammatically [3, 21–23] and, in doing so, an accurate approximation to the centroid density [3].

The first set of diagrams to be studied contains only one vertex, that is,

$$\bullet \; = \; \circ \; + \; \bigcirc \; + \; \bigcirc\!\!\!\!\times\!\!\!\!\bigcirc \; + \cdots$$

$$= -\beta \, \Delta V - \frac{1}{2} \beta \, \Delta V^{(2)} \alpha(0) - \frac{1}{2^2 2!} \beta \, \Delta V^{(4)} \alpha^2(0) + \cdots, \tag{2.14}$$

where the superscripts "(i)" denote the order of the spatial derivative, and all terms $\Delta V^{(i)}$ are evaluated at the centroid position q_c. The various terms in Eq. (2.14) correspond to the corrections due to local quantum path fluctuations $\tilde{q}(\tau)$. By summing the series, one obtains a closed expression for Eq. (2.14) in the form of a Gaussian average over a single variable having a Gaussian width factor $\alpha(0)$ such that [3]

$$\langle \Delta V \rangle_\alpha = \frac{1}{\sqrt{2\pi\alpha(0)}} \int d\tilde{q} \, \Delta V(q_c + \tilde{q}) \exp[-\tilde{q}^2/2\alpha(0)] . \tag{2.15}$$

This form of the effective potential incorporates some degree of quantum effects as well as the anharmonicity of the potential. In the case of the free-particle reference system, the centroid-constrained propagator is given by [3]

$$\alpha_{fp}(\tau) = \frac{\lambda^2}{2} [3(1 - 2u)^2 - 1] , \tag{2.16}$$

where $u = \tau/\hbar\beta$ and $\lambda^2 = \hbar^2\beta/12m$. Substitution of this expression for $\alpha(0)$ in Eq. (2.15), and using the fact that for the free-particle reference system $V_{\text{ref}} = 0$, one recovers the well-known Feynman–Hibbs quasi-classical theory [1] for the effective centroid potential from Eq. (2.15). As shown in Paper I, however, any general quadratic reference system for the propagator $\alpha(\tau)$ can be used in the expression (2.16), which leads to greater accuracy than the simple Feynman–Hibbs approach.

The next order of diagram is a ring diagram with two vertices and two lines:

$$\bigcirc\!\!\!\!\bigcirc = \frac{1}{4\hbar^2} (\Delta V^{(2)})^2 \int_0^{\hbar\beta} d\tau_1 \int_0^{\hbar\beta} d\tau_2 \, \alpha^2(\tau_1 - \tau_2) . \tag{2.17}$$

For example, substituting Eq. (2.16) into Eq. (2.17) yields the first correction term to the Feynman–Hibbs approximation [3]:

$$\frac{1}{4\hbar^2} (\Delta V^{(2)})^2 \int_0^{\hbar\beta} d\tau_1 \int_0^{\hbar\beta} d\tau_2 \, \alpha_{fp}^2(\tau_1 - \tau_2) = \frac{\beta^2 \lambda^4}{20} (V^{(2)})^2 . \tag{2.18}$$

Again, any quadratic reference propagator $\alpha(\tau)$ could be used here instead of the free-particle one.

The next level of complexity is to add local quantum fluctuations to the diagrams with two vertices. The latter diagrams are given by [3]

$$\text{(2.19)}$$

This procedure is equivalent to replacing the potential ΔV in Eq. (2.17) with the effective potential $\langle \Delta V \rangle_\alpha$ of Eq. (2.15), and this correction can be introduced in the same fashion for all higher-order diagrams. Before doing this, however, it is advantageous to include more diagrams in the vertex corrections. In fact, the set of local fluctuation diagrams [Eq. (2.14)] is only the simplest correction, so one should extend the analysis to incorporate all ring corrections, given by [3]

$$= -\frac{1}{2\hbar} \Delta V^{(2)} \int_0^{\hbar\beta} d\tau_1 \, \alpha(\tau_1)$$

$$+ \frac{1}{2\hbar^2} \Delta V^{(2)} \int_0^{\hbar\beta} d\tau_1 \int_0^{\hbar\beta} d\tau_2 \, \alpha(\tau_1 - \tau_2)\alpha(\tau_2 - \tau_1)\langle \Delta V^{(2)} \rangle_\alpha$$

$$+ \cdots .$$

$$\text{(2.20)}$$

After expressing the convolution integrals in the powers of α_n, one obtains the simple closed form for the ring diagrams [3]:

$$-\beta \left[\sum_{n \neq 0} \frac{1}{2} \frac{\alpha_n^2}{1 + \beta \langle \Delta V^{(2)} \rangle_\alpha \alpha_n^2} \partial^2 \right] \Delta V = -\beta \mathscr{D} \, \Delta V , \qquad \text{(2.21)}$$

where $\mathscr{D}$ is defined as an operator. One can also include diagrams with multiple rings hanging on the same vertex, leading symbolically to [3]

$$= -\beta \left(1 + \mathscr{D} + \frac{1}{2}\mathscr{D} + \frac{1}{2^2 2!} \mathscr{D}^2 + \cdots \right) \Delta V . \qquad \text{(2.22)}$$

Equation (2.22) now defines an effective potential which has the same form as Eq. (2.15), except that the Gaussian width factor is given by

$$\bar{\alpha} = \sum_{n \neq 0} \frac{\alpha_n}{1 + \beta \langle \Delta V^{(2)} \rangle_\alpha \alpha_n} . \qquad \text{(2.23)}$$

This form of the equation for the effective potential involves the second derivative of the potential averaged about the centroid using the *unoptimized* reference propagator width factor α.

As suggested earlier, ΔV in Eq. (2.23) can be replaced by an effective potential difference $\Delta \bar{V}$. This procedure is a form of renormalization [3]. Analytically, the renormalized element appears as an unknown variable to be solved from self-consistent equations which relate the unrenormalized quantities to the renormalized ones. Usually, it is possible to discover such a relationship by analyzing the topology of diagrams. In this case, as a result of the renormalization of the vertices, one arrives at the expression for the renormalized potential difference $\Delta \bar{V}$ by replacing the Gaussian width factor α in the right-hand side of Eq. (2.23) by the effective width factor $\bar{\alpha}$. Note that the resulting transcendental equation must be solved self-consistently, that is, the resulting expression for $\Delta \bar{V}$ is given by [3]

$$\Delta \bar{V} = \langle \Delta V(q_c + \tilde{q}) \rangle_{\bar{\alpha}} , \tag{2.24}$$

with

$$-\beta \, \Delta \bar{V} \equiv \bullet \tag{2.25}$$

and $\langle \cdots \rangle_{\bar{\alpha}}$ denotes a Gaussian average with a width factor

$$\bar{\alpha} = \sum_{n \neq 0} \frac{\alpha_n}{1 + \beta \, \Delta \bar{V}^{(2)} \alpha_n} . \tag{2.26}$$

The notation "$\Delta \bar{V}^{(2)}$" means that the second derivative of $\Delta V(q_c + \tilde{q})$ is taken with respect to $\tilde{q}$. Renormalized quantities are now denoted by an overbar [3]. Diagrammatically, a black vertex stands for a renormalized potential [cf. Eq. (2.25)] and a bold line stands for a renormalized centroid-constrained propagator. Again, the general procedure of diagram renormalization is not new and has been used many times in, for example, Green's function theory, mean field theory, and so on. However, Paper I was the first place such a procedure was applied in a general way to the centroid density. A multidimensional generalization of the formalism is also discussed in the Appendix of Paper I.

The equations given above have a straightforward interpretation in the context of quadratic reference systems. By substituting the centroid-constrained propagator for the LHO into Eq. (2.26), and using a general LHO frequency ω such that $\Delta V(q_c + \tilde{q}) = V(q_c + \tilde{q}) - \frac{1}{2}m\omega^2\tilde{q}^2$ with $\alpha_n^{-1} = m\beta(\Omega_n^2 + \omega^2)$, the renormalized LHO frequency $\bar{\omega}$ is defined by the expression [3]

$$m\bar{\omega}^2 \equiv \langle V^{(2)}(q_c + \tilde{q}) \rangle_{\bar{\alpha}}$$

$$= \frac{1}{\sqrt{2\pi\bar{\alpha}}} \int d\tilde{q} \, V''(q_c + \tilde{q}) \exp(-\tilde{q}^2/2\bar{\alpha}) . \tag{2.27}$$

The value of the renormalized LHO frequency is determined from the solution to this transcendental equation, where the renormalized Gaussian width factor in Eqs. (2.26) and (2.27) is given by

$$\bar{\alpha} = \sum_{n \neq 0} \frac{1}{m\beta(\Omega_n^2 + \bar{\omega}^2)}$$

$$= \frac{1}{m\beta\bar{\omega}^2} \left[\frac{\hbar\beta\bar{\omega}/2}{\tanh(\hbar\beta\bar{\omega}/2)} - 1 \right] . \tag{2.28}$$

It should be noted that these equations are to be solved for each position of the centroid q_c. The frequency in Eq. (2.27) is the same as the effective frequency obtained for the optimized LHO reference system using the path-integral centroid density version of the Gibbs–Bogoliubov variational method [1, pp. 303–307; 2, pp. 86–96]. Correspondingly, Eqs. (2.27) and (2.28) are exactly the same as those in the quadratic effective potential theory [1, 21–23]. The derivation above does not make use of the variational principle but, instead, is the result of the vertex renormalization procedure. The diagrammatic analysis thus provides a method of systematic identification and evaluation of the corrections to the variational theory [3].

To improve on the optimized LHO theory for the effective centroid potential, one considers the contribution from higher-order diagrams by diagrammatically imposing the condition $\langle \Delta V^{(2)} \rangle_{\bar{\alpha}} = 0$ [cf. Eq. (2.27)]. This condition specifies that all vertices linked to two lines must vanish, that is [3],

$$\longrightarrow\!\!\bullet\!\!\longrightarrow \; = 0 . \tag{2.29}$$

All diagrams containing this element will therefore vanish, giving the result $\alpha = \bar{\alpha}$. The leading corrections in the centroid density expansion in Eq. (2.13) are then found to be [3]

$$\bigcirc\!\!=\!\!\bigcirc \; = \frac{1}{2! \, 3!} \beta \langle \Delta V^{(3)} \rangle_{\bar{\alpha}}^2 \frac{1}{\hbar} \int_0^{\hbar\beta} d\tau \, \alpha^3(\tau) \tag{2.30}$$

and

$$\text{[diagram]} = \frac{1}{2!\,4!}\,\beta\langle\Delta V^{(4)}\rangle^2_{\bar\alpha}\,\frac{1}{\hbar}\int_0^{\hbar\beta}d\tau\,\alpha^4(\tau)\,, \qquad (2.31)$$

where the centroid-constrained propagator for the LHO is given by

$$\bar\alpha(\tau) = \frac{1}{m\beta\bar\omega^2}\left[\frac{b/2}{\sinh(b/2)}\cosh[(1-2u)b/2]-1\right], \qquad (2.32)$$

in which $b = \hbar\beta\bar\omega$ and $u = \tau/\hbar\beta$. These two terms provide an improvement on the optimized harmonic reference centroid density approximation, giving [3]

$$\rho_c(q_c) = \rho_{c,\text{ref}}(q_c)\exp\left\{-\beta\left[\langle\Delta V\rangle_{\bar\alpha} - \frac{1}{2!\,3!}\frac{f_3(b)}{\hbar\bar\omega}\left(\frac{\hbar}{m\bar\omega}\right)\langle\Delta V^{(3)}\rangle^2_{\bar\alpha}\right.\right.$$
$$\left.\left.- \frac{1}{2!\,4!}\frac{f_4(b)}{\hbar\bar\omega}\left(\frac{\hbar}{m\bar\omega}\right)^4\langle\Delta V^{(4)}\rangle^2_{\bar\alpha} + \cdots\right]\right\}. \qquad (2.33)$$

Here $f_n(b)$ is a dimensionless coefficient, defined by

$$f_n(b) = \left(\frac{1}{b}\right)^{n-1}\int_0^1 du\left[\frac{b/2}{\sinh(b/2)}\cosh(1-2u)b/2 - 1\right]^n, \quad (2.34)$$

which becomes a constant in the limit of large b. Additional corrections can be included by adding more complicated diagrams, and this procedure is discussed below within the context of the renormalization of the lines (i.e., the centroid-constrained imaginary time propagator).

The line is the other essential element of the diagrams, and it represents the centroid-constrained propagator. While the renormalization of vertices leads to optimization of the effective potential, the renormalization of lines leads to optimization of the Euclidean centroid-constrained propagator defined in Eq. (2.1). This propagator can be obtained formally by the appropriate functional differentiation of Eq. (2.12) [3].

Following the diagrammatic analysis of Paper I, all the leading diagrams for $C_c(\tau, q_c)$ are obtained from Eq. (2.13) [3]:

$$C_c(\tau, q_c) = \begin{array}{l}\text{[diagram]} + \text{[diagram]} + \text{[diagram]} + \cdots \\ + \text{[diagram]} + \text{[diagram]} + \cdots\end{array}$$

$$+ \ -\!\!-\!\!\bigominus\!\!-\!\!- \ + \ -\!\!-\!\!\bigominus\!\!-\!\!- \ + \ \cdots \ . \tag{2.35}$$

The collection of diagrams above represents all possible contributions to the centroid-constrained correlation function. In fact, all the decorations attached to the intermediate vertices can be removed if the vertex is renormalized. This operation is carried out by replacing all the ΔV's by $\Delta \bar{V}$'s, giving

$$\Delta \bar{V} = \langle \Delta V(q_c + \tilde{q}) \rangle_{C_c(0, q_c)} , \tag{2.36}$$

where the Gaussian width factor is now $C_c(0, q_c)$ instead of $\alpha(0)$. In the case of fully renormalized vertices, $C_c(0, q_c)$ is equivalent to the renormalized reference centroid-constrained propagator $\bar{a}$ of Eq. (2.26). These two notations are hereafter taken to be equivalent.

The simplest set of lines is the chain collection, given by [3]

$$\boldsymbol{-\!\!\!-} \ = \ -\!\!- \ + \ -\!\!\!-\!\!\bullet\!\!-\!\!- \ + \ -\!\!\bullet\!\!-\!\!\bullet\!\!- \ + \ \cdots \ , \tag{2.37}$$

where the bold line stands for a renormalized line. This diagram can also be expressed in the compact form

$$\boldsymbol{-\!\!\!-} \ = \ -\!\!- \ + \ -\!\!\!-\!\!\bullet\!\!\boldsymbol{-\!\!\!-} \tag{2.38}$$

which leads to the self-consistent equation,

$$\bar{\alpha}_n = \alpha_n - \beta \, \Delta \bar{V}^{(2)} \bar{\alpha}_n \alpha_n . \tag{2.39}$$

After noting that $\bar{\alpha} = \Sigma_{n \neq 0} \, \bar{\alpha}_n$, it is seen that Eqs. (2.36) and (2.39) are the same as the optimized LHO reference equations (2.24) and (2.26). In fact, differentiating a ring collection of the generating functional always produces a chain collection in the correlation diagrams [53, 54]. The multidimensional generalization of (2.39) is described in the Appendix of Paper I.

The next stage in the diagrammatic analysis is to include all the two-line-loop corrections in the renormalization scheme, given by [3]

$$-\!\!\bigominus\!\!\boldsymbol{-} \ + \ -\!\!\bigominus\!\!\bigominus\!\!\boldsymbol{-} \ + \ \cdots$$

$$= \tfrac{1}{2} \, \alpha_n \bar{\alpha}_n \bar{\alpha}_n^2 (\beta \, \Delta \bar{V}^{(3)})^2 - \tfrac{1}{4} \, \alpha_n \bar{\alpha}_n (\overline{\alpha_n^2})^2 (\beta \, \Delta \bar{V}^{(3)})^2 \beta \, \Delta \bar{V}^{(4)} + \cdots . \tag{2.40}$$

Because of the imaginary-time convolutions in the expression, the analytical expressions for the diagrams above are written in Fourier space, where $\overline{\alpha}_n^2$ is the contribution from the two-line loop, given by

$$\overline{\alpha}_n^2 = \sum_{m \neq 0} \bar{\alpha}_{n-m} \bar{\alpha}_m . \tag{2.41}$$

Since $\overline{\alpha}^2$ is a convolution expression, the self-consistent equation for $\bar{\alpha}$ is not local in Fourier space, and therefore one can no longer seek a single effective frequency solution as in Eq. (2.27). Therefore, this diagrammatic analysis demonstrates that the optimized LHO reference system is the best possible quadratic potential with which to approximate an anharmonic potential, a fact reached independently from the GB variational perspective. Further corrections to the centroid density are thus beyond an effective potential description [3].

The infinite summation in Eq. (2.41) can be carried out to yield a closed equation that is given by

$$\bar{\alpha}_n = \alpha_n - \alpha_n \bar{\alpha}_n \beta \, \Delta \bar{V}^{(2)} + \frac{\frac{1}{2} \alpha_n \bar{\alpha}_n \overline{\alpha}_n^2 (\beta \, \Delta \bar{V}^{(3)})^2}{1 + \frac{1}{2} \overline{\alpha}_n^2 (\beta \, \Delta \bar{V}^{(4)})} . \tag{2.42}$$

This equation can be solved numerically. It is important here to incorporate infinite terms corresponding to the same class of diagrams so that at low temperature and high anharmonicity the self-consistent equation will not diverge. Even higher-order corrections to the propagator diagrams will consist of multiline loops and their combinations [3].

C. Averaging Formalism

In this subsection the general averaging formalism in terms of the centroid density is discussed [3]. The term *general* means here that the rules are presented by which to calculate operator averages and imaginary time correlation functions at different levels of approximation. Such a formulation is not completely trivial, due to the mathematical differences between the centroid approach and the usual rules of quantum statistical mechanics. The emphasis in this section is primarily an analytical one intended to explore the quantum–classical correspondence in statistical mechanics. Moreover, the rules outlined here for calculating averages and correlation functions are formulated with the assumption that a good analytic expression for the centroid density is in hand (cf. Section IIB), thereby bypassing the need for a numerical approach.

1. Operator Averages

In the normal path-integral perspective [2, 17–19] the equilibrium average of a coordinate-dependent operator A is given by the expression

$$\langle A \rangle = Z^{-1} \int \cdots \int \mathcal{D}q(\tau) A[q(0)] \exp\{-S[q(\tau)]/\hbar\} . \qquad (2.43)$$

From the cyclic invariance of the trace, the operator can be evaluated at any point along the cyclic imaginary-time path $q(\tau)$. This is depicted schematically in Fig. 2, where the centroid variable, which is off the thermal loop, is also shown. It is nontrivial, therefore, to reformulate the rules for calculating an operator average in order that the centroid density be used as the statistical distribution [34, 45]. To illustrate this principle, the case of an operator average is described here in some detail. As the first step, Eq. (2.43) is rewritten as

$$\langle A \rangle = \int dq_c \frac{\int \cdots \int \mathcal{D}q(\tau) \delta(q_c - q_0) A[q(0)] e^{-S[q(\tau)]/\hbar}}{\int dq_c \, \rho_c(q_c)} . \qquad (2.44)$$

The operator A is next represented in Fourier space, that is,

$$A(q) = \int \frac{dk}{2\pi} \hat{A}(k) \exp(ikq) , \qquad (2.45)$$

where $\hat{A}(k)$ is the Fourier transform of the operator. Equation (2.43) can

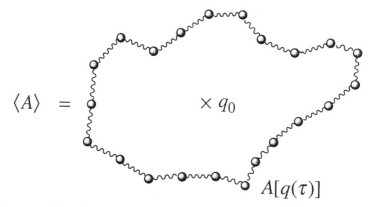

Figure 2. Calculation of an operator equilibrium average $\langle A \rangle$ in the discretized Feynman path-integral picture. The operator $A[q(\tau)]$ can be evaluated anywhere on the path-integral ring, or thermal loop, $0 \le \tau \le \hbar\beta$, due to the cyclic invariance of the trace operation. The centroid variable, which is off the loop, is also shown.

then be written as

$$\langle A \rangle = Z^{-1} \int dq_c \, \rho_c(q_c) \int \frac{dk}{2\pi} \hat{A}(k) e^{ikq_c}$$

$$\times \left[\frac{\int \cdots \int \mathcal{D}\tilde{q}(\tau) e^{ik\tilde{q}(0)} e^{-S[q_c+\tilde{q}(\tau)]/\hbar}}{\int \cdots \int \mathcal{D}\tilde{q}(\tau) e^{-S[q_c+\tilde{q}(\tau)]/\hbar}} \right], \qquad (2.46)$$

where the notation for the action functional "$S[q_c + \tilde{q}(\tau)]$" denotes path fluctuations around a centroid that is fixed at q_c. A cumulant average of the term $\exp[ik\tilde{q}(0)]$ in the brackets in Eq. (2.46) can then be performed over the path fluctuation variable $\tilde{q}(\tau)$ [3]. This average is truncated at second order, but it need not be, and the variable $\tilde{q}(\tau)$ is assumed to exhibit symmetric fluctuations about the centroid. After performing the inverse Fourier transforms in Eq. (2.46), the final result is given by [3]

$$\langle A \rangle \simeq \langle A_c(q_c) \rangle_{\rho_c}, \qquad (2.47)$$

where the effective centroid-dependent quasiclassical function A_c is given by [3]

$$A_c(q_c) = \langle A(q_c + \tilde{q}) \rangle_{C_c(0,q_c)}$$

$$= \frac{1}{\sqrt{2\pi C_c(0, q_c)}} \int d\tilde{q} \, A(q_c + \tilde{q}) \exp[-\tilde{q}^2/2C_c(0, q_c)]. \qquad (2.48)$$

Here the variable $\tilde{q}$ is clearly a Gaussian variable with a width factor $C_c(0, q_c)$. The effective classical function $A_c(q_c)$ depends on the centroid variable q_c and to calculate the equilibrium average $\langle A \rangle$ it to be averaged in a classical-like fashion in Eq. (2.47) over the normalized centroid density $\rho_c(q_c)/Z$. It should be noted that the result above from Paper I was obtained earlier and independently by Cuccoli et al. [34] and Voth [45] using somewhat more specialized perspectives.

2. Imaginary-Time Correlation Functions

A general imaginary-time correlation function for coordinate-dependent operators is defined as [17–19]

$$C_{AB} = \langle A[q(\tau)]B[q(0)] \rangle$$

$$= Z^{-1} \int \cdots \int \mathcal{D}q(\tau) A[q(\tau)] B[q(0)] \exp\{-S[q(\tau)]/\hbar\}, \qquad (2.49)$$

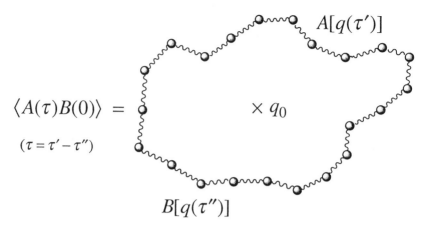

Figure 3. Calculation of the imaginary-time correlation function $\langle A(\tau)B(0)\rangle$ in the discretized Feynman path-integral picture. The operators $A[q(\tau')]$ and $B[q(\tau'')]$ can be evaluated at any two points on the thermal loop subject to the constraint $0 \leq \tau = \tau' - \tau'' \leq \hbar\beta$. The centroid variable, which is off the loop, is also shown.

where the operators $A[q(\tau)]$ and $B[q(0)]$ can be evaluated anywhere along the path such that $0 \leq \tau = \tau' - \tau'' \leq \hbar\beta$. This calculation is depicted schematically in Fig. 3 and, as in Section II.C.1, the challenge is to reformulate the calculation so that the centroid distribution, which is off the loop, can be utilized in the averaging [3]. To accomplish this, one uses the fact that the correlation function $C_{AB}(\tau)$ can always be expressed in terms of the centroid-constrained propagator $C_c(\tau, q_c)$ in Eq. (2.1) at the level of a second-order cumulant approximation [3]. The latter analysis is more complicated than in the case of an operator average, so the reader is referred to Paper I for the details. The final result is given by

$$C_{AB}(\tau) \simeq \langle p_{AB}(\tau, q_c)\rangle_{\rho_c}, \qquad (2.50)$$

where the centroid-dependent imaginary-time-correlated operator product $p_{AB}(\tau, q_c)$ is defined in the centroid density picture by the double-Gaussian average

$$p_{AB}(\tau, q_c) = \langle A[q_c + 2^{-1/2}(q_+ + q_-)]B[q_c + 2^{-1/2}(q_+ - q_-)]\rangle_{C_c^+, C_c^-}. \qquad (2.51)$$

Here, q_+ and q_- are Gaussian variables with width factors $C_c^+(\tau, q_c)$ and

$C_c^-(\tau, q_c)$, given by

$$C_c^+(\tau, q_c) = C_c(0, q_c) + C_c(\tau, q_c),$$
$$C_c^-(\tau, q_c) = C_c(0, q_c) - C_c(\tau, q_c).$$
$$(2.52)$$

Again, this formula is most useful for cases in which one has an analytical expression for the centroid density and the Gaussian width factor [3].

D. Phase-Space Perspective

The formal analysis in Papers I and II was based on the Feynman path integral formulation in coordinate space [1, 2]. Although it can be argued that momentum-dependent quantities can be obtained by taking time derivatives of the position coordinate, complications may arise due to the time ordering of the quantum operators, especially when mixed position and momentum terms appear in such operators. A better treatment, therefore, requires a path integral centroid formulation in phase space which not only generalizes the earlier position centroid-based formulation [3, 4, 8] but also provides a remedy to the time-ordering problem [59]. The first steps toward such a formulation are contained in Paper III (see also Ref. 40 for a related effective phase-space perspective). In a fashion similar to Paper I, the results were formulated in terms of a statistical averaging over the centroid density but instead, the *phase-space* centroid density. These results are described below [5]. While the resulting formulas resemble those in coordinate space, they must be defined in phase space through a compact vector-matrix notation. Unless specified otherwise, the results will therefore be presented for an N-dimensional system with the position and momentum variables described by the N-dimensional column vectors $\mathbf{q}$ and $\mathbf{p}$, respectively. The generalized vector $\boldsymbol{\zeta}$ is defined as the collection of the $2N$ degrees of freedom in phase space [i.e., $\boldsymbol{\zeta} \equiv (\mathbf{p}, \mathbf{q})$].

The phase-space centroid density is defined as [5, 40]

$$\rho_c(\boldsymbol{\zeta}_c) = \int \cdots \int \mathcal{D}\boldsymbol{\zeta}(\tau)\delta(\boldsymbol{\zeta}_c - \boldsymbol{\zeta}_0) \exp\{-S[\boldsymbol{\zeta}(\tau)]/\hbar\}, \qquad (2.53)$$

where the path centroid vector in phase space is given by

$$\boldsymbol{\zeta}_0 = \frac{1}{\hbar\beta} \int_0^{\hbar\beta} d\tau\, \boldsymbol{\zeta}(\tau). \qquad (2.54)$$

The action functional for the imaginary time phase-space path integral is given by [11]

$$S[\zeta(\tau)] = \int_0^{\hbar\beta} d\tau \, \{ \tfrac{1}{2}\mathbf{p} \cdot (\tau)\mathbf{m}^{-1} \cdot \mathbf{p}(\tau) - i\mathbf{p}(\tau) \cdot \dot{\mathbf{q}}(\tau) + V[\zeta(\tau)] \} \, , \quad (2.55)$$

where $\mathbf{m}$ is the diagonal particle mass matrix and $\dot{\mathbf{q}}(\tau)$ is understood as the imaginary time velocity vector. The quantum partition function is related to Eq. (2.53) such that $Z = \int d\zeta_c \rho_c(\zeta_c)$. The phase-space centroid-constrained correlation functions are defined by the $2N \times 2N$ matrix [5]

$$\mathbf{C}_c(\tau, \mathbf{q}_c) = \frac{\int \cdots \int \mathcal{D}\zeta(\tau)\delta(\zeta_c - \zeta_0)(\zeta(\tau) - \zeta_0)(\zeta(0) - \zeta_0) \exp\{-S[\zeta(\tau)]/\hbar\}}{\int \cdots \int \mathcal{D}\zeta(\tau)\delta(\zeta_c - \zeta_0) \exp\{-S[\zeta(\tau)]/\hbar\}} \, .$$

$$(2.56)$$

Each element of this matrix with the indices "i, j" is given explicitly by the 2×2 block

$$[\mathbf{C}_c(\tau, \mathbf{q}_c)]_{ij} = \begin{pmatrix} C_c(\tilde{p}_i(\tau)\tilde{p}_j(0), \mathbf{q}_c) & C_c(\tilde{p}_i(\tau)\tilde{q}_j(0), \mathbf{q}_c) \\ C_c(\tilde{q}_i(\tau)\tilde{p}_j(0), \mathbf{q}_c) & C_c(\tilde{q}_i(\tau)\tilde{q}_j(0), \mathbf{q}_c) \end{pmatrix}, \quad (2.57)$$

where $\tilde{\zeta}(\tau)$ is the quantum path fluctuation with respect to the centroid position ζ_c [i.e., $\zeta(\tau) = \zeta_c + \tilde{\zeta}(\tau)$]. The elements of the centroid-constrained correlation function matrix in Eq. (2.57) can also be obtained by adding linear external field terms to the action functional in Eq. (2.55) and then taking the appropriate second-order functional derivatives [5]. The centroid-constrained correlation function matrix in Eq. (2.57) is independent of the momentum centroid $\mathbf{p}_c$ if the potential V is independent of the momentum variable.

It proves informative first to revisit the expression for the equilibrium average of a general operator but now in the phase-space centroid perspective. This simple analysis identifies the centroid-constrained correlation function matrix in Eqs. (2.56) and (2.57) as providing the effective centroid "width" factors in phase space. In the phase-space path integral perspective [11], the equilibrium average of an operator A is given by the expression

$$\langle A \rangle = Z^{-1} \int \cdots \int \mathcal{D}\zeta(\tau)A[\zeta(0)] \exp\{-S[\zeta(\tau)]/\hbar\} \, . \quad (2.58)$$

From the cyclic invariance of the phase-space trace, the operator can be evaluated at any point along the cyclic imaginary time path $\zeta(\tau)$. The average in Eq. (2.58) is first reexpressed so that the centroid variable appears explicitly in the statistical averaging [3, 5, 34, 40, 45]:

$$\langle A \rangle = Z^{-1} \int d\boldsymbol{\zeta}_c \left[\int \cdots \int \mathcal{D}\boldsymbol{\zeta}(\tau)\delta(\boldsymbol{\zeta}_c - \boldsymbol{\zeta}_0)A[\boldsymbol{\zeta}(0)] \exp\{-S[\boldsymbol{\zeta}(\tau)]/\hbar\} \right].$$

(2.59)

The operator $A(\boldsymbol{\zeta})$ is then represented in $2N$-dimensional Fourier space and the cumulant average over the phase-space path fluctuation variables $\tilde{\boldsymbol{\zeta}}(\tau)$ is carried out through second order [5]. After performing the inverse Fourier transformation, the final result for the operator average in the phase-space centroid picture is approximated by the classical-like form [5]

$$\langle A \rangle \simeq \langle A_c(\boldsymbol{\zeta}_c) \rangle_{\rho_c},$$

(2.60)

where the effective centroid-dependent quasiclassical operator A_c is given by

$$A_c(\boldsymbol{\zeta}_c) = \langle A(\boldsymbol{\zeta}_c + \tilde{\boldsymbol{\zeta}}) \rangle_{\mathbf{C}_c(0, \boldsymbol{\zeta}_c)}$$

$$= \frac{1}{\sqrt{\det[2\pi\mathbf{C}_c(0, \mathbf{q}_c)]}} \int d\tilde{\boldsymbol{\zeta}}\, A(\boldsymbol{\zeta}_c + \tilde{\boldsymbol{\zeta}}) \exp[-\tilde{\boldsymbol{\zeta}} \cdot \mathbf{C}_c^{-1}(0, \mathbf{q}_c) \cdot \tilde{\boldsymbol{\zeta}}/2].$$

(2.61)

Here the vector variable $\tilde{\boldsymbol{\zeta}}$ is a Gaussian vector with associated width matrix given by $\mathbf{C}_c(0, \mathbf{q}_c)$. The result above from Paper III is the simplest generalization of the expression obtained in Paper I for coordinate-dependent operators. The expression reveals the role played by the centroid-constrained correlation function matrix [Eq. (2.57)] in defining the effective width factor in phase space for the centroid quasiparticle. A more careful treatment of the operator ordering problems demonstrates that the derivation of the equations above involves additional approximation beyond second-order truncation of the cumulant expansion [59].

In Paper I, general imaginary-time correlation functions were expressed in terms of an averaging over the coordinate-space centroid density $\rho_c(q_c)$ and the centroid-constrained imaginary-time-position correlation function $C_c(\tau, q_c)$. This formalism was extended in Paper III to the phase-space centroid picture so that the momentum could be treated as an independent variable. The final result for a general imaginary-time correlation function is found to be given approximately by [5, 59]

$$C_{AB}(\tau) \simeq \langle p_{AB}(\tau, \boldsymbol{\zeta}_c) \rangle_{\rho_c},$$

(2.62)

where the centroid-dependent imaginary-time-correlated operator product $p_{AB}(\tau, \zeta_c)$ is defined as the multiple Gaussian average

$$p_{AB}(\tau, \zeta_c) = \langle A(\zeta_c + \zeta_1)B(\zeta_c + \zeta_2)\rangle_{\mathbf{C}_c^+, \mathbf{C}_c^-} . \qquad (2.63)$$

Here, the vectors ζ_1 and ζ_2 are related to the two Gaussian vectors ζ_+ and ζ_-, such that

$$\zeta_{1,2} = \frac{1}{\sqrt{2}}(\zeta_+ \pm \zeta_-) , \qquad (2.64)$$

where ζ_+ and ζ_- have Gaussian width matrices given by

$$\begin{aligned}
\mathbf{C}_c^+(\tau, \mathbf{q}_c) &= \mathbf{C}_c(0, \mathbf{q}_c) + \mathbf{C}_c(\tau, \mathbf{q}_c) , \\
\mathbf{C}_c^-(\tau, \mathbf{q}_c) &= \mathbf{C}_c(0, \mathbf{q}_c) - \mathbf{C}_c(\tau, \mathbf{q}_c) .
\end{aligned} \qquad (2.65)$$

Due to the compact notation adopted in the preceding analysis, the final expressions in Eqs. (2.62) and (2.63) resemble those appearing in Paper I. Nonetheless, they are more complicated, due to the cross terms which appear because of the noncommutation of position and momentum operators. In fact, the two off-diagonal terms in the matrix $\mathbf{C}_c(\tau, \zeta_c)$ are complex functions and, in turn, complex conjugates of each other. In the expression for operator averages in Eq. (2.60) and for the time-correlated operator product in Eq. (2.63), different operator orderings may lead to different values. The final expression in terms of the Gaussian averages should always be consistent with the original choice of the operator ordering. A more detailed (and more complicated) analysis of the operator-ordering problem is given in Ref. 59.

E. Numerical Examples

In this subsection, representative results from calculations designed to probe the accuracy of preceding equilibrium theory are presented. The main results given here are summarized as (1) a test of the centroid-based formulation for imaginary-time correlation functions (Section II.C.2), (2) a test of the analytic diagrammatic approach for the centroid density (Section II.A), and (3) a test of the analytic diagrammatic approach for the imaginary time propagator (Section II.A). Other numerical results on equilibrium properties are given in Paper I.

The numerical calculations reported here have been performed for a completely *nonquadratic* potential, given by

$$V(q) = q^3 + q^4/2 , \qquad (2.66)$$

where the mass m and $\hbar$ are taken to be unity. The inverse temperature $\beta = 1/k_B T$ is thus the same as the value as the dimensionless parameter $\beta\hbar\omega$. All of the analytical results for this potential are compared with PIMC simulation results, the details of which are given in Paper I. It should be noted that these calculations represent a serious test of the accuracy of the centroid-based formulation because the potential in Eq. (2.66) contains no intrinsic quadratic term.

In Fig. 4 the imaginary-time correlation function $\langle q^3(\tau)q^3(0)\rangle$ is plotted for a temperature of $\beta = 5$ as a function of the dimensionless variable $u = \tau/\hbar\beta$. The centroid-based formalism is seen to be in very good agreement with the numerically exact result, confirming the validity of the various approximations for this completely nonquadratic example.

As a direct test of the analytic theory for the centroid density, the quantum correction factor for the thermal rate constant of an Eckart barrier potential was calculated within the context of path-integral quantum transition-state theory [42–44, 49]. The results are tabulated in

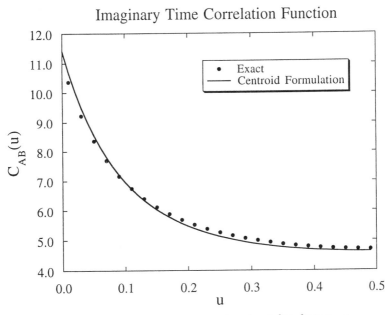

Figure 4. Plot of the imaginary-time correlation function $\langle q^3(\tau)q^3(0)\rangle$ for the nonquadratic potential described in Section II.E [Eq. (2.66)]. The correlation function is plotted as a function of the dimensionless variable $u = \tau/\hbar\beta$ with $\beta = 5$. The solid circles are the numerically exact results, while the solid line is for the optimized LHO theory in Eqs. (2.24)–(2.28) used in the centroid-based formulation of the correlation function in Eq. (2.50).

G.A. VOTH

TABLE I
Quantum Correction Factors[a] for the Thermal Rate Constant of an Eckart Barrier
Potential[b]

u	Γ_1	Γ_2	Γ_{MC}
6	4.4	4.4	4.4
8	15.0	17.0	17.0
10	73.0	110.6	105.0
12	514.0	1278.0	1240.0

[a] The quantum corrections Γ_1 are based on the optimized LHO reference potential approximation for the centroid density in Eqs. (2.24)–(2.28), while Γ_2 are the results including the higher-order terms in Eqs. (2.33). The quantum correction Γ_{MC} is the path-integral Monte Carlo result reported in Ref. 42.

[b] The Eckart barrier potential is given by $V(q) = V_0 \cosh^{-2}(q/a_0)$, with the parameter values $2\pi V_0/\hbar\omega_b = 12.0$ and $u = \beta\hbar\omega_b$ in the present calculations, and ω_b is the magnitude of the classical barrier frequency.

Table I. The difference between the numerical path-integral results and the optimized LHO approximation for the centroid density increases as the temperature is lowered, so the higher-order analytic corrections in Eq. (2.33) become important in the extreme quantum limit. As seen from Table I, the renormalized analytic theory with corrections through quartic order is essentially exact to within the MC error (ca. 5%) of the numerical results.

To probe the effect of the higher-order correction terms in Eq. (2.42) for the imaginary time propagator, the centroid-constrained propagator was first evaluated in the optimized LHO approximation from Eq. (2.39) (Fig. 5). Although the effective LHO approximation provides a good approximation to the exact centroid-constrained correlation function, the correction from the two-line-loop diagram renormalization improves the agreement with the PIMC data significantly. Other numerical examples for equilibrium properties are given in Paper I.

III. DYNAMICAL PROPERTIES

In Papers II and III, the centroid-based theory was significantly extended to treat perhaps one of the most challenging problems in condensed matter theory—the computation of general real-time quantum correlation functions $\langle A(t)B(0) \rangle$. Consistent with the general theme of this research, the properties of dynamical correlation functions were explored using the centroid-based perspective of quantum statistical mechanics. To be more specific, in one approach, real-time dynamical information was extracted with the help of the centroid-constrained formalism for imaginary-time

Centroid-Constrained Imaginary Time Correlation Function

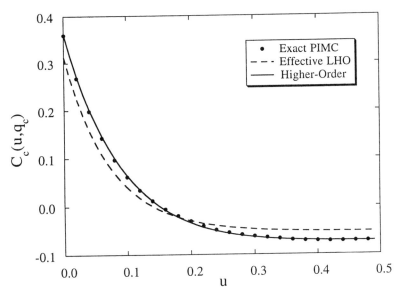

Figure 5. Plot of the centroid-constrained correlation function $C_c(\tau, q_c)$ defined in Eq. (2.1) for the nonquadratic potential described in Section II.E [Eq. (2.66)]. The numerically exact results are shown by the solid circles. The correlation function obtained from the optimized LHO approximation in Eq. (2.39) is shown by the dashed line, while the solid line depicts the results obtained by including the two-line-loop correction from Eq. (2.42). The correlation functions are plotted as a function of $u = \tau/\hbar\beta$ and for $\beta = 10$ and $q_c = 0.0$.

correlation functions developed in Paper I. As summarized in Section III.A, there is a dual outcome from this exercise. First, an approximate, but mathematically direct analytical approach for real-time position correlation functions can be derived which is a direct outgrowth of the centroid-based formalism developed in Paper I. This procedure involves the analytic continuation of imaginary time position correlation functions $\langle q(\tau)q(0)\rangle$ via the inverse Wick rotation $\tau \rightarrow it$. The analytically continued effective harmonic result expresses the time correlation as the centroid density-weighted superposition of locally optimized linear harmonic oscillator (LHO) time correlation functions and coincidentally provides the basis for a quantum mechanical generalizations [7] of the classical instantaneous normal mode perspective (see, e.g., Refs. 60 and 61). As shown below, this theory has its strengths and weaknesses.

The second and much more important outcome of the centroid-based dynamical analysis was the identification [4, 8] and subsequent mathematical justification [5] of the conceptually and computationally promising centroid molecular dynamics (CMD) perspective for computing real-time quantum correlation functions. Of the two approaches to time correlation functions described here, the CMD approach is the better one because it can be applied to general systems for long time scales and the numerical effort in such an application scales with the number of particles in the same way as a classical molecular dynamics calculation. Centroid MD is motivated by the results of mathematical analysis and is further supported by several reasonable physical arguments [4, 5, 8].

A.   Effective Harmonic Theory

1.   *Position Time Correlation Functions*

By definition, the centroid variable occupies a central role in the behavior of the centroid-constrained imaginary-time correlation function in Eq. (2.1). However, it is even more interesting to analyze the role of the centroid variable in the *real-time* quantum position correlation function [4, 8]. This information can in principle be extracted from the exact centroid-constrained correlation function $C_c(\tau, q_c)$ through the analytic continuation $\tau \rightarrow it$. Such a procedure, however, is generally not tractable unless there is some prior simplification of the problem. One such simplification is achieved [4, 8] through use of the optimized reference quadratic action functional, given by [3, 21–23]

$$S_{\text{ref}}[q(\tau)] = \int_0^{\hbar\beta} d\tau \left\{ \frac{m}{2} \dot{q}(\tau)^2 + \frac{1}{2} m\bar{\omega}^2 [q(\tau) - q_c]^2 \right\}, \qquad (3.1)$$

where $\bar{\omega}$ is the centroid-dependent optimized effective LHO frequency. The value for the optimized frequency $\bar{\omega}$ is determined from the solution of the transcendental equation in Eq. (2.27) with the effective thermal width factor $\bar{\alpha}$ for a particular position of the path centroid given by Eq. (2.28).

The real-time expression for the centroid-constrained position correlation function in Eq. (2.1) can be determined analytically for the optimized LHO reference system via the inverse Wick rotation $\tau \rightarrow it$. Then, by virtue of Eq. (2.3), the real-time position correlation function is found in the effective harmonic theory to be given by [4, 8]

$$C_{\text{ref}}(t) = \langle C_c(t, q_c) + q_c^2 \rangle_{\rho_c}$$

$$= \frac{1}{Z} \int dq_c \, \rho_c(q_c) \frac{\hbar}{2m\bar{\omega}} \left[\frac{1}{\tanh(\hbar\beta\bar{\omega}/2)} \cos \bar{\omega}t - i \sin \bar{\omega}t \right] + \langle q \rangle^2. \qquad (3.2)$$

The real-time correlation function at this level of theory is seen to be the superposition of centroid correlation functions for effective harmonic oscillators defined at each centroid position q_c. Each centroid correlation function is then statistically weighted by the centroid density for the given value of q_c. The expression above for the correlation function in Eq. (3.2) represents a quantum mechanical instantaneous normal-mode perspective for condensed-phase correlation functions [7], the multidimensional details of which are given in Paper V.

One significant feature of Eq. (3.2) is the factorization of the expression into the centroid density (i.e., the centroid statistical distribution) and the dynamical part, which depends on the centroid frequency $\bar{\omega}$. It is not obvious that such a factorization should occur in general. For example, a rather different factorization occurs when the conventional formalism for computing time correlation functions is used [i.e., a double integration in terms of the *off-diagonal* elements of the thermal density matrix and the Heisenberg operator $q(t)$ is obtained]. This result sheds light on the dynamical role of the centroid variable in real-time correlation functions (cf. Section III.B) [4, 8].

2. General Time Correlation Functions

The centroid-based effective harmonic theory for position time correlation functions can be extended to calculate general time correlation functions of the form $\langle A(t)B(0)\rangle$ [4]. This approach begins with the results from Papers I and III for general imaginary-time correlation functions in the centroid-based perspective. These expressions are given by Eq. (2.50) or (2.62) and, in turn, involve the centroid-dependent imaginary-time-correlated operator product $p_{AB}(\tau, q_c)$ [Eq. (2.51) or (2.63)]. In principle, the double-Gaussian average in Eq. (2.51) or (2.63) can be performed for certain forms of the general operators A and B. The resulting expression will then involve an average of functions that depend on the elements of $C_c(\tau, q_c)$ over the normalized centroid density. For example, the coordinate-dependent imaginary time correlation function $\langle q^3(\tau)q^3(0)\rangle$ is given in this approach by

$$\langle q^3(\tau)q^3(0)\rangle = \langle q_c^6 + 3q_c^4[2C_c(0, q_c) + 3C_c(\tau, q_c)]$$
$$+ 9q_c^2[C_c^2(0, q_c) + 2C_c^2(\tau, q_c) + 2C_c(\tau, q_c)C_c(0, q_c)]$$
$$+ 3C_c(\tau, q_c)[3C_c^2(0, q_c) + 2C_c^2(\tau, q_c)]\rangle_{\rho_c}. \qquad (3.3)$$

Note that this function is calculated for each value of the centroid variable and averaged over the centroid density. This equation is also exact in the case of a globally harmonic potential.

After obtaining the closed-form imaginary-time expression [e.g., Eq. (3.3)], one can continue the optimized LHO expression analytically for the centroid-constrained propagator $C_c(\tau, q_c)$ by replacing τ with it and then use the resulting expression in, for example, Eq. (3.3). In the effective harmonic theory, the real-time version of $C_c(\tau, q_c)$ for each value of the centroid is given by

$$C_c(t, q_c) \simeq \frac{\hbar}{2m\bar{\omega}} \left[\frac{1}{\tanh(\hbar\beta\bar{\omega}/2)} \cos \bar{\omega}t - i \sin \bar{\omega}t \right] - \frac{1}{m\bar{\omega}^2\beta}, \quad (3.4)$$

where the optimized frequency $\bar{\omega}$ depends on the centroid position via Eq. (2.27). One replaces $C_c(\tau, q_c)$ with the effective harmonic $C_c(t, q_c)$ at each centroid value in the closed-form average for the correlation function [e.g., Eq. (3.3)], resulting in the effective harmonic approximation to the real-time correlation function [4, 8].

The results obtained from the effective harmonic theory [4, 5] are not completely satisfactory for two reasons:

1. The accuracy of the analytically continued version of Eq. (3.2) is only as good as the accuracy of the analytically continued centroid-constrained propagator $C_c(t, q_c)$. While the effective LHO version of this function given in Eq. (3.4) is accurate at short times, anharmonicities in the real potential will cause it to deviate from the exact behavior at long times even in the classical limit.

2. The operator representations in Eq. (2.50) or (2.62) are expressed at the level of second-order cumulant expansions. Although this approximation is an excellent one for imaginary-time calculations, real-time correlation functions are more sensitive to nonlinear interactions and hence less predictable in their behavior. In principle, however, the cumulant averages could be carried out to higher order.

B. Centroid Molecular Dynamics Method

In this subsection the more accurate CMD method [4–6, 8] is described and analyzed in some detail. The method holds great promise for the study of quantum dynamics in condensed matter because systems having nonquadratic many-body potentials can be simulated for relatively long times. The numerical effort in this approach scales with system size as does a classical MD simulation, although the total overall computational cost will always be larger. Here the CMD method is first motivated by further analysis of the effective harmonic theory. This discussion is an abbreviated form of the historical line of development contained in Paper

II and an early Communication [8]). The initial proposal [4, 8] for the CMD method was an ad hoc one reinforced by numerical results, but the method was subsequently justified mathematically in Paper III. The latter justification [5] is also described in some detail below.

To begin, it is useful to revisit the effective harmonic result of Section III.A. In doing this it is informative to introduce a different effective harmonic real-time correlation function, given in terms of the centroid variable by [4, 8]

$$C_{\text{ref}}^*(t) = \frac{1}{Z} \int dq_c \, \rho_c(q_c) \frac{1}{m\bar{\omega}^2\beta} \cos \bar{\omega}t + \langle q_c \rangle_{\rho_c}^2 . \tag{3.5}$$

This correlation function is the exact analog of a classical-like superposition of effective LHO position correlation functions with centroid frequencies $\bar{\omega}$. As shown below, it is related to the quantum position correlation function in Eq. (3.2) by a Fourier relation [4, 8]. By noting that the first term on the right-hand side of Eq. (3.5) describes the correlation of fluctuations about the mean value $\langle q_c \rangle_{\rho_c} = \langle q \rangle$, Eq. (3.5) can be rewritten as

$$C_{\text{ref}}^*(t) = \frac{1}{Z} \int dq_c \, \rho_c(q_c) \langle q_c(t)q_c(0) \rangle_{\bar{\omega}} , \tag{3.6}$$

where $\langle q_c(t)q_c(0) \rangle_{\bar{\omega}}$ is a position correlation function calculated for *quasiclassical centroid dynamics* on the optimized LHO centroid potential. The dynamical centroid trajectories for this correlation function are thus given by the classical-like equations

$$q_c(t) = q_c(0) \cos\{\bar{\omega}[q(0)]t\} + \frac{p_c(0)}{m\bar{\omega}[q(0)]} \sin\{\bar{\omega}[q(0)]t\} . \tag{3.7}$$

The symbol $\langle \cdots \rangle_{\bar{\omega}}$ in Eq. (3.6) denotes initial condition averaging using the optimized LHO approximation to the phase-space centroid density. It should be noted that the momentum centroid $\tilde{p}_0$ is always decoupled from the position coordinates and is described by the *classical* Boltzmann momentum distribution. This result is both interesting and significant for the role of the centroid variable in dynamics [4, 5, 8].

The relationship of $C_{\text{ref}}^*(t)$ to the optimized LHO representation of the quantum position correlation function $C_{\text{ref}}(t)$ is found to be [4, 8]

$$\tilde{C}_{\text{ref}}(\omega) = (\hbar\beta\omega/2)[\coth(\hbar\beta\omega/2) + 1]\tilde{C}_{\text{ref}}^*(\omega) . \tag{3.8}$$

Accordingly, it can also be shown [4] that Eq. (3.5) is the optimized LHO representation of the Kubo-transformed [62] position correlation function, given by

$$\psi(t) = \frac{1}{\hbar\beta} \int_0^{\hbar\beta} d\tau \langle q(0)q(t+i\tau) \rangle . \qquad (3.9)$$

The latter connection is important for the mathematical justification [5] of CMD (cf. Section III.B.1) and for one particular CMD-based approach [4, 5] for computing general time correlation functions (cf. Section III.B.3).

For general systems, the optimized LHO result for the centroid position correlation function $C^*(t)$ in Eq. (3.6) can only be viewed as a short-time approximation [4, 5, 8]. It is then reasonable to assume that Eq. (3.6) must actually be an approximation to some more accurate expression. In two of the earlier centroid papers [4, 8], it was suggested and argued that, in fact, a more accurate representation of the centroid correlation function should be the classical-like expression

$$C^*(t) = \langle q_c(t)q_c(0) \rangle_{\rho_c} , \qquad (3.10)$$

where the centroid trajectories are instead generated by the effective classical-like equations of motion for the *exact centroid effective potential* defined in Eq. (1.8) [4–6, 8]:

$$m\ddot{q}_c(t) = -\frac{dV_c(q_c)}{dq_c} . \qquad (3.11)$$

The centroid force $-dV_c(q_c)/dq_c$ defined on the right-hand side of Eq. (3.11) is the quantum mechanical potential of mean force for the centroid, given by

$$-\frac{dV_c(q_c)}{dq_c} = -\frac{\int \cdots \int \mathcal{D}q(\tau)\delta(q_c - q_0)\{dV[q(0)]/dq\}\exp\{-S[q(\tau)]/\hbar\}}{\int \cdots \int \mathcal{D}q(\tau)\delta(q_c - q_0)\exp\{-S[q(\tau)]/\hbar\}} .$$

$$(3.12)$$

In the CMD correlation function [Eq. (3.10)], the notation $\langle \cdots \rangle_{\rho_c}$ means that the exact (normalized) phase-space centroid density is used to average the initial conditions of the centroid trajectories in the usual classical sense. Once the centroid correlation function $C^*(t)$ in Eq. (3.10) is calculated, the exact quantum position correlation function $C(t)$ can be estimated through a Fourier transform relationship similar to Eq. (3.8), but now for the general CMD case [4, 5, 8],

$$\tilde{C}(\omega) \simeq (\hbar\beta\omega/2)[\coth(\hbar\beta\omega/2) + 1]\tilde{C}^*(\omega) \, . \qquad (3.13)$$

An alternative justification for this relation is that $C^*(t)$ is an approximation [4] to the Kubo-transformed position correlation function $\psi(t)$ in Eq. (3.9) and that the relationship of the latter function to $C(t)$ is always given by the expression in Eq. (3.13).

In contrast to the effective harmonic prescription for centroid-based dynamics, in CMD the force is a *unique* function of the system. That is, the force on a centroid trajectory at some time and position in space is not different from the force experienced by a different centroid trajectory at that same point in space but at a different time. Furthermore, the centroid trajectories are derived from the same effective potential as the one giving the *exact* centroid statistical distribution so that a *centroid ergodic theorem* will hold. The CMD approach satisfies this condition, while the analytically continued optimized LHO theory may not. Finally, CMD recovers the exact limiting expressions for globally harmonic potentials and for general classical systems.

1. Justification of Centroid Molecular Dynamics

In the early papers [4, 8], the development of the CMD method was guided in part by the effective harmonic analysis and, in part, by physical reasoning. In Paper III, however, a mathematical justification of CMD was provided. In the latter analysis, it was shown that (1) CMD always yields a mathematically well-defined approximation to the quantum Kubo-transformed position or velocity correlation function, and (2) the *equilibrium* path centroid variable occupies an important role in the time correlation function because of the nature of the preaveraging procedure in CMD. Critical to the analysis of CMD and its justification was the phase-space centroid density formulation of Paper III, so that the momentum could be treated as an independent dynamical variable. The relationship between the centroid correlation function and the Kubo-transformed position correlation function was found to be unique if the centroid is taken as a dynamical variable. The analysis of Paper III will now be reviewed. For notational simplicity, the equations are restricted to a two-dimensional phase space, but they can readily be generalized.

In CMD, the centroid variable evolves according to the classical-like equation of motion [4, 5, 8]

$$\dot{q}_c(t) = p_c(t)/m \, ,$$
$$\dot{p}_c(t) = F_c(t) \, , \qquad (3.14)$$

where the centroid force is defined compactly as

$$F_c(q) = \langle f(q_c + \tilde{q}) \rangle_c . \tag{3.15}$$

Here the symbol $\langle \cdots \rangle_c$ denotes the centroid-constrained average with the phase-space imaginary time path integral and $f = -dV/dq$. With the centroid trajectories in hand from the equations above, the CMD time correlation function is given by [4–6, 8]

$$C^*(t) = \langle q_c(t)q_c(0) \rangle_{\rho_c} , \tag{3.16}$$

where the subscript ρ_c means that the initial conditions of the centroid trajectories are averaged with the phase-space centroid density.

After expanding $C^*(t)$ as a Taylor series in terms of t, one obtains

$$C^*(t) = \sum_{n=0}^{\infty} \frac{t^{2n}}{(2n)!} C^{[2n]} , \tag{3.17}$$

where the expansion coefficients are expressed as

$$C^{[2n]} = \langle q_c \mathcal{L}_c^{2n} q_c \rangle_{\rho_c} . \tag{3.18}$$

The operator $\mathcal{L}_c$ is the classical Poisson bracket

$$\mathcal{L}_c A = \{A, H_c\} \tag{3.19}$$

for a classical-like centroid Hamiltonian defined by $H_c = p_c^2/2 + V_c(q_c)$.

One can similarly expand the Kubo-transformed position correlation function, given by [5]

$$\psi(t) = \frac{1}{\hbar \beta} \int_0^{\hbar \beta} d\tau \langle q(0)q(t + i\tau) \rangle , \tag{3.20}$$

which is related directly to the quantum response function as well as the quantum dynamical position, momentum, and cross-correlation functions. The Taylor expansion of Eq. (3.20) yields

$$\psi(t) = \sum_{n=0}^{\infty} \frac{t^{2n}}{(2n)!} \psi^{[2n]} , \tag{3.21}$$

where the expansion coefficients are given by

$$\psi^{[2n]} = \frac{1}{\hbar \beta} \int_0^{\hbar \beta} d\tau \langle q(\tau)\mathcal{L}^{2n} q(0) \rangle . \tag{3.22}$$

Here $\mathcal{L}$ is the commutator operator

$$\mathcal{L}A = \frac{1}{i\hbar}[A, H].$$ (3.23)

After making use of the definition of the centroid variable and the cyclic invariance of trace, one obtains

$$\psi^{[2n]} = \langle q_c \langle \mathcal{L}^{2n} q \rangle_c \rangle_{\rho_c}.$$ (3.24)

Upon inspection of the two expansions, it is seen that the centroid correlation function and the Kubo-transformed correlation function take a similar analytical form [cf. Eqs. (3.18) and (3.24)], the difference being between the terms $\mathcal{L}_c^{2n} q_c$ and $\langle \mathcal{L}^{2n} q \rangle_c$ in Eqs. (3.18) and (3.24), respectively. The expansions can be determined explicitly and compared term by term. The first few terms are:

$n = 0$:

$$\langle q \rangle_c = q_c ;$$ (3.25)

$n = 1$:

$$\langle \mathcal{L}^2 q \rangle_c = \langle f/m \rangle_c = F_c/m ;$$ (3.26)

$n = 2$:

$$\langle \mathcal{L}^4 q_c \rangle = \frac{1}{4m^3} \langle (f^{(2)} p^2 + 2pf^{(2)} p + p^2 f^{(2)}) \rangle_c + \frac{1}{m^2} \langle f^{(1)} f \rangle_c$$ (3.27)

for the Kubo correlation function [Eq. (3.24)], but for the centroid correlation function [Eq. (3.18)],

$$\mathcal{L}_c^4 q_c = \frac{1}{m^3} F_c^{(2)} p_c^2 + \frac{1}{m^2} F_c^{(1)} F_c .$$ (3.28)

Therefore, the two leading terms of the Taylor expansions for the centroid and Kubo correlation functions are the same, the difference between them beginning with the third term (i.e., at order t^4). The latter term can be taken as an example of how to evaluate the leading correction term to the centroid correlation function (and thereby demonstrate that the centroid correlation function is a well-defined approximation to the Kubo correlation function). The Gaussian representation of operators in phase space [Eq. (2.61)] proves to be useful but not essential in this analysis.

It is first noted that the centroid average of a product of operators can be written as a product of the centroid averages of operators and a leading correction term [4, 5]:

$$\langle AB \rangle_c \simeq A_c B_c + A'_c B'_c C_c , \qquad (3.29)$$

where terms higher order in the phase space width factor C_c [Eq. (2.56)] have been ignored. By virtue of the cyclic invariance of the trace, it can be demonstrated that

$$C_c(\tilde{p}(0)\tilde{q}(0), q_c) + C_c(\tilde{q}(0)\tilde{p}(0), q_c) = 0 . \qquad (3.30)$$

It is also seen that the application of the commutator $\mathscr{L}$ of Eq. (3.23) four times in Eq. (3.27) leads to a symmetrized arrangement of momentum and coordinate operators. Combining this fact with Eqs. (3.29) and (3.30), it is found that there are no centroid-constrained correlation functions that mix momentum and coordinate operators to the order of the leading correction term. Therefore, Eq. (3.27) becomes

$$\langle \mathscr{L}^4 q \rangle_c = \frac{1}{m^3} f_c^{(2)} \langle p^2 \rangle_c + \frac{1}{m^2} \langle f^{(1)} f \rangle_c , \qquad (3.31)$$

where $f_c^{(n)}$ stands for the centroid-constrained operator average $f_c^{(n)} \equiv \langle f^{(n)} \rangle_c$. Equation (3.29) is useful when comparing the second term on the right-hand side of Eq. (3.31) with the second term of Eq. (3.28) (i.e., $\langle f^{(1)} f \rangle_c \simeq F_c^{(1)} F_c$). The correction factor for this term contains the width factor and second- or higher-order derivatives of the force. The focus of the effort can thereby shift to a comparison of the first term on the right-hand sides of Eqs. (3.28) and (3.31).

Because the centroid force F_c is a function and f is quantum operator, the nth-order derivative of the centroid force, $F_c^{(n)}$, and the centroid-constrained average of the nth-order derivative of the force, $f_c^{(n)}$, are different even though the centroid force is equal to the centroid-constrained average of the force (i.e., $F_c = \langle f \rangle_c$). Application of the chain rule to $F_c^{(n)}$ reveals the difference between the force derivative terms as being given by

$$\delta f_c^{(n)} = f_c^{(n)} - F_c^{(n)} = -\tfrac{1}{2} f_c^{(1+n)} C_c^{(1)}(\tilde{q}(0)\tilde{q}(0), q_c) + \cdots , \qquad (3.32)$$

where the derivative of the width factor appears here instead of the width factor itself. This observation is quite significant because it indicates that an expression in terms of the centroid force or its derivatives agrees with its quantum counterpart $f_c^{(n)}$ to *all orders* in the width factor C_c, the corrections coming only in the spatial derivatives of the latter factor.

From Eqs. (3.18) and (3.24) it is seen that the latter correction will also be averaged over the centroid density, so large deviations occur only if the width experiences large fluctuations that persist in the *average* sense.

Returning to the first term on the right-hand side of Eq. (3.31), the quantum fluctuations in momentum contribute a further deviation from the similar term in the centroid correlation function [Eq. (3.28)]. The difference between the two terms for all powers of n is found to be [5]

$$\delta p_c^n = \langle p^n \rangle_c - p_c^n = \frac{n}{2} p_c^{n-2} C_c(\tilde{p}(0)\tilde{p}(0), q_c) + \cdots, \qquad (3.33)$$

where $C_c(\tilde{p}(0)\tilde{p}(0), q_c)$ is the width factor for the momentum path fluctuations. The terms associated with this correction have a value of n no smaller than 2. The average of the term p_c^{n-2} over the phase-space centroid density will simply factorize and give a constant since the distribution of the momentum centroid is the classical Boltzmann distribution and independent of the higher-order-path Fourier modes [4, 5, 8, 63].

Taking into account all of the preceding considerations and generalizing them to terms of higher order, the difference between the Kubo-transformed position correlation function [Eq. (3.22)] and the CMD correlation function [Eq. (3.16)] can be summed to give [5]

$$\psi(t) - C^*(t) = \sum_{n=2}^{\infty} \frac{t^{2n}}{(2n)!} \Delta^{[2n]}, \qquad (3.34)$$

with $\Delta^{[2n]}$ being the difference between the two sets of coefficients at order t^{2n}. The explicit expression for $\Delta^{[2n]}$ is given by [5]

$$\Delta^{[2n]} = \psi^{[2n]} - C^{[2n]} \propto \Big\langle q_c C_c \prod_l F_c^{(n_l)} \Big\rangle_{\rho_c}, \qquad (3.35)$$

where at least one n_l is no smaller than 2 and terms that involve the spatial derivatives of the width factor C_c have been neglected. The result above can also be confirmed by a dimensional analysis.

The preceding analysis proves that the numerical success of CMD [4–6, 8, 48] is not accidental and, in fact, is both physically and mathematically understandable. Some comments on the analysis are as follows:

(a) At first glance, Eq. (3.35) highlights two properties of CMD that are clear [4, 5, 8]. Namely, the method is exact if the potential contains no global anharmonicity (i.e., $F_c^{(n_l)} = 0$) or if the system is near the classical limit (i.e., C_c is small). However, upon closer inspection Eq. (3.35) also reveals why CMD is a good approximation in highly quantized situations. In particular, the centroid force F_c [Eq. (3.15)] at the centroid position q_c is computed by *averaging* the classical force over the imaginary time Feynman paths around the centroid. This averaging tends to smooth out significantly any "kinks" in the classical potential energy function, at least over the length scale of the particle's thermal width. The more "quantum" the particle, the more this averaging occurs. This behavior is very important to the success of CMD since the higher-order derivatives of the centroid force in Eq. (3.35) will tend to be small, giving a small correction factor in Eq. (3.35) even though the quantum thermal width C_c may be sizable. For CMD to become inaccurate, the locally nonlinear features of the effective *centroid* potential must be large on a length scale greater than the thermal width factor of the centroid quasiparticle. Moreover, since any correction terms are then *averaged* in Eq. (3.35) over the centroid distribution, the effective anharmonic behavior must also be present to a significant degree in the regions of greatest centroid density. As the system becomes more classical, the effective anharmonicities (i.e., the higher derivatives of the centroid force) will indeed become larger, but this effect is compensated by an even larger reduction in the thermal width factor C_c in Eq. (3.35). (Recall that CMD always yields the exact classical limit.)

(b) The definition of the centroid force captures the major contribution of quantum fluctuations and predicts the right quantum dynamics to within a tolerance proportional to the *averaged* higher-order derivatives of the centroid force and the centroid thermal width factor C_c. The CMD correlation function therefore contains terms that are infinite order in $\hbar$. The leading corrections to the centroid dynamics depend on the thermal width factor for the potential at hand (i.e., not just the free particle width, which is of order $\hbar^2$).

(c) In general, the Kubo-transformed position correlation function [Eq. (3.20)] is an ideal candidate for any type of approximation because the integration over the imaginary time τ eliminates the imaginary part of the correlation function and also averages out certain quantum fluctuations.

(d) To shed some light on the unique role of the centroid in formulating a classical-like approach to the position correlation function, one can apply the Taylor expansion to the symmetrized position correlation function $\bar{C}(t)$, that is,

$$\bar{C}(t) = \frac{1}{2} \langle [q(t), q(0)]_+ \rangle = \sum_{n=0}^{\infty} \frac{t^{2n}}{(2n)!} \langle q \mathcal{L}^{2n} q \rangle, \qquad (3.36)$$

where $[\cdots]_+$ is the anticommutator. If one attempts to carry out a term-by-term analysis as was done for the Kubo-transformed position correlation function, complications arise even in the first few terms because it is unclear how to define the effective momentum distribution and mass in the general case [64]. (Note again that the *centroid* momentum distribution is simply the Boltzmann distribution.) Focusing instead on the Kubo-transformed position correlation function [Eq. (3.20)] reveals the factorization of the centroid variable, which, in turn, leads one to the factorization of the centroid-constrained average in Eq. (3.24). The subsequent identification of the centroid force in the Taylor expansion terms of the correlation function supports the conclusion that the centroid variable can indeed be viewed as a *dynamical* variable at a well-defined level of approximation.

(e) To improve its accuracy, CMD should be augmented by an additional quantum factor. The correction to the centroid force begins at the t^4 term in the Taylor series expansion, so such a factor will not add linearly to the deterministic centroid force, but it might instead be in the form of a time convolution. Although only speculation at this point, such a *quantum memory function* should depend locally on the width factor and the anharmonicity, but still yield a time-reversible dynamics.

2. Position and Velocity Time Correlation Functions

The CMD position correlation function $C^*(t)$ has been shown to be an approximation to the exact Kubo-transformed position correlation function [5]. It is therefore relatively straightforward to obtain a CMD approximation for the quantum velocity and cross-correlation (position-velocity) functions. We first rewrite the Fourier relation in Eq. (3.13) as

$$\tilde{C}_{qq}(\omega) \simeq (\hbar\beta\omega/2)[\coth(\hbar\beta\omega/2) + 1]\tilde{C}^*(\omega) . \tag{3.37}$$

In turn, the Fourier transform of the position correlation function $\tilde{C}_{qq}(\omega)$ can be related to Fourier transforms of the exact velocity and cross-correlation functions:

$$\tilde{C}_{vv}(\omega) = \omega^2 \tilde{C}_{qq}(\omega) ,$$
$$\tilde{C}_{vq}(\omega) = i\omega \tilde{C}_{qq}(\omega) . \tag{3.38}$$

A CMD velocity correlation function $C_{vv}^*(t)$ can then be defined in a classical-like manner such that

$$C_{vv}^*(t) = \langle v_c(t)v_c(0)\rangle_{\rho_c} = -\frac{\partial^2}{\partial t^2}\langle q_c(t)q_c(0)\rangle_{\rho_c} , \tag{3.39}$$

where $v_c(t)$ and $q_c(t)$ are, respectively, the centroid velocity and position variables, and these variables obey the classical-like equations of motion of CMD [4–6, 8]. Since the Fourier relation between the centroid position and velocity time correlation functions is given by

$$\tilde{C}_{vv}^*(\omega) = \omega^2 \tilde{C}^*(\omega) \,, \tag{3.40}$$

one obtains from Eqs. (3.38) and (3.40) the CMD approximation for the Fourier transform of the exact velocity time correlation function:

$$\tilde{C}_{vv}(\omega) \simeq (\hbar\beta\omega/2)[\coth(\hbar\beta\omega/2) + 1]\tilde{C}_{vv}^*(\omega) \,. \tag{3.41}$$

Note that since the diffusion constant is the zero-frequency component of the Fourier transform of the velocity correlation functions [65], Eq. (3.41) immediately implies for a "tagged" quantum particle i in three-dimensional space that

$$D \simeq \frac{1}{3} \int_0^\infty dt \langle \mathbf{v}_{c,i}(t) \cdot \mathbf{v}_{c,i}(0) \rangle_{\rho_c} \,. \tag{3.42}$$

This expression is a Green–Kubo-like expression for the quantum self-diffusion constant in CMD. Working backwards from its usual derivation shows that the self-diffusion constant can also be obtained from the long-time behavior of the slope of the centroid mean-squared displacement:

$$D \simeq \frac{1}{6} \lim_{t \to \infty} \frac{d}{dt} \langle |\mathbf{q}_{c,i}(t) - \mathbf{q}_{c,i}(0)|^2 \rangle_{\rho_c} \,. \tag{3.43}$$

3. General Time Correlation Functions

The CMD method is based on the propagation of quasiclassical centroid trajectories $q_c(t)$ derived from the mean force on the centroid as a function of position [cf. Eqs. (3.11) and (3.12)]. This method, combined with Eq. (3.13), generally provides an accurate representation of the exact quantum real-time position or velocity correlation functions. There is a significant theoretical question, however, on how CMD might be used to compute time correlation functions of the form $\langle A(t)B(0) \rangle$, where A and B are general quantum operators [4, 5]. Two approaches to this problem will now be described.

Since CMD provides an accurate approximation to the quantum position or velocity correlation function, one approach to the general correlation function problem is to introduce these correlation functions directly into an approximate expression for the correlation function [4, 5].

The case of coordinate-dependent operators A and B will first be described for simplicity. In this approach, the general *imaginary-time* correlation function $C_{AB}(\tau) = \langle A(\tau)B(0)\rangle$ is expressed as

$$C_{AB}(\tau) = \int \frac{dk_1}{2\pi} \int \frac{dk_2}{2\pi} \, \hat{A}(k_1)\hat{B}(k_2)\langle \exp[ik_1 q(\tau) + ik_2 q(0)]\rangle , \quad (3.44)$$

where

$$\langle \cdots \rangle \equiv \frac{\int \cdots \int \mathscr{D}q(\tau)(\cdots)e^{-S[q(\tau)]/\hbar}}{\int \cdots \int \mathscr{D}q(\tau)e^{-S[q(\tau)]/\hbar}} . \quad (3.45)$$

The cumulant average of the exponential term in Eq. (3.44) can be performed and truncated at second order, giving

$$\langle \exp[ik_1 q(\tau) + ik_2 q(0)]\rangle$$
$$\simeq \exp\{ik_1\langle q\rangle + ik_2\langle q\rangle - \tfrac{1}{2}[k_1^2 C_\delta(0) + 2k_1 k_2 C_\delta(\tau) + k_2^2 C_\delta(0)]\} , \quad (3.46)$$

where the imaginary-time position fluctuation correlation functions are defined by

$$C_\delta(\tau) = \langle \delta q(\tau)\delta q(0)\rangle , \quad (3.47)$$

with $\delta q = q - \langle q\rangle$. To perform the integrals over k_1 and k_2 in Eq. (3.46), new imaginary-time correlation functions are defined as

$$C_\delta^\pm(\tau) = C_\delta(0) \pm C_\delta(\tau) . \quad (3.48)$$

After performing the k integrals in Eq. (3.46), the expression for the general imaginary time correlation function is given by the double-Gaussian average

$$C_{AB}(\tau) = \langle A(\langle q\rangle + \tilde{q}_1)B(\langle q\rangle + \tilde{q}_2)\rangle_{C_\delta^+, C_\delta^-} , \quad (3.49)$$

where the variables $\tilde{q}_1$ and $\tilde{q}_2$ are related to the two Gaussian variables q_+ and q_- such that

$$\tilde{q}_{1,2} = \frac{1}{\sqrt{2}}(q_+ \pm q_-) . \quad (3.50)$$

The Gaussian variables q_+ and q_- have width factors given by $C_\delta^+(\tau)$ and $C_\delta^-(\tau)$, respectively, defined in Eq. (3.48).

The final step is to continue the foregoing imaginary-time expression for $C_\delta^\pm(\tau)$ analytically via the inverse Wick rotation $\tau \to it$. The resulting expression for the real-time correlation function $C_{AB}(t)$ is thus given by [4]

$$C_{AB}(t) \simeq \langle A(\langle q \rangle + \tilde{q}_1)B(\langle q \rangle + \tilde{q}_2) \rangle_{C_\delta^+(t), C_\delta^-(t)}, \tag{3.51}$$

where the time-dependent Gaussian width factors are given by

$$C_\delta^\pm(t) = C_\delta(0) \pm C_\delta(t). \tag{3.52}$$

Since $C_\delta(t)$ equals $\langle q(t)q(0) \rangle - \langle q \rangle^2$, the correlation functions in Eq. (3.52) can be calculated using the CMD method for real-time position correlation functions [4–6, 8]. It should be noted that both the real and imaginary parts of $C(t) = \langle q(t)q(0) \rangle$ are required in Eq. (3.52) and the CMD position correlation function $C^*(t)$ defined in Eq. (3.10) can be used to estimate these terms through inversion of the Fourier transform relation in Eq. (3.13).

Equation (3.51), which is the central result of this subsection, has two particularly appealing characteristics: (1) it involves only the real-time position correlation function $\langle q(t)q(0) \rangle$, which can be determined from CMD, and (2) the double-Gaussian average in Eq. (3.51) does not have to be expressed in closed form [i.e., it can be evaluated numerically at each time point using the values of the complex Gaussian width factors $C_\delta^\pm(t)$ computed from the CMD algorithm]. On the negative side, Eq. (3.51) is exact only in the globally harmonic limit due to the truncation of the cumulant expansion in Eq. (3.46). Furthermore, Eq. (3.51) may disagree with the exact classical limit for highly anharmonic systems at elevated temperatures. Nevertheless, Eq. (3.51) represents a promising step toward the calculation of general real-time correlation functions using the CMD method. As an example, this method is expected to be useful in the study of nonlinear high-frequency molecular vibrations in condensed phases.

The cumulant expansion with CMD approach for general correlation functions can be extended to calculate correlation functions having momentum-dependent operators via the phase-space perspective of Paper III. Only the final expressions will be given here, so the reader is referred to the original paper for the details [5]. The approximate result for the real-time correlation function $C_{AB}(t)$ in this approach is given by [5]

$$C_{AB}(t) \simeq \langle A(\langle \zeta \rangle + \zeta_1)B(\langle \zeta \rangle + \zeta_2) \rangle_{C_\delta^+(t), C_\delta^-(t)}, \tag{3.53}$$

where the real-time-dependent Gaussian width matrices are given by

$$\mathbf{C}_\delta^\pm(t) = \mathbf{C}_\delta(0) \pm \mathbf{C}_\delta(t), \tag{3.54}$$

and

$$[\mathbf{C}_\delta(t)]_{ij} = \begin{pmatrix} \langle \delta p_i(t)\delta p_j(0) \rangle & \langle \delta p_i(t)\delta q_j(0) \rangle \\ \langle \delta q_i(t)\delta p_j(0) \rangle & \langle \delta q_i(t)\delta q_j(0) \rangle \end{pmatrix}. \qquad (3.55)$$

Here $\delta\zeta(t) = \zeta(t) - \langle \zeta \rangle$. The correlation function elements of Eq. (3.54) can be calculated using the CMD position correlation function $C^*(t)$ and inversion of the Fourier transform relations in Eqs. (3.37) and (3.38). It should be noted that Eq. (3.53) simplifies considerably if the operators A and B depend on only one phase-space coordinate (i.e., most of the Gaussian integrals can be integrated out of the expression).

A second CMD algorithm for calculating general correlation functions was proposed in Papers II and III. This algorithm employs CMD along with a semiclassical centroid-based representation of the quantum operators. It is therefore called *CMD with semiclassical operators*. Although this approach is less mathematically direct, the goal is one of utility. That is, the algorithm is easy to use and capable of recovering the exact classical limit for general correlation functions. This approach, therefore, may be the method of choice for studying many problems in chemical physics which are nearly classical in nature. The basic equations underlying the algorithm are simply stated below—the reader is referred to the appendix of Paper II for the supporting mathematical analysis.

The CMD with semiclassical operators approach centers around the computation of a classical-like correlation function $C^*_{AB}(t)$, given by

$$C^*_{AB}(t) = \langle A_c(t)B_c(0) \rangle_{\rho_c}, \qquad (3.56)$$

where the initial condition averaging is performed with the (normalized) phase-space centroid density defined in Eq. (2.53). The semiclassical operators $O_c(q_c(t))$ in Eq. (3.56) are given by the time-dependent analog of Eq. (2.61):

$$O_c(\zeta_c(t)) = \langle O(\zeta_c(t) + \tilde{\zeta}) \rangle_{C_c(0,\mathbf{q}_c(t))}. \qquad (3.57)$$

Here $\zeta_c(t)$ is the phase-space centroid trajectory which obeys the CMD equation of motion in Eq. (3.16), and the time-dependent Gaussian width matrix $\mathbf{C}_c(0, \mathbf{q}_c(t))$ for the vector $\tilde{\zeta}$ is given by the centroid-constrained correlation function matrix in Eq. (2.56) with the position centroid $\mathbf{q}_0$ located at $\mathbf{q}_c(t)$. As shown in the appendix of Paper II, the general centroid correlation function in Eq. (3.56) is an approximation to the Kubo-transformed version of the exact correlation function $C_{AB}(t)$. Therefore, to calculate $C_{AB}(t)$ one makes use of the Fourier relationship

[4]

$$\tilde{C}_{AB}(\omega) \simeq (\hbar\beta\omega/2)[\coth(\hbar\beta\omega/2) + 1]\tilde{C}^*_{AB}(\omega) . \tag{3.58}$$

The expression in Eq. (3.56) is intended to maximize the utility of the CMD method in a transparent, though approximate, fashion for general correlation functions. However, it can be somewhat lacking in its description of certain features of low-temperature nonlinear vibrational correlation functions. The reader is referred to Paper II for further analysis and comments. It should also be noted that Eq. (3.56) reduces to the CMD expression in Eq. (3.10) in the case of a position correlation function. As will be shown in Section III.D.3, the numerical results for general correlation functions indicate that for such problems the Gaussian "smeared" representation of the operators in Eq. (3.57) will be superior to a simpler, but incorrect, classical representation of the operators in terms of the centroid variables. Schenter et al. [66] and Doll [67] have employed the latter procedure to estimate the dynamical corrections to the path-integral QTST rate constant [42–44], although their approach will be exact in the globally quadratic limit for the barrier potential and bath modes [43].

C. Centroid Molecular Dynamics Algorithms

In the implementation of CMD for a realistic many-body system, the $3N$-dimensional centroid trajectories are generated by the effective classical equations of motion [4–6, 8]

$$\mathbf{m} \cdot \frac{d^2}{dt^2} \mathbf{q}_c(t) = \mathbf{F}_c(\mathbf{q}_c) , \tag{3.59}$$

where $\mathbf{q}_c(t)$ is the $3N$-dimensional column vector of centroid positions, $\mathbf{m}$ is the diagonal particle mass matrix, and $\mathbf{F}_c(\mathbf{q}_c)$ is the quantum mechanical centroid mean force vector. The latter quantity is expressed as the operation of a $3N$-dimensional gradient vector in centroid configuration space, $\boldsymbol{\nabla}_c$, on the mean centroid potential $V_c(\mathbf{q}_c)$:

$$
\begin{aligned}
\mathbf{F}_c(\mathbf{q}_c) &= -\boldsymbol{\nabla}_c V_c(\mathbf{q}_c) \\
&= -\frac{\int \cdots \int \mathscr{D}\mathbf{q}(\tau)\delta(\mathbf{q}_c - \mathbf{q}_0)\{\boldsymbol{\nabla}V[\mathbf{q}(0)]\} \exp\{-S[\mathbf{q}(\tau)]/\hbar\}}{\int \cdots \int \mathscr{D}\mathbf{q}(\tau)\delta(\mathbf{q}_c - \mathbf{q}_0) \exp\{-S[\mathbf{q}(\tau)]/\hbar\}} .
\end{aligned} \tag{3.60}
$$

Here $S[\mathbf{q}(\tau)]$ is the imaginary-time action functional and the imaginary-time position centroid vector defined in Eq. (3.60) is

$$\mathbf{q}_0 = \frac{1}{\hbar\beta} \int_0^{\hbar\beta} d\tau\, \mathbf{q}(\tau)\,. \tag{3.61}$$

In the many-dimensional case, the effective temperature-dependent centroid potential $V_c(\mathbf{q}_c)$ in Eq. (3.60) is defined by

$$V_c(\mathbf{q}_c) = -k_B T \ln\left\{ \rho_c(\mathbf{q}_c) \Big/ \left[\prod_{i=1}^{N} (m_i/2\pi\hbar^2\beta) \right]^{3/2} \right\}\,, \tag{3.62}$$

where the multidimensional position centroid density $\rho_c(\mathbf{q}_c)$ is given by [1, 3]

$$\rho_c(\mathbf{q}_c) = \int \cdots \int \mathscr{D}\mathbf{q}(\tau)\delta(\mathbf{q}_c - \mathbf{q}_0) \exp\{-S[\mathbf{q}(\tau)]/\hbar\}\,. \tag{3.63}$$

As described earlier, once the CMD trajectories are in hand the classical-like correlation function in Eq. (3.10) is calculated and related to the quantum correlation function through the Fourier relation in Eq. (3.13).

Although CMD is a substantial breakthrough in the approximate computation of quantum time correlation functions, the determination of the centroid force in Eq. (3.59), as defined by Eq. (3.60), represents an algorithmic challenge for realistic many-body simulations. Equation (3.60) shows that the centroid force is given by the average of the local potential gradient over quantum path fluctuations about the constrained centroid variables. Of course, all numerical path-integral simulation techniques [17–19] can be adapted to compute this average, but the real issue is one of *computational efficiency* so that the centroid force can be readily computed "on the fly" during the real-time integration of the CMD equations [Eq. (3.59)]. This computation is not trivial for quantum many-body systems. (It will, however, be much easier than a numerical frontal assault on the many-body time-dependent Schrödinger equation or real-time Feynman path integrals [1].)

In Paper IV, several algorithms were developed and explored for the efficient computation of the centroid force in the CMD equations for many-body systems. Two of the algorithms are "direct" numerical path-integral approaches [17–19] in which the path averaging explicit in Eq. (3.60) is performed on the fly during the time integration of the CMD equations. One approach is based on a normal-mode path-integral MD algorithm [68] and has recently been extended and applied by Cao and Martyna [48]. A second direct algorithm is based on the staging path-integral Monte Carlo (MC) algorithm [69, 70], with several numerical tricks introduced to speed up the computations. The latter algorithm is described in detail below. A rather different and even more efficient approach also proposed in Paper IV approximates the centroid force within the framework of the locally optimized effective harmonic perspective [3, 21–23, 41]. This approach is described below together with two independent algorithms developed to optimize the variational parameters while propagating the centroid variables in time. The third CMD algorithm outlined below is perhaps the simplest: It involves the calculation of the excess centroid free energy as a function of the separation between a pair of particles, modeling this free energy by pairwise centroid pseudopotentials, and then running CMD simulations for an effective many-body system that interacts only through such pairwise interactions. It should be noted that all of these CMD algorithms could be synthesized to treat different parts of a physical problem within a single CMD simulation.

1. Direct Centroid Molecular Dynamics Approaches

In the discretized version of the Feynman path integral [17–19], the centroid force in Eq. (3.60) is written as

$$F_c(q_c) =$$

$$-\frac{\int \cdots \int \Pi_{i=1}^{P} dq_i \delta(q_c - q_0) V'(q_1, \ldots, q_P) \exp[-S_P(q_i, \ldots, q_P)/\hbar]}{\int \cdots \int \Pi_{i=1}^{P} dq_i \, \delta(q_c - q_0) \exp[-S_P(q_i, \ldots, q_P)/\hbar]},$$

$$(3.64)$$

where from the cyclic invariance of the trace,

$$V'(q_1, \ldots, q_P) \equiv \frac{1}{P} \sum_{i=1}^{P} \frac{dV(q)}{dq} \bigg|_{q=q_i} \qquad (3.65)$$

and the centroid variable in discrete notation is given by Eq. (1.6). The discretized path-integral action functional $S_P(q_i, \ldots, q_P)$ is given by $\hbar \beta V_P(q_i, \ldots, q_P)$, where $V_P(q_i, \ldots, q_P)$ is given by Eq. (1.4). (For simplicity here, one-dimensional notation has again been employed.)

In a path-integral Monte Carlo (PIMC) calculation, the centroid force can readily be calculated from Eq. (3.64) by using the importance sampling function $\exp[-S_P(q_i, \ldots, q_P)/\hbar]$ and pairwise MC moves to enforce the centroid constraint. For a path-integral molecular dynamics (PIMD) calculation [71], one defines fictitious momenta p_i for each of the quasiparticles q_i and then runs an MD simulation with Hamilton's equations based on the fictitious Hamiltonian

$$H_P = \sum_{i=1}^{P} \frac{p_i^2}{2m'} + V_P(q_i, \ldots, q_P) , \tag{3.66}$$

where, in the case of CMD, m' must equal m/P so that the centroid variable has the physical particle mass m [4, 5, 8]. If the PIMD trajectory adequately samples the canonical ensemble, a particular time average over that trajectory will yield the centroid force, that is,

$$F_c(q_c) = -\frac{1}{\hbar\beta} \int_0^{\hbar\beta} d\tau \langle V'[q(\tau)] \rangle_c$$

$$= -\lim_{T \to \infty} \frac{1}{T} \int_0^T dt \frac{1}{P} \sum_{i=1}^{P} \frac{dV(q)}{dq} \bigg|_{q=q_i(t), q_0 = q_c} , \tag{3.67}$$

where $\langle \cdots \rangle_c$ denotes a centroid-constrained path-integral average [cf. Eqs. (3.60) and (3.64)]. In a PIMD calculation, the centroid constraint must be enforced through, for example, a holonomic constraint. In a PIMC calculation, the average over time in Eq. (3.67) is replaced by an average over centroid-constrained Monte Carlo moves.

A direct numerical path-integral computation of the centroid force as outlined above will undoubtedly provide an accurate value of the centroid potential surface. For low-dimensional systems, the centroid force might indeed be calculated for each point in space and stored on a grid in computer memory and later recalled in the CMD calculation. As the dimensionality of the system increases, however, this straightforward procedure is no longer feasible, due to the exponential growth of the computational effort. The real issue, therefore, is how to carry out such a computation efficiently within the context of the time integration of the CMD equations [Eq. (3.59)]. Consequently, more specialized approaches are required [6].

One approach [6, 48] for the computation of the centroid force during a CMD calculation is based on the normal-mode path-integral molecular dynamics (NMPIMD) algorithm [68]. In the CMD/NMPIMD algorithm, the centroid variable naturally separates from the other Feynman path modes. This feature provides an important simplification of the computation of the centroid force [Eq. (3.60)] and in distinguishing the sampling inherent in that computation from the real-time propagation of the centroid variable. The computational effort implicit in NMPIMD relies on the efficiency of the path-integral algorithm, which, in turn, depends on the accuracy of the propagator used in the sampling. It is therefore advantageous to include the quadratic part of the potential energy function in the definition of the NMPIMD propagator [6]. In this approach, one of the normal modes is the centroid variable q_0, which must propagate according to the CMD equations. The other modes are canonically sampled by assigning them a set of fictitious momenta and (light) masses and attaching a Nosé–Hoover chain to each of the mode variables [48, 72]. The NMPIMD algorithm can be incorporated into a CMD calculation in one of two ways. The first is to calculate the centroid force using Eq. (3.67) at each time step in the CMD calculation. In this approach, the centroid force computation is algorithmically distinct from the CMD time integration. The second approach is, in effect, to compute the centroid force on the fly within the CMD algorithm [48]. This calculation can, in principle, be accomplished by making the fictitious masses of the normal-mode variables small enough so that the centroid force is convergently computed on the (longer) time scale of the centroid motion. In this case the centroid force on that time scale is given essentially by

$$F_c(q_c) \simeq \frac{1}{\Delta t_c} \int_0^{\Delta t_c} dt \, \frac{1}{P} \sum_{i=1}^{P} \frac{dV(q)}{dq} \bigg|_{q=q_i(t), q_0=q_c}, \qquad (3.68)$$

where Δt_c is the time step needed to integrate the slow centroid motion accurately. Clearly, the average must converge on the time scale Δt_c. The algorithm is a kind of extended Lagrangian technique, like the Car–Parrinello algorithm [73, 74], but there is an important difference. In the latter case, the goal is to have the parameters (e.g., plane-wave co-efficients) oscillate quickly and "tightly" around the minimum (i.e., the adiabatic ground-state Born–Oppenheimer surface). In a CMD calcula-tion, however, the idea is to have the path fluctuation modes sample their full *canonical equilibrium distribution* (and to do so quickly). Other "tricks" can also be implemented to improve the convergence of the NMPIMD calculation of the centroid force in the CMD calculation. These are described in Paper IV. One drawback of the CMD/NMPIMD algorithm, however, is the computational overhead associated with the normal-mode transformation back to the quasiparticle coordinates at each MD time step to transform. This transformation can become time consuming if the number of path-integral discretizations P is large.

An alternative direct CMD algorithm [6] which becomes particularly advantageous if the number of discretizations P is large is based on a combination of MD with the staging PIMC method [69, 70]. The par-ticular implementation of the staging algorithm in the CMD method is similar to that discussed at length in Ref. 75, but an effective harmonic reference system is incorporated into the definition of the transition probability distribution function for selecting trial configurations of the path-integral quasiparticle chain [69, 70, 75]. The details of the latter are given in Paper IV, while its implementation in the CMD algorithm is described here in detail.

The basic approach of the CMD/PIMC algorithm is to incorporate staging PIMC smoothly into the CMD time integration in order to compute the centroid force. To accomplish this, a single CMD time step Δt can be divided into N_{MC} segments. As a new set of MC chain configurations are sampled for a fixed centroid, the centroid can then be moved according to a smaller time step $\delta t = \Delta t / N_{MC}$ using the averaged centroid force over the MC path integral configurations. The N_{MC} staging PIMC samplings are thus evenly distributed on a finer MD grid of spacing δt so that the PIMC evolves smoothly with the CMD motion. Such an approach is more numerically effective than simply computing the centroid force at the beginning of each time step. Although the fluctuations of the centroid force can be observed on the fine time scale δt, the centroid trajectory becomes smooth on the larger time scale Δt. This behavior is similar to the stochastic "Brownian" motion of a heavy (centroid) particle solvated in a light (path fluctuation) bath. In the extreme limit of $N_{MC} \to \infty$, one recovers the mean force average according to the PIMD case of $T \to \infty$ in Eq. (3.67). In this case the force fluctuations on the centroid will become completely smoothed out so that the centroid feels the mean centroid force.

The complete CMD/PIMC algorithm is summarized as follows:

1. Generate a MC chain segment (i.e., moves of a subset of quasiparticles) by the staging PIMC method.

2. Uniformly adjust the positions of all quasiparticles to fix the centroid position for each move.

3. Accept or reject the new centroid-constrained configurations using the MC algorithm.

4. Average the centroid force over a pass of such centroid-constrained PIMC moves and then move the centroid accordingly for a small time step $\delta t = \Delta t / N_{MC}$ using the velocity Verlet algorithm.

5. Repeat steps 1–4 N_{MC} times to complete one CMD time step Δt within the framework of the CMD integration.

The complete CMD simulation should be carried out for different values of N_{MC} so that convergence of the mean centroid force becomes apparent. Moreover, in the staging part of the algorithm [part 1], the number of quasiparticles moved in each chain segment should be adjusted to achieve a reasonable MC acceptance ratio in part 3. Other considerations for the direct calculation of the centroid force within the CMD/PIMC algorithm are given in Paper IV.

2. *Effective Harmonic Computation of the Centroid Force*

An alternative and efficient algorithm [6] for computing the centroid force in CMD makes use of the variational effective quadratic theory for the centroid potential [3–5, 8, 21–23, 41]. This approach, which yields an approximate centroid force, represents the centroid potential as a variationally optimized quadratic potential centered at the centroid position. The effective potential is then constantly updated as the centroid propagates in the CMD time evolution.

A central algorithmic challenge in using the effective harmonic theory is the computation of the many-dimensional Gaussian average inherent in the variational expression for the centroid potential [3–5, 21–23, 41]. This average is given by

$$\langle V(\mathbf{q}_c + \tilde{\mathbf{q}}) \rangle_c$$

$$= \frac{1}{\sqrt{\det[2\pi \mathbf{C}_c(0, \mathbf{q}_c)]}} \int d\tilde{\mathbf{q}} \, V(\mathbf{q}_c + \tilde{\mathbf{q}}) \exp[-\tilde{\mathbf{q}} \cdot \mathbf{C}_c^{-1}(0, \mathbf{q}_c) \cdot \tilde{\mathbf{q}}/2] \,. \qquad (3.69)$$

From this expression one can obtain the effective harmonic approximation to the centroid force, given by

$$\mathbf{F}_{c,\mathrm{eff}}(\mathbf{q}_c) = -\nabla_c \langle V(\mathbf{q}_c + \tilde{\mathbf{q}}) \rangle_c \big|_{\mathbf{C}_c} \,, \qquad (3.70)$$

where the notation "$\cdots |_{\mathbf{C}_c}$" indicates that the centroid derivatives in the gradient do not act on the matrix $\mathbf{C}_c$. The Gaussian width factor matrix $\mathbf{C}_c(0, \mathbf{q}_c)$, in this case, is the position subblock of the generalized centroid-constrained correlation function matrix in Eq. (2.57). The elements of the centroid-dependent effective force constant matrix can be similarly expressed as

$$[\mathbf{K}_c(\mathbf{q}_c)]_{ij} = \frac{\partial^2}{\partial q_i \, \partial q_j} \langle V(\mathbf{q}_c + \tilde{\mathbf{q}}) \rangle_c \big|_{\mathbf{C}_c} \,. \qquad (3.71)$$

It should be pointed out that the Gaussian-averaged potential at the centroid $\mathbf{q}_c$ in Eq. (3.69) is different from the effective centroid potential introduced in Eq. (3.62). However, the centroid force in Eq. (3.70) can be shown to be the effective harmonic approximation to the exact centroid force in Eq. (3.60).

At the level of the effective quadratic approximation, the Gaussian width matrix is formally expressed as [6, 7]

$$C_c(0, \mathbf{q}_c) = \sum_{n \neq 0} [\beta \mathbf{m}\Omega_n^2 + \beta \mathbf{K}_c(\mathbf{q}_c)]^{-1}, \tag{3.72}$$

where $\mathbf{m}$ is the $3N$-dimensional particle mass matrix and $\Omega_n = 2\pi n/\hbar\beta$. A centroid-dependent unitary matrix $U(\mathbf{q}_c)$ can be found to diagonalize the mass-scaled centroid force constant matrix $\bar{\mathbf{K}}_c(\mathbf{q}_c)$, giving the eigenfrequencies

$$U^\dagger(\mathbf{q}_c)\bar{\mathbf{K}}_c(\mathbf{q}_c)U(\mathbf{q}_c) = \mathbf{I} \cdot \boldsymbol{\omega}_c^2, \tag{3.73}$$

where $\boldsymbol{\omega}_c^2$ is the column vector of centroid-dependent eigenvalues and $\mathbf{I}$ is the $3N$-dimensional identity matrix. The Gaussian width factor matrix in Eq. (3.72) can be determined from the relation

$$C_c(0, \mathbf{q}_c) = \bar{U}(\mathbf{q}_c)[\mathbf{I} \cdot \bar{\boldsymbol{\alpha}}(\mathbf{q}_c)]\bar{U}^\dagger(\mathbf{q}_c), \tag{3.74}$$

where the individual elements of the normal-mode thermal width factor vector are given by

$$\bar{\alpha}_l(\mathbf{q}_c) = \frac{1}{\beta\omega_{c,l}^2} \left\{ \frac{\hbar\beta\omega_{c,l}/2}{\tanh(\hbar\beta\omega_{c,l}/2)} - 1 \right\} \tag{3.75}$$

and the mass-scaled unitary matrix $\bar{U}(\mathbf{q}_c)$ is given by

$$\bar{U}(\mathbf{q}_c) = \mathbf{m}^{-1/2}U(\mathbf{q}_c). \tag{3.76}$$

Thus the set of optimized frequencies $\{\omega_{c,l}\}$ are variationally obtained as the self-consistent solution to the transcendental matrix equations (3.71)–(3.75) for a given centroid position.

Many systems can be well described by pairwise (or site–site) potentials. If such is the case, the Gaussian average in Eq. (3.69) simplifies considerably and can be expressed in terms of the summation over all pair interactions [6]:

$$\langle V(\mathbf{q}_c + \tilde{\mathbf{q}})\rangle_c = \sum_{i=1}^{N} \sum_{j>i}^{N} \langle v(\mathbf{r}_{ij})\rangle_c, \tag{3.77}$$

where $\mathbf{r}_{ij}$ is the three-dimensional vector connecting the ith and jth particles. For a specific pair of interactions, the other degrees of freedom can be integrated out of the average, leading to a Gaussian average in a lower-dimensional space [6].

The Gaussian average in Eq. (3.69) can be carried out analytically only for polynomials, Gaussian potentials, exponentials, and their combinations. Gaussian averages of other functions cannot be expressed in closed analytical forms. The numerical integration of even the three-dimensional Gaussian averages for pairwise interactions can be time consuming and would reduce the efficiency of the effective harmonic algorithm for CMD. These difficulties are magnified when several iterations are required to achieve a convergent solution to the variational transcendental equations. The best strategy, therefore, is to represent the physical potential by a set of functions that can be Gaussian averaged analytically. Several such approaches were presented in Paper IV.

Since the optimal parameters for the effective harmonic approximation of the centroid potential $V_c(\mathbf{q}_c)$ [Eq. (3.62)] are determined by a variational minimization of $V_{c,\mathrm{eff}}(\mathbf{q}_c)$ with respect to those parameters, [3, 21–23, 41] an extended Lagrangian technique [73, 74] can be employed to minimize the effective centroid potential and calculate the resulting centroid force while propagating the centroid variables in time [6]. The similarity between CMD with the effective harmonic theory for the centroid force and the Car–Parrinello (CP) extended Lagrangian method [73, 74] is clear: The centroid motion is the analog to the CP nuclear motion, the variational parameters in the effective harmonic theory are the analog to the CP plane-wave coefficients, and the centroid potential is similar to the ground-state CP adiabatic energy functional. To outline the extended Lagrangian method as it applies to CMD, for simplicity the following discussion is restricted to a one-dimensional problem. The relevant formulas for multidimensional systems are described at the end.

In the effective harmonic approximation, [3, 21–23, 41] the centroid potential in Eq. (3.62) is given by

$$V_c(q_c; \xi)$$

$$= -k_B T \ln[(b/2)/\sinh(b/2)] + \langle V(q_c + \tilde{q})\rangle_c - m\omega_c^2 C_c(0, q_c)/2 , \quad (3.78)$$

where $b = \hbar\beta\omega_c$, with ω_c being the frequency of the effective harmonic potential; the effective harmonic expression for $C_c(0, q_c)$ is given in Eq. (3.72). The parameter ξ is the variational parameter that is optimized to minimize the centroid potential. This parameter can be chosen to be the centroid frequency ω_c or the width factor C_c.

At this point one can introduce the extended Lagrangian [6],

$$L_{\mathrm{ex}}(\dot{q}_c, \dot{\xi}, q_c, \xi) = \tfrac{1}{2}m\dot{q}_c^2 + \tfrac{1}{2}m_\xi\dot{\xi}^2 - V_c(q_c; \xi) , \quad (3.79)$$

where $\dot{\xi}$ is the velocity associated with the variational parameter ξ and m_ξ is its fictitious mass. The Lagrangian above forms the basis for the CMD equations in which the variational parameter ξ and, in turn, the centroid force are computed simultaneously with the centroid motion.

If the variable ξ in the extended Lagrangian is taken to be the Gaussian width factor C_c, one obtains the fictitious force [6]

$$F_{C_c}(q_c, C_c) \equiv -\frac{\partial V_c(q_c; C_c)}{\partial C_c}$$

$$= -\tfrac{1}{2}[\langle \partial_c^2 V(q_c + \tilde{q})\rangle_c - m\omega_c^2]. \tag{3.80}$$

Since the width factor C_c is in this case the dynamical variable, the centroid frequency can be obtained from C_c through the one-dimensional versions of Eqs. (3.74) and (3.75). The optimal variational value for C_c is obtained by setting Eq. (3.80) equal to zero and self-consistently solving for C_c. For a given centroid initial condition, this value of C_c would provide the initial condition for the variable C_c in the extended Lagrangian simulation.

The generalization of Eq. (3.80) multidimensional systems is given by

$$-\frac{\partial V_c(\mathbf{q}_c; \mathbf{C}_c)}{\partial \mathbf{C}_c} = -\tfrac{1}{2}[\mathbf{K}_c(\mathbf{q}_c) - \bar{\mathbf{U}}(\mathbf{q}_c)\mathbf{I} \cdot \boldsymbol{\omega}_c^2 \bar{\mathbf{U}}^\dagger(\mathbf{q}_c)] \tag{3.81}$$

where the relevant quantities here are defined in Eqs. (3.71)–(3.76). For a system of $3N$ degrees of freedom, there are $3N(3N + 1)/2$ independent variational parameters (i.e., elements of the matrix $\mathbf{C}_c$). The multidimensional trajectory from the extended Lagrangian generates the width factor matrix $\mathbf{C}_c(t)$ as a function of time. To compute the force in Eq. (3.81), the matrix trajectory $\mathbf{C}_c(t)$ is used in Eq. (3.71) to compute the first term on the right-hand side of Eq. (3.81). The second term is extracted from $\mathbf{C}_c(t)$ by inverting Eq. (3.74) to get the elements $\alpha_l(\mathbf{q}_c)$ and, in turn, by solving for the vector $\boldsymbol{\omega}_c^2$ from Eq. (3.75). The latter vector, along with the transformation matrix $\bar{\mathbf{U}}(\mathbf{q}_c)$, is then used in the second term on the right-hand side of Eq. (3.81).

The essence of the extended Lagrangian method [73, 74] is to choose small enough fictitious masses $\{m_\xi\}$ in the multidimensional version of Eq. (3.79) that the fictitious variables $\{\xi\}$ rapidly oscillate around the minimum of the centroid free-energy surface [Eq. (3.78)]. The centroid variables should then exhibit an adiabatic, conservative motion because there is little energy exchange between them and the fictitious degrees of freedom. One useful technique might be to associate a Nosé–Hoover chain [72] to each fictitious variable to keep it from heating up.

As an alternative to the extended Lagrangian method discussed above, it turns out that a simple iterative solution [6] is quite feasible for the multidimensional effective harmonic variational equations [i.e., Eq. (3.81) set equal to zero]. This algorithm is rather straightforward: Solve the transcendental equation iteratively at a fixed centroid until convergence, calculate the centroid force, and move the centroid accordingly in the MD integrator; repeat the procedure for each time step. Many fewer iterations are required if the optimal parameters from the preceding time step are used as the initial guess at the current time step. Provided that the centroid displacement is small, the convergence of the iterative scheme should be fast.

One difficulty associated with the iterative method occurs when negative curvatures are present in the classical Hessian matrix for a given centroid configuration (i.e., at inflection points or in a barrier region). If the initial guess of parameters is poor, the iterative scheme may not converge, even though the Gibbs–Bogoliubov variational principle [1, pp. 303–307; 2, pp. 86–96] ensures that the centroid potential is finite and that a solution must therefore exist. If such a problem arises in the iterative algorithm at a given time step, the iterative process can be restarted with a smaller value of $\hbar$, which can then be adjusted until the neighborhood of a fixed point is located. For $\hbar = 0$, the classical force is obtained, which, of course, is well defined. It should be noted that the effective quadratic theory is being implemented in this CMD algorithm within the context of the *local* centroid force for each time step in the time integration. The local instantaneous centroid potential is *not* being used to extrapolate the dynamics to long times as in a quantum instantaneous normal-mode theory [7]. If such were the case, unphysical instabilities in the centroid dynamics might indeed occur.

3. Pairwise Centroid Pseudopotentials

The exact excess free energy of the centroids for a many-body system can be approximated as the superposition of the excess free energy of the pair centroid interactions [6]:

$$V_c(\mathbf{q}_c) \simeq \sum_{i=1}^{N} \sum_{j>i}^{N} v_{\text{eff}}(r_{c,ij}) \,, \tag{3.82}$$

where the effective centroid pair potential [76] $v_{\text{eff}}(r_{c,ij})$ can, for example, be computed from the expression for two particles:

$$\exp[-\beta v_{\text{eff}}(r_{c,12})] =$$

$$\frac{\int \cdots \int \mathscr{D}\mathbf{q}_1(\tau)\mathscr{D}\mathbf{q}_2(\tau)\delta(r_{c,12} - \tilde{r}_{12}) \exp\{-S[\mathbf{q}_1(\tau), \mathbf{q}_2(\tau)]/\hbar\}}{\int \cdots \int \mathscr{D}\mathbf{q}_1(\tau)\mathscr{D}\mathbf{q}_2(\tau)\delta(r_{c,12} - \tilde{r}_{12}) \exp\{-S_{fp}[\mathbf{q}_1(\tau), \mathbf{q}_2(\tau)]/\hbar\}} . \quad (3.83)$$

Here $\mathbf{q}_1$ and $\mathbf{q}_2$ are the three-dimensional vectors for particles 1 and 2, respectively, $S[\mathbf{q}_1(\tau), \mathbf{q}_2(\tau)]$ is the action functional for the interacting particles, $S_{fp}[\mathbf{q}_1(\tau), \mathbf{q}_2(\tau)]$ is the free-particle action functional, and

$$\tilde{r}_{12} = \frac{1}{\hbar\beta} \int_0^{\hbar\beta} d\tau |\mathbf{q}_1(\tau) - \mathbf{q}_2(\tau)| . \quad (3.84)$$

The effective centroid pair potential $v_{\text{eff}}(r_{c,ij})$, which yields a central force and is a function of temperature as well as mass, can be calculated accurately by the direct PIMC sampling methods discussed in Section III.C.1 and saved on a one-dimensional grid. The CMD equations (3.59) can then be integrated using the centroid force, computed from the pairwise centroid pseudopotential. Obviously, this scheme will be most effective for nearly classical systems where many-body quantum correlations are negligible or for weakly interacting systems where there is a low probability for three particles to be mutually within some relevant interaction range. For highly condensed phases of strongly quantized particles, the compact structure ensures that there will be several neighbors for each quantum particle, and thus the contribution from many-body quantum correlations becomes significant. Nevertheless, the overall centroid potential might still be represented as a pairwise form, but the simple algorithms outlined above to compute $v_{\text{eff}}(r_{c,12})$ will be inadequate. More detailed approaches remain to be developed, but the motivation is great since, once the centroid potential is pairwise represented, a CMD study of quantum dynamical time correlations *is no more difficult than a classical MD calculation.*

D. Numerical Examples and Applications

Various numerical tests and applications of the CMD method are presented in this subsection. These examples range from one-dimensional anharmonic potentials to actual liquid simulations. In these calculations, various CMD algorithms [6] were used to obtain the results.

1. Position Correlation Functions

To test the CMD approach on a system for which numerically exact quantum dynamical results can be obtained, a one-dimensional nonlinear oscillator model was employed, given by [4, 5]

$$V(q) = \tfrac{1}{2}q^2 + cq^3 + gq^4 , \qquad (3.85)$$

with the parameters $c = 0.10$, $g = 0.01$, $\hbar = 1.0$, and $m = 1.0$. In these units the inverse temperature β is given in terms of values of the dimensionless parameter $\beta\hbar\omega$. The potential in Eq. (3.85) has a single minimum at $q = 0$, exhibiting a cubic anharmonicity for small deviations from the minimum and a quartic anharmonicity for larger displacements. At low temperatures where mode quantization effects are operational, the cubic anharmonicity is the dominant perturbation. The anharmonicity of the potential in Eq. (3.85) shifts the energy gap between the ground and first excited vibrational states downward by 6% compared to the harmonic limit of the energy spectrum. In molecular terms, this relatively large anharmonic shift is like a 180-cm^{-1} shift for a 3000-cm^{-1} C–H stretching mode or a 60-cm^{-1} shift of a 1000-cm^{-1} C–C stretching mode. A temperature of $\beta = 10$ was employed in this calculation, which is typical of a C=C double bond at 300 K. The details of the exact and CMD calculations are given in Papers II and III. In Fig. 6 the real part of the position correlation function is shown for this model. The CMD formalism is clearly the most accurate approach. The effective harmonic result is accurate for short times, but it dephases too rapidly, while the classical correlation function is in very poor agreement with the exact quantum result.

As another example of a CMD calculation, a three-dimensional particle in a nonseparable potential well was studied [6]. The potential in this case was given by

$$V(q_1, q_2, q_3) = \frac{1}{2} \sum_{i=1}^{3} (q_i^2 + gq_i^4) + cq_1 q_2 q_3 , \qquad (3.86)$$

where $g = 0.1$, $\hbar = 1.0$, $m = 1.0$, and $\beta = 5.0$. This particular example was chosen [6] as a test of the effective harmonic method for CMD [cf. Section III.C.2] using the extended Lagrangian approach. The extended Lagrangian calculation was carried out for the three centroid variables and the six dynamical elements of the Gaussian width factor matrix with a time step of $\Delta t = 0.01$ and a fictitious mass of $m_\xi = 0.1$. This approach will be a good one for CMD if the fictitious mass is sufficiently small so that the associated variational parameters (i.e., the Gaussian width factors $\mathbf{C}_c$

Position Time Correlation Function

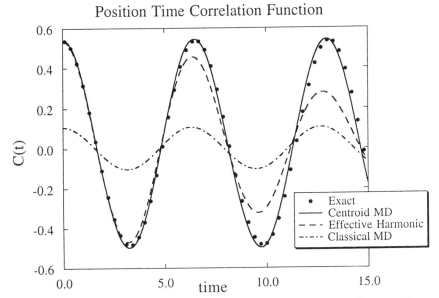

Figure 6. Plot of the real-time position correlation function for the nonlinear oscillator described in Eq. (3.85) at a temperature of $\beta = 10$. The solid circles are the exact quantum results, the solid line is the CMD result from Eq. (3.10), the dashed line is the analytically continued effective harmonic result from Eq. (3.2), and the dot-dashed line is the classical MD result.

in the present case rapidly average the dynamics about the minimized centroid potential surface. The rule of thumb is to choose m_ξ small enough so that the fictitious kinetic energy is less than 1% of the total energy and to make Δt small enough so that the total energy does not drift more than a few percent at the end of a single trajectory. Since for this example the Gaussian averages could be expressed in closed form, the extended Lagrangian calculation took very little CPU time. The quantum position correlation function (not the centroid correlation function) is shown in Fig. 7 for $c = 0.1$ along with the classical MD result. The extended Lagrangian simulation for this example was found to be both stable and efficient, and the quantum effects were seen to be quite significant [6].

A much more realistic set of CMD calculations is presently under way at Penn [46] which involves the dynamics of excess proton migration in water. To this end, a valence bond type of model has been developed for the stable states in the H_3O^+–H_2O dimer. This approach has proven reasonable for path-integral modeling of other, simpler kinds of proton transfer processes [77] and it can also be generalized to a multistate

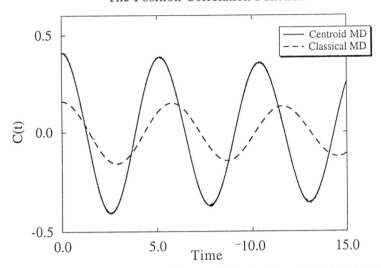

Figure 7. Plot of the real-time position correlation function calculated by the extended Lagrangian CMD method of Section III.C.2 for the three-dimensional potential given in Eq. (3.86) at a temperature of $\beta = 5$. Also shown is the classical MD result.

model so that proton translocation involving multiple water molecules can be studied. In this method, two distinct diabatic states are used where the proton is bound to one water molecule in the first state (H_{11}) and to the other water molecule in the second state (H_{22}) (cf. Fig. 8). Preliminary CMD results have been obtained for the isolated H_3O^+–H_2O dimer. In the top panel of Fig. 9 is shown a classical trajectory for the dimer, while the lower panel is the CMD trajectory obtained by running the direct CMD/PIMC algorithm. When the proton coordinate has a value around 0.5, the system is in the H_3O^+–H_2O configuration, while a value of -0.5 indicates that the proton has hopped and the system has transformed to the H_2O–H_3O^+ configuration. The classical barrier for hopping along the proton coordinate in this particular model was around $1.0\ kcal/mol$, while

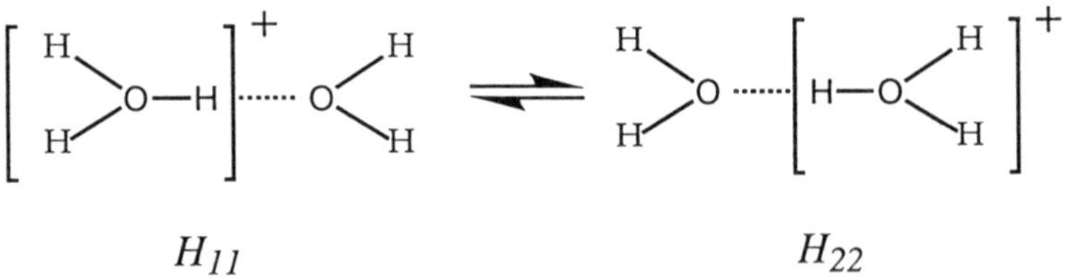

$$\left[\begin{matrix} H \\ \\ H \end{matrix} \!\! \diagdown\!\! O\!-\!H \right]^{+} \cdots\cdots O \!\!\diagup\!\!\! \begin{matrix} H \\ \\ H \end{matrix} \quad \rightleftharpoons \quad \begin{matrix} H \\ \\ H \end{matrix} \!\!\diagdown\!\! O \cdots\cdots \left[H\!-\!O \!\!\diagup\!\!\! \begin{matrix} H \\ \\ H \end{matrix} \right]^{+}$$

$$H_{11} \qquad\qquad\qquad\qquad\qquad\qquad H_{22}$$

Figure 8. Two-state valence-bond picture for the H_3O^+–H_2O dimer.

Classical MD Trajectory

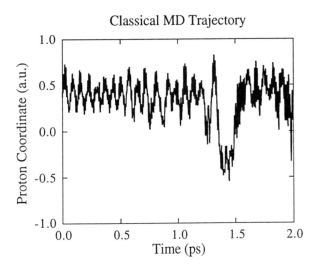

Centroid MD Trajectory

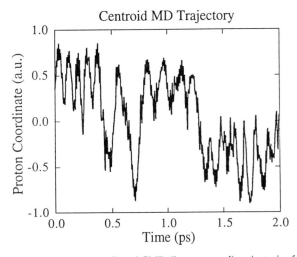

Figure 9. Classical MD (*top panel*) and CMD (*bottom panel*) trajectories for the proton transfer asymmetric stretch coordinate in the H_3O^+–H_2O dimer (cf. Fig. 8).

$k_B T$ is 0.7 kcal/mol at 300 K. The classical proton therefore hops only infrequently since it does not, on average, have enough thermal energy to surmount the barrier. On the other hand, it is seen that the quantum proton hops in the dimer much more frequently, with a larger amplitude, since the effective barrier felt by the quantum centroid "particle" is much lower than the classical case, due to tunneling and quantization effects (cf. Section IV.B.1).

2. Velocity Correlation Functions

It is of considerable interest to calculate quantum velocity correlation functions using CMD.* In Fig. 10, the velocity (actually momentum) correlation function is shown for the nonlinear oscillator in Eq. (3.85) at $\beta = 10$. This simple test was again designed to compare the exact result with the CMD estimate. The latter is clearly very accurate.

Preliminary results from a much more complex CMD simulation of liquid water have recently been obtained. Shown in Fig. 11 are the quantum and classical hydrogen velocity correlation functions computed, in the former case, by CMD and, in the latter case, by classical MD. This

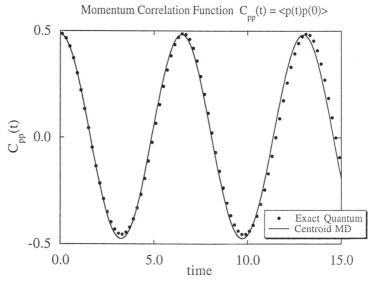

Figure 10. Plot of the real-time momentum autocorrelation function for the nonlinear oscillator described in Eq. (3.85) at a temperature of $\beta = 10$. The solid circles are the exact quantum results, while the solid line is the CMD result [5].

* The momentum correlation function can be calculated with CMD in two ways. The first method is to compute the centroid position correlation function and to use the usual Fourier relationships between position and velocty correlation functions to obtain the momentum correlation function. The second route is to calculate the centroid momentum correlation function and then use its Fourier relation with the momentum correlation function directly (see Ref. 5). It can be shown that the two methods are equivalent, although the latter is numericaly preferable because the former has a tendency to amplify the high-frequency noise in the transform.

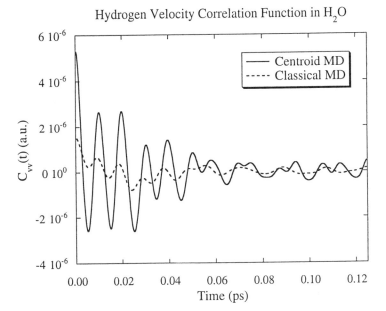

Figure 11.   Classical and quantum hydrogen velocity time correlation function for liquid water at 300 K. The classical result is obtained from classical MD and the quantum result is from CMD. The difference between the classical and quantum results is due primarily to the large zero-point energy in the O—H bonds. The simulation consisted of 125 periodically replicated flexible water molecules represented by site–site point-charge Coulomb and O–O Lennard-Jones interactions [47].

calculation involved 125 periodically replicated flexible point charge-like water molecules, quantized with $P = 25$, at 300 K. This simulation employed the CMD/PIMC algorithm [6]. Clearly, the quantum effects are significant for the hydrogen velocity correlation function of water, coming primarily from the large zero-point quantization of the O–H bond. These quantum effects are important, for example, for the accurate prediction of molecular vibrational spectra using simulation methods. The complete details of this simulation and a more complete set of results will be presented in a forthcoming publication [47].

3.   General Correlation Functions

To test the methods outlined in Section III.A.2 for calculating general correlation functions, the correlation function $\langle q^3(t)q^3(0) \rangle$ was calculated for the nonlinear potential defined in Eq. (3.85). This correlation function presents a nontrivial test for the approximate methods because of its nonlinearity and the fact that its amplitude is essentially completely

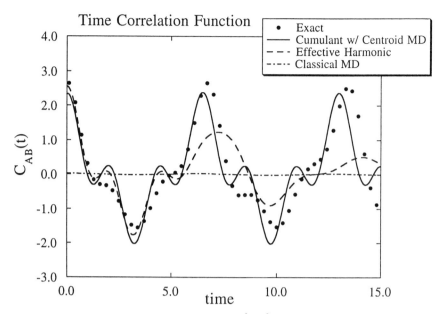

Figure 12. Plot of the correlation function $\langle q^3(t)q^3(0)\rangle$ for the nonlinear potential in Eq. (3.85) at a temperature of $\beta = 10$. The solid circles show the exact quantum results, the solid line is the cumulant expansion with CMD theory of Section III.B.3, the dashed line is the analytically continued effective harmonic result from Section III.A.2, and the dot-dashed line is the classical MD result.

quantum mechanical in origin at lower temperatures. In Fig. 12, the results for a temperature of $\beta = 10$ are shown. The correlation function from the cumulant expansion theory is in best agreement with the exact result, although there is too much structure and symmetry in the oscillations. Again the effective harmonic result dephases much too quickly, being accurate only at short times, while the classical result is extremely inaccurate.

In Fig. 13 the CMD with semiclassical operators approach of Section III.B.3 is compared to the exact result and to the classical result at $\beta = 10$ for the potential in Eq. (3.85). Also shown is a CMD result but one calculated without using the Gaussian-averaged operators defined in Eq. (3.57), that is, by instead using the classical limit of the operators expressed in terms of the centroid variables such that

$$C^*_{AB}(t) \approx \langle A_{\rm cl}(q_c(t))B_{\rm cl}(q_c(0))\rangle_{\rho_c}, \qquad (3.87)$$

where $A_{\rm cl}$ and $B_{\rm cl}$ are the classical limits of the operators A and B (cf. Section II.C.1). The CMD with semiclassical operators result does not

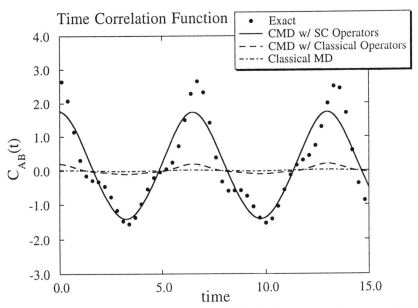

Figure 13. Plot of the correlation function $\langle q^3(t)q^3(0)\rangle$ for the nonlinear potential in Eq. (3.85) at a temperature of $\beta = 10$. The CMD with semiclassical operators method of Section III.B.3 is shown by the solid line, the exact result is given by the solid circles, and the classical result shown by the dot-dashed line. Also shown by the dashed line is a CMD result, but one obtained by using the classical form of the operators as in Eq. (3.87).

reproduce well the high-frequency oscillation of the exact result, but it is clearly superior to both the classical MD and CMD with classical operator results. The Gaussian "smearing" of the effective operators is evidently an important factor in the accuracy of the CMD with semiclassical operators approach

In Fig. 14 the correlation function $\langle q^2(t)q^2(0)\rangle$ is shown for the nonlinear potential in Eq. (3.85) at $\beta = 10$. This correlation function presents another nontrivial test of the various approximate methods because, classically, it can have no negative values while, quantum mechanically, it be negative due to interference effects. Clearly, only the cumulant method can describe the latter effects. The classical result is extremely poor for this low-temperature correlation function. The CMD with semiclassical operators method also cannot give a correlation function with negative values in this case. This feature of the latter method arises because the correlation of the two operators at different times is ignored when the Gaussian averages are performed. Consequently, the semiclassical operator approximation underestimates the quantum real-time interference of the two operators and thus fails to

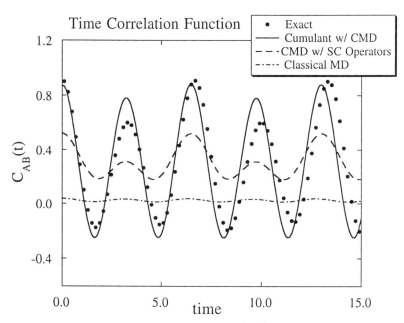

Figure 14. Plot of the correlation function $\langle q^2(t)q^2(0) \rangle$ for the nonlinear potential in Eq. (3.85) at a temperature of $\beta = 10$. The solid circles are the exact quantum results, the solid line is the cumulant expansion with CMD theory of Section III.B.3, the dashed line is the CMD with semiclassical operators result from Section III.B.3, and the dot-dashed line is the classical MD result.

provide the finer dynamical details of some general time correlation functions at low temperatures. On the other hand, the accuracy of the CMD/semiclassical operator method is already superior to a classical calculation and will improve as the temperature is increased. It also provides a simple and intuitive semiclassical algorithm to evaluate general quantum time correlation functions.

To test the methods outlined in Section III.B.3 for calculating general correlation functions in the phase-space centroid perspective, the correlation function $\langle A(t)B(0) \rangle$, where $A = pq$ and $B = qp$, was studied [5]. The results of this calculation are shown in Fig. 15 for the nonlinear potential in Eq. (3.85) at $\beta = 10$. The classical MD result is, as expected, extremely inaccurate for this low temperature. The CMD with semiclassical operators method does not reproduce the amplitude and negative values of this correlation function as well. On the other hand, the cumulant method can describe the quantum interference effects for this correlation function, and it appears to do so quite well.

Although the CMD results are quite encouraging and far superior to

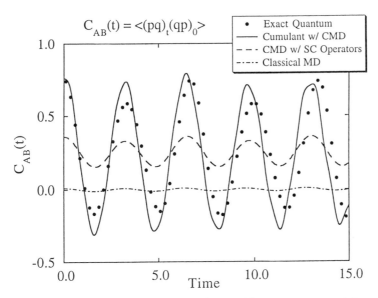

Figure 15. Plot of the correlation function $\langle A(t)B(0)\rangle$, where $A = pq$ and $B = qp$, for the nonlinear potential in Eq. (3.85) at a temperature of $\beta = 10$. The solid circles show the exact quantum results, the solid line is the cumulant expansion with CMD theory of Section III.B.3, the dashed line is the CMD with semiclassical operators result from Section III.B.3, and the dot-dashed line is the classical MD result.

classical MD, it seems evident that none of the CMD approaches developed in Section III.B.3 for general correlation functions are completely satisfactory under all circumstances. Future research must be devoted to this important issue.

4. Quantum Self-Diffusion Constants

The self-diffusion constant is a key property of condensed-matter systems. Obviously, it is desirable to be able to estimate the quantum effects for this quantity. With CMD, such a thing can be accomplished for highly nontrivial systems. Specifically, if a system of quantum particles is assumed to evolve under the laws of centroid dynamics, the diffusion constant can be calculated from either Eq. (3.42) or (3.43).

In an early test case [6], the self-diffusion process of a quantum particle in a classical solvent was studied. The solvent in this case was a Lennard-Jones fluid with parameters given in Paper IV, while the solute–solvent pair potential had the simple form [75] $v(r) = [B/(C + r^6) - 1])(A/r^4)$, where $A = 0.665$, $B = 89,099$, $C = 12,608$ (in atomic units). The simulation consisted of 512 total particles at a temperature $T = 309\ K$

and a reduced density of $\rho^* = 0.3$. The solute was a quantum particle having the electron mass m_e, but Planck's constant $\hbar$ was reduced by a factor of 10, so that a discretization number of $P = 100$ was adequate in the simulation. The details of the CMD simulation, in this case the direct CMD/PIMC method, are described in Paper IV. The mean-squared displacement of the quantum solute for this system is plotted in Fig. 16 from the CMD calculation and compared with the classical MD result. The quantum effect of diminished diffusion is clearly manifested, although the quantum solute is not in the fully localized limit [75, 78].

In Paper IV, the self-diffusion process in fluid neon was also studied with CMD using the pairwise pseudopotential method. In Fig. 17 the centroid velocity time correlation function is plotted for quantum neon using the pseudopotential method and for classical neon. When the quantum mechanical nature of the Ne atoms is taken into account, the diffusion constant is reduced by a small fraction. In the gas phase and to some degree in liquids, the diffusion process can be viewed as a sequence of two-body collisions, the frequency of which depends on the collision cross section. Because the quantum centroid cross section is larger than the corresponding classical value, the quantum diffusion constant is found

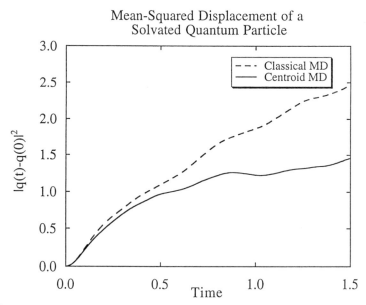

Figure 16. Plot of the mean-squared displacement for a quantum particle solvated in a classical Lennard-Jones fluid. The solid line is the CMD/PIMC algorithm described in Section III.C.1, while the dashed line is for classical MD.

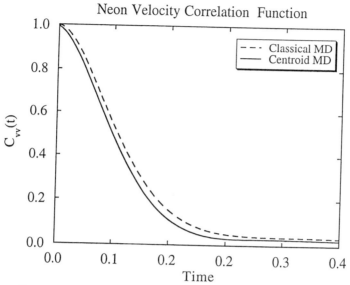

Figure 17. Plot of the centroid velocity correlation function for liquid neon. The solid line is the CMD result calculated with the centroid pseudopotential approximation, while the dashed line is the classical MD result. The self-diffusion constant is proportional to the time integral of the centroid velocity correlation functions.

to be smaller (although this is not *always* the case; cf. the next paragraph).

Perhaps the best supporting evidence for the viability and validity of the CMD approach comes from a recent study of self-diffusion in liquid p-H_2 [48]. In this study the computed CMD value for the self-diffusion constant at $T = 25$ K and $V = 31.7$ cm^3/mol was found to be $D = 1.5$ Å^2/ps, compared to the experimental value of $D_{exp} = 1.6$ Å^2/ps and the classical limit of $D_{cl} = 0.5$ Å^2/ps. Clearly, very good agreement with the experimentally measured diffusion constant was obtained from the quantum CMD calculation, while the classical MD result was too small by a factor of 3, due to the quantum "softness" (higher diffusivity) of the p-H_2 fluid. Similar quality of agreement was obtained at other densities (e.g., at the triple point of liquid hydrogen, $T = 14$ K, the experimental self-diffusion constant is $D = 0.4$ Å^2/ps compared to the CMD value of $D = 0.35$ Å^2 ps). It should also be noted that equilibrium quantities such as the radial distribution function, the mean kinetic energy, and the compressibility were also computed in Ref. [48], independently establishing the accuracy of the empirical H_2–H_2 pair potential. The agreement between the CMD and experimental results is therefore not likely to be accidental.

IV. ACTIVATED DYNAMICS AND QUANTUM TRANSITION-STATE THEORY

Quantum mechanical transition-state theory (QTST) is described from the path integral centroid perspective in this subsection. Interestingly, it has proven particularly difficult to reconcile the classical and quantum mechanical perspectives in the formulation of transition-state theory. The centroid-based path-integral quantum transition-state theory (PI-QTST) approach has proven to be a significant step forward in that regard [42–44, 49]. Not only does the theory retain many of the appealing aspects of classical TST, including the feature that only equilibrium information is necessary to estimate the thermally activated rate constant, it also includes the effects of quantum mechanical tunneling and mode quantization on the rate constant. Therefore, The PI-QTST approach provides a consistent and computationally powerful generalization of classical TST [79–84]. Since PI-QTST has been reviewed previously [44], the discussion below will largely be confined to the centroid perspective, in light of the more recent developments described in previous sections and in Refs. [4–8].

A. Formalism

The exact quantum mechanical activated rate constant expression can be written as [85, 86]

$$k = x_R^{-1} \frac{1}{\hbar\beta} \int_0^{\hbar\beta} d\tau \langle \dot{h}_P(-i\tau) h_P(t_{\mathrm{pl}}) \rangle , \tag{4.1}$$

where x_R is the reactant equilibrium mole fraction, h_P is the product state operator (usually taken to be a step function in the coordinate representation), and t_{pl} is the plateau time, at which the correlation function assumes an essentially constant value [87–94]. This quantity is obviously very difficult to evaluate for a many-body system since one must solve the exact quantum dynamics for a many-dimensional nonlinear potential. Fortunately, the path-integral centroid perspective allows one to formulate [42–44, 49] an accurate and computationally useful approximation to Eq. (4.1).

By virtue of a number of complementary perspectives [42–44, 49], it turns out that the reduced centroid density at the dividing surface along some reaction coordinate q is the central quantity in determining the value of the rate constant. The latter quantity is given by

$$Q_c^* = \int d\mathbf{x}_c \, \rho_c(q^*, \mathbf{x}_c) , \qquad (4.2)$$

where q^* is the value of q that defines the planar dividing surface in the centroid variables. The exact rate constant in Eq. (4.1) can then be expressed in terms of Q_c^* as

$$k = \frac{1}{2} \bar{v} \frac{Q_c^*}{Q_R} , \qquad (4.3)$$

and the velocity factor $\bar{v}$ is formally given by

$$\frac{1}{2} \bar{v} = \left(\frac{m}{2\pi\hbar^2\beta} \right)^{-1/2} \frac{1}{\hbar\beta} \int_0^{\hbar\beta} d\tau \, \mathrm{Tr}[e^{-\beta(H - F_c^*)} \dot{h}_P(-i\tau) h_P(t_{pl})] , \qquad (4.4)$$

where the excess path-integral centroid free energy at the dividing surface is defined as [43, 44]

$$F_c^* = -k_B T \ln[Q_c^*/(m/2\pi\hbar^2\beta)^{1/2}] . \qquad (4.5)$$

In general, the exact solution for the factor $\bar{v}$ is not known since it is expressed in terms of the exact Hamiltonian. A *dynamical model* for the correlation function can therefore be adopted which provides an estimate of the velocity factor $\bar{v}$ in the rate constant expression in Eq. (4.3). The most reasonable model one can adopt for the dynamics in the velocity factor [Eq. (4.4)] is for a *free particle* along the q direction [42–44] (The corresponding models for the other modes of the system can be arbitrary.) The resulting expression for $\bar{v}$ is given by [42]

$$\bar{v}_{\mathrm{FP}} = \sqrt{2/\pi m\beta} . \qquad (4.6)$$

Within the context of Eq. (4.3), the approximation above leads to the PI-QTST rate constant expression, given by [42–44]

$$k^{\mathrm{QTST}} = (2\pi m\beta)^{-1/2} \frac{Q_c^*}{Q_R} , \qquad (4.7)$$

or, equivalently [43, 44],

$$k^{\mathrm{QTST}} = \frac{k_B T}{h Q_R} \exp(-\beta F_c^*) . \qquad (4.8)$$

This formula may be readily applied to a variety of problems because, in the spirit of classical transition-state theory [79–84], only equilibrium information is required in order to estimate the rate constant. Additionally, the excess free energy in Eq. (4.5) can be calculated directly from PIMC techniques with umbrella sampling (this has now been done many times; see, e.g., Sections IV.B.1 and IV.B.2). Equation (4.7) has also been rederived by Stuchebrukhov [95] using a different analysis.

The formulation in Eq. (4.7) assumes that a planar dividing surface along q has been used and that the coordinate system is a simple rectilinear one. The dividing surface can be variationally rotated to maximize the centroid barrier free energy (i.e., bottleneck) in a classical-like fashion [43, 96, 97], or even a nonplanar dividing surface can be used [63, 97]. Although the well-known variational bound property of classical TST does not exist in the PI-QTST formulation [44], analytical [43, 96, 97], and numerical calculations [96], as well as physical insight, strongly suggest such a variational procedure is sensible and will, in fact, improve the absolute accuracy of the theory.

Before proceeding to some applications of PI-QTST, it is worthwhile to consider the relation of the latter theory to the CMD method described in previous sections. The essential feature of CMD is that classical-like trajectories can be propagated on the centroid potential to study quantum dynamical correlations in condensed matter. In terms of elementary arguments, therefore, if such trajectories encounter bottlenecks that impede their progress along the many-dimensional centroid potential (i.e., barriers), the average flow of the nonstationary centroid phase-space distribution will be reduced below the barrierless limit by a factor of $\exp(-\beta \Delta F_c^*)$, where ΔF_c^* is the classical-like centroid free energy for surmounting the barrier. Remarkably, this simple Arrhenius-like argument in the context of CMD *exactly* yields the PI-QTST formula [Eq. (4.8)] for the quantum activated rate constant. In fact, the PI-QTST rate constant follows naturally from the CMD picture if one defines it to be given by

$$k^{\mathrm{QTST}} \equiv x_R^{-1} \lim_{t \to 0^+} \langle \dot{h}_{P,\mathrm{cl}}[q_c(0)] h_{P,\mathrm{cl}}[q_c(t)] \rangle_{\rho_c}, \qquad (4.9)$$

where x_R is the equilibrium mole fraction of reactants, $h_{P,\mathrm{cl}}$ is the classical product population function (i.e., step function), and $q_c(t)$ is the reaction coordinate CMD trajectory. The expression above cannot be completely correct, of course, because the quantum flux operators have been treated in the classical limit [4, 66]. Nevertheless, it is known that the formula is quite accurate [66], so the errors are likely to arise from the subtleties in treating general quantum operators within the CMD context [4, 5]. These errors appear to be confined to the preexponential factor of the PI-QTST expression [44, 98–101].

B. Selected Applications

A few relatively recent applications of PI-QTST are summarized in this subsection. For other applications and extensions of the theory, the reader is referred to the growing list of PI-QTST papers in such areas as electron transfer theory [102–105] and simulation [50, 98–100, 102, 106], proton transfer theory [107] and simulation [46, 77, 107–111], hydrogen diffusion in [112] and on [113–116] metals, molecular diffusion [117] and adsorption [118, 119] on metals, and in the theory of condensed-phase effects in quantum activated dynamics [43, 63, 66, 96, 97, 120–122].

1. Proton Transfer in Polar Solvents

Proton transfer (PT) reactions in condensed phases are of considerable importance in chemistry and particularly interesting because they can involve large quantum mechanical effects. In addition, since proton transfers involve a redistribution of solute electronic charge density, a substantial contribution to the activation free energy may come from solvent reorganization effects (i.e., solvent fluctuations are necessary to create a degeneracy between the two proton binding states to the proton can transfer). It has been suggested that intramolecular vibrations may also play a crucial role in modulating the PT process. Proton transfers in polar solvents thereby provide fascinating systems to study the interplay between solvent activation dynamics, intramolecular mode coupling, and quantum tunneling effects. Not surprisingly, they have attracted a considerable degree of experimental and theoretical attention in recent years.

The benefit of the PI-QTST approach for PT reactions is that it includes both the solvent and intramolecular contributions to the PT activation free energy while treating any number of particles in the system quantum mechanically (including, of course, the transferring proton). The method can also bridge the adiabatic and nonadiabatic limits of PT reactions and is applicable to situations where the proton itself is thermally activated. To cast the path integral QTST equation into a form particularly well suited for computer simulation, the partition function for the system in the reactant configuration is first rewritten rigorously in terms of an effective quantum reactant population operator [34, 45] averaged over the centroid density in the vicinity of the reactant well. For the system in the reactant configuration, the latter quantity may then be accurately approximated to be of a Gaussian form with a width factor given by $(\beta m \omega_{c,w}^2)^{-1/2}$. After some straightforward manipulations, the

PI-QTST PT rate constant from Eq. (4.8) may be reexpressed as [108]

$$k_{PT}^{QTST} = \frac{\omega_{c,w}}{2\pi} \exp(-\beta \, \Delta F_c^*) \,, \tag{4.10}$$

where the difference in centroid free energies between the reactant and transition state is given by [77, 108]

$$\begin{aligned} \Delta F_c^* &= -k_B T \ln[Q_c(q^*)/Q_c(q_r)] \\ &= -k_B T \ln[P_c(q_r \to q^*)] \,. \end{aligned} \tag{4.11}$$

The probability $P_c(q_r \to q^*)$ to move the reaction coordinate centroid variable from the reactant configuration to the transition state is readily calculated [108] by PIMC or PIMD techniques [17–19] combined with umbrella sampling [77, 108, 123] of the reaction coordinate centroid variable. In the latter computational technique, a number of "windows" are set up which confine the path centroid variable of the reaction coordinate to different regions. These windows connect in a piecewise fashion the possible centroid positions in going from the reactant state to the transition state. A series of Monte Carlo calculations are then performed, one for each window, and the centroid probability distribution in each window is determined. These individual window distributions are then smoothly joined to calculate the overall probability function in Eq. (4.11). An equivalent approach is to calculate the centroid mean force and integrate it from the reactant well to barrier top (i.e., a "reversible work" approach for the calculation of the quantum activation free energy [109, 124]).

One application of PI-QTST to PT has been to study a model A–H–A′ PT solute in a polar solvent [77]. This computational study provided a detailed examination of the specific features of PT, including the competition between proton tunneling and solvent activation, the influence from intramolecular vibrational modulation of the PT barrier, and the role of electronic polarizability of both the solute and the solvent. Changes in the total quantum activation free energy, and hence the reaction probability, due to these different effects were calculated (cf. Fig. 18). By virtue of these studies, it was found that to fully understand the rate of a given PT reaction, one must deal with a number of complex, nonlinear interactions. Examples of such interactions include the nonlinear dependence of the solute dipole on the position of the proton, the coupling of the solute dipole to both the proton coordinate and to other vibrational modes, and the intrinsically nonlinear interactions arising from both solute and solvent polarizability effects. Perhaps the most important conclusion

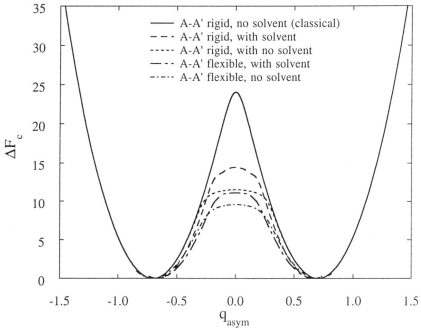

Figure 18. PI-QTST activation free-energy curves as a function of the proton asymmetric stretch coordinate for a A–H–A′ proton transfer model (see Ref. 77). The solid line depicts the classical free-energy curve for the solute in isolation with a rigid A–A′ distance, while the dotted line is the quantum free energy for the rigid, isolated solute with a fully quantized proton. The long-dashed line is the quantum free-energy curve for the isolated solute in which the A–A′ distance is allowed to fluctuate. The dot-dashed and short-dashed lines depict the quantum free-energy curves for the rigid and flexible solutes, in the polar solvent.

reached in Ref. 77 was that the influence of solvent electronic polarization cannot be neglected and must be treated quantum mechanically if one wishes to quantify proton tunneling behavior accurately in polar solvents (cf. Fig. 19). More details of the calculations and discussion of the results can be found in Ref. 77.

Another example of a PI-QTST PT calculation is shown in Fig. 20 for the activation free energy along the asymmetric PT coordinate in the aqueous $H_3O^+–H_2O$ dimer solvated in 125 periodically replicated water molecules at 300 K [46]. Using valence-bond modeling for the reaction (cf. Fig. 8) to reproduce the gas-phase dimer potential energy surface and charge distribution, it was found that the classical barrier in aqueous solution is quite small (ca. 0.5 kcal/mol). However, when the hydrogen nuclei are quantized with $P = 25$, the barrier along the centroid reaction

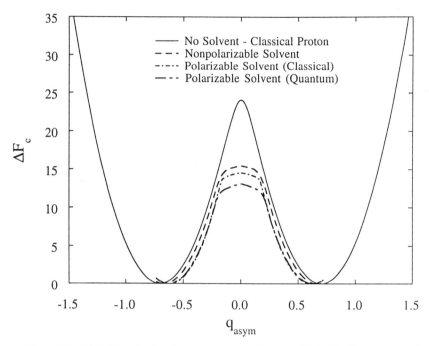

Figure 19. PI-QTST activation free-energy curves for a model A–H–A′ proton transfer system which demonstrate the effect of solvent electronic polarizability (see Ref. 77). The solid line depicts the classical free-energy curve for the solute in isolation with a rigid A–A′ distance. The short-dashed line is for the rigid solute in an electronically nonpolarizable solvent. The long-dashed line depicts the quantum free-energy curve for the rigid solute solvated in a quantum polarizable solvent, while the dot-dashed line depicts the quantum free energy for the same solute but in a classically polarizable solvent.

coordinate completely disappears (cf. Fig. 20). This PI-QTST result therefore suggests that the nuclear quantization will be important in the aqueous proton transport problem. A forthcoming publication will address this issue in more detail [46].

2. Heterogeneous Electron Transfer

Another significant application of the centroid-based PI-QTST has been in the area of heterogeneous electron transfer (ET) across the electrode-electrolyte interface [50]. In the latter research, a computer simulation method was developed for the study of such reactions in the adiabatic limit while allowing for the full quantization of all water solvent modes (i.e., collective dipole librations, molecular bends and stretches, etc.). The particular system studied was the Fe^{2+}/Fe^{3+} ion reacting with a Pt(111) electrode. The solvent activation free energy ΔF_c^* for the ET

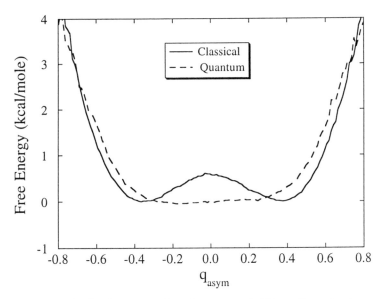

Figure 20. Activation free energy along the asymmetric PT coordinate for the aqueous-phase H_3O^+–H_2O dimer at 300 K. The result for classical hydrogen nuclei is shown by the solid curve, while the quantum centroid free energy is given by the dashed curve.

reaction was cast as the difference in free energy obtained by moving the path centroid of a collective solvation reaction coordinate from the reactant configuration to the top of the barrier in the quantum centroid free-energy curve [50]. In this case the reaction coordinate centroid was defined as

$$\Delta E_0 = \frac{1}{\hbar \beta} \int_0^{\hbar \beta} d\tau \, \Delta E(\tau)$$

$$\simeq \frac{1}{P} \sum_{i=1}^{P} \Delta E(\mathbf{x}_i) \, ,$$

(4.12)

where ΔE is the instantaneous diabatic energy gap, or solvation difference "coordinate," between the Fe^{2+} and Fe^{3+} diabatic states, $\mathbf{x}_i$ are all system coordinates at imaginary time slice i, and the second equation is written in the discretized path integral notation [17–19]. Specific numerical values of the energy gap centroid are denoted by Δe.

Shown in Fig. 21 are the quantum and classical solvent curves calculated for the same set of potential parameters. The water model was a flexible point-charge model, the specifics of which are described in

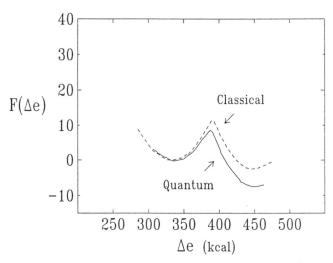

Figure 21. Adibatic solvent activation free-energy curves for the Fe^{2+}/Fe^{3+} electron transfer reaction with a platinum electrode at 300 K calculated obtained using the model of Ref. 50, path-integral quantum transition-state theory, and umbrella sampling. The solid line depicts the quantum adiabatic free-energy curve, while the dashed line depicts the curve in the classical limit. In both cases, the left-hand well corresponds to the Fe^{2+} stable state, while the right-hand well is the Fe^{3+} stable state.

detail in Ref. 50 (although the harmonic limit of the intramolecular water potential was used here). The differences in classical and quantum free energies have quite a large effect on the ET rate (i.e., a quantum increase of a factor of about 50 for the oxidation of Fe^{2+} over the classical limit with a decrease by a factor of about 1000 for the reduction of Fe^{3+}). Moreover, the overall thermodynamic driving force is substantially different for the two limits, reflecting the differences in solvation of the ions by the classical and quantized water models. This PI-QTST study uncovered the importance of quantized water modes in heterogeneous electron transfer processes. More details and discussion can be found in Ref. 50 and in forthcoming publications.

V. CONCLUDING REMARKS

In this article a perspective on quantum statistical mechanics and dynamics has been reviewed that is based on the path centroid variable in Feynman path integration [1, 3–8, 21–23]. Although significant progress has been achieved in this research effort to date, much remains to be done. For example, in terms of the calculation of equilibrium properties it remains to be seen whether the centroid perspective will offer any significant advantage over more established numerical and analytical path-integral approaches [17–19]. Although it is too early to know the answer to this equation, the formal properties and physical insight inherent in the centroid-based analysis seem compelling. Moreover, the analytic theory for the centroid density will itself be valuable for the determination and analysis of the centroid potential, which is the central quantity in both CMD and PI-QTST.

In terms of dynamics it seems clear, however, that the centroid perspective, specifically CMD and PI-QTST, represent significant theoretical and computational advances. Indeed, the more mature PI-QTST approach has now been widely employed in the theory and simulation of quite a few condensed matter activated rate processes (cf. Section IV). Within the context of a transition-state theory, the errors in PI-QTST have been found to be small and confined largely to the preexponential factor. Yet the theory can certainly be improved by focusing on the origins of these errors. In terms of dynamical correlation functions, the CMD approach certainly holds promise for many interesting applications. In fact, such applications are already under way to simulate proton and electron transport in liquids and biomolecular systems, vibrational relaxation phenomena in condensed phases, the dynamics of high-frequency vibrations in complex quantum systems, and hydrogen bond dynamics. Additional fundamental research will need to be carried out to characterize and control the errors in the method even better, as well as to develop numerical procedures to systematically *improve upon* the CMD result for the time correlation function. Moreover, the problem of treating general operators in CMD time correlation functions has not been solved completely and requires further research. More work is also required to develop increasingly efficient CMD algorithms, perhaps exploiting the advantages of parallel computing, as well as to formulate accurate centroid pseudopotentials for general many-body systems.

As the research on centroid theory evolves in the future, additional far-reaching questions will probably arise. For example, can electronically nonadiabatic transitions be readily included in the CMD method? Can Bose–Einstein and Fermi–Dirac statistics be included? Correspondingly, is Bose condensation related to the coalescence of centroid momenta? Can the Pauli exclusion principle be described by effective repulsive or imaginary terms in the centroid potential to simulate real-time Fermion dynamics? These are questions with unknown answers, but the path centroid perspective is clearly a different and promising way to describe complex quantum systems which should continue to yield new and interesting results for some time to come.

ACKNOWLEDGMENTS

The author expresses his sincere gratitude to Jianshu Cao for his many invaluable and essential contributions to this research. Acknowledgements are also in order to past and present members of the author's group who have contributed to various phases of this research: John Lobaugh, George Haynes, Jay Straus, Diane Sagnella, August Calhoun, Rigoberto Hernandez, Daohui Li, Huadong Gai, Indrani Bhattacharya-Kodali, Ying-Chieh Sun, Yuri Boroda, Marc Pavese, Amir Karger, Charles Ursenbach and Kurt Kistler. This research was supported by grants from the National Science Foundation and the Office of Naval Research. The author is a recipient of a National Science Foundation Presidential Young Investigator Award, a David and Lucile Packard Fellowship in Science and Engineering, an Alfred P. Sloan Foundation Research Fellowship, a Camille Dreyfus Teacher–Scholar Award, and a Dreyfus Foundtion New Faculty Award.

REFERENCES

1. R. P. Feynman and A. R. Hibbs, *Quantum Mechanics and Path Integrals*, McGraw-Hill, New York, 1965.

2. R. P. Feynman, *Statistical Mechanics*, Addison-Wesley, Reading, Mass., Chap. 3.

3. J. Cao and G. A. Voth, *J. Chem. Phys.* **100**, 5093 (1994); this is referred to as "Paper I."

4. J. Cao and G. A. Voth, *J. Chem. Phys.* **100**, 5106 (1994); this is referred to as "Paper II."

5. J. Cao and G. A. Voth, *J. Chem. Phys.* **101**, 6157 (1994); this is referred to as "Paper III."

6. J. Cao and G. A. Voth, *J. Chem. Phys.* **101**, 6168 (1994); this is referred to as "Paper IV."

7. J. Cao and G. A. Voth, *J. Chem. Phys.* **101**, 6184 (1994); this is referred to as "Paper V."

8. J. Cao and G. A. Voth, *J. Chem. Phys.* **99**, 10070 (1993).

9. L. S. Schulman, *Techniques and Applications of Path Integration*, Wiley, New York, 1986).

10. H. Kleinert, *Path Integrals in Quantum Mechanics, Statistics, and Polymer Physics*, 2nd ed., World Scientific, Singapore, 1995; see Chaps. 5 and 17 for path-integral centroid-related material.

11. M. S. Swanson, *Path Integrals and Quantum Processes*, Academic Press, San Diego, 1992).

12. G. Roepstorff, *Path Integral Approach to Quantum Physics*, Springer-Verlag, Berlin, 1994.

13. F. Weigel, *Phys. Rep.* **16**, 57 (1975).

14. C. DeWitt-Morette, A. Maheshwari, and B. Nelson, *Phys. Rep.* **50**, 255 (1979).
15. M. S. Marinov, *Phys. Rep.* **60**, 1 (1980).
16. D. C. Khandekar and S. V. Lawande, *Phys. Rep.* **137**, 115 (1986).
17. B. J. Berne and D. Thirumalai, *Ann. Rev. Phys. Chem.* **37**, 401 (1986).
18. D. Chandler, in *Liquides, cristillisation et transition vitreuse les Houches, Session LI*, D. Levesque, J. Hansen, and J. Zinn-Justin, Eds., Elsevier, New York, 1991.
19. J. D. Doll, D. L. Freeman and T. L. Beck, *Adv. Chem. Phys.* **78**, 61 (1990).
20. J. D. Doll and J. E. Gubernatis, *Quantum Simulations of Condensed Matter Phenomena*, World Scientific, Singapore, 1990.
21. R. Giachetti and V. Tognetti, *Phys. Rev. Lett.* **55**, 912 (1985).
22. R. Giachetti and V. Tognetti, *Phys. Rev. B* **33**, 7647 (1986).
23. R. P. Feynman and H. Kleinert, *Phys. Rev. A* **34**, 5080 (1986).
24. H. Kleinert, *Phys. Lett. A* **118**, 267 (1986).
25. W. Janke and H. Kleinert, *Chem. Phys. Lett.* **137**, 162 (1987).
26. D. Chandler and P. G. Wolynes, *J. Chem. Phys.* **80**, 860 (1981).
27. S. Liu, G. K. Horton, and E. R. Cowley, *Phys. Rev. B* **44**, 11714 (1991).
28. S. Liu, G. K. Horton, E. R. Cowley, A. R. McGurn, A. A. Maradudin, and R. F. Wallis, *Phys. Rev. B* **45**, 9716 (1992).
29. A. Cuccoli, A. Macchi, M. Neumann, V. Tognetti, and R. Vaia, *Phys. Rev. B* **45**, 2088 (1992).
30. A. Cuccoli, A. Macchi, V. Tognetti, and R. Vaia, *Phys. Rev. B* **47**, 14923 (1993).
31. A. Cuccoli, V. Tognetti, P. Verrucchi, and R. Vaia, *Phys. Rev. B* **46**, 11601 (1992).
32. A. Cuccoli, V. Tognetti, P. Verrucchi, and R. Vaia, *J. Appl. Phys.* **75**, 5814 (1994), and references cited therein.
33. A. R. Völkel, A. Cuccoli, M. Spicci, and V. Tognetti, *Phys. Lett. A* **182**, 60 (1993), and references cited therein.
34. A. Cuccoli, V. Tognetti, and R. Vaia, *Phys. Rev. A* **44**, 2734 (1991).
35. A. Cuccoli, V. Tognetti, A. A. Maradudin, A. R. McGurn, and R. Vaia, *Phys. Rev. B* **46**, 8839 (1992).
36. A. Cuccoli, M. Spicci, V. Tognetti, and R. Vaia, *Phys. Rev. B* **45**, 10127 (1992).
37. A. Cuccoli, M. Spicci, V. Tognetti, and R. Vaia, *Phys. Rev. B* **47**, 7859 (1993).
38. A. Cuccoli, V. Tognetti, A. A. Maradudin, A. R. McGurn, and R. Vaia, *Phys. Rev. B* **48**, 7015 (1993).
39. A. Cuccoli, V. Tognetti, A. A. Maradudin, A. R. McGurn, and R. Vaia, *Phys. Lett. A* **196**, 285 (1994).
40. A. Cuccoli, V. Tognetti, P. Verrucchi, and R. Vaia, *Phys. Rev. A* **45**, 8418 (1992).
41. J. Cao and B. J. Berne, *J. Chem. Phys.* **92**, 7531 (1990).
42. G. A. Voth, D. Chandler, and W. H. Miller, *J. Chem. Phys.* **91**, 7749 (1989).
43. G. A. Voth, *Chem. Phys. Lett.* **170**, 289 (1990).
44. G. A. Voth, *J. Phys. Chem.* **97**, 8365 (1993). For a review of path integral quantum transition state theory, see this paper.
45. G. A. Voth, *Phys. Rev. A* **44**, 5302 (1991).
46. J. Lobaugh and G. A. Voth, *J. Chem. Phys.* (in press, 1996).
47. J. Lobaugh, A. Calhoun, M. Pavese, and G. A. Voth, to be published.
48. J. Cao and G. J. Martyna, *J. Chem. Phys.* (in press, 1996).

49. M. J. Gillan, *J. Phys. C* **20**, 3621 (1987). This paper is related to the path integral quantum transition state theory work of Refs. 42–44.

50. J. B. Straus, A. Calhoun, and G. A. Voth, *J. Chem. Phys.* **102**, 529 (1995).

51. H. Kleinert and H. Meyer, *Phys. Lett. A* **184**, 319 (1994). This work is closely related to the diagrammatic formulation of the centroid density in Ref. 3. However, in the latter paper the locally optimized quadratic reference potential approximation was explicitly derived from the perspective of the diagrammatic analysis, thus enabling the introduction of the higher-order corrections without having to subtract the over-counted diagrams resulting from the quadratic optimization. Moreover, it was shown how the lines (i.e., the centroid-constrained imaginary time propagators) could be renormalized along with the vertices in a consistent fashion to improve substantially on the accuracy of the effective quadratic variational method.

52. D. Chandler, *J. Phys. Chem.* **88**, 3400 (1984).

53. T. Morita and K. Hiroike, *Prog. Theor. Phys.* **25**, 537 (1961).

54. C. D. Dominicis, *J. Math. Phys.* **3**, 183 (1962).

55. J. E. Mayer, *J. Chem. Phys.* **10**, 629 (1942).

56. R. D. Mattuck, *A Guide to Feynman Diagrams in the Many-Body Problem*, McGraw-Hill, New York, 1976.

57. A. L. Fetter and J. D. Walecka, *Quantum Theory of Many-Particle Systems*, McGraw-Hill, New York, 1971.

58. E. N. Economou, *Green's Function in Quantum Mechanics*, Springer-Verlag, Berlin, 1983.

59. R. Hernandez, J. Cao, and G. A. Voth, *J. Chem. Phys.* **103**, 5018 (1995).

60. G. Seeley and T. Keyes, *J. Chem. Phys.* **91**, 5581 (1989).

61. B. Xu and R. M. Stratt, *J. Chem. Phys.* **92**, 1923 (1990).

62. R. Kubo, N. Toda, and N. Hashitsume, *Statistical Physics II*, Springer-Verlag, Berlin, 1985.

63. M. Messina, G. K. Schenter, and B. C. Garrett, *J. Chem. Phys.* **98**, 8525 (1993).

64. G. J. Martyna, *J. Chem. Phys.* (in press, 1996). In this paper, an effective set of molecular dynamics equations are specified that provide an alternative path-integral approach to the calculation of position and velocity time correlation functions. This approach is essentially based on the Wigner phase-space function. For general nonlinear systems, the appropriate MD mass in this approach is not the physical mass, but it must instead be a position-dependent effective mass.

65. B. J. Berne and R. Pecora, *Dynamic Light Scattering*, Wiley-Interscience, New York, 1976.

66. G. K. Schenter, M. Messina, and B. C. Garrett, *J. Chem. Phys.* **99**, 1674 (1993).

67. J. D. Doll, *J. Chem. Phys.* **81**, 3536 (1984). In Appendix A of this paper, it was speculated that the quantum-activated rate constant might be computed from classical dynamics on the Feynman–Hibbs effective potential energy surface.

68. J. Cao and B. J. Berne, *J. Chem. Phys.* **99**, 2902 (1993).

69. E. L. Pollock and D. M. Ceperley, *Phys. Rev. B* **30**, 2555 (1984).

70. M. Sprik, M. L. Klein, and D. Chandler, *Phys. Rev. B* **31**, 4234 (1985).

71. M. Parrinello and A. Rahman, *J. Chem. Phys.* **80**, 860 (1984).

72. G. J. Martyna, M. Tuckerman, and B. J. Berne, *J. Chem. Phys.* **98**, 1990 (1992).

73. R. Car and M. Parrinello, *Phys. Rev. Lett.* **55**, 2471 (1985).

74. D. K. Remler and P. A. Madden, *Mol. Phys.* **70**, 921 (1990). This reference contains a

review and discussion of the Car–Parrinello extended Lagrangian technique for first-principles molecular dynamics simulations.

75. D. F. Coker, B. J. Berne, and D. Thirumalai, *J. Chem. Phys.* **86**, 5689 (1987).

76. R. Vaia and V. Tognetti, *Int. J. Mod. Phys.* **4**, 2005 (1990).

77. J. Lobaugh and G. A. Voth, *J. Chem. Phys.* **100**, 3039 (1994).

78. D. Chandler, Y. Singh, and D. M. Richardson, *J. Chem. Phys.* **81**, 1975 (1984).

79. H. Eyring, *J. Chem. Phys.* **3**, 107 (1934).

80. E. Wigner, *J. Chem. Phys.* **5**, 720 (1937).

81. W. H. Miller, *Acc. Chem. Res.* **9**, 306 (1976).

82. P. Pechukas, *Annu. Rev. Phys. Chem.* **32**, 159 (1981).

83. D. G. Truhlar, W. L. Hase, and J. T. Hynes, *J. Phys. Chem.* **87**, 2664 (1983).

84. P. Hänggi, P. Talkner, and M. Borkovec, *Rev. Mod. Phys.* **62**, 250 (1990).

85. G. A. Voth, D. Chandler, and W. H. Miller, *J. Phys. Chem.* **93**, 7009 (1989).

86. T. Yamamoto, *J. Chem. Phys.* **33**, 281 (1960).

87. D. Chandler, *J. Chem. Phys.* **68**, 2959 (1978).

88. J. J. Montgomery, Jr., D. Chandler, and B. J. Berne, *J. Chem. Phys.* **70**, 4056 (1979).

89. R. O. Rosenberg, B. J. Berne, and D. Chandler, *Chem. Phys. Lett.* **75**, 162 (1980).

90. J. Keck, *J. Chem. Phys.* **32**, 1035 (1960).

91. J. B. Anderson, *J. Chem. Phys.* **58**, 4684 (1975).

92. C. H. Bennett, in *Algorithms for Chemical Computation*, ACS Symposium Series 46, R. Christofferson, Ed., American Chemical Society, Washington, D.C., 1977.

93. J. T. Hynes, in *The Theory of Chemical Reactions*, M. Baer, Ed., CRC Press, Boca Raton, Fla., 1985.

94. B. J. Berne, in *Multiple Timescales*, J. Brackbill and B. Cohen, Eds., Academic Press, New York, 1985.

95. A. A. Stuchebrukhov, *J Chem. Phys.* **95**, 4258 (1991).

96. M. Messina, G. K. Schenter, and B. C. Garrett, *J. Chem. Phys.* **99**, 8644 (1993).

97. E. Pollak, *J. Chem. Phys.* **103**, 973 (1995).

98. C. H. Mak and J. N. Gehlen, *Chem. Phys. Lett.* **206**, 103 (1993).

99. C. H. Mak and J. N. Gehlen, *J. Chem. Phys.* **98**, 7361 (1993).

100. R. Egger and C. H. Mak, *J. Chem. Phys.* **99**, 2541 (1993).

101. M. Topaler and N. Makri, *J. Chem. Phys.* **101**, 7500 (1994).

102. J. S. Bader, R. A. Kuharski, and D. Chandler, *J. Chem. Phys.* **93**, 230 (1990).

104. P. G. Wolynes, *J. Chem. Phys.* **87**, 6559 (1987). This author derived a connection between the imaginary-time-path-integral method and the golden rule limit of electron transfer dynamics. This formulation is basically equivalent to the result of centroid theory for the case of nonadiabatic electron transfer.

105. J. N. Gehlen, D. Chandler, H. J. Kim, and J. T. Hynes, *J. Phys. Chem.* **96**, 1748 (1992).

106. A. Warshel and Z. T. Chu, *J. Chem. Phys.* **93**, 4003 (1990).

107. D. Li and G. A. Voth, *J. Phys. Chem.* **95**, 10425 (1991).

108. J. Lobaugh and G. A. Voth, *Chem. Phys. Lett.* **198**, 311 (1992).

109. H. Azzouz and D. Borgis, *J. Chem. Phys.* **98**, 7361 (1993).
110. D. Laria, G. Ciccotti, M. Ferrario, and R. Kapral, *Chem. Phys.* **180**, 181 (1994).
111. J.-K. Hwang, Z. T. Chu, Y. Yadav, and A. Warshel, *J. Phys. Chem.* **95**, 8445 (1991).
112. M. J. Gillan, *Phys. Rev. Lett.* **58**, 563 (1987).
113. Y.-C. Sun and G. A. Voth, *J. Chem. Phys.* **98**, 7451 (1993).
114. S. L. Rick, D. L. Lynch, and J. D. Doll, *J. Chem. Phys.* **99**, 8183 (1993).
115. T. R. Mattsson, U. Engberg, and G. Wahnström, *Phys. Rev. Lett.* **71**, 2615 (1993).
116. T. R. Mattsson and G. Wahnström, *Phys. Rev. B* **51**, 1885 (1995).
117. A. Calhoun and D. Doren, *J. Chem. Phys.* **97**, 2251 (1993).
118. G. Mills and H. Jónsson, *Phys. Rev. Lett.* **72**, 1124 (1994).
119. G. Mills, H. Jónsson, and G. K. Schenter, *Surf. Sci.* **324**, 305 (1995).
120. G. A. Voth and E. V. O'Gorman, *J. Chem. Phys.* **94**, 7342 (1991).
121. G. A. Voth, *Ber. Bunsenges. Phys. Chem.* **95**, 393 (1991).
122. G. R. Haynes and G. A. Voth, *Phys. Rev. A* **46**, 2143 (1992).
123. J. P. Valleau and G. M. Torrie, in *Statistical Mechanics, Part A*, B. Berne, Ed., Plenum, New York, 1977.
124. G. K. Schenter, G. Mills, and H. Jónsson, *J. Chem. Phys.* **101**, 8964 (1994).

MULTICONFIGURATIONAL PERTURBATION THEORY: APPLICATIONS IN ELECTRONIC SPECTROSCOPY

BJÖRN O. ROOS, KERSTIN ANDERSSON,
MARKUS P. FÜLSCHER, PER-ÅKE MALMQVIST,
and LUIS SERRANO-ANDRÉS

Department of Theoretical Chemistry, Chemical Centre, Lund, Sweden

KRISTIN PIERLOOT

Department of Chemistry, University of Leuven, Heverlee-Leuven, Belgium

MANUELA MERCHÁN

Departamento de Química Física, Universitat de València, Spain

CONTENTS

Advances in Chemical Physics, Volume XCIII, Edited by I. Prigogine and Stuart A. Rice.
ISBN 0-471-14321-9 © 1996 John Wiley & Sons, Inc.

ABSTRACT

Applications of the complete active space (CAS) SCF method and multiconfigurational second-order perturbation theory (CASPT2) in electronic spectroscopy are reviewed. The CASSCF/CASPT2 method was developed five to seven years ago and the first applications in spectroscopy were performed in 1991. Since then, about 100 molecular systems have been studied. Most of the applications have been to organic molecules and to transition metal compounds. The overall accuracy of the approach is better than 0.3 eV for excitation energies except in a few cases, where the CASSCF reference function does not characterize the electronic state with sufficient accuracy.

Some of the more important aspects of the theory behind the method are described in the review. In particular, the choice of the zeroth-order Hamiltonian is discussed together with the intruder-state problem and its solution. A generalization of the method to a multistate perturbation approach is suggested. Problems specifically related to spectroscopic applications are discussed, such as the choice of the active space and the treatment of solvent effects.

The spirit is to show some of the results, but also to guide users of the approach by pointing to the problems and limitations of the method. The review covers some of the newer applications in the spectroscopy of organic molecules: acetone, methylenecyclopropene, biphenyl, bithiophene, the protein chromophores indole and imidazole, and a series of radical cations of conjugated polyenes and polyaromatic hydrocarbons. The applications in transition metal chemistry include carbonyl, nitrosyl, and cyanide complexes, some dihalogens, and the chromium dimer.

I. INTRODUCTION

The improvement in recent years of the methods and techniques of *ab initio* quantum chemistry has considerably increased the possibilities of obtaining accurate theoretical information about spectroscopic properties of molecular systems and of their photochemistry (see, e.g., articles in Ref. 1). It is in particular the developments of the multiconfigurational approaches that have had a profound impact on the possibilities for large-scale applications in spectroscopy. The electronic structure for most ground-state molecules is rather simple, with the wavefunction dominated by a single (Hartree–Fock) configuration. The situation is different for the excited states. Here the energy separation between different electronic configurations is smaller, which results in larger mixing. Singly and doubly excited configurations close in energy are not uncommon. A typical case in organic systems results from the interaction of two C=C double bonds. The triplet excitations in each of them can couple to a singlet with an energy close to that of a singly excited singlet state in one of them. This has profound consequences for the lower part of the electronic spectrum of such systems, for example, in pyrrole, thiophene, and related system (for details, see the article by Roos et al. in Ref. 1).

Situations such as the one described above are normally treated using multiconfigurational (MC) SCF theories. Most MCSCF calculations are today performed within the framework of the complete active space (CAS) SCF scheme [2]. The generality of this approach makes it especially suitable for studies of excited states. The wavefunction is defined by selecting a set of active orbitals and is constructed as a linear expansion in the set of configuration functions (CFs) that can be generated by occupying the active orbitals in all ways consistent with an overall spin and space symmetry [full configuration interaction (CI)]. No assumptions are made regarding the character of the excited states (singly or multiply excited) or the shape of the molecular orbitals. The latter are optimized with the only limitations set by the chosen basis set. This can be done independently for each state but is more commonly done as an average over a set of excited states. The CASSCF method has been used to study a large number of spectroscopic problems for small and medium-sized molecules.

However, while the CASSCF method can handle the near-degeneracy problem in a balanced and effective way, it does not include the effects of dynamic (external) correlation. For small molecules it is possible to treat this problem using multireference (MR) CI or similar techniques. Two recent review articles in the electronic spectroscopy of diatomic [3] and triatomic [4] molecules provide good illustrations of the present state of the art. The MRCI method is in principle capable of very high accuracy. It is also very general and can treat any type of electronically excited state. However, the computational effort increases steeply with the number of correlated electrons. It is therefore not a method that can be used for larger molecules. An illustration is given by a recent MRCI study of the electronic spectrum of the pyrimidine molecule, which failed due to the size bottleneck [5].

An alternative approach for the treatment of dynamic correlation effects within the multiconfigurational framework was developed six years ago [6, 7], based on an earlier suggestion [8]. The idea was to use second-order perturbation theory. As all configurations with large co-efficients were included into the CASSCF wavefunction, the presumption was that it should be possible to obtain an accurate estimate of the contributions from the remainder of the full CI space by means of low-order perturbation theory, where the CASSCF wavefunction acts as the zeroth-order approximation (the reference function). A perturbation method is also size consistent and can therefore be applied to systems with many electrons without loss of accuracy (in practice, it turns out that the approach is only "almost size consistent," but this has no practical consequences, since the nonadditive terms give very small contributions to the correlation energy [9]). The algorithm for the solution of the first-order equations has a complicated structure, and an effective implementation is possible only if one can compute effectively up to fourth-order density matrices for the CASSCF wavefunction. This prob-lem was solved using the split unitary group approach [10]. The resulting method was named CASPT2 and has been added to the MOLCAS quantum chemistry software [11]. All results that will be presented here have been obtained using this program. The CASPT2 method has been applied to a large body of chemical problems. Provided that extended basis sets are used and the active space of the CASSCF reference function is properly chosen, it gives highly accurate relative energies for a broad range of chemical problems (for a recent review, see Ref. 12). Due to the general nature of the reference function, the method works equally well for ground and excited states and for electronic states that are not well characterized by a single configuration. The applications in electronic spectroscopy have been especially rewarding, since alternative methods exist only for small molecules. The results obtained for excited states of organic molecules have recently been reviewed [13].

The present review is concerned with general aspects of applications of the CASSCF/CASPT2 method in molecular spectroscopy. We first introduce the reader to the method. This is brief; we refer readers to the original papers and the thesis of Andersson [14] for details. The pitfalls of the method are discussed with special emphasis on the intruder-state problem. The applications in spectroscopy are divided into two parts. One deals with recent results obtained for organic systems and the other with transition metal compounds. For each of these types of application, we discuss the general requirements that have to be fulfilled to obtain accurate results. These include the choice of active space, basis set, and so on. The final discussion emphasizes the limitations of the method and where future work should be concentrated to increase the range of applications and the effectiveness of the approach.

II. MULTICONFIGURATIONAL PERTURBATION THEORY

The (hopefully converging) series of results from any systematic iterative method defines the partial sums of a perturbation series. We consider the approximate solution to the electronic Schrödinger equation. In this context a Taylor series expansion in one perturbation parameter is almost always implied:

$$\hat{H} = \hat{H}_0 + \lambda\hat{H}_1 \ , \tag{1}$$

where $\hat{H}_0$ is a simplified operator, which can be chosen on grounds of expediency, and λ is the formal perturbation parameter.

In particular, state-specific methods produce a state function and an energy:

$$\hat{H}(\lambda)\Psi(\lambda) = E(\lambda)\Psi(\lambda)$$

$$\Psi(\lambda) = \Psi_0 + \sum_1^\infty \lambda^k \Psi_k \tag{2}$$

$$E(\lambda) = E_0 + \sum_1^\infty \lambda^k E_k$$

as functions of λ, expressed as Taylor expansions, and we are usually interested only in the value $\lambda = 1$.

By contrast, effective Hamiltonian methods produce a stable subspace and an effective Hamiltonian matrix:

$$\hat{H}(\lambda)\mathbf{\Psi}(\lambda) = \mathbf{\Psi}(\lambda)\mathbf{H}^{\text{eff}}(\lambda) \ ,$$

$$\mathbf{\Psi}(\lambda) = \mathbf{\Psi}_0 + \sum_1^\infty \lambda^k \mathbf{\Psi}_k \ , \tag{3}$$

$$\mathbf{H}^{\text{eff}}(\lambda) = \mathbf{H}_0 + \sum_1^\infty \lambda^k \mathbf{H}_k \ ,$$

where $\mathbf{\Psi}$ is now a row array of wavefunctions. The space spanned by $\mathbf{\Psi}_0$ is called the model space. For $\lambda = 1$, the exact energies are obtained as the eigenvalues of the matrix $\mathbf{H}^{\text{eff}}$. The state-specific perturbation series is thus a special case of the effective Hamiltonian method, with a one-dimensional model space.

If any of the functions of λ above is not analytic for all arguments $|\lambda| \leq 1$, the perturbation series of that function does not converge. We will be concerned only with the conventional situation where the wavefunction is contained in a finite (but potentially huge) space spanned by Slater determinant functions constructed by a finite set of orbitals, or equivalently, such determinants precombined into configuration functions (CFs). From the point of view of this article, the Hamiltonian $\hat{H}$ is thus not the exact electronic Hamiltonian but is its projection in the CF space. Our aim is thus to construct a perturbation theory that approximates the full CI result in this space. The FCI equations are algebraic equations, and lack of convergence of the perturbation series is then caused by near coincidence of energies of model functions with functions outside the model space, which are in this context called intruders.

In this article we deal exclusively with the CASPT2 method, where the series expansions are truncated to first order in wavefunctions, and second order in energy. Intruders are tolerable if they do not contribute to the energy at this level. We assume that the reader has some familiarity with the CASSCF method, which is used to obtain the root function. More information can be obtained in earlier review articles [2, 12, 13]. The equations to be solved, to the second order, are simple:

$$\hat{H}_0 \Psi_0 = E_0 \Psi_0 \,,$$

$$E_1 = \langle \Psi_0 | \hat{H}_1 | \Psi_0 \rangle \,,$$

$$(\hat{H}_0 - E_0) \Psi_1 = -(\hat{H}_1 - E_1) \Psi_0 \,, \tag{4}$$

$$E_2 = \langle \Psi_1 | \hat{H}_1 | \Psi_0 \rangle \,.$$

The Ψ_1 wavefunction is restricted to be orthonormal to Ψ_0, which gives the equation system a unique and well-conditioned solution if there is no intruder. Obviously, the first equation is normally not to be solved for Ψ_0. Instead, that wavefunction is given, and $\hat{H}_0$ is designed to have it as an eigenfunction.

There are a number of important differences between single-determinant root functions and multiconfigurational root functions. One of the most important is that a one-electron operator cannot in general have a multiconfigurational Ψ_0 as an eigenfunction. Consider first the ordinary Møller–Plesset perturbation series: The $\hat{H}_0$ operator is chosen to be $\hat{H}_0 = \hat{F}$, the Fock operator of Hartree–Fock theory. The orbitals are usually chosen to diagonalize $\hat{F}$. The solution of the perturbation equations will preserve the number of excited electrons, so the first-order wavefunction can involve only single and double excitations. Similarly, the nth order involves $1-2n$ excited electrons. A zero-electron excitation is never produced, because of orthogonality.

From a formal point of view, the simplest multiconfigurational extension again uses a diagonal, one-electron $\hat{H}_0$. However, this is simple only when the *space of interacting determinants* is small enough. This contains the set of determinants that are produced by single and double excitations from those determinants that are used in the root wavefunction. Moreover, it requires that $\hat{F}$ has Ψ_0 as eigenfunction, which is possible for a multiconfigurational wavefunction only if all the active orbitals have the same orbital energy (quasidegenerate perturbation theory). For general purposes it is advantageous to work with CASSCF wavefunctions. There are then typically 10^{10} interacting determinants, and also the quasidegenerate $\hat{H}_0$ is rarely adequate.

By contrast, CASPT2 uses orbital excitation operators applied to the root wavefunction to express the perturbation function. Apart from the root function itself, this generates the singly excited wavefunctions $\{\hat{E}_{pq}\Psi_0\}$, the doubly excited wavefunctions $\{\hat{E}_{pqrs}\Psi_0\}$, and so on. These lists are here written in a general way—in practice, many of the terms listed are zero, and the rest may be linearly dependent. Furthermore, perhaps surprising at first, in contrast to the ordinary single-configurational case, the space spanned by the doubly excited wavefunctions contains the singly excited ones. This set of functions, orthonormalized against the CAS-CI space, will be called the first-order interacting (FOI) space. In CASPT2, the number of independent variables is roughly, as for MP2, equal to the number of occupied pair of orbitals, times the number of virtual pairs. This small number has been obtained at the price of complexity: In order to make Ψ_0 an eigenfunction of $\hat{H}_0$, we must use a projection operator $\hat{P}_0 = |\Psi_0\rangle\langle\Psi_0|$. To have Ψ_1 contained in the doubly excited space, there is a further projection. Due to the linear dependencies, the equation systems are overcomplete. Finally, the equation system does not have a diagonal equation matrix, but a dense one, often with a poor condition number and important nondiagonal elements.

Those complications are technical. For details, we refer to the original work [6, 7, 14], and to a recent review article [12]. Understanding the program from a functional point of view requires merely these pertinent facts: It is an ordinary MP perturbation theory, taken to second order in energy. Due to its use of a multiconfigurational root function, the equations may take a longer time to solve, but the rewards are a general applicability (regardless of multiplicity and symmetry, also for excited states) and often a higher accuracy as well since important correlation is already built into the root function. Two important issues follow immediately from the wider range of applications. The definition of $\hat{H}_0$ will be suitably extended to multiconfigurational cases in the next section. Then the intruders must be dealt with—they are rare in near-closed-shell applications but are common among excited states, and occasionally appear during bond-breaking and similar near-degenerate situations.

A. Zeroth-Order Hamilton

In single-determinant perturbation theory, all formulas can be evaluated either by diagram techniques or by simple algebra. The energy, or any other quantity, at any level of perturbation theory, is obtained by a short sequence of operations, either of matrix multiplication type, or else division by energy denominators. This simplicity arises from the fact that every matrix element in the formulas will be a product of basic integrals, divided by such denominators, times a product of Kronecker deltas in the orbital indices. In the CASPT2 case, this remains true for those orbital indices that refer to either inactive or secondary orbitals in the CASSCF: those that are always doubly occupied, or not occupied at all, in the root (or reference) function. However, all formulas that involve active orbitals, which are those with different occupancy in different terms in the reference function, will be quite complicated. This is true already for the simplest possible zeroth-order Hamiltonian: the Møller–Plesset-like ansatz, which uses a Fock matrix (suitably generalized). The special case of a quasidegenerate Fock matrix is simpler still but is useful only for very simple references of the ROHF type, where all active orbitals are singly occupied and symmetry equivalent.

The $\hat{H}_0$ operator must be simple, but it must also be a reasonable approximation to $\hat{H}$ for wavefunctions in the interacting space. If $\hat{H}$ had, in fact, *been* expressible as a one-electron operator $\hat{f} = \Sigma f_{pq}\hat{a}_p^\dagger\hat{a}_q$, with matrix elements unknown to us, the matrix elements could be obtained by the formula $f_{pq} = -\langle\Phi|[[\hat{H}, \hat{a}_q^\dagger], \hat{a}_p]_+|\Phi\rangle$, for any arbitrary normalized wavefunction. If it is not an exact one-electron operator, this formula can be used for a specific Φ—in our case we will choose the root function Ψ_0—or an average of several Φ functions or an ensemble of wavefunctions, to obtain an approximation. In particular, if Ψ_0 is a component of a set of symmetry-equivalent degenerate states, and if we require $\hat{H}_0$ to preserve this symmetry, we must use the average over the members of this degenerate set.

In our previous work [6, 7], the operator defined above was evaluated for the average over the spin of Ψ_0, since we require that $\hat{H}_0$ does not break spin symmetry. The result is

$$f_{pq} = h_{pq} + \sum_{rs} D_{rs}[(pq|rs) - \tfrac{1}{2}(pr|sq)] , \qquad (5)$$

where spatial orbitals and the spin-summed density matrix $\mathbf{D}$ are used, as a consequence of the spin summations. $\mathbf{h}$ is the core Hamiltonian, and for a closed-shell wavefunction, this Fock matrix is identical to the conventional one used in Hartree–Fock theory. The extension of the Fock matrix to general correlated wavefunctions has been used before in other contexts. It interpolates between two matrices: the eigenvalues (with opposite sign) of one are ionization energies, and the eigenvalues of the other are electron affinities, both in the sense of the extended Koopmans theorem. We have for some time used the term *standard Fock matrix* for this matrix and suggest this as a suitable and highly needed name, since a host of other named matrices have been suggested for various purposes. We thus regard CASPT2 as an extension of the Møller–Plesset method, which allows CASSCF reference functions.

Multiconfigurational perturbation theory as outlined above has been applied successfully to a large number of chemical problems (see Refs. 12 and 13 and references therein). The zeroth-order Hamiltonian has thus turned out to be a good choice for a broad range of problems. However, in a systematic test of geometries and binding energies of 32 molecules containing first-row atoms [15], it was noticed that wavefunctions dominated by an open-shell configuration were favored over wavefunctions dominated by a closed-shell configuration. This led to dissociation energies underestimated with between 3 and 6 kcal/mol times the number of extra electron pairs formed. It would be desirable to give a more balanced treatment of these two kinds of wavefunctions for a more accurate determination of dissociation energies and excitation energies.

One deficiency can be seen in the case of a high-spin open-shell wavefunction. If the condition that $\mathbf{f}$ should not be symmetry breaking is relaxed, by replacing $\mathbf{D}$ with the spin density matrix and using spin orbitals, we arrive at

$$f_{pq}^{\alpha} = h_{pq}^{\alpha} + \sum_{r=1}^{N^{\beta}} [(\phi_p^{\alpha}\phi_q^{\alpha}|\phi_r^{\alpha}\phi_r^{\alpha}) + (\phi_p^{\alpha}\phi_q^{\alpha}|\phi_r^{\beta}\phi_r^{\beta}) - (\phi_p^{\alpha}\phi_r^{\alpha}|\phi_r^{\alpha}\phi_q^{\alpha})]$$

$$+ \sum_{r=N^{\beta}+1}^{N^{\alpha}} [(\phi_p^{\alpha}\phi_q^{\alpha}|\phi_r^{\alpha}\phi_r^{\alpha}) - (\phi_p^{\alpha}\phi_r^{\alpha}|\phi_r^{\alpha}\phi_q^{\alpha})],$$

(6)

$$f_{pq}^{\beta} = h_{pq}^{\beta} + \sum_{r=1}^{N^{\beta}} [(\phi_p^{\beta}\phi_q^{\beta}|\phi_r^{\beta}\phi_r^{\beta}) + (\phi_p^{\beta}\phi_q^{\beta}|\phi_r^{\alpha}\phi_r^{\alpha}) - (\phi_p^{\beta}\phi_r^{\beta}|\phi_r^{\beta}\phi_q^{\beta})]$$

$$+ \sum_{r=N^{\beta}+1}^{N^{\alpha}} (\phi_p^{\beta}\phi_q^{\beta}|\phi_r^{\alpha}\phi_r^{\alpha}),$$

(7)

where $\mathbf{h}^\alpha$ and $\mathbf{h}^\beta$ are the matrix representations of $\hat{h}$ in the spin α and spin β orbital spaces, the N^α electrons of α spin are described by the spatial orbitals ϕ_p^α, and the N^β electrons of β spin are described by the spatial orbitals ϕ_p^β. Just as the spin-summed standard Fock matrix coincides with the conventional one in the closed-shell case, the matrix displayed above coincides with the conventional spin-orbital Fock matrix when evaluated for a high-spin ROHF wavefunction. This Fock matrix is different for α and β spin, and since a systematic component in the CASPT2 error is associated with open shells, it could be hypothesized that this difference could be important to preserve. However, it turns out that a spin-dependent Fock matrix cannot be used with properly spin-coupled wavefunctions.

Consider, for instance, the dissociation of N_2 into two ground-state (e.g., ^4S) N atoms. The local effects of the exact Hamiltonian on one of the N atoms (its spectrum, for instance) is the same also when it is considered as one part of a molecule with large enough interatomic distance. This locality property must be shared by the zero-order Hamiltonian if the perturbation theory is to be size consistent. But the local state of the N atom, during dissociation of the molecule from the ground ($^1\Sigma_g^+$) state, is not described by a wavefunction: It is a mixed state, containing four different spin projections. This mixed state is what any size-consistent Hamiltonian will see, locally. It must treat each spin component in an equal way. The naïve use of a spin-dependent Fock matrix would actually have almost no effect on the ground-state dissociation of the N_2 molecule. It would have the intended effect on the high-spin $^7\Sigma$ case, which could be used at the dissociation limit since it has the same energy, but that is not a possible way out for intermediate interatomic distances.

However, there is another important difference between the α and β electrons in the high-spin case: The *open* α orbitals are populated, the *open* β are not; thus the open α orbitals should use orbital energies appropriate for ionization, while the β ones should be appropriate for electron affinity. To use distinct orbital energies for active orbitals, depending on whether an electron is added or removed, is a principle that would work for arbitrary spin-coupled cases. It is not possible for a strictly one-electron operator. Several ways to approach this problem have been suggested:

1. Two-electron operators can be added to $\hat{H}_0$. A small number of important terms were suggested by Murray and Davidson for open-shell perturbation theory [16] and later by Kozlowski and Davidson for more general multiconfiguration perturbation theory [17]. Such formulas, which

we will call *OPT2-like*, can all be written as $\hat{H}_0 = \Sigma \; \varepsilon_p \hat{n}_p + \eta_p \hat{n}_p (2 - \hat{n}_p)$ with somewhat different definitions of the parameters ε_p and η_p ($\hat{n}_p$ is the number operator for orbital p). The OPT2-like schemes lack orbital invariance. This is problematic in low-spin ROHF cases, since it is not invariant even to rotations among symmetry-equivalent orbitals. It is also problematic for more general MCSCF root functions, since near-degeneracy gives the orbital set an unlimited sensitivity to small changes in geometry or to conditions at distant parts of a molecule, and the variations in orbitals representation affect the second-order energy.

2. Because of the orbital dependence of the OPT2 scheme, Kozlowski and Davidson have also suggested a scheme called IOPT [18], where the distinction between added and removed active orbitals is achieved by a term involving the *sum* of active orbital occupation. However, this scheme has very large size consistency errors [9].

3. A more comprehensive set of two-electron operators can be included. Inclusion of terms involving four active orbitals gives a Hamiltonian that is exact within the CAS-CI space. As a side effect, it has Ψ_0 as eigenfunction, without projections. It turns out that while the preliminary stages of the CASPT2 method become much more complicated by the extra terms, the final iterative equation solution is simplified. This method is thus quite reasonable for small active spaces and is presently investigated by Dyall [19]. However, it is not invariant to orbital rotation between, for example, an inactive and a fully occupied active orbital.

4. We said earlier that simplicity demands a one-electron $\hat{H}_0$. This is not quite true, of course: Corrections of a very general nature can be defined operationally, by manipulating the matrix elements or the energy denominators, rather than by first defining operators and then computing their matrix elements. This is straightforward in a determinant or CF formulation, but using the FOI space, it is no longer trivial to ensure that the operator defined is Hermitian, or that a "small" correction gives small effects on the correlation energy. We have made a large number of experiments with this kind of correction but have failed to find one that combines simplicity, orbital invariance, and size consistence while taking due account to the strong nonorthogonality and overcompleteness of the set of excited wavefunction terms.

5. The average implied in the standard Fock matrix is not necessarily the best compromise between ionization energy/electron affinity of active orbitals. In a number of applications, the contribution to the second-order correlation energy is much larger for a few excitations from open shells to low-lying virtual orbitals than the contributions from exciting

into the open shells. A better compromise is then to keep closer to the ionization-energy-type Fock matrix, which in our high-spin example is the f^α. For ROHF perturbation theory, this has been recognized for a long time. Modified Fock matrices suitable for CASPT2 have been studied by Andersson [20].

In CASSCF theory, with changes in conformation or perturbations (including also the perturbation of one group in a molecule due to distant groups), there may be large changes in orbitals due to rotations between inactive and heavily occupied (at that conformation) orbitals, or between weakly occupied (at that conformation) correlating orbitals and virtual ones. Also, while rotations within the active space are irrelevant to the CASSCF wavefunction, attempts to make them more strictly defined via canonicalization schemes fail because these schemes do not in general yield orbitals that vary slowly enough with geometry or perturbation. Because of this, we feel that representation invariance is an issue of decisive importance. Dyall's scheme is attractive and has many important invariance properties, but it does depend on the subdivision into inactive/ active orbitals, also for fully occupied orbitals. We have tried the last of the alternatives listed above, since this can be made representation independent.

In Ref. 20 several choices of a Fock operator for a general CASSCF reference state were discussed. They were all constructed such that for our high-spin example, the f^α Fock matrix is obtained for the open orbitals. These modified Fock operators were constructed with the aid of a matrix $\mathbf{K}$, defined as

$$K_{pq} = \sum_{rs} (\mathbf{Dd})_{rs} (pr|sq) , \tag{8}$$

where $\mathbf{d} = 2 \cdot \mathbf{1} - \mathbf{D}$ is a hole density matrix. In the classical open-shell cases, with occupation numbers only $0, 1$, or 2, this matrix is called the *open-shell exchange matrix* and denoted $\mathbf{K}$ or $\mathbf{K}^\circ$. The more general extension defined above has many applications, and we suggest that the term *open-shell exchange matrix* can be used for it as well. In terms of this matrix, the three choices of *open-shell corrected* Fock matrices discussed in Ref. 20 are

$$\mathbf{g}_1 = -\tfrac{1}{4}(\mathbf{DKd} + \mathbf{dKD}) , \tag{9}$$

$$\mathbf{g}_2 = -\tfrac{1}{2}(\mathbf{Dd})^{1/2}\mathbf{K}(\mathbf{Dd})^{1/2} , \tag{10}$$

$$\mathbf{g}_3 = -\tfrac{1}{2}(\mathbf{Dd})\mathbf{K}(\mathbf{Dd}) . \tag{11}$$

The major effect of introducing a correction to the Fock matrix is an enlarged energy gap between the active and secondary orbitals. The enlargement is small for CASSCF wavefunctions dominated by a closed-shell configuration and larger for other kinds of wavefunctions. Since the correction does not affect the inactive–inactive and secondary–secondary subblocks of the Fock matrix, the energy gap between the inactive and active orbitals will consequently be reduced. The enlarged energy gap between the active and secondary orbitals will lead to larger energy denominators for the most relevant contributions to the second-order energy. The result is smaller absolute values of the second-order energy for reference functions with many open shells. However, if inactive orbitals high in energy are present, the decreased energy gap between them and the active orbitals may lead to contributions to the second-order energy with small energy denominators.

It has often been assumed that the MCSCF Fock matrix can be used as $\hat{H}_0$ for multiconfiguration root functions. This works for a few simple open-shell cases but is in general wrong. The MCSCF Fock matrix, which is used to find energy-optimized orbitals in MCSCF (such as CASSCF), is

$$f_{pq}^{\mathrm{MCSCF}} = \sum_r h_{pr} D_{rq} + \sum_{rst} (pr|st) P_{rq,st} , \qquad (12)$$

where $P_{qr,st}$ is the two-electron spin-free density matrix. The optimization criterion is that this matrix is symmetric. First, there is an obvious occupation number dependence that must be eliminated: The orbital energy of a strongly occupied orbital becomes twice as large as that of the standard Fock matrix. The correlating orbitals, which should be very high in energy, are multiplied with low occupation numbers and have thus numerically small, and in fact, *negative* eigenvalues. The naïve interpretation of this phenomenon is that $\mathbf{f}^{\mathrm{MCSCF}}$ is a valid one-electron Hamiltonian matrix but in a nonorthogonal basis set with overlap matrix $\mathbf{D}$, so that a more reasonable eigenvalue equation would be $\mathbf{f}^{\mathrm{MCSCF}}\psi = \varepsilon \mathbf{D}\psi$. However, any realistic experiment shows that the orbital energies so obtained, which may be regarded as approximate ionization energies (with negative sign) in the sense of the extended Koopmans theorem, will behave as expected only for orbitals with high occupation number. For weakly occupied ones, the energy starts to *go down*. It never reaches positive values, and has in fact (if the EKT is valid) an upper limit that is the exact negative of the lowest ionization energy, in the basis-set limit.

For the weakest occupied orbitals, it makes sense to interchange particles and holes. The resulting analog to the MCSCF Fock matrix is then $2\mathbf{f} - \mathbf{f}^{\mathrm{MCSCF}}$, and the hole density matrix is, as before, $\mathbf{d} = 2 \cdot \mathbf{1} - \mathbf{D}$.

A physically meaningful eigenvalue equation is $(2\mathbf{f} - \mathbf{f}^{\mathrm{MCSCF}})\psi = \varepsilon(2 \cdot \mathbf{1} - \mathbf{D})\psi$, which gives Koopmans type of *electron affinities* (with reversed sign) as eigenvalues, but only for weakly occupied orbitals. For the strongly occupied orbitals, the energies go *up* to high values.

The counterintuitive behavior of the electron orbital energies for high occupation number, and of hole orbital energies for low occupation number, are simple examples of a fairly general phenomenon: the selection effect. When adding, removing, or exciting electron(s), the result is a new unnormalized wavefunction with another energy. The energy difference is not just an approximate sum of orbital energies, since the new wavefunction is not obtained by simply changing orbital occupations up and down. In addition, we must select that part of the correlated wavefunction that gives a nonzero result. As an example, consider $\Psi_0 = |\cdots(c_1\psi_t\bar{\psi}_t - c_2\psi_u\bar{\psi}_u)\rangle$, where a bonding orbital is correlated by double excitations to an antibonding $(c_1 \gg c_2)$. The excitation operator $\hat{E}_{ut}$ provides a wavefunction term $\alpha|\cdots(\bar{\psi}_t\psi_u - \psi_t\bar{\psi}_u)\rangle$ with an estimated excitation energy $\varepsilon_u - \varepsilon_t$, just as expected. But *so does* $\hat{E}_{tu}$. Instead of the expected energy difference, we get its negative. In fact, the expected result would imply that the excitation changes the occupation numbers as $n_t \leftarrow n_t + 1$, $n_u \leftarrow n_u - 1$, which is nonsense in this case since $n_t \approx 2$ and $n_u \approx 0$. This is never observed for single-determinant root functions. Of course, the complete picture is complicated by non-diagonal and two-electron terms, but we wish only to give a simple qualitative explanation to a counterintuitive selection effect in the use of multiconfiguration root functions. The same phenomenon also explains why $\mathbf{f}^{\mathrm{MCSCF}}$ fails to give proper orbital energies for weakly occupied orbitals, even if the occupation number dependence has been divided off.

Adding the two eigenvalue equations removes the **D** matrix dependence. Divided by 2, the result is the standard Fock matrix eigenvalue equation. This has sometimes been taken to mean that the *eigenvalues* of this are just the arithmetic mean of approximate negative ionization potentials (IPs) and electron affinities (EAs). However, it is obvious that in making this average, the first equation contains negative IPs weighted with the occupation number n, and the second, negative EAs weighted by $(2 - n)$. Thus when n goes toward either 0 or 2, the equation remains well behaved and the eigenvalues have the proper Koopmans interpretation.

B. Multistate CASPT2

As already mentioned, there is a so-called effective Hamiltonian approach to multiconfiguration perturbation theory. This approach is often applied with the complete active space as the model space. We have avoided this formulation and use a single-root perturbation theory, primarily because of the advantages in using the FOI space, and also to minimize complications involving intruders. However, the relative amplitude of the configurations in the CAS-CI space is then unalterable. A slight different CASSCF wavefunction would have been the optimal choice for the root function. In principle, the CASSCF wavefunction can be reoptimized, but this is not without complications and is probably not a convergent procedure for excited states. An alternative is then to use the effective Hamiltonian approach but with only a small space of CASSCF wavefunctions rather than a large space of configuration functions.

One of the major advantages of the CASSCF/CASPT2 approach is that excited electronic states of molecular systems can be obtained almost as easy as ground electronic states. By changing the nuclear coordinates, one can generate entire potential surfaces that contain information of chemical interest such as equilibrium structure, reaction paths, and so on. Such potential surfaces do not normally cross each other, but they do at points of high symmetry and also along the conical intersection paths, which are curves or hypercurves of dimension $N - 2$, where N is the total number of internal coordinates. When the energy surfaces cross or come near each other, this is often highly significant to spectroscopic properties and to reaction probabilities. In a region of conformations where a state crossing occurs, there is no bound to the derivative of the state function with respect to coordinates. Rapid changes occur in all properties and transition moments.

It is thus of importance that crossings and avoided crossings are treated as accurately a possible. Single-state perturbation methods are not generally applicable, due to strong divergence. They can still be useful if truncated at very low order in perturbation theory. CASPT2 will still give useful results in such situations, provided that the crossing states are included in the CAS-CI, so that they do not interact with the root function. However, the accuracy of the result is often impaired. Rapid changes in the root function occur where the potential surfaces cross at the CASSCF level. This is rarely at the same geometry as for the correlated wavefunction. The rather abrupt changes in second-order correlation energy can give rise to multiple minima, unphysical, repeated curve crossing, and other unwanted phenomena [21].

Apart from the qualitative errors that are obvious in near-crossing situations, there is also the question of precision. As an example, among the many molecular spectra that have been computed by CASPT2, there are a handful of energies with untypically large errors. These seem to be caused by erroneous mixing, at the CASSCF level, of wavefunction terms with different dynamic correlation.

The effective Hamiltonian formalism avoids these drawbacks by treating several zeroth-order functions at once. These functions span the model space, and they are the basis of a matrix Hamiltonian. The energies are obtained as eigenvalues of this matrix, so if the matrix elements are smooth functions of geometry, the drawbacks above can be avoided. In particular, different energy surfaces do not normally cross, but when they do, this is dealt with in a stable manner by the final diagonalization of the effective Hamiltonian matrix, not by the series expansion. Actually, the reference functions in CASPT2 are CASSCF wavefunctions, which are not smooth functions of geometry everywhere but are subject to large and rapid changes where the CASSCF energy surfaces cross. However, these changes amount to rotations within the model space, which can be shown not to affect the result.

Assume that we have decided on some particular zeroth-order Hamiltonian, as simple as possible, for which we require that

$$\hat{H}_0 \Psi_0^i = E_0^i \Psi_0^i , \qquad i = 1, \ldots, n , \qquad (13)$$

where n is the dimension of the model space. Here Ψ_0^i are CASSCF wavefunctions, and $\hat{H}_0$ is similar to the one used for single-state CASPT2. From the reference functions, we form new functions by adding a perturbation part. In the intermediate normalization, the perturbation added to a Ψ_0^i function is orthonormal to *all* the reference functions. This has the consequence that we cannot require that the new functions solve individual Schrödinger equations. Instead, we require that the interesting eigenfunctions can all be obtained as linear combinations of the perturbed functions. This indirect formulation allows the perturbation expansions to converge also in curve-crossing situations.

The general procedure is most easily described in terms of a few operators. Define the projectors $\hat{P} = \Sigma_1^n |\Psi_0^i\rangle\langle\Psi_0^i|$ and $\hat{Q} = \hat{1} - \hat{P}$, and use a *wave operator* $\hat{\Omega}$ to write the perturbed functions $\Psi^i = \hat{\Omega}\Psi_0^i$. Instead of the Schrödinger equation, we write

$$\hat{H}\hat{\Omega}\Psi_0^i = \hat{\Omega}\hat{H}^{\text{eff}}\Psi_0^i , \qquad (14)$$

where $\hat{H}^{\text{eff}}$ is an operator acting entirely *within* the model space (i.e., only its action on functions in the model space is relevant), and it maps any such function into the model space. For any Ψ that can be expressed as a linear combination of reference functions, it now follows from linearity that if $\hat{H}^{\text{eff}}\Psi = E\Psi$, then also $\hat{H}\hat{\Omega}\Psi = E\hat{\Omega}\Psi$. A set of n solutions to the complete Schrödinger equation can thus be obtained by solving the eigenvalue problem for the operator $\hat{H}^{\mathrm{eff}}$. A matrix representation is used in practice, but the operator formulation is practical for working out and concisely denote perturbation schemes and shows explicitly that the procedure is invariant to rotations among the CASSCF wavefunctions. The matrix $\mathbf{H}^{\mathrm{eff}}$ obeys

$$H_{ij}^{\mathrm{eff}} = \langle \Psi_0^i | \hat{H}^{\mathrm{eff}} \Psi_0^j \rangle ,$$

$$\hat{H}^{\mathrm{eff}} \Psi_0^j = \sum_i \Psi_0^i H_{ij}^{\mathrm{eff}} . \tag{15}$$

The wave operator and $\hat{H}^{\mathrm{eff}}$ are not completely specified. For our purposes it is simplest to set $\hat{\Omega}\hat{Q} = \hat{H}^{\mathrm{eff}}\hat{Q} = \hat{0}$. The perturbation equations are obtained by the ansatz $\hat{\Omega} = \hat{P} + \Sigma_1^\infty \lambda^k \hat{\Omega}_k$. A convenient relation is the *Bloch* equation

$$[\hat{\Omega}, \hat{H}_0]\hat{P} = \hat{H}_1 \hat{\Omega}\hat{P} - \hat{\Omega}\hat{P}\hat{H}_1\hat{\Omega}\hat{P} , \tag{16}$$

which gives the perturbation equations to all orders by substituting the expansion of $\hat{\Omega}$, replacing $\hat{H}_1$ by $\lambda\hat{H}_1$, and collecting terms of equal order. In computation this will give a matrix commutator equation at each order. It is most easily solved in a basis where $\hat{H}_0$ is diagonal, where it simplifies to division by energy denominators. Each denominator is the energy difference between one $\hat{H}_0$ eigenfunction in the model space, and one outside it. Any degeneracy in the model space is thus irrelevant.

Multiplying Eq. (14) from the left by $\hat{P}$ gives $\hat{P}\hat{H}\hat{\Omega}\hat{P} = \hat{H}^{\mathrm{eff}}$. Inserting the $\hat{\Omega}$ expansion gives

$$\hat{H}^{\mathrm{eff}} = \sum_{k=0}^\infty \lambda^k \hat{H}_k^{\mathrm{eff}} = \hat{P}\hat{H}_0\hat{P} + \lambda\hat{P}\hat{H}_1\hat{P} + \sum_{k=2}^\infty \lambda^k \hat{P}\hat{H}_1\hat{\Omega}_{k-1}\hat{P} . \tag{17}$$

For CASPT2, the equations to second order are

$$\langle \Psi_0^i | \hat{H}_0^{\mathrm{eff}} | \Psi_0^j \rangle = \langle \Psi_0^i | \hat{H}_0 | \Psi_0^j \rangle = E_0^i \delta_{ij} , \tag{18}$$

$$\langle \Psi_0^i | \hat{H}_1^{\mathrm{eff}} | \Psi_0^j \rangle = \langle \Psi_0^i | \hat{H}_1 | \Psi_0^j \rangle , \tag{19}$$

$$\langle \Psi_0^i | \hat{H}_2^{\mathrm{eff}} | \Psi_0^j \rangle = \langle \Psi_0^i | \hat{H}_1 \hat{\Omega}_1 | \Psi_0^j \rangle . \tag{20}$$

The zeroth-order eigenfunctions are identical to $\{\Psi_0^i\}_{i=1}^n$ and their energies are equal to E_0^i. At the first-order level the matrix $\langle \Psi_0^i | \hat{H}_0^{\mathrm{eff}} + \hat{H}_1^{\mathrm{eff}} | \Psi_0^j \rangle = \langle \Psi_0^i | \hat{H} | \Psi_0^j \rangle$ should be diagonalized. If Ψ_0^i are CASSCF wavefunctions with a common orbital set, this matrix is already diagonal and the first-order energies are identical to the CASSCF energies (E^i_{CAS}). Finally, at the second-order level the matrix $\langle \Psi^i_0|\hat{H}^{eff}_0 + \hat{H}^{eff}_1 + \hat{H}^{eff}_2|\Psi^j_0\rangle = E^i_{CAS}\delta_{ij} + \langle \Psi^i_0|\hat{H}_1\hat{\Omega}_1|\Psi^j_0\rangle$ should be diagonalized. For obtaining the second-order energy we have to determine $\hat{\Omega}_1$ from the lowest-order Bloch equation:

$$[\hat{\Omega}_1, \hat{H}_0]\hat{P} = \hat{H}_1\hat{P} - \hat{P}\hat{H}_1\hat{P} . \tag{21}$$

To make the solution of this equation as simple as possible, we define the zeroth-order Hamiltonian in a similar way as for the single-state case (see Refs. 6 and 7) using various projection operators. This will then allow us to express

$$\hat{\Omega}_1\Psi^i_0 = \sum_{I=1}^{M} C^i_I|I\rangle , \qquad |I\rangle \in V_{SD} , \tag{22}$$

where V_{SD} is the space spanned by single and double replacement states generated from all the n states Ψ^i_0 and M its dimension. Equations (21) and (22) together give the following system of equations:

$$\sum_{J=1}^{M} \langle I|E^i_0 - \hat{H}_0|J\rangle C^i_J = \langle I|\hat{H}_1|\Psi^i_0\rangle \equiv V^i_I , \qquad I = 1, \ldots , M , \tag{23}$$

from which the expansion coefficients C^i_I can be determined. Equation (23) is similar to the one for single-state CASPT2. The difference lies in the space V_{SD}. For a multistate CASPT2 this space will be approximately n times larger than the space obtained in single-state CASPT2. Using the definitions above, the matrix representation of the second-order effective Hamiltonian can be expressed as

$$\langle \Psi^i_0|\hat{H}^{eff}_0 + \hat{H}^{eff}_1 + \hat{H}^{eff}_2|\Psi^j_0\rangle = E^i_{CAS}\delta_{ij} + \sum_{I=1}^{M} C^j_I V^{i*}_I = E^i_{CAS}\delta_{ij} + \mathbf{V}^{i\dagger}\mathbf{C}^j .$$

$$\tag{24}$$

The final step in a multistate CASPT2 calculation is a diagonalization of the matrix in Eq. (24), which gives the energy of the states under consideration up to second order. We have used here a formulation where $\hat{H}^{eff}$ and $\mathbf{H}^{eff}$ are non-Hermitian. However, this is of no computational importance as long as $\mathbf{H}^{eff}$ is a small matrix. The increase in computational cost will be due primarily to the increased number of variables: the size of the system of linear equations in CASPT2 will be proportional to the number of states. Since in many cases two reference functions are sufficient, this does not dramatically increase the computational time or storage. We are presently coding a CASPT2 version with an effective Hamiltonian. It will be of special interest for studies of avoided crossings but also for cases where one can expect that a single optimized CASSCF reference function has the wrong shape due to lack of dynamic correlation (e.g., for negative ions).

C. Intruder-State Problem

The success of a low-order perturbation treatment is vitally dependent on the strength of the perturbation. Fast convergence is obtained only for weak perturbations. In addition, E_0—the eigenvalue of $\mathbf{H}_0$ corresponding to Ψ_0—should not be close to any other eigenvalue. If this happens for a function in the FOI space, the second-order energy will become divergent. Such intruder states seldom appear in the normal Møller–Plesset perturbation theory, which is used primarily for molecules in their ground state. The CASPT2 approach is, however, frequently used also for excited states. Here it is not uncommon to find intruder states. To realize why they appear, one has to understand the structure of the CI space in which the wavefunction is expanded. The initial partitioning of the occupied molecular orbitals into an inactive and an active part defines the CAS-CI space as all the CFs that are obtained by distributing the active electrons among the active orbitals in all possible ways consistent with a given space and spin symmetry. The CASSCF wavefunction is expanded in this CI space. The orthogonal complement does not interact with the CAS wavefunction and will consequently not appear in the first order wavefunction. The CFs in the FOI space will have at least one electron in the external (virtual) orbital space and/or at least one hole in the inactive space,

Assume now that one orbital, which is important for the description of the electronic spectrum of the system, is missing from the active space (it may be inactive or external). Excitations involving this orbital will then belong to the FOI space. It is not unlikely that the zeroth-order energy for one or several of the CFs where this orbital is excited (to or from) will have low energies and may act as intruder states in a CASPT2 calculation. Clearly, the remedy to the problem is in this case to include the orbital in the active space.

It is, however, not always possible to solve the problem so easily. The active space may already be at the limit of what the method can handle (normally, 12–14 orbitals) and further extension is then not possible. A typical case are calculations with extended basis sets, which include diffuse or semidiffuse functions. Such basis sets give rise to a number of

Rydberg-like states. Calculations, which aim also at describing excited states of Rydberg type, will have most of the diffuse functions in the active space. However, it often happens that some diffuse electronic states are left in the FOI space. The energy lowering due to dynamic correlation is usually smaller for diffuse states than for the more compact valence excited states. At the zeroth-order level, the diffuse states will therefore be artificially stabilized with respect to other excited states, and accidental degeneracy may occur, even if these states are much higher in energy when dynamic correlation corrections have been added. This situation is rather common in CASPT2 calculations on excited states. However, the interaction element coupling the intruder state with Ψ_0 is often very small. Thus the contribution to the second-order correlation energy may be without importance even if the energy denominator is small. In such cases it may be allowed to disregard the contribution of the intruder state. Another possibility is to delete the diffuse orbital that causes the problem. It has been shown in a number of cases that virtually identical results are obtained whether the orbital is deleted from the virtual or added to the active space [22].

Even if intruder states most commonly appear in calculations on excited states, there are a few cases where it also happens for ground-state energy surfaces. An especially pathological system is the chromium dimer. With an active space of 12 orbitals (Cr $3d$ and $4s$) and 12 active electrons, several intruder states appear in the region around the minimum on the ground-state potential curve (for details, see Refs. 14 and 23). They have their origin in orbitals of mainly $4p$ character. The best solution to the problem would have been to include these orbitals in the active space since they are rather close in energy to the $4s$ orbitals. However, such a procedure would have resulted in an active space consisting of 18 orbitals, which is far beyond the capability of the present computer implementation. It was shown in a recent study of Cr_2 that a useful potential curve could be constructed by excluding the rather narrow regions where the intruder states appeared [23], but this is clearly an unsatisfactory solution to the problem. There are two main reasons why intruder states show up in the potential curve for Cr_2. One is the low energy of the $4p$ orbitals. The second is the strongly degenerate character of the CASSCF wavefunction with six weak chemical bonds. The occupation numbers of several of the active orbitals are as a consequence close to 1. As discussed in Section II.B, such a situation leads to a diminished energy gap between the active and virtual orbitals and thus increases the possibility for intruder states. For Cr_2 it was possible to reduce the problem by introducing a modification in $\hat{H}_0$, which stabilized the active orbitals [23, 24]. As discussed earlier, the modification introduced by Andersson [25] cannot be adopted generally. It stabilizes the active orbitals, which makes the energy difference to the inactive orbitals smaller. Thus the method will work only when all valence electrons are active (as in Cr_2). It will not work in cases where one can expect that the energy difference between active and inactive orbitals is small. In such cases the modification will increase the probability for intruder states.

A very simple way to avoid near-degeneracies in the zeroth-order Hamiltonian is to introduce a level shift [26]. $\hat{H}_0$ is then replaced with $\hat{H}_0 + \varepsilon \hat{P}_I$, where ε is a small positive number and $\hat{P}_I$ is the projection operator for the FOI space. The first-order equation with the shifted Hamiltonian is

$$(\mathbf{H}_0 - (E_0 - \varepsilon)\mathbf{S})\tilde{\mathbf{C}} = -\mathbf{V} , \tag{25}$$

where $\tilde{\mathbf{C}}$ is the new first-order CI vector (the expansion vector for Ψ_1 in the FOI space). It is easy to derive a formula for the level-shifted second-order energy in terms of $\tilde{\mathbf{C}}$ and $\mathbf{C}$,

$$\tilde{E}_2 = E_2 + \varepsilon \mathbf{C}^\dagger \mathbf{S} \tilde{\mathbf{C}} . \tag{26}$$

$\mathbf{C}$ can be written as $\tilde{\mathbf{C}} + o(\varepsilon)$ provided that ε is small compared to the eigenvalues of $\mathbf{H}_0$, which will be the case if there are no intruder states. If the intruder states have been removed by the level shift, it is possible to make an approximate *a posteriori* correction of the second-order energy to an assumed unshifted calculation without the intruder state. The correction is obtained from Eq. (26) by replacing $\mathbf{C}$ with $\tilde{\mathbf{C}}$,

$$E_2 \approx \tilde{E}_2 - \varepsilon \tilde{\mathbf{C}}^\dagger \mathbf{S} \tilde{\mathbf{C}} , \tag{27}$$

which can be written as

$$E_2 \approx \tilde{E}_2 - \varepsilon \left(\frac{1}{\tilde{\omega}} - 1 \right) , \tag{28}$$

where $\tilde{\omega}$ is the weight of the CASSCF reference function in a normalized zeroth plus first-order function. The correction is given by $E_2 - \tilde{E}_2$, which we shall call the LS (level-shift) correction.

The procedure above can work only if the effect of the level shift is negligible in the stable case when no intruder state appears. Otherwise, we have introduced a new parameter into the model over which we have no control. The behavior of the level-shifted CASPT2 approach has recently been tested in a series of calculations on the N_2 and Cr_2 molecules [26]. The first test was performed on the potential curve for the

$X^1\Sigma_g^+$ ground state of N_2. An ANO-type basis set of size $5s4p3d2f$ was used [27]. The active space consisted of the eight orbitals generated from the nitrogen valence orbitals with 10 active electrons. Calculations were performed at 19 points spread around the minimum of the potential curve, and the spectroscopic constants were obtained by numerical solution of the vibrational Schrödinger equation. The results obtained with different level shifts ε are illustrated in Table I. This table contains two entries, one where the shift was applied [Eq. (25)] without LS correction and one where this correction was added to $\tilde{E}_2$. In the first case we see an almost linear increase of the second-order energy with the level shift (the slope is about 0.05). The computed spectroscopic constants are, however, affected surprisingly little. Between a shift of zero and 0.2 hartree, the bond distance decreases 0.0009 Å, the stretching frequency increases $8\,\mathrm{cm}^{-1}$, and the bond energy increases with 0.106 eV. The LS corrected second-order energy removes almost all these changes. The total second-order energy only changes 0.6 millihartree, the bond energy 0.016 eV, the frequency $1\,\mathrm{cm}^{-1}$, and the bond distance 0.0001 Å. Considering the general accuracy of a second-order perturbation treatment,

TABLE I

Effect of the Level Shift in the Zeroth-Order Hamiltonian on the Spectroscopic Properties of the Ground-State Potential of N_2

Shift (H)	r_e (Å)	ω_e (cm^{-1})	D_e (eV)	Δ_e[a]
		Without the LS Correction		
0.00	1.1036	2324	9.21	—
0.05	1.1033	2326	9.24	2.7
0.10	1.1031	2328	9.27	5.3
0.20	1.1027	2332	9.32	10.3
0.50	1.1017	2342	9.42	23.7
		With the LS Correction		
0.00	1.1036	2324	9.21	—
0.05	1.1036	2324	9.21	0.041
0.10	1.1035	2324	9.22	0.158
0.20	1.1035	2325	9.23	0.589
0.50	1.1031	2329	9.28	3.029
		Experimental[b]		
—	1.0977	2359	9.91	—

[a] Shift in millihartrees with respect to the CASPT2 energy without a level shift.
[b] From Ref. 212.

these effects of the level shift are clearly negligible. Even a level shift as large as 0.5 hartree gives rise to very small shifts in computed properties. The conclusion is that a small level shift will not affect computed properties if the second-order energy is LS corrected and there are no intruder states. The computed bond energy for N_2 is 0.7 eV smaller than the experimental value, which corresponds to 0.23 eV per electron pair. This is an illustration of the systematic error in the CASPT2 approach discussed earlier: The bond energy is underestimated with 4–6 kcal/mol per electron pair. Notice that application of the level shift (without LS correction) decreases this error. That is because the shift will have a larger effect when the eigenvalues of $\mathbf{H}_0$ are smaller, as is the case for the nitrogen atom compared to the nitrogen molecule.

To study what happens if there is an intruder state, we performed level-shifted calculations on an excited $(A^3\Sigma_u^+)$ of N_2, for which an intruder state crosses the potential curve close to the minimum [7]. The details of the calculations were the same as for the ground state. The result is shown in Table II and is also illustrated in Fig. 1, which shows the potential curve for three values of the level shift: 0.00, 0.05 and 0.10

TABLE II
Effect of the Level Shift in the Zeroth-Order Hamiltonian on the Spectroscopic Properties of the $(A^3\Sigma_u^+)$ Excited State of N_2

Shift (H)	r_e (Å)	ω_e (cm^{-1})	D_e (eV)	$\Delta_e{}^a$
		Without the LS Correction		
0.00	1.3095	1340	3.36	—
0.05	1.2917	1427	3.38	3.0
0.10	1.2931	1433	3.38	6.9
		With the LS Correction		
0.00	1.3095	1340	3.36	—
0.05	1.2874	1378	3.40	−1.3
0.10	1.2931	1427	3.38	−0.35
		$2\pi_g$ *Orbitals Added to the Active Space*^{b}		
0.00	1.2953	1409	3.30	—
		Experimental^c		
—	1.2866	1461	3.68	—

^{a} Shift in millihartrees with respect to the CASPT2 energy without a level shift.
^{b} Data from Ref. 7.
^{c} From Ref. 212.

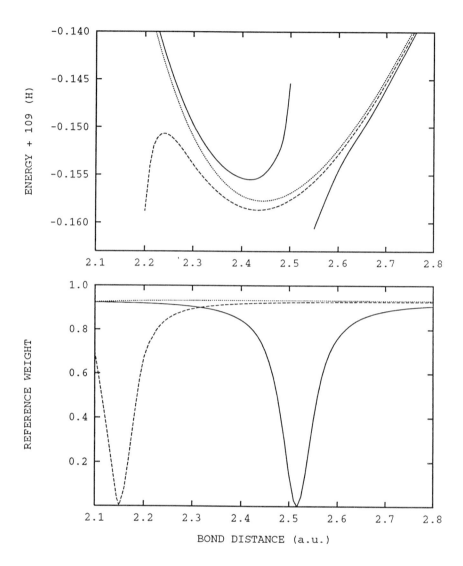

Figure 1. Potential curves for the $(A^3\Sigma_u^+)$ excited state of N_2 for three values of the level shift: 0.00 (solid line), 0.05 (dashed line) and 0.10 a.u. (dotted line). The LS correction has been applied. The lower diagram gives the corresponding weight of the CASSCF reference function.

hartree and the weight, ω, of the corresponding CASSCF reference function. The appearance of the intruder state is clearly seen for the unshifted potential as a singularity at a bond distance of about 1.36 Å. The weight ω drops to zero at this point. The spectroscopic constants presented in Table II have been obtained by fitting the potential only to the points for which the reference weight is acceptable. It is clear from the figure that such a procedure is rather meaningless when intruder states appear in the potential close to the minimum. It must be regarded as fortuitous that the results presented in Table II do not have larger errors.

A level shift of 0.05 hartree (with LS correction) moves the intruder state to shorter internuclear distances. It will, however, still affect the region around the minimum. A level shift of 0.10 hartree will remove the intruder entirely and the potential curve is now normal in the minimum region with constant weight of the reference function, which indicates a balanced description of the dynamic correlation effects. Away from the intruder-state region, the effect of the level shift on the total energy is less than 1 millihartree. The spectroscopic constants obtained with the level-shifted potential are of the same quality as those obtained for the ground state. We conclude that intruder states can be removed by applying a level shift together with the LS correction. The result of such a calculation will be similar to those obtained by incorporating the intruder state into the active space, which is the ideal (but often impossible) solution to the problem. It must, however, be emphasized that the level-shifted CASPT2 method will work only in cases were the interaction between the intruder state and the reference function is weak. Strongly interacting states should be included into the CAS space by increasing the number of active orbitals. In the N_2 case it was possible to remove the intruder state by adding the two $2\pi_g$ orbitals to the active space [7]. The resulting spectroscopic parameters are very similar to those obtained by the level-shift method (cf. Table II).

A severe test of the level-shifted CASPT2 procedure is Cr_2. Previous studies of this molecule have shown that a number of intruder states appear along the ground-state potential curve [14]. Modifying the zeroth-order Hamiltonian only partially solved the problem, and it was only by avoiding the intruder state regions that approximate spectroscopic constants could be computed for Cr_2 [23]. Not surprisingly, intruder states also appeared frequently for the excited-state potential curves [24]. The chromium dimer has been studied here using the same basis set as was used in Ref. 23: ANO 8s7p6d4f. The active space comprised the chromium $3d$ and $4s$ orbitals and corresponding electrons. The $3s$ and $3p$ electrons were included in the CASPT2 treatment, and relativistic corrections were added as in the earlier work. About 25 points on the potential curve were computed with a spacing of 0.05 a.u. around the minimum. Calculations were performed for three values of the level shift: 0.05, 0.10, and 0.20 a.u.

The resulting potential curves are shown in Fig. 2 and the computed spectroscopic constants are given in Table III. Corresponding results without level shift have been presented in the earlier work (cf. Fig. 5 and Table 2 in Ref. 23). As can be seen in the figure, there are at least three intruder states in the region around the minimum. A level shift of 0.05 a.u. does not remove them. A level shift of 0.10 a.u. gives a potential curve without intruder states, but the dip in the CASSCF weight in the minimum region shows that they still influence the potential. A constant weight is obtained with a level shift of 0.20 a.u. As in previous work, the spectroscopic constants for the case with intruders present have been obtained by using only those parts of the potential curve where the CASSCF weight is large. The bond distance computed in this way is reasonable, as is the bond energy. The harmonic frequency is, however, $100 \, \text{cm}^{-1}$ too large. Increasing the level shift from 0.05 to 0.10 brings the computed frequency to the experimental value (cf. Table III). The bond distance has, however, increased. The weight is still low in the region around minimum, which makes the LS correction large. Using 0.20 a.u. instead makes the calculation stable. The computed bond energy and the frequency are now close to experiment. The excellent agreement with experiment may, however, be somewhat fortuitous, as indicated by preliminary results obtained with a large basis set containing g-type functions. We refer to the original article for details [26].

The examples above indicate that a level-shifted CASPT2 method with LS correction is a promising approach to deal with the intruder state problem. The results obtained for the challenging system Cr_2 are surprisingly accurate. They show once again that the CASPT2 approach is capable of giving accurate structural and energetic data for complex electronic states once the intruder state problem has been removed.

III. Applications in Spectroscopy

The CASPT2 approach has been used in a large variety of applications ranging from molecular structure determinations in ground and excited states [15] to calculations of binding energies of transition metal compounds [28, 29]. There is, however, little doubt that the most spectacular success of the new approach has been in electronic spectroscopy and photochemistry. Up until now, a lot of effort has been put into the refinement of computational methods for molecular ground states and it

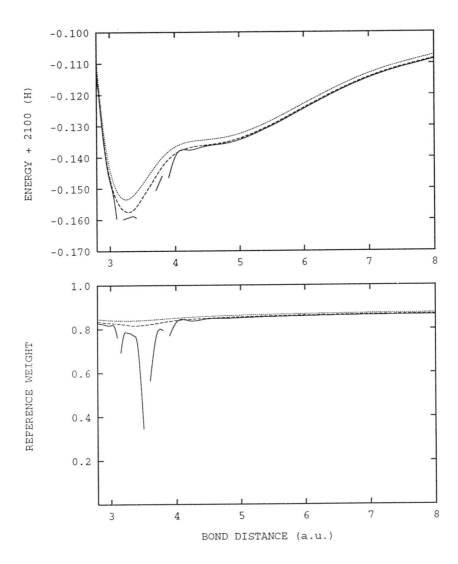

Figure 2. Potential curves for the ground state of Cr_2 for three values of the level shift: 0.05 (solid line), 0.10 (dashed line) and 0.20 a.u. (dotted line). The LS correction has been applied. The lower diagram gives the corresponding weight of the CASSCF reference function.

TABLE III
Effect of the Level Shift in the Zeroth-Order Hamiltonian on the Spectroscopic Properties of the Ground State of Cr_2 (Including Relativistic Corrections)

Shift (H)	r_e (Å)	ω_e (cm^{-1})	$\Delta G_{1/2}$ (cm^{-1})	D_0 (eV)	$\Delta_e{}^a$
		Without the LS Correction			
0.05	1.713	496	478	1.38	—
0.10	1.717	470	457	1.31	10.2
0.20	1.712	459	448	1.16	30.6
		With the LS Correction			
0.05	1.707	597	531	1.56	—
0.10	1.736	488	475	1.53	2.5
0.20	1.714	482	467	1.42	6.6
		Experimental			
—	1.679[b]	481[c]	452	1.44[d]	—

[a] Shift in millihartrees with respect to the CASPT2 energy with a level shift of 0.05 a.u.
[b] From Ref. 213.
[c] From Ref. 214.
[d] From Ref. 215.

is today possible to obtain accurate results for many molecular properties. Most of the approaches are based on the Hartree–Fock approximation as the reference for the estimation of correlation contributions. In contrast, much less effort has been put into the development of efficient computational tools for the excited states. The CI singles approach (singly excited configurations with respect to a HF reference function) has been advocated as a simple and systematic tool for larger molecules [30]. The method is, however, too approximate and limited to be of any practical value. Attempts to include correlation effects by means of perturbation theory have not been successful. The only general tool is multireference CI, but this approach has severe limitations, as pointed out above. Recent attempts to extend the coupled cluster methods to excited states are promising, but so far they are limited to states dominated by single excitations with respect to a HF reference [31].

Early applications of the CASPT2 method to the electronic spectrum of the nickel atom [32] and the benzene molecule [33] were promising and showed that the method gave the same accuracy for excited states as for ground-state properties. This was expected, since the generality of the CASSCF reference function makes the second-order approximation equally valid for all states. The positive surprise was the high accuracy.

Excitation energies were computed with errors less than 0.2 eV in both cases. This was a higher accuracy than any of the earlier calculations on Ni and benzene had achieved. The CASPT2 method has during the last three years been applied to a large number of electronic spectra. Some of the earlier applications in organic chemistry have already been reviewed [13]. Here we discuss additional applications in organic systems but also the electronic spectroscopy of transition metal compounds, where the demands on the approach are different.

A. Electronic Spectroscopy for Organic Molecules

As mentioned above, one of the first applications of the CASPT2 method was the electronic spectrum of the benzene molecule. This study was successful and gave results of higher quality than had been obtained using large-scale MRCI techniques [34]. It was quite clear that the reason for the larger errors obtained in the earlier studies were the difficulties these *ab initio* methods had in treating the dynamic correlation effects. Benzene and molecules of similar size contain many valence electrons, all of which must be included in the correlation treatment. Extensive basis sets are necessary to account for the correlation effects. In a conventional configuration interaction (CI) approach the size of the CI expansion consequently becomes large. Of special importance are the $\sigma-\pi$ correlation effects, which manifest themselves as a dynamic polarization of the σ framework. These correlation effects are state dependent and omitting them leads to erratic predictions of relative energies.

A recent multireference (MR) CI study of the electronic spectrum of pyrimidine [5] showed that large CI expansions were needed to converge the results for the excitation energies. Convergence could actually not be achieved without extending the length of the CI expansion beyond the limits set by the MRCI program [about 3 million configuration functions (CFs)]. The remaining error in the computed excitation energies was in some cases 0.4 eV. Although this error might be acceptable, the test showed that the MRCI method is not a practical tool for studies of electronic spectra of larger organic molecules.

Why is the more approximate CASPT2 method more successful in electron spectroscopy than the more elaborate MRCI method? One answer is that the correlation treatment is based on a full CASSCF reference instead of a (often small) selected set of configurations. The result is a more balanced treatment of different excited states. Another difference is that the simpler CASPT2 method can be used with larger basis sets, which gives a wider area of application and increased accuracy. It is also a size-extensive approach, which is necessary in application to larger systems where many electrons have to be correlated (there are small size-extensivity violating terms in $\hat{H}_0$, but they are of no practical importance). The method has been used to estimate correlation contributions to excitation energies in as large molecules as porphin with 114 correlated electrons [35]. The general experience is that the CASPT2 method, when applied to electronic spectroscopy of organic molecules, yields excitation energies with an accuracy of 0.3 eV or better. A few exceptions to this rule exist [36]. They are usually due to an inappropriate description of the CASSCF reference function. An example is valence–Rydberg mixing, which may be incorrectly described by the CASSCF wavefunction, due to errors in relative energies at this level of approximation.

The CASPT2 applications in organic spectroscopy have recently been reviewed [13]. The applications discussed included linear polyenes (from ethene to octatetraene), five-membered rings and norbornadiene (systems with two interacting double bonds), benzene and benzene derivatives, the porphin molecule, charge transfer in aminobenzonitriles, and the nucleic acid base monomers cytosine, thymine, uracil, and guanine. Here we review continued applications, including dimeric systems such as biphenyl and bithiophene, one more example of interacting double bonds: the methylenecyclopropene system, formaldehyde and acetone, imidazole and indole, and a series of aromatic cations and anions (Fig. 3). However, before describing these applications, we discuss some of the general features and problems that have to be considered. They include the choice of basis set and active orbitals, problems with intruder state, valence–Rydberg mixing, and so on. One section deals with the solvent effects on the electronic spectra.

1. Basis Sets

The basis set, inevitably, represents a compromise between the accuracy of a calculation and the computational effort. Recent improvements in calculating two-electron repulsion integrals [37, 38] combined with the atomic natural orbital (ANO) basis sets introduced by Almlöf and Taylor [39] now allow the use of large primitive sets routinely. Two such libraries of basis sets have been constructed by Widmark et al. [27, 40] and Pierloot et al. [41] and are derived from van Duijneveldt's [42] primitive basis sets. The larger set, for convenience called ANO-L, includes $(8s4p3d)$ primitives for hydrogen and $(14s9p4d3f)$ for the atoms Li–Ne. The smaller primitive set, hereafter referred to as ANO-S, has the following size: $H(7s3p)$ and $Li–Ne(10s6p3d)$. Unless stated otherwise, the ANO-L basis sets are applied in the present calculations.

ANO-type basis sets are based on a general contraction scheme and thus offer a route to high-quality one-particle basis sets without making

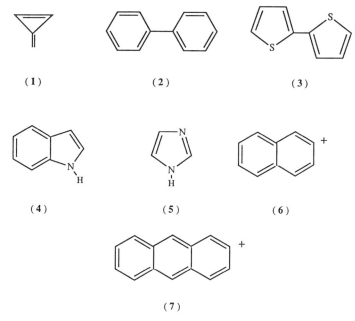

Figure 3. Structure formulas for methylenecyclopropene (**1**), biphenyl (**2**), bithiophene (**3**), indole (**4**), imidazole (**5**), naphthalene$^+$ (**6**), and anthracene$^+$ (**7**).

the wavefunction calculation unfeasible. Moreover, they can be improved systematically by progressively increasing the number of contracted functions. In general, if the basis sets are contracted to TZP (triple-zeta plus polarization) quality in the valence space (e.g., for first-row atoms the contracted size would be $4s3p1d$), the resulting basis sets are optimal to treat relative correlation and polarization effects in the valence space, as has been shown in a number of applications.

It is, however, also clear that quantitative results can be obtained only when the basis sets are large enough to treat the different types of excited states on a comparable level of accuracy. This means that the basis set must be able to describe equally well valence and Rydberg states, and flexible enough to cover states of an ionic nature at the same time as less diffuse covalent states.

A test of the basis-set effects on the excitation energy and transition moments of vertically excited singlet states in pyrazine has been published [43]. Table IV shows some recent data on the basis-set dependency of the excitation energy for the lowest vertically excited singlet states of neutral butadiene and its positive and negative ions. In accord with the previous results, we conclude that basis sets of the contracted size ($4s3p1d/2s$)

TABLE IV
Basis-Set Effects on the Excitation Energies (eV) of Butadiene and Its Ions[a]

State	ANO-S $(3s2p1d/2s)$	ANO-S $(4s3p1d/2s)$	ANO-S $(4s3p2d/2s)$	ANO-S $(4s3p2d/2s1p)$	ANO-L $(4s3p2d/2s1p)$	ANO-L $(4s3p2d1f/2s1p)$
			Neutral Butadiene			
1^1B_u	6.07	6.06	6.00	5.98	6.02	6.07
1^1A_g	6.34	6.16	6.11	6.09	6.10	6.07
			Butadiene Radical Cation			
1^2A_u	2.42	2.41	2.41	2.40	2.40	2.39
2^2A_u	4.18	4.14	4.14	4.12	4.11	4.09
			Butadiene Radical Anion			
1^2B_g	2.50	2.23	2.20	2.21	2.24	2.24
2^2B_g	3.26	2.87	2.84	2.85	2.72	2.67

[a] The geometry of butadiene in the different electronic states has been computed at the ROHF/MP2 level of approximation using 6-31G* basis sets. Excitation energies have been computed with the CASSCF/CASPT2 method.

appear to be an optimal compromise between the quality and size of a calculation. The novel aspects are that it hardly makes any difference whether ANO-L or ANO-S sets are used. Moreover, for the butadiene cation the valence description can even be reduced to DZP (double-zeta plus polarization) quality.

If one is about to study the electronic spectrum of a neutral molecule in the ultraviolet region, say up to 8 eV, a large number of states will be included of which some may be Rydberg states. As the excited electron approaches the ionization limit it will move in very diffuse orbitals, resembling the hydrogenic functions and with radial maximum several times larger than the size of the molecule. Obviously, standard basis sets need to be supplemented with an adequate set of Rydberg ANO basis functions. In numerous applications it has been found that it is advantageous to add such basis sets routinely, even if one is not interested in the Rydberg states. Recently, we demonstrated [44] that the neglect of Rydberg states may lead to a significant artificial mixing of Rydberg and valence states.

The present method to obtain the Rydberg basis functions is based on the universal Gaussian basis sets devised by Kaufmann et al. for representing Rydberg and continuum wave functions [45], and the contraction coefficients are obtained in the following way: A CASSCF wavefunction for the cation lowest in energy is determined with the uncontracted, extra basis set placed at the center of charge. From the resulting set of MOs, the lowest virtual orbitals of each angular momentum component are selected and used to construct an averaged density matrix, which is then diagonalized. The resulting natural orbitals are used as basis functions for the Rydberg states.

The uncontracted Rydberg basis sets typically include eight Gaussian functions per angular function. This large number of primitives is needed to describe properly the region of space where the excited electron penetrates the charge distribution of the remaining molecular core.

Finally, in calculations of electronic spectra of negatively charged systems, the extra electron may orbit in a large number of diffuse states before dissociation. However, the potential felt by the electron at large distances will not be a central field. Consequently, these diffuse orbitals cannot be approximated by hydrogen atom–like functions. Therefore, the most promising way to obtain accurate results for negatively charged systems is, apparently, to augment the atomic basis sets and to test for saturation of one-particle space.

2. Problems and Limitations

CASPT2 is not a black-box technique. It is demanding on the user, who has to know enough chemistry and quantum chemistry to be able to choose an appropriate active space. It may sometimes be necessary to experiment with different active spaces to check that the results are converged with respect to this parameter. The output of the calculations has to be inspected carefully for intruder states. When intruder states appear, appropriate steps must be taken to remove them. There are limitations to what applications are possible and the user must be aware of that and must be able to judge the quality of the calculation. In this section we discuss some of the problems and limitations of the method and give some illustrations that might be helpful for other users of the CASPT2 method.

One of the severe bottlenecks of the CASPT2 approach is the size of the active space. It is very hard to perform calculations with more than 12 active orbitals when the system has no symmetry. With symmetry it is possible to extend the practical limit to 13 or 14, but above that the calculations become too heavy. When studying only one energy surface, it is rare that more than 12 active orbitals are needed to describe the near-degeneracies in the system. Calculations in spectroscopy are, however, more demanding. Different orbitals are active in different excited states. Most electronic spectra have Rydberg states mixed in with the valence states. Thus Rydberg orbitals have to be included. As an example, consider the molecule methylenecyclopropene (MCP) [(**1**) in Fig. 3], the spectrum of which is discussed in more detail below. A minimum set of valence active orbitals is 4. To this must be added nine Rydberg orbitals ($3s$, $3p$, and $3d$), which gives a total of 13. More orbitals may have to be added if intruder states appear in the calculations. The calculation would hardly be possible without symmetry. However, since the molecule has C_{2v} symmetry, it is possible to divide up the Rydberg orbitals such that calculations in different symmetries are made with the appropriate subset that contributes to excited states within that symmetry. The selection is illustrated later in Table V, which shows that it was in no case necessary to use more than nine active orbitals.

The problem was solved in a somewhat different way in a recent study of the phenol molecule [46]. The calculations of the excited states in this molecule were subdivided into three groups: the valence $\pi \rightarrow \pi^*$ states, Rydberg $\pi \rightarrow \pi^*$ states, and $\pi \rightarrow \sigma^*$ states. The molecule was assumed to be planar. For the valence states, the orbitals and the CASSCF states were obtained by minimizing the average energies of the lowest seven A' states. Three Rydberg orbitals had been obtained in earlier pilot studies and were deleted from the MO basis. The 21 occupied sigma orbitals were inactive, and the active space consisted of nine π orbitals: the oxygen lone pair, the six-ring π, and two extra-valence orbitals, which were needed to avoid intruder state problems (see also the earlier study of the benzene molecule [33]). The key to the problem is here the identification of the Rydberg orbitals and the exclusion of them from the one-electron basis. A similar approach was used in a recent study of the valence excited states of uracil, thymine, and cytosine [44, 47].

The $\pi \rightarrow \pi^*$ Rydberg states were computed by starting out from

TABLE V

CASSCF Wavefunctions (Number of Active Electrons) Used for Valence and Rydberg Transitions in Methylenecyclopropene

Wavefunction[a]	State	Number of Configurations[b]	N_{states}[c]
CASSCF(0301)(4)	$^3B_2(2b_1 \rightarrow 1a_2^*)$	9	1
CASSCF(0301)(4)	$^3A_1(2b_1 \rightarrow 3b_1^*)$	6	1
CASSCF(0501)(4)	$^1A_1(2b_1 \rightarrow 3p_x, 3d_{xz}, 3b_1^*)$	65	4
CASSCF(5301)(4)	$^1B_1(2b_1 \rightarrow 3s, 3p_z, 3d_{z^2}, 3d_{x^2-y^2})$	175	5
CASSCF(0311)(6)	$^1B_1(4b_2 \rightarrow 1a_2^*)$	9	1
CASSCF(0302)(4)	$^1B_2(2b_1 \rightarrow 1a_2^*, 3d_{xy})$	22	2
CASSCF(0331)(4)	$^1A_2(2b_1 \rightarrow 3p_y, 3d_{yz})$	57	3
CASSCF(1301)(6)	$^1A_2(8a_1 \rightarrow 1a_2^*)$	9	1

[a] Within parentheses the number of active orbitals of symmetry a_1, b_1, b_2, and a_2 of the point group C_{2v}.

[b] Number of configurations in the CASSCF wavefunction.

[c] States included in the average CASSCF calculation.

orbitals obtained in a calculation on the cation. First, a CASSCF calculation was performed, where the 21 occupied sigma orbitals were frozen, and the oxygen lone-pair orbital inactive, with 12 π-orbitals active, for the average of the 12 lowest A' states. This resulted in an optimized active orbital set containing the six valence π orbitals, three correlating orbitals, and three well-formed Rydberg orbitals. Six Rydberg states (excitations out of the two highest occupied π-orbitals to the $3p\pi$ and the two $3d\pi$ orbitals) were identified, and for each the natural orbitals were extracted and used as a *state-specific* orbital set for a set of CASCI calculations (CASSCF without orbital reoptimization) again using nine active orbitals. The type of orbitals in the active space is now different: Each orbital set contains only that state-optimized Rydberg orbital which must be present, and there are minor variations of the correlating orbitals, depending on which of the two ion rests the state contains.

The $\pi \to \sigma^*$ Rydberg states were obtained from a CASSCF with the 21 occupied sigma orbitals, and also the oxygen lone pair, inactive. The active space comprised six sigma orbitals and the six ring valence π orbitals, and the average energy of the lowest 12 states was minimized. The orbital optimization produced six Rydberg orbitals of σ symmetry. The 12 states were each used as the root state for subsequent CASPT2 calculations, without reoptimization. In all, 25 electronic states were determined for phenol, 13 of A' and 12 of A'' symmetry.

The validity of the CASSCF/CASPT2 method for computing excitation energies and properties of the excited states relies on the possibility of obtaining a valid reference function for the perturbation treatment. This is not always a trivial task. The reason is that the dynamic correlation energy is normally different in different excited states. Its effect on the excitation energy is usually small and positive for Rydberg states, it is negative and sometimes large for valence excited states. The ordering of the states is, as a result, different at the CASSCF level before the dynamic correlation energy has been added. At this level states may appear which are not going to remain in the final list, since they are not well characterized (higher Rydberg states, for example, for which there are no appropriate basis functions). A good example is given in the study of the furan molecule [48]. It was decided to include four $^1A_1$ states in the study, three valence states and one Rydberg state. A four-state average CASSCF calculation gave, however, only two valence states (the ground state and one excited state). The two other states were of Rydberg character. It was not until eight states were included in the averaging that the third valence state was found, with an energy of 10.6 eV relative to the ground state. The calculation was then repeated, but now with a weight of the four desired states nine times larger than that of the other states. The final CASPT2 calculation gave three valence excited states and one Rydberg state in the energy range 0–7.8 eV. The decrease in the excitation energy was 2.9 eV for the fourth state, while it was only 0.2 eV for the Rydberg state. This and similar examples show that it is important to know in advance the general structure of the excited states of interest. The CASSCF calculations must be set up such that they are included among the resulting wavefunctions. It may be necessary to perform several CASSCF calculations in order to arrive at a state-averaged optimization emphasizing only those states for which the CASPT2 calculations will be performed.

Sometimes it may be difficult to know in advance how to construct the active orbital space, because the near-degeneracies in the system are not known and/or the space that would be satisfactory is too large to be possible. Guidance can then be obtained by performing restricted active space (RAS) SCF calculations on the excited states of interest (the RASSCF method is described in Refs. 10 and 49). The RASSCF method is characterized by a greater flexibility in the CF space used to build the wavefunction. Instead of a single active space, three subspaces are distinguished, RAS-1, RAS-2, and RAS-3. A certain number of active electrons are distributed among the three orbital spaces but now with the added restriction that at most a specified number of holes are allowed in the RAS-1 space and at most a specified number of electrons in RAS-3. One way to use this in spectroscopic applications is to diminish the size of the RAS-2 space to an absolute minimum and, instead, add orbitals to the RAS-1 and RAS-3 spaces. An extreme but often useful calculation is to have an empty RAS-2 space and to allow only single and double excitations from RAS-1 to RAS-3 (SDCI). The point is, of course, that it is possible to use many active orbitals in this way. Inspection of the natural orbital occupation numbers will then give information about the important near-degeneracies, which have to be included into the CASSCF calculation that precedes CASPT2.

The procedure above was used in a recent study of the lower excited states of free-base porphin [35]. This molecule has 24 π orbitals and 24 π electrons. It is clearly impossible to have all these orbitals and electrons active. Therefore, an SDCI-type RASSCF calculation was first performed with the 24 π orbitals active. The occupation numbers were then used as a guidance in a series of CASSCF/CASPT2 calculations on the excited states. It was possible to increase the active space in a systematic way until the computed excitation energies had converged. There is no guarantee that this is always possible, however. If not, the CASSCF/CASPT2 method cannot be used to study the electronic spectrum. One such example is C_{60}. The minimum active space that can be used contains 15 orbitals and the calculation is thus outside the limits of the method. Other difficult cases are discussed in section III.B.

3. Solvation Effects

For isolated molecules of modest size, electronic excitation spectra can be predicted accurately by ab initio methods. On the other hand, such spectra may be of only limited relevance to bench chemists, who usually perform their experiments in a solvent. In this section we describe how solute–solvent interactions are included in the theoretical treatment of the electronic spectra within the present computational approach.

A microscopic description of solvation effects can be obtained by statistical mechanical simulation techniques or by a supermolecule approach. Through the advent of femtosecond time-resolution spectroscopy, this viewpoint has gained attraction. In the supermolecule approach, the surrounding solvent molecules are included in the quantum mechanical system. For studies of electronic excitations by statistical mechanics models, the solvent molecules are treated classically, but the quantum mechanical equations for the solute includes the solvent–solute interactions, resulting in a hybrid method [50, 51]. However, these methods are computationally demanding. In contrast, at the phenomenological level the solvent is regarded as a dielectric continuum, and there are a number of approaches [52–58] based on a classical reaction field concept. Typically, these methods are not more complicated than corresponding calculations for molecules *in vacuo*, except that self-consistency may require iteration. Here we use a modified version of the self-consistent reaction field (SCRF) approach of Bernhardsson et al. [59] to predict the shifts in band positions upon solvation.

Assuming that the solute is placed into a spherical cavity of radius a, Kirkwood derived a closed expression for the energy of the system,

$$E = E^0 - \frac{1}{2} \sum_{l,m} c_l M_l^m M_l^m . \tag{29}$$

Here M_l^m is the real component m of a 2^l-pole moment of order l. The reaction field factor c_l is

$$c_l = -\frac{l! \, (l+1)(\varepsilon - 1)}{(l+1)\varepsilon + l} \frac{1}{a^{2l+1}} \tag{30}$$

and depends on the dielectric permittivity ε of the solvent. Similar formulas may be derived for more general cavity shapes, but with much larger number of multipoles and with coefficients to be determined for each case.

In the derivation of Eq. (29) and (30), equilibrium between the electronic state of the solute and the reaction field has been assumed. This condition is not fulfilled for an electronic excitation. The time dependence is approximated by partitioning the reaction field factor into two parts, which give a slow and fast component, respectively, of the reaction field. The reaction field is coupled to the average charge density of the quantum mechanical system, and the coupled equations are iterated to self-consistency. For a relaxed state such as the ground state in an absorption spectrum, the total dielectric constant is used, and the full reaction field is determined by the solute density. For excited states, however, that fraction of the reaction field which is due to slow relaxation processes is held fixed, and only the remainder is determined by the excited-state density, using the dielectric constant ε^{∞}. The slow reaction field is obtained as the fraction $(c_l - c_l^{\infty})/c_l$ of each individual 2^l-pole component. c_l^{∞} is the reaction field factor using $\varepsilon^{\infty} \approx n^2$ and n is the refraction index for some suitable frequency in the infrared or visible range.

A repulsive potential between the solute and the cavity walls is added to the reaction field(s). This repulsion is due to exchange interaction between the solute and the solvent molecules, which is modeled as

$$E_{ex} = \int \int \int \rho(r, \theta, \phi) f(r) \, d\phi \, d\theta \, dr , \qquad (31)$$

where ρ denotes the electron density of the solute and $f(r)$ a penalty function of the form

$$f(r) = \sum_i \beta_i \exp[\alpha_i(r - R_i)^2] \qquad (32)$$

and r, θ, and ϕ are polar coordinates relative to the cavity center. The $f(r)$ is only a function of the distance to the center and helps to confine the electron density to the cavity. Matrix elements of this function over a CGTO basis are easily programmed and are part of the standard repertoire of our integral program [11]. For our purposes we found four terms sufficient.

The excitation $2^1A_1 \leftarrow 1^1A_1$, from the ground state to the next singlet valence state of aminobenzonitrile (ABN) and dimethyl-ABN (DMABN) was used as test cases. Due to the larger dipole moment in the excited state, the transitions shift to the red in dipolar media for ABN [60] as well as for DMABN [61]. The latter has a larger shift, measured in several solvents with ε up to 38.8 (Acetonitrile), which makes it a better example.

The geometries of ABN and DMABN were optimized at the CASSCF level of approximation [62]. ANO-S basis sets contracted as C, N[3s2p1d]/H[2s] were used. The cavity radius was optimized to minimize the absolute energy of the ground state at the CASSCF level of theory. This gave $a = 13.0a_0$ for ABN in diethylether, and $14.2a_0$, $13.8a_0$, and $13.5a_0$, for DMABN in cyclohexane, n-butylchloride, and acetonitrile, respectively.

The left side of Fig. 4 shows the CASPT2 energy diagram for the isolated molecule. We have found no experimental gas-phase excitation energy. The right-hand side shows the energy levels in a butyl chloride solution ($\varepsilon = 9.65$). There, the computed excitation energy is 4.32 eV, in excellent agreement with experiment (4.3 eV; see Ref. 61). Upon solvation, all energy levels are lowered. The lowering is larger for the larger dielectric constants, and is larger for the excited state than for the ground state. For acetonitrile, with $\varepsilon = 38.8$, the ground state is stabilized by 0.04 eV, but the excited state by 0.14 eV, giving a total red shift of 0.10 eV. The experimental shift is somewhat larger, 0.2 eV, in this case, but the experimental shifts are not very precise, and we conclude that the results are in agreement with experiment. The dipole moments of the ground and the excited states, 6.6 D and 14.2 D, respectively, are increased by the reaction field and reach 7.0 D and 15.2 D, respectively, in acetonitrile. They all have the same direction, parallel to the C_2 axis of the molecules.

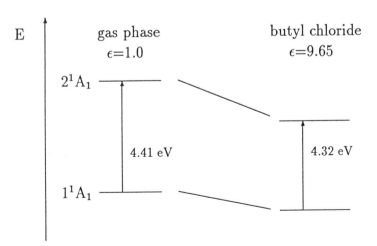

Figure 4. Solvation effects in the DMABN molecule.

This method was also used to study the electronic spectrum of imidazole. Section III.A.7 is devoted to the excited states of chromophores in proteins, and the imidazole results are presented there.

4. Spectroscopy of Carbonyl Compounds: Formaldehyde and Acetone

Aldehydes and ketones are important chromophoric groups, which play a central role in many different areas of chemistry. Formaldehyde is the prototype molecule for these kinds of compounds. Its electronically excited states have therefore been investigated extensively both experimentally and theoretically (see Refs. 63–65 and references cited therein). Acetone is the simplest aliphatic ketone. It is probably the best experimentally studied system of this group of important organic systems. The interpretation of its electronic spectrum has been and remains a subject of experimental interest [66–73]. In contrast to formaldehyde, acetone has been much less studied theoretically, undoubtedly due to the larger size of the molecule. To our knowledge there exist only two previous *ab initio* studies [74, 75]. Formaldehyde, on the other hand, is frequently used for testing new theoretical methods developed to treat excited states, because of its apparent simplicity and the numerous studies available.

At present there is consensus in the assignments of the vertical transitions of the low-energy part of the formaldehyde and acetone spectra, which are supported by CASSCF/CASPT2 calculations [65, 76]. However, the location and nature of all but the lowest singlet excited states of valence character is still one of the unresolved key problems of small carbonyl compounds. Qualitatively, the interpretation of the valence electronic transitions of the carbonyl chromophore involves a bonding π orbital of CO, a nonbonding $2p_y$ orbital on oxygen (hereafter named n_y), and an antibonding π^* orbital between C and O. Within the C_{2v} symmetry, with the z axis along the C–O bond, the lowest-energy electron promotion is from the HOMO, the b_2 nonbonding n_y orbital, to the b_1 antibonding π^* orbital resulting in the $^{1,3}A_2(n_y \rightarrow \pi^*)$ states. The $^{1,3}A_1 (\pi \rightarrow \pi^*)$ states for the carbonyl group are expected to appear at energies similar to those found for the corresponding states in ethylene. However, the position of the $^1A_1$ valence state has not yet been determined unambiguously. At the CASPT2 level, the vertical transition energy for the valence $^1A_1 (\pi \rightarrow \pi^*)$ state was computed to be above the $3s, 3p$, and $3d$ members of the lowest Rydberg series, and the oscillator strengths were large for both formaldehyde and acetone. It is assumed that the $(\pi \rightarrow \pi^*)$ transition probably takes the form of a broad, undetected feature, underlying the somewhat congested Rydberg region between 7 and 12 eV, in the same way as the Rydberg transitions of ethylene are superimposed on the broad singlet valence transition. Indeed, inspection of the electron energy-loss spectra of formaldehyde shows that the spectral peaks above ≈ 8 eV, corresponding to Rydberg series converging on the first ionization limit, are positioned on a nonzero background [77]. The band origin of the $(\pi \rightarrow \pi^*)$ state is therefore expected to appear close to 8 eV. The CASPT2 results indicate a more complex structure in this region, although it may be noted that the computed lowest excitation energy for the planar geometry is 7.84 eV [65]. For acetone, two-photon photoacoustic spectroscopy gives evidence of a coupling between this valence state and the $^1A_1$ $(n_y \rightarrow 3p_y)$ Rydberg state [71]. Similar evidence is obtained from two- and three-photon resonantly enhanced multiphoton ionization (REMPI) spectra of the $3p$ Rydberg transitions [73].

Knowledge about the vertical transitions is not enough for a full understanding of the spectroscopy of formaldehyde and acetone. The complicated photochemistry of formaldehyde has been described in recent papers by Hachey et al. [63, 64]. The interaction between the Rydberg and valence excited states in acetone has recently been studied with the CASSCF/CASPT2 method [76]. In principle, this can only be fully achieved with access to the full energy hypersurfaces, an obviously difficult task. We did not attempt to perform such a study. However, one of the main features expected for the equilibrium geometry of the $^1A_1$ $(\pi \rightarrow \pi^*)$ excited state is a considerably increased CO distance compared to the ground state, since the excitation is from a bonding to an antibonding orbital. The energy was computed as a function of the CO bond length for the four lowest electronic states of $^1A_1$ symmetry, the ground state, the $n_y \rightarrow 3p_y$ and the $n_y \rightarrow 3d_{yz}$ Rydberg states, and the $\pi \rightarrow \pi^*$ valence state, which are placed vertically at 7.26, 7.91, and 9.16 eV, respectively. An active space (2230) with six active electrons was employed. It comprised the CO σ, σ^* (2000) and π, π^* (0200) orbitals, the lone pair, n_y, and $3p_y$ and $3d_{yz}$ Rydberg orbitals.

The computation of the CO-stretch potential energy curves is not straightforward with the present theoretical approach. The reason is the interaction of Rydberg and valence excited states. The reference functions for the perturbation calculation of the correlation energy are determined at the CASSCF level of approximation. However, at this level the interaction of the various states are grossly in error, since the dynamic correlation effects for the valence excited states are normally larger than for the Rydberg states. In the region of the crossing, the reference functions will therefore not describe correctly the mixing of the various states. This error cannot be corrected by a second-order treatment, since higher-order terms that mix the different reference functions will be important. Only a multistate theory, which includes the mixing in the perturbation treatment, can solve this problem. The problem does not arise for the vertical transitions since there is no Rydberg–valence mixing in acetone at the ground-state equilibrium geometry.

The interaction between Rydberg and valence excited states can be assumed to be weak and only important close to the crossing points of the corresponding potential curves. One solution is to neglect the coupling by performing the calculations in such a way that the two types of states do not appear together. This approach was used in the calculation of the potential curves for the $^1A_1$ excited states. For CO distances shorter than 1.5 Å, $\pi \rightarrow \pi^*$ is the fourth excited state at the CASSCF level. Calculations with the active space (2230) and six active electrons were performed in this region. The CASSCF state average calculation included three roots (the ground state and the $n_y \rightarrow 3p_y$ and the $n_y \rightarrow 3d_{yz}$ states). The $3p_x$ and $3d_{xz}$ orbitals were deleted from the MO space to avoid intruder state problems from excitations to these orbitals from the occupied π orbital. The $\pi \rightarrow \pi^*$ excited-state potential was obtained with the active space (2200), which does not include the n_y and Rydberg orbitals. Only the four electrons of the CO bond were active. The state average calculation included only two roots. The same two Rydberg orbitals of π symmetry were deleted. The two results give an indication of the stability of the approach. First, the two potential curves for the ground state were virtually identical. Second, the computed vertical excitation energy for the $\pi \rightarrow \pi^*$ state was 9.20 eV, which is only 0.04 eV larger than the value obtained with the larger active space, which was used for all the vertical transitions. This illustrates again that the CASPT2 results are stable with respect to modifications in the active space as long as they do not grossly change the structure of the CASSCF reference function.

The resulting potential curves are shown in Fig. 5. The computed curves cross each other, but in the figure the avoided crossings have been indicated by introducing a small separation. The minimum of the ground-state curve is found for $r(CO) = 1.226$ Å with $\nu_3(CO) = 1698 \text{ cm}^{-1}$, which can be compared to the experimental results $(1.222 \text{ Å}, 1731 \text{ cm}^{-1})$ [73, 78]. The 2^1A_1 state has a double minimum potential. The inner minimum is dominated by the $3p_y$ Rydberg state with an equilibrium bond length slightly shorter than that of the ground state. The outer minimum is the $\pi \rightarrow \pi^*$ state. It is located 7.42 eV above the ground state. The outer minimum is the $\pi \rightarrow \pi^*$ state. It is located 7.42 eV above the ground state at a CO distance of 1.58 Å. The excitation energy has thus decreased 1.78 eV compared to the vertical value.

The next state also has a double minimum potential along this coordinate. It contains contributions from all three diabatic curves. The

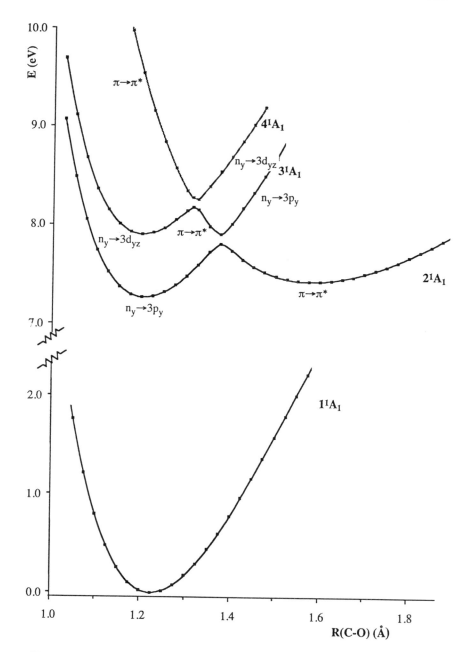

Figure 5. Potential curves for the $^1A_1$ states in acetone as a function of the CO bond length.

inner minimum is dominated by the $3d_{yz}$ Rydberg state, while the outer minimum is due to the avoided crossing between the $\pi \to \pi^*$ and the $3p_y$ states. The upper state is a combination of the wings of the $\pi \to \pi^*$ state and the $3d_{yz}$ state. The result is one sharp minimum at 1.32 Å, 8.26 eV above the ground state.

The transition moments for the three excitations from the ground state are smoothly varying functions of the CO distance. For the diabatic $3p_y$ Rydberg state the values vary from about 0.11 at short distances to 0.00 at $r(CO) = 1.5$ Å. The corresponding values for the $3d_{yz}$ Rydberg state are 0.59 to 0.45, and for the $\pi \to \pi^*$ state 1.26 to 0.75 [at $r(CO) = 1.95$ Å]. We used these data to perform a vibronic analysis, treating the system as a diatomic molecule. This is obviously a crude approximation, since it neglects all couplings to other modes and treats only the CO stretch frequency. Nevertheless, such a study might give some insight into the intensity distribution for excitations from the ground state to the three excited states. Since this calculation neglects all vibronic coupling between the three excited states, only the lowest vibrational quantum numbers were considered. The rovibrational Schrödinger equation was solved and the intensities were computed by numerical integration. Only excitations out of the $v'' = 0$ ground-state level were considered. This analysis shows that the bands of the 2^1A_1 system should be weak. The largest intensity is obtained for the 0–0 transition at 7.25 eV. The two lowest vibrational modes are entirely confined to the Rydberg part of the potential. It is only the third state that has any contribution from the $\pi \to \pi^*$ potential. But here the Franck–Condon overlap is small. The largest intensity for bands below 8.0 eV is found for the 0–0 transition to the 3^1A_1 state, which occurs at 7.9 eV. This is the $3d_{yz}$ transition. Appreciable intensity contributions from the $\pi \to \pi^*$ excited state appear only for the 4^1A_1 potential and start at 8.35 eV.

Recent theoretical studies of formaldehyde by Hachey et al. [64] and by the present authors [65] have revealed more features of the potential surfaces. The similarity between formaldehyde and acetone makes it possible to draw conclusions for the latter molecule based on the results obtained for formaldehyde, especially since the results obtained here for the $^1A_1$ states are similar in quality to those obtained by Hachey et al. The $^1B_1$ ($\sigma \to \pi^*$) potential has a minimum at $r(CO) = 1.47$ Å and crosses the $^1A_1 (\pi \to \pi^*)$ surface close to this point. Bending the CH_2 group out of the plane causes them to mix. The optimum bending angle was computed to be 46°. At shorter distances the $^1B_1$ surface mixes strongly with the $(n_y \to 3d_{xy})$ transition. It is likely that a similar situation obtains in acetone. Thus out-of-plane modes in combination with the CO stretch will allow for mixing of the $^1A_1$ and $^1B_1$ states. This will complicate the vibrational spectrum further, compared to the simplified analysis performed above.

In addition, the bent $2^1A'$ ($\sigma \to \pi^*$, $\pi \to \pi^*$) surface can perturb the $^1B_2$ ($n_y \to 3s$) state [63]. The equilibrium geometry for the $^1A_2$ ($n_y \to \pi^*$) state is likewise expected to be bent (in formaldehyde the bending angle is 31° [65]). The results obtained by Hachey et al. for H_2CO shows that the 2^1A_2 ($n_y \to 3p_x$) potential is crossed by the $^1B_1$ ($\sigma \to \pi^*$) potential. Thus the dipole-forbidden transition to the $^1A_2$ state can become vibrationally allowed, borrowing intensity from the interfering $^1B_1$ state via modes of b_2 symmetry.

Assuming the two molecules to be similar, we conclude that $n_y \to$ Rydberg states may mix through vibronic interactions with the valence excited states of $^1A_1$ and $^1B_1$ symmetry. The large transition moment of the $^1A_1$ ($\pi \to \pi^*$) state can consequently be used to give intensity to Rydberg states of $^1A_1$, $^1B_1$, and $^1B_2$ symmetry. Direct observation of the $\pi \to \pi^*$ state is, on the other hand, difficult, due to the small Franck–Condon factors for the lower vibrational levels.

5. Interacting Double Bonds: Methylenecyclopropene

We have shown [13, 79, 80] that the electronic spectra of cis-1,3-buta-diene [81], cyclopentadiene, aromatic five-membered heterocycles [48, 79], and norbornadiene [80] can be understood on the basis of a model with two interacting double bonds. Cyclopentadiene (CP) is the prime example of a ring-shaped molecule with a conjugated π-electron system, and its structure can be related to that of short polyenes such as cis-1,3-butadiene (CB) and the simplest heterocycles, such as pyrrole (PY), furan (FU), and thiophene (TP). In the series, cis-1,3-butadiene, cyclopentadiene, norbornadiene (NB), the latter is the most complex system, with the two ethylenic units coupled through indirect conjugation and π, σ interaction. One more system will be added here to the set of molecules with two interacting double bonds: methylenecyclopropene (MCP) [(1) in Fig 3] the simplest cross-conjugated π-electron system.

MCP, the archetype of nonalternant hydrocarbon known as the fulvenes, is a molecule of considerable theoretical and experimental interest [82–90]. It was synthesized and characterized for the first time in 1984 Billups et al. [84] and Staley and Norden [85] independently. An ultraviolet (UV) spectrum was included in the latter study. The electric dipole moment and heavy-atom molecular structure were reported two years later [88]. The experimental value of the dipole moment, $\mu = 1.90 \pm 0.02$ D, remarkably large for such a small molecule, which has been related to the strong transfer of π-electron density out of the ring due to the two double bonds interacting in a T-type arrangement. The lowest adiabatic (8.15 ± 0.03 eV) and vertical (8.41 ± 0.05 eV) ionization potentials have been determined by photoelectron spectroscopy [89]. Most of the *ab initio* work has focused on the ground state of the molecule [86–89]. In the early study performed by Johnson and Schmidt [82] on the sudden polarization effect of the system, the low-lying excited states of planar MCP were characterized at the MCSCF level. A limited atomic basis set of double-zeta quality was used.

For the qualitative description of the electronic transitions it is helpful to consider the interaction of two independent ethylenic moieties. The interaction of the degenerate bonding, π_b, and antibonding, π_a, MOs of two ethylenic units results in the following four MOs: $\pi_1 = \pi_b + \pi_b$, $\pi_2 = \pi_b - \pi_b$, $\pi_3 = \pi_a + \pi_a$, and $\pi_4 = \pi_a - \pi_a$. With the geometrical arrangement in NB as reference, they lead to the orbital ordering $a_1(\pi_1)$, $b_1(\pi_2)$, $b_2(\pi_3)$, and $a_2(\pi_4)$ of the C_{2v} point group. Based on this structure the following valence singlet states can be expected and were actually obtained for NB. The one-electron promotions $\pi_2 \rightarrow \pi_3$ and $\pi_1 \rightarrow \pi_4$ yield two states of A_2 symmetry, while $\pi_1 \rightarrow \pi_3$ and $\pi_2 \rightarrow \pi_4$ give two states of B_2 symmetry. The corresponding electronic configurations are nearly degenerate and mix to form B_2^- and B_2^+ states. The minus state has the lowest energy and low intensity, while the plus state is pushed up and carries most of the intensity. There is an additional state of A_1 symmetry formed by two electron replacements, mainly the $(\pi_2)^2 \rightarrow (\pi_3)^2$ doubly excited configuration. Similar arguments can be used to rationalize the electronic transitions found in related systems as the five-atom rings (CP, PY, FU and TP) and *cis*- and *trans*-butadiene (CB and TB).

Due to the perpendicular arrangement of the double bonds in MCP, the two antibonding π_a ethylenic MOs cannot interact, since they have different symmetry. Interaction is, however, possible between the π_b MOs. In increasing order of orbital energies, the π-valence orbitals are: $1b_1(\pi_1)$, $2b_1(\pi_2)$, $1a_2(\pi_3)$, and $3b_1(\pi_4)$. The SCF orbital energies show that configurations involving the first occupied π orbital will not play a significant role in the description of the low-lying excited states. The $1b_1$ MO is around 5.8 eV below the HOMO, $2b_1$, which has a Koopmans' ionization potential of 8.1 eV (C[$4s3p1d$]/H[$2s1p$] results). Electronic states dominated by excitations out of the $1b_1$ MO will consequently not appear in the low-energy part of the spectrum. Two π-derived low-lying singlet and triplet valence excited states can therefore be expected in MCP: $2b_1 \rightarrow 1a_2$ gives rise to a state of B_2 symmetry and the excitation $2b_1 \rightarrow 3b_1$ results in a state of A_1 symmetry. Due to the localization of the $1a_2$ MO on the two carbon atoms of the ring, the transition from the ground state to the first singlet excited state, $^1B_2$ ($2b_1 \rightarrow 1a_2$), is expected with low intensity, contrary to the case of short polyenes in which the corresponding feature is the most prominent. With both the $2b_1$ and $3b_1$ MOs spread on the four carbon atoms, the valence transition to the $^1A_1$ ($2b_1 \rightarrow 3b_1$) state is expected to be most intense. In addition, because of the high ring strain inherent in its molecular structure, the σ and σ^* orbitals of MCP are expected to be shifted significantly relative to those of unstrained hydrocarbons. Indeed, the HOMO-1 ($4b_2$) describes the σ C–C bonds adjacent to the double bond in the cyclic part of the molecule. Valence features involving σ MOs therefore had to be included in description of the electronic spectrum of MCP. Rydberg states arising from excitations out of the HOMO are predicted to be interleaved among the valence excited states.

The ground-state geometry determined in gas phase by microwave spectroscopy [88] was employed. Calculations were carried out within C_{2v} symmetry, placing the molecule in the yz plane with z axis bisecting the HCH angle. The ANO-L type C [$4s3p1d$]/H[$2s1p$] basis set was used, supplemented with a $1s1p1d$ set of Rydberg functions, placed in the charge centroid of the $^2B_1$ state of the MCP cation. The active spaces used and the type of states computed are listed in Table V.

The active space comprising the π-valence MOs (0301) was used in the computation of the low-lying $^3B_2$ and $^3A_1$ states. For the study of the singlet excited states the active space was extended to include the corresponding Rydberg orbitals as appropriate. The results of the study are presented in Table VI. *Ab initio* transition energies obtained in previous calculations at the MCSCF level [82] are also included. The computed dipole moment for the ground state at the single-state optimized CASSCF (0301) level, 1.80 D, is in agreement with experiment, 1.90 D [88]. Three valence singlet states, 1^1B_2, 3^1B_1, and 4^1A_1, were located below 7 eV. As expected, the lowest singlet excited state is of B_2 symmetry. The computed intensities for the two lowest valence transitions (both involving excitations to the $1a_2$ MO) are relatively low. The third valence transition represents the most intense feature of the system. The expectation values of x^2, y^2, and z^2 reveal the valence nature of the 1^1B_2 and 3^1B_1 states, while the 4^1A_1 state is somewhat more diffuse. The 1^1B_2 and 3^1B_1 states have opposite polarity to that of the ground state, due to the promotion of an electron into the $1a_2$ orbital entirely localized on the three-membered ring. Due to the ionic nature of the 4^1A_1 valence state, its computed dipole moment is high, 4.81 D, with the same polarity as the ground state [a carbanion center (CH_2^-) attached to a cyclopropenyl cation]. Comparison of the CASSCF and CASPT2 results demonstrates the importance of dynamic correlation effects to the excitation energies. The largest dynamic correlation contribution corresponds to the 4^1A_1 valence state, 1.22 eV. It is therefore not surprising

TABLE VI
Calculated Excitation Energies (eV) and Other Properties of the Vertical Excited States of Methylenecyclopropene

State	CASSCF	CASPT2	μ^a	Total[b] $\langle x^2 \rangle$	$\langle y^2 \rangle$	$\langle z^2 \rangle$	Oscillator Strength	Other Results MCSCF[c]
Ground state ($1\,^1A_1$)	—	—	-1.80	-20.0	-15.2	-16.9	—	—
Singlet states								
$1\,^1B_2(2b_1 \rightarrow 1a_2^*)$	4.71	4.13	$+2.07$	-22.6	-18.8	-14.4	0.0093	4.71
$1\,^1B_1(2b_1 \rightarrow 3s)$	5.04	5.32	$+2.19$	-41.7	-36.5	-32.8	0.0028	—
$1\,^1A_2(2b_1 \rightarrow 3p_y)$	5.28	5.83	$+1.23$	-34.7	-66.2	-25.6	Forbidden	—
$2\,^1B_1(2b_1 \rightarrow 3p_z)$	5.52	5.84	$+1.58$	-36.7	-32.2	-64.0	0.0275	—
$2\,^1A_1(2b_1 \rightarrow 3p_x)$	5.00	5.85	$+0.78$	-71.7	-30.7	-26.7	0.0670	—
$3\,^1B_1(4b_2 \rightarrow 1a_2^*)$	7.29	6.12	$+0.01$	-22.2	-15.8	-15.8	0.0064	—
$2\,^1A_2(2b_1 \rightarrow 3d_{yz})$	5.94	6.54	-1.86	-37.6	-74.4	-70.1	Forbidden	—
$4\,^1B_1(2b_1 \rightarrow 3d_{z^2})$	6.14	6.55	-5.78	-42.0	-54.3	-86.6	0.0012	—
$5\,^1B_1(2b_1 \rightarrow 3d_{x^2-y^2})$	6.19	6.62	-0.31	-82.1	-64.5	-30.5	0.0005	—
$3\,^1A_1(2b_1 \rightarrow 3d_{xz})$	5.77	6.71	-0.46	-84.2	-34.8	-74.5	0.0062	—
$2\,^1B_2(2b_1 \rightarrow 3d_{xy})$	6.53	6.75	-0.58	-80.4	-75.7	-29.6	0.0035	—
$4\,^1A_1(2b_1 \rightarrow 3b_1^*)$	8.04	6.82	-4.81	-32.2	-17.7	-21.6	0.5840	9.36
Triplet states								
$1\,^3B_2(2b_1 \rightarrow 1a_2^*)$	3.50	3.24	$+0.97$	-20.3	-17.2	-15.0	—	3.32
$1\,^3A_1(2b_1 \rightarrow 3b_1^*)$	4.72	4.52	-0.85	-20.8	-15.6	-18.2	—	4.87

[a] Dipole moment (CASSCF) in Debye (experimental value for the ground state is -1.90 a.u. [88]).
[b] Expectation value (CASSCF) of x^2, y^2, and z^2 (in a.u.).
[c] MCSCF results employing a double-zeta basis set from Ref. 82.

that previous MCSCF calculations using a small atomic basis set [82] placed this state energetically too high, at 9.36 eV.

The calculated vertical transition energy from the ground to the 1^1B_1 ($2b_1 \rightarrow 3s$) Rydberg state is 5.32 eV. It is a weak transition. The $3p$ Rydberg states, $1^1A_2, 2^1B_1$, and 2^1A_1, have similar excitation energies. The 2^1A_1 state is placed at 5.85 eV above the ground state. The oscillator strength for the corresponding transition is around 0.07. It is actually the second most intense feature of the spectrum. The five members of the $3d$ series have been computed to lie in the energy range 6.54–6.75 eV. The dipole-allowed $3d$ transitions have lower intensities than the $3p$ transitions. In contrast to the $3s$ and $3p$ Rydberg states (cf. Table VI), the $3d$ states have the same polarity as the ground state. The singlet–triplet gaps are computed to be 0.89 and 2.30 eV for the 1^3B_2 and 1^3A_1 states, respectively. MCP has the smallest singlet–triplet splitting for the HOMO$\rightarrow$LUMO excited state (0.89 eV), as compared to the related systems NB (1.86 eV), CP (2.12 eV), and CB (2.77 eV). It is an additional reflection of the different (localized) nature of the LUMO in MCP.

As far as we are aware, no gas-phase spectroscopic data are available for comparison. The UV spectrum of MCP in n-pentane at $-78°C$ is, however, available [85]. It displays a broad low-intensity band at 4.01 eV, a more narrow low-intensity band at 5.12 eV, and a strong band at 6.02 eV. The position of the first band is strongly solvent dependent, 4.49 eV in methanol at $-78°C$. The solvation effects are smaller for the other absorption bands; 5.00 and 5.90 eV in methanol, respectively. Therefore, comparison of the calculated (gas-phase) results with the available experimental data in solution has to be done with caution. The presence of a condensed medium perturbs the electronic transitions in several ways. Rydberg states are precluded and bands are usually broad with low resolution. The observed bands of the UV spectrum reported by Staley and Norden [85] should therefore be related to the computed valence transitions. The first band can be identified with the lowest vertical singlet–singlet transition, computed at 4.13 eV in gas phase. The fact that the energy of the excitation increases in a more polar solvent is consistent with the change of sign of the dipole moment (from -1.80 D to $+2.07$ D), which makes the interaction with the solvent repulsive in the upper state during the time scale of the excitation. A broad band is also expected due to strong relaxation effects in the solvent. A bathochromic shift is expected for the $1^1A_1 \rightarrow 4^1A_1$ transition since the dipole moment of the 4^1A_1 state is 3 D larger than that of the ground state and with the same sign. Finally, comparison between the observed and calculated intensities of the three valence transitions located below 7 eV supports the identification of the weak band centered at 5.12 eV in n-pentane, with the

second valence transition, which has a computed vertical energy of
6.12 eV in the gas phase. The dipole moment is, however, small and one
would not expect a solvent red shift of 1 eV. It is, however, possible that
other effects than the solvent are important for the differences between
the spectrum of MCP in gas phase and solution. One possibility is the
nonvertical nature of the transitions. Johnson and Schmidt [82] have
calculated the low-lying diradical and zwitterionic states of 90° twisted
MCP. The exocyclic π-bond twisting leads to a pronounced stabilization
with respect to the correlated states of planar MCP. The equilibrium
geometries of the excited states are, in general, expected to be twisted.
The importance of this phenomenon has been demonstrated in ethylene,
and it is by now well established that the maximum of the π–π^* valence
band, which occurs at 7.66 eV [91], does not correspond to the vertical
transition, which has a best theoretical estimate of about 8 eV [92], but
instead, to a somewhat twisted molecule. Another possible source of
error in the description of the 4^1A_1 state is related to the increased spatial
diffuseness of this state compared to the other valence excited states. In
cases where there is an erroneous valence–Rydberg mixing at the
CASSCF level, the CASPT2 does not lead to an improvement of the
excitation energy, and larger errors are obtained (about 0.4 eV for the V
state of ethylene and 0.3 eV for the lowest singlet–singlet transition of TB
[36]). This possibility was checked in a series of calculations using
different active spaces and different basis sets. Additional MRCI calculations were also performed to check the behavior of the CASPT2
approach itself. The result of all these studies (which will be reported in
detail elsewhere [93]) unambiguously show that the vertical transition
energy of the 4^1A_1 state is located around 7 eV.

To summarize, the main features of the electronic spectrum of
methylenecyclopropene can be rationalized within the scheme of two
interacting double bonds. The lowest triplet state is mainly described by
the HOMO$\rightarrow$LUMO excitation. The energy difference between the π_3
and π_2 orbitals is about the same in MCP, NB, CP, and CB. This is
reflected in similar values for the lowest singlet–triplet transition, 3.24,
3.42, 3.15, and 2.81 eV, respectively. The orbital energy model is too
crude to allow a more detailed analysis of the small relative energy
differences. The corresponding singlet excited state, 1^1B_2, located at
4.13 eV, lies around 1 eV below the corresponding state in the related
systems NB $(1^1A_2: 5.28\,\text{eV})$, and $CP(1^1B_2: 5.27\,\text{eV})$. Orbital energy
differences are about the same as in the other diene systems [79]. Thus
the explanation lies in the electron repulsion terms. The excitation moves
the electron from an orbital, which is delocalized over all four atoms to
an orbital entirely localized on the ring atoms. This greatly reduces the

size of the corresponding exchange integral, with a corresponding down shift of the energy for the singlet state and a small singlet–triplet separation.

The strongest vertical transition corresponds to the 4^1A_1 valence state, computed in the high-energy side of the spectrum above the $n = 3$ Rydberg series, at 6.82 eV, is described primarily by the $\pi_2 \rightarrow \pi_4$ configuration, with a small valence–Rydberg mixing. No near degeneracy occurs with excitations involving the lowest π orbital, which can be rationalized considering the large orbital energy spacing with the remaining π orbitals. The T-type interaction between the two double bonds and the intrinsic strain of the molecule makes the system unique, with a valence state described primarily by one-electron promotion from the highest σ orbital (HOMO-1) to the LUMO present at a moderate energy, between the $3p$ and $3d$ Rydberg series. The shifts noted between the computed spectrum and the available UV spectrum in n-pentane and methanol are probably due to both the nonvertical nature of the valence bands and to solvent effects.

Taking into account that methylenecyclopropene is a highly reactive compound, the experimentally determined gas-phase spectrum might not be easily accessible for some time. The interplay between experiment and theory therefore has importance for gaining further insight into the properties of this fundamental nonalternant hydrocarbon.

6. Interacting Fragments: Biphenyl and Bithiophene

With the studies of the electronic spectra of basic systems as benzene [33] and thiophene [79] performed successfully, it was tempting to try to relate the electronic states of two such interacting fragments to those of the corresponding monomer. Two cases may occur: two weakly coupled π systems such as in biphenyl [(2) in Fig. 3] or stronger coupling as in bithiophene [(3) in Fig. 3]. In the former situation it was possible to identify the excited states with the excited states in benzene. The orbitals are only weakly perturbed and retain their benzene identity to a large extent. Biothiopene is different. The shorter CC link distance leads to dimer orbitals that are considerably modified. The HOMO and HOMO-1 orbitals in thiophene are close in energy and mix strongly when perturbed. It is thus not possible to identify the excitation pattern in the dimer from the orbitals of thiophene. The main features in the energy range up to 6 eV of both biphenyl [94] and bithiophene [95] have been analyzed from CASSCF wavefunctions and CASPT2 energies. The results have been used to assign the experimental spectra, resulting in agreement between computed and measured excitation energies.

Apart from the intrinsic interest of biphenyl (BP) and bithiophene

(BT), they also act as models of the conjugated organic polymers polyparaphenylene (PPP) and polythiophene (PT), respectively. As is discussed in detail elsewhere [96], the results for the electronic excited states of the cation and anion of BP can be used for interpreting the polaron states in doped PPP. The main features of the spectra are reviewed briefly below.

Biphenyl Molecule. Most calculations were performed for a planar molecule with D_{2h} symmetry. The molecule is placed in the yz plane with z as the long molecular axis. A structure optimized at the CASSCF level was used [97]. The active space comprised the 12 π-valence orbitals, with 12 π electrons active. The computed excitation energies and oscillator strengths are collected in Table VII. In this table the observed bands are also related to the calculated vertical excitation energies. When comparing the experimental and theoretical data, one has to keep in mind that the geometry of the ground state of BP depends on the phase. A planar structure with large oscillations has been derived for solid biphenyl at

TABLE VII

Calculated and Experimental Singlet–Singlet Vertical Excitation Energies and Oscillator Strengths in Planar Biphenyl

	Excitation Energy (eV)			Oscillator Strength	
States	CASSCF	CASPT2	Exp.	Calc.	Exp.
1^1B_{3g}	4.66	4.04	4.11(0–0),[a,b] 4.17,[c] 4.37(0–0)[d]	Forbidden	—
1^1B_{2u}	4.92	4.35	4.20,[a] 4.26(0–0),[e] 4.59,[f] 4.59[g]	0.0005	0.02,[f] 0.01[g]
1^1B_{1u}	6.62	4.63[h]	5.21,[f] 4.80,[g] 4.95,[i] 4.92,[j] 5.02[k]	0.6200	0.36,[f] 0.28[g]
2^1B_{3g}	7.98	5.07	—	Forbidden	—
$1^1B_{2g}(3s)$	5.65	5.60	—	Forbidden	—
2^1B_{2u}	8.87	5.69	6.41,[f] 5.85,[g] 5.96,[j]	0.4300	0.65,[f] 0.43[g]
2^1B_{1u}	8.51	5.76	6.41,[f] 6.16,[g] 6.14[j]	0.8355	0.65,[f] 0.45[g]
2^1A_g	6.40	5.85	5.64[c]	Forbidden	—
3^1A_g	8.49	5.85	5.64[c]	Forbidden	—

[a] Crystal absorption spectrum at 4.2 K [98].
[b] Two-photon excitation spectrum of crystal biphenyl [99].
[c] Two-photon excitation spectrum in ethanol [103].
[d] Supersonic jet laser spectroscopy [217].
[e] Absorption spectrum in argon matrix [100].
[f] Vapor spectrum [102].
[g] Crystal transitions derived from the reflection spectra [102].
[h] The value obtained with the true (twisted) ground-state geometry is 5.09 eV.
[i] Electron energy-loss spectrum for biphenyl deposited on a thin film of solid argon at 20 K [106].
[j] The UV spectrum in stretched film [105].
[k] Absorption spectrum in cyclohexane [104].

temperatures above 40 K. In solution, in the crystal below 40 K, and in the gas phase, nonplanar structures with different twist angles are found (see Ref. 97 for details). All this shows that the ground-state energy surface is flat with respect to the twist angle. Simple arguments based on the orbital shapes show that the excited states have a more profound minimum at the planar structure. The assumption of a planar ground state may therefore give too small vertical excitation energies. In one such a case $(1\,^1B_{1u})$ this was illustrated in the recent study by computing also the excitation energy at the true ground-state equilibrium geometry [97]. The difference was found to be 0.46 eV.

As is shown in Table VII, experimental data in crystal, in solution, and in the gas phase give evidence of two low-energy transitions. This is reproduced by the calculations. The symmetry assignments are in agreement with the experimental data for the biphenyl crystal [98, 99]. The transition energies obtained are consistent with the crystalline data [98, 99] and biphenyl trapped in argon matrix [100]. Correlation with experimental data in the gas phase is also good considering that BP adopts a twisted conformation in this phase. In the energy interval 4.52–4.58 eV, two long progressions of the torsional vibration have been observed in the one- and two-photon absorption spectra of biphenyl in a supersonic jet. The fact that the intensities of these progressions are reversed in one- and two-photon spectroscopies supports assignment to the $1\,^1B_{3g}$ and $1\,^1B_{2u}$ states, as discussed by Murakami et al. [101]. A previous vapor spectrum showed a weak band at 4.59 eV with polarization along the short molecular axis as determined from an analysis of the reflection spectra in the crystal [102]. This band is therefore assigned to the $1\,^1B_{2u}$ state. The small value computed for the oscillator strength of this state is supported by the low intensity of the observed band, which in addition is obscured by the next band of the vapor spectrum. In solution, the experimental information about the two lowest singlet excited states comes primarily from two-photon spectroscopy [103]. The lowest band, located at 4.17 eV in ethanol solution, has been found to correspond to a final state of B symmetry. Dick and Hohlneicher [103] have assigned this band to a $^1B_3$ state based on the similarity to the lowest bands of the two-photon spectrum of fluorene. The band at 4.17 eV is then attributed to our $1\,^1B_{3g}$ state. It should be noted that in D_{2h} symmetry the $1\,^1B_{2u}$ state is two-photon forbidden, but in D_2 symmetry, in which the state becomes B_2, it is allowed.

The third singlet excited state is of B_{1u} symmetry with a CASPT2 computed excitation energy of 4.63 eV and an oscillator strength of 0.62. The vapor spectrum shows a band of medium intensity at 5.21 eV [102]. This band has been found to be polarized along the long molecular axis and is located at 4.80 eV in crystalline biphenyl [102]. An energy difference of 0.4 eV is therefore observed for the excitation energies measured in the gas and crystal phases. Values around 5 eV have been reported for this band in the absorption spectrum measured in solution [104] and for biphenyl deposited on film [105, 106].

The variation in the position of the maximum of the band depending on the media can be related to the twist angle of the molecule in the different media. Thus, in the gas phase, where biphenyl has the largest twist angle, the excitation energy acquires its maximum value, while it is minimum in crystalline biphenyl, which has a planar geometry at room temperature. These experimental facts seem to suggest that the $1\,^1B_{1u}$ excited state is planar, considering the low barrier to rotation determined for the ground state [97, 107]. It is therefore clear why the computed value is smaller than all the experimental data and closer to the value in the crystal phase. To get more insight into the influence of the geometry on the computed excitation energy to the $1\,^1B_{1u}$ state, calculations were carried out using a twisted geometry for the ground state of biphenyl. These calculations place the $1\,^1B_{1u}$ (or $1\,^1B_1$ in D_2) state at 5.09 eV ($f = 0.58$), which is in agreement with the gas-phase value [102].

Dick and Hohlneicher have suggested that the bands observed at energies higher than 4.7 eV are related to final states of A symmetry [103]. Thus a sharp maximum at 4.7 eV and a shoulder at about 5.0 eV were assigned to a low-lying ^1A (or ^1A_g) state [103]. An exhaustive investigation of the ^1A_g states has been undertaken to explain the above-mentioned features of the two-photon spectrum. We have been unable to find a low-lying ^1A_g state, despite increasing the number of roots in the average CASSCF calculation to seven. The first state of this symmetry appears at 5.85 eV. The previous assignment of the two-photon spectrum should therefore be revised. The $2\,^1B_{2u}$ and $2\,^1B_{1u}$ states are calculated to be close in energy, with computed values of 5.69 and 5.76 eV, respectively, which are in agreement with experimental findings [102, 105, 108] considering the fact that the upper states are planar.

Two optically forbidden states of A_g symmetry are found at 5.85 eV. These states clearly correspond to the intense two-photon absorption starting at 5.64 eV, which was related to states of A symmetry on the basis of the value of the two-photon polarization parameter [103]. The CASSCF wavefunction of the $2\,^1A_g$ state is described mainly by doubly excited configurations which have a total weight of 43%, with predominance of the doubly excited configuration HOMO → LUMO. By contrast, the CASSCF wavefunction of the $3\,^1A_g$ state is described mainly by four singly excited configurations, involving the four highest occupied and four lowest unoccupied MOs.

To summarize, there is agreement between the spectrum computed and the large amount of spectral information. The theoretical data confirm the sensitivity of the excitation energies to the structure of the ground state, which varies from nearly planar in crystals to 44°C twisted in the gas phase. The appearance of the first Rydberg transition has been predicted. It has been shown that no excited state of ^1A_g symmetry appears below 5.85 eV. As far as we are aware, this represents the first *ab initio* study of BP. Comparison with earlier semiempirical studies shows that they are not able to account for all features of the electronic spectrum of biphenyl (see the discussion in Ref. 94).

Bithiophene Molecule. Calculations were performed at the planar *trans* geometry (C_{2h} symmetry) and at the twisted *trans*-bithiophene conformation (C_2 symmetry) obtained at the MP2/6-31G* level [109]. In most calculations the active space comprised the 10 valence π orbitals. Because of intruder states, two orbitals (one of each π symmetry) had to be added for the 2^1A_g excited state in the twisted conformation. The 12 π electrons were active. The computed excitation energies and transition intensities for the planar and twisted forms are presented in Table VIII together with relevant experimental information.

Four excited states of ^1B_u and two of ^1A_g symmetry (^1B and ^1A in the twisted form) have been studied. The first ^1B_u state is dominated by the HOMO→LUMO excitation, but contains in addition a sizable contribution from HOMO $- 2$→LUMO. The same is true for the next ^1B_u state even if there are now larger contributions from higher excitations. The first state of ^1A_g symmetry corresponds mainly to the HOMO $- 1$→LUMO excitation but contains 23% double excitations. The character of being doubly excited is even larger in the next state of this symmetry (42%), and here it is the $(\text{HOMO})^2$→$(\text{LUMO})^2$ that dominates the wavefunction.

The fact that the ground-state potential is flat [109] implies that variations in computed excitation energies for different geometries will reflect mostly the shape of the upper state potential. Comparison with experiment is complicated, since measured bands depend strongly on temperature and medium. It is not unlikely that the *cis* conformation is more stable than *trans* in polar media, since there is a large difference in dipole moment. It is only low-temperature gas-phase data that can be compared directly to the present results.

The potential curves for the excited states are different from those of the ground state. In the region of the *trans* conformer they have a distinct minimum at the planar geometry. The energy difference to the twisted geometry varies from 0.14 eV for the 2^1B_u state to 0.45 eV for the 2^1A_g

TABLE VIII
Calculated and Experimental Singlet–Singlet Excitation Energies (eV) and Calculated
Oscillator Strengths in Planar and Twisted *trans*-Bithiophene

	State				
	1^1B_u	2^1B_u	2^1A_g	3^1A_g	3^1B_u
Planar Conformation					
CASSCF	5.64	6.13	5.92	5.31	7.82
CASPT2	3.88	4.15	4.40	4.71	5.53
Oscillator strength	0.062	0.014	Forbidden	Forbidden	0.253
Twisted Conformation					
CASSCF	5.76	6.29	6.07^a	5.56	7.86
CASPT2	4.36	4.22	4.90^a	4.99	5.79
Oscillator strength	0.049	0.034	0.013	0.0014	0.172
Experimental Energies					
	$3.86(0–0),^b 3.67(0–0)^c$		$4.96,^d 5.02^{e,f}$		6.46^e
	$4.13,^d 4.11,^e 4.09^f$		$4.48(0–0)^g$		5.93^f

[a] State computed using the active space (66) due to intruder-state problems.
[b] Fluorescence excitation spectrum of bithiophene seeded into a supersonic helium expansion [110].
[c] Fluorescence excitation spectrum of solid solutions of bithophene in *n*-hexane at 4.2 K [218].
[d] Gas-phase absorption spectrum at room temperature [110].
[e] UV absorption spectrum in cyclohexane [104].
[f] UV absorption spectrum in methanol [111].
[g] Two-photon fluorescence excitation spectrum of a dilute solution of bithiophene in crystalline *n*-hexane at 77 K [112].

state. Since the potential for the ground state is flat, these energy differences are almost identical to the difference between the two sets of excitation energies given in Table VIII. As a result, the computed excitation energies for the planar conformation are close to the 0–0 transition energies. The computed "vertical" energies (those for the twisted form) are more difficult to relate directly to experimental data, again because the potential for the ground state is flat while the upper potential is not. It should be noted that the transition moments vary strongly with geometry.

Four excited states were found at low energies (below 5.0 eV). A third ^1B_u state is found with energies around 5.5 eV (0–0 transition). The first excited state is of ^1B_u symmetry with a computed 0–0 energy of 3.88 eV. This is in agreement with the recent fluorescence excitation spectrum of Chadwick and Kohler [110], who place the 0–0 transition at 3.86 eV. The corresponding spectrum in solid solution of n-hexane places the band at 3.67 eV, corresponding to a solvent shift of -0.19 eV (this is almost exactly the red shift expected if the molecule is forced to be planar in the condensed phase). The second state is also of ^1B_u symmetry. This possibility was not discussed earlier. The 0–0 energy is 4.15 eV. It has considerably lower intensity than the first state in planar geometry, but they become more similar when the molecule is twisted. Gas-phase absorption spectroscopy finds the first band with a peak at 4.13 eV [110]. The band is broad with several features on the low-energy side, and it is not unlikely that it contains more than one transition. Also ultraviolet (UV) absorption spectra in solution show the same band [104, 111].

The first band of ^1A_g symmetry appears at 4.40 eV in the planar conformation. It is shifted to 4.90 eV when the molecule is twisted. The two-photon excitation spectrum of Birnbaum and Kohler [112] locates the 0–0 band of a state of ^1A_g symmetry at 4.48 eV, which is again in agreement with the present result. It is discussed whether they see the first excited state of this symmetry or the second. The present results strongly indicate that it is the first, but the 3^1A_g state is only 0.31 eV higher in energy. Considering the error bars of the computed excitation energies (0.3 eV), it cannot be ruled out completely that it is the third state that is observed. Both states are forbidden for planar geometry but become allowed when the system is twisted. UV absorption spectroscopy finds a second band with a peak at about 5.0 eV [104, 110, 111]. This band probably contains both ^1A_g transitions, which in the twisted conformation are almost degenerate.

The UV absorption spectra contain a third band with maximum at about 6.0 eV [104, 111]. The calculations assign this band to the third state of ^1B_u symmetry, which is computed to have a 0–0 transition energy of 5.53 eV, which increases to 5.79 eV at the twisted geometry. The UV absorption spectra have been recorded in solution and direct comparison is therefore not possible, since the ground-state conformation of the molecule is not known. It is probable that it is cis, at least in polar solvents. But the assignment of the band is clear. A fourth ^1B_u state has actually been calculated with an adiabatic energy of 6.02 eV (not included in Table VIII). The band at 6 eV is probably composed of both these transitions. The computed intensity for the latter state is, however, small.

The main features of the low-energy part of the electronic spectrum of bithiophene are understood. Corresponding studies of the ter-thiophene molecule are under way. Together, the results may be used to calibrate data obtained with more approximate methods. Extrapolation to larger oligomers becomes possible by using data computed with simpler semiempirical methods adjusted to fit *ab initio* results obtained for the dimer and trimer.

7. Spectroscopy of Protein Chromophores

The main features of the near- and far-ultraviolet spectra of the proteins are related to the absorption properties of the aromatic amino acids phenylalanine (Phe), tyrosine (Tyr), tryptophan (Trp), and histidine (His) [113, 114]. The peaks observed in the absorption spectra up to 185 nm (6.70 eV) can be assigned to the excited states of the chromophore-acting molecules benzene, phenol, indole, and imidazole, respectively. In the present section we focus on the theoretical description of the most representative valence singlet excited states of the aromatic amino acid chromophores. As the results for benzene and phenol have been recently described [13, 46], only the results for indole, [(4) in Fig. 3] and imidazole [(5) in Fig. 3] are reviewed here [115, 116]. The theoretical results support the assignment of four $\pi \to \pi^*$ valence singlet states as responsible for the most relevant observed peaks: two low-lying and weaker states usually named ^1L_b and ^1L_a, the splitting of which depends on the specific system, and two higher ^1B_b and ^1B_a states with larger intensities. Imidazole (and therefore histidine) is different, however, with only two transitions below 7 eV.

It is the configurations related to the HOMO, HOMO − 1 and LUMO, LUMO + 1 orbitals that are responsible for the four singlet (and four triplet) states: the HOMO → LUMO state, two states as a mixture of the nearly degenerate HOMO − 1 → LUMO and HOMO → LUMO + 1 configurations, and a final HOMO − 1 → LUMO + 1 state [13, 117]. The relative order of the HOMO → LUMO state and the antisymmetric combination of the two nearly degenerate configurations will depend on the relative energies of the corresponding orbitals and the strength of the interaction. Even in highly symmetric systems, such as naphthalene, the HOMO − 1 → LUMO and HOMO → LUMO + 1 excitations belong to the same symmetry and are able to interact; the other two excitations, HOMO → LUMO and HOMO − 1 ← LUMO + 1, belong to another symmetry. In this case the large difference in energy prevents strong mixing of the two configurations. The intensities give a good measure of the strength of the interaction. The two low-lying states have low intensities, one because it is the antisymmetric combination of two configurations [118], the other because of the orbital structures and often because it contains a large fraction of doubly excited states. These are the L states as they were labeled by Platt [119]. The complementary states are pushed up in energy and have larger intensity. They were labeled B states [119]. Obviously, in large systems, more combinations are available and the structure becomes more complex. Configurations involving other orbital excitations mix together with doubly excited states, leading to a complex multiconfigurational character of the states. In addition, lowering the symmetry permits further mixing among the configurations. This happens, for example, in indole and imidazole, and modifies the simple model.

Indole Molecule. Tryptophan is the most important emissive source in proteins, with the indole chromophore, [(**4**) in Fig. 3] responsible for its low-energy absorption bands [120]. The studies of the electronic spectrum of indole have focused almost exclusively on the position and nature of the two low-lying valence singlet states, ^1L_b and ^1L_a, due to their apparent degeneracy and their different behavior in solvents. While the sharp ^1L_b band origin and maximum in gas phase were established at 4.37 eV [121, 122], the maximum (4.77 eV [122, 123] in gas phase) and the origin (estimated from 4.42 to 4.70 eV [123–126]) of the ^1L_a state have been disputed, because of the strong overlap between the two bands. The properties of the bands are sensitive to solvent effects and to substitutions in the indole ring [120]. The less studied group of intense transitions in the domain from 216 to 192 nm (5.74 to 6.45 eV) has been related to the strongly allowed $^1B_{a,b}$ manifold in polyhexacenes [114, 122]. The data concerning the triplet and Rydberg states are not discussed here.

The ground-state gas-phase equilibrium geometry was optimized at the CASSCF level employing an ANO basis set [41] contracted to N, C/$3s2p1d$, H/$2s$. The molecule is planar and was placed in the xy plane. The calculations of the vertical electronic spectrum used a larger ANO basis set [27] contracted to C,N/$4s3p1d$, H/$2s$ and supplemented with $1s1p1d$ Rydberg-type functions in the cation charge centroid. The active space included the eight valence π orbitals plus the nitrogen lone-pair orbital and appropriate Rydberg orbitals.

Naphthalene [127] can be used as a model for the general structure of the excited states. A previous CASSCF/CASPT2 study showed that the low-lying ^1L_b state (1^1B_{3u}) at 4.03 eV is a mixture of the above-mentioned singlet excited configurations, leading to a weak band ($f = 0.0004$). The opposite mixture of the same configurations gives the ^1B_b (2^1B_{3u}) state at 5.54 eV, which has the largest intensity in the spectrum ($f = 1.34$). The interaction is not as strong in the ^1L_a (1^1B_{2u}) state ($f = 0.050$) at 4.56 eV and the ^1B_a (2^1B_{2u}) state ($f = 0.31$) at 5.93 eV. Both are well represented by a single excitation, HOMO$\rightarrow$LUMO and HOMO $- 1 \rightarrow$ LUMO $+ 1$, respectively.

Table IX presents the results for the valence π states in indole. The four states discussed, ^1L_b, ^1L_a, ^1B_b, and ^1B_a, computed at 4.43, 4.73, 5.84, and 6.44 eV, respectively (see also Table XII), are in agreement both in energies and intensities with the experimental data. The ^1L_b state, at 4.43 eV ($f = 0.05$), has the expected composition of the wavefunction: HOMO $- 1 \rightarrow$ LUMO, 44% and HOMO $\rightarrow$ LUMO $+ 1$, 22%. As pointed out, it is the antisymmetric combination of the configurations that gives the low value of the oscillator strength. Both energy and intensity are in agreement with the experimental values, 4.37 eV [121, 122] and 0.05 [128]. The ^1L_a state is composed mainly of the HOMO $\rightarrow$ LUMO configuration, 54%. The computed excitation energy, 4.73 eV, agrees with experiment, 4.77 eV.

The most intense feature in the spectrum was computed at 5.84 eV ($f = 0.46$). It is the ^1B_b state, which has a wavefunction composed mainly of the HOMO $- 1 \rightarrow$ LUMO, 11%, and HOMO $\rightarrow$ LUMO $+ 1$, 42%, configurations. Several peaks and shoulders in the region 5.80–6.02 eV [122, 129] have been assigned to the most intense band. Finally, the ^1B_a state is computed at 6.44 eV ($f = 0.26$), in agreement with the band observed at 6.35 eV [122, 129]. Both ^1B_b and ^1B_a bands have lower intensities here than they have in naphthalene, a consequence of the larger configurational mixing in the wavefunction due to the lowering of the symmetry. The computed CASSCF dipole moments give us some hints about the expected behavior of the various bands in the presence of polar solvents. The ground-state dipole moment is computed to be 1.86 D (experiment: 2.09 D [130]). The dipole moment for the ^1L_b state, 0.85 D, differs, however, from experiment 2.3 D [131]. The difference is outside the error limits of the theoretical value. The computed dipole moment for the ^1L_a state, 5.7 D, is in better agreement with experiment, 5.4 D [132]. For the ^1B_b and ^1B_a states the values computed are 3.8 and 1.8 D, respectively. The ^1L_a and ^1B_b can be expected to be more affected by environmental effects because of the larger changes in their dipole moments compared to the ground state. This is seen clearly in the experimental spectra [129].

A less intense state at 6.16 eV with 16% double excitations in the wavefunction, and two higher and intense states, also with important contributions from doubly excited configurations, complete the computed valence spectrum. The Rydberg states series are not going to be described here, but the beginning of two series were calculated at 4.85 eV ($5a'' \rightarrow 3s$ state) and 5.33 eV ($4a'' \rightarrow 3s$ state).

In addition to the energies and intensities we have also included the transition moment directions in Table IX. These quantities, obtained in UV or infrared (IR) linear dichroism experiments, have been used

TABLE IX

Calculated and Experimental Excitation Energies, Oscillator Strengths, Dipole Moments (μ), and Transition Moment Directions for the $\pi \to \pi^*$ Excited Valence Singlet States in Indole

State	Excitation Energy (eV)			μ (D)	Oscillator Strength		TM Direction[c] (deg)	
	CAS	PT2	Expt.[a]		This Work	Expt.[b]	This Work	Expt.[d]
1¹A'	—	—	—	1.86	—	—	—	—
2¹A'	4.83	4.43	4.37	0.85	0.050	0.045	+37	+42 ± 5
3¹A'	6.02	4.73	4.77	5.69	0.081	0.123	−36	−46 ± 5
4¹A'	6.97	5.84	6.02	3.77	0.458	—	+16	0 ± 15
5¹A'	6.74	6.16	—	0.93	0.003	—	+5	—
6¹A'	7.35	6.44	6.35	1.80	0.257	—	−55	> ± 30
7¹A'	7.52	6.71	—	2.34	0.138	—	−10	—
8¹A'	7.98	6.75	—	1.12	0.245	—	−12	—

[a] Experiments from Refs. [121–123 and 129].
[b] From Ref. 128.
[c] The molecule is placed on the xy plane. The ring-shared bond is the y axis. The x axis is defined to be the pseudosymmetry long axis of indole, which joins the end carbon of the pyrrole moiety with the midpoint of the ring-shared bond. Angles are defined counterclockwise from the x axis. The nitrogen is placed in the fourth quadrant.
[d] Experimental data from Ref. 133.

extensively for identifying excited states. Among all the experimental data the most recent measurements [133] have been selected in Table IX to show the agreement between theory and experiment.

Imidazole Molecule. The absorption spectra of imidazole [(5) in Fig. 3] in various solvents exhibit a broad band with maximum around 207 nm (6.00 eV) [134–136], which is usually attributed to the first $\pi \to \pi^*$ transition. A second band was observed in the 195–178 nm (6.38–6.97 eV) region, with a modest dependency upon solvent and pH [136]. Several weak bands have been reported at low energies. Our calculations have assigned them as triplet states and they were described and discussed in the original paper [116]. The recent study by Caswell and Spiro [135] reported a broad absorption band envelope at 207 nm (6.00 eV) with two overlapping peaks at 218 nm (5.69 eV) and 204 nm (6.08 eV), which were assumed to correspond to two different $\pi \to \pi^*$ transitions.

The theoretical studies on imidazole are scarce. A recent MRCI study by Machado and Davidson [137] considered valence and Rydberg states together for the first time. On the other hand, the available experimental spectra of imidazole are limited to methanol, ethanol, and aqueous solutions and it is difficult to establish the effects of the solvent on the different transitions. The structure of the computed gas-phase spectrum differ from the spectra in a solvated medium. An unambiguous assignment of the main valence bands in the spectrum of imidazole thus requires consideration of such effects. The reaction field method, explained above, was employed with the CASSCF/CASPT2 method to analyze the influence of the solvent on the valence $\pi \to \pi^*$ states of imidazole and assign the most important features.

The experimental ground-state geometry [138] was used with the molecule in the xy plane. ANO basis functions contracted as C, N/ $4s3p1d$, H/$2s1p$, were used, supplemented with $1s1p1d$ Rydberg functions in the cation charge centroid. The study of the electronic spectrum of imidazole involved $\pi \to \pi^*$, $n \to \pi^*$, and Rydberg states (in the gas phase). The optimal radius for the cavity in the reaction field calculations was computed to be 7.0 bohr in both water and ethanol.

Table X compiles the computed gas-phase excitation energies, oscillator strengths, and dipole moments for the valence singlet excited states of the imidazole molecule. Previous MRCI results by Machado and Davidson [137] are also included. Table XI lists the results obtained in the reaction field model. Three valence $\pi \to \pi^*$ and two valence $n \to \pi^*$ states have been found in the gas-phase spectrum up to 8.5 eV. In addition, the beginning of three Rydberg series have been computed from

TABLE X
Calculated Excitation Energies, Oscillator Strengths, and Dipole Moments (μ) for the Excited Valence Singlet States in Gas-Phase Imidazole

| | Excitation energy (eV) | | | μ | Oscillator Strength | |
State	CAS	PT2	MRCI[a]	(D)	This Work	MRCI[a]
$1^1A'$	—	—	—	3.70	—	—
$1^1A''(n\pi^*)$	7.02	6.52	—	0.22	0.011	—
$2^1A'(\pi\pi^*)$	7.51	6.72	7.71	4.61	0.126	0.080
$3^1A'(\pi\pi^*)$	8.43	7.15	8.18	3.00	0.143	0.070
$2^1A''(n\pi^*)$	8.42	7.56	—	2.79	0.013	—
$4^1A'(\pi\pi^*)$	10.06	8.51	—	3.85	0.594	—

[a] CIR6 calculations from Machado and Davidson [137].

TABLE XI
Calculated and Experimental Excitation Energies, Oscillator Strengths, (f) and Dipole Moments (μ) for the Valence Singlet States of Imidazole in Gas Phase, Ethanol, and Water

| | Gas phase | | | | Ethanol | | | Water | | |
State	PT2 (eV)	f	μ (D)	Expt.[a]	PT2 (eV)	f	μ (D)	PT2 (eV)	f	μ (D)
$1^1A'$(ground state)	—	—	3.70	—	—	—	4.09	—	—	4.07
$2^1A'(\pi\pi^*)$	6.72	0.126	4.61	6.0	6.32	0.036	3.61	6.32	0.024	3.57
$3^1A'(\pi\pi^*)$	7.15	0.143	3.00	6.5	6.53	0.307	3.42	6.53	0.275	3.41
$4^1A'(\pi\pi^*)$	8.51	0.594	3.85	—	7.48	0.600	4.76	7.48	0.561	4.75

[a] Experimental excitation energies are the same in ethanol and water [135, 136].

the highest occupied orbitals $3a''(\pi)$, $2a''(\pi)$, and $15a'(n)$ to the $3s$ Rydberg orbital at 5.71, 7.10, and 7.10 eV, respectively. The details of this part of the spectrum are discussed elsewhere [116]. The gas-phase results place a $n \rightarrow \pi^*$ state at 6.52 eV as the lowest singlet excited valence state with low intensity, while the two valence $\pi \rightarrow \pi^*$ singlet states appear at 6.72 and 7.15 eV with moderate intensities. The most intense transition is the third valence $\pi \rightarrow \pi^*$ state at 8.52 eV. Analysis of the wavefunction shows strong mixing of different configurations in the low-lying states. One of the most important consequences is the increase in intensity in both states and a general increase in the energies with respect to the other protein chromophores. The indole model can be related to the interacting double-bond model explained in the methylenecyclopropene section. In imidazole, however, the general model should be only considered an approximation. As an illustration, the ^1L_b state wavefunction is composed of not only the two configurations expected but three: HOMO $\rightarrow$ LUMO, 37%; HOMO $-1 \rightarrow$ LUMO, 23%, and

HOMO $\rightarrow$ LUMO + 1, 25%, while the corresponding weights for the ^1L_a state are 53%, 24%, and 12%.

As the experimental spectra for imidazole are available only in the polar solvents water and ethanol, there is no knowledge of the solvent effects on the spectrum. The addition of the effects of the environment by using the reaction field method, as explained in an earlier section, emphasizes the sensitivity of the imidazole states to the presence of a polar solvent. The theoretical results are presented in Table XI. No important differences were found for the solvents water and ethanol within the present model. The calculated excitation energies were 6.32, 6.53, and 7.48 eV in both solvents. The experimental spectra have peaks at 6.0 and around 6.5 eV [135, 136] for both water and ethanol as solvents.

As can be inferred from the data in Table XI, solvation leads to large changes in dipole moments. For the $2^1A'$ state, the dipole moment drops by more than 1 D. The solvent effects already appear to be saturated for ethanol. We also note that these changes are much larger for imidazole than they were for the ABN molecule (see Section III.A.3). It should be pointed out that the theoretical model used is crude and is not expected to provide extremely accurate shifts in excitation energies and other properties. Some expectation values, as for example the dipole moments computed at the CASSCF level, turn out to be sensitive to various parameters, such as the basis set, the active space, and the number of roots selected in the CASSCF state average calculations. Test calculations [116] have shown, however, that in this specific case the large solvent shifts are an artefact of the CASPT2 method. Too large excitation energies are obtained in the gas phase, due to an incorrect mixing of Rydberg and valence excited states, similar to what has been observed in ethene and butudiene [36].

Electronic Spectra of Proteins. The chromophore groups in a protein form a complex system of molecular fragments differing in structure and position and contributing as a whole to the spectroscopic properties of the protein. The absorption spectrum of a protein can be related to the superposition of the spectra of its chromophore groups. The most characteristic features have been found in the region 220–190 nm (5.64–6.53 eV), where the peptide or the amide chromophores absorb, and the range around 280 nm (4.43 eV), in which tyrosine and tryptophan absorption occur [114]. When aromatic amino acids are present, the energy region typical of peptide and amide absorption shows intense features corresponding to the higher states of the aromatic chromophores. Table XII compiles our theoretical and the experimental results

TABLE XII

Theoretical and Experimental Energies (eV) and Oscillator Strengths (in parentheses)[a] for the Most Representative Valence Singlet Excited States of the Aromatic Amino Acids[b] and Their Chromophores

| State | Benzene | | Phenylalanine, | Phenol | | Tyrosine, |
	Theory	Expt.	Expt.	Theory	Expt.	Expt.
^1L_b	4.84 (forb.)	4.90 (—)	4.85 (0.01)	4.53 (0.007)	4.51 (0.02)	4.51 (0.02)
^1L_a	6.30 (forb.)	6.20 (—)	6.00 (0.16)	5.80 (0.005)	5.77 (0.13)	5.70 (0.15)
^1B_b	—	6.94 (1.05)	6.60 (1.00)	6.50 (0.68)	—	—
^1B_a	7.02 (0.82)	—	—	6.56 (0.78)	6.60 (1.1)	6.46 (0.71)

| State | Indole | | Tryptophan, | Imidazole | | Histidine, |
	Theory	Expt.	Expt.	Theory[c]	Expt.	Expt.
^1L_b	4.43 (0.05)	4.37 (0.05)	4.31 (—)	6.32 (0.02)	6.0 (—)	5.7–6.3 (0.11)
^1L_a	4.73 (0.08)	4.77 (0.12)	4.60 (0.10)	6.53 (0.28)	6.5 (—)	5.7–6.3 (0.11)
^1B_b	5.84 (0.46)	6.02 (—)	5.69 (0.62)	7.48 (0.56)	>7.0 (—)	>7.0 (—)
^1B_a	6.44 (0.26)	6.35 (—)	6.35 (0.38)	—(—)	—(—)	—(—)

[a] Oscillator strength values in amino acids have been normalized to unity for the most intense band of phenylalanine. They are used only for comparison.

[b] Amino acids experimental data in neutral aqueous phase from Refs. 113 and 114. See also Refs. 13, 115, and 116 and the text.

[c] Theoretical CASPT2 energies include the solvent effects in imidazole.

for the aromatic amino acids and their chromophores. There is agreement between the theoretical (see Refs. 13, 46, 115, and 116) and experimental results on the molecules benzene, phenol, indole, and imidazole for energies and intensities. Experimentally, a general lowering of the energies can be observed in the amino acids, due to substitutions in the rings and solvent effects. All experimental data for the amino acids have been obtained in aqueous solutions. In some cases, increased intensities are observed due to symmetry breaking, for instance, in going from benzene to phenylalanine. The basic structure of the excited states, as discussed above, is, however, easy to observe: two low-lying weak ^1L_b and ^1L_a states, with the second usually more intense and more affected by the solvent environment, and two high-lying, more intense ^1B_b and ^1B_a states, actually overlapped in phenylalanine and tyrosine. Energies and intensities reflect the nature of the states and the configurational mixing observed in their wavefunction. The most important deviation from this scheme appears in imidazole and also in histidine. Here the solvent effects have been included in the theoretical results due to the importance of the environment for the energy of the low-lying states. The observed spectrum has only a single broad absorption band between 5.7 and 6.3 eV, while the next band appears at 7.0 eV [113, 114]. The theoretical results suggest the presence of both ^1L_b and ^1L_a bands in the 5.7–6.3 eV band envelope.

8. Electronic Spectra of Radical Cations of Linear Conjugated Polyenes and Polycyclic Aromatic Hydrocarbons

Linear conjugated polyenes (LCP) and polycyclic aromatic hydrocarbons (PAH) as well as their ions are key reactants in many chemical processes or can be identified as transient species. They have rich and broad electronic spectra, ranging from the ultraviolet to the visible and near-infrared [139, 140]. The electronic spectra of linear conjugated polyenes have served as tests for computational chemistry ever since the first semiempirical calculations were performed and have been discussed extensively in literature. However, it was not until recently that accurate, theoretical predictions of excited-state properties were attained by *ab initio* calculations [36, 141].

The principal experimental techniques to characterize the LCP radical cations have been photoelectron spectroscopy (PES) and electron absorption (EA) in matrices [142–150]. The production of radicals in matrices is a slow process and hence EA probes the electronic structure of the chromophore at the equilibrium geometry of the radical ion. In PES the chromophore has no time to relax. However, even ionization results in a remarkable attenuation of the bond-length alternation in comparison with the neutral polyenes, the difference in the excitation energies derived from PES and EA are marginal.

Recently, Fülscher et al. [151] presented calculations on excited states in LCP radical cations using the CASSCF/CASPT2 method and ANO-S basis sets contracted as C, N[$3s2p1d$]/H[$2s$]. The geometries of the LCPs were optimized at the RHF/MP2 and ROHF/MP2 level of approximation, respectively, using 6-31G* basis sets. The calculated excitation energies shown in Table XIII were accurate to 0.2 eV or better. For the 2^2A_u state of octatetraene the shift upon going from the cation to the neutral geometry has been calculated to 0.16 eV, in agreement with the experimental value 0.2 eV [145].

The wavefunctions of the LCP radical cations are of rather simple structure and are dominated by single excitations. The two lowest π-excited states of equal symmetry are characterized by configuration mixing of two competing single excited configurations: HOMO − 1 → HOMO and HOMO → LUMO. Since the LCP radical cations are electron deficient, the wavefunctions are compact and well described by basis sets of modest size. Rydberg states do not appear in the low-energy part of the spectrum.

TABLE XIII
π Excitation Energies of Linear Conjugated Polyene Radical Cations

State	Photoelectron Spectrum			Electron Absorption Spectrum		
	ΔE_{calc} [a]	ΔE_{expt} [b]	f_{calc} [c]	ΔE_{calc} [a]	ΔE_{expt} [b]	f_{calc} [c]
			All-*trans*-1,3-butadiene			
1^2A_u	2.36	2.40	0.049	2.47	2.30	0.014
2^2A_u	4.56	—	0.644	4.23	4.20	0.651
			All-*trans*-1,3,5-hexatriene			
1^2B_g	1.93	1.96	0.073	1.98	1.96	0.013
2^2B_g	3.59	—	0.910	3.37	3.30	1.001
2^2A_u	3.26	—	Forbidden	3.45	—	Forbidden
			All-*trans*-1,3,5,7-octatetraene			
1^2A_u	1.63	1.67	0.116	1.66	1.67	0.0158
2^2A_u	3.04	3.00	1.238	2.86	2.77	1.370
2^2B_g	2.88	3.16	Forbidden	2.99	—	Forbidden
3^2A_u	3.72	—	0.001	3.88	—	0.001

[a] Calculated excitation energy in eV using the CASSCF/CASPT2 method.
[b] Experimental excitation energy in eV taken from Ref. 145 and references therein.
[c] Calculated oscillator strength.

It is suspected that 0.2% of interstellar carbon may exist in the form of PAH radical cations and suggested that they may be possible carriers of the visible, diffuse interstellar absorption band (DIB), which extends from 4000 Å into the near IR [152]. Encouraged by the results on the LCP radical cations, Schütz et al. [153] undertook to calculate the electronic spectra of PAH radical cations starting with naphthalene [(6) in Fig. 3] and anthracene [(7) in Fig. 3].

The naphthalene cation has been studied by absorption spectroscopy in solutions [154, 155], glasses [156–159], and matrices [160]. Gas-phase data stem from PES [161, 162] and multiphoton dissociation [163, 164] spectra. Recently, the 180- to 900-nm absorption spectrum of naphthalene$^+$ formed by photoionization in a neon matrix has been reinvestigated [165]. These experiments show seven progressions, with peak maxima at 1.84, 2.72, 3.29, 4.06, 4.49, 5.07, and 5.57 eV. The excitation energies calculated by Schütz et al. for the cation geometry are shown in Table XIV. They are in agreement with experiment. The largest deviation amounts to 0.13 eV (2^2B_{2g}). Table XIV also shows the predicted PES, which matches experiment.

Unfortunately, the experimental information for anthracene$^+$ is not as rich as for naphthalene$^+$. Transitions to the lowest π-excited state occur at 1.1 eV and are dipole forbidden. In addition, PES [161, 162] predicts transitions at 1.76, 2.73, and 2.81 eV. The EA spectrum [166] of anthracene$^+$ recorded in argon matrices agrees with the absorption maxima in organic glasses [140], and the vibronic fine structure is in accord with the Raman spectrum of polycrystalline anthracene [167] at room temperature. The CASSCF/CASPT2 calculations presented by

TABLE XIV
π-Excited States of the Naphthalene Radical Cation

State	Photoelectron Spectrum			Electron Absorption Spectrum		
	ΔE_{calc} [a]	ΔE_{expt} [b]	f_{calc} [c]	ΔE_{calc} [a]	ΔE_{expt} [d]	f_{calc} [c]
1^2B_{1u}	0.73	0.73	Forbidden	0.98	—	Forbidden
1^2B_{2g}	1.78	1.90	0.052	1.90	1.84	0.044
1^2B_{3g}	2.63	2.68	0.017	2.64	2.72	0.001
2^2B_{1u}	3.47	—	0.068	3.28	3.29	0.087
2^2B_{2g}	4.01	—	0.078	3.94	4.07	0.051
2^2B_{3g}	3.99	—	Forbidden	3.98	—	Forbidden

[a] Calculated excitation energy in eV using the CASSCF/CASPT2 method.
[b] Experimental excitation energy in eV from Ref. 161.
[c] Calculated oscillator strength.
[d] Experimental excitation energy in eV from Ref. 165.

TABLE XV
π-Excited States of the Anthracene Radical Cation

State	Photoelectron Spectrum			Electron Absorption Spectrum		
	$\Delta E_{calc}{}^a$	$\Delta E_{expt}{}^b$	$f_{calc}{}^c$	$\Delta E_{calc}{}^a$	$\Delta E_{expt}{}^d$	$f_{calc}{}^c$
1^1B_{3g}	1.13	1.12	Forbidden	1.30	—	Forbidden
1^2A_u	1.48	1.76	0.084	1.56	1.72	0.079
1^2B_{1u}	2.16	—	0.002	2.03	2.19	0.010
2^2B_{2g}	2.56	2.73	Forbidden	2.66	—	Forbidden
2^2B_{1u}	2.66	2.81	0.070	2.75	2.89	0.063
2^2A_u	3.15	—	0.091	3.08	—	0.114
2^1B_{3g}	3.38	—	Forbidden	3.31	—	Forbidden

^{a} Calculated excitation energy in eV using the CASSCF/CASPT2 method.
^{b} Experimental excitation energy in eV from Ref. 161.
^{c} Calculated oscillator strength.
^{d} Experimental excitation energy in eV from Ref. 166.

Schütz et al. (cf. Table XV) agree with experimental data, although the deviations are somewhat larger than for naphthalene$^+$.

As indicated by Schütz et al., the CASSCF/CASPT2 calculations on the larger PAH radical cations are at the limit of the current implementation of the method and available resources. Presently, it is not possible to compute the high-energy spectrum of large PAH radical cations, because of the demands on the size of the active space. On the other hand, these calculations illustrate that the CASSCF/CASPT2 method is capable of yielding reliable, quantitative results in areas not accessible to experiment, for example, molecules or molecular ions appearing in low concentrations in stellar atmospheres or in interstellar matter.

B. Electronic Spectroscopy for Transition Metal Compounds

The CASPT2 method has been applied successfully in a number of studies of properties of transition metal atoms and compounds containing one or two metal atoms. Most of these applications have dealt with ground-state properties, such as structure and binding energies [23, 28, 29, 168, 169], but a number of studies of electronic spectra and excited states have also been performed [22, 24, 32, 170, 171]. The first transition metal system for which the CASPT2 approach was applied was the nickel atom. The reason for the interest in this system was an earlier MRCI study, which had shown that it was necessary to include two sets of d-type orbitals ($3d$ and $4d$) into the reference space for the SD-MRCI wavefunction in order to get accurate results for the splitting between the d^8s^2, d^9s, and d^{10} states [172]. The CASPT2 study showed that it was indeed possible to obtain accurate relative energies, but only with an active space comprising 14 orbitals ($3d$, $4d$, $4s$, and $4p$). The results were actually more accurate than those obtained in the earlier MRCI study. The $3d$ *double-shell effect* was clear: Adding a second $3d$ shell to the active space changes the $d^9s \rightarrow d^{10}$ excitation energy from 0.42 to 1.87 eV (the final result obtained by adding core–valence correlation corrections is 1.77 eV, only 0.03 eV from the experimental value). While it is gratifying that the CASPT2 method is able to compute this notoriously difficult energy difference, it is less satisfactory that such a large active space was needed. The study also pointed to the importance of including $3s$, $3p$ correlation terms. Their effect on the relative energies were on the order of 0.1 eV. As was well known from earlier studies, it is necessary to add an estimate of the relativistic effects by means of first-order perturbation theory.

Below we illustrate how the experience obtained in the nickel study has been transferred to the molecular case. It has been shown in studies of structure and bonding that the $3d$ double-shell effect is important when the number of $3d$ electrons changes. Thus the extra $3d$ shell plays a crucial role for the electronic structure in NiH and CuH, while bonding in transition metal dimers can be described without the second $3d$ shell [23, 168]. However, there is an important modification compared to the atomic case. It was found that a $4d$ orbital is needed only when the doubly occupied $3d$ orbital does not interact directly with ligand orbitals. If it does, it is necessary to have the corresponding ligand orbital active, and the addition of one more orbital does not affect the result. Instead, the active orbital describing the interaction with the ligand will acquire a certain amount of $4d$ character; how much depends on the covalency of the metal ligand bond. Thus one is led to an active space that will comprise 10 orbitals with up to 10 active electrons for all compounds involving a transition metal ion and closed-shell ligands. We shall see that such a binding model will work for many systems and will give an accurate description of the spectroscopy, provided that extra active orbitals are added in the case of charge transfer excitation to ligand orbitals not included in the 10-orbital space. It is actually the first consistent and quantitative model for the bonding in transition metal complexes that is capable of describing the ground-state structure, the metal ligand bond strength, and the spectroscopy simultaneously. The examples given in the following sections illustrate this. There are cases where the model does not work. They are characterized by the need for larger active spaces than the 14-orbital limit set by the present technology. The chromium dimer is an example of a system with a complex metal–metal bond. It is shown that the spectroscopy of this system can be well described within the present model and with an active space comprising only the valence orbitals $3d$ and $4s$. The nickel and copper dimers have been treated successfully with a similar active space [168] and the same approach has been used for larger di-metal complexes such as cyclobutane–diazadivanadium [169] and dichromium–tetraformate [173].

1. Special Considerations

The accurate calculation of molecules containing transition metal atoms or ions requires some special considerations, which are absent (or at least less important) for the lighter elements in most organic molecules. First, care has to be taken in constructing the CASSCF reference wavefunction. For all but the smallest systems it is not possible simply to include all valence electrons into the active space; most often a careful choice of active electrons and orbitals has to be made, based on the knowledge of electronic structure properties. Given the wide variety of different bonding types that can be formed between a transition metal and its surroundings, this choice is not always obvious a priori, and different options may have to be considered. In this section we present an overview of some important near-degeneracy effects that have been encountered during our recent applications of the CASSCF/CASPT2 method in transition metal chemistry.

Another feature specific to the correlation problem in transition metals and their compounds concerns the role played by the $3s, 3p$ semicore electrons. It has been known for a long time [174] that the $3s, 3p$ orbitals in the first-row transition metal atoms cannot be considered truly core orbitals. The $3s, 3p$ core orbitals are confined to the same region of space as the $3d$ valence orbitals and their role in the correlation treatment can therefore not simply be ignored if accurate results are to be obtained. It has been shown [174] that $3s, 3p$ correlation thoroughly affects the energy separation between states, with a different number of $4s$ electrons in transition metal atoms, but more so for the elements at the left side of the series than for Ni and Cu, where its importance becomes small. In transition metal compounds, the importance of the $4s$ orbital decreases as the metal gets surrounded by more ligands, and the lowest electronic transitions most often occur between terms belonging to the same $3d$ configuration. A detailed investigation of the importance of $3s, 3p$ correlation for this type of transitions was performed in a recent CASSCF/CASPT2 study [171] of the lowest excited states in the transition metal ions Ti^{2+}–Ni^{2+} and V^{3+}–Cu^{3+}. It was shown in this systematic study that the core–core correlation contributions were as important as the core–valence contributions. Of special importance were double excitations from the $3p$ shell to the $3d$ shell. The total effect could amount to up to $0.5\,eV$ for relative energies within a given $(3d)^n$

configuration. We refer to the original paper for details [171]. The role of $3s$, $3p$ correlation has been investigated further in all molecular applications presented in this review, including the charge-transfer transitions in $Cr(CO)_6$ and $Ni(CO)_4$ (Section III.B.2).

Up to now, applications of the CASSCF/CASPT2 method to transition metal systems have been confined to compounds containing transition metals belonging to the first transition series. Although relativistic effects are most certainly less important here than for the members of the second and third series, they can often not be ignored if accuracy is required. For example, it is well known that including the mass-velocity and Darwin terms from the Pauli equation has a profound effect (up to 0.35 eV) on the relative energy of the $3d^n4s^2$–$3d^{n+1}4s^1$–$3d^{n+2}$ states in the heaviest elements Ni–Cu, where they act to stabilize "s-electron rich" configurations [175]. Since, on the other hand, correlation effects generally favor $3d$ over $4s$ occupation, both effects tend to cancel. Ignoring relativistic effects may, therefore, in combination with an incomplete treatment, easily lead to the right result for the wrong reasons. Relativistic effects are quite often also important for the bonding in transition metal compounds, where they may alter the shape of potential curves, bond distances, and binding energies. The only case where the Darwin and mass-velocity terms may safely be ignored is in the calculation of $3d \rightarrow 3d$ transitions in the spectra of transition metal compounds. However, in this case one may have to deal with the effect of spin-orbit coupling, at least in cases where different states have close-lying energies. In this section we present the results of a recent CASSCF/CASPT2 study on $ScCl_2$, $CrCl_2$, and $NiCl_2$, which may be considered as a case study of relativistic effects in first-row transition metal compounds. Indeed, due to the simultaneous occurrence of a strong $3d$–$4s$ mixing and a weak ligand field splitting of the M^{2+} ground state, both the mass-velocity and Darwin terms and spin-orbit coupling effects turned out to play a crucial role in the determination of the ground state and the relative energy of low-lying excited states in these systems.

Near-Degeneracy Effects and Choice of the Active Space. Although the emphasis of the present review is on spectroscopy, this section is concerned primarily with the ground state. Indeed, in most organic systems the HF configuration dominates the ground-state and near-degeneracy effects become important only for excited states. The same is no longer true when entering the field of transition metal chemistry. It is well known, for example, that ground-state properties of molecules such as $Cr(CO)_6$ and ferrocene are hard to describe using single reference methods, due to the presence of large nondynamical correlation effects.

Thus the most elaborate study performed on $Cr(CO)_6$ using the single-reference CCSD(T) method [176] still captured only 71% of the total binding energy.

One way to look at the problem of nondynamical correlation is to consider the occupation numbers of the natural orbitals resulting from a CASSCF calculation. Such a procedure is not foolproof of course, since it does not reveal any near-degeneracy effects that have not already been included in the CASSCF active space. Additional confirmation of the fact that the chosen active space is indeed adequate for the problem under consideration must therefore be obtained from the results of subsequent CASPT2 treatment, more specifically from the absence of any important intruder states in the final first-order wavefunction. Most of the molecules have been treated successfully with the CASPT2 method, using the active spaces presented in this section [28, 29, 170, 177, 178]. We will, however, also present two examples, CrF_6 and CrO_4^{2-}, for which the CASPT2 method turned out to be less successful, since all nondynamical correlation effects in these molecules could not be captured with an active space of at most 14 orbitals, the limit of the present CASPT2 code.

In Tables XVI and XVII we have collected the natural orbital occupation numbers from different CASSCF calculations on a selection of octahedral and tetrahedral molecules of first-row transition metals. In all cases a CASSCF calculation was performed using a "basic" active space of 10 orbitals. For some of the molecules additional calculations with larger active spaces (up to 14 orbitals) are also presented. Only ground-state results are presented, except for the molecules CrF_6^{3-}, $Cr(CN)_6^{3-}$, CoF_6^{3-}, and $Co(CN)_6^{3-}$, for which we have included a selected number of excited states.

For all octahedral molecules presented in Table XVI, the basic 10-orbital active space consists of the metal $3d$ orbitals, residing in the representations $t_{2g}(3d_\pi)$ and $e_g(3d_\sigma)$ as well as their bonding or antibonding counterparts within the same representations. All systems are characterized by a $t_{2g}(3d_\pi)^n$ ground state (where $n = 0, 3, 6$), with a formally empty antibonding $e_g(3d_\sigma)$ shell. Within the e_g representation, the corresponding bonding σ ligand orbitals were therefore included. Within t_{2g}, the included orbitals depend on the character of the ligand concerned: for the π-donor ligands F^-, H_2O, and NH_3 a doubly occupied ligand π shell was added, whereas for the π acceptors CN^- and CO, the included orbitals are the low-lying unoccupied π^* orbitals of t_{2g} symmetry.

For the Cr^{3+} and Co^{3+} compounds, the natural orbital occupation numbers in Table XVI reveal some clear trends, which can be related to the nature of the metal–ligand interaction. First we notice that for the π donors F^-, H_2O, and NH_3, correlation effects within the t_{2g} representa-

TABLE XVI

Natural Orbital Occupation Numbers Resulting from Different CASSCF Calculations on Some Representative Octahedral Transition Metal Compounds

Complex	State	Ligand			Metal $3d$		Ligand π^*	
		Other	t_{2g}	e_g	t_{2g}	e_g	t_{2g}	Other
CrF_6^{3-}	$^4A_{2g}$	—	5.988	3.998	3.002	0.012	—	—
	$^4T_{2g}$	—	5.997	3.994	2.002	1.006	—	—
$Cr(H_2O)_6^{3+}$	$^4A_{2g}$	—	5.998	3.987	3.002	0.013	—	—
$Cr(NH_3)_6^{3+}$	$^4A_{2g}$	—	6.000	3.969	3.000	0.031	—	—
$Cr(CN)_6^{3-}$	$^4A_{2g}$	—	—	3.947	2.971	0.053	0.029	—
	$^4T_{2g}$	—	—	3.969	1.984	1.029	0.018	—
CoF_6^{3-}	$^5T_{2g}$	—	6.000	3.998	4.001	2.001	—	—
	$^1T_{1g}$	—	5.999	3.989	4.957	1.056	—	—
	$^1A_{1g}$	—	6.000	3.982	5.901	0.118	—	—
$Co(H_2O)_6^{3+}$	$^1A_{1g}$	—	6.000	3.971	5.915	0.082	—	—
$Co(NH_3)_6^{3+}$	$^1A_{1g}$	—	6.000	3.936	5.923	0.096	—	—
$Co(CN)_6^{3-}$	$^1A_{1g}$	—	—	3.915	5.905	0.099	0.081	—
	$^1T_{1g}$	—	—	3.939	4.936	1.006	0.059	—
	$^5T_{2g}$	—	—	3.975	4.970	2.021	0.034	—
$Cr(CO)_6$	$^1A_{1g}$	—	—	3.958	5.763	0.049	0.230	—
	$^1A_{1g}$	—	—	3.945	5.738	0.050	0.226	$0.041(t_{1u})$
	$^1A_{1g}$	—	—	3.953	5.751	0.050	0.226	$0.020(t_{2u})$
	$^1A_{1g}$	—	—	3.953	5.756	0.051	0.230	$0.009(t_{1g})$
CrF_6	$^1A_{1g}$	—	5.819	3.860	0.183	0.139	—	—
	$^1A_{1g}$	$5.861(t_{1g})$	5.796	3.857	0.344	0.142	—	—
	$^1A_{1g}$	$5.888(t_{1u})$	5.825	3.862	0.214	0.211	—	—
	$^1A_{1g}$	$5.901(t_{2u})$	5.803	3.862	0.297	0.137	—	—

TABLE XVII

Natural Orbital Occupation Numbers Resulting from Different CASSCF Calculations on Some Representative Tetrahedral Transition Metal Compounds

Complex	State	Ligand				Metal 3d		Ligand π^*		
		a_1	t_1	t_2	e	t_2	e	t_2	e	t_1
Ni(CO)$_4$	$^1A_1$	—	—	—	—	5.856	3.915	0.145	0.084	—
	$^1A_1$	—	—	—	—	5.855	3.915	0.143	0.084	0.004
Cr(NO)$_4$	$^1A_1$	—	—	—	—	0.297	3.773	5.697	0.233	—
	$^1A_1$	—	—	—	—	0.349	3.579	5.264	0.290	0.519
CrF$_4$	$^3A_2$	—	—	5.896	3.984	0.106	2.013	—	—	—
	$^3A_2$	1.988	5.990	5.896	3.984	0.124	2.020	—	—	—
CrO$_4$$^{2-}$	$^1A_1$	—	—	5.757	3.880	0.243	0.120	—	—	—
	$^1A_1$	1.943	5.885	5.727	3.879	0.360	0.205	—	—	—

tion are unimportant. For the Co^{3+} systems this is self-evident, since in this case both the bonding and antibonding t_{2g} shells are doubly occupied. Yet for Cr^{3+} also, with a half-occupied $t_{2g}(3d_\pi)$ shell, the occupation numbers of the corresponding ligand orbitals are close to 2 in all cases. The same is not, however, true for the e_g σ lone-pair ligand orbitals. Here we find a strongly decreasing occupation number within the series $F^- > H_2O > NH_3$, or with a deincreasing polarizability of the ligand. For the metals the trend is also clear: Except for CoF_6^{3-} in its $^5T_{2g}$ ground state, the Co^{3+} compounds are characterized systematically by a significant lower ligand e_g occupation number than the Cr^{3+} compounds.

The trends observed can also be connected to the covalency of the metal–ligand interaction. Indeed, although the subdivision in either metal or ligand orbitals in Table XVI is essentially correct, some mixing between both types of orbitals does occur in all cases. This mixing reflects the covalent contribution to the bonding—since no mixing would indicate a purely ionic interaction—and looking at the composition of the molecular orbitals, we notice that the mixing increases rather steeply within the series $F^- < H_2O < NH_3$ and $Cr^{3+} < Co^{3+}$. For example, at the RHF level, the ligand e_g orbitals contain a (derived from Mulliken gross populations) $3d$ contribution of 10.9% for CrF_6^{3-}, 12.7% for $Cr(H_2O)_6^{3+}$, and 20.5% for $Cr(NH_3)_6^{3+}$. For the cobalt compounds the corresponding numbers are 11.4%, 16.3%, and 24.5%. However, it is also clear that the covalency of the bonding is not yet captured to its full extent by the RHF wavefunction. Indeed, as indicated by the natural orbital occupation numbers in Table XVI, the CASSCF wavefunctions contain a significant contribution from excitations out of the predominantly ligand e_g orbitals into the antibonding metal $e_g(3d_\sigma)$ orbitals. These excitations go together with a transfer of charge density from the ligands into the metal, thus enforcing the covalency of the bonding. Furthermore, from the trends observed in the occupation numbers it is clear that the importance of this effect increases with the inherent covalency of the bonds. In other words, nondynamical correlation effects tend to become more important as the covalency of the metal–ligand bonds increases, while at the same time they serve to reinforce the covalency even further.

The cyanide compounds in Table XVI are no exception to this rule. On the contrary, of all Cr^{3+} and Co^{3+} complexes included, the cyanide complexes are definitely characterized by the most covalent bonds. In these complexes, σ donation from the carbon lone pairs into the formally empty $e_g(3d_\sigma)$ orbitals is counteracted by π backdonation from the filled $t_{2g}(3d_\pi)$ orbitals into the CN π^* orbitals. Both bonding types contain a significant covalent contribution. It should therefore come as no surprise that nondynamical correlation effects are more important for the cyanide compounds than for the other π-donor Cr^{3+} and Co^{3+} complexes in Table XVI. Correlation effects now serve to increase both σ donation and π backdonation. The effect is again largest for the cobalt complex, with 0.085 additional electron transferred from the carbon lone pair into the $3d_\sigma$ orbitals and 0.081 electron from $3d_\pi$ into the CN π^* orbitals.

The Cr^{3+} and Co^{3+} compounds in Table XVI belong to a group usually referred to as Werner complexes. They constitute a group of complexes treated by spectroscopy [179] and photochemistry [180] with considerable success in the 1970s and the beginning of the 1980s by means of simple semiempirical ligand field models, based on an ionic view of the metal–ligand interaction. The success of these models can be traced back to the fact that the bonding in these systems is indeed primarily ionic. Yet even within the framework of ligand field theory, the presence of covalent effects could not be ignored. They were included into the model by means of the *nephelauxetic effect* [181]. The term *nephelauxetic* is based on the Greek term "cloud expansion," which indicates its meaning. It was found that the interelectronic repulsion within the metal $3d$ shell (usually expressed in terms of the Racah parameters A, B, C) is reduced significantly in a transition metal complex compared to the free metal ion, as a result of delocalization of the metal $3d$ electrons into the ligand orbitals, and vice versa. The extent of the reduction is dependent both on the nature of the ligands and of the metal, and from a comparison of the ligand field spectra of an extended range of complexes two nephelauxetic series were constructed, classifying either the ligands or the metal ions with respect to their tendency to form covalent bonds. Interestingly, the sequence of covalency/nondynamical correlation found for the metals and ligands included in this study corresponds exactly to their position in the nephelauxetic series [179]: $F^- > H_2O > NH_3 > CN^-$ and $Cr^{3+} > Co^{3+}$. It is tempting to assume that this correspondence can be extrapolated to the entire range of metals and ligands included in these series and that the nephelauxetic series could actually be used to predict the extent of nondynamical correlation effects to be expected for a specific metal–ligand combination. A more definite confirmation of this surmise can be given only after a larger variety of metals and ligands have been studied.

Carbonyl does not appear in the nephelauxetic series. Indeed, organometallic compounds fall outside the scope of ligand field theory, due to the fact that these systems are usually characterized by stronger covalent bonds than those of the Werner complexes. Yet CO is closely related to CN^-. As CN^- it is a σ donor and a π acceptor, although the relative importance of both contributions to bond formation is different for both ligands, CN^- being the stronger σ donor and CO the stronger π acceptor. This is also reflected by the natural orbital occupation numbers in Table XVI. Looking at the isoelectronic $Co(CN)_6^{3-}$ and $Cr(CO)_6$ molecules, we notice that both the e_g carbon lone pair and the t_{2g} π^* population are significantly higher in $Cr(CO)_6$ than in $Co(CN)_6^{3-}$. The most covalent bonding mechanism creates the largest correlation effect also in this case. For $Cr(CO)_6$ we have included in Table XVI the results from a number of additional CASSCF calculations performed with an active space of 13 orbitals, obtained by extending the original 10-orbital active space with one additional CO π^* shell of different symmetries. The occupation numbers obtained from these calculations clearly confirm the fact that the orbitals that are actually allowed to mix with the metal $3d$ orbitals are by far the most important in the correlation treatment. The occupation numbers of the t_{1u}, t_{2u}, and t_{1g} CO π^* shells all turn out to be manifestly smaller than the t_{2g} occupation number.

We now turn to Table XVII for a comparison of $Ni(CO)_4$ with $Cr(CO)_6$. Since $Ni(CO)_4$ is a formal d^{10} system, CO-to-Ni σ donation is unimportant. The carbon lone pairs were therefore omitted from the $Ni(CO)_4$ basic 10-orbital active space. Instead, a virtual e and t_2 shell were included, since the CO π^* orbitals now end up in the representations e, t_1, and t_2, of which both e and t_2 are allowed to mix with the $3d$ orbitals. However, although the correlating e and t_2 orbitals are designated as CO π^* orbitals in Table XVII, they in fact turn out to contain a considerable amount of Ni $4d$ character. This observation reflects the importance of the $3d$ *double-shell effect* in transition metals containing a large number of d electrons. It has already been illustrated that the introduction of a second d shell has a major impact on the accuracy of the computed CASPT2 excitation energies in the Ni atom [32] and on the quality of the description of the bonding in the NiH and CuH molecules [168]. Here we find that the second d shell is actually at least as important as the CO π^* orbitals for the electron correlation in $Ni(CO)_4$. To illustrate this point further, we have plotted the difference of the total density between the CASSCF and the SCF wavefunction for $Ni(CO)_4$ compared to $Cr(CO)_6$ (cf. Fig. 6). For $Cr(CO)_6$ this plot nicely illustrates the effect of correlation on the bonding, with an increased σ donation and π backdonation in the CASSCF treatment compared to SCF (although some $3d-4d$ correlation can also be spotted). On the other hand, the $Ni(CO)_4$ plot clearly shows the combined effect of radial extension of the $3d$ electrons in the CASSCF treatment and the correlation effects connected with the Ni–CO bonding.

The present results obtained for $Cr(CO)_6$ and $Ni(CO)_4$ are a clear

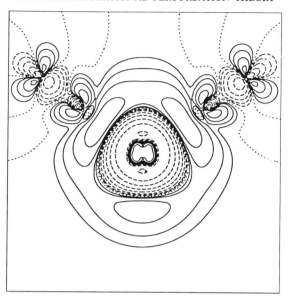

Figure 6. Total density difference plot between the CAS(10/10) wavefunction and the Hartree–Fock wavefunction for $Ni(CO)_4$ (*upper part*) and $Cr(CO)_6$ (*lower part*). The figure describes the density shifts in a plane containing the central nickel and two CO ligands. Full contours correspond to an increase in electron density and dashed contours to a decrease. At the dotted lines $\Delta\rho = 0$. The values of the $\Delta\rho$ contours are ± 0.00025, ± 0.0005, ± 0.001, ± 0.002, ± 0.004, ± 0.008, ± 0.016, ± 0.032, and 0.0 a.u.

indication of the difficulty in treating organometallic compounds with single-reference methods. The presence of strong nondynamical correlation effects in these systems, both the $3d$ double-shell effect and correlation effects on the bonding, severely limits the accuracy of any calculation that does not include these effects from the start. The CASPT2 method has therefore turned out to be more than a feasible alternative for the description of the bonding in these systems. Starting from the 10-orbital active space described here, a CASPT2 treatment of the bonding in $Ni(CO)_4$ and $Cr(CO)_6$ [29] has produced results which are far more accurate than the results obtained up to now with any single-reference method. Furthermore, this is also the case for the molecules $Ni(C_2H_4)$ [28], $Ni(CO)_x$ ($x = 1-3$) [29], $Fe(CO)_5$ [29], and ferrocene [28], which have not been included here. In all cases, an active space of only 10 orbitals proved to be large enough as a starting point for the CASPT2 treatment. An active space of this size is perfectly manageable for the present CASPT2 code, leading to CPU times that are smaller (less than 1 CPU hour on an IBM RS6000/580) than can possibly be expected from a single-reference configuration interaction or coupled-cluster calculation, considering the large number of electrons [up to 74 for $Cr(CO)_6$] included in the correlation treatment.

One may therefore ask if CASPT2 has conquered the problems in the difficult field of *ab initio* transition metal chemistry. The answer to this question critically depends on whether or not the 10-orbital-active-space rule observed here turns out to be applicable to all transition metal systems in general. Unfortunately, this is not the case. A few recent CASPT2 studies have indicated that in some special cases it is unavoidable to extend the active space with ligand valence orbitals that are of different symmetries than those of the metal $3d$ orbitals. All cases met up to now have one thing in common: They concern systems that no longer fit nicely into the picture of a transition metal complex being built from the interaction between a metal ion with an incomplete $3d$ shell, on the one hand, and a number of ligands with closed shells, on the other hand. A notorious example is $Cr(NO)_4$. $Cr(NO)_4$ is isoelectronic with $Ni(CO)_4$, and both molecules have an identical ground-state electronic configuration. Yet, trying to fit the bonding pattern in $Cr(NO)_4$ into the picture of $Ni(CO)_4$, one ends up with a formal d^{10} Cr^{4-} ion and four NO^+ ligands. Such a picture is obviously unrealistic, and the natural orbital occupation numbers in Table XVII nicely illustrate this fact. First we have to remark that while subdivision of the molecular orbitals into either ligand or metal orbitals is essentially correct for all other systems in Tables XVI and XVII, it has become completely blurred in case of $Cr(NO)_4$. The orbitals designated as metal $3d$ in fact each contain only

54% Cr $3d$ character, while we find 37% and 40% $3d$ character in NO π^* orbitals of t_2 and e symmetry, respectively. This already indicates that the bonding in $Cr(NO)_4$ is purely covalent and should better be viewed as being built from neutral Cr and four NO radicals. Furthermore, looking at the occupation numbers of the natural orbitals resulting from the 10-orbital CASSCF calculation, we notice that the t_2 orbitals with predominant $3d$ character have now become the weakly occupied orbitals, with a population of only 0.297 electron, while most electrons, 5.697, are found in the predominantly NO π^* orbitals. However, it is clear that the remaining NO π^* shell of t_1 symmetry can in this case most certainly not be ignored in the CASSCF treatment. Adding it to the active space gives it an occupation number of as much as 0.519 electron. A CASSCF treatment of $Cr(NO)_4$ with only 10 active orbitals, as presented recently by Bagus and Nelin [182], simply does not catch all important electronic structure features of this molecule.

In fact, the importance of the $t_1 \pi^*$ shell in $Cr(NO)_4$ was noted several years ago by Bauschlicher and Siegbahn [183], who called its contribution to the bonding a purely ionic contribution. Although this definition may seem spurious at first in view of the present considerations, it becomes clear if one realizes that the presence of 0.519 electron in the t_1 shell is indeed the result of a purely ionic charge transfer between orbitals which are, for symmetry reasons, not allowed any covalent interaction. An important consequence of this fact is also that any calculation on $Cr(NO)_4$ using single-reference methods is doomed to fail. Indeed, where an SCF treatment at least partially describes the covalency of the bonding within the t_2 and e representations through mixing at the orbital level, it does not even begin to capture the ionic component of the bonding associated with the t_1 shell. The same remark holds for any multireference treatment with a limited reference space. However, in a recent CASSCF/CASPT2 study [178] on the series $Ni(CO)_4$, $Co(CO)_3(NO)$, $Fe(CO)_2(NO)_2$, $Mn(CO)(NO)_3$, $Cr(NO)_4$, we have shown that results of similar accuracy can be obtained for all these molecules, on the condition that the size of the CASSCF active space is gradually increased from 10 to 13 within the series. Although the calculations with 13 active orbitals were, of course, more demanding in terms of computer time (with a maximum of 4 CPU hours), the CASSCF/CASPT2 method also turned out to be successful for $Cr(NO)_4$, with no further important intruder states appearing in the first-order wavefunction.

However, for two other molecules, CrF_6 and CrO_4^{2-}, the CASSCF/CASPT2 method turned out to be less successful. Textbooks on inorganic chemistry usually give the central chromium in CrO_4^{2-} an oxidation number $(+6)$, thus describing the molecule as consisting of d^0 Cr^{6+} and four O^{2-} closed-shell ligands. The same reasoning describes CrF_6 as d^0 Cr^{6+} + six F^-. However, this simple picture is, of course, again far from realistic. The results of a Mulliken population analysis point instead to a charge of only $+1.7$ in CrO_4^{2-} and 1.6 in CrF_6. Again, this charge reduction is the result of a considerable covalent mixing between the metal $3d$ and ligand $2p$ orbitals as well as from an additional charge transfer from the ligands to the metal at the CASSCF level. However, although the observed charge reorganizations are significantly less severe than for $Cr(NO)_4$, they are no longer constrained to a limited set of orbitals. This is again illustrated by Tables XVI and XVII, where we have included the results of several CASSCF calculations for both molecules, in which the basic 10-orbital active space was extended with different sets of ligand $2p$ orbitals. As one can see, the additional orbitals do not have occupation numbers close to 2. For comparison, we have also added CrF_4 to Table XVII. As opposed to CrO_4^{2-}, CrF_4 is still a 10-orbital-active-space molecule, even with a formal charge of $(+4)$ on Cr. However, the results obtained for CrF_6 and CrO_4^{2-} indicate that an accurate CASPT2 calculation on these molecules would have to be based on an active space including the Cr $3d$ as well as all ligand $2p$ orbitals, resulting in 17 orbitals for CrO_4^{2-} and 23 orbitals for CrF_6. Both calculations are obviously impossible at this moment. CASSCF/CASPT2 calculations using smaller active spaces have been tried on both molecules. For CrF_6, calculation of the barrier for pseudorotation to a trigonal prism using a 10-orbital active space [184] resulted in a barrier height more than twice as high than the result obtained from both a CCSD(T) [185] and DFT study [186] and which should probably be considered as less reliable. On the other hand, calculation of the spectrum of CrO_4^{2-} completely failed, even with 14 active orbitals, due to the appearance of severe intruder states even for the lowest excited states [187].

We end this section with some considerations on spectroscopy. To do so, we return to the Werner systems discussed earlier, in particular to the fluoride and cyanide compounds, for which we have included in Table XVI the results of a few excited states. The excited states were chosen such that each corresponds to a configuration with a different $e_g(3d_\sigma)$ occupation number. Although the different ligand field states are subject to the same type of nondynamical correlation effects, we notice that in some cases the occupation numbers of the ligand orbitals are significantly different for different states. This is especially true for the cyanide complexes, for which correlation effects actually tend to lose importance with each electron transferred from the $t_{2g}(3d_\pi)$ into the $e_g(3d_\sigma)$ orbitals. The ligand field spectrum of these systems was described successfully in a recent CASSCF/CASPT2 study [170], using an active space of 10 orbitals. In Section III.B.2 we present the results obtained for $Co(CN)_6{}^{3-}$ and for the isoelectronic $Fe(CN)_6{}^{4-}$ and $Cr(CO)_6$ systems. The fluoride molecule $CrF_6{}^{3-}$ is the most ionic system considered, with few or no nondynamical correlation effects on the bonding. A quantitative description of the spectrum of this ion has been obtained in a CASSCF/ACPF study based on an active space containing only the five $3d$ orbitals [188, 189]. The same is true for $CrF_6{}^{2-}$ and $CrF_6{}^{4-}$, the spectra of which have been the subject of a recent CASPT2 study [190]. On the other hand, for the slightly more covalent $CoF_6{}^{3-}$ it was shown [191] that a significant improvement in transition energies could be obtained by extending the active space with the ligand e_g orbitals.

When considering organometallic systems, a calculation of the charge transfer states may become more of interest. Obviously, an extension of the basic 10-orbital active space can in this case most often not be avoided, since the charge transfer states included in this active space may not be the ones of interest. For example, the active space of 10 orbitals used for the calculation of the ground state of $Cr(CO)_6$ only includes the CO π^* orbitals of t_{2g} symmetry, while the lowest charge transfer bands in its spectrum are due to transitions to the CO π^* t_{1u} and t_{2u} shells. On the other hand, adding both the t_{1u} and the t_{2u} shells on top of the basic 10 orbitals would result in an active space of 16 orbitals, which is too large. In calculating charge transfer spectra, one may therefore have to compromise by using different active spaces for different excited states. This does not present any serious drawback, as long as there are no strong interactions between excited states treated with different active spaces. CASSCF/CASPT2 calculations on the charge transfer spectra of $Cr(CO)_6$, $Fe(CO)_5$, and $Ni(CO)_4$ have recently been performed with considerable success [29, 177]. The results obtained for $Cr(CO)_6$ and $Ni(CO)_4$ are presented in Section III.B.2. More details concerning the active spaces used are given there.

Relativistic Effects: ScCl₂, CrCl₂, and NiCl₂. A nice illustration of the importance of core correlation and relativistic effects to the bonding properties of first-row transition metal compounds was found in a recent systematic CASPT2 study of transition dihalides MF_2 and MCl_2 (M = Sc– Zn). Here we review some important results obtained for $ScCl_2$, $CrCl_2$, and $NiCl_2$. The calculations were performed using different active spaces, depending on the metal. Thus for $SrCl_2$ and $CrCl_2$, only the five $3d$ orbitals were included, while for $NiCl_2$ the σ_g and π_g chlorine valence orbitals and a second d shell were added to account both for the increased covalency of the metal–chlorine bonds and the $3d$ double-shell effect, thus obtaining 13 active orbitals. Two sets of CASPT2 calculations were performed, one set including all valence electrons (M $3d$, Cl $3s$, $3p$) and a second set including in addition also the metal $3s$, $3p$ electrons. Relativistic effects were taken into account by including the mass-velocity and Darwin terms using first-order perturbation theory at the CASSCF level. Furthermore, the effect of spin-orbit coupling was estimated using an effective one-electron spin-orbital coupling operator, together with a scaling of the metal charge to an effective value that reproduces the spectroscopically observed atomic levels [192].

Fairly large ANO type basis sets were used in this study: $(17s12p9d4f)/[7s6p4d2f]$ for the metal [41] and $(17s12p5d4f)/[6s5p2d1f]$ for chlorine [40]. Even with these basis sets, basis-set superposition errors on the M–Cl bond distances are significant. They were examined in detail for $MnCl_2$ by means of the full counterpoise method. For the valence-only calculations, a correction of 0.005 Å was found, due exclusively to the incompleteness of the chlorine basis set. Including the $3s$, $3p$ electrons increases the metal contribution, but not too drastically. A total correction of 0.010 Å was now obtained. It was assumed that these results can be extrapolated to the entire series of MCl_2 compounds, and they have been included in all results reported here.

Transition metal dihalides are generally characterized by a weak ligand field splitting of the free M^{2+} ground state. In a $D_{\infty h}$ point group, the $3d$ shell is split into δ_g, π_g, and σ_g orbitals. Simple ligand field arguments indicate the d-orbital energy level sequence $\delta_g < \pi_g < \sigma_g$. This would then give $ScCl_2$ a $^2\Delta_g(\delta_g^1)$ and $CrCl_2$ a $^5\Sigma_g^+(\delta_g^2\pi_g^2)$ ground state, while for $NiCl_2$ a $^3\Pi_g(\delta_g^4\pi_g^3\sigma_g^1)$ ground state should be obtained. However, the results from our CASPT2 study suggest that none of these states in fact correspond to the actual ground state in the molecules considered. In Table XVIII we have collected the results of a CASPT2 geometry optimization of all possible molecular states originating from the ground state in the free M^{2+} ion. In all three molecules, the ground state as predicted from ligand field theory actually shows up as the first excited state, although in the case of $CrCl_2$ the energy difference with the calculated $^5\Pi_g$ ground state is small and becomes positive only when both $3s$, $3p$ correlation and relativistic effects are taken into account. Obviously, qualitative ligand field arguments are unable to capture the actual bonding situation in these molecules. In $ScCl_2$, the single d electron clearly prefers the σ_g orbital (which is supposed to be highest in energy) over δ_g and π_g, and also in the other two systems there is a slight preference for σ_g over π_g. This preference for occupying the σ_g orbital can easily be understood when considering the shape of this orbital,

TABLE XVIII

Ligand Field Splitting of the Sc^{2+} ^2D Ground State, the Cr^{2+} ^5D Ground State, and the Ni^{2+} ^3F Ground State in the Corresponding Dichlorides

| | | | Only Valence Electrons Correlated | | | | Also Metal 3s, 3p Correlated | | | |
| | | | Nonrelativistic | | Relativistic[a] | | Nonrelativistic | | Relativistic[a] | |
	Main Conf.		R_e (Å)	T_e (cm^{-1})	R_e (Å)	T_e (cm^{-1})	R_e (Å)	T_e (cm^{-1})	R_e (Å)	T_e (cm^{-1})
$ScCl_2$	$^2\Sigma_g^+$	$\delta_g^0\pi_g^0\sigma_g^1$	2.324	0	2.318	0	2.297	0	2.292	0
	$^2\Delta_g$	$\delta_g^1\pi_g^0\sigma_g^0$	2.404	336	2.401	1126	2.371	1905	2.367	2710
	$^2\Pi_g$	$\delta_g^0\pi_g^1\sigma_g^0$	2.434	6159	2.431	7023	2.402	8170	2.400	9055
$CrCl_2$	$^5\Pi_g$	$\delta_g^2\pi_g^1\sigma_g^1$	2.205	0	2.198	0	2.183	0	2.175	0
	$^5\Sigma_g^+$	$\delta_g^2\pi_g^2\sigma_g^0$	2.244	−634	2.240	−247	2.230	−11	2.226	408
	$^5\Delta_g$	$\delta_g^1\pi_g^2\sigma_g^1$	2.233	4883	2.224	4875	2.215	5066	2.206	5062
$NiCl_2$	$^3\Sigma_g^-$	$\delta_g^4\pi_g^2\sigma_g^2$	2.075	0	2.060	0	2.069	0	2.055	0
	$^3\Pi_g$	$\delta_g^4\pi_g^3\sigma_g^1$	2.087	800	2.077	1048	2.091	679	2.080	960
	$^3\Phi_g$	$\delta_g^4\pi_g^2\sigma_g^2$	2.091	3187	2.076	3076	2.086	2713	2.072	2605
	$^3\Delta_g$	$\delta_g^3\pi_g^4\sigma_g^1$	2.122	3130	2.111	3535	2.118	3760	2.108	4177

[a] Including the Darwin and mass-velocity terms from the Pauli equation.

A   B   C

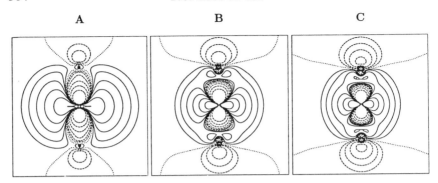

Figure 7. The $9\sigma_g^+$ orbital in $ScCl_2$ (A), $CrCl_2$ (B), and $NiCl_2$ (C). The figure shows the orbitals in a plane containing the central metal and the two chlorine atoms. Values of the contours are ±0.128, ±0.064, ±0.032, ±0.016, ±0.008, ±0.004, and 0.0 a.u.

shown in Fig. 7. These plots clearly indicate a large $3d_\sigma$–$4s$ mixing in the σ_g orbitals, which serves to construct a $3d$–$4s$ hybrid in such a way that the maximum density is located in the plane perpendicular to the metal–chlorine bonds, thus minimizing the exchange repulsion with the Cl^- σ electron pair. The mixing is extreme in $ScCl_2$ and decreases strongly when going to $CrCl_2$ and further to $NiCl_2$.

Apart from leading to a breakdown of the ligand field model, the strong $3d$–$4s$ mixing in the σ_g orbitals also induces a relativistic contribution to the relative energies of the low-lying states in Table XVIII. Looking at the configurations corresponding to the various states, we notice that in all cases the introduction of Darwin and mass-velocity terms leads to a stabilization of configurations with the highest σ_g occupation number. The effect is small but definitely present. It is largest for $ScCl_2$, about $800\,cm^{-1}$, and decreases when going to $CrCl_2$ and $NiCl_2$, in harmony with the fact that the $4s$ contribution to the σ_g orbital decreases in the same order. Notice also that in $CrCl_2$ the relativistic contribution to the energy of the transition between the $^5\Pi_g$ and $^5\Delta_g$ states is negligibly small. Both are characterized by the same σ_g occupation number, and the transition between them corresponds to a $\delta_g \rightarrow \pi_g$ transition. This is a confirmation of the fact that relativistic effects are negligible for "pure" $3d$–$3d$ transitions.

Table XVIII also includes the effect of correlation of the semicore $3s$, $3p$ electrons on the transition energies. As one can see, the effect is substantial, especially for $ScCl_2$, where it increases the energy difference between the different states with up to $2000\,cm^{-1}$. A significantly smaller effect is found for $CrCl_2$ and $NiCl_2$.

Both $3s$, $3p$ correlation and relativistic effects also affect the value of the bond distances in Table XVIII. These distances were obtained from an individual optimization at the different computational levels. Bond-angle deformations were also considered, and in fact the $^5\Pi_g$ state in $CrCl_2$ was found to be slightly bent, with a Cl–Cr–Cl angle varying between 155 and 165° for the various levels of approximation. For simplicity, the $D_{\infty h}$ notation has been maintained for this state. The effect of $3s$, $3p$ correlation is again largest for $ScCl_2$, where it results in a shortening of the bond lengths by around 0.03 Å for all states. Smaller effects are found for $CrCl_2$, while for $NiCl_2$ the effect is almost absent. On the other hand, relativistic corrections to the bond distances are clearly largest for $NiCl_2$ while almost absent for $ScCl_2$.

On the whole it is clear that the combined effect of $3s$, $3p$ correlation and relativistic (mass-velocity and Darwin) corrections has important consequences for both the structure and the relative energy of the low-lying states in the considered molecules. Up to this point the effects of spin-orbit coupling have not yet been considered. However, for both $CrCl_2$ and $NiCl_2$ substantial effects can be expected, since in both molecules the first excited state is close in energy to the ground state but is characterized by a different M–Cl bond distance. For example, looking at $CrCl_2$ we notice that the final bond distances obtained for the $^5\Sigma_g^+$ and $^5\Pi_g$ states differ by as much as 0.051 Å, while the energy difference between them is only 408 cm^{-1}. It is clear that the actual ground state in this molecule is a mixture of both states, with a flexible structure. The Cr–Cl bond distance in $CrCl_2$ reported from a gas-phase electron diffraction study [193] is 2.207 Å, in between the values calculated for the two states. In fact, the same study reports $CrCl_2$ as a highly bent molecule, which is not reproduced by our calculations. However, determination of the $CrCl_2$ structure was greatly hindered by the presence of dimeric species in the sample. This may explain the discrepancy between the experimental and calculated bond angles.

The effect of spin-orbit coupling was considered in more detail for $NiCl_2$. Using the final CASPT2 energies (with $3s$, $3p$ correlation and including the Darwin and mass-velocity terms) as the diagonal elements of the spin-orbit coupling matrix, the resulting Σ_g^+ ground state was found to be composed of 76% $^3\Sigma_g^-$ and 24% $^3\Pi_g$ configurations. A reoptimization of the Ni–Cl bond distance for this state resulted in a value of 2.062 Å, 0.007 Å longer than the result obtained for the pure $^3\Sigma_g^-$ state, and 0.014 Å shorter than the experimental Ni–Cl bond distance of 2.076 Å [194]. The latter agrees with the results obtained for all other MCl_2 systems studied in Ref. 195. It was found that CASPT2 systematically underestimates the M–Cl bond distance in these systems, by 0.01–0.02 Å.

Using the optimized distance for the Σ_g^+ ground state, a calculation of the spectrum of $NiCl_2$ was performed, both with and without spin-orbit coupling. $3s$, $3p$ correlation and the Darwin and mass-velocity terms were included in both sets of calculations. The calculations now included all states corresponding to the four lowest ^3F, ^1D, ^3P, and ^1G states in Ni^{2+}. The results are shown in Fig. 8. The calculated spectrum with spin-orbit coupling is in agreement with the experimental absorption spectrum of $NiCl_2$ [196]. In the region $0-7000\,cm^{-1}$, a broad and complex absorption band is found, corresponding to the series of closely spaced levels originating from the Ni^{2+} ^3F ground state. The Ni^{2+} ^3P state is split into $^3\Pi_g$ and $^3\Sigma_g^-$, calculated without spin-orbit coupling at $11,589\,cm^{-1}$ and $21,845\,cm^{-1}$, respectively. The highest $^3\Sigma_g^-$ state is clearly unaffected by spin-orbit coupling. This is confirmed by the experimental spectrum, containing a single band at $21,180\,cm^{-1}$. On the other hand, the lowest $^3\Pi_g$ state is surrounded by two singlet states, $^1\Delta_g$ and $^1\Pi_g$. Spin-orbit coupling therefore results in a series of states with considerable $^3\Pi_g$ character, calculated between $10,667$ and $13,316\,cm^{-1}$, in perfect agreement with the appearance of a series of sharp bands in the region $11,000-14,000\,cm^{-1}$ of the experimental spectrum. Furthermore, low-intensity absorption bands occur in the region $14,000-19,000\,cm^{-1}$ of the experimental spectrum, agreeing with the fact that all states calculated in this region have predominantly singlet character.

As a conclusion, accurate results are obtained with the CASSCF/CASPT2 method for the metal dichlorides. However, this accuracy can be achieved only with a full correlation treatment, including all valence electrons and the metal semicore $3s$, $3p$ electrons. Relativistic effects can also become quite important in some cases and should therefore be included if accuracy is required.

2. Transition Metal Complexes with Cyanide and Carbonyl Ligands

The accurate calculation of excited states and electronic spectra of organometallic systems still remains a major challenge to *ab initio* methods. The simultaneous occurrence of important nondynamical correlation effects and a large number of electrons make these systems virtually impossible to handle with configuration-interaction or coupled-cluster methods unless severe restrictions are imposed on either the size of the reference space or the number of electrons actually included in the correlation treatment. Even if these restrictions are calibrated carefully, it is most often impossible to avoid a substantial loss of accuracy. The fact that the CASSCF/CASPT2 method does not suffer from these restrictions a priori makes it a promising alternative for the calculation of electronic spectra of organometallics and other large transition metal

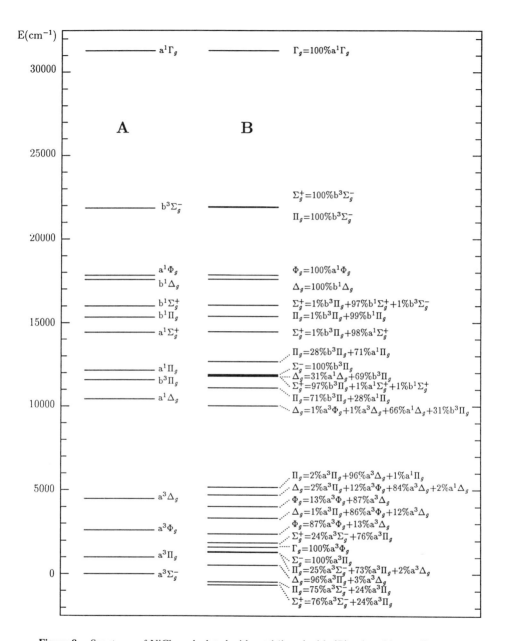

Figure 8. Spectrum of NiCl$_2$, calculated without (A) and with (B) spin-orbit coupling.

systems, provided that an appropriate reference active space can be constructed, including all near-degeneracy effects as well as the excited states of interest.

The first application of the CASSCF/CASPT2 method to the spectra of large transition metal systems concerned a systematic study of the ligand field spectra of $M(CN)_6^{n-}$ systems, where $M = V$, Cr, Mn, Fe, and Co and $n = 3$ or 4 [170]. It was found there that a balanced description of the ground state and all excited ligand field states could be obtained with an active space of 10 orbitals, consisting of the bonding and antibonding combinations of the metal $3d$ and the cyanide σ and π^* valence orbitals (see also Section III.B.1). With this reference space, all calculated CASPT2 transition energies were in excellent agreement with the experimental data available, with errors in the range 0.0–0.35 eV. More recently, the method was applied in a study of the ground-state properties and the electronic spectra of $Cr(CO)_6$, $Fe(CO)_5$, and $Ni(CO)_4$ [29, 177], where the emphasis was now on the calculation of charge transfer states. Since not all relevant excited charge transfer states are contained in the active space of 10 orbitals, appropriate for the ground-state description, it had to be extended for the calculation of the spectra. However, it was found that with an active space of at most 13 orbitals, satisfactory results could be obtained, with an accuracy comparable to the ligand field state calculations.

Some of the most important results obtained from the studies above are reviewed below, including a comparison of the ligand field spectra of the isoelectronic systems $Co(CN)_6^{3-}$, $Fe(CN)_6^{4-}$, and $Cr(CO)_6$, the charge transfer states in $Cr(CO)_6$, and the spectrum of $Ni(CO)_4$. Due to the size of the systems considered, rather limited ANO basis sets [41] were used in these studies: $(17s12p9d4f)/[6s4p3d1f]$ for the metal and $(10s6p3d)/[3s2p1d]$ for C, N, and O, resulting in 208 contracted functions for the octahedral systems and 152 contracted functions for $Ni(CO)_4$. All calculations were performed using experimental geometries. The calculations on the octahedral $Cr(CO)_6$, $Co(CN)_6^{3-}$, and $Fe(CN)_6^{4-}$ systems were done using D_{2h} symmetry, while for the tetrahedral $Ni(CO)_4$ molecule D_2 symmetry was used. However, in all cases additional symmetry restrictions were imposed in the CASSCF step to prevent mixing between molecular orbitals belonging to different representations in the actual point group (O_h or T_d) of the molecules.

The role of $3s$, $3p$ correlation was investigated in all cases by comparing the results obtained from two sets of CASPT2 calculations, denoted as CASPT2(v) and CASPT2(c-v), respectively. The CASPT2(v) calculations include the metal $3d$ and all valence electrons of the ligands originating from the $2s$, $2p$ orbitals on C, N, and O, while the metal

$3s$, $3p$ electrons are included in the CASPT2(c-v) calculations. Thus up to 74 electrons were correlated in $Co(CN)_6^{3-}$, $Fe(CN)_6^{4-}$, and $Cr(CO)_6$ and 58 electrons in $Ni(CO)_4$. The role of relativistic effects was not investigated but is expected to be small, at least for the ligand field transitions. The effect of spin-orbit coupling on the ligand field spectrum of $Cr(CO)_6$ was investigated in a recent study by Daniel and Ribbing [197], but turned out to be very small.

In addition to calculated CASSCF/CASPT2 transition energies, results for the oscillator strengths of the spin and orbitally allowed charge transfer transitions in $Cr(CO)_6$ and $Ni(CO)_4$ are presented. They were obtained using transition dipole moments obtained by the CASSCF state interaction method (CASSI). CASPT2(c-v) values were used for the excitation energies, appearing in the expression for the oscillator strengths.

Ligand Field Spectrum of $Co(CN)_6^{3-}$, $Fe(CN)_6^{4-}$, and $Cr(CO)_6$. The ligand field spectra of the isoelectronic species $Co(CN)_6^{3-}$, $Fe(CN)_6^{4-}$, and $Cr(CO)_6$ show an analogous structure. All three systems are characterized by a $^1A_{1g}$ $(2t_{2g}^6)$ ground state, and their ligand field spectrum therefore consists of excitations from the π bonding $2t_{2g}$ orbitals to the σ antibonding $6e_g$ orbitals, giving rise to the following states: $^3T_{1g}$, $^3T_{2g}$, $^1T_{1g}$, $^1T_{2g}$. The spectra of the cyanides $Co(CN)_6^{3-}$ and $Fe(CN)_6^{3-}$ are simple and well understood [198]. In both cases three bands were located in the ligand field part of the spectrum, well separated from the charge transfer bands, which do not appear below 5.5 eV. They were assigned as transitions to the $^3T_{1g}$, $^1T_{1g}$, and $^1T_{2g}$ states, respectively. However, the same is no longer true for $Cr(CO)_6$. Due to the stronger π-acceptor character of CO compared to CN^-, the $3d \rightarrow CO$ π^* charge transfer transitions appear at lower wavenumbers, thus obscuring the weaker ligand field transitions. The spectrum of $Cr(CO)_6$ is dominated by two intense absorption bands at 4.43 eV and 5.41 eV, corresponding to the two orbitally and spin allowed $^1A_{1g} \rightarrow {}^1T_{1u}$ transitions [199–201]. Three weak shoulders appearing at 3.60, 3.91, and 4.83 eV were obtained from a Gaussian analysis of the spectrum by Beach and Gray in 1968 [199]. The shoulder at 4.83 eV was assigned as the second spin-allowed transition to $^1T_{2g}$, while the two weak shoulders at the low-energy side of the spectrum were assigned as vibrational components of the $^1A_{1g} \rightarrow {}^1T_{1g}$ transition. Although this assignment was made rather tentatively, it became widely accepted over the years [179, 200].

Somewhat remarkably, the Beach and Gray assignment was also reproduced almost exactly in a ROHF study of the $Cr(CO)_6$ spectrum

[202] with calculated transition energies of 3.86 eV and 4.91 eV for the $^1A_{1g} \rightarrow {}^1T_{1g}$ and $^1A_{1g} \rightarrow {}^1T_{2g}$ transitions, respectively. This is unexpected, because it is known that the Hartree–Fock method systematically tends to underestimate seriously the ligand field strength in first-row transition metal systems. For example, an ROHF calculation on $Co(CN)_6^{3-}$ [203] resulted in a value of 3.08 eV for the first spin-allowed transition $^1A_{1g} \rightarrow {}^1T_{1g}$, which is 0.95 eV lower than the experimental value. Up to now, no other studies including correlation have been reported for the spectrum of $Cr(CO)_6$ nor for the hexacyanometalate complexes. An INDO/S CI study on $Cr(CO)_6$ was reported recently by Kotzian et al. [201], who also remeasured the spectrum but made no attempt to locate the ligand field transitions. In this study both spin-allowed ligand field transitions were calculated below the charge transfer bands, at 3.64 eV ($^1T_{1g}$) and 4.14 eV ($^1T_{2g}$), and assigned to the shoulders at 3.60 and 3.91 eV in the Beach and Gray spectrum [199].

CASSCF/CASPT2 results obtained for the ligand field transitions in the three complexes are collected in Table XIX. Both CASSCF (with 10

TABLE XIX

Comparison of the CASSCF and CASPT2 Excitation Energies (eV) for the d-d Transitions in $Co(CN)_6^{3-}$, $Fe(CN)_6^{4-}$, and $Cr(CO)_6$

State	$Co(CN)_6^{3-}$					
	CASSCF	CASPT2(v)	$\omega_v{}^a$	CASPT2(c-v)	$\omega_{c-v}{}^a$	Expt.[b]
$^3T_{1g}$	3.90	3.38	0.569	3.14	0.559	3.22
$^3T_{2g}$	4.55	3.89	0.567	3.63	0.557	
$^1T_{1g}$	4.91	4.23	0.567	3.98	0.558	4.03
$^1T_{2g}$	6.30	5.22	0.564	4.80	0.554	4.86
	$Fe(CN)_6^{4-}$					
	CASSCF	CASPT2(v)	$\omega_v{}^a$	CASPT2(c-v)	$\omega_{c-v}{}^a$	Expt.[b]
$^3T_{1g}$	3.37	2.91	0.564	2.67	0.552	2.94
$^3T_{2g}$	4.02	3.42	0.563	3.14	0.551	
$^1T_{1g}$	4.38	3.85	0.562	3.60	0.551	3.84
$^1T_{2g}$	5.73	4.77	0.560	4.33	0.549	4.59
	$Cr(CO)_6$					
	CASSCF	CASPT2(v)	$\omega_v{}^a$	CASPT2(c-v)	$\omega_{c-v}{}^a$	
$^3T_{1g}$	5.07	4.46	0.572	4.28	0.558	
$^3T_{2g}$	5.57	4.86	0.553	4.64	0.537	
$^1T_{1g}$	5.66	5.05	0.576	4.85	0.561	
$^1T_{2g}$	6.48	5.43	0.568	5.13	0.553	

[a] Weight of the CASSCF wavefunction in the first-order wavefunction.
[b] From Ref. 198.

active orbitals) and CASPT2 results are reported. The weights of the CASSCF wavefunction in the final CASPT2 first-order wavefunctions are almost exactly the same for three complexes and stable for the different excited states. This is an indication that the chosen 10-orbital active space is large enough to include the essential electronic structure features in all three complexes. It also shows that a balanced treatment of all excited states has been achieved.

Calculated transition energies for $Co(CN)_6^{3-}$ and $Fe(CN)_6^{4-}$ at the CASSCF level are clearly too high, with errors up to 1 eV and more. This is a general observation for the CASSCF results obtained for all hexacyanometalate complexes studied in Ref. 170, where it could be traced back to the fact that CASSCF tends to overestimate the occupation of the antibonding $3t_{2g}$ and $6e_g$ orbitals included in the active space, but does so more for the ground state than for the excited states. A considerable improvement is obtained at the CASPT2(v) level. All excited states are drastically lowered with respect to the CASSCF results by the perturbational correlation treatment of the valence electrons. However, the contribution of the metal semicore $3s$, $3p$ electrons is most certainly not negligible: The difference between the CASPT2(v) and CASPT2(c-v) results amounts to 0.4 eV for the $^1A_{1g} \rightarrow {}^1T_{2g}$ transition in both $Co(CN)_6^{3-}$ and $Fe(CN)_6^{4-}$. At the CASPT2(c-v) level, excellent results are obtained. For $Co(CN)_6^{3-}$, the transition energies calculated agree with the experimental values to within 0.1 eV. The error is in each case slightly negative, implying that the splitting of the various $2t_{2g}^5 6e_g^1$ states is described even more accurately. The same remark holds for $Fe(CN)_6^{4-}$. The transition energies calculated are slightly less accurate in this case, with a maximum error of 0.27 eV. However, the relative energies of the various $2t_{2g}^5 6e_g^1$ states again agree to within 0.05 eV with the experimental splitting.

In view of the close relationship between both cyanide complexes and $Cr(CO)_6$, similar accuracy is expected at the CASPT2(c-v) level for the latter molecule. The results for $Cr(CO)_6$ in Table XIX do show similar trends, in that the transition energies generally decrease as the correlation treatment is improved. However, the CASPT2(c-v) results obtained for the energies of the spin-allowed transitions to $^1T_{1g}$ and $^1T_{2g}$ do not compare at all well with the values of 3.60, 3.91, and 4.83 eV reported from Gaussian analysis of the spectrum by Beach and Gray [199]. For the second transition, the difference between the calculated result and the band position according to Beach and Gray is 0.30 eV, which is acceptable, but for the first transition a discrepancy of 0.94 eV or more is found. It is highly unlikely that the present CASPT2 method would lead to such a large error for this transition. Furthermore, this would also imply that the error for the splitting between both states is at least 0.61 eV, which is even more unlikely. Considering the quality of the CASPT2 results obtained for $Co(CN)_6^{3-}$ and $Fe(CN)_6^{4-}$, we therefore believe that the weak shoulders at 3.60 eV and 3.91 eV in the spectrum of $Cr(CO)_6$, if present, should not be assigned as ligand field transitions. The presence of the second band at 4.83 eV conforms better to the present results. Given that the Beach and Gray assignment of this band is correct and that the CASPT2(c-v) result for the $^1T_{2g}-^1T_{1g}$ splitting in $Cr(CO)_6$ is as accurate as in the case of the cyanide compounds, we can predict the position of the first transition to $^1T_{1g}$ at 4.55 eV, right below the first intense $^1A_{1g} \rightarrow {}^1T_{1u}$ charge transfer transition. In the next section, where we present the charge transfer spectrum of $Cr(CO)_6$, we will show that there are, in fact, plenty of other candidates to take over the assignment of the shoulders at both 3.60 and 3.91 eV.

The results from the present CASPT2 study also explain the abnormal "accuracy" of the results obtained using ROHF for the ligand field states. When comparing the ROHF result for the $^1A_{1g} \rightarrow {}^1T_{1g}$ transition to the present CASPT2(c-v) result, we now find a difference of 0.99 eV, a perfectly normal ROHF error, of the same magnitude as the ROHF error obtained for $Co(CN)_6^{3-}$. We also note the large discrepancy (more than 1 eV) between the present CASPT2 results and the INDO/S CI from Kotzian et al. [201] for the spin-allowed ligand field transitions.

Charge Transfer States in $Cr(CO)_6$. The electronic spectrum of metal carbonyl compounds is usually dominated by charge transfer (MLCT) transitions from the metal $3d$ orbitals into the CO π^* orbitals. In the case of $Cr(CO)_6$, the 12 CO π^* orbitals are found in the molecular orbitals $9t_{1u}$, $2t_{2u}$, $2t_{1g}$, and $3t_{2g}$. The region 3.5–7 eV in its spectrum is therefore built from a broad range of charge transfer transitions [200], of which, however, only two $^1A_{1g} \rightarrow {}^1T_{1u}$ transitions are both spin and symmetry allowed. They give rise to two intense and broad absorption bands, with a maximum at 4.43 eV and 5.41 eV and an intensity ratio of 1:9.

Only the $3t_{2g}$ CO π^* shell is included in the active space used for the calculation of the ligand field states. Thus the 10 orbitals should have to be extended with the other three CO π^* shells to be able to calculate the full spectrum using the same active space for all states, ending up with an impossible number of 19 active orbitals. The only alternative is to use different active spaces for different excited states. Denoting the basic $(5,6)e_g$, $(2,3)t_{2g}$ 10-orbital active space as active space A, we decided in favor of the following options:

1. All charge transfer states of gerade symmetry were calculated with an active space of 13 orbitals, constructed by extending space A with the $2t_{1g}$ shell. This active space will be denoted as B.

2. To include all charge transfer states of ungerade symmetry, active space A would have to be extended with both the $9t_{1u}$ and $2t_{2u}$ shells, leading to a still unmanageable number of 16 orbitals. In a first calculation, both shells were therefore added in turn, thus constructing active space C1 by adding the $9t_{1u}$ shell and active space C2 by adding $2t_{2u}$. However, this selection of orbitals prevents interaction in the CASSCF step among states belonging to the configurations $2t_{2g}^5 9t_{1u}^1$ and $2t_{2g}^5 2t_{2u}^1$, an interaction that could be important. Therefore, a second set of calculations was performed with an active space of 12 orbitals, denoted as D, in which now both the $9t_{1u}$ and $2t_{2u}$ shells were included, in addition to $(2.3)t_{2g}$. The $(5,6)e_g$ σ orbitals were no longer included, but looking at Table XVI we notice that they are less important than $(2,3)t_{2g}$.

Excitations to all singlet states were included in the calculations, as well as the two $^1A_{1g} \rightarrow {}^3T_{1u}$ excitations, which are spin forbidden but orbitally allowed. The $^1T_{1g}$ and $^1T_{2g}$ ligand field states were now also recalculated using active space B. The results are shown in Table XX. We have included in this table the composition of the CASSCF wavefunction in terms of the different singly excited configurations. For the ungerade states, the composition was taken from the calculations performed with active space D.

The general trends observed in the calculated results are the same as for the ligand field states. The CASSCF transition energies are again definitely too high, with errors amounting up to 2 eV and more. The transition energies are strongly reduced at the CASPT2(v) level. The effect of semicore $3s$, $3p$ correlation turns out to be the least important for the transitions to ungerade charge transfer states, for which the difference between the CASPT2(c-v) and CASPT2(v) results is never larger than 0.1 eV. Significantly larger differences, up to 0.3 eV, are found for some of the gerade states, again including the $^1T_{2g}$ ligand field state. At the CASPT2(c-v) level, all charge transfer transitions in $Cr(CO)_6$ are calculated in the region between 3.5 and 7 eV, which agrees nicely with the experimental spectrum. From the composition of the wavefunctions, we find the following overall energy sequence for the CO π^* orbitals: $9t_{1u} < 2t_{2u} < 3t_{2g} < 2t_{1g}$. The transition energies calculated for the two spin and orbitally allowed $^1A_{1g} \rightarrow {}^1T_{1u}$ transitions are satisfactory, with a maximum error of 0.34 eV. There can be no doubt that these two transitions are responsible for the two intense bands in the spectrum. Both singlet ligand field states are now found between the two strong

TABLE XX
Absorption Spectrum of $Cr(CO)_6$: CASSCF and CASPT2 Results

Final State	Composition[a]	Active Space	Transition Energy (eV)				Oscillator Strength	
			CASSCF	CASPT2(v)	CASPT2(c-v)	Expt.[b]	Calc.	Expt.[b]
a^1E_u	85%($2t_{2g} \rightarrow 9t_{1u}$)	C_1-D	5.11–5.28	3.49–3.67	3.41–3.59	—	—	—
a^1A_{2u}	85%($2t_{2g} \rightarrow 9t_{1u}$)	C_1-D	5.14–5.32	3.65–3.66	3.58–3.58	—	—	—
a^1T_{2u}	76%($2t_{2g} \rightarrow 9t_{1u}$), 9%($2t_{2g} \rightarrow 2t_{2u}$)	C_1-D	5.18–5.26	3.77–3.64	3.70–3.56	—	—	—
a^3T_{1u}	79%($2t_{2g} \rightarrow 9t_{1u}$), 10%($2t_{2g} \rightarrow 2t_{2u}$)	C_1-D	5.25–5.36	3.97–3.78	3.90–3.69	—	—	—
b^1E_u	2%($2t_{2g} \rightarrow 2t_{1u}$), 82%($2t_{2g} \rightarrow 2t_{2u}$)	C_2-D	5.86–6.16	4.05–4.14	3.97–4.05	—	—	—
a^1A_{1u}	84%($2t_{2g} \rightarrow 2t_{2u}$)	C_2-D	5.92–6.29	4.23–4.20	4.15–4.10	—	—	—
b^1T_{2u}	4%($2t_{2g} \rightarrow 9t_{1u}$), 82%($2t_{2g} \rightarrow 2t_{2u}$)	C_2-D	6.21–6.57	4.39–4.52	4.32–4.43	—	—	—
a^1T_{1u}	66%($2t_{2g} \rightarrow 9t_{1u}$), 21%($2t_{2g} \rightarrow 2t_{2u}$)	C_1-D	6.15–5.97	4.61–4.19	4.54–4.11	4.43	1.33–0.20	0.25
b^3T_{1u}	4%($2t_{2g} \rightarrow 9t_{1u}$), 83%($2t_{2g} \rightarrow 2t_{2u}$)	C_2-D	6.23–5.59	4.59–4.61	4.51–4.51	—	—	—
a^1E_g	86%($2t_{2g} \rightarrow 3t_{2g}$), 2%($2t_{2g} \rightarrow 2t_{1g}$)	B	7.17	4.78	4.58	—	—	—
a^1T_{1g}	5%($2t_{2g} \rightarrow 6e_g$), 83%($2t_{2g} \rightarrow 3t_{2g}$)	B	6.80	5.02	4.82	—	—	—
b^1T_{1g}	88%($2t_{2g} \rightarrow 6e_g$), 1%($2t_{2g} \rightarrow 3t_{2g}$)	B	5.66	5.04	4.85	—	—	—
a^1T_{2g}	74%($2t_{2g} \rightarrow 6e_g$), 14%($2t_{2g} \rightarrow 3t_{2g}$)	B	6.42	5.38	5.08	—	—	—
b^1T_{1u}	13%($2t_{2g} \rightarrow 9t_{1u}$), 77%($2t_{2g} \rightarrow 2t_{2u}$)	C_2-D	7.16–7.75	5.16–5.30	5.07–5.20	5.41	1.63–2.58	2.30
b^1E_g	8%($2t_{2g} \rightarrow 3t_{2g}$), 75%($2t_{2g} \rightarrow 2t_{1g}$)	B	9.09	5.52	5.42	—	—	—
b^1T_{2g}	30%($2t_{2g} \rightarrow 6e_g$), 55%($2t_{2g} \rightarrow 3t_{2g}$), 1%($2t_{2g} \rightarrow 2t_{1g}$)	B	7.47	5.74	5.43	—	—	—
a^1A_{2g}	85%($2t_{2g} \rightarrow 2t_{1g}$)	B	8.01	5.71	5.62	—	—	—
c^1T_{1g}	2%($2t_{2g} \rightarrow 3t_{2g}$), 83%($2t_{2g} \rightarrow 2t_{1g}$)	B	8.41	6.00	5.91	—	—	—
c^1T_{2g}	1%($2t_{2g} \rightarrow 6e_g$), 11%($2t_{2g} \rightarrow 3t_{2g}$), 72%($2t_{2g} \rightarrow 2t_{1g}$)	B	8.41	6.03	5.92	—	—	—
b^1A_{1g}	11%($2t_{2g} \rightarrow 3t_{2g}$), 9%($5e_g \rightarrow 6e_g$)	B	10.81	7.12	6.89	—	—	—

[a] For the ungerade charge transfer states, the composition in terms of an excitation to either $9t_{1u}$ or $2t_{2u}$ is obtained from the CASSCF calculation with the active space D.

[b] From Ref. 199.

charge transfer bands, at 4.85 eV (b^1T_{1g}) and 5.08 eV (a^1T_{2g}). Comparing these results to the results in Table XIX, we notice that while the $^1T_{1g}$ state is calculated at exactly the same position with active spaces A and B, a small difference between the results is obtained for the $^1T_{2g}$ state. This can be traced back to the composition of the a^1T_{2g} state in Table XX, which now contains 14% $2t_{2g} \to 3t_{2g}$ charge transfer character. However, the difference is small, only 0.05 eV, giving further credit to the 10-orbital active space A originally chosen for calculation of the ligand field states. With active space B, the difference between the result calculated for the $^1T_{2g}$ state and the shoulder observed at 4.86 eV in the Beach and Gray spectrum is only 0.26 eV. However, from the results in Table XX it seems just as likely that this shoulder is actually due to the first $^1A_{1g} \to {}^1T_{1g}$ ligand field transition, or to a superposition of both. As for the shoulders at 3.60 eV and 3.91 eV, there seems to be little doubt that these should actually be assigned as either the spin-forbidden but orbitally allowed $^1A_{1g} \to {}^3T_{1u}$ transition or to one of the orbitally forbidden singlet–singlet transitions, of which there are plenty in the region 3.6–4.1 eV.

For states with ungerade symmetry, the agreement between the transition energies calculated with active spaces C1–C2 and D is in most cases acceptable, with a maximum difference of 0.2 eV. An exception is the first $^1A_{1g} \to {}^1T_{1u}$ transition. In this case the difference between the results calculated with the active spaces C1 and D amounts to 0.43 eV. We notice, however, that the $^1T_{1u}$ state under consideration is subject to a considerable interconfigurational mixing between the configurations with either $9t_{1u}$ or $2t_{2u}$ occupied. This interaction is significantly larger for this state than for any of the other ungerade states. This may explain the larger discrepancy between the results obtained with active space D, where the interaction is treated variationally at the CASSCF level, and active space C1, where it is treated perturbationally in the CASPT2 step. However, considering that both active spaces C1 and D were in fact chosen as a compromise for the the 16-orbital space that one would really prefer to use for the ungerade states, the results obtained should still be considered as satisfactory.

Neglect of the interaction between both $^1T_{1u}$ states when using the active spaces C1–C2 has more serious consequences for the oscillator strengths calculated, due to the fact that CASSCF wavefunctions are used for this purpose. Indeed, with the active space C1–C2, rather similar transition moments (−2.00 a.u.; −2.09 a.u.) are calculated for the pure $2t_{2g} \to 9t_{1u}$ and $2t_{2g} \to 2t_{2u}$ transitions, leading to an intensity ratio of only 1:1.2 for the two orbitally allowed transitions, much smaller than the experimental value of about 1:9. On the other hand, active space D contains both $^1T_{1u}$ states as a combination of the pure transitions, differing only in sign. As a result, the transition moment for the first, negative, combination is weakened, while it is enforced for the second, positive, combination. Although the calculated intensity ratio, 1:12.9, is now slightly too high, the absolute values of the oscillator strengths calculated with active space D are in good agreement with the values given from experiment.

As a final note we add that the recently reported INDO/S CI calculations of the $Cr(CO)_6$ spectrum resulted in an intensity ratio of 1:400 for the two transitions allowed, due to the fact that with this method the interconfigurational mixing in both $^1T_{1u}$ states is grossly overestimated. Although the calculated transition energies reported, 4.60 eV and 5.79 eV, conformed with experiment, both $^1T_{1u}$ states were in fact calculated as an almost equal mixture of $2t_{2g}^5 9t_{1u}^1$ and $2_{2g}^5 2t_{2u}^1$.

Spectrum of Ni(CO)$_4$. Since $Ni(CO)_4$ is formally a d^{10} system, no ligand field transitions appear in its electronic absorption spectrum, which consists solely of Ni $3d \rightarrow CO\ \pi^*$ charge transfer bands. Within the point group T_d, the CO π^* orbitals transform as e, t_1, and t_2, and one would a priori expect five bands in the spectrum, corresponding to five spin and orbitally allowed $^1A_1 \rightarrow {}^1T_2$ transitions. However, all experimental spectra of $Ni(CO)_4$, recorded either in solution [204], in a matrix environment [205], or in the gas phase [201], reveal at most three bands. The solution spectrum exhibits a main peak at 6.0 eV with two shoulders at 5.5 eV and 5.2 eV. Of these, only the band at 5.2 eV was detected in a matrix isolation spectrum, where instead a new shoulder appeared at 4.5 eV. Finally, in the most recent recording of the spectrum in the gas phase, the main peak at 6.0 eV and two shoulders at 5.4 eV and 4.6 eV could be discriminated. No experimental oscillator strengths have been reported, but in all cases the intensity of the bands increases steeply with an increasing wavenumber.

The theoretical interpretation of the $Ni(CO)_4$ spectrum has always been hampered by the absence of a more resolved structure in the experimental spectra. Up to now, a detailed assignment has only been provided in the most recent study of the gas-phase spectrum, where it was based on calculations performed with the INDO/S CI method [201]. To our knowledge, no previous *ab initio* results have been reported for the spectrum of $Ni(CO)_4$.

As for $Cr(CO)_6$, a calculation of the complete charge transfer spectrum of $Ni(CO)_4$, including singlet excited states of all possible symmetries, has been performed in a recent CASSCF/CASPT2 study [177]. Here the results obtained for the symmetry-allowed $^1A_1 \rightarrow {}^1T_2$

transitions are reported. Transition energies obtained at the CASSCF and CASPT2 level as well as oscillator strengths obtained using the CASSI method are presented in Table XXI, where we have included the INDO/S CI results for comparison.

Since the number of CO π^* orbitals in $Ni(CO)_4$ is limited to eight, no selection of different active spaces for different excited states has to be made in this case. Instead all states were calculated using the same active space of 13 orbitals, consisting of the $2e$, $9t_2$ Ni $3d$, and the $3e$, $10t_2$, and $2t_1$ CO π^* orbitals, and including 10 electrons. However, we remember that ground-state CASSCF calculations on $Ni(CO)_4$ indicate a considerable contribution of Ni $4d$ character in the $3e$ and $10t_2$ orbitals, accounting for the $3d$ double-shell effect (see also Table XVII and Fig. 6). Part of the Ni $4d$ character inevitably gets lost in the CASSCF calculations on excited states containing either a singly occupied $3e$ or $10t_2$ orbital, since the corresponding orbital now turns into an exclusively CO π^* orbital. This situation introduces a slight unbalance in the calculations, which would only be removed at the expense of adding an additional virtual e and t_2 shell. The unbalance is most strongly reflected in the CASSCF results in Table XXI. As usual, all transition energies are calculated too high at the CASSCF level. We notice, however, the large error, up to 3 eV and more, for the three lowest $^1T_2$ states with either a singly occupied $10t_2$ or $2e$ orbital. Both shells are left unoccupied in the highest two $^1T_2$ states, which are therefore more in balance with the ground state.

Fortunately, the CASPT2 results do not seem to suffer from the unbalanced CASSCF situation. Indeed, the weights of the CASSCF wavefunction in the first-order CASPT2(v) wavefunction are about the same for all states, indicating a balanced set of calculations. The CASPT2 results are again affected by the inclusion of a $3s$, $3p$ correlation, although the effect is almost negligible for the two highest charge transfer states. The largest effect is found for the a^1T_2 state, for which the difference between CASPT2(v) and CASPT2(c-v) amounts to 0.43 eV. Obviously, correlation of the semicore $3s$, $3p$ electrons cannot be neglected if one wants to get an accurate description of the relative energy of different charge transfer states in organometallic compounds.

Based on the transition energies obtained at the CASPT2(c-v) level, together with the calculated oscillator strengths, the following assignment of the spectral bands in the experimental $Ni(CO)_4$ spectra is proposed:

1. There can be little or no doubt that the lowest and least intense band observed at 4.5–4.6 eV in the matrix isolation and gas-phase spectra is due to the $9t_2 \rightarrow 10t_2$ transition. According to the

TABLE XXI

Absorption Spectrum of $Ni(CO)_4$: CASSCF and CASPT2 Results

Final State	Composition	Transition Energy (eV)				Oscillator Strength	
		CASSCF	CASPT2(v)	CASPT(c-v)	INDO/S	CASSI	INDO/S
a^1T_2	$92\%(9t_2 \rightarrow 10t_2)$	7.49	4.72	4.29	4.15	0.29	0.14
b^1T_2	$92\%(9t_2 \rightarrow 3e)$	7.57	5.45	5.19	4.36	0.38	0.11
c^1T_2	$94\%(2e \rightarrow 10t_2)$	7.34	5.75	5.54	4.91	0.29	0.02
d^1T_2	$93\%(9t_2 \rightarrow 2t_1)$	7.67	6.20	6.17	5.36	0.47	0.45
e^1T_2	$88\%(2e \rightarrow 2t_1), 6\%(9t_2 \rightarrow 2t_1)$	8.16	6.90	6.89	6.20	0.83	1.74

calculations, this transition is well separated from the other transitions. The calculated transition energy, 4.29 eV agrees well with the experimental band position, and the oscillator strength calculated indicates that this is indeed the least intense band in the spectrum.

2. The band positions calculated for the b^1T_2 ($9t_2 \rightarrow 3e$) and c^1T_2 ($2e \rightarrow 10t_2$) states, 5.19 eV and 5.54 eV, are in agreement with the experimental band positions of the two shoulders observed at 5.2 eV and 5.5 eV in the solution spectrum, which should therefore be assigned as such. On the other hand, only one band is observed in this region in the gas-phase spectrum, at 5.4 eV. This band should therefore be related to both the $9t_2 \rightarrow 3e$ and $2e \rightarrow 10t_2$ transitions.

3. According to the results, the most intense band at 6.0 eV in the solution- and gas-phase spectra should be assigned as the $9t_2 \rightarrow 2t_1$ transition. The only possible alternative could be that it is in fact a superposition of both the transitions from $9t_2$ and $2e$ to $2t_1$. However, the calculated splitting between both transitions is more than 0.6 eV, and one would therefore expect them to appear as distinct bands in the spectrum. We also notice that the highest charge transfer transition could just as well fall outside the range measured by the spectra, which never reached below 200 nm (above 6.2 eV).

The present assignment is different from the assignment proposed by Kotzian et al. [201] based on the INDO/S CI results, shown in Table XXI. With that method all allowed charge transfer transitions in $Ni(CO)_4$ are predicted at significantly lower energies. Consequently, the three bands appearing in the gas-phase spectrum were assigned as follows: the band at 6.0 eV as e^1T_2, the band at 5.4 eV as d^1T_2, and the band at 4.6 eV as a superposition of the lowest three transitions. The latter assignment seemed to be justified by the low oscillator strengths calculated for these transitions. However, the relative oscillator strengths of the allowed charge transfer states are quite different with the two methods. As indicated earlier by the results presented for $Cr(CO)_6$, this observation must be connected to the extent of interconfigurational mixing observed for the different $^1T_2$ states. Actually, almost no mixing occurs in the CASSCF wavefunctions (see Table XXI), while the INDO/S CI method gives a large mixing. In view of the excellent results obtained with the CASSI method for the oscillator strengths in $Cr(CO)_6$, we believe that the present CASSI results for the oscillator strengths should also be considered as being closer to the truth than the INDO/S CI results. This then puts some serious doubt on the assignment given by Kotzian et al. [201] for the 4.6-eV band. In view of the low intensity of this band, it could never be composed of three transitions with a considerable oscillator strength.

Finally, we want to refer to the results of a recent CASSCF/CASPT2 study of the spectrum of Fe(CO)$_5$ [29]. A comparison with the INDO/S CI results showed that in this case also, CASSCF/CASPT2 gives significantly higher transition energies and quite different oscillator strengths. It is clear that the CASSCF/CASPT2 method is highly superior to INDO/S CI, which, at this moment, seems to represent the only method alternative for calculation of charge transfer spectra of organometallic compounds.

3. Cr$_2$ Molecule and Its Spectroscopy

As for all transition metal quantum chemistry, the primary difficulty in describing the chromium dimer concerns electron correlation. This is demonstrated clearly by the ground-state properties computed at the RHF-SCF and CASSCF levels of theory. At the experimental bond length the molecular energies given by the two methods are far above that of two separated atoms (more than 20 eV with the RHF-SCF method) [206]. Another difficulty concerns the multiconfigurational character of the wavefunction for the ground state. The weight of the Hartree–Fock configuration in a CASSCF wavefunction based on an active space comprising 12 active orbitals (3d and 4s) is less than 50%. It is clear that single-configurational correlation methods are inappropriate for this particular molecule [207].

In some very recent calculations on the electronic spectrum of the chromium dimer [23, 25] we were able to show how well suited the CASSCF/CASPT2 approach is for this molecule. The plenitude of configurations needed for its description at the zeroth-order level does not lead to prohibitively large expansions at the first-order level. This is one of the advantages with CASPT2 over MRCI techniques. Further, the CASSCF/CASPT2 method allows us to calculate excited electronic states almost as easy as ground electronic states. However, the calculations were not straightforward because of problems with intruder states for some electronic states (cf. Section II.C). The intruder states appear at short internuclear distances (typically between 2.8 and 4.4 a.u.) and they give rise to rather weak singularities in the potential curves. In Fig. 9 the potential energy curves of 18 electronic states of Cr$_2$ are displayed. Here, only points with reference weights larger than 0.7 were taken into account when constructing the curves. Although the intruder states constitute a problem, in particular for calculating accurate values of vibrational frequencies, they do not destroy the overall picture. With the level shift

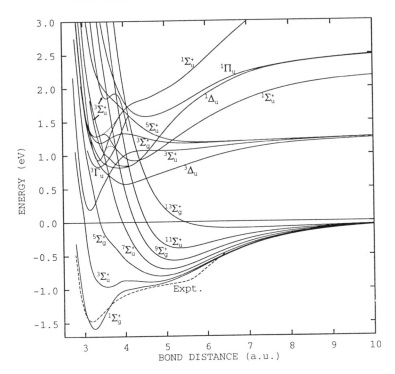

Figure 9. Potential curves for Cr_2 computed using the CASSCF/CASPT2 method. The dashed curve represents the experimental potential for the ground state. The dotted curves indicate avoided crossings.

introduced in Section II.C we will hopefully be able to obtain accurate spectroscopic constants also for the excited states.

An additional difficulty concerns the crossing of the two $^1\Sigma_u^+$ states at the CASPT2 level of theory (see Fig. 9). At the CASSCF level of theory these two states are dominated primarily by two configurations. The weight of the two configurations changes heavily with bond distance. At equilibrium, the two configurations are weighted equally in each state, while at somewhat larger distances one configuration dominates. This is a typical situation where a multistate CASPT2 method should be used, treating the two $^1\Sigma_u^+$ states simultaneously (see Section II.C).

Despite the problems stated above, we were able to give explanations to some of the experimental data about Cr_2. Using an ANO basis set of size $8s7p6d4f$ and including $3s$, $3p$ correlation effects and relativistic corrections, the following results were obtained (experimental results in parentheses):

1. The intensity profile of the rotational spectrum for the $A^1\Sigma_u^+ \leftarrow X^1\Sigma_g^+$ 0–0 band is probably due to strong nonadiabatic interactions near the inner walls of the $A^1\Sigma_u^+$ state and another $^1\Sigma_u^+$ state lower in energy.

2. The location of the $A^1\Sigma_u^+$ state is 2.86 eV (2.70 eV) above the ground state, and that of the lower $^1\Sigma_u^+$ state is 2.51 eV (2.40 eV) above the ground state.

3. Spectroscopic constants for the $^3\Sigma_u^+$ state located 1.82 eV (1.76 eV) above the ground state are $r_e = 1.67$ Å (1.65 ± 0.02 Å) and $\Delta G_{1/2} = 680$ cm^{-1} (574 cm^{-1}).

4. The $^5\Sigma_g^+$ state in the ground-state manifold is a plausible candidate for the metastable state detected in 1985 by Moskovits et al. [216]. The vibrational frequency $\omega_e = 148$ cm^{-1} (79 cm^{-1}) does differ from the experimental value. The difference can be attributed, at least partially, to matrix effects. After populating the metastable state Moskovits et al. [216] observed an intense absorption at 2.11 eV. This absorption may be the vertical transition $^5\Sigma_u^+ \leftarrow {}^5\Sigma_g^+$ calculated at 2.07 eV.

In summary, there is agreement between the spectroscopic data computed and available experimental information. The CASSCF/CASPT2 method has thus shown that it is capable of treating not only the ground state but also a large number of excited states for a molecule that has been a frustrating challenge for quantum chemistry for more than 10 years. With the newly implemented level-shift technique, the intruder state problem is also solved, which makes the calculations less involved and the results more stable. The entire field of spectroscopy of transition metal dimers and their ions is therefore open for an accurate theoretical analysis.

IV. SUMMARY

We have in this review presented a number of applications of the CASSCF/CASPT2 method in electronic spectroscopy. The general conclusion to be drawn from these and a large body of other applications is that if the method can be applied, it yields accurate results. The limitations of the approach is set by the possibility to select a small number of active orbitals, which describe the nondynamical correlation. This is almost always possible, but we have seen cases where the smallest acceptable active space is already outside the limits of the program. The CrF_6 molecule is one example, C_{60} is another. The application to polyaromatic hydrocarbons discussed earlier is an example which shows how the size of the active space can easily outgrow the capacity of the present technology. Extensions of the method to widen the range of applications are, however, possible:

1. The number of active orbitals can be increased by introducing direct calculations of the third-order density matrices without intermediate storing, combined with a new iterative solution to the first-order CI equations, which does not require diagonalization of large matrices (cf. Ref. 7).

2. It is also necessary to increase the capacity of the CASSCF program such that larger CI expansions can be handled. The limit today is about 10^6 CFs. Techniques to do this exist. An alternative is to use a RASSCF instead of a CASSCF reference function, which would allow a considerably larger number of active orbitals. However, the generalization of CASPT2 to RASPT2 is not trivial.

3. A CASPT2 gradient code is under development. It will enable more detailed studies of potential surfaces for excited states.

4. A multistate CASPT2 method, which is also under development, will improve results in cases where several energy surfaces of the same symmetry are close in energy and interact strongly. The most important situations where this occurs are probably close avoided crossings as illustrated for some excited states in Cr_2 [23] and close encounters between valence and Rydberg excited states. These situations are characterized by large modifications of the CASSCF reference function due to dynamic correlation.

5. The search for an improved zeroth-order Hamiltonian continues with the goal to remove the systematic error in binding energies. The methods presented in a previous section work well for cases with no inactive orbitals close in energy to the active orbitals. The correction terms move the orbital energies of the active orbitals closer to the ionization energies, which restores the balance for excitations out of them. To correct simultaneously for the imbalance of excitations into the active orbitals, it seems necessary to introduce two-electron terms in the zeroth-order Hamiltonian, which leads to considerable technical complications for large active spaces.

It is thus clear that there is room for considerable improvement of the CASSCF/CASPT2 approach. However, even at its present level of development it is a very useful and accurate method. We have presented examples from electronic spectroscopy, but the same method can be used to study energy surfaces for chemical reactions, including geometries of equilibria and transition states. Good examples are recent studies of the Bergmann reaction (autoaromatization of hex-3-ene-1,5-diyne) [208] and the Cope rearrangement [209]. The CASPT2 method has been used to compute vibration energies of ozone and the ozonide anion [210, 211]. It is thus a general method. It has limitations in accuracy, but this drawback is balanced by the generality of the method and the size of the molecules that may be studied.

ACKNOWLEDGMENTS

The research reported in this paper has been supported by a grant from the Swedish Natural Science Research Council (NFR), by IBM Sweden under a joint study contract, by the Cooperación Científico-Técnica (Ministerio de Asuntos Exteriores) of Spain, by projects PB91-0634 and PB94-0986 of Spanish DGICYT, and by the Belgian National Science Foundation (NFWO) together with the Belgian government (DPWB). Luis Serrano acknowledges a postdoctoral grant from the DGICYT of the Ministerio de Educación y Ciencia of Spain. Important contributions to the MOLCAS system, of which CASPT2 is one module, have been given by Roland Lindh, Jeppe Olsen, and Per-Olof Widmark. The authors would like to acknowledge fruitful discussions with Gunnar Karlström and Andrzej Sadlej. The following persons are acknowledged for their contributions to the work on transition metal compounds: Luc G. Vanquickenborne, Erik Van Praet, Eftimios Tsokos, Ralf Åkesson, Carl Ribbing, Kristine Ooms, Birgit Dumez, Rosendo Pou-Amérigo, Joakim Persson, and Nazzareno Re. Apart from the present authors, the following persons have contributed to the work on organic systems: Johan Lorentzon, Mercedes Rubio, Enrique Ortí, and Martin Schütz.

REFERENCES

1. S. R. Langhoff, Ed., *Quantum Mechanical Electronic Structure Calculations with Chemical Accuracy*, Kluwer, Dordrecht, The Netherlands, 1995.

2. B. O. Roos, in *Advances in Chemical Physics*; *Ab Initio Methods in Quantum Chemistry, II*, K. P. Lawley, Ed., Wiley, Chichester, West Sussex, England, 1987.

3. H. Partridge, S. R. Langhoff, and C. W. Bauschlicher, Jr., in *Quantum Mechanical Electronic Structure Calculations with Chemical Accuracy*, S. R. Langhoff, Ed., Kluwer, Dordrecht, The Netherlands, 1995.

4. M. Perić, B. Engels, and S. D. Peyerimhoff, in *Quantum Mechanical Electronic Structure Calculations with Chemical Accuracy*, S. R. Langhoff, Ed., Kluwer, Dordrecht, The Netherlands, 1995.

5. P.-Å. Malmqvist, B. O. Roos, M. P. Fülscher, and A. Rendell, *Chem. Phys.* **162**, 359 (1992).

6. K. Andersson, P.-Å. Malmqvist, B. O. Roos, A. J. Sadlej, and K. Wolinski, *J. Phys. Chem.* **94**, 5483 (1990).

7. K. Andersson, P.-Å. Malmqvist, and B. O. Roos, *J. Chem. Phys.* **96**, 1218 (1992).

8. B. O. Roos, P. Linse, P. E. M. Siegbahn, and M. R. A. Blomberg, *Chem. Phys.* **66**, 197 (1982).

9. P.-Å. Malmqvist, to be published, 1995.

10. P.-Å Malmqvist, A. Rendell, and B. O. Roos, *J. Phys. Chem.* **94**, 5477 (1990).

11. K. Andersson, M. P. Fülscher, G. Karlström, R. Lindh, P.-Å Malmqvist, J. Olsen, B. O. Roos, A. J. Sadlej, M. R. A. Blomberg, P. E. M. Siegbahn, V. Kellö, J. Noga, M. Urban, and P.-O. Widmark, *MOLCAS Version 3*, University of Lund, Lund, Sweden, 1994.

12. K. Andersson and B. O. Roos, In *Modern Electron Structure Theory*, Vol. 1, R. Yarkony, Ed., World Scientific, New York, 1995.

13. B. O. Roos, M. P. Fülscher, Per-Åke Malmqvist, M. Merchán, and L. Serrano-Andrés, In *Quantum Mechanical Electronic Structure Calculations with Chemical Accuracy*, S.R. Langhoff, Ed., Kluwer, Dordrecht, The Netherlands, 1995.

14. K. Andersson, Ph.D. thesis, University of Lund, Lund, Sweden, 1992.

15. K. Andersson and B. O. Roos, *Int. J. Quantum Chem.* **45**, 591 (1993).

16. C. W. Murray and E. R. Davidson, *Chem. Phys. Lett.* **187**, 451 (1991).

17. P. M. Kozlowski and E. R. Davidson, *J. Chem. Phys.* **100**, 3672, (1994).

18. P. M. Kozlowski and E. R. Davidson, *Chem. Phys. Lett.* **226**, 440 (1994).

19. K. G. Dyall, *J. Chem. Phys.*, **102**, 4909 (1995).

20. K. Andersson, *Theor. Chim. Acta* **91**, 31 (1995).

21. J.-P. Malrieu, J.-L. Heully, and A. Zaitsevski, *Theor. Chim. Acta* **90**, 167 (1995).

22. B. O. Roos, in *New Challenges in Computational Quantum Chemistry*, P. J. C. Aerts, R. Broer, and P. S. Bagus, Eds., University of Groningen, Groningen, The Netherlands, 1994.

23. K. Andersson, B. O. Roos, P.-Å. Malmqvist, and P.-O. Widmark, *Chem. Phys. Lett.*, **230**, 391 (1994).

24. K. Andersson, *Chem. Phys. Lett.* **237**, 212 (1995).

25. K. Andersson, *Theor. Chim. Acta* **91**, 31 (1995).

26. B. O. Roos and K. Andersson, *Chem. Phys. Lett.*, **245**, 215 (1995).

27. P.-O. Widmark, P.-Å. Malmqvist, and B. O. Roos, *Theor. Chim. Acta* **77**, 291 (1990).

28. K. Pierloot, B. J. Persson, and B. O. Roos, *J. Phys. Chem.* **99**, 3465 (1995).

29. B. J. Persson, B. O. Roos, and K. Pierloot, *J. Chem. Phys.* **101**, 6810 (1994).

30. J. B. Foresman, M. Head-Gordon, J. A. Pople, and M. J. Frisch, *J. Phys. Chem.* **96**, 135 (1992).

31. A. Balková and R. J. Bartlett, *Chem. Phys. Lett.* **193**, 364 (1992).

32. K. Andersson and B. O. Roos, *Chem. Phys. Lett.* **191**, 507 (1992).

33. B. O. Roos, K. Andersson, and M. P. Fülscher, *Chem. Phys. Lett.* **192**, 5 (1992).

34. J. M. O. Matos, B. O. Roos, and P.-Å. Malmqvist, *J. Chem. Phys.* **86**, 1458 (1987).

35. M. Merchán, E. Ortí, and B. O. Roos, *Chem. Phys. Lett.* **226**, 27 (1994).
36. L. Serrano-Andrés, M. Merchán, I. Nebot-Gil, R. Lindh, and B. O. Roos, *J. Chem. Phys.* **98**, 3151 (1993).
37. R. Lindh, U. Ryu, and B. Liu, *J. Chem. Phys.* **95**, 5889 (1991).
38. R. Lindh and B. Liu, *manuscript* (1991).
39. J. Almlöf and P. R. Taylor, *J. Chem. Phys.* **86**, 4070 (1987).
40. P.-O. Widmark, B.J. Persson, and B. O. Roos, *Theor. Chim. Acta* **79**, 419 (1991).
41. K. Pierloot, B. Dumez, P.-O. Widmark, and B. O. Roos, *Theor. Chim. Acta* **90**, 87 (1995).
42. F. B. van Duijneveldt, *Technical Report RJ945*, IBM Research, 1970.
43. M. P. Fülscher and B. O. Roos, *Theor. Chim. Acta* **87**, 403 (1994).
44. M. P. Fülscher and B. O. Roos, *J. Am. Chem. Soc.* **117**, 2089 (1995).
45. K. Kaufmann, W. Baumeister, and M. Jungen, *J. Phys. B* **22**, 2223 (1989).
46. J. Lorentzon, P.-Å. Malmqvist, M. P. Fülscher, and B. O. Roos, *Theor. Chim. Acta* **91**, 91 (1995).
47. J. Lorentzon, M. P. Fülscher, and B. O. Roos, *J. Am. Chem. Soc.*, **117**, 9265 (1995).
48. L. Serrano-Andrés, M. Merchán, I. Nebot-Gil, B. O. Roos, and M. P. Fülscher, *J. Am. Chem. Soc.*, **115**, 6184 (1993).
49. J. Olsen, B. O. Roos, P. Jørgensen, and H. J. Aa. Jensen, *J. Chem. Phys.* **89**, 2185 (1988).
50. S. Tenno, F. Hirata, and S. Kato, *Chem. Phys. Lett.* **214**, 391 (1993).
51. P. L. Muiño and P. R. Callis, *J. Chem. Phys.* **100**, 4093 (1994).
52. J. Hylton-McCreery, R. E. Christofferson, and G. G. Hall, *J. Am. Chem. Soc.* **98**, 7191, 7198 (1976).
53. O. Tapia and O. Goscinski, *Mol. Phys.* **29**, 1653 (1975).
54. S. Miertuš, E. Scrocco, and J. Tomasi, *Chem. Phys.* **55**, 117 (1981).
55. K. V. Mikkelsen, Ågren, H. J. Aa. Jensen, and T. Helgaker, *J. Chem. Phys.* **89**, 3086 (1988).
56. G. Karlström, *J. Phys. Chem.* **92**, 1315 (1988).
57. C. J. Cramer and D. G. Truhlar, *J. Am. Chem. Soc.* **113**, 8305 (1991).
58. M. M. Karelson and M. C. Zerner, *J. Phys. Chem.* **96**, 6949 (1992).
59. A. Bernhardsson, G. Karlström, R. Lindh, and B. O. Roos, to be published.
60. K. A. Zachariasse, Th. von der Haar, A. Hebecker, U. Leinhos, and W. Kühnle, *Pure Appl. Chem.* **65**, 1745 (1993).
61. J. Lipinski, H. Chojnacki, Z. R. Grabowski, and K. Rotkiewicz, *Chem. Phys. Lett.* **70**, 449 (1980).
62. L. Serrano-Andrés, M. Merchán, B. O. Roos, and R. Lindh, *J. Am. Chem. Soc.* **117**, 3189 (1995).
63. M. Hachey, P. J. Bruna, and F. Grein, *J. Chem. Soc. Faraday Trans.* **90**, 683 (1994).
64. M. R. J. Hachey, P. J. Bruna, and F. Grein, *J. Phys. Chem.* **99**, 8050 (1995).
65. M. Merchán and B. O. Roos, *Theor. Chim. Acta*, **92**, 227 (1995).
66. E. E. Barnes and W. T. Simpson, *J. Chem. Phys.* **39**, 670 (1963).
67. R. McDiarmid and A. Sabljić, *J. Chem. Phys.* **89**, 6086 (1988).
68. A. Gedanken and R. McDiarmid, *J. Chem. Phys.* **92**, 3237 (1990).
69. J. G. Philis, J. M. Berman, and L. Goodman, *Chem. Phys. Lett.* **167**, 16 (1990).

70. R. McDiarmid, *J. Chem. Phys.* **95**, 1530 (1991).

71. S. N. Thakur, D. Guo, T. Kundu, and L. Goodman, *Chem. Phys. Lett.* **199**, 335 (1992).

72. J. G. Philis and L. Goodman, *J. Chem. Phys.* **98**, 3795 (1993).

73. X. Xing, R. McDiarmid, J. G. Philis, and L. Goodman, *J. Chem. Phys.* **99**, 7565 (1993).

74. V. Galasso, *J. Chem. Phys.* **92**, 2495 (1990).

75. B. Hess, P. J. Bruna, R. J. Buenker, and S. D. Peyerimhoff, *Chem. Phys.* **18**, 267 (1976).

76. M. Merchán, B. O. Roos, R. McDiarmid, and X. Xing, *J. Chem. Phys.*, in press (1995).

77. S. Taylor, D. G. Wilden, and J. Comer, *Chem. Phys.* **70**, 291 (1982).

78. R. Nelson and L. Pierce, *J. Mol. Spectrosc.* **18**, 344 (1965).

79. L. Serrano-Andrés, M. Merchán, M. Fülscher, and B. O. Roos, *Chem. Phys. Lett.* **211**, 125 (1993).

80. B. O. Roos, M. Merchán, R. McDiarmid, and X. Xing, *J. Am. Chem. Soc.* **116**, 5927 (1994).

81. L. Serrano-Andrés, B. O. Roos, and M. Merchán, *Theor. Chim. Acta* **87**, 387 (1994).

82. R. P. Johnson and M. W. Schmidt, *J. Am. Chem. Soc.* **103**, 3244 (1981).

83. J. P. Malrieu and D. Maynau, *J. Am. Chem. Soc.* **104**, 3021 (1982).

84. W. E. Billups, L.-J. Lin, and E. W. Casserly, *J. Am. Chem. Soc.* **106**, 3698 (1984).

85. S. W. Staley and T. D. Norden, *J. Am. Chem. Soc.* **106**, 3699 (1984).

86. B. A. Hess, Jr., D. Michalska, and L. J. Schaad, *J. Am. Chem. Soc.* **107**, 1449 (1985).

87. D. Michalska, B. A. Hess, Jr., and L. J. Shaad, *Intern. J. Quantum Chem.* **29**, 1127 (1986).

88. T. D. Norden, S. W. Staley, W. H. Taylor, and M. D. Harmony, *J. Am. Chem. Soc.* **108**, 7912 (1986).

89. S. W. Staley and T. D. Norden, *J. Am. Chem. Soc.* **111**, 445 (1989).

90. M. Bogey, M. Cordonnier, J.-L. Destombes, J.-M. Denis, and J.-C. Guillemin, *J. Mol. Spectrosc.* **149**, 230 (1991).

91. C. Petrongolo, R. J. Buenker, and S. D. Peyerimhoff, *J. Chem. Phys.* **76**, 3655 (1982).

92. R. Lindh and B. O. Roos, *Intern. J. Quantum Chem.* **35**, 813 (1989).

93. R. González-Luque, M. Merchán, and B. O. Roos, *Z. Physik*, in press (1995).

94. M. Rubio, M. Merchán, E. Ortí, and B. O. Roos, *Chem. Phys. Lett.* **234**, 373 (1995).

95. M. Rubio, M. Merchán, E. Ortí, and B. O. Roos, *J. Chem. Phys.* **102**, 3580 (1995).

96. M. Rubio, M. Merchán, E. Ortí, and B. O. Roos, *J. Phys. Chem.*, in press (1995).

97. M. Rubio, M. Merchán, and E. Ortí, *Theor. Chim. Acta*, **91**, 17 (1995).

98. R. M. Hochstrasser, R. D. McAlpine, and J. D. Whiteman, *J. Chem. Phys.* **58**, 5078 (1973).

99. R. M. Hochstrasser and H. N. Sung, *J. Chem. Phys.* **66**, 3265 (1977).

100. A. Baca, R. Rossetti, and L. E. Brus, *J. Chem. Phys.* **70**, 5575 (1979).

101. J. Murakami, M. Ito, and K. Kaya, *J. Chem. Phys.* **74**, 6505 (1981).

102. T. G. McLaughlin and L. B. Clark, *Chem. Phys.* **31**, 11 (1978).

103. B. Dick and G. Hohlneicher, *Chem. Phys.* **94**, 131 (1985).

104. B. Nordén, R. Håkansson, and M. Sundbom, *Acta Chem. Scand.* **26**, 429 (1972).

105. J. Sagiv, A. Yogev, and Y. Mazur, *J. Am. Chem. Soc.* **99**, 6861 (1977).

106. P. Swiderek, M. Michaud, G. Hohlneicher, and L. Sanche, *Chem. Phys. Lett.* **187**, 583 (1991).

107. O. Bastiansen and S. Samdal, *J. Mol. Struct.* **128**, 115 (1985).

108. L. O. Edwards and W. T. Simpson, *J. Chem. Phys.* **53** 4237 (1970).

109. E. Ortí, P. Viruela, J. Sánchez-Marín, and F. Tomás, *J. Phys. Chem.*, **99**, 4955 (1995).

110. J. E. Chadwick and B. E. Kohler, *J. Phys. Chem.* **98**, 3631 (1994).

111. R. H. Abu-Eittah and F. A. Al-Sugeir, *Bull. Chem. Soc. Japan* **58**, 2126 (1985).

112. D. Birnbaum and B. E. Kohler, *J. Chem. Phys.* **96**, 2492 (1992).

113. D. B. Wetlaufer, in *Advances in Protein Chemistry*, Vol. 17, C. B. Anfinsen, M. L. Anson, K. Bailey, and J. T. Edsall, Eds., Academic Press, New York, 1962.

114. A. P. Demchenko, *Ultraviolet Spectroscopy of Proteins*. Springer-Verlag, Berlin, 1986.

115. L. Serrano-Andrés and B. O. Roos, *J. Amer. Chem. Soc.*, in press (1995).

116. L. Serrano-Andrés, M. P. Fülscher, B. O. Roos, and M. Merchán, *J. Phys. Chem.*, in press (1995).

117. L. Serrano-Andrés, Ph.D. thesis, University of Valencia, Valencia, Spain, 1994.

118. R. Pariser, *J. Chem. Phys.* **24**, 250 (1956).

119. J. R. Platt, *J. Chem. Phys.* **17**, 489 (1949).

120. A. Weissberger and E. C. Taylor, Eds., *The Chemistry of Heterocyclic Compounds*, Vol. 25, Wiley-Interscience, New York, 1972–1983.

121. J. M. Hollas, *Spectrochim. Acta* **19**, 753 (1963).

122. P. Ilich, *Can. J. Spectrosc.* **67**, 3274 (1987).

123. E. H. Strickland, J. Horwitz, and C. Billups, *Biochemistry* **4914**, 25 (1970).

124. A. A. Rehms and P. R. Callis, *Chem. Phys. Lett.* **140**, 83 (1987).

125. D. M. Sammeth, S. Yan, L. H. Spangler, and P. R. Callis, *J. Phys. Chem.* **94**, 7340 (1990).

126. T. L. O. Bartis, L. I. Grace, T. M. Dunn, and D. M. Lubman, *J. Phys. Chem.* **97**, 5820 (1993).

127. M. Rubio, M. Merchán, E. Ortí, and B. O. Roos, *Chem. Phys.* **179**, 395 (1994).

128. A. Z. Britten and G. Lockwood, *Spectrochim. Acta* **32A**, 1335 (1976).

129. H. Lami, *Chem. Phys. Lett.* **48**, 447 (1977).

130. W. Caminati and S. Di Bernardo, *J. Mol. Struct.* **240**, 253 (1990).

131. C. T. Chang, C. Y. Wu, A. R. Muirhead, and J. R. Lombardi, *Photochem. Photobiol.* **19**, 347 (1974).

132. H. Lami and N. Glasser, *J. Chem. Phys.* **84**, 597 (1986).

133. B. Albinsson and B. Nordén, *J. Phys. Chem.* **96**, 6204 (1992).

134. T. G. Fawcett, E. R. Bernarducci, K. Krogh-Jespersen, and H. J. Schugar, *J. Am. Chem. Soc.* **102**, 2598 (1980).

135. D. S. Caswell and T. G. Spiro, *J. Am. Chem. Soc.* **108**, 6470 (1986).

136. P. E. Grebow and T. M. Hooker, *Biopolymers* **14**, 871 (1975).

137. F. B. C. Machado and E. R. Davidson, *J. Chem. Phys.* **97**, 1881 (1992).

138. D. Christen, J. H. Griffiths, and J. Sheridan, *Z. Naturforsch.* **37A**, 1378 (1982).

139. E. Clar, *Polycyclic Hydrocarbons*, Academic Press, New York, 1964.

140. T. Shida, *Electronic Absorption Spectra of Radical Ions*, Elsevier, Amsterdam, 1988.

141. L. Serrano-Andrés, R. Lindh, B. O. Roos, and M. Merchán, *J. Phys. Chem.* **97**, 9360 (1993).

142. R. C. Dunbar, *Chem. Phys. Lett.* **32**, 508 (1975).

143. R. C. Dunbar, *Anal. Chem.* **46**, 723 (1976).

144. R. C. Dunbar and H. H.-I. Teng, *J. Am. Chem. Soc.* **100**, 2279 (1978).

145. T. Bally, S. Nitsche, K. Roth, and E. Haselbach, *J. Am. Chem. Soc.* **106**, 3927 (1984).

146. T. Bally, S. Nitsche, K. Roth, and E. Haselbach, *J. Chem. Phys.* **89**, 2528 (1985).

147. T. Bally, S. Nitsche, and K. Roth, *J. Chem. Phys.* **84**, 2577 (1986).

148. M. Beez, G. Bieri, H. Bock, and E. Heilbronner, *Helv. Chim. Acta* **56**, 1028 (1973).

149. M. Allan, J. Dannacher, and J. P. Maier, *Chem. Phys.* **73**, 3114 (1980).

150. T. B. Jones and J. P. Maier, *Intern. J. Mass. Spectrom. Ion Phys.* **31**, 287 (1979).

151. M. P. Fülscher, S. Matzinger, and T. Bally, *Chem. Phys. Lett.* **236**, 167 (1995).

152. F. Salama and L. J. Allamandola, *Nature* **358**, 42 (1992).

153. M. Schütz, M. P. Fülscher, and B. O. Roos, to be published.

154. R. Gschwind and E. Haselbach, *Helv. Chim. Acta* **62**, 941 (1979).

155. M. O. Delcourt and M. J. Rossi, *J. Phys. Chem.* **86**, 3233 (1982).

156. T. Shida and W. H. Hamill, *J. Chem. Phys.* **44**, 2375 (1966).

157. T. Shida and S. Iwata, *J. Am. Chem. Soc.* **95**, 3473 (1973).

158. A. Kira, M. Imamura, and T. Shida, *J. Phys. Chem.* **80**, 1445, (1980).

159. A. Kira, I. Nakamura, and M. Imamura, *J. Phys. Chem.* **81**, 511 (1977).

160. L. Andrews, B. J. Kelsall, and T. A. Blankenship, *J. Phys. Chem.* **86**, 2916 (1982).

161. P. A. Clark, F. Brogli, and E. Heilbronner, *Helv. Chim. Acta* **55**, 1410 (1972).

162. W. Schmidt, *J. Chem. Phys.* **66**, 828 (1977).

163. R. C. Dunbar and R. Klein, *J. Am. Chem. Soc.* **98**, 7994 (1976).

164. M. S. Kim and R. C. Dunbar, *J. Chem. Phys.* **72**, 4405 (1980).

165. F. Salama and L. J. Allamandola, *J. Chem. Phys.* **94**, 6964 (1991).

166. L. Andrews, R. S. Friedman, and B. J. Kelsall, *J. Phys. Chem.* **89**, 4016 (1985).

167. N. Otha and M. Ito, *Chem. Phys.* **20**, 71 (1977).

168. R. Pou-Amérigo, M. Merchán, I. Nebot-Gil, P.-Å. Malmqvist, and B. O. Roos, *J. Chem. Phys.* **101**, 4893 (1994).

169. N. Re, A. Sgamellotti, B. J. Persson, and B. O. Roos, *Organometallics* **14**, 63 (1995).

170. K. Pierloot, E. Van Praet, L. G. Vanquickenborne, and B. O. Roos, *J. Phys. Chem.* **97**, 12220 (1993).

171. K. Pierloot, E. Tsokos, and B. O. Roos, *Chem. Phys. Lett.* **214**, 583 (1993).

172. C. W. Bauschlicher Jr., P. E. M. Siegbahn, and L. G. M. Pettersson, *Theor. Chim. Acta* **74**, 479 (1988).

173. K. Andersson, B. J. Persson, and B. O. Roos, to be published.

174. C. W. Bauschlicher, S. P. Walch, and H. Partridge, *J. Chem. Phys.* **76**, 1033 (1982).

175. R. L. Martin and P. J. Hay, *J. Chem. Phys.* **75**, 4539 (1981).

176. L. A. Barnes, B. Liu, and R. Lindh, *J. Chem. Phys.* **98**, 3978 (1993).

177. K. Pierloot, E. Tsokos, and L. G. Vanquickenborne, to be published.
178. R. Åkesson, K. Pierloot, and L. G. Vanquickenborne, to be published.
179. A. B. P. Lever, *Inorganic Electronic Spectroscopy*, Elsevier, Amsterdam, 1984.
180. L. G. Vanquickenborne and A. Ceulemans, *Coord. Chem. Rev.*, **48**, 157 (1983).
181. C. K. Jørgensen, *Modern Aspects of Ligand Field Theory*, North-Holland, Amsterdam, 1971.
182. P. S. Bagus and C. J. Nelin, in *New Challenges in Computational Quantum Chemistry*, P. J. C. Aerts, R. Broer and P. S. Bagus, Eds., University of Groningen, Groningen, The Netherlands, 1994.
183. C. W. Bauschlicher and P. E. M. Siegbahn, *J. Chem. Phys.* **85**, 2802 (1986).
184. K. Pierloot and B. O. Roos, *Inorg. Chem.* **31**, 5353 (1992).
185. C. J. Marsden, D. Moncrieff, and G. E. Quelch, *J. Phys. Chem.* **98**, 2038 (1994).
186. L. G. Vanquickenborne, A. E. Vinckier, and K. Pierloot, *Inorg. Chem.*, in press (1995).
187. K. Pierloot, unpublished results, 1995.
188. K. Pierloot and L. G. Vanquickenborne, *J. Chem. Phys.* **93**, 4154 (1990).
189. K. Pierloot, E. Van Praet, and L. G. Vanquickenborne, *J. Chem. Phys.* **96**, 4163 (1992).
190. K. Pierloot, E. Van Praet, and L. G. Vanquickenborne, *J. Chem. Phys.* **102**, 1164 (1995).
191. L. G. Vanquickenborne, K. Pierloot, and E. Duyvejonck, *Chem. Phys. Lett.* **224**, 207 (1994).
192. C. Ribbing and C. Daniel, *J. Chem. Phys.* **100**, 6591 (1994).
193. M. Hargittai, O. V. Dorofeeva, and J. Tremmel, *Inorg. Chem.* **24**, 3963 (1985).
194. M. Hargittai, *Inorg. Chim. Acta* **180**, 5 (1991).
195. K. Pierloot, B. Dumez, K. Ooms, C. Ribbing, and L. G. Vanquickenborne, to be published.
196. C. W. DeKock and D. M. Gruen, *J. Chem. Phys.* **44**, 4387 (1966).
197. C. Daniel, private communication, 1995.
198. J. J. Alexander and H. B. Gray, *J. Am. Chem. Soc.* **90**, 4260 (1968).
199. N. A. Beach and H. B. Gray, *J. Am. Chem. Soc.* **90**, 5713 (1968).
200. C. F. Koerting, K. N. Walzl, and A. Kupperman, *J. Chem. Phys.* **86**, 6646 (1987).
201. M. Kotzian, N. Rösch, H. Schröder, and M. C. Zerner, *J. Am. Chem. Soc.* **111**, 7687 (1989).
202. K. Pierloot, J. Verhulst, P. Verbeke, and L. G. Vanquickenborne, *Inorg. Chem.* **28**, 3059 (1989).
203. L. G. Vanquickenborne, M. Hendrickx, I. Hyla-Kryspin, and L. Haspeslagh, *Inorg. Chem.* **25**, 885 (1986).
204. A. F. Schreiner and T. L. Brown, *J. Am. Chem. Soc.* **90**, 3366 (1968).
205. A. B. P. Lever, G. A. Ozin, A. J. L. Hanlan, W. J. Power, and H. B. Gray, *Inorg. Chem.* **18**, 2088 (1979).
206. A. D. McLean and B. Liu, *Chem. Phys. Lett.* **101**, 144 (1983).
207. G. E. Scuseria, *J. Chem. Phys.* **94**, 442 (1991).

208. R. Lindh and B. J. Persson, *J. Am. Chem. Soc.* **116**, 4963 (1994).

209. D. A. Hrovat, K. Morokuma, and W. Thatcher Borden, *J. Am. Chem. Soc.* **116**, 1072 (1994).

210. P. Borowski, K. Andersson, P.-Å Malmqvist, and B. O. Roos, *J. Chem. Phys.* **97**, 5568 (1992).

211. P. Borowski, B. O. Roos, S. C. Racine, T. J. Lee, and S. Carter, *J. Chem. Phys.* **103**, 266 (1995).

212. K. P. Huber and G. Herzberg, *Molecular Spectra and Molecular Structure*, Vol. 4, *Constants of Diatomic Molecules*, Van Nostrand Reinhold, New York, 1979.

213. V. E. Bondybey and J. H. English, *Chem. Phys. Lett.* **94**, 443 (1983).

214. S. M. Casey and D. G. Leopold, *J. Phys. Chem.* **97**, 816 (1993).

215. K. Hilpert and K. Ruthardt, *Ber. Bunsenges, Phys. Chem.* **91**, 724 (1987).

216. M. Moskovits, W. Limm, and T. Mejean, *J. Chem. Phys.* **82**, 4875 (1985).

217. H. Im and E. R. Bernstein, *J. Chem. Phys.* **88**, 7337 (1988).

218. D. Birnbaum and B. E. Kohler, *J. Chem. Phys.* **95**, 4783 (1991).

ELECTRONIC STRUCTURE CALCULATIONS FOR MOLECULES CONTAINING TRANSITION METALS

PER E. M. SIEGBAHN

Department of Physics, University of Stockholm, Stockholm, Sweden

CONTENTS

I. INTRODUCTION

Molecules containing transition metals are by tradition considered to be difficult to treat computationally. The fractional occupation of the d shells frequently gives rise to a manifold of close-lying electronic states, each with different chemical behavior. The near-degeneracy problem for transition metal systems is furthermore enhanced by the fact that d bonds are often formed with nonoptimal overlap. This leads to the appearance of nonproper dissociation behavior at the equilibrium geometry. The problem in treating these sometimes severe nondynamical near-degeneracy problems is for transition metals coupled with the problem to treat the unusually strong dynamical correlation problem for the tightly packed

Advances in Chemical Physics, Volume XCIII, Edited by I. Prigogine and Stuart A. Rice.
ISBN 0-471-14321-9 © 1996 John Wiley & Sons, Inc.

electrons in the d shell. The main problem in the treatment of nondynamical correlation effects is that very flexible configuration expansions may be required, while the main problem in the treatment of the dynamical correlation problem is that basis functions with high l quantum numbers are in principle required. Transition metal atoms are also heavy enough that relativistic effects play an important chemical role. This is particularly true for the third-row transition metals, but the relativistic effects cannot be neglected even for the first-row transition metals if high accuracy is required. It is this computational challenge to treat transition metal systems combined with the very large experimental and technical interest that has led to a constantly growing theoretical interest in these systems. There is no doubt that application to transition metal complexes is going to be a major area of quantum chemical methods in the coming decade. The present review is intended to cover what has been learned so far in general terms as to treatment of these systems, with an emphasis on the calculation of energetics such as thermodynamic stabilities, bond strengths, and barrier heights.

Even though the above-mentioned computational problems are present for many systems containing transition metals, recent progress in an understanding of these systems has led to a much more optimistic outlook for the future in this area. This improved situation is not only due to the development of better methods but is to a large extent due to the fact that these problems are n^t as difficult to treat as anticipated 10–20 years ago. For example, it was believed that the near-degeneracy problems mentioned above would definitely require the use of multiconfigurational self-consistent field (MCSCF) and multireference configuration interaction (MR-CI) methods in almost all cases for transition metal systems. In contrast to this expectation, the practical and economical single-reference methods based on orbitals from a simple SCF calculation have turned out to handle much more difficult problems than was believed possible. This is probably the most important finding the last 10–20 years in computational quantum chemistry. It should be stressed that this development is to a large extent a result of the vast experience from calculations on transition metal complexes and only to a minor extent due to the development of better single-reference methods than was available 20 years ago. This is a good example of the quite different situation for quantum chemical method development at present from that present during the 1970s. Method development now relies entirely on the experience from large numbers of calculations and is not meaningful without a close knowledge of the results from these calculations. Another important and extremely useful development that has recently taken place is also directly a result of the experience from a large number of calculations rather than a purely theoretical insight. This is the development of schemes capable of removing the major part of the basis-set problem. As mentioned above, to treat the dynamical correlation problem of the transition metal d shell, basis functions with high l quantum numbers are otherwise required. These schemes make use of one or a few semiempirical parameters in combination with modest basis sets. The treatment of relativistic effects has also turned out to be much simpler than what was expected. The use of simple perturbation theory, including the Darwin and mass-velocity terms, an almost trivial computational effort, leads to surprisingly accurate results, at least for the first- and second-row transition metal systems. There are, in fact, at present no known cases where this simple treatment fails for these systems. The use of relativistic effective core potentials (RECPs) are also of the same accuracy if they are carefully calibrated. For systems containing transition metals of the third-row or higher elements, the situation is less clear due to the limited amount of experience. It appears at the moment that RECPs combined with a simple, effective one-electron spin-orbit treatment might be sufficient for these systems.

In parallel to the development and understanding of the use of conventional quantum chemical methods there has also been a development of density functional theory (DFT)–based methods. Progress in this area has been particularly successful during the last five years. Two new features lie behind this progress. The one that is almost exclusively mentioned is the use of gradient corrections of the density to improve the description of nonlocal effects. The other development, which may be even more important in this area, is that semiempirical parameters have been incorporated into the methods (compare the development of quantum chemical methods described above). Taken together, these developments have led to highly competitive DFT methods. In the long run this development may turn out to be the most important since these methods also have a higher potential for treating much larger systems than do the other methods. Several examples are given in this review, where DFT and standard quantum chemical methods are compared.

Several excellent reviews have been written during the past five years on the application of electronic structure calculations of systems containing transition metal atoms. In the book edited by Dedieu on transition metal hydrides [1], several chapters relate to this subject, and the same is true of the book on the treatment of d and f electrons edited by Salahub and Zerner [2]. A book edited by van Leeuwen et al. on theoretical aspects of homogeneous catalysis [3] covers important areas of application. Furthermore, the review by Veillard [4] has given an overview of the range of systems that have been studied by *ab initio* calculations, and Koga and Morokuma [5] have reviewed studies of reactions involving large organometallic systems. Finally, in a book edited by Yarkony [6] on electronic structure methods, Bauschlicher and others have recently reviewed application on molecules containing transition metals covering ionic and neutral systems ranging from small to rather large systems. For work prior to 1990, references can be found in these reviews. The present review only complements these other reviews and concentrates on the more recent development.

This review is organized as follows. First, an overview is given on relevant method development during the last 10 years. This development not only concerns applications on transition metals but is quite general, covering both standard quantum chemical methods and DFT methods. A subsection is related to the determination of geometries, which is a key aspect of all modern applications. Section III the particular aspects of these methods that relate to different groups of systems containing transition metals are described. There is one subsection each for the three rows of the transition metals. Section III.A is not on the first row as might have appeared most logical, but is, instead, concerned with the treatment of the second transition row, which turns out to be easiest to handle. In Section III.B we discuss the methods relevant for the treatment of systems containing third-row transition metals, where there is so far rather limited experience of accurate applications. In Section III.C the first transition row is discussed. This row is in many ways the hardest to treat. For example, severe near-degeneracy effects lead to sometimes very poor geometries at the SCF level and even at some correlated levels. In Section IV, examples from typical applications are given, showing what can be obtained in terms of accuracy and results. In Section V, some more difficult applications are described. One of these is for the chromium dimer, which used to be considered a typical transition metal system but is now considered to be an extreme case of mostly methodological interest. Unlike the expectations 10 years ago, the solution of the chromium dimer problem is nowadays not considered to have any significant impact on the treatment of transition metal systems in general. The review is summarized in Section VI.

II. GENERAL METHODS

Every modern quantum chemical application consists of two separate parts. First, the geometry should be determined by an optimization of all degrees of freedom in most cases. The time when geometries were assumed or taken from experiments is past. The only situations where geometry optimization of all degrees of freedom are not performed are where a freezing of variables is desired for some reason or where the optimization is likely not to change the assumed structure in any significant way. Examples of the first situation is where a cluster of metal atoms are used to model part of a metal surface. An example of the second situation is where the geometry optimization is performed at such a high level of treatment that programs that generate gradients are not available. In those cases it is practical to fix an unimportant C–H distance, for example. The second step of an application is to calculate the energy for the optimized geometry. It is another major finding based on the large number of applications the past 20 years that the energy step and the geometry step should optimally be done at quite different levels of computational accuracy. The geometry can normally be obtained at a rather low level of treatment. The energy will in most cases change only insignificantly if a geometry obtained at a higher level is used. Since a high-level geometry optimization is expensive, it will in the long run reduce the number of systems that is possible to treat within a given computer time, and this will therefore lead to a smaller amount of useful information generated. It is therefore always a task of major importance optimally to design a calculation not to be unnecessarily expensive. In the present section the experience obtained for the energy step and the geometry step is discussed in quite general terms in two subsections. The particular experience for transition metal systems is described in Section III.

A.   Accurate Energies

When the development of *ab initio* methods for molecules started about 25 years ago, the dream was to obtain methods that would allow a consistent convergence toward exact results for chemically interesting problems. Although this dream has been possible to fulfill for very small systems of mainly academic interest, it now appears that this will continue to be a dream forever for more interesting systems. This has been realized by most quantum chemists, and the 1990s can therefore be considered to constitute a new era of quantum chemistry, termed the *pragmatic era* [7]. The term *pragmatic* indicates that information based on general experience or experiments is allowed to enter into the results when the final answer is given. Even though the pragmatic era in this way contrasts the pure *ab initio* era from 1970 to 1990, it is also quite different from the previous semiempirical era dominating quantum chemistry up to 1970. In a typical semiempirical calculation, like an extended Hückel calculation, the result is normally not intended for quantitative use, or alternatively a very large number, sometimes 100, parameters were used to obtain quantitative accuracy. In contrast, in a typical modern quantum chemical application very high quantitative accuracy, with errors of only a few kcal/mol on bond energies, is obtained and only one to three semiempirical parameters are used.

To illustrate the accuracy typically obtained in a pure *ab initio* calculation, some results from a benchmark test consisting of 32 simple first-row molecules [8] are collected in Table I. This table contains atomization energies obtained at the Hartree–Fock and a few different correlated levels as well as empirically corrected results to be discussed below. The atomization energy is the energy required to dissociate a molecule into atoms. Zero-point vibrational effects are included. Representative results of a standard correlation treatment can be seen in the

TABLE I

Atomization Energies (kcal/mol) for the First-Row Benchmark Test Obtained at Various Levels of Theory Using Various Basis Sets

Method:	SCF	SCF	ACPF	CCSD(T)	ACPF	CCSD(T)	Expt.
Basis Set:	DZP	W6532	DZP	W6532	DZP	W6532	
X^a:			100.0	100.0	77.9	94.1	
H_2	75.6	77.9	99.5	102.9	105.6	104.5	103.3
CH	50.3	53.3	66.2	77.1	81.3	81.1	79.9
$CH_2(^3B_1)$	140.4	144.6	162.8	175.8	179.4	180.3	179.6
$CH_2(^1A_1)$	114.9	119.9	150.7	166.3	170.9	171.7	170.6
CH_3	220.1	226.0	266.3	284.6	289.2	290.8	289.2
CH_4	298.0	304.7	364.8	386.8	393.4	394.5	392.5
NH	40.5	44.3	69.7	76.3	77.0	78.3	79.0
NH_2	99.0	105.6	153.9	166.9	167.7	170.7	170.0
NH_3	173.8	182.8	253.2	271.9	273.4	277.5	276.7
OH	60.0	64.2	93.4	100.0	101.8	102.2	101.3
OH_2	139.2	147.5	203.4	216.1	219.5	220.4	219.3
FH	90.6	94.6	127.9	133.9	137.5	136.4	135.2
HCCH	271.2	283.6	344.2	376.5	384.4	387.3	388.9
H_2CCH_2	392.7	404.1	485.2	520.5	531.3	532.8	531.9
CN	74.8	84.4	145.8	166.9	174.9	174.5	176.6
HCN	184.3	195.5	267.7	292.3	299.8	300.8	301.8
CO	167.8	177.0	233.3	248.1	260.9	255.1	256.2
HCO	168.0	177.3	242.2	261.5	272.3	269.3	270.3
H_2CO	237.6	247.4	326.4	348.4	360.5	357.2	357.2
N_2	102.8	115.4	198.9	216.3	222.9	222.6	225.1
NO	43.6	53.5	127.9	141.4	149.2	146.9	150.1
O_2	14.9	24.6	99.9	112.2	121.5	117.7	118.0
CO_2	235.5	249.6	346.0	369.6	385.1	379.6	381.9
Δ^b	86.2	78.7	22.2	5.4	1.9	1.2	

^{a} Optimized scale parameter defined in the text.
^{b} Mean absolute, deviation compared to experiments.

column under ACPF (DZP) without any empirical corrections ($x = 100.0$) (DZP represents a "standard double-zeta plus polarization basis"). As seen on these results using the ACPF (averaged coupled pair functional) method [9], the accuracy is not very high. The ACPF method is a coupled cluster type of method with approximately the same accuracy as the CCSD (coupled cluster singles and doubles) method [10]. If some simple molecules containing bonds to hydrogen are considered first, it can be seen that for methane the error is 28 kcal/mol, which is 7 kcal/mol for each bond. The accuracy is similar for ammonia and water. Even the hydrogen molecule has an error of 4 kcal/mol. For multiply bonded systems the accuracy is correspondingly lower, with an error of 23 kcal/mol for CO and 26 kcal/mol for N_2, which again leads to an error of 7–9 kcal/mol for each bond. It is clear that with an error of this magnitude for each bond, the accuracy for a reaction energy where several bonds are formed and broken cannot be high either unless there is a fortuitous or systematic cancellation of errors.

To see the overall picture of the benchmark test, the mean absolute deviations Δ are given for several methods and basis sets in Table II. These methods include the previously mentioned ACPF and CCSD methods, but also the MCPF (modified coupled pair functional) method [11], the MP2 (Møller–Plesset second-order perturbation theory) method and the CCSD(T) method [12], where a perturbational estimate of the triple excitations has been added. The basis sets include the DZP basis discussed above and the nearly equivalent VDZ basis set, a DZ basis set

TABLE II

Mean Absolute Deviations Δ (kcal/mol) Obtained After Scaling the Correlation Error Using Various Basis Sets and Methods for the First-Row Benchmark Test

Method	Basis Set	Unscaled Δ	X^a	Δ
MCPF	DZ	40.0	64.1	15.8
MP2	DZP	25.1	74.1	7.9
MCPF	DZP	22.2	77.8	2.4
ACPF	DZP	22.2	77.9	1.9[b]
MCPF	VDZ	23.0	77.6	2.4
CCSD	VDZ	21.0	76.8	2.2
CCSD(T)	VDZ	17.6	80.0	2.6
MCPF	W321	23.1	76.8	4.0
MCPF	W432	19.5	81.4	2.8
MCPF	W5521	12.7	87.3	2.9
CCSD	W6532	10.6	88.2	3.2
CCSD(T)	W6532	5.4	94.1	1.2

[a] Optimized scale parameter defined in the text.
[b] Two reference states for F_2.

without polarization functions and a set of ANO (approximate natural orbital) contracted [13] basis sets from Widmark et al. [14], the Wxyzt basis sets. These basis sets are based on very large $14s,9p,4d,3f$ primitive basis sets and Wxyzu means that there are x contracted s functions, y contracted p functions, and so on. For further details of these calculations, see Ref. 8. The pure *ab initio* results are given under the column termed "unscaled." A few conclusions can be drawn from these results. First, the deviation Δ obtained at the double zeta plus polarization level (DZP, VDZ, and W321 basis sets) are quite stable around 20 kcal/mol, irrespective of which infinite-order correlation method is used. The second and most striking observation from the results in Table II is the slow convergence of Δ with basis-set size. At the MCPF level the Δ value goes from 23.1 kcal/mol using the W321 basis, down to 19.5 kcal/mol using the much larger W432 basis, and down to 12.7 kcal/mol using the extended W5521 basis set. The result for the atomization energies at this rather high level is thus still rather poor. Going to the larger W6532 basis at the CCSD level improves the result slightly, down to 10.6 kcal/mol. However, the only result in Table II that can be considered to be of quantitative absolute accuracy is obtained using the same W6532 basis at the CCSD(T) level, where Δ is 5.4 kcal/mol. It should be added that the CCSD(T) method is probably the most accurate method available today for this type of molecules. Also, a CCSD(T) calculation using the W6532 basis set is a very ℯxpensive calculation and it was in fact not even possible to do a simple molecule such as ethane at this level using the present set of programs. These results give a very pessimistic outlook for pure *ab initio* methods for applications on mainstream chemical problems.

The only way to proceed further from a pure *ab initio* scheme is to apply some type of correction on the results. The simpler this correction and the more well defined the scheme, the better. The parametrized configuration interaction with parameter X (PCI-X) scheme [8, 15] is built on the assumption that the correlation energy (ΔE) can be written as a Taylor expansion of the calculated correlation energy ΔE_c as

$$\Delta E = \Delta E_c + X' \, \Delta E_c + \tfrac{1}{2} X'' \, \Delta E_c^2 + \cdots .$$

To a first approximation the correlation energy can be taken to be linear in ΔE_c, and with $X = 100/(1 + X')$, X becomes the percentage of the correlation effect obtained in the calculation. With only the linear term retained, the extrapolation obtained in the PCI-X scheme becomes equal to a scaling of the correlation energy. The assumption of linearity in ΔE_c is a very good approximation. The inclusion of the quadratic term leads to an improvement in the average absolute deviation Δ compared to experiments for the benchmark test in Table I of less than 0.1 kcal/mol using a DZP basis set at the MCPF level. The PCI-X scaling procedure can be termed a multiplicative correction scheme. This is not the only possibility for correcting *ab initio* energies in a simple and well-defined way. In the G1 and G2 theories of Pople and co-workers [16, 17], an additive correction scheme is used. It has been shown in Ref. 8 that unless very large basis sets are used, the PCI-X multiplicative correction scheme is far superior. There are a few minor points where the PCI-X scheme deviates from a simple scaling procedure (see details in Refs. 8 and 12). The most important of these points is that Hartree–Fock limit corrections should in principle be added since only the correlation energy is corrected. The Hartree–Fock limit correction is very important to explain some of the results where scaling otherwise fails, but in many areas of application, such as for most transition metal systems, it is not actually needed.

The results for the benchmark test using the PCI-X scaling scheme are shown in Tables I and II. The X values given in this table are optimized for the benchmark to minimize Δ. The main conclusion that can be drawn from these results is that scaling always leads to large, sometimes dramatic, improvements. This is particularly true for the double-zeta plus polarization quality basis sets. For the DZP basis the MCPF Δ value goes from 22.2 kcal/mol without scaling to 2.4 kcal/mol with scaling. The success of the scaling is surprisingly insensitive to the details of the basis sets. This is a major advantage for actual applications since any standard basis set can be used and the results after scaling are much more stable and still equally well defined as the *ab initio* results without scaling. A few further comments are worth making on the results in Tables I and II. First, for the present benchmark the results when scaling is applied are equally good at the simpler MCPF or ACPF levels as at the more advanced CCSD or CCSD(T) levels, which is another advantage since the MCPF and ACPF methods are much cheaper to use. Second, unpolarized basis sets do not give high accuracy even after scaling, and the MP2 method is also clearly inferior to methods such as MCPF after the scaling correction. It is furthermore interesting to note that even for the most advanced calculations done here, using the CCSD(T) method with the W6532 basis, scaling significantly improves the results.

In a survey of modern computational chemistry, the recent rapid progress of density functional (DFT) methods must also be mentioned. As noted earlier, the combination of more advanced functionals giving a better description of nonlocal effects and the use of a few semiempirical parameters have led to very powerful DFT methods. In fact, Becke [18] has shown that with only three empirical parameters and with density gradient and Hartree–Fock exchange corrections, results for the benchmark in Table I (and other similar benchmarks) are as accurate as the PCI-X results based on MCPF and using DZP basis sets. This DFT method, hereafter termed B3LYP, has also been shown to give very accurate results for transition metal systems. The B3LYP method has been selected here as a representative modern DFT method, and several comparisons between this method and standard quantum chemical methods are made in subsequent sections. The B3LYP functional can be written as

$$F^{\text{B3LYP}} = (1 - A)F_x^{\text{Slater}} + AF_x^{\text{HF}} + BF_x^{\text{Becke}} + CF_c^{\text{LYP}} + (1 - C)F_c^{\text{VWN}},$$

where F_x^{Slater} is the Slater exchange, F_x^{HF} is the Hartree–Fock exchange, F_x^{Becke} is the exchange functional of Becke [18], F_c^{LYP} is the correlation functional of Lee et al. [19] and F_c^{VWN} is the correlation functional of Vosko et al. [20]. A, B, and C are the coefficients determined by Becke [18] using a fit to experimental heats of formation. Since the method uses Hartree–Fock exchange and has to compute some additional integrals, it is slightly slower than Hartree–Fock but normally slightly faster than MP2.

Dissociation curves of diatomic molecules are frequently used to assess the quality of computational methods. To illustrate different computational aspects of the dissociation of a molecule, curves for the HF molecule are collected in Figs. 1 and 2. The curve termed EXP in these figures is not an actual experimental curve but should be very close to the exact one. It was obtained by scaling a MR-ACPF(W6532) curve to give the exact experimental dissociation energy. The other curves shown in the figures are all obtained using DZP quality basis sets. In Fig. 1 the entire dissociation out to atomic fragments is shown. The main aspect that can be seen here is whether or not a method is capable of handling proper dissociation of a molecule. It can be seen in the figure that the PCI-80 scheme does not give proper dissociation since it is based on a single reference state. Similarly, curves not shown in the figure, like those from the SCF method, the single-reference MCPF method, and the CCSD(T) method based on restricted Hartree–Fock (RHF) orbitals, all fail to give proper dissociation of HF. So does the B3LYP-DFT method if RHF orbitals are used (not shown in the figure). If proper dissociation is the main criterion of a good calculation, these methods are not to be recommended. Proper dissociation is, on the other hand, seen clearly for the curve based on the MR-ACPF method, where orbitals from complete active space SCF (CASSCF) calculations were used. The active space

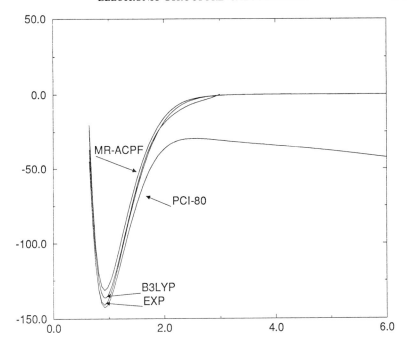

Figure 1. Dissociation curves for the HF molecule using various methods. DZP basis sets were used throughout.

used consisted of σ and σ^* orbitals. The B3LYP-DFT curve based on unrestricted HF (UHF) orbitals also dissociates properly. These methods are therefore superior to the previous methods in the sense of proper dissociation.

A question that remains, however, is of what consequence improper dissociation is for an actual application on a given chemical problem. It is, for example, commonly argued that a method without proper dissociation cannot be used where breaking and forming bonds occur and should thus not be useful for calculations on transition states of chemical reactions. To show that this argument is too simplified, another aspect of the dissociation curves is emphasized in Fig. 2. In this figure the curves obtained from the same methods are shown in a more narrow region between 0.7 and 1.5 Å around the equilibrium geometry. These curves were selected to answer the following question: If the calculation for economic reasons is restricted to the use of a limited basis set, what method should be used for the calculation of a barrier? The relevance of the deviation of a curve from the exact curve in connection with calculations of reactions concerns at what energy (X kcal/mol) above the

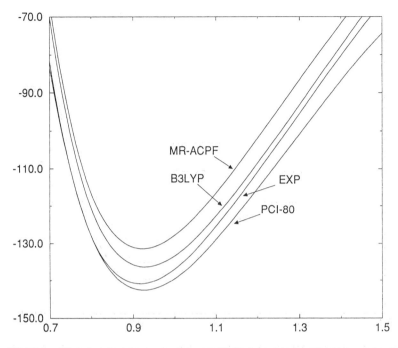

Figure 2. Dissociation curves around the equilibrium for the HF molecule using various methods. DZP basis sets were used throughout. The curve marked "EXP" is based on MR-ACPF calculations using the W6532 basis set scaled to give the experimental dissociation energy.

minimum the deviation becomes significant. A useful criterion is then to consider a method as useful for calculating reaction barriers when the barriers are lower than X kcal/mol. In this context it is important to consider the overall quality of the calculation, where the size of the basis set is at least as important as the intrinsic dissociation behavior of a method. For the curves in Fig. 2, the following conclusions can be drawn. Around the minimum the PCI-80 curve based on MCPF(DZP) calculations is energetically closest to the exact curve, as expected. The deviation of the curve termed MR-ACPF around the minimum is quite significant, 9.4 kcal/mol. As the bond is stretched, the improper dissociation behavior of the PCI-80 curve starts to be noticed. At equilibrium the deviation is 1.7 kcal/mol, at 20 kcal/mol above the minimum the deviation is 3.4 kcal/mol, and at 40 kcal/mol above the minimum the deviation is 5.4 kcal/mol. At this point the MR-ACPF(DZP) curve is still farther away from the exact curve, with a deviation of 9.1 kcal/mol. It can therefore be concluded that even for barriers as high as 40 kcal/mol, the error obtained from improper dissociation is actually smaller than the basis-set error from using a DZP basis set. Unfortunately, most applications on chemically significant systems will, at least in the near future, be restricted to basis sets of this size.

The important implication from the foregoing discussion related to Figs. 1 and 2 concerns the treatment of transition states. It can be concluded that as long as bond breaking and bond formation occur at an energy that is not very high above the energy for the equilibrium geometry, the use of single-reference methods is normally quite adequate for treating transition states. For example, based solely on energy arguments, it is not more dangerous to use a single-reference method for a transition state with a barrier less than 20 kcal/mol than it is to use a DZP basis set (as opposed to the exact basis set) for an equilibrium geometry (see Figs. 1 and 2). Since barriers for most of the interesting catalytical reactions involving transition metals are quite low, single-reference methods have been used both at equilibria and transition states in most of the examples discussed below. It can finally be added that if there should be any uncertainty concerning the adequacy of the single-reference-state treatment of a particular transition state, it is always possible to look at the CI wavefunction to ascertain whether any large coefficients have appeared.

B. Geometries

There are few areas of quantum chemistry where the development has been as successful as in the area of determination of structures. It is probably true today that there is no experimental technique available for general structures that is more accurate than even a simple quantum chemical treatment. Examples where failures are encountered obviously exist, but this is also true of any experimental technique. The time is therefore nearly gone when the sole purpose of a theoretical paper was to compare a structure to an experimental structure determined earlier, just to become confident in the theoretical approach. Instead, the time should be close when the reverse situation becomes more common. A clear advantage of theoretical methods is also that geometries of short-lived species and transition states can be obtained with nearly the same ease as those of stable structures. In addition to this success, theory has added a new dimension to the study of geometric structures, and this is the energy aspect on structures. A calculation cannot only determine the structural changes of a system but also how important this change is for the energy. It should be remembered that the structure is normally just a tool for obtaining insight into the energetics of a system. This energetic information is quite difficult to obtain from experiments and can be inferred only indirectly. In a theoretical study, typical conclusions could, for example, be that a bond angle changes by a large number of degrees but that this is not very important (for the energy), whereas in other cases, the conclusion could be that the bond distance changes only by so much but that this is quite important, which is an almost new perspective on structures. With these positive aspects of theoretical structure determinations in mind, the main important drawback should still be remembered: It is quite expensive to determine the geometry of a molecule, which means that the systems to be studied are severely limited in size from a general chemical viewpoint. This situation is changing rather rapidly with the development of faster workstations, but it is clear that this will continue to be the main drawback of theoretical methods. The present practical limit for a transition metal–containing system is 20–30 atoms, and this will perhaps improve to 100 atoms in the coming few years. This means that models of actual systems will continue to play a very important role in theoretical studies. This is not only a situation forced by limitations on computational speed but actually belongs to the nature of theoretical studies, where more general information than just the structure of a particular system is normally requested.

In this section some examples are given of geometries obtained for first-row systems of the kind shown in the benchmark test in Table I. Geometries for transition metal systems are discussed in Section III. The first main conclusion that can be drawn concerning first-row systems is that very simple computational methods are sufficient in most cases. Almost all systems in the benchmark in Table I were in fact done at the SCF level using DZ basis sets without polarization functions. Still, the mean absolute deviation of the atomization energies compared to experiments for the benchmark test is only 2.4 kcal/mol for the MCPF(DZP) method after scaling is applied. The CCSD(T)(W6532) deviation after scaling using the same geometries is as low as 1.2 kcal/mol. This deviation thus represents an upper limit of what can be obtained by using better geometries. Probably, the improvement would be only a few tenths of a kcal/mol on average, which is an entirely insignificant amount compared to the other errors in the calculations, at least when transition metals are present in the molecules. This result is very important, and it is interesting that it has taken 25 years to quantify this error. Of course, the reason for this is that very accurate energies are required to draw this conclusion.

It was stated above that almost all geometries for systems in the benchmark were determined at the SCF level using DZ basis sets. There is one marked exception and that is ammonia, for which an SCF(DZ) optimization can almost be claimed to fail. The scaled MCPF(DZP) energy for the SCF(DZ) geometry is thus as much as 3.5 kcal/mol worse than that obtained for an SCF(DZP) geometry. Since the total atomization energy of ammonia is 276.7 kcal/mol, 3.5 kcal/mol may not appear to be much, but this is still significant compared to the errors for the other molecules in the benchmark. The reason ammonia needs d polarization functions on nitrogen is that the umbrella angle is sensitive to angular polarization. Ethylene is another molecule where a geometric optimization at a higher level is not entirely unimportant. For ethylene the SCF(DZ) geometry is 1.0 kcal/mol worse than, for example, the MP2(DZ) and B3LYP(DZ) geometries. The energies for the MP2(DZP) and MP2(DZ) geometries are almost identical. It should also be remembered in this context that the total atomization energy is an extreme example; normally, the interest if focused on an energy difference between reasonably similar systems, where there will be a large error cancellation. In those cases an accurate geometry determination will be even less important.

It is clear that with such a good performance for the simple SCF(DZ) geometries, geometries determined at higher levels will also be adequate. A few detailed examples from recent investigations could be of some interest. The examples above from ethylene give the following geometries at different levels. At the SCF(DZ) level the C–C distance is 1.33 Å and the C–H distance is 1.07 Å, at MP2(DZ) they are 1.37 and 1.09 Å, at MP2(DZP) they are 1.35 and 1.08 Å, and at B3LYP(DZ) they are 1.35 and 1.09 Å. These are typical variations obtained at different levels and apparently do not matter for the energy. A slightly more complicated system has recently been studied, $C_6H_5-CH_2$ (Ph–Et), containing a benzene ring. The geometry for the lowest triplet excited state was obtained at the MP2(DZ) and B3LYP(DZ) levels, with the following results: For MP2(DZ) the C–C distances in the ring are four distances of 1.50 Å and two distances of 1.38 Å, while for B3LYP(DZ) they are 1.49 and 1.36 Å. The C–C distances in the ethyl group are for MP2(DZ) 1.52 and 1.58 Å and for B3LYP(DZ) 1.52 and 1.55 Å. These C–C distances, particularly in the ring, are expected to be sensitive to the treatment, and as seen from these results, they still vary by at most 0.02–0.03 Å between the methods. The total energy effect of all these changes at the B3LYP(DZ) level is 2.4 kcal/mol, which is a very small amount for such a large system. It will therefore not be of any consequence for the chemistry, at which of these levels the geometry is optimized. This is typical for systems not containing transition metals.

The oxygen molecule is an interesting and rare case where the geometric optimizations at different levels give significantly different results. In this case it is the MP2(DZ) optimization that fails. The PCI-80 energy for the MP2(DZ) optimized geometry is as much as 6.2 kcal/mol worse (higher) than for the SCF(DZ) geometry. The B3LYP(DZ) geometry is 0.6 kcal/mol better than the SCF(DZ) geometry. If the MP2 method is used, at least DZP basis sets have to be employed for the oxygen molecule.

In summary, from the examples above it can be concluded that sufficiently reliable geometries normally can be obtained from any one of the SCF, MP2, or B3LYP methods using DZ basis sets. Since the B3LYP method appears less basis set sensitive and it is slightly cheaper than MP2 and only slightly more expensive than SCF, there is seldom any reason not to obtain the geometries at the B3LYP level for these first-row systems.

III. METHODS FOR TRANSITION METALS

There are different requirements for the methods that should be used to treat the different rows of the transition metals. Two factors are of primary importance in this context. The first is the amount of near-degeneracy that is present, which in turn is strongly related to the overlap between the d orbitals and the ligand orbitals. In this context the relative size of the metal d and s orbitals is a key factor. The difference in size between these orbitals is largest for the first transition row, and near-degeneracy effects are therefore most critical for this row. The second factor for selecting an appropriate method is the importance of relativistic effects. In particular, the size of the spin-orbit effects play a dominant role, but the adequacy of using perturbation theory for treating the Darwin and mass-velocity terms is also significant. These questions naturally become more important the heavier the atoms are and are therefore most critical for the treatment of the third transition row. Below, methods found to be adequate for the three different rows are described and examples from calculations are given.

A. Second-Row Transition Metals

The second transition row is the easiest to treat computationally, so the discussion of methods requirements will start with this row. The principal reasons that this row is easiest to treat are that the $4d$ and $5s$ orbitals are of equal size and that spin-orbit effects are not yet important. For a certain distance to a ligand the $4d$ and $5s$ orbitals will have similar overlap with the ligand orbitals. Bonds can therefore be formed with optimal overlap for both these metal orbitals simultaneously. This is not true for systems where the size of the s orbital is much larger than the size of the d orbitals. When a d bond in those systems is formed at the distance for optimal overlap of the s orbital, the d bond has to be stretched. When a bond is stretched, nonproper dissociation behavior of the single determinant description sets in, and MCSCF and multireference methods are required (see Section III.C). A related effect is that when the s and d orbitals are of similar size as they are for the second transition row, the important sd hybridization can occur at the orbital level without significant configuration mixing. When metal complexes of the second transition row are treated, it is therefore quite unusual that multireference methods are required. In this context it is important to note that this is the case not only for equilibria but also in most cases for transition states (see Section II). This is true at least if the barriers for the reactions are not very high, which means that it is likely to hold for the catalytically most important reactions.

The other principal reason that the second transition row is easiest to treat is that spin-orbit effects are not yet significant. The situation is quite different for the third transition row (see below). For the second transition row, perturbation theory for the treatment of the Darwin and mass-velocity terms [21] is adequate. This can be seen on the results for the metal hydrides of this row shown in Table III. In this table the dissociation energies computed using perturbation theory for the relativistic effects are compared to results obtained by Eriksson and Wahlgren [22] using the more sophisticated no-pair theory of Hess [23]. In no-pair theory a transformation of the Dirac equation to two-component form is performed allowing a self-consistent (spin-free) treatment of relativistic effects to all orders. Finally, results obtained using relativistic ECPs (RECPs) [22] are also given in the table. As can be seen, the results show very small deviations in the three different treatments. It should also be remembered that slightly different basis sets have to be used in the three

TABLE III

Dissociation Energies (kcal/mol) of Second-Row Transition Metal Hydrides Obtained Without Relativistic Effects and Obtained Using Perturbation Theory, Relativistic ECPs (RECPs), or No-Pair Theory for These Effects[a]

System	Nonrelativistic	Perturbation Theory	No-Pair Theory	RECP
YH	64.8	67.9	68.5	67.5
ZrH	54.0	55.5	56.4	55.6
NbH	55.8	60.5	61.2	59.7
MoH	47.8	50.4	52.0	51.0
TcH	49.1	43.1	46.1	45.6
RuH	53.0	54.0	56.1	55.1
RhH	63.5	63.5	65.4	64.6
PdH	38.5	50.9	51.1	50.2

[a] The MCPF method was used with DZP quality basis sets.

treatments, and some of the small deviations obtained toward the middle of the row are more likely to be due to this difference in basis sets than to the actual treatment of relativistic effects.

Another aspect of the relativistic effects should also be mentioned in the context of the test of different methods shown in Table III, and this is that the relativistic effects are by far largest for small systems. The reason for this is that the relativistic effects are largest when there is a change of atomic state on the metal. The result for PdH is a good illustration of this. The bond in PdH is formed with the palladium atom in a $4d^9 5s^1$ configuration. At long distance the ground state of the palladium atom has a $4d^{10}$ configuration. This means that when the bond in PdH is formed, most of the atomic relativistic effect for the splitting between these states enters into the bond strength. This is verified by the results in Table III. For the palladium atom this effect is 15.4 kcal/mol, and for PdH the effect is 12.4 kcal/mol. In contrast, for a large palladium complex the optimal $4d^9 5s^1$ bonding state is likely to be present both before and after the bond formation, and the relativistic effect on the bond strength will therefore be small. The large relativistic effect in PdH is also in contrast to the almost absent effect in RhH. Clearly, the bonding Rh state in RhH is the same state as the ground state for the free atom. This also shows that the relativistic effects are normally quite easy to predict since they are basically atomic in origin.

As mentioned above, the $4d$ and $5s$ orbitals are similar enough in size that MCSCF and multireference techniques are generally not needed. This does not mean that these orbitals are precisely equal in size. The $4d$ orbitals are in fact still substantially smaller than the $5s$ orbitals. This is particularly true to the right, where the ratio of $\langle 5s \rangle / \langle 4d \rangle$ is 2.57 for palladium [6]. For yttrium to the left, the ratio is only 1.61.

An important consequence of the absence of near-degeneracy effects for the second transition row is that the geometries can be determined at a low level. In a systematic investigation of the requirements for geometric optimizations for this row [24], the $M(CH_3)H_x$ systems were studied from yttrium to palladium (except molybdenum). For simplicity, the number of hydrogen atoms (x) was chosen to make the systems have closed-shell ground states. Different methods were used to optimize the geometries of these systems. At the simplest level an SCF optimization was performed, at the next level the MP2 method was used, and at the highest level the quadratic configuration interaction with singles and doubles (QCISD) method [25] was used. Standard DZ basis sets without polarization were used throughout. For each optimized geometry the energy was evaluated using the same method—the MCPF method with a DZP basis set, which is known to give accurate relative energies. The

TABLE IV
Calculated MCPF(DZP) Energy Differences (kcal/mol)
Relative to the Energy Obtained for the SCF Structure for
Selected Second-Row Transition Metal Complexes[a]

System	SCF	MP2	QCISD
YH_2CH_3	0.0	+0.1	+0.4
ZrH_3CH_3	0.0	+0.4	+0.8
NbH_4CH_3	0.0	+0.3	+0.8
TcH_4CH_3	0.0	+1.2	+1.4
RuH_3CH_3	0.0	+0.9	+1.4
RhH_2CH_3	0.0	+1.8	+1.4
$PdHCH_3$	0.0	+1.1	+0.8

[a] In the various columns the results using different levels of geometry optimization are given. A positive energy difference means that the lowest energy is obtained for the SCF structure.

results are shown in Table IV, where each entry is the MCPF(DZP) energy difference compared to the energy obtained at the SCF(DZ) optimized geometries. The most important conclusion that can be drawn from these results is that it is normally sufficient to optimize the geometries at the SCF level even though there are very large correlation effects on the binding energies. The origin of this behavior is that the correlation energy is quite constant in the molecular region. One reason for this is that structure-sensitive near-degeneracy effects are seldom present for second-row transition metal complexes (see above).

As can be seen in Table IV, the SCF geometries are actually slightly better [they yield lower MCPF(DZP) energies] than when higher-level methods are used. This small and not very important effect has to be due to a cancellation of errors. A few additional test calculations were therefore performed with polarization functions added on carbon and hydrogen in the geometric optimization. For the yttrium system the MCPF(DZP) energy is now exactly the same for the SCF and MP2 geometries and 0.1 kcal/mol better for the QCISD geometry. For the palladium complex the MCPF energy for the QCISD geometry is 0.5 kcal/mol lower than for the MP2-optimized geometry using the polarized basis set, and it is 0.1 kcal/mol lower for the QCISD geometry using the polarized basis set than it is for the SCF geometry using the small basis set. It is clear that as the basis sets are enlarged, the results using the QCISD method for the geometric optimization must give better geometries than using the simple SCF method. For a double-zeta plus polarization quality basis set, this appears to be true but by a very small amount. It is important to note that the SCF(DZ) geometry is not a

Pauling point in the sense that an improvement in the level of geometric optimization does not lead to any significant worsening of the geometry. On the contrary, very small effects are found as the level of geometric optimization is improved, which indicates convergence in the treatment of the geometry.

It has been argued above that as long as a barrier for a reaction is low, it should be sufficient in general to optimize the geometry also for a transition state at a low level. To test the quality of the geometric optimization for a transition state, the case of the oxidative addition of the O–H bond in water to a palladium atom was selected. The PdHOH system can be considered to be a representative example where the barrier is rather low but where the correlation effects still are quite large for the reaction energetics, about 40 kcal/mol. The results are given in Table V. At the highest level of treatment, with QCISD-optimized geometries using a basis set with polarization functions, the insertion barrier is 26.6 kcal/mol. At the lowest level, with SCF-optimized geometries using a basis set without polarization functions, the barrier is 26.8 kcal/mol. If polarization functions are used in the SCF geometric optimization, the barrier becomes 26.7 kcal/mol. The barrier height is thus remarkably stable to the level of treatment used in the geometric optimization. The situation is quite similar for the binding energy of the insertion product, also given in the table. These results thus support the arguments given above that it is normally not more difficult to obtain a transition-state geometry than the geometry of an equilibrium.

From the results above it can thus be concluded that a simple SCF(DZ) geometric optimization should be adequate in most cases for second-row transition metal complexes. Another even more important question is at what level the energy has to be determined for obtaining quantitative accuracy. It was shown in Section II that a simple scaling

TABLE V

Results Obtained for the Transition State and Insertion Product of the Oxidative Addition of Water to a Palladium Atom[a]

Structure	Method	Basis Set	ΔE
Transition state	SCF	No polarization	−26.8
	SCF	With polarization	−26.7
	QCISD	With polarization	−26.6
Insertion product	SCF	No polarization	+13.9
	QCISD	With polarization	+14.0

[a] The columns on method and basis set refer to the geometry optimization step. The relative energies ΔE (kcal/mol) are obtained at the MCPF(DZP) level and are given with respect to the asymptotic MCPF(DZP) energy for free water and a metal atom.

using the PCI-X scheme is extremely successful in yielding reliable energies for first-row systems. The optimal scale parameter X is close to 80 for a coupled cluster type of method such as MCPF using DZP basis sets. This PCI-80 scheme has been tested for a large number of second-transition-row systems where experimental dissociation energies are available [15]. First, the method was tested on available bond strengths from experiments on cations (Table VI) and then on essentially all bond strengths for second-row transition metal systems listed by Huber and Herzberg [26]. Some of the latter systems show very large near-degeneracy effects and have to be treated with multireference methods (the results for those systems are described in Section V). The results for the other systems, where near-degeneracy effects are less severe, are given in Table VII. A major problem of testing the present method against

TABLE VI

Bond Strengths (D_0 in eV) in Various Cationic and Neutral Second-Row Transition Metal Systems[a]

System	SCF	MCPF	PCI-80	Exp.
MoH	0.90	1.98	2.33	2.30 ± 0.22^{b}
RuH	1.08	2.31	2.70	2.43 ± 0.22^{b}
RhH	1.19	2.53	2.95	2.56 ± 0.22^{b}
PdH	1.31	2.00	2.39	2.43 ± 0.26^{b}
YH^{+}	2.16	2.32	2.74	2.65 ± 0.04^{b}
ZrH^{+}	1.30	2.24	2.56	2.39 ± 0.13^{b}
NbH^{+}	1.21	2.10	2.45	2.34 ± 0.13^{b}
MoH^{+}	0.27	1.39	1.91	1.82 ± 0.13^{b}
RuH^{+}	0.73	1.49	1.68	$1.78 \pm 0.13^{c}(1.65 \pm 0.09^{d})$
RhH^{+}	−0.16	1.59	2.13	$1.56 \pm 0.13^{b}(1.69 \pm 0.09^{d})$
PdH^{+}	−0.07	1.83	2.49	$2.04 \pm 0.13^{b}(2.08 \pm 0.09^{d})$
YCH_3^{+}	2.01	2.29	2.78	2.56 ± 0.04^{b}
$RuCH_3^{+}$	0.50	1.53	1.83	$2.34 \pm 0.22^{c}(1.69 \pm 0.09^{d})$
$RhCH_3^{+}$	−0.56	1.38	2.02	$2.04 \pm 0.22^{c}(1.30 \pm 0.09^{d})$
$PdCH_3^{+}$	−0.12	1.83	2.56	$2.56 \pm 0.22^{c}(1.91 \pm 0.22^{d})$
YCH_2^{+}	1.57	3.10	3.90	4.03 ± 0.13^{b}
$NbCH_2^{+}$	0.62	3.28	4.12	$4.64 \pm 0.30^{e}(4.25^{d})$
$RuCH_2^{+}$	−0.39	2.69	3.51	$-(3.70 \pm 0.09^{d})$
$RhCH_2^{+}$	−0.45	2.81	3.78	$3.95 \pm 0.22^{e}(3.66 \pm 0.13^{d})$
$PdCH_2^{+}$	0.53	2.41	3.12	$-(2.97 \pm 0.17^{d})$
$\overline{\lvert \Delta E \rvert}^{f}$	1.88	0.41	0.25	

[a] The energies are calculated relative to ground-state metal atoms and ligands.
[b] From Ref. 27.
[c] From Ref. 28.
[d] From Ref. 29 (preliminary results).
[e] From Ref. 30.
[f] $\lvert \Delta E \rvert$ is the average absolute deviation compared to experiments.

TABLE VII

Bond Strengths (D_0 in eV) in Various Neutral Second-Row Transition Metal Systems[a]

System	SCF	MCPF	PCI-80	Expt.
Y_2	0.07	1.00	1.83	1.6^b
Zr_2	−2.15	1.68	3.05	3.052 ± 0.001^c
Pd_2	0.01	0.40	0.82	1.02 ± 0.2^d
YO	4.03	6.31	7.19	$7.2_9{}^b$
ZrO	3.61	7.03	7.88	7.85^b
NbO	3.21	6.80	7.70	7.8^b
MoO	0.43	4.30	5.27	5.0^b
PdSi	0.00	1.88	2.75	$3.2_1{}^b$
RhSi	−1.13	2.84	3.84	4.05^b
RuSi	−1.03	2.79	3.76	4.08^b
RhC	−0.03	4.73	5.90	6.01^b
RuC	0.52	5.02	6.14	6.68^b
RhB	0.72	4.12	4.95	4.89^b
RuB	0.21	3.55	4.37	4.60^b
PdB	1.17	2.72	3.24	$3.3_7{}^b$
ZrN	1.44	4.88	5.88	$5.8_1{}^b$
PdAl	0.56	1.87	2.24	2.60^b
YS	3.66	4.72	5.29	5.45^b
$\overline{\|\Delta E\|}^e$	3.83	0.97	0.18	

[a] The energies are calculated relative to ground-state metal atoms and ligands.
[b] From Ref. 26.
[c] From Ref. 31.
[d] From Ref. 32.
[e] $|\Delta E|$ is the average absolute deviation compared to experiments.

experiments for transition metals is that the experiments can be very uncertain. In fact, it is probably true that if all the experimental bond strengths listed in Ref. 26 were exchanged with the present theoretical values, the overall accuracy would probably not be lower. Two more comments are needed before the results are discussed. First, Hartree–Fock limit corrections have not been added, since they were found to be rather small in the cases tested. Second, the dissociation energies are calculated with respect to atomic s^1 states of the atoms and then adjusted to ground-state dissociation by using experimental excitation energies. The arguments for this procedure are discussed in detail in Ref. 15.

The 20 entries in Table VI refer to comparisons to experimental gas-phase measurements on transition metal cations [27–30]. The average absolute deviation from these experiments is 0.25 eV at the PCI-80 level. It can be noted that in general the agreement between the PCI-80 values and experiments is quite good in particular when the difficulty of treating these systems, both experimentally and theoretically, is taken into account. It must be emphasized again that the experimental results cannot be considered as the final answer in all cases. As shown in Table VI, new experimental results differing from the old results by as much as 0.7 eV have recently appeared. Because of this, it is sometimes difficult to draw definite conclusions concerning the PCI-80 accuracy. However, the situation becomes more clear in the comparison for the 18 entries in Table VII. For these systems the average absolute deviation from experiments is 0.18 eV at the PCI-80 level. This is a very large improvement compared to the unparametrized MCPF deviation of 0.97 eV. The average absolute deviation at the SCF level of 3.83 eV also illustrates the difficulty in treating these systems theoretically. A few results in the table are particularly noteworthy, showing very large improvements from the parametrization. In particular, Zr_2 is the only transition metal dimer for which the dissociation energy is accurately known experimentally through the presence of a predissociation [31]. For this multiply bonded system the correlation effects are dramatic. Even the MCPF value is as far as 1.5 eV away from experiment. The PCI-80 value is fortuitously close to experiment, with an agreement to 0.01 eV, which should, of course, not be taken too seriously. The most interesting aspect of this result is that it shows that Zr_2 is not qualitatively different from first-row systems such as CH_4 in that about 80% of the correlation effect is obtained in a standard DZP treatment. In this respect the transition metal dimers are thus not more difficult than any other molecules. However, a major difference is that the absolute correlation effect on the bond strength (about 5 eV) is very much larger for Zr_2.

Since accurate experimental results are often difficult to find for transition metal complexes, an alternative when a method is tested is to make comparisons to as accurate calculations as possible. The most accurate calculations available for the second transition row are those by Bauschlicher et al. [33] for the MCH_2^+ cations. These calculations are of multireference type using very large basis sets, and include error estimates. The comparison between the PCI-80 results and these accurate results [34] are shown in Table VIII. As seen on most of the entries in the table, the PCI-80 results and the MRCI results from Ref. 33 are in almost perfect agreement. The exceptions occur for niobium and molybdenum, where it is argued that the PCI-80 results are the ones that should be trusted. For more details, see Ref. 34.

Another set of results in Table VIII is also of some interest. These results are labeled "DFTG," which stands for "DFT method including gradient correction." The gradient corrections by Perdew and Wang [35] were used for the exchange and by Perdew for the correlation functionals [36]. This DFT method differs from the B3LYP method mainly in that no

TABLE VIII
Bond Strengths (D_0 in kcal/mol) for Second-Row Transition Metal $MCH_2{}^+$ Systems Using Various Methods[a]

M	MCPF	PCI-80	MRCI[b]	DFTG	Expt.[c]
Y	76.6	89.9	90	100.2	93 ± 3[d]
Zr	86.4	101.4	101	101.1	—
Nb	78.0	95.0	89	97.0	107 ± 7[e](98[f])
Mo	62.1	81.6	71	82.0	—
Tc	65.9	84.4	83	100.5	—
Ru	59.5	80.9	80	91.5	85.4 ± 2[f]
Rh	64.3	87.1	84	95.1	84.5 ± 3[f](89.6 ± 5[e])
Pd	56.6	71.9	70	73.5	68.6 ± 4[f]
$\|\Delta D_0\|$[g]	17.8	0.0	3.1	6.7	

[a] The energies are calculated relative to ground-state CH_2 and the bonding s^1 state of the metal cations, with the experimental excitation energy used to relate to ground-state ions.
[b] The MRCI results, which include error estimates, are from Ref. 33.
[c] Experimental values adjusted to 0 K by subtracting 1.5 kcal/mol following the procedure of Ref. 33.
[d] From Ref. 27.
[e] From Ref. 30.
[f] From Ref. 29 (preliminary results).
[g] $\|\Delta D_0\|$ is the average absolute deviation from the PCI-80 results.

empirical parameters are adjusted and that no Hartree–Fock exchange is included, which means that this DFTG method is substantially faster than the B3LYP method. The binding energies in the table are calculated with respect to the bonding atomic s^1 asymptote and adjusted to ground-state atoms by using experimental excitation energies. Although this procedure leads to only marginal improvements at the PCI-80 level, it is very important for obtaining reasonable results at the present DFTG level. This is because the splittings between the atomic states are not very well reproduced at the DFTG level [34]. As seen in Table IV, the DFTG results always indicate a higher binding energy than with PCI-80 and MRCI and in several cases give rather large overestimates compared to the PCI-80 (or MRCI) results. The results for zirconium, niobium, molybdenum, and palladium are quite acceptable, with deviations between 0 and 4 kcal/mol, while for the other metals the DFTG results are between 8 and 16 kcal/mol larger than the PCI-80 results. Overall, the average absolute deviation for the DFTG results with 6.7 kcal/mol is about a third of the corresponding deviation for the MCPF results of 17.8 kcal/mol. Even though these DFTG results are useful and promising, it appears worthwhile to proceed to introduce empirical parameters to improve the results in this scheme also.

There are two main conclusions to be drawn from the examples above. First, a single-reference scheme based on a coupled cluster type method followed by an empirical scaling, such as that in the PCI-80 scheme, is sufficient for obtaining quantitative accuracy for the energy in most cases for the second transition row. The second main conclusion is that a simple SCF(DZ) optimization is usually sufficient for the geometry. Exceptions to both these rules exist. The most obvious exceptions occur when near-degeneracy effects are present. These are rare cases, but molecules with multiple metal–metal bonding are often of this type. Another set of systems for which severe near-degeneracy is present is the metal oxides to the right. For these systems a mixing of ionic and covalent contributions to the bonding is a complication. Both metal dimers where metal–metal multiple bonding is present and metal oxides can be handled with some success using a multireference treatment followed by scaling (discussed in Section V).

For the geometry optimization there are other cases where near-degeneracy effects are not very important and where a simple SCF(DZ) treatment is still not sufficient. These cases have been found by trial and error and not purely by theoretical insight. There are essentially only two groups of systems where the SCF(DZ) geometric optimization will fail for the second transition row. Complexes to the right where ligands with strong donation–backdonation bonding are present belong to the first of these groups. The classical example of a ligand like this is ethylene, but phosphine can also be problematic. The carbonyl ligand is less of a problem, and water and ammonia ligands where electrostatics determine the bonding are well handled by an SCF(DZ) optimization. $Pd(C_2H_4)$ is an example of what happens (Table IX). At the SCF(DZ) level the metal–ligand distance becomes much too long, by more than 0.3 Å. The reason for this is that the system can dissociate properly into closed-shell

TABLE IX

Geometries and Energies for the $Pd(C_2H_4)$ and $Rh(C_2H_4)$ Obtained at Various Levels of Geometry Optimization[a]

System	Method	M–L	C–C	ΔE
$Pd(C_2H_4)$	SCF	2.34	1.35	0.0
	MP2	2.08	1.43	−8.9
	B3LYP	2.03	1.42	−10.0
$Rh(C_2H_4)$	SCF	2.09	1.44	0.0
	MP2	2.16	1.38	+5.6
	B3LYP	1.99	1.44	−3.4

[a] ΔE (kcal/mol) is the total PCI-80 energy difference compared to the energy at the SCF(DZ) geometry, M–L and C–C are the metal–ligand and C–C distances (Å).

fragments at the SCF level, combined with the fact that the bonding is strongly correlation dependent. The correlation dependence is common, but proper dissociation at the SCF level is quite unusual. For olefin complexes to the left, for example, nonproper dissociation will hold the system together and give rather accurate geometries. This may appear as a fortuitous artifact of the calculations but can also be regarded as a description of the bonding which is better balanced. There are many such cases also for lighter elements. The F_2 molecule is, for example, energetically unbound at the SCF level, but the geometry obtained at this level is still quite good. This is not normally regarded as completely fortuitous but rather as a result of a reasonable single determinant description around the equilibrium geometry. For $Pd(C_2H_4)$ a simple MP2(DZ) treatment gives a reliable geometry with a PCI-80 energy that is much lower, by 8.9 kcal/mol, than the one at the SCF(DZ) geometry (see Table IX). The DFT-B3LYP geometry is even better, with a PCI-80 energy 1.1 kcal/mol lower than the one at the MP2 geometry. For comparison the results for $Rh(C_2H_4)$ are also shown in Table IX. For this system it is the MP2 geometry optimization that gives the worst results. This is due to an exaggerated spin polarization at the UHF level. $Pd(C_2H_4)$ is a closed-shell system and will therefore not experience the same problem. The rhodium system is again well handled at the DFT-B3LYP level and this is therefore the only treatment that is sufficient for both these systems. The PCI-80 energy at the DFT-B3LYP geometry is as much as 9.0 kcal/mol lower than at the MP2 geometry. With the present experience the DFT-B3LYP method thus appears very promising, and there are in fact to our knowledge no known examples from the second transition row where this level does not give a sufficiently accurate geometry.

The second group of systems where an SCF(DZ) geometry optimization is not adequate is where the structure has *agostic bonding*, a term used for the attractive interaction between metals and C–H bonds. Examples of this type of interaction are the bond between a metal center and a beta C–H bond, and the bond between an essentially undistorted methane molecule and a metal complex. This type of bonding has a large amount of van der Waals character, which will obviously require a treatment of correlation effects. Some examples from a recent study of precursors for C–H activation of methane [37] are shown in Table X, where the results obtained using the SCF(DZ) optimized geometries are clearly seen to be quite erratic. Another problem that is a consequence of the bad SCF(DZ) description of the precursor can be mentioned in this context. For the RhCl(CO) system the reaction with methane leads to an unbound precursor if SCF(DZ) geometries are used. This fact, combined

TABLE X
Binding Energies for the Precursor of the Oxidative Addition
Reaction $RhXL + CH_4 + \Delta E \rightarrow RhXL(CH_4)^a$

Reactant	Method	ΔE
RhClCO	MP2	-3.4
	SCF	$+3.8$
RhHCO	MP2	-10.8
	SCF	-8.6
RhHNH$_3$	MP2	-14.5
	SCF	-7.1

^{a} Energies at the PCI-80 level are calculated relative to
ground-state RhXL systems and methane; "Method" indicates the geometry optimization method.

with the fact that at the SCF(DZ) geometry for the transition state for
the C–H bond breaking the PCI-80 energy is below the reactants, leads to
results which appear to indicate that there is no true transition state for
C–H bond breaking. This is not true. At the MP2 level of geometry
optimization the PCI-80 energy first goes down to -3.4 kcal/mol for the
precursor and then passes over a transition state at an energy of
-0.7 kcal/mol before the bond is broken. This is thus an example where
the potential surface is distorted at the SCF level so that a qualitatively
wrong picture emerges at this level. It should be noted that this behavior
is not related to a poor description in the bond-breaking region but to a
poor description in the precursor region.

B. Third-Row Transition Metals

The present discussion on methods has so far concentrated on the second
transition row, for which by far the largest number of systematic studies
of transition metal systems have been performed. In the present subsection the much more limited experience on the third transition row is
discussed. The first general point that should be emphasized concerns the
requirement on the correlation treatment. It was argued above that
single-reference methods are adequate for treating the second transition
row in most cases since near-degeneracy effects are seldom present. The
main reason for this is that the $4d$ and $5s$ orbitals are reasonably similar in
size, which leads to optimal overlap with ligand orbitals simultaneously
for these metal orbitals. Also, sd hydridization can occur easily at the
orbital level. These arguments should be even more relevant for the third
row since the ratio $\langle 6s \rangle / \langle 5d \rangle$ is smaller than the ratio $\langle 5s \rangle / \langle 4d \rangle$.
Therefore, near-degeneracy effects are not expected to be severe in
general for the third transition row. This means both that single-reference

methods should be adequate for the correlation treatment and that SCF geometries should be sufficiently accurate in general. It must be remembered that the exceptions mentioned above for the second transition row for cases of bonds with strong donation–backdonation and for agostic bonds should also be present for the third transition row. The experience available at present from calculations on the third transition row supports these general conclusions.

The main difference from a computational point of view between the second and third transition rows is that the relativistic effects are much larger for the third row. Therefore, an investigation of methods for the third transition row must start with a discussion of these effects. The first question in this context is how the spin-free effects, like those from the mass-velocity and Darwin terms, have to be treated. The second, much more difficult question is how the spin-orbit effects should be treated. For the first two transition rows spin-orbit effects can normally be neglected, but this is not generally true for the third transition row.

To investigate different treatments of the spin-free relativistic effects, Wahlgren [38] performed some comparative calculations for Hg_2^{2+}, and the results are shown in Table XI. The calculations were for simplicity done at the SCF level, but the conclusions drawn should also be valid when correlation effects are added. To read Table XI correctly, it is important to note that Hg_2^{2+} is only quasibound in a local potential minimum, with a large negative binding energy at the SCF level. The first comment that should be made on the results in Table XI is that the inclusion of relativistic effects is very important for both the geometry

TABLE XI

Comparison of Various Methods to Treat the Spin-Free Relativistic Effects for Hg_2^{2+} [a]

Treatment	Method	r_e	D_e
All-electron	Nonrelativistic	3.05	−71.8
	First order	2.79	−78.6
	No-pair	2.74	−83.7
ECP	Nonrelativistic	3.07	−72.8
	First order	3.06	−72.9
RECP	Conventional	2.74	−83.6
RECP[b]	Conventional	2.77	−84.0

Source: Ref. 38.

[a] The equilibrium distance R_e is given in Å and the dissociation energy D_e (note that the molecule is unbound) in kcal/mol. Calculations are done at the SCF level.

[b] Relativistic ECP of Hay and Wadt [39].

and the energies. It should be added that these effects are also important, but much less so, for the first and second transition rows. Since the spin-free effects are trivially included in the calculations [21], there is no reason to leave them out, even in the treatment of molecules of lighter elements. For Hg_2^{2+} the effect on the geometry is a shortening by 0.31 Å, and the effect on the binding energy is a decrease by 11.9 kcal/mol at the all-electron no-pair level. A second important conclusion is that perturbation theory for these relativistic effects does not work very well. Although most of the shortening of the bond distance, 0.26 Å, is obtained, only about half of the decrease in binding energy, 6.8 kcal/mol, is found at this level. The next stage in the comparison is to use a nonrelativistic ECP. This ECP reproduces the nonrelativistic all-electron results well, but when perturbation theory for the relativistic effects is used together with this ECP, almost no effect is obtained. The difference between the nonrelativistic ECP results and the first-order ECP results is only 0.01 Å for the distance and 0.1 kcal/mol for the binding energy. The reason for this total failure of perturbation theory at the ECP level is that the inner nodes of the $5s$ and $6s$ orbitals are poorly described [38]. At the next stage of the comparison relativistic ECPs (RECPs) were tried. As seen on the results in the table, this approach works very well. The results using the RECP developed by Wahlgren is within 0.01 Å of the all-electron bond distance and within 0.2 kcal/mol of the all-electron binding energy. The results using the relativistic ECP of Hay and Wadt [39] are almost as good. Since the use of RECPs is very much cheaper than the all-electron no-pair calculations, the conclusion to be drawn based solely on the results for Hg_2^{2+} is that RECPs should be used for these effects.

The results for Hg_2^{2+} thus indicate that the use of RECPs is a reliable procedure. However, since mercury is not a transition metal, it is still not clear that the procedure will work for cases with open d shells. To see if this is the case, results from a recent study by Wittborn and Wahlgren [40] on the third-row transition metal hydrides are shown in Table XII. The main conclusion from these results is that the RECPs work excellently. The PCI-80 results using RECPs are almost indistinguishable from the results at the all-electron no-pair level. The deviation for the bond distance is at most 0.02 Å, and the deviation for the dissociation energy is at most 2.8 kcal/mol and usually smaller.

Even though the appropriateness of using RECPs for the spin-free relativistic effects is useful to know, the most critical question in the treatment of the third transition row concerns the spin-orbit effects. Ideally, these effects should be treated explicitly, but a generally available well-established procedure to include these effects combined with a high level of correlation treatment does not yet exist. Hess et al. have recently

PER E.M. SIEGBAHN

TABLE XII

Results from All-Electron (AE) No-Pair and Relativistic ECP (RECP) Calculations on the Third-Row Transition Metal Hydrides

Molecule	Method	$r_e{}^a$	$D_e{}^a$
LaH	AE	2.07	68.3
	RECP	2.08	67.9
HfH	AE	1.86	70.3
	RECP	1.85	69.6
TaH	AE	1.82	54.3
	RECP	1.82	56.0
WH	AE	1.72	69.0
	RECP	1.73	67.9
ReH	AE	1.64	52.7
	RECP	1.64	53.1
OsH	AE	1.59	69.6
	RECP	1.59	70.9
IrH	AE	1.54	82.5
	RECP	1.56	85.3
PtH	AE	1.49	87.0
	RECP	1.51	86.3

Source: Ref. 40.

^{a} Distances are given in Å and PCI-80 energies in kcal/mol.

written a comprehensive review of the present status of the treatment of these effects [41], where the most important references can be found. For heavier transition metals the work by Balasubramanian and co-workers [42] is particularly worth mentioning. In the meantime, before this type of method has been well established and tested, the simplest assumption that can be made is that the main spin-orbit effects are quenched when the chemical bonds are formed. The basis for this assumption is the chemical situation before spin-orbit effects are added. An example can be taken from a recent study of the reaction between excited mercury atoms and different reactants [43]. If the reaction between $Hg^{*3}P(6s^16p^1)$ and a methyl radical is considered, it is clear that before the chemical reaction the three p directions of the Hg^* $6p$ orbital will be equivalent and lead to degeneracy for the three different states without inclusion of spin-orbit effects. When the spin-orbit operator is introduced, the interaction between these degenerate states will be large. However, when a chemical interaction has taken place, the degeneracy is lifted. With a strong interaction as between Hg^* and CH_3, it is not unreasonable to assume that the degeneracy is almost entirely gone when the bond is formed. The interaction between the p_σ orbital and the methyl radical orbital will be strongly attractive, while the interaction between the p_π orbital and methyl will be repulsive and the degenerate states will thus move apart. When the spin-orbit operator is introduced, the interaction between these states will then be small and the energy gain therefore also small. This model was tested in Ref. 53 on different mercury systems where experimental results are available [44, 26], and the results are shown in Table XIII. In practice, the assumption of total quenching of spin-orbit effects after the reaction means that for the unperturbed atom, spin-orbit effects will lead to a lowering of the energy by 6.8 kcal/mol, which is the difference between the calculated average value and the Hg* $^3P_1$ component. For the products of the reaction and for the transition state, spin-orbit effects will not lead to any lowering at all. As seen on the results in the table, the results are quite good. The most surprising result is that even for a rather weak interaction, such as the one between Hg* and lone-pair ligands such as water and ammonia, it appears that the spin-orbit effects are still quenched. This result should be seen in the context of the quite large splittings of 5.1 and 13.2 kcal/mol between the three different components of the ^3P_J states. It is interesting to note that the present conclusion about the quenching of spin-orbit effects is perfectly in line with the conclusion drawn from experiments [44]. It is stated there that spin-orbit interaction must be reduced by at least a factor of 10 near the potential minima, and it is also noted that this is a surprising result.

From the examples above it is thus clear that the total quenching of spin-orbit effects is a good approximation for mercury even in cases where the interaction is not very strong. Again, this conclusion is not obviously applicable to transition metals with open d shells. In fact, it is clear that there must be cases where there is state degeneracy also for the

TABLE XIII
Various Hg–X and Hg*–X Bond Strengths ΔE (kcal/mol) Obtained at the PCI-80 Level[a]

	ΔE	
	Calc.	Expt.
Hg–H	6.7	8.6[b]
Hg*–H$_2$O	6.3	8.1[c]
Hg*–NH$_3$	15.5	17.4[c]

[a] Spin-orbit effects are assumed to be totally quenched for the molecular systems.
[b] From Ref. 26.
[c] From Ref. 44.

molecule after the reaction, and a simple picture of total spin-orbit quenching must then be less accurate. This type of effect has not been studied thoroughly in general, but several studies on third-row transition metal hydrides have been performed. In particular, there is a study on PtH_2 by Dyall [45], who used a full four-component Dirac–Hartree–Fock procedure. For PtH_2, which is a singlet state, the spin-orbit effects are zero to first order. The higher-order spin-orbit effects for PtH_2 were estimated by Dyall to be 4.3 kcal/mol based on a comparison between the fully relativistic treatment and a no-pair calculation where spin-orbit effects are not included. This is a rather uncertain procedure where other effects are included, but it gives an idea of the order of the effect. It is therefore already clear that the error in the total quenching approach appears to be larger for PtH_2 than for the mercury systems discussed above.

In a recent study of the reactivity of Pt and Ir atoms with methane [46], some other useful rules were recommended if the total quenching model should be used for transition metals. First, for platinum, bond strengths should be related to the $^1S(d^{10})$ state since this state is low lying and has only one component. Results referring to the lowest J component of the ground-state platinum atom can then be obtained from experimental spectra [47]. For reactions with the iridium atom, the situation is more complicated. When the proper atomic state of the Ir atom is selected as a reference point for the calculations, it is important to use a state where the Landé interval rule is fulfilled. This leads to a well-defined average value which can then be related to the calculated value. For iridum the best state using this criterion turns out to be the $^4F(d^8s^1)$ state. For other third-row transition metal atoms, states should be selected in the same manner. It can be added that for lighter elements such as those of the second transition row, the Landé interval rule is normally well fulfilled in most cases.

In summary, for the methodological aspects of treating the third transition row, much more remains to be investigated. However, it is expected that reliable energies should be possible to obtain using single-reference methods, due to the essential absence of severe near-degeneracy effects. Similarly, it should be possible to obtain reliable geometries at a low level, such as the SCF, MP2, or B3LYP levels, using small basis sets. It has also been shown that the spin-free relativistic effects are well handled by RECPs. For the treatment of spin-orbit effects, two conclusions can be drawn. First, for larger saturated systems it is probably reasonably accurate simply to ignore these effects, since the effects are likely to be important mainly in cases when there is a change of the degeneracy. For reactions where atoms are involved, a combination of total quenching and a selection of a reference atomic state where the Landé interval rule is fulfilled is a useful model. These approaches are recommended until well-established procedures to treat the spin-orbit effects explicitly in the calculations become generally available, which should happen fairly soon.

C. First-Row Transition Metals

Even though the first transition row represents the lightest of the transition metals, molecules containing these metals are usually the most difficult to treat. As mentioned earlier, the reason for this is the large difference in size between the $3d$ and $4s$ orbitals. The largest difference in size occurs for nickel, where $\langle 4s \rangle / \langle 3d \rangle$ is as large as 3.36. To the left the ratio decreases down to 2.03 for scandium. This leads to severe near-degeneracy effects when the d bonds are formed since the bonds have to be stretched from the position of optimal overlap. This effect complicates substantially both geometry optimization and the calculation of reliable energies for molecules with elements of this transition row. These problems cannot be claimed to have been solved entirely at the present stage. In this subsection several examples of tests of methods are given that illustrate the experience obtained so far of methods that have been applied to this transition row.

The first problem encountered is for the geometry optimization. In Table XIV an example is given from a study of the O–H insertion reaction between H_2O and the nickel atom [48]. For this type of reaction with a second-row transition metal atom, an SCF geometry optimization works well both for the equilibrium geometry of the insertion product and for the transition state (see Table V). The results in Table XIV show that for the nickel atom, an SCF geometry optimization works very poorly. The MCPF energy is almost 15 kcal/mol worse (higher) at the SCF optimized geometry of the product than at geometries obtained at higher levels. For the transition state an SCF treatment happens to be better than for the product with an MCPF energy only 6 kcal/mol worse than for more accurate geometries. When correlation is included in the geometry optimization at the MP2 level, the geometry of the product appears to be reasonably accurate, but the MCPF energy for the MP2 transition-state geometry is actually 2.4 kcal/mol worse than for the SCF geometry. The situation is rather similar for a CASSCF geometry optimization, even though the main near-degeneracy effects have been taken care of in the active space. The active space contains four electrons in four active orbitals. It is only at the highest level of treatment, at the MCPF or QCISD levels, that accurate geometries are obtained for both the product equilibrium and for the transition state. Polarization func-

TABLE XIV

Optimized Geometries (Å) and Energies (kcal/mol) for the Singlet NiH(OH) Insertion Product and Transition State[a]

Method	Ni–H	Ni–O	H–Ni–O	Ni–O–H$_1$	ΔE(MCPF)
			Product		
SCF	1.60	1.74	131.4	151.5	+1.3
MP2	1.39	1.71	89.6	122.2	−12.6
CASSCF	1.51	1.75	109.2	136.7	−11.0
MCPF	1.40	1.71	103.8	116.1	−13.2
			Transition State		
SCF	1.65	1.78	48.7	131.8	+9.0
MP2	1.26	1.74	66.4	110.2	+11.4
CASSCF	1.69	1.89	48.5	128.2	+9.8
QCISD	1.57	1.85	49.6	112.8	+2.9
MCPF	1.57	1.82	46.2	113.1	+3.1

[a] The MCPF energies are given relative to the triplet ground-state asymptote. The method concerns the level of geometry optimization. H$_1$ is the hydroxyl hydrogen atom.

on both hydrogen and oxygen. When these MCPF or QCISD geometries are used, energies obtained with large basis sets at a high level, such as the CCSD(T) level give a potential surface that is consistent with experimental observations for this reaction [48].

From the results in Table XIV it might be concluded that the MP2 method is reliable for equilibrium geometries but not for transition states. Another example will show that failures can also be obtained for equilibria. The MP2 geometry obtained for the seemingly simple insertion product of the methane reaction with nickel, NiH(CH$_3$), is severely wrong. A correct geometry obtained at the CI level with polarized basis sets for this system has an angle of 94.0° between the Ni–H and Ni–methyl bonds [49]. The B3LYP geometry is quite accurate, with a bond angle of 95.2° and a low MCPF energy. At the CASSCF level with four active electrons in four active orbitals a reasonable bond angle of 96.6° is obtained, but due to too-long bond distances, the MCPF energy at this point is still 6.3 kcal/mol worse than for the B3LYP geometry. At the SCF level a much too large angle of 112.0° is obtained, with a total MCPF energy 11.2 kcal/mol worse than for the B3LYP geometry. At the MP2 level using polarized basis sets, the optimized angle is only 44.9°, which means that the C–H bond is halfway dissociated, with a distance of 1.38 Å. The MCPF energy for this MP2 geometry is as much as 20.2 kcal/mol worse than for the B3LYP geometry. Near-degeneracy effects are quite severe for this system, with a large CI coefficient of 0.30 appearing in the wavefunction. At the MP2 level these near-degeneracy effects will be exaggerated, and when the geometry is optimized at this level, the energy will become lower as these effects become larger. Therefore, at the MP2 level the system will start to dissociate into a nickel atom and methane to maximize near-degeneracy, but it is not possible to dissociate all the way since then more is lost due to improper dissociation than is gained by near-degeneracy. The MP2 geometry will therefore stay halfway dissociated.

The problems mentioned above do not mean that the MP2 method cannot be useful for geometry optimization of other types of systems of the first transition row. Ricca et al. [50] have, for example, shown that quite reliable geometries are obtained at the MP2 level for certain cationic iron systems such as $FeCO^+$ and FeH_2O^+ and several states of $FeCH_4^+$ where SCF optimizations give very poor results in most cases.

The near-degeneracy problem that affects the accuracy of geometry optimization of first-row transition metal systems also clearly affects the requirements on the methods used to calculate the final energies. In Table XV a few different methods have been used to calculate the bond strengths of first-row MCH_2^+ systems. The experimental results are taken from Armentrout et al. [51]. Most of the calculated results are taken from Bauschlicher et al. [33], who used very large basis sets with f functions on

TABLE XV
Metal–Ligand Bond Strengths in First-Row Transition Metal MCH_2^+ Systems (kcal/mol) for Various Methods[a]

M	DFT B3LYP	MCPF[b]	MR-ACPF[b]	Best Estimate[b]	Expt.[c]
Sc	83.6	75.3	80.1	85	96.9
Ti	81.9	74.8	78.1	83	91.9
V	78.4	69.5	73.5	79	78.5
Cr	59.6	42.4	50.7	57	52.3
Mn	66.9	43.0	55.9	61	69.3
Fe	79.2	59.1	67.2	74	81.5
Co	83.4	72.2	76.2	79	76.0
Ni	81.1	70.4	73.5	77	73.7
Cu	69.9	56.5	58.3	61	62.4
Δ^d	4.9	11.5	5.5	4.2	—

[a] Very large basis sets were used in Ref. 33 and a DZP basis set was used for the DFT-B3LYP calculations.
[b] From Ref. 33.
[c] From Ref. 51.
[d] Δ is the average absolute deviation compared to experiments excluding Sc and Ti.

carbon and d functions on hydrogen. Their MR-ACPF calculations based on internal contraction [52] must be considered state of the art at present for pure *ab initio* treatments. Based on their results, they suggest that the experimental results for scandium and titanium are too high, and these results are therefore excluded in the following discussion of average absolute deviations Δ compared to experiments. The Δ value obtained from the MR-ACPF calculations is 5.5 kcal/mol, where it should be remembered that the error bars for this type of mass-spectrometric experiments are between 2 and 5 kcal/mol. Based on the experience on deficiencies of their one- and many-particle basis sets, Bauschlicher et al. also arrive at the best estimates given in the table. These estimates should probably be regarded as equal in accuracy to the experimental values. The Δ value from the best estimates is 4.2 kcal/mol. As expected, the MCPF values using very similar large basis sets are of lower accuracy than the MR-ACPF results, with a Δ value of 11.5 kcal/mol. In this context the results at the DFT-B3LYP level using DZP basis sets have to be regarded as very promising, with a Δ value of only 4.9 kcal/mol. It should be added that the DFT-B3LYP calculations require only a few percent of the computation time taken by the MR-ACPF calculations. Whereas the large-basis-set MR-ACPF calculations will be restricted to small systems for a very long time to come, DFT-B3LYP calculations with a DZP basis set can already be applied to very large systems.

It was shown above for the second transition row that scaling the correlation effects as in the PCI-80 scheme gives very accurate energies. Even though the severe near-degeneracy effects for the first transition row are expected to complicate the situation substantially, it is still worthwhile to test a scaling procedure also for this transition row. In Table XVI, results of scaling using different methods are shown for the same set of MCH_2^+ systems as discussed above and shown in Table XV. The scale parameter X has been optimized for this set of molecules. As seen in Table XVI, the scaling appears to be rather successful in most cases, with Δ values of 3.7 kcal/mol at the MCPF level with scale parameter 83, 4.9 kcal/mol at the CCSD level with scale parameter 76, and 3.5 kcal/mol at the CCSD(T) level with scale parameter 87. However, a severe problem in this context is that the scale parameter is much larger than the optimal values for the same-quality basis sets for the first-row molecules in Table II, at least for the MCPF and CCSD(T) methods. This will lead to problems when reactions are studied where, for example, a C–H bond is broken and an M–C bond is formed, since a common X value has to be used. The origin of the large X values in Table XVI is that near-degeneracy effects lead to overestimates of the unscaled correlation energies. The severity of the near-degeneracy effects

TABLE XVI

Metal–Ligand Bond Strengths in First-Row Transition Metal MCH_2^+ Systems (kcal/mol) for Various Methods with Scaling, where X is the Scale Parameter[a]

M	MCPF $X = 83$	CCSD $X = 76$	CCSD(T) $X = 87$	Best Estimate[b]	Expt.[c]
Sc	89.0	91.7	89.5	85	96.9
Ti	92.6	92.5	89.8	83	91.9
V	85.0	82.9	79.9	79	78.5
Cr	66.3	62.0	59.2	57	52.3
Mn	69.8	62.1	61.2	61	69.3
Fe	77.8	73.6	74.4	74	81.5
Co	76.3	75.6	75.6	79	76.0
Ni	73.4	74.2	73.3	77	73.7
Cu	61.8	65.9	62.7	61	62.4
Δ[d]	3.7	4.9	3.5	4.2	—

[a] DZP basis sets were used throughout.
[b] From Ref. 33.
[c] From Ref. 51.
[d] Δ is the average absolute deviation compared to experiments excluding Sc and Ti; from Ref. 33.

can be seen particularly well for systems in the middle of the row, where the $CrCH_2^+$ system has an error compared to experiments that is as large as 14.0 kcal/mol at the MCPF-83 level. The $CrCH_2^+$ system has the strongest near-degeneracy effects of all these systems, with the largest norm of the $t2$ vector in the CCSD calculation and one of the largest CI coefficients in the MCPF calculation, and this is what leads to the large error for chromium. However, near-degeneracy does not always lead to severe overestimates of the bond strengths. The $MnCH_2^+$ system, for example, has almost as strong near-degeneracy as that of the $CrCH_2^+$ system and is still in almost perfect agreement with experiments using the MCPF-83 scheme. Until a better balanced scaling scheme has been devised, the appropriate conclusion is that results using scaling should not be trusted when there are severe near-degeneracy effects.

The set of results using the CCSD(T) method shown in Table XVI may not be the last word concerning this method for the first transition row. In a set of calculations for the same MCH_2^+ systems, Bauschlicher et al. [53] have shown that the CCSD(T) method can work very well for these systems. They used the same basis set as for the MR-ACPF calculations and obtained almost identical binding energies at the CCSD(T) level. The largest deviation for any of these systems compared to the MR-ACPF results is only 1.5 kcal/mol. This is a remarkable result considering the strong near-degeneracy effects present. The effect on the binding energies of the triple excitations is quite large, varying between 5 and 12 kcal/mol. In light of these results, the large-scale parameter required for the results at the CCSD(T) level in Table XVI is quite disappointing. This problem is probably not related to the unscaled CCSD(T) results by themselves but is connected to the scaling procedure used. At present the entire correlation energy is scaled, which appears to be an accurate procedure when the near-degeneracy effects are small, as for the systems in the benchmark test in Table I. However, when a large part of the correlation energy has a nondynamical origin, this type of scaling scheme works less well. This is relatively easy to understand since the basis set dependence, for example, of a small CASSCF calculation is much more similar to the dependence at the Hartree–Fock level than at the fully correlated level. The multiple excitation dependence within the active space is also normally taken care of, in contrast to the case for the excitations to the external space. The nondynamical part of the correlation energy should therefore, to a first approximation, not be scaled at all. Work is in progress to obtain a better definition of the part of the correlation energy that should be scaled.

The DFT-B3LYP results shown in Table XV are not the only test results obtained using this method for first-row transition metals. Ricca and Bauschlicher [54, 55] have also tested the accuracy of the energies calculated using this method. In particular, they showed that the successive binding energies of $Fe(CO)_5^+$ obtained by the DFT-B3LYP method are in extremely good agreement with the most reliable experiments and with other results obtained by high-level *ab initio* methods.

The final method suitable for treatment of the first transition row is the CASPT2 method [56], which uses second-order perturbation theory based on a CASSCF reference function. The method is designed to be able to treat efficiently very large reference spaces containing several thousand reference states. This is actually required quite often since the method does not work very well if a limited reference space is used. There is in this context a very large difference between the MR-CI and CASPT2 methods. For the same accuracy, one or two orders of magnitude more reference states are needed for the CASPT2 method. This does not mean that the CASPT2 method is less efficient. On the contrary, a CASPT2 calculation is generally much faster than a MR-CI circulation. These differences, both in the number of reference states and in the speed, are obviously due to the use of second-order perturbation theory in the CASPT2 method. The CASPT2 method is probably the best method today for treating very difficult cases with severe near-degeneracy effects. Situations where this method is particularly useful are treatment of excited states and spectra and for metal–metal multiple bonding. The method is not as useful for less complicated problems, which constitute the majority of the chemical problems encountered in transition metal chemistry. Examples of CASPT2 applications are given in Section V.

IV. APPLICATIONS

The main emphasis of this review is on methods used to treat transition metal systems, so this section on applications will be rather brief and far from complete. Instead, we mainly give illustrations of what can be done by the methods described in previous sections. For a more comprehensive coverage of the applications that have been done, some reviews were mentioned in the introduction. For example, very accurate *ab initio* calculations have recently been covered by Bauschlicher et al. [6], and reactions of large organometallic systems are covered by Koga and Morokuma [5]. For a survey of methods and applications in the very interesting and promising area of density functional theory, the reader is referred to the reviews by Ziegler [57] and Salahub et al. [58].

A majority of the applications on transition metal complexes have dealt with their geometric structures. Several systematic studies of the accuracy of the geometries have been done, some of which have already been mentioned. In Section III.A it was concluded that SCF geometries are quite accurate in general for the second transition row. Examples taken from Ref. 24 and shown in Tables IV and V illustrate this point. Benson et al. [59] have studied 150 different transition metal chalcogenido complexes and optimized the geometries at the SCF level. Transition metals from all three rows were studied, mainly those to the left and in the middle of the rows. With few exceptions agreement between calculated and experimental geometries is excellent. Frenking and co-workers also studied a large number of different transition metal complexes (see, e.g., Refs. 60 and 61), again with very good results in most cases. For the transition metal hexacarbonyls they used the MP2 method and obtained very good structures for molybdenum (second transition row) and tungsten (third transition row). For chromium (first transition row) a somewhat less accurate geometry was obtained, in line with what was discussed above about difficulties in treating this row.

The study by Ehlers and Frenking [60] mentioned above for transition metal hexacarbonyls also contain an interesting discussion of bond strengths. They used the CCSD(T) method to calculate these energies and compared their results both to experiments and to previous calculations. Sometimes rather large discrepancies were found compared to previous DFT calculations which did not use gradient corrections [62]. For molybdenum and tungsten, Ehlers and Frenking obtained excellent agreement with experiment (within 2 kcal/mol) for the first CO dissociation energy, but for chromium the discrepancy is quite large, 8–10 kcal/mol. Ehlers and Frenking argue that this discrepancy for chromium in part could be due to deficiencies in the optimized geometries. Part of the discrepancy should also come from the difficulty in describing the near-degeneracy effects of this row. Ziegler et al. [63] used density functional methods with gradient corrections and obtained results very similar to those of Ehlers and Frenking for all systems, including the chromium system, but argue that the discrepancy to experiment for chromium is due to a problem in the analysis of the experiment.

Impressive calculations for several sequences of reaction steps of the catalytic hydroboration of ethylene by $HB(OH)_2$ using a model Wilkinson catalyst, $RhCl(Ph_3)_2$, have been done by Musaev et al. [64]. Different mechanisms were compared involving more than 30 intermediates and transition states. Full-geometry optimizations were performed at the MP2 level. This study shows that it is now possible to understand catalytic reactions at a very detailed level. The hydroboration reaction is just one example, although a particularly spectacular one, of the type of detailed studies that have been performed by Morokuma et al. For earlier work along similar lines, see Ref. 5.

A very large number of systematic studies has recently been made for small and medium-sized second-row transition metal complexes (see Refs. 65–86). These studies use essentially the same computational procedure. The geometries are mostly optimized at the SCF level and correlation energies were obtained using the MCPF method, more recently within the PCI-80 scheme. About 1200 systems have up until now been studied in this way. A few examples of the types of results obtained are given here. In Fig. 3 the results obtained for some simple diatomic molecules are given for the entire second transition row [72]. In this figure the main effects responsible for covalent and ionic bonding in transition metal complexes can be seen. For covalently bound transition metal hydrides the bond strength varies from metal to metal but stays approximately constant across the row. The variations seen can be explained in quantitative detail by loss of exchange energy and by promotion energies [27, 72, 87]. The exchange energy loss appears because the metal atoms, in particular toward the middle of the row, have high spins in their ground states and the spin is reduced when the covalent bond is formed. The promotion energies involved are those sometimes needed to reach the bonding state, which in the case of the hydrides is the s^1 state of the metal. In contrast to the hydride curve, the curves in Fig. 3 for the more ionically bound halides rise sharply to the left in the row. The two factors primarily responsible for the large

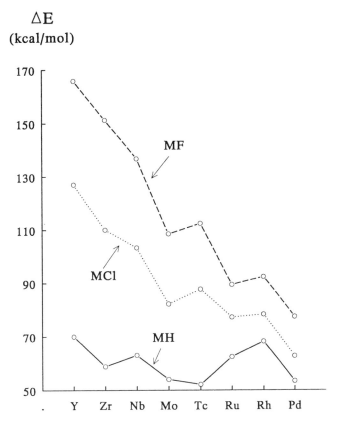

Figure 3. Bond strengths obtained at the PCI-80 level for second-row transition metal diatomic hydrides, fluorides, and chlorides.

increase in halide binding energies to the left are the decreasing ionization energies to the left and the attractive interactions between the halide lone pairs and the empty metal $4d$ orbitals. To the right, lone-pair interaction with the filled $4d$ orbitals is repulsive. For fluorine these interactions are strongest, leading to bond strengths that are almost 90 kcal/mol stronger to the left than to the right, where the Pd–F bond strength is only 77.4 kcal/mol.

In Fig. 4 another use of trends of calculated bond strengths is shown. In this case the bond strengths between the metal atoms and the methyl radical are compared to the bond strengths to the acetyl radical. For the acetyl radical, interaction between the metal and the lone pairs of the oxygen atom can be either attractive or repulsive, and an interesting question is how strong this interaction is. The metal–methyl bond

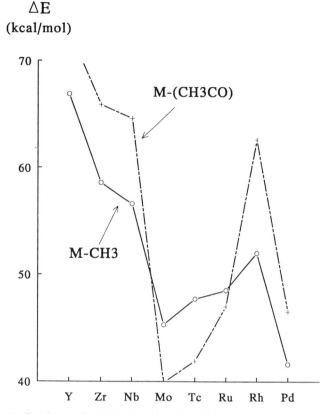

Figure 4. Bond strengths obtained at the PCI-80 level for second-row transition metal methyl and acetyl systems.

strengths in this case serve as reference values for a situation where lone-pair interaction is not present. As seen in Fig. 4, the metal–acetyl bonds are stronger than the metal–methyl bonds to the left for yttrium, zirconium, and niobium, where there are empty d orbitals on the metal. This attractive interaction leads to an η^2 bonding, shown in Fig. 5 for the yttrium system. From Fig. 4 this attraction gives about 5 kcal/mol additional bonding to the metal. In the middle of the row, where all the d orbitals are occupied (mostly singly occupied), the interaction with the oxygen lone pairs becomes repulsive, as expected. The rhodium system to the right is an interesting example showing that interaction with the lone pair can also be very attractive in this region of the periodic table. In this case the attraction appears due to a change of ground state for the acetyl system. The rhodium–methyl system is a triplet state, whereas the acetyl

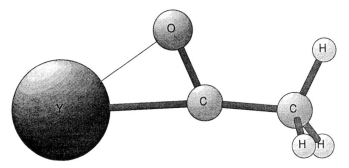

Figure 5. Structure of the yttrium–acetyl system showing the attractive interaction between the metal and the oxygen lone pairs.

system is a singlet. By this closed-shell coupling, the rhodium–acetyl system can create a hole in the d shell where the oxygen lone pair can fit and form an attractive interaction to a partly unshielded metal nucleus. This attraction gives about 10 kcal/mol additional binding. Use of this type of trend is thus an important way to break down interaction energies in quantitative detail.

The final example from the large number of systematic studies mentioned above will be taken from a recent collaborative effort between experiment and theory for reactions between the second-row transition metals and different hydrocarbons [88]. This is a very comprehensive study and serves to illustrate what it is now possible to do with theoretical methods. PCI-80 calculations of the accuracy demonstrated in Tables VI and VII were performed for the reactions of the entire sequence of second-row transition metals with methane, ethylene, and cyclopropane, following the reactions through intermediate and over transition states. An example of a C–C insertion transition state for the cyclopropane reaction is shown in Fig. 6. Figure 7 illustrates the type of results obtained for the cyclopropane reactions. The transition-state energies for C–H and C–C insertion are shown and also the stabilities of the product, which is just a fraction of the results in the paper. It is interesting to note the development in this area, where 10 years ago a study of the C_{2v}-restricted C–C insertion reaction of a Pd atom with cyclopropane was a major effort [89]. The results obtained in that study represent only two points in Fig. 7 and the accuracy obtained at that time was not even close to that obtainable today. It should be added that it is concluded in Ref. 88 that the theoretical energetic results obtained at the PCI-80 level are normally consistent to within 2–3 kcal/mol with the experimental rate measurements of the same study.

The final examples given in this section illustrate both what can

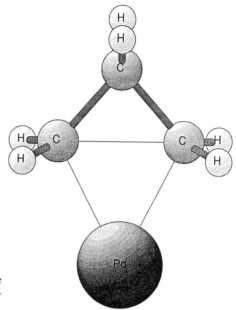

Figure 6. Transition state for the C–C insertion reaction between the palladium atom and cyclopropane.

presently be done using the methods described above and some of the problems not solved at present. These results concern the reactions between $M(C_5H_5)(CO)$ and methane, where the metals M are cobalt, rhodium, and iridium of the same column in the periodic table. The transition state for the iridium reaction is shown in Fig. 8. In Table XVII results obtained at four points along the reaction paths for these reactions are presented. The first point is the reactant, which is either a singlet or a triplet state; the second point is the η^2 precursor, with agostic bonding to an essentially undistorted methane molecule; the third point is the transition state for C–H bond breaking; and the fourth point is the product of the insertion reaction. At the PCI-80 level this is quite a large system, and the basis set therefore had to be reduced from DZP, which is normally used, to DZ without polarization for the cyclopentadienyl and carbonyl ligands. Still, the PCI-80 calculation is close to the limit of what can presently be done. At the B3LYP level this is not true for the energy evaluation, which is quite fast, but geometry optimization at this level is rather tedious. For applications on larger systems, DFT methods without Hartree–Fock exchange [57, 58] are more practical. If the results for iridium in Table XVII are first considered, it can be concluded that there is extremely good agreement between the PCI-80 and B3LYP results. Counting from the ground-state triplet asymptote, the energies are within 1 kcal/mol of each other. This is not so for rhodium or cobalt. For

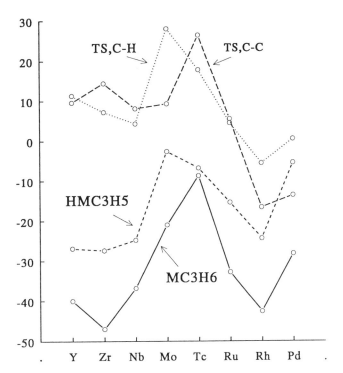

Figure 7. Energies obtained at the PCI-80 level for the reaction between the second-row transition metals and cyclopropane. MC_3H_6 are the C–C inserted metallacycles and HMC_3H_5 are the C–H inserted products. TS,C–H and TS,C–C are the transition states for C–H and C–C insertion, respectively.

rhodium the PCI-80 results are in general agreement with what has been inferred based on experiments [90]. The reaction exothermicity should, for example, be larger than 15 kcal/mol. The PCI-80 result is 17.1 kcal/mol. The B3LYP result, on the other hand, is only 6.3 kcal/mol using the largest basis set, and it is clear that this method has a problem to describe the rhodium reaction. The origin of this problem is at present unknown. The situation is different for the cobalt reaction. In this case it is known (see above) that the near-degeneracy effects for the first transition row make results using the PCI-80 scheme sometimes unreliable. For the cobalt reaction in Table XVII, the CI coefficients are rather large, and the PCI-80 results might therefore not be trusted. The accuracy of the B3LYP results is not completely clear at this stage, but the results are at

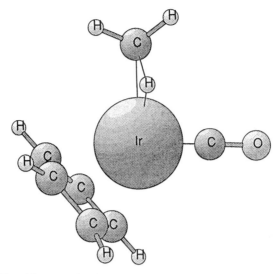

Figure 8. Transition state for the reaction between $Ir(C_5H_5)(CO)$ and methane.

TABLE XVII
Energies (kcal/mol) Obtained at Various Levels for the Reaction Between $M(C_5H_5)(CO)$
and $CH_4{}^a$

System		PCI-80 DZPF	B3LYP DZ	B3LYP DZP	B3LYP DZPF
$Co(^3A)$	Reactant	0.0	—	0.0	—
$Co(^1A)$	Reactant	8.7	—	28.1	—
	η^2	−13.2	—	17.5	—
	Product	−26.8	—	−3.9	—
$Rh(^3A)$	Reactant	5.9	−1.9	−1.5	−0.7
$Rh(^1A)$	Reactant	0.0	—	0.0	—
	η^2	−14.5	−7.1	−7.6	−7.9
	Transition state	−5.9	6.0	3.8	3.1
	Product	−17.1	−4.9	−5.7	−6.3
$Ir(^3A)$	Reactant	0.0	—	0.0	—
$Ir(^1A)$	Reactant	1.1	—	4.6	—
	η^2	−7.4	—	−7.2	—
	Transition state	−6.5	—	−6.4	—
	Product	−28.0	—	−29.0	—

^{a} The DZPF basis set uses polarization functions on both the metal and ligands. In the DZP basis the f function on the metal is removed, and in the DZ basis set the polarization functions on the ligands are also removed. The systems marked η^2 are the precursors of the reaction with an essentially undistorted methane.

least in line with the qualitative information available from experiments [91]. This information says that in contrast to the rhodium and iridium complexes, the cobalt complex does not form bound precursor complexes and does not insert into C–H bonds of alkanes. With the quite good experience available for the B3LYP method for the first transition rows (see, e.g., Table XV), it is therefore possible that the B3LYP results in Table XVII the cobalt reaction are actually reasonable. More experience with this method is needed to prove this point. What these examples show is that it is unlikely that a method can be found that can be trusted in all cases. It is therefore usually only with a large amount of experience for similar systems that correct decisions can be made concerning the accuracy of a certain method.

V. DIFFICULT APPLICATIONS

In the introduction it was stated that transition metal–containing systems have often been considered as unusually difficult to treat by theoretical methods. Even though respect for these systems has been exaggerated in many ways, there are definitely cases for which it is still very difficult to obtain results of even qualitative accuracy, and this does not refer to obvious cases of very large molecules. The classically difficult case is the chromium dimer, which has challenged method-oriented theoreticians for at least a decade. However, as solution to the chromium dimer problem is now in sight, and some of the more recent development are described briefly below.

Experimentally, the dissociation energy of Cr_2 has been determined to 1.44 ± 0.05 eV [92], the bond distance to 1.68 Å [93], and the potential curve has recently been shown to have an unusual shape, with a broad shoulder on the outer wall [94]. The ground state is a singlet having a sextuple bond. SCF calculations give a too short bond and a dissociation energy more than 20 eV above the energy of two separated atoms [95]. CASSCF calculations with an active space allowing for proper dissociation, on the other hand, yield a weakly bound system (0.35 eV) with a much too long bond distance (3.25 Å) [96]. This solution corresponds to a single s-s bond and an antiferromagnetic singlet coupling of the d electrons. The most ambitious approach until recently was made by Werner and Knowles, who used their internally contracted CI method with a large reference space allowing for proper dissociation [97]. However, even this method failed to give a short bond length. Recently, Bauschlicher and Partridge [98] have looked at the Cr_2 problem again using a variety of CCSD(T) and hybrid DFT methods of B3LYP type.

None of the methods investigated gave a qualitatively correct potential curve. Most of the methods, including the B3LYP method, gave much too long bond distances, corresponding to the single s-s bond solution. The only approach considered that gave a reasonable bond distance was the UBLYP procedure, which is another hybrid DFT variant. However, the binding energy was severely overestimated using this method.

The method available that should be most suited to treating the severe near-degeneracy effects present for Cr_2 is the CASPT2 method [56], mentioned in Section III.C. With this method it is straightforward to use a reference space that allows for proper dissociation, but as shown by Knowles and Werner [97], this is not enough. A large basis set, $3s,3p$ correlation, and preferably an even larger reference space should also be used to obtain a qualitatively correct binding. When this was done by Andersson et al. using the CASPT2 approach, a bond distance of 1.71 Å and a dissociation energy of 1.54 eV were obtained [99], both in very good agreement with experiment. However, these calculations still failed to give a reasonable potential curve. The reason for this was that singularities were encountered in the perturbation energy expression due to the presence of intruder states, a classical problem in multireference perturbation theory. A simple ad hoc procedure was then adopted to construct a potential curve, where the points along the potential curve that experienced these singularities were simply ignored. When this was done, a curve in qualitative agreement with the experimental curve [94] could be constructed, including the shoulder on the outer wall. From a theoretical point of view this procedure is not entirely satisfactory, but quite recently a more rigorous approach has been adopted successfully to remove the singularities [100]. The long-standing problem of generating an accurate theoretical potential curve for Cr_2 thus appears finally to be solved.

A few more examples of applications on transition metals of the CASPT2 method can serve to illustrate the potential of the method and some of the limitations. Highly accurate bond energies have, for example, recently been determined with this method for $Ni(CO)_n$ complexes [101]. Results in better agreement with experiment than for any other method that has been applied to these systems were obtained. Very large basis sets and $3s,3p$ correlation and basis set superposition errors were found to be important. The reference space used consisted of the $3d$ and $3d'$ correlating orbitals. It was concluded that this is the absolute minimum requirement for obtaining accurate results. This is a much larger reference space than that required for similar accuracy using an ordinary multireference CI calculation (see further above). It is interesting to note that the carbonyl π and π^* orbitals are not included in the active space in the CASPT2 calculations, even though $\pi \rightarrow \pi^*$ excitations on CO are among the most important excitations in this system. Normally, the optimal active space for a CASPT2 application has to be found after considerable experimentation, and this is one of the major problems in the use of this method for routine chemical applications. If another chemical problem involving $Ni(CO)_n$ had been studied where a CO ligand had participated in a reaction, for example, it is more than likely that the CO π and π^* orbitals would have had to be in the active space. Another problem for accurate applications is that the only procedure available to reach the basis-set limit is to saturate the basis set, which is very costly. No scaling procedure is available at present. However, the most severe problem with the method is that the active space quickly becomes very large when chemical reactions are studied. The most important applications of the CASPT2 method in the context of ground-state reactions of transition metal complexes will therefore be for particular cases known to be difficult, where other methods fail. Another area where the CASPT2 method is very useful is in the case of spectra for transition metal complexes. In fact, the CASPT2 method is at present the only method, together with the very costly MR-CI method, that gives reliable spectra for these systems. Recent studies of hexacyanometalate complexes [102] and iron pentacarbonyl [101] are examples of successful applications of the use of the CASPT2 method for spectra. For other examples, see the chapter by Roos.

The second-row transition metal dimers are easier than those of the first row but are still challenging systems. Due to the metal–metal multiple bonding, these systems sometimes have near-degeneracy effects that require a multireference treatment. These systems were therefore selected as test cases for a scaling scheme based on a multireference treatment. An immediate question in this context concerns the definition of the correlation energy, since in the PCI-80 scaling scheme only the correlation energy is scaled. If a system has near-degeneracy and requires a multireference treatment, the most natural extrapolation of the PCI-80 scheme to the multireference case is to scale only the energy difference between the multireference CI energy and the reference energy, which means that the nondynamical correlation energy is not scaled. If this is done, it is clear that at the limit of a very large reference space this scheme will not work since dynamical correlation effects will appear in the reference space and they will therefore not be scaled. Therefore, as small reference spaces as possible are necessary. The results for the second-row transition metal dimers are shown in Table XVIII. The technetium dimer was not done since no experimental value is available for comparison, and for the molybdenum dimer no meaningful result

TABLE XVIII
Binding Energies (eV) for the Second-Row Transition Metal Dimers

System	State	Reference State[a]	MR-ACPF[b]	PCI-80[c]	Expt.[d]
Y_2	$^5\Sigma_u^-$	1	1.00	1.83	1.6[e]
Zr_2	$^1\Sigma_g^+$	1	1.68	3.05	3.052[f]
Nb_2	$^3\Sigma_g^-$	1	3.00	4.85	5.21[e]
Ru_2	$^7\Delta_u$	7	2.40	3.15	3.39[g]
Rh_2	$^5\Delta_g$	10	1.82	2.40	2.92[e]
Pd_2	$^3\Sigma_u^+$	2	0.45	0.85	1.02[e]

[a] Number of reference states used in the multireference calculations.
[b] Unscaled results.
[c] Results are those obtained after scaling.
[d] The experimental error bars are ± 0.2 eV except for Zr_2, where it is ± 0.001 eV.
[e] Ref. 103.
[f] Ref. 31.
[f] Ref. 104.

could be obtained using the present scaling procedure. As seen in the table, the results after scaling constitute a major improvement over the unscaled results. In general, rather good agreement with experiment is found. For Rh_2, where the smallest meaningful reference space contained 10 configurations, the computed dissociation energy of 2.40 eV is somewhat below the experimental value of 2.92 ± 0.2 eV. This could be due to the use of a too large reference space, which contains some dynamical correlation effects. For the other systems it is not clear whether the experimental or theoretical ones are most accurate.

The transition metal oxides to the right in the periodic table are other systems where near-degeneracy effects are strong. For PdO, for example, the largest coefficient is 0.29 in the MCPF wavefunction, and there are a large number of coefficients between 0.10 and 0.25. The experimental dissociation energy is 66.2 kcal/mol [26]. At the MCPF level a dissociation energy of 39.5 kcal/mol is found, which improves up to 50 kcal/mol at the multireference level [73]. Despite the large coefficients, the straightforward PCI-80 value of 66.0 kcal/mol is in almost perfect agreement with experiment, which must be partly fortuitous. A PCI-80 scaling of a two-reference-state ACPF treatment, along the same lines as described above for the metal dimers, gives 59.3 kcal/mol. With four reference states the value improves to 65.0 kcal/mol. The scaling procedure on top of a multireference treatment is thus quite promising for these very difficult cases, and more work will be devoted to the development of such schemes in the near future.

VI. SUMMARY

Recent progress in the understanding and treatment of molecules containing transition metals have been discussed. Methods appropriate for
the treatment of the various rows of the transition metals have been
described. The second transition row is the easiest to treat by standard
quantum chemical methods. The reason for this has been analyzed, and it
was shown that the key factor in this context is the relative size of the
metal s and d orbitals. When these orbitals are very different in size, as
they are for the first transition row, severe near-degeneracy effects will
frequently appear. For the second and third transition rows, these orbitals
are similar enough in size that bonds with optimal overlap can be formed
simultaneously for these metal orbitals, and sd hybridization can also
occur at the orbital level without significant configuration mixing. The
absence of near-degeneracy effects for complexes of the second and third
transition rows has two important consequences: (1) the geometries can
in general be obtained at the SCF or MP2 level, and (2) single-reference-
state methods can be used for the energies. For the first transition row
these simple procedures frequently break down. For example, for the
seemingly simple $NiHCH_3$ molecule, a qualitatively wrong geometry is
obtained at the MP2 level and a very poor geometry at the SCF level. For
the third transition row, the main complication is that the spin-orbit
effects can sometimes not be neglected.

There are two very important recent developments in the area of
electronic structure calculations for transition metal complexes. The first
of these concerns the development of DFT methods capable of achieving
quantitative accuracy. The second development concerns improvement in
the accuracy of standard quantum chemical methods. For both the DFT
and standard quantum chemical methods, the introduction of one or a
few empirical parameters has been very important and has lead to a
dramatic improvement in accuracy at no additional cost. A simple scaling
of the correlation energy using the PCI-X scheme has led to an accuracy
comparable to that from experiments for the second transition row. A
similar accuracy is probably also possible to achieve for the third
transition row, provided that spin-orbit effects are not important. For the
first transition row, on the other hand, DFT hybrid methods, as ex-
amplified here by the B3LYP method, have given quite promising results
both for geometries and energetics. For this row, the most accurate
standard quantum chemical methods are the CASPT2 and CCSD(T)
methods. However, very large basis sets are required, and some type of
scaling scheme for these methods is urgently needed to make them
routinely applicable to large systems. In the case of the CASPT2 method,

which has proven extremely useful for special cases, a standard procedure to select a reference space that is not too big is also needed. For the future, the most important development will be toward larger systems, where the DFT methods have a much higher potential than that of standard quantum chemical methods. In particular, methods that do not require the calculation of all two-electron integrals can already be applied to very large systems. What is required is that these methods be carefully calibrated against experiments so that an accuracy similar to that of the hybrid DFT methods can be achieved. The coming years will probably see a large amount of work in this direction.

REFERENCES

1. A. Dedieu, Ed., *Transition Metal Hydrides*, VCH Publishers, New York, 1992.

2. D. R. Salahub and N. C. Zeruer, Eds., *Metal–Ligand Interactions: From Atoms, to Clusters, to Surfaces*, Kluwer, Dordrecht, The Netherlands, 1991.

3. P. W. N. M. van Leeuwen, J. H. van Lenthe, and K. Morokuma, Eds., *Theoretical Aspects of Homogeneous Catalysis, Applications of Ab Initio Molecular Orbital Theory*, Kluwer, Dordrecht, The Netherlands, 1994.

4. A. Veillard, *Chem. Rev.* **91**, 743 (1991).

5. N. Koga and K. Morokuma, *Chem. Rev.* **91**, 823 (1991).

6. C. W. Bauschlicher, S. R. Langhoff, and H. Partridge, in *Modern Electronic Structure Theory*, D. R. Yarkony, Ed., World Scientific, London, 1995.

7. H. F. Schaefer III, interviewed by P. Preuss, *Discovery*, November 1993, p. 137.

8. P. E. M. Siegbahn, M. Svensson, and P. J. E. Boussard, *J. Chem. Phys.*, **102**, 5377 (1955).

9. R. J. Gdanitz and R. Ahlrichs, *Chem. Phys. Lett.* **143**, 413 (1988).

10. G. D. Purvis and R. J. Bartlett, *J. Chem. Phys.* **76**, 1910 (1982).

11. D. P. Chong and S. R. Langhoff, *J. Chem. Phys.* **84**, 5606 (1986).

12. K. Raghavachari, G. W. Trucks, J. A. Pople, and M. Head-Gordon, *Chem. Phys. Lett.* **157**, 479 (1989).

13. J. Almlöf and P. R. Taylor, *J. Chem. Phys.* **86**, 4070 (1987).

14. P.-O. Widmark, P.-Å. Malmqvist, and B. O. Roos, *Theor. Chim. Acta* **77**, 291 (1990).

15. P. E. M. Siegbahn, M. R. A. Blomberg, and M. Svensson, *Chem. Phys. Lett.* **223**, 35 (1994).

16. J. A. Pople, M. Head-Gordon, D. J. Fox, K. Raghavachari, and L. A. Curtiss, *J. Chem. Phys.* **90**, 5622 (1989).

17. L. A. Curtiss, K. Raghavachari, G. W. Trucks, and J. A. Pople, *J. Chem. Phys.* **94**, 7221 (1991).

18. A. D. Becke, *Phys. Rev.* **A38**, 3098 (1988); A. D. Becke, *J. Chem. Phys.* **98**, 1372 (1993); A. D. Becke, *J. Chem. Phys.* **98**, 5648 (1993).

19. C. Lee, W. Yang, and R. G. Parr, *Phys. Rev. B* **37**, 785 (1988).

20. S. H. Vosko, L. Wilk, and M. Nusair, *Can. J. Phys.* **58**, 1200 (1980).

21. R. L. Martin, *J. Phys. Chem.* **87**, 750 (1983); see also R. D. Cowan and D. C. Griffin, *J. Opt. Soc. Am.* **66**, 1010 (1976).

22. L. A. Eriksson and U. Wahlgren, to be published.

23. B. A. Hess, *Phys. Rev. A* **33**, 3742 (1986).

24. P. E. M. Siegbahn and M. Svensson, *Chem. Phys. Lett.* **216**, 147 (1993).

25. J. A. Pople, M. Head-Gordon, and K. Raghavachari, *J. Chem. Phys.* **87**, 5968 (1987).

26. K. P. Huber and G. Herzberg, *Molecular Spectra and Molecular Structure*, Vol. 4, Van Nostrand Reinhold, New York, 1979.

27. P. B. Armentrout, in *Selective Hydrocarbon Activation: Principles and Progress*, J. A. Davies, P. L. Watson, A. Greenberg, and J. F. Liebman, Eds., VCH, New York, 1990.

28. M. L. Mandich, L. H. Halle, and J. L. Beauchamp, *J. Am. Chem. Soc.* **106**, 4403 (1984).

29. Y.-M. Chen and P. B. Armentrout, private communication.

30. R. L. Hettich and B. S. Freiser, *J. Am. Chem. Soc.* **108**, 5086 (1986).

31. C. A. Arrington, T. Blume, M. D. Morse, M. Doverstål, and U. Sassenberg, *J. Phys. Chem.* **98**, 1398 (1994).

32. I. Shim and K. A. Gingerich, *J. Chem. Phys.* **80**, 5107 (1984).

33. C. W. Bauschlicher, Jr., H. Partridge, J. A. Sheehy, S. R. Langhoff, and M. Rosi, *J. Phys. Chem.* **96**, 6969 (1992).

34. L. A. Eriksson, L. G. M. Pettersson, P. E. M. Siegbahn, and U. Wahlgren, *J. Chem. Phys.* **102**, 872 (1995).

35. J. P. Perdew and Y. Wang, *Phys. Rev. B* **33**, 8800 (1986).

36. J. P. Perdew, *Phys. Rev. B* **33**, 8822 (1986); **34**, 7406 (1986).

37. P. E. M. Siegbahn and M. Svensson, *J. Am. Chem. Soc.* **116**, 10124 (1994).

38. U. Wahlgren, unpublished.

39. P. J. Hay and W. R. Wadt, *J. Chem. Phys.* **82**, 299 (1985).

40. C. A. M. Wittborn and U. Wahlgren, *Chem. Phys.*, in press.

41. B. Hess, C. M. Marian, and S. D. Peyerimhoff, to be published.

42. K. Balasubramanian and D. Dai, *J. Chem. Phys.* **93**, 7243 (1990).

43. P. E. M. Siegbahn, M. Svensson, and R. H. Crabtree, submitted to *J. Am. Chem. Soc.*

44. M.-C. Duval, B. Soep, and W. H. Breckenridge, *J. Phys. Chem.* **95**, 7145 (1991).

45. K. J. Dyall, *J. Chem. Phys.* **98**, 2191 (1993).

46. J. J. Carroll, J. C. Weisshaar, P. E. M. Siegbahn, C. A. M. Wittborn, and M. R. A. Blomberg, *J. Phys. Chem.* **99**, 14388 (1995).

47. C. E. Moore, *Atomic Energy Levels*, U.S. Department of Commerce, National Bureau of Standards, U.S. Government Printing Office, Washington, D.C., 1952.

48. S. A. Mitchell, M. A. Blitz, P. E. M. Siegbahn, and M. Svensson, *J. Chem. Phys.* **100**, 423 (1994).

49. M. R. A. Blomberg, U. Brandemark, and P. E. M. Siegbahn, *J. Am. Chem. Soc.* **105**, 5557 (1983).

50. A. Ricca, C. W. Bauschlicher, and M. Rosi, *J. Phys. Chem.* **98**, 9498 (1994).

51. P. B. Armentrout, L. S. Sunderlin, and E. R. Fisher, *Inorg. Chem.* **28**, 4436 (1989), and references therein.

52. H.-J. Werner and P. J. Knowles, *J. Chem. Phys.* **82**, 5053 (1985); P. J. Knowles and H.-J. Werner, *Chem. Phys. Lett.* **115**, 259 (1985).

53. C. W. Bauschlicher, H. Partridge, and G. E. Scuseria, *J. Chem. Phys.* **97**, 7471 (1992).

54. A. Ricca and C. W. Bauschlicher, *J. Phys. Chem.* **98**, 12899 (1994).

55. A. Ricca and C. W. Bauschlicher, *Theor. Chim. Acta* **92**, 123 (1995).

56. K. Andersson, P.-Å. Malmqvist, B. O. Roos, A. J. Sadlej, and K. Wolinski, *J. Phys. Chem.* **94**, 5483 (1990).

57. T. Ziegler, *Chem. Rev.* **91**, 651 (1991).

58. D. R. Salahub, M. Castro, R. Fournier, P. Calaminici, N. Godbout, A. Goursot, C. Jamorski, H. Kobayashi, A. Martinez, I. Papai, E. Proynov, N. Russo, S. Sirois, J. Ushio, and A. Vela, in *Theoretical and Computational Approaches to Interface Phenomena*, H. Sellers and J. Olab, Eds., Plenum Press, New York, 1995, p. 187.

59. M. T. Benson, T. R. Cundari, S. J. Lim, H. D. Nguyen, and K. Pierce-Beaver, *J. Am. Chem. Soc.* **116**, 3955 (1994).

60. A. W. Ehlers and G. Frenking, *J. Am. Chem. Soc.* **116**, 1514 (1994).

61. R. Stegmann, A. Neuhaus, and G. Frenking, *J. Am. Chem. Soc.* **115**, 11930 (1993).

62. T. Ziegler, V. Tschinke, and C. Ursenbach, *J. Am. Chem. Soc.* **109**, 4825 (1987).

63. J. Li, G. Schreckenbach, and T. Ziegler, *J. Am. Chem. Soc.* **117**, 486 (1995).

64. D. G. Musaev, A. M. Mebel, and K. Morokuma, *J. Am. Chem. Soc.* **116**, 10693 (1994).

65. M. R. A. Blomberg, P. E. M. Siegbahn, and M. Svensson, *J. Am. Chem. Soc.* **114**, 6095 (1992).

66. M. R. A. Blomberg, P. E. M. Siegbahn, and M. Svensson, *J. Phys. Chem.* **96**, 9794 (1992).

67. P. E. M. Siegbahn and M. R. A. Blomberg, *J. Am. Chem. Soc.* **114**, 10548 (1992).

68. P. E. M. Siegbahn, M. R. A. Blomberg, and M. Svensson, *J. Am. Chem. Soc.* **115**, 1952 (1993).

69. P. E. M. Siegbahn, M. R. A. Blomberg, and M. Svensson, *J. Am. Chem. Soc.* **115**, 4191 (1993).

70. M. R. A. Blomberg, P. E. M. Siegbahn, and M. Svensson, *J. Inorg. Chem.* **32**, 4218 (1993).

71. M. R. A. Blomberg, P. E. M. Siegbahn, and M. Svensson, *J. Phys. Chem.* **97**, 2564 (1993).

72. P. E. M. Siegbahn, *Theor. Chim. Acta* **86**, 219 (1993).

73. P. E. M. Siegbahn, *Chem. Phys. Lett.* **201**, 15 (1993).

74. P. E. M. Siegbahn, *Theor. Chim. Acta* **87**, 441 (1994).

75. P. E. M. Siegbahn, *J. Am. Chem. Soc.* **115**, 5803 (1993).

76. P. E. M. Siegbahn, *Theor. Chim. Acta* **87**, 277 (1994).

77. P. E. M. Siegbahn, *Chem. Phys. Lett.* **205**, 290 (1993).

78. M. R. A. Blomberg, C. A. M. Karlsson, and P. E. M. Siegbahn, *J. Phys. Chem.* **97**, 9341 (1993).

79. P. E. M. Siegbahn, *J. Phys. Chem.* **97**, 9096 (1993).

80. P. E. M. Siegbahn and M. R. A. Blomberg, *Organometallics* **13**, 354 (1994).

81. M. R. A. Blomberg, P. E. M. Siegbahn, and M. Svensson, *J. Phys. Chem.* **98**, 2062 (1994).

82. P. E. M. Siegbahn, *Theor. Chim. Acta* **88**, 413 (1994).

83. P. E. M. Siegbahn, *J. Organomet. Chem.* **478**, 83 (1994).

84. P. E. M. Siegbahn, *J. Organomet. Chem.* **491**, 231 (1995).

85. P. E. M. Siegbahn, *Organometallics* **13**, 2833 (1994).

86. P. E. M. Siegbahn, *J. Am. Chem. Soc.* **116**, 7722 (1994).

87. E. A. Carter and W. A. Goddard III, *J. Phys. Chem.* **92**, 5679 (1988).

88. J. J. Carroll, J. C. Weisshaar, M. R. A. Blomberg, P. E. M. Siegbahn, and M. Svensson, *J. Phys. Chem.* **99**, 13955 (1995).

89. J.-E. Bäckvall, E. E. Björkman, L. Pettersson, P. Siegbahn, and A. Strich, *J. Am. Chem. Soc.* **107**, 7408 (1985).

90. E. P. Wasserman, C. B. Morse, and R. G. Bergman, *Science* **255**, 315 (1992).

91. A. A. Bengali, R. G. Bergman, and C. B. Moore, *J. Am. Chem. Soc.* **117**, 3879 (1995).

92. K. Hilpert and K. Ruthardt, Ber. Bunsenges. *Phys. Chem.* **91**, 724 (1987).

93. V. E. Bondybey and J. H. English, *Chem. Phys. Lett.* **94**, 443 (1983).

94. S. M. Casey and D. G. Leopold, *J. Phys. Chem.* **97**, 816 (1993).

95. A. D. McLean and B. Liu, *Chem. Phys. Lett.* **101**, 144 (1983).

96. M. M. Goodgame and W. A. Goddard, *Phys. Rev. Lett.* **48**, 135 (1982); *J. Phys. Chem.* **85**, 215 (1981).

97. H.-J. Werner and P. J. Knowles, *J. Chem. Phys.* **89**, 5803 (1988).

98. C. W. Bauschlicher and H. Partridge, *Chem. Phys. Lett.* **231**, 277 (1994).

99. K. Andersson, B. O. Roos, P.-Å. Malmqvist, and P.-O. Widmark, *Chem. Phys. Lett.* **230**, 391 (1994).

100. B. O. Roos and K. Andersson, private communication.

101. B. J. Persson, B. O. Roos, and K. Pierloot, *J. Chem. Phys.* **101**, 6810 (1994).

102. K. Pierloot, E. Van Praet, L. G. Vanquickenborne, and B. O. Roos, *J. Phys. Chem.* **97**, 12220 (1993).

103. K. A. Gingerich, *Curr. Top. Mater. Sci.* **6**, 345 (1980).

104. A. R. Miedema and K. A. Gingerich, *J. Phys. B* **12**, 2081 (1979).

THE INTERFACE BETWEEN ELECTRONIC STRUCTURE THEORY AND REACTION DYNAMICS BY REACTION PATH METHODS

MICHAEL A. COLLINS

Research School of Chemistry, Australian National University, Canberra, Australia

CONTENTS

Advances in Chemical Physics, Volume XCIII, Edited by I. Prigogine and Stuart A. Rice.
ISBN 0-471-14321-9 © 1996 John Wiley & Sons, Inc.

I. INTRODUCTION

Theories of chemical reaction dynamics and molecular spectroscopy require a knowledge of the molecular potential energy surface (PES) [1, 2]. This surface describes the variation in the total electronic energy of the molecule as a function of the nuclear coordinates, within the Born–Oppenheimer approximation, and hence it determines the forces on the atomic nuclei.

The nature of the PES can be probed either experimentally or by using *ab initio* quantum chemical calculations. As the name implies, *ab initio* electronic structure methods represent the least biased approach; deriving the form of a PES from experimental data is model dependent because we have no exact means of solving the quantum equations for reaction dynamics or inverting the resulting scattering data to obtain the PES.

For a given configuration of the atomic nuclei, the total electronic energy can now be evaluated *ab initio* to within chemical accuracy for small to medium-sized molecules, using a hierarchy of quantum chemistry methods. Further information about the shape of the energy surface may be obtained by evaluating derivatives of the energy with respect to the molecular coordinates. Of course, derivatives of the molecular energy with respect to the atomic Cartesian coordinates (say) may be obtained by finite difference of the energy. Higher-order derivatives may be derived similarly. However, for some levels of theory, the computational expense of calculating energy gradients or higher derivatives has been dramatically reduced by the derivation of analytic formulas for such derivatives [3–8]. These analytic formulas for Hartree–Fock wavefunctions and for some correlated levels of theory have been incorporated in a number of quantum chemistry programs [9]. The availability of analytic gradients and higher derivatives has substantially changed the practice of quantum chemistry; exploration of the PES to determine the molecular geometry at various critical features is now relatively commonplace.

However, the energy and energy derivative values at a small number of molecular geometries do not in themselves constitute a molecular potential energy surface. A single Born–Oppenheimer PES for a chemically reacting system is a smooth function which accurately describes the molecular energy as interatomic distances change by up to several angstroms. An important contemporary challenge for chemical theory is the development of methods that use the output of quantum chemistry to construct such "global" potential energy surfaces for use in chemical reaction dynamics (and spectroscopy). The purpose of this paper is to review how this challenge has been taken up using the concept of a reaction path.

We begin by outlining the task and suggesting why a reaction path approach to evaluating a PES is plausible. Three important technical questions must then be addressed: Exactly how do we define a reaction path, how do we calculate it, and how do we use it to describe a PES? Because this task lies along the interface between *ab initio* quantum chemistry and reaction dynamics, even these three technical questions cannot be answered completely without reference to how we intend to evaluate the reaction dynamics. So when we go on to describe how the reaction dynamics is to be studied, we have to return from time to time to the questions of what path and what PES. Without claiming to cover the area exhaustively, we hope to arrive at a reasonable overview of how the simple reaction path approach works in practice. This is an active area of research, and some very useful reviews have already appeared [10–12].

However, we need to go beyond the simpler ideas. Chemical reactions are more complex than the first few sections of this article would suggest. For example, there may be multiple reaction paths (or mechanisms, if you like) for a single reaction; there may be multiple symmetry related paths (versions of the same mechanism) for a reaction; there may be different reaction paths for different reactions in close proximity on the PES (competing reactions). These macroscopic difficulties for the reaction path concept have echoes even in the simple case of a single reaction path when we consider in detail the shape of the valley walls. In later sections we explore a way in which the reaction path approach forms the nucleus of a more general approach to interfacing *ab initio* quantum chemistry and reaction dynamics in these more complex scenarios.

II. MOLECULAR POTENTIAL ENERGY SURFACES AND REACTION PATHS

It is easy to see just how "big" is the task of constructing a molecular PES. At first glance, the total energy of a molecule of N atoms appears to be a function of all $3N$ Cartesian coordinates that specify the positions of the atoms. However, the energy is actually only a function of the relative positions of the atoms; it does not depend on the three coordinates that specify the position of the molecule as a whole or on the three coordinates needed to specify the orientation of the molecule. Thus the energy is only a function of $3N - 6$ "internal" coordinates, $\{R\}$. Nonetheless, this is a lot of dimensions: three for a triatomic molecule (well, we are used to that space), but six for a tetraatomic, and absurdly large thereafter.

Exactly where in the huge $(3N - 6)$-dimensional space do we start to calculate the electronic energy (keeping in mind that each single calculation might be quite expensive)? If we wanted to know about something relatively simple such as the low-energy vibrational motion of a stable molecule, the answer is obvious: Calculate the energy, energy gradients, and higher derivatives of the energy at the equilibrium geometry, determined by the internal coordinates R^{eq}. That is, express the PES, $V(R)$, as a Taylor series about the equilibrium configuration:

$$
\begin{aligned}
V(\mathbf{R}) = V(\mathbf{R}^{eq}) + \sum_{n=1}^{3N-6} (R_n - R_n^{eq})\left(\frac{\partial V}{\partial R_n}\right)^{eq} \\
+ \frac{1}{2!} \sum_{n=1}^{3N-6} \sum_{m=1}^{3N-6} (R_n - R_n^{eq})(R_m - R_m^{eq})\left(\frac{\partial^2 V}{\partial R_m \partial R_m}\right)^{eq} + \cdots .
\end{aligned}
\tag{2.1}
$$

Of course, at equilibrium the gradients, $\partial V/\partial R_n$, vanish, so that Eq. (2.1) simplifies somewhat. This expansion is useful because a molecule vibrating at low energy may not "sample" geometries that are much different from the equilibrium one, so that the series in Eq. (2.1) may be truncated accurately even after the second-order term. However, during a chemical reaction, several interatomic distances change by many angstroms while others change by at least a few tenths of an angstrom.

How can we deal with this? Suppose that we considered representing a PES as a table that might be interpolated to give the energy at any relevant molecular configuration. For a molecule of N atoms, requiring $3N - 6$ coordinates to determine its configuration, a table with K rows for each coordinate would contain K^{3N-6} energy values. For N greater than 3, such a table would be prohibitively expensive to evaluate, since K may be on the order of 10 or more. So brute force will not work. Broadly speaking, two alternative approaches have been employed.

Most commonly, functional forms of the $3N - 6$ coordinates, which contain adjustable parameters, have been developed to describe the PES for particular types of molecules and molecular motions (e.g., bond stretching). These parameters are then varied to ensure that the PES "fits" some combination of empirical or *ab initio* data (see, e.g., Refs. 1 and 13–18). For chemically reacting molecules, such a PES may take the form of an expansion in two-body, three-body, and higher interactions [1]. These fitting methods have been very successful in many cases, particularly for small molecules, but are restricted by the choice of functional form and may not easily transfer to different molecules. One cannot be sure that a particular choice of functional form is sufficiently flexible, that it does not preclude the description of some important aspect of the PES, or that it is very inaccurate in some important domain of the coordinates. Moreover, the larger the range of configurations that the function must encompass, the more flexible and complicated the form must be (and hence the more difficult to construct).

The second approach to this problem of constructing a PES for chemical reaction dynamics is based on the concept of a reaction path. This concept goes back a long way [19] and is well known to many readers. However, at the risk of boring some, we'll go over the simple ideas involved.

To begin with, we may not have to calculate the PES throughout the whole of the $(3N - 6)$-dimensional coordinate space. Suppose that we only want to study the chemical reaction when the reactants have less than some maximum amount of energy available to them; call it E_{max}. Then to study the reaction dynamics using classical mechanics would only require us to know the PES, $V(\mathbf{R})$, at those molecular geometries, $\mathbf{R}$, for which

$$V(\mathbf{R}) \le E_{max} . \qquad (2.2)$$

If quantum dynamics were proposed, we could still restrict the values of $\mathbf{R}$ in a similar way, simple using a higher cutoff E_{max} to ensure that the quantum wavefunction at $\mathbf{R}$ had exponentially decayed to a negligible value when $V(\mathbf{R}) = E_{max}$.

Equation (2.2) does not appear to be very helpful, since we would have to know the answer, $V(\mathbf{R})$, to begin with. However, keeping Eq. (2.2) in mind, let's look at what we know qualitatively about $V(\mathbf{R})$ in a simple case: the collinear collision of an atom, A, with a diatomic, BC:

$$A + BC \rightarrow AB + C . \qquad (2.3)$$

Here $V(\mathbf{R}) = V(R_{AB}, R_{BC})$. We can graph such a function of two variables, and it must look rather like that shown in Fig. 1. When R_{AB} is large, the PES depends only on R_{BC} and must look something like a Morse or Lennard-Jones function, $U_{BC}(R_{BC})$ (similarly, when R_{BC} is large). As R_{AB} becomes smaller, the molecular energy changes, at first by

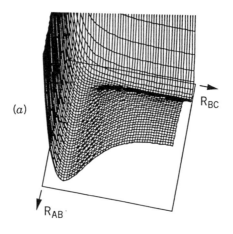

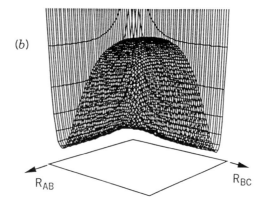

Figure 1. Schematic representation of the PES for a collinear $A + BC \rightarrow AB + C$ reaction as a function of the AB and BC interatomic distances: (a) viewed from above, showing the reaction valley between the repulsive walls at short bond lengths and the plateau region corresponding to fragmentation; (b) viewed from the side, showing the rise and fall of the valley floor.

the nonbonded interaction of A with B, so that the energy may fall at fixed R_{BC}; then the energy rises as electron repulsion dominates. Reducing R_{AB} further may cause a new bond to form between A and B. This will become a very high energy state unless the BC bond breaks; that is, R_{BC} becomes large. As R_{BC} increases while R_{AB} is short, we have a similar variation of the PES due to the nonbonded $B \cdots C$ interaction superimposed on a diatomic potential, $U_{AB}(R_{AB})$. The picture then is as shown in Fig. 1, essentially a valley that bends from the R_{AB} direction to the R_{BC} direction, whose valley floor rises and falls.

The very flat plateau region in Fig. 1 corresponds to fragmentation, A + B + C. If the energy here is greater than E_{max}, this part of the PES is irrelevant, and the interesting part of the PES looks like that shown in Fig. 2: a valley with roughly parabolic walls whose floor rises to an intermediate maximum and whose direction bends in going from the entrance valley to the exit valley. The intermediate maximum is a simple topological feature known as a saddle point on the PES. A saddle point is a stationary point,

$$\frac{\partial V(\mathbf{R})}{\partial R_\alpha} = 0 , \qquad \alpha = 1, \ldots, 3N - 6 . \tag{2.4}$$

At stable stationary points, energy minima on the PES, all the eigenvalues of the Hessian or matrix of second derivatives (also called the force constant matrix), $\partial^2 V / \partial R_\alpha \, \partial R_\beta$, are positive. At a saddle point, however, one (and only one) eigenvalue is negative (the energy falls as

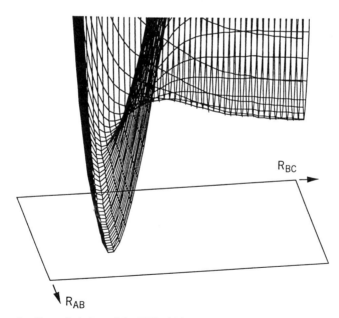

Figure 2. Expanded view of the PES of Fig. 1, showing only energies up to and a little above the energy at the saddle point.

we move from the saddle point down the valley floor toward either reactants or products). The saddle point may be "early" if it occurs in the entrance valley before the corner, or "late" if it occurs in the product valley after the corner [20–22].

While Figs. 1 and 2 apply strictly to a collinear A + BC reaction, we know that the picture of a potential energy "valley" confined (more or less) by repulsive walls also holds for the PES of polyatomic reactions in three dimensions.

Since the advent of analytic gradient methods, it has become commonplace for quantum chemists to investigate the basic structure of the PES for a chemical reaction in terms of topological features such as saddle points, mountaintops, and energy minima. Quantum chemistry programs now have searching algorithms to help locate such features [23–27]. Typically, to estimate the enthalpy and activation energy for a chemical reaction, one would begin by locating and characterizing the energy minima corresponding to the reactants and products. Next, using an educated guess for an initial geometry and a saddle point searching algorithm, one locates the saddle point where one eigenvalue of the force constant matrix is negative [let's denote the eigenvector by $\mathbf{u}(1)$]. Or rather, one locates a saddle point. To verify that this saddle point connects the reactant and product minima, a gradient-following algorithm is used to follow the PES downhill from the saddle point to a minimum [28–32]. By stepping away from the saddle point in the directions given by $\pm\mathbf{u}(1)$, one can show which minima are connected by this saddle point.

Once we have located the important stationary geometries on the PES, how do we proceed to form an accurate approximation for the PES? Even the PES in Fig. 2 looks far too complicated to be described neatly in terms of a simple Taylor expansion [e.g., Eq. (2.1)] about a single geometry. In the $(3N - 6)$-dimensional configuration space, a single geometry is represented by a point, a topological object of zero dimension. The next simplest type of object is one-dimensional: a curve or path in the space. So it seems natural, looking at Fig. 2, to suggest that we try to express the PES as some form of expansion about a path that follows the valley from reactants to products. Which path? Well without getting technical, it seems natural to choose a path that follows the middle of the valley in some sense, or the valley floor, or some such choice which does not unnaturally prejudice the shape of the valley by being closer to one side or the other. We have used some rather ill-defined terms in the last sentence, and part of our task will be to define the necessary concepts precisely.

III.   WHAT REACTION PATH?

Let me say at the outset that there is *not* just one and only one useful definition of a reaction path. Rather, the concept of a reaction valley or *reaction swath* [11] has real utility in those simple cases where the reaction dynamics takes place in a restricted region such as the valleys of Figs. 1 and 2. Depending on how we wish to perform dynamical calculations, for example, various definitions of a reaction path might be appropriate.

A.   Paths of Steepest Decent

However, to rationalize a popular choice of reaction path, we might begin by noting that Figs. 1 and 2 are only informative graphs of a PES if the coordinates employed are mass-weighted Cartesian coordinates (MWC); for we can only relate the changing height of the PES *directly* to forces on the atoms if all the coordinates move in straight lines when the PES is flat. If $\mathbf{x}(n)$ denotes the Cartesian coordinate vector (x_n, y_n, z_n) of the nth atom (of mass m_n) in a molecule of N atoms, we define $\boldsymbol{\xi}(n)$ as

$$\boldsymbol{\xi}(n) = \sqrt{m_n}\, \mathbf{x}(n) . \tag{3.1}$$

The classical Lagrangian is then

$$
\begin{aligned}
L &= \sum_{n=1}^{N} \frac{1}{2} m_n \left| \frac{d\mathbf{x}(n)}{dt} \right|^2 - V[\mathbf{x}(1), \mathbf{x}(2), \dots, \mathbf{x}(N)] \\
&= \sum_{n=1}^{N} \frac{1}{2} \left| \frac{d\boldsymbol{\xi}(n)}{dt} \right|^2 - V[\boldsymbol{\xi}(1), \boldsymbol{\xi}(2), \dots, \boldsymbol{\xi}(N)] .
\end{aligned}
\tag{3.2}
$$

Thus all MWC coordinates experience classical forces opposite to the potential gradient:

$$\frac{d^2 \boldsymbol{\xi}(n)}{dt^2} = -\frac{\partial V}{\partial \boldsymbol{\xi}(n)} , \qquad n = 1, \dots, N . \tag{3.3}$$

This is just Newton's law of classical motion. If $\boldsymbol{\zeta}$ represents the $3N$ dimensional vector $[\boldsymbol{\xi}(1), \boldsymbol{\xi}(2), \dots, \boldsymbol{\xi}(N)]$, Eq. (3.3) is written as

$$\frac{d^2 \boldsymbol{\zeta}}{dt^2} = -\frac{\partial V}{\partial \boldsymbol{\zeta}} . \tag{3.4}$$

Now, let's consider what would be the result of solving these classical equations of motion if the initial position $\zeta(t=0)$ is at the saddle point and the initial velocities contain a component toward either the reactant or product valley. Clearly, $\zeta(t)$ will trace out a trajectory something like that of Fig. 3; the system will accelerate down the valley while oscillating (vibrating) from side to side. Changing the initial atomic velocities will generate similar but different oscillating trajectories or paths. None will pass directly down the middle of the valley. However, if we subject the atoms to heavy friction, so strong as to reduce the velocities to almost zero, a little algebra shows that the path traced out is given by

$$\frac{d\zeta}{dt} = -\frac{\partial V}{\partial \zeta}. \tag{3.5}$$

This is an equation for the *path of steepest decent*. There are two paths of steepest descent from the saddle point, one toward reactants and the other toward the products. The combined path from reactants to products is called the *intrinsic reaction path* (IRP) [33–39]. Friction has eliminated the oscillatory motion from this path, so that it resembles the dashed path shown in Fig. 3.

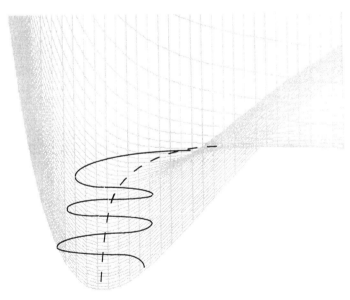

Figure 3. Another expanded view of the PES of Fig. 1, sketching the projection of a classical trajectory onto the PES (solid line) and a similar projection of the path of steepest descent (dashed line).

The arbitrary parameter t in Eq. (3.4) no longer corresponds to time and can be eliminated in favor of the distance along the path. If $\zeta(t)$ and $\zeta(t + dt)$ are neighboring points on the IRP, they are separated by a distance ds:

$$
\begin{aligned}
ds^2 &= |\zeta(t + dt) - \zeta(t)|^2 \\
&= \left| \frac{d\zeta(t)}{dt} \right|^2 dt^2 \\
&= \left| \frac{\partial V}{d\zeta(t)} \right|^2 dt^2 .
\end{aligned}
\tag{3.6}
$$

From Eqs. (3.5) and (3.6), we have an equation for the IRP in terms of the distance, s, along the path:

$$
\frac{d\zeta}{ds} = - \frac{\partial V / \partial \zeta}{|\partial V / d\zeta|} .
\tag{3.7}
$$

The distance, s (which has units of length mass$^{1/2}$), is called the *intrinsic reaction coordinate* (IRC) and is clearly a measure of progress along the reaction path. By convention, we take $s = 0$ at the saddle point and $s < 0$ (>0) in the reactant (product) valley.

Equation (3.7) is an equation for the path of steepest descent in the $3N$ mass-weighted Cartesian coordinates. In fact, we can write similar equations for the path of steepest descent in any set of coordinates, say some complete set of $(3N - 6)$-internal coordinates, such as interatomic distances, $\mathbf{R}$:

$$
\frac{d\mathbf{R}}{ds'} = - \frac{\partial V / \partial \mathbf{R}}{|\partial V / d\mathbf{R}|} ,
\tag{3.8}
$$

where s' is a new reaction coordinate, the distance along a path of steepest descent in interatomic distance coordinates (say). In fact, one can show that any such path of steepest descent which begins near a saddle point will lead to an energy minimum in the reactant or product valleys. Note that the energy is unchanged to first order in any small displacement of the coordinate vector $\mathbf{R}$ (or ζ) away from the path in a direction orthogonal to the path,

$$
[\mathbf{R} - \mathbf{R}(s')] \cdot \frac{d\mathbf{R}}{ds'} = 0 \Rightarrow [\mathbf{R} - \mathbf{R}(s')] \cdot \frac{\partial V}{\partial \mathbf{R}} = 0 .
\tag{3.9}
$$

In this sense, all such paths are called *minimum energy paths* (MEPs). So the steepest descent path on a PES in any well-defined, complete set of coordinates will connect the saddle point with the appropriate energy minima and is a minimum energy path. All such paths are plausible choices as a reaction path.

B. How to Calculate a Minimum Energy Path

A first glance at Eq. (3.5) or (3.8) suggests that we can simply obtain an MEP by stepwise solution of the first-order differential equation:

$$\boldsymbol{\zeta}(\delta t) = \boldsymbol{\zeta}(t = 0) - \frac{\partial V}{\partial \boldsymbol{\zeta}}\bigg|_{t=0} \delta t \,;$$

$$\boldsymbol{\zeta}(2\delta t) = \boldsymbol{\zeta}(\delta t) - \frac{\partial V}{\partial \boldsymbol{\zeta}}\bigg|_{t=\delta t} \delta t \,; \quad \text{etc.}$$

(3.10)

There are two problems here. One is easily disposed of: The gradient is zero at $t = 0$ (at the saddle point), so the scheme in Eq. (3.10) does not progress away from the saddle point. One can show, however [40], that a MEP must approach a stationary point along the direction of the eigenvector of the force constant matrix, **F**, with lowest eigenvalue. At the saddle point, there is one negative eigenvalue of **F**, so we can simply replace the first step in Eq. (3.10) by

$$\boldsymbol{\zeta}(\delta t) = \boldsymbol{\zeta}(t = 0) \pm \mathbf{u}(1) \, \delta t \,,$$

(3.11)

where $\mathbf{u}(1)$ is the eigenvector of **F** with negative eigenvalue, and $\pm$ is chosen to fall into either the reactant or product valleys. The second and subsequent steps can then be taken according to Eq. (3.10), however at high computational cost, because the steps, δt, must be very small to solve Eq. (3.5) accurately. This is the second problem with Eq. (3.10). It arises from the fact that the $3N$ equations in Eq. (3.5) are a *stiff* system of first-order differential equations [41]. The step size, δt, is determined by the rate at which the gradient vector $\partial V / \partial \boldsymbol{\zeta}$ changes as the geometry changes (as determined by the force constant matrix, **F**). In a molecule, **F** has a large range of eigenvalues (proportional to the square of the vibrational frequencies). While the step size must be small to describe motion accurately in the stiffest directions (the largest eigenvalues are associated with the stretching of strong bonds), motion along the reaction path direction can be dominated by the lowest-frequency motions (including the asymptotic stretching of broken bonds). So small steps are needed merely to ensure that the numerical approximation to the reaction path does not erroneously involve distorting strong bonds. This problem can be partly overcome with the use of more sophisticated numerical procedures that still only require a knowledge of the gradient $\partial V / \partial \boldsymbol{\zeta}$

[42, 43]. Alternatively, if one has analytic second derivatives, Eq. (3.5) can be expanded near $t = t_0$ locally as

$$\frac{d\zeta}{dt} = -\left\{\frac{\partial V}{\partial \zeta}\bigg|_{t=t_0} + \mathbf{F}(t_0)[\zeta - \zeta(t_0)]\right\}. \tag{3.12}$$

This equation can be solved in terms of the eigenvectors of $\mathbf{F}(t_0)$ to advance one step forward from $\zeta(t_0)$ on the MEP. Much larger steps can be taken using this local quadratic approximation method [41, 44]. Even more substantial increases in the step size have been achieved by accounting approximately for third derivatives of the energy along the path (although only *ab initio* second derivatives are actually computed [45]).

A great deal of effort has been expended on efficient calculation of the MEP [12, 30–32, 41–50]. At the moment, if relatively inexpensive second derivatives are available, the cubic corrected local quadratic method is most efficient; otherwise, reasonably efficient gradient-only methods are available [49]. This is an area where methods are still advancing and recently proposed methods may prove to be better still when these methods have been used with dynamical calculations on real *ab initio* PES [12, 43, 48, 50]. At this stage some examples of steepest descent reaction paths might be informative.

C. Some Examples

Figure 4 shows the energy profile for the reaction [51]

$$\text{NH}(^3\Sigma^-) + \text{H}_2(^1\Sigma_g^+) \rightarrow \text{NH}_2(^2B_1) + \text{H}(^2S). \tag{3.13}$$

This reaction shows a classically simple variation of the energy along a reaction path. The NH_3 molecule has only a plane of symmetry on the path, with a *cis* structure at the saddle point. Figure 5 shows how the six interatomic distances change as a function of the reaction coordinate. The geometry clearly evolves very smoothly and simply as the reaction proceeds. Figure 6 shows how the molecular vibrational frequencies change along the reaction path (for MWC coordinates, the eigenvalues of $\mathbf{F}$ are the squares of the frequencies).

Figure 7 shows the energy profile along the reaction path for the reaction [41, 52, 53]

$$\text{NH}_3^+ + \text{H}_2 \rightarrow \text{NH}_4^+ + \text{H}. \tag{3.14}$$

This reaction profile is slightly more complicated than that of Fig. 4 due to the weak complexes in the reactant and product valleys. The saddle

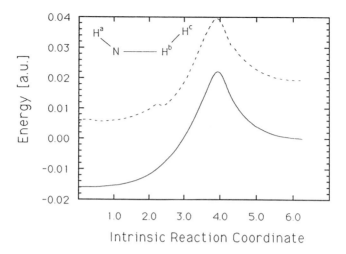

Figure 4. Energy profile along the IRP (solid line) for the reaction $NH(^3\Sigma^-) + H_2(^1\Sigma_g^+) \rightarrow NH_2(^2B_1) + H(^2S)$, calculated at the MCSCF level of theory (see Ref. 51). The dashed curve presents the energy profile along the IRP corrected for the zero-point energy of the molecular vibrations (see Section V.A.3). The intrinsic reaction coordinate is given in atomic units ($amu^{1/2}$ bohr); the energy (in atomic units) is shown relative to that of the products. The geometry of the saddle point is sketched with labeled atoms.

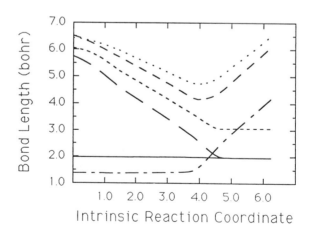

Figure 5. Internuclear distances (in bohr) for $^3NH_3$ along the IRP of Fig. 4. Bond lengths: $N-H^a$, ——; $N-H^b$, —— ; $N-H^c$ – – –; H^a-H^b, - - - -; H^a-H^c, · · · ·; H^b-H^c, —·—·.

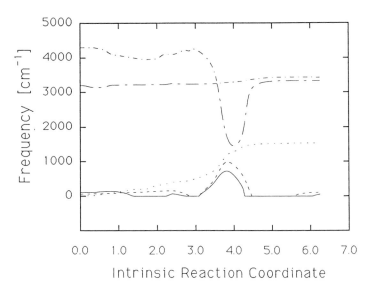

Figure 6. Projected (see Section IV.A) harmonic vibrational frequencies for $^3NH_3$ along the IRP of Fig. 4. The assignment of the vibrational modes is given in Ref. 51.

point and product complex have C_{3v} symmetry, while the complex in the reactant valley has C_{2v} symmetry.

Figure 8 presents an example of a very complicated set of reaction paths on the PES for the dissociation reaction [54–57]:

$$HCOH^+ \rightarrow HCO^+ + H . \qquad (3.15)$$

There are two distinct reaction pathways for dissociation of the most stable *trans*-HCOH$^+$ structure, denoted as routes A and B. Only the structure at the intermediate minimum for the formaldehyde radical cation has more than a plane of symmetry. These three examples give some impression of the variation in the behavior observed in intrinsic reaction paths.

D. Distinguished Coordinate and Other Reaction Paths

While various higher-order methods allow large step sizes in the integration of Eq. (3.5) or (3.8); nevertheless, there are often many steps required to connect the saddle point with reactants or products. Hence, particularly if second-order derivatives are required, the computational effort is substantial.

One method of generating a type of reaction path at somewhat

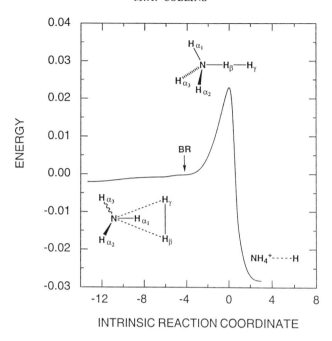

Figure 7. Energy profile along the IRP for reaction (3.14) (see Ref. 41). The path was calculated at the Hartree–Fock/6-31G(d, p) level and the energy (in a.u.) is shown relative to that of the separated reactants. The intrinsic reaction coordinate has units of amu$^{1/2}$ bohr. BR denotes the location of a branching point. The geometries at the minimum in the reactant valley and at the saddle point are sketched.

reduced cost uses a *distinguished reaction coordinate*. One simply chooses some internal coordinate, a *distinguished coordinate*, such as the length of a bond that breaks during the reaction, or the distance between the product fragments (or a combination of both) as the reaction coordinate (see, e.g., Refs. 58–61). At a number of fixed values of this coordinate [ranging from its value at the saddle point to that at reactants (and products)], the molecular geometry is determined by a constrained energy minimization. This produces points on a *distinguished reaction path*. The computational cost of this procedure is not simply proportional to the number of points originally chosen; the more widely spaced are the values of the distinguished coordinate, the more computation is required in the geometry optimization.

   Another proposal to limit the computation required to determine a reaction path involves minimizing a line integral of the energy. The line integral is approximated as a finite sum of the energies at equidistantly spaced points along a path, where the first and last points are the

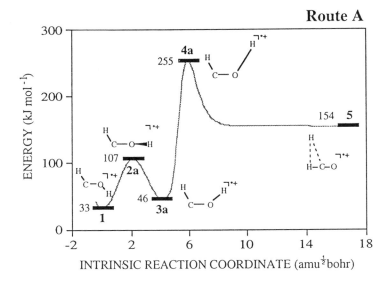

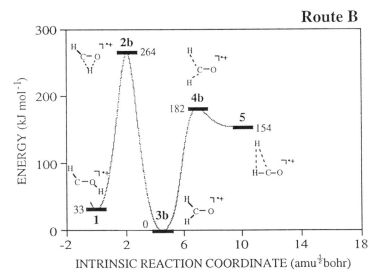

Figure 8. Energy profile along the intrinsic reaction paths for reaction (3.15), shown relative to the energy of the formaldehyde radical cation (see Refs. 56 and 57). There are two separate routes for the dissociation of structure **1** through to a weak complex in the product valley, structure **5**. The paths were calculated at the Hartree–Fock/6-31G** level of theory. The molecular structures at the stationary points on the paths are sketched.

reactants and products, respectively (and the saddle point is a fixed intermediate point). First proposed by Elder and Karplus for use in complex biological reactions, this method has been modified and applied to *ab initio* PES [47, 48, 62]. The line integral is minimized by optimizing the geometries corresponding to all the variable points along the path. This is achieved by nonlinear least-squares fit, with the coordinates of the points as the unknown parameters, using a conjugate gradient technique [48]. The optimal set of geometries define the reaction path.

Much simpler still is the rectilinear reaction path proposed by Ruf and Miller. To study the transfer of the hydroxyl hydrogen in malonaldehyde, they suggested a reaction path that is just a straight line connecting the reactant and product configurations in Cartesian coordinates rather than an IRP [40, 63, 64]. To understand completely their objection to using an IRP, we would have to consider the PES and indeed the total Hamiltonian in terms of coordinates based on a reaction path. However, for the moment, we recall that Ruf and Miller showed that an IRP must always approach a stationary point on the PES from the direction of the eigenvector of F with lowest eigenvalue. They suggested that this implies that an IRP path may have unphysical bends near the ends, as the lowest-frequency mode in a stable molecule may be unrelated to the reaction; for example, it may be the weakly hindered rotation of a methyl group. Curves or bends in a reaction path can have dynamical consequences which we explore below. In fact, such "unphysical bends" can be quite substantial for other reasons as well. As an example, route A of Fig. 8 presented the variation of the energy along an *ab initio* IRP that describes one mechanism for the dissociation of the hydroxyl hydrogen from $HCOH^+$. From the *cis*-structure minimum (**3a** of Fig. 8) onward, this reaction mechanism is inherently simple; the OH bond stretches while the other OCH angle becomes linear. However, the IRP displays a long tail where the energy falls very slowly. Here the dissociating H atom swings around to the other (HC) end of the molecule. This results from the fact that the absolute energy minimum configuration for $HCO^+ \cdots H$ has the H nearer the carbon end of the HCO^+ molecule. This long tail section of the IRP is irrelevant to the dissociation dynamics, as the translational energy available to the departing H atom dwarfs the small energy ridges on either side of the path as its bends around to the carbon side of the molecule. Any realistic evaluation of the dynamics would show the H atom dissociating more or less directly away from the oxygen end of the HCO^+ molecule [56].

This example raises an important general point: A reaction path is not dynamically significant if the potential energy walls surrounding it are not significantly high by comparison with the energy available. A reaction path is significant only to the extent to which the surrounding walls limit the dynamics.

E.   Bifurcations and Other Paths

There have been some related investigations of other types of paths that might characterize a PES. A *gradient extremal path* is defined as a path on which the gradient vector at each point is an eigenvector of the Hessian or force constant matrix [65, 66]. Gradient extremal paths often, but not always, follow valley floors or ridge crests on a PES; they can run up and down hill; they can terminate abruptly; and they do not necessarily connect saddle points with minima. Hence they are not useful as reaction paths [46, 65, 67, 68]. However, gradient extremal paths do have the interesting feature that they pass through *valley ridge inflection points*.

In the simple case represented in Fig. 1, there is a single continuous valley floor. However, it is possible for a valley floor to bifurcate into two or more valleys, as represented in Fig. 9. Such bifurcations are associated

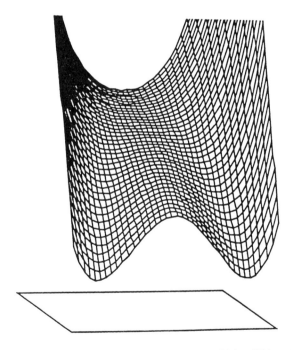

Figure 9.   Energy surface for two degrees of freedom which exhibits a bifurcation of a single valley. The data were produced using a modified form of model functions presented by Valtazanos and Ruedenberg [67].

with a breaking of symmetry [69, 70]. A path of steepest descent cannot bifurcate except at a stationary point, because the right-hand side of Eq. (3.8) is otherwise well defined and single valued [46]. Hence the steepest descent path in Fig. 9 changes from being a valley floor above the valley bifurcation to being a *ridge crest* below it. This bifurcation point is a valley ridge inflection point. Locating such points on the PES may be useful in some contexts, and an algorithm for finding them has been developed [71].

Figure 9 raises an important practical point. If the IRP is a ridge crest below a bifurcation point, it may be of less interest to us than the path(s) that follow the lower valleys. In practice, unless one takes special care to follow the path of steepest descent precisely, numerical error will displace the calculated path from the precise ridge crest and lead to one or other of the bifurcated valleys. This is a case where precisely following the steepest descent path may be contraindicated; indeed, methods to ensure that the calculated path diverges from the path of steepest descent as close as possible to the valley ridge inflection point may be useful [43]. Alternatively, this again looks like a situation where the available energy is in excess of the height of the walls separating the (bifurcated) valleys; no single path is dynamically significant by itself.

IV. COORDINATES AND PATH-BASED SURFACES

A. Intrinsic Reaction Paths and Natural Collision Coordinates

Let's assume for the moment that we have evaluated a simple IRP that contains no bifurcations. How do we base a PES on this path? The information we have derived is actually a table of values of the molecular coordinates, $\zeta(s_n)$, the potential energy, $V(s_n)$, and the gradient vector, $\partial V / \partial \zeta(s_n)$, at N_{path} discrete values of the reaction coordinate, s_n, along the path. If we have used Hessian-based methods to derive the IRP, we also have the force constant matrix $F(s_n)$ at each of these points. We cannot represent the walls of Fig. 2 without values of F; so we must at least calculate $F(s)$ at a subset of the points to have $F(s_n)$, $n = 1, \ldots, N_{PES}$. The PES will be given by an interpolation procedure, so that N_{PES} must be large enough, and the points must be sufficiently evenly distributed, for the interpolation procedure to be sufficiently accurate. Numerical trials have been carried out in test cases to determine how large N_{PES} must be [42, 49]. Clearly, the higher the ratio, N_{PES}/N_{path}, the more efficient are the Hessian-based methods for determining the IRP.

In the vicinity of any one of these discrete points, s_n, we can expand the PES as a Taylor series like that of Eq. (2.1):

$$V(\zeta) = V(s_n) + [\zeta - \zeta(s_n)]^T \left.\frac{\partial V}{\partial \zeta}\right|_{\zeta=\zeta(s_n)} + \frac{1}{2}[\zeta - \zeta(s_n)]^T \mathbf{F}|_{\zeta=\zeta(s_n)}[\zeta - \zeta(s_n)],$$

$$(4.1)$$

where ζ^T is a $3N$-dimensional row vector.

Quite generally, the expansion in Eq. (4.1) may be based on the geometry at any value of s by interpolation: using one-dimensional spline interpolation, for example, to give $\zeta(s)$, $V(s)$, $\partial V/\partial \zeta(s)$, and $\mathbf{F}(s)$ from the tables $\{s_n, \zeta(s_n)\}$, $\{s_n, V(s_n)\}$, $\{s_n, \partial V/\partial \zeta(s_n)\}$, and $\{s_n, \mathbf{F}(s_n)\}$. Equation (4.1) then provides the PES at any geometry ζ *if* we have some means of deciding what value of s to use for that ζ, that is, $s(\zeta)$. The traditional approach to this is to use *natural collision coordinates* [34–36, 72–74].

Figure 10 shows the IRP on a contour diagram of the PES of Fig. 1. For each geometry ζ off the path, we find a point on the path, $\zeta(s)$, which minimizes (with respect to s) the distance, $|\zeta - \zeta(s)|$. A little algebra implies that the vector $\zeta - \zeta(s)$ must be orthogonal to the direction of the

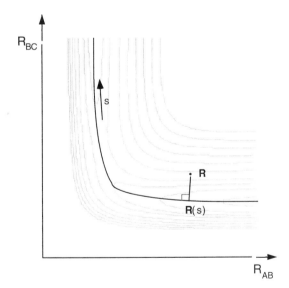

Figure 10.   Reaction path on a contour diagram of the PES of Fig. 1. The reaction coordinate s is the distance along the path. Points **R** off the path are related to s in natural collision coordinates, as indicated [see Eq. (4.2) for the corresponding formula in MWC coordinates].

path:

$$[\zeta - \zeta(s)]^T \frac{d\zeta(s)}{ds} = 0 . \tag{4.2}$$

This determines $s(\zeta)$ [as long as Eq. (4.2) has a unique solution for s]. Now because the path follows the local gradient vector [see Eq. (3.8)], $\zeta - \zeta(s)$ is orthogonal to the gradient, and there is no linear term in $\zeta - \zeta(s)$ for a Taylor expansion of the PES about $\zeta(s)$:

$$V(\zeta) = V(s) + \tfrac{1}{2} [\zeta - \zeta(s)]^T F(s)[\zeta - \zeta(s)] . \tag{4.3}$$

Now Eq. (4.3) should be simplified further since it logically reads: Given ζ, find s from Eq. (4.2), evaluate $V(s)$ and $F(s)$ by interpolation, then evaluate the quadratic term (a double sum over $3N$ MWC coordinates at fixed s). For fixing s implies that the $3N$ MWC coordinates are constrained and only $3N - 1$ coordinates are free. In addition, we know that the PES is invariant to three coordinates that describe overall translation of the molecule and three coordinates that describe overall rotation of the molecule. Thus only $3N-7$ coordinates are actually significant in the quadratic term in Eq. (4.3). Miller et al. [75] eliminate seven coordinates by imposing the condition (4.2) and requiring that the center of mass remain at the origin:

$$\sum_{i=1}^{N} \sqrt{m_i}[\xi(i) - \xi(i, s)] = 0 , \tag{4.4}$$

and that the displacements from the path do not contribute to the angular momentum (the Eckart condition):

$$\sum_{i=1}^{N} [\xi(i) - \xi(i, s] \times \frac{d\xi(i, s)}{ds} = 0 . \tag{4.5}$$

Avoiding as much detail as possible (see Ref. 75), we define $\chi(3N, s)$ as the $3N$-dimensional unit vector corresponding to displacement of ζ along the path at s, $\chi(3N - 1, s)$, $\chi(3N - 2, s)$, and $\chi(3N - 3, s)$ as unit vectors corresponding to translation of ζ in the x, y, and z directions, and $\chi(3N - 4, s)$, $\chi(3N - 5, s)$, and $\chi(3N - 6, s)$ as unit vectors corresponding to rigid rotation of the molecule with structure $\zeta(s)$. Then the projected force constant matrix $\mathbf{F}^P$,

$$\mathbf{F}^P(s) = \left[\mathbf{I} - \sum_{j=0}^{6} \chi(3N-j)\chi^T(3N-j) \right] \mathbf{F}(s) \left[\mathbf{I} - \sum_{j=0}^{6} \chi(3N-j)\chi^T(3N-j) \right],$$

(4.6)

has seven zero eigenvalues. We can locally transform our coordinates from $\zeta - \zeta(s)$ to the normal coordinates of $\mathbf{F}^P$ $\{Q_k, k = 1, \ldots, 3N\}$:

$$\zeta_i = \zeta_i(s) + \sum_{k=1}^{3N-1} L_{i,k}Q_k.$$

(4.7)

Note that there is no component for Q_{3N} that would represent a displacement in s:

$$L_{i,3N} = \frac{d\zeta_i(s)}{ds}.$$

(4.8)

The PES expanded about the IRP can then be written in very simple form:

$$V_{IRP} = V_{IRP}(s, \mathbf{Q})$$

$$= V(s) + \sum_{k=1}^{3N-7} \tfrac{1}{2} \, \omega_k^2(s)Q_k^2.$$

(4.9)

Thus our IRP-based PES is one in which the potential rises and falls along the IRP with harmonic walls orthogonal to the IRP in $3N - 7$ dimensions, with frequencies that change with the reaction coordinate (as in Fig. 6). Equation (4.9) is precisely the simple picture we had in mind from the minute we drew Fig. 2.

However, the general law of conservation of difficulty usually applies in spades to molecular dynamics. When we want to do dynamics, we must deal with the Hamiltonian for the system: that is, the PES plus the kinetic energy. Invariably, when the PES is simple in one set of coordinates, the kinetic energy will be complicated, and vice versa. Thus, for the coordinates $\{s, Q_1(s), \ldots, Q_{3N-7}(s)\}$, the kinetic energy (for total angular momentum equal to zero, and neglecting the vibrational angular momentum) is given by T_{IRP} [75]:

$$T_{IRP} = \frac{1}{2} \frac{\left[p_s - \sum_{j=1}^{3N-7} \sum_{k=1}^{3N-7} Q_j P_k B_{k,j}(s) \right]^2}{\left[1 + \sum_{j=1}^{3N-7} Q_j B_{j,3N}(s) \right]^2} + \sum_{k=1}^{3N-7} \frac{1}{2} P_k^2.$$

(4.10)

Here p_s is the momentum conjugate to s, P_k is the momentum conjugate to Q_k, $B_{k,3N}$ is a kinematic coupling of the kth orthogonal vibration to the reaction coordinate (s is the $3N$th coordinate and this coupling term is written more transparently as $B_{k,s}$), and the $B_{j,k}$ in the numerator of Eq. (4.10) are kinematic couplings of the vibrational modes [see Eqs. (4.7) and (4.8)]:

$$B_{j,k}(s) = \sum_{n=1}^{3N} \frac{dL_{n,j}}{ds} L_{n,k} . \tag{4.11}$$

These coupling effects arise because the IRP is not a straight line in MWC coordinates. As s changes, the character of the motion with coordinate Q_k changes ($dL_{i,j}/ds \neq 0$, in general) and the direction of the motion along the path changes ($d^2\zeta_i/ds^2 \neq 0$) because the IRP curves, so the conjugate momenta are coupled. By analogy with the definition of natural collision coordinates in two dimensions [35] the total curvature of the IRP is $\kappa(s)$:

$$\kappa(s) = \left(\sum_{k=1}^{3N-7} B_{k,3N}^2 \right)^{1/2} . \tag{4.12}$$

It is important to remember that the kinetic energy of Eq. (4.10) applies only to the case of zero angular momentum ($\mathbf{J} = 0$). For the moment, we simply note that the complete kinetic energy for nonzero angular momentum is complicated significantly by vibration–rotation coupling [75]. Together, Eqs. (4.9) and (4.10) define the reaction path Hamiltonian (RPH) (for zero angular momentum), H_{IRP}:

$$H_{IRP} = T_{IRP} + V_{IRP}(s, \mathbf{Q}) . \tag{4.13}$$

All the parameters in H_{IRP} can be determined from the molecular geometry along the IRP, $\zeta(s)$, together with the energy profile $V(s)$ and the force constant matrices, $\mathbf{F}(s)$ [12, 44].

An advantage of the RPH approach is that if only one degree of freedom, the reaction coordinate, undergoes large-amplitude motion, and if the couplings between the reaction coordinate and other degrees of freedom are weak, approximate quantum treatments of the dynamics can be carried out [76–80].

B. Variations on the Same Theme

It is a typical ploy for a theoretician to try and simplify a problem by using a more convenient set of coordinates. Which set of coordinates is convenient will depend on the system studied, and even on what is to be measured, as various dynamical approximations are best implemented in terms of different coordinates. Thus the reaction path Hamiltonian of Eq. (4.13) can be modified in a number of ways.

First, following earlier work by Hougen et al. [81], Miller [82] has pointed out that a very similar path-based Hamiltonian can be derived for any path, not just the IRP, using natural collision coordinates, including a special coordinate to describe motion along the path (orthogonal to overall rotation and translation) and coordinates for vibrational motion orthogonal to the path. For an arbitrary path, there are linear terms in the PES, as the path is not locally aligned with the gradient, and there are coupling terms between the vibrational modes in both the kinetic and potential energies.

As we have seen, the use of natural collision coordinates with an IRP leads to kinematic coupling between the momenta, as in Eq. (4.10), with separable contributions to the PES. However, if it is convenient, one can carry out a succession of canonical transformations (changes of coordinates and consequent changes to the conjugate momenta) along the path to produce a separable kinetic energy operator. However, the coupling between motions is not eliminated, merely transferred to the PES, which now contains quadratic terms involving products like $Q_i Q_j$ [83].

As we noted above for a general path, natural collision coordinates give rise to coupling terms in both the kinetic and potential energies. By choosing the path "appropriately" one might hope to effectively separate some orthogonal motions by having these two sources of coupling effectively cancel.

C. Potentials for Distinguished Coordinate Paths

As we also noted above, a reaction path is often determined in internal coordinates (rather than MWC coordinates) using a *distinguished coordinate*. By construction, the distinguished coordinate is monotonic along the path, and is taken to be the reaction coordinate. For example, the distinguished coordinate may be defined as the difference of the lengths of a bond which forms and a bond which breaks during the reaction. Also by construction, the PES is stationary with respect to the other $3N - 7$ internal coordinates. Denoting the internal reaction coordinate by R_1, and the remaining internal coordinates by $R_2, \ldots, R_{3N-6}$, one can write the PES as

$$V(\mathbf{R}) = V_{\text{path}}(R_1) + \frac{1}{2} \sum_{j=2}^{3N-6} \sum_{k=2}^{3N-6} [R_j - R_j(R_1)]$$

$$\times [R_k - R_k(R_1)] \frac{\partial^2 V}{\partial R_j \, \partial R_k} \bigg|_{\substack{R_j = R_j(R_1) \\ R_k = R_k(R_1)}}. \tag{4.14}$$

Here $V_{\mathrm{path}}(R_1)$ is the energy profile along the reaction path, viewed as a function of R_1. Evaluating $V(\mathbf{R})$ from a discrete set of points along the path is accomplished by interpolation in the same way as for an IRP (see Section IV.A). The force constant matrix in internal coordinates, contained in Eq. (4.14), is obtained from the Cartesian gradient and force constant matrix by standard methods [84].

D. IRP Potentials in Internal (Valence) Coordinates

Although the intrinsic reaction path and associated potential energy derivatives are obtained in MWC coordinates, the PES may be expressed in internal valence coordinates if desired [85], as for a distinguished coordinate path. By expressing the PES as a function of internal valence coordinates, we ensure that the PES is explicitly independent of the orientation of the molecule. [The IRP PES of Eq. (4.9) is expressed in normal coordinates, linear combinations of the MWC coordinates, and is independent of orientation by virtue of a continuous change of variables along the IRP, as given by Eq. (4.7).] By expressing the PES in internal coordinates, we can use the PES independently of the reaction path Hamiltonian approach. Moreover, by eliminating the intrinsic reaction coordinate, s, we ensure that the potential is always well defined. For, unfortunately, natural collision coordinates are not always well defined [86]: there may be more than one solution for s in Eq. (4.2), *due to the fact that the reaction path is curved.* This ambiguity arises on the inside of the reaction path as it curves, as indicated in Fig. 11. An IRP potential in natural collision coordinates will always be undefined in some region. As long as we can identify an internal coordinate which is monotonic along the IRP (and well defined everywhere), we can use this coordinate in place of s and avoid this problem. In simple cases this coordinate will be

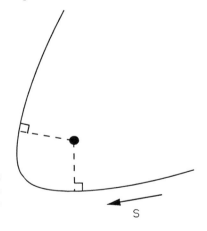

Figure 11. In natural collision coordinates the relation between points off the path and the reaction coordinate s may be ambiguous due to curvature of the path.

obvious when we graph the usual valence coordinates (bond lengths, angles, etc.) corresponding to $\zeta(s)$ along the IRP. Sometimes, as in Fig. 5, a monotomic coordinate is not obvious at first but might be found easily by taking linear combinations of bond lengths, for example [58]. A more general procedure for finding a monotonic internal reaction coordinate has been reported [56].

E. Reaction Surfaces

The concept of a reaction path is based on the idea that there is only one essential degree of freedom, which undergoes very large amplitude motion during the reaction. It was understood rather early in the development of PES based ón a reaction path that cases may arise where there are two or more quite independent motions that undergo large-amplitude motion. Carrington and Miller [87, 88] therefore developed a new Hamiltonian, a *reaction surface Hamiltonian*, which is completely analogous to the RPH but has two reaction coordinates (see also Ref. 89). The motivation there was to bypass problems associated with natural collision coordinates for very highly curved paths. However, a reaction surface is generally necessary when some of the orthogonal vibrational modes undergo large-amplitude motion [90], *a fortiori* when the reaction path bifurcates [70]. We discuss such problems below when we consider the transition to more nearly global surfaces.

V. DYNAMICS ON PATH-BASED SURFACES

We have now established the basic forms of PES, which are based on a reaction path. Let's now look at how these surfaces are used in dynamics. The study of chemical reaction dynamics has been limited by a number of factors in addition to the paucity of reliable PES [91]. The cutting edge of exact quantum reaction dynamics is still to be found at the level of tetraatomic systems in three dimensions [92]. The reaction dynamics of polyatomic molecules have generally been studied using approximate dynamics, like classical and semiclassical dynamics, or using quantum dynamics for a small subset of the actual degrees of freedom or by using statistical theories. Even within these general categories of approximate dynamics, various approximations place different requirements on our knowledge of the PES. Since this review concentrates on the role of a reaction path, we discuss the dynamics in the context of the type of PES required for their implementation.

A. Near-Quadratic Surfaces

By a near-quadratic surface, we mean a PES like that of Eq. (4.9) or (4.14), where the PES is taken to be a second-order Taylor expansion in directions orthogonal to the reaction path, although we will include cases where some higher-order terms are included.

1. Reaction Path Hamiltonian Dynamics

A natural starting place for this level of PES is the reaction path Hamiltonian (RPH) of Eq. (4.13). As we noted in Section IV, all the frequencies and kinematic parameters required for the RPH can be determined from the path $\zeta(s)$, the energy $V(s)$, and the force constant matrix $\mathbf{F}(s)$ at discrete points along the path. The dynamics is complicated by rotation–vibration coupling and the kinematic coupling due to the curvature of the path. To my knowledge, no full quantum treatment of the RPH has been attempted for nonzero total angular momentum ($\mathbf{J} \neq 0$). I am aware of only one RPH calculation for $\mathbf{J} \neq 0$: Billing [89, 93] has derived a perturbation expansion of the full RPH and used this to perform quantum–classical calculations in which the classical trajectories are allowed to tunnel to the product valley from their turning points in the reactant valley.

The approximate RPH for $J = 0$ (presented above) has been used in a number of studies. The principal question in applying the RPH is to decide how to approximate the effect of the kinematic coupling elements. Miller and co-workers have used a semiclassical perturbation approximation to the scattering ($\mathbf{S}$) matrix [94], together with an infinite-order sudden approximation to the RPH dynamical coupling. This gives an explicit expression for state-to-state scattering matrix elements for both reactive and inelastic events [77]. This approach has also been used to study tunneling contributions to energy transfer and reaction rates [95], and the rate of a unimolecular isomerization [96]. Tunneling is discussed in more detail below.

A qualitatively appealing feature of the RPH is that the kinematic coupling between motion along the reaction coordinate and the vibrational modes is made explicit in the $B_{j,s}$ matrix elements. From the early work of Polanyi and co-workers on collinear triatomic reactions, we know that from the position of the saddle point relative to the corner in the reaction path, qualitative conclusions can be drawn about the effect of the energy distribution in the reactants on the rate and about the distribution of energy in the products [20, 22]. The $B_{j,s}$ appear to contain quite analogous and quite detailed information about curvature effects for polyatomic systems, identifying which modes should couple to (and exchange energy with) the reaction coordinate. If the $B_{j,s}$ can be neglected for some modes throughout the RP, these modes should be purely "spectators" to the chemical reaction and should behave adiabatically [97]. That is, the frequency of such modes may change with s, but their quantum numbers will tend to be unaffected by the reaction. Similarly, if only some $B_{j,s}$ are large while others are always small, the vibrational modes can be categorized into strongly coupled modes and bath modes. The strongly coupled modes can be treated fully dynamically along with the reaction coordinate, while the bath modes can be treated as perturbations [76, 78, 79, 98, 99]. Only a few *ab initio* studies [80, 100–103] have used the $B_{j,s}$ elements to make even qualitative statements about energy distribution effects in reactants or products, despite the fact that these quantities can be obtained directly from the IRP. *Ab initio* studies of reactions that report data like Figs. 4–6 are becoming more frequent [104–108] and might provide further qualitative insight if the coupling elements $B_{j,s}$ were also presented and discussed in terms of possible mode-specific effects. Of course, if the IRP contains "unphysical" bends, as has been suggested in some cases [82], the $B_{j,s}$ would not be physically relevant. However, very few data are available from which we could form a general view of the qualitative utility of examining these coupling elements. The total reaction path curvature, as measured by $\kappa(s)$ in Eq. (4.12), has mostly been considered in terms of its effect on "corner cutting" or tunneling through the potential barrier in the vicinity of the saddle point on the reaction path [11, 109–115].

2. Variational Transition-State Theory

A near-quadratic reaction path PES can be applied directly to the statistical theory of reaction rates known as *variational transition-state theory* (VTST). This theory has been reviewed thoroughly previously [11, 116–121], so that here we merely try to give a simple descriptive view with references to recent progress.

In transition-state theory (TST) one divides the configuration space of the system into two parts by defining a surface (transition state) that separates reactants from products. Classically, the rate of reaction at energy E (and angular momentum **J**) is taken to be proportional to the flux of trajectories across this surface. If all trajectories that begin in the reactant region and proceed on to the product region cross the surface *only once*, then (assuming that the dynamics in the reactant region is ergodic) TST gives the exact classical rate [122]. The TST rate is a classical upper bound to the exact rate if the "no recrossing" assumption is violated, as we will have overcounted the reactive trajectories.

Variational transition-state theory (VTST) seeks to find the best lower bound by varying the position of the dividing surface to find the surface that minimizes the rate.

This theory originated with the work of Wigner and Keck [123]. Beginning in the 1970s, a number of authors, including Truhlar, Garrett and co-workers in particular, have developed variational transition-state theory using *ab initio* quantum chemistry to evaluate reaction path–based potentials and to calculate bimolecular reaction rate constants as a function of energy and temperature. They have produced a number of practical methods for polyatomic systems which have been made generally available in the POLYRATE program [124].

The transition state is taken to be the $(3N - 7)$-dimensional plane orthogonal to the reaction path at that value of s which minimizes the TST rate. For reactions at given energy [microcanonical variational transition-state theory (μVT)], the bimolecular rate constant is given by [125]

$$k^{\mu \mathrm{VT}}(E) = \min_s \frac{N^{\mathrm{GT}}(E, s)}{h \phi^R(E)}. \qquad (5.1)$$

Here $\phi^R(E)$ is the density of states per unit volume and energy for the reactants at energy E, and $N^{\mathrm{GT}}(E, s)$ is the number of energetically available states of the generalized transition state at reaction coordinate s. While Eq. (5.1) is an upper bound in classical mechanics, the usual quasiclassical approach is to replace $N^{\mathrm{GT}}(E, s)$ and $\phi^R(E)$ with the corresponding quantum values. The quantum vibrational density of states can be evaluated using an efficient direct-counting algorithm [126].

Why should we bother finding the optimum value of s rather than siting the transition state at the saddle point as in conventional transition-state theory? Because $N^{\mathrm{GT}}(E, s)$ may depend strongly on s. In part, $N^{\mathrm{GT}}(E, s)$ may depend strongly on s because the harmonic frequencies, $\omega_k(s)$, may vary with s (see, e.g., Fig. 6). For a single oscillator, the number of accessible states up to an energy E is determined by the relative height of the oscillator energy minimum with respect to E and by the spacing of the energy levels. The height of the oscillator minimum at s is $V(s)$; the spacing of the levels is $\hbar\omega(s)$. If, for example, some of the $\omega_k(s)$ have maximum values away from the saddle point, the number of available states at energy E may be smaller there than at the saddle point (see, e.g., Ref. 127). We envisage the valley in Fig. 2 being narrower at this value of s, so that trajectories cannot "get through" from reactants to products. Such narrowings of the reaction path valley are termed *dynamical bottlenecks*. Basically, variational transition-state theory accounts for the possibility that the width of the reaction path valley may be as important as the barrier height in determining the reaction rate constant.

Implicitly, the microcanonical theory for $k^{\mu VT}$ applies to the reaction path Hamiltonian of Eq. (4.14), in which $J = 0$. Below we consider a microcanonical theory that accounts for rotation.

Canonical variational transition-state theory (CVT) gives the rate constant at temperature T as [125]

$$k^{\mathrm{CVT}}(T) = \min_s \frac{kT}{h} \frac{Q^{\mathrm{GT}}(T, s)}{\Phi^R(T)} e^{-\Delta V(s)/kT}, \qquad (5.2)$$

where $\Delta V(s)$ is the energy on the reaction path (relative to the reactants), $\Phi^R(T)$ is the reactant partition function per unit volume at temperature T, and $Q^{\mathrm{GT}}(T, s)$ is the partition function of the generalized transition state at s. The partition functions are generally taken to be products of the separated electronic, vibrational, and rotational partition functions. The rotational partition function is usually taken to be that for a rigid rotor with geometry $\zeta(s)$ and is often evaluated classically. An optimum value of s is found for each value of the temperature. Truhlar, Garrett, and co-workers commonly employ an improved version of canonical variational transition-state theory (ICVT) [128], in which the classical contributions to the rate from energies below threshold are set to zero.

The partition functions and numbers of accessible states can all be calculated from the reaction path data, $\zeta(s)$ and $\mathbf{F}(s)$, assuming harmonic vibrations and separation of vibration and rigid-body rotation. The vibrational partition function may be improved in accuracy by accounting for anharmonicity in some modes. This has been done simply in a separable mode approximation (e.g., Morse stretches and quartic terms in the bending potentials) [129–131]. There are now a number of examples of applications of various forms of canonical variational transition-state theory using *ab initio* reaction path calculations [15, 106, 108, 112, 131–142].

For reactions involving light atoms, particularly hydrogen atom or proton transfer reactions, a major correction to the reaction rate constant is due to tunneling [11, 12, 58, 64, 76, 96, 106, 111, 112, 114–116, 129, 132, 133, 143–152].

3. Tunneling

A first glance at Fig. 4 might suggest that tunneling corrections, to a VTST rate constant, for example, could easily be evaluated in terms of a transmission coefficient for tunneling through a one-dimensional barrier. Garrett et al. have shown that tunneling corrections to VTST can be incorporated consistently as a multiplicitive transmission coefficient [132]. However, some complications must be taken into account.

In terms of the RPH (for $J = 0$) of Eq. (4.14), the simplest approximation to tunneling assumes that the process is vibrationally adiabatic; that is, the $3N - 7$ orthogonal vibrational modes have constant quantum numbers. Even then, several levels of further simplification have been considered [76, 114, 115, 143]. First, we might ignore all kinematic coupling elements, including $B_{j,s}$. Tunneling is then one-dimensional and subject to the effective Hamiltonian, H_{Ad}:

$$H_{Ad} = \tfrac{1}{2} \, p_s^2 + V(s) + \sum_{k=1}^{3N-7} \hbar\omega_k(s)(n_k + \tfrac{1}{2}) \, . \tag{5.3}$$

The only complication here is that tunneling occurs on an effective potential that depends on the excitation of the vibrational modes. Note that even if the vibrational modes are in their ground states, the zero-point energy of the modes may make a substantial contribution to the height and shape of the barrier [82]. For example, Fig. 4 shows the energy profile of the adiabatic ground state [all n_k equal to zero in Eq. (5.3)] along the IRP for the $NH(^3\Sigma^-) + H_2(^1\Sigma_g^+) \rightarrow NH_2(^2B_1) + H(^2S)$ reaction.

More complex treatments of tunneling take account of the kinematic coupling of the vibrational motion with motion along the reaction path. By transforming $\{Q_k, P_k\}$ variables in the RPH of Eq. (4.13) to action-angle variables, Gray et al. [76] could apply classical perturbation theory to this coupling (for assumed small $B_{j,k}$). With this method, one can derive new action variables which are more nearly constants of the classical motion. Tunneling is then evaluated via a semiclassical transmission coefficient at constant values of the perturbationally improved action variables. Colwell and Handy have applied these methods to the decomposition of the methoxy radical [102, 153]. As we discussed above, Miller and Shi [77] derived a semiclassical infinite-order sudden approximation to the coupling induced by the $B_{j,k}$; at its simplest level, this gives the tunneling transmission coefficient in terms of the reaction path curvature, $\kappa(s)$. Cerjan et al. [95] have shown how this approximation can be applied to a variety of dynamical processes. In a different approach, Carrington et al. [96] employed a Feshbach model for the $B_{j,k}$ coupling in considering the lifetime of vinylidene for rearrangement to acetylene. All these tunneling approximations only require knowledge of the PES in terms of the path, $\zeta(s)$, plus $V(s)$ and $\mathbf{F}(s)$.

An appealing way to view the complications caused by the kinematic coupling elements is presented in Fig. 12. The IRP path is shown on a

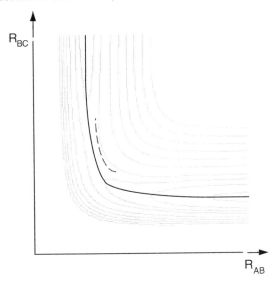

Figure 12. Reaction path and PES of Fig. 10, indicating the location of the Marcus–Coltrin tunneling path (dashed line). This is the line of classical turning points (on the inner side of the path) for the vibration(s) orthogonal to the reaction path. The Marcus–Coltrin path is shown only in the region of the saddle point assuming that the PES lies above the total energy of the reactants in that region.

contour diagram of a simple PES in the vicinity of the saddle point. Tunneling on the IRP has been shown to give transmission coefficients which are much too small. Marcus and Coltrin [111] showed that the kinematic coupling leads, for energies below the barrier, to a reverse "bobsled" effect and preferential tunneling along a path inside the bend of the IRP. The Marcus–Coltrin tunneling path is the line of classical turning points for the vibrations orthogonal to the IRP (on the "inner side" of the bending path). This path is shorter than the corresponding classically forbidden section of the reaction path, so that the tunneling transmission coefficient may be larger than that calculated for tunneling on the reaction path. At least when the curvature of the IRP path, $\kappa(s)$, is not too large, this path leads to the maximum transmission probability. Skodje et al. [114, 115] have derived semiclassical adiabatic ground-state (SAG) transmission coefficients based on the Marcus–Coltrin path for the small curvature limit. The detailed kinematic couplings are approximated by an effective mass for the tunneling "particle" whose value depends on the curvature, $\kappa(s)$. Given that the curvature is not too large and the Marcus–Coltrin path is not too far from the IRP, all these tunneling corrections only require knowledge of the PES in terms of $V(s)$ and $\mathbf{F}(s)$.

However, it may be that the IRP is very strongly curved in the vicinity of the variational transition state [109]. In that case, the optimum tunneling path may sharply cut the corner of the IRP; indeed, it may even penetrate the inside of the corner to regions where the natural collision coordinates are undefined. Note that in the large-curvature case, the tunneling is not adiabatic: The tunneling path implicitly passes outside the region occupied by the vibrational ground state and is associated with energy transfer from motion on the IRP to the orthogonal vibrational modes. Liu et al. [110] describe the transmission coefficient in this case as a convolution of contributions from multiple straight-line tunneling paths which connect the turning point of the vibrational ground state on the reactant side of the barrier with all accessible vibrational states on the product side. Since these paths may deviate significantly from the reaction path, *additional calculations of the energy* in this inside corner region may be required to evaluate accurately the semiclassical tunneling integral.

Most recently, Truhlar, Garrett, and co-workers employ a combination of various small-curvature SAG and large-curvature (straight-line tunneling) approximations [106, 110, 124, 154, 155].

4. More on How the Ab Initio Data Are Used

Before the later 1980s, *ab initio* studies based on a reaction path (including the IRP) usually involved fitting the *ab initio* data to some functional form for the PES. That is, one performed *ab initio* calculations of the energy and possibly the energy gradient and force constant matrix at significant points on the PES, including the stationary points, and adjusted the parameters of some analytic function of the molecular coordinates to obtain a fitted PES [14, 17, 18, 156–160]. Often, some scaling of this *ab initio* surface was performed to ensure that the relative energies of the reactants, saddle point, and products agreed with experimental measurements. The surface for the reaction

$$OH + H_2 \rightarrow H_2O + H, \qquad (5.4)$$

derived by Schatz and Elgersma [161] from *ab initio* calculations by Walch and Dunning [162], is an excellent example of this procedure. The IRP and associated force constant data were subsequently obtained by solving Eq. (3.5), as discussed in Section III.B, using the fitted functional form for the PES.

This fitting of a more or less global surface is not only unnecessary for applications such as VTST but very time consuming. "Direct" use of the *ab initio* data by interpolation methods promises to facilitate the application of *ab initio* calculations to reaction path dynamics [12, 44, 56, 57,

106, 112, 139, 141, 142, 151, 152, 157, 163–165]. If a large number of *ab initio* calculations of the geometry, energy, energy gradient, and force constant matrix have been performed along the reaction path, direct interpolation of this data is quite straightforward, as discussed in Section IV.A. It may well be, however, that very high level correlated *ab initio* calculations are required for a reasonable description of the reaction energetics, and such a complete set of data is prohibitively expensive to calculate. Approximate schemes to employ a smaller number of high-level calculations to scale or otherwise correct lower-level data have been reported [106, 112, 139, 141, 142, 152, 163, 164, 166]. In some cases the method used is not so much an interpolation of a small number of data distributed along the reaction path as the fitting of a simple functional form of the energy profile on the path, $V(s)$ [112].

How the *ab initio* data are used is associated with the question of what coordinates are employed. Naturally, in distinguished coordinate paths, internal coordinates are employed. Earlier we noted that some authors have investigated the suitability of using internal coordinates even when the reaction path is an IRP (originally calculated in MWC coordinates) [56, 138, 167–169]. The transformation of the energy derivatives from Cartesian to internal coordinates is carried out by solving the simultaneous equations that come from application of the chain rule for differentiation of the energy [84]. Perhaps the major motivation for this work comes from the limitations of natural collision coordinates. A related subtlety leads to a variation in the frequencies of the force constant matrix $F(s)$ in internal coordinates: At a nonstationary point on the IRP, $F(s)$ depends on the way in which s is associated with points off the IRP. Equation (4.2) is just one of a number of equally plausible ways of choosing how geometries off the path are related to points on the reaction path. $N^{GT}(E, s)$, the number of energetically available states of the generalized transition state, will therefore depend on our choice of coordinates [169]. This arbitrary variation of $N^{GT}(E, s)$, and hence the reaction rate constant, is really a reflection of the limitations of the reaction path–based approach to variational transition-state theory when the PES is a second-order truncated Taylor expansion.

B. Very Anharmonic Surfaces

So far we have discussed statistical and dynamical theories that have largely been based on a quadratic expansion of the PES about the reaction path (albeit with a limited account of anharmonicity in some vibrational modes). Still within that same spirit, very anharmonic vibrations such as intramolecular hindered rotations (torsional motion) have been treated more realistically [56, 110, 138]. However, in many cases, even when only a single reaction path is relevant, much more "global" descriptions of some aspects of the PES are required.

1. Statistical Studies

This is especially true for certain unimolecular and recombination reactions [121, 170]. For many such reactions there is little or no barrier. The position of the generalized transition state is very dependent on the variation of the frequencies along the reaction path and on the angular momentum of the fragments or reactants. As a polyatomic molecule splits into two fragments, the fragments may rotate and have orbital angular momentum relative to one another. Six vibrational degrees of freedom of the entire (nonlinear) molecule cease to be vibrations and instead describe the relative orientation and the relative translation of the (nonlinear) fragments. If one of these six coordinates is taken to be the reaction coordinate (say, the distance between the centers of mass of the fragments), there must be at least five degrees of freedom which completely change their character during the reaction. In addition, there may be other vibrational motions of the molecule whose frequencies and characters change radically (e.g., because steric hindrance to a torsional motion is removed on fragmentation). Thus there is a set of transitional modes whose energetically accessible amplitudes of motion change dramatically along the reaction path. To consistently determine the location of any dynamical bottleneck or generalized transition state, we will need to consider essentially global motion in these modes. Moreover, we need to take into account that the orbital angular momentum of the fragments will produce an effective centrifugal potential for the radial or reaction coordinate motion. Thus we need to consider explicitly the reaction rate as a function of both total energy and total angular momentum, $k(E, J)$, something that was neglected in the simple cases above of a "tight" transition state which occurs in the vicinity of a high-energy saddle point on the PES.

We will not discuss the relative merits of various theories of unimolecular reactions. Here we simply point out that the concept of a reaction path is still useful in understanding the PES and dynamics, even for "loose" transition states. Even though the dynamics in the transitional modes is inherently of large amplitude, the remaining vibrational modes may still be treated as above, and one can still label the position of a generalized transition state by a reaction coordinate, say R. Following Wardlaw and Marcus [121, 171–173], the number of accessible states at energy E and total angular momentum J at reaction coordinate R is N_{EJ}:

$$N_{EJ}(R) = \int_0^E N_V(E - \varepsilon)\Omega_J(\varepsilon) \, d\varepsilon \ . \tag{5.5}$$

This is a convolution of the density of states accessible for the transitional modes, Ω_J, with the number of states accessible to the other vibrational modes, N_V (calculated as a quantum sum in a harmonic approximation or with separable anharmonic corrections as above). The difficult task then is to calculate Ω_J.

Since the transition state in such reactions generally occurs when the fragments are separated by 2 Å or more, the transition state is relatively "loose," so that the energy levels for the transitional modes are closely spaced. Hence Wardlaw and Marcus, and later authors, have evaluated Ω_J classically (with perhaps a quantum correction [174]). The classical evaluation of the density of states, Ω_J, is achieved by a Monte Carlo integration over the phase space of the transitional modes, either in terms of action-angle variables [121] or in terms of Euler angles (describing the relative orientation of the fragments) and their conjugate momenta [175]. Since this Monte Carlo integration is a computationally expensive procedure, improvements in the efficiency of the method have been pursued [176].

Ab initio calculations have entered this area in a typical sequence. To begin with, the PES for this Monte Carlo integration has been an analytic functional form derived by fitting to experimental data and some *ab initio* calculations [121]. A little later, Klippenstein and co-workers produced reaction path potentials specifically for this purpose: again analytic functional forms for the PES, but fitted to the energy and force constant matrices evaluated on the reaction path [59, 177, 178]. Most recently, Klippenstein et al. [179] have dispensed with a functional form for the PES and calculated N_{EJ} "directly," that is, by evaluating the *ab initio* energy at each step in the Monte Carlo integration. In the application studied ($CH_2CO \rightarrow CH_2 + CO$), this required 4000 calculations of the energy (at the MP2/6-31G* level of theory).

2. Classical Trajectory Studies

Classical mechanics has been used to study bimolecular reactions for many years [180]. The major advantage of classical mechanics lies in its ease of application; the major disadvantage is that it is basically wrong. However, whenever tunneling does not make a significant contribution to the reaction dynamics, we can expect classical mechanics to be semiquantitatively accurate for a process that is classically allowed. To justify using classical mechanics, we can often invoke the gambler's defence: Asked why he played roulette in the local casino when he knew the wheel was rigged, the inveterate gambler replied: "It's the only wheel in town."

In the context of reaction path studies, Billing has employed classical trajectories directly to simulate the dynamics of an approximation to the reaction path Hamiltonian (including rotational motion) [89, 93]. Classical trajectories have also been used on model PES for the case where the reaction path bifurcates [181].

However, the real power of classical trajectory studies comes from the ease with which polyatomic reactions involving many degrees of freedom can be studied using Cartesian coordinates. If one can write the PES in a set of internal coordinates, $\mathbf{R}$, Hamilton's equations of motion are given by

$$\frac{dx_n}{dt} = \frac{p_n}{m_n},$$

$$\frac{dp_n}{dt} = -\sum_{k=1}^{3N-6} \frac{\partial V}{\partial R_k} \frac{\partial R_k}{\partial x_n}, \qquad n = 1, \dots, 3N. \tag{5.6}$$

These are quite simple first-order coupled differential equations that can be solved using any of a myriad of numerical methods. Bimolecular collisions at given energy (and angular momentum) or at given temperature can be simulated by using Eq. (5.6) with the appropriate initial conditions [180]. One only needs to have a PES in internal coordinates (there are some difficulties here, which are discussed below).

PES based on distinguished coordinates or similar reaction surface potentials [70] are directly applicable in Eq. (5.6). For *ab initio* reaction path potentials, we discussed the conversion of IRP-based potentials to internal coordinates in Section IV.D. Our group has used trajectory studies to follow the decomposition of $HCOH^+$, as in Eq. (3.15) [56, 57], and the competition between exchange and abstraction reactions such as Eq. (3.14). These studies have used IRP based potentials converted to internal coordinates.

Note that a path-based PES is unlikely to have the correct asymptotic form; that is, a Taylor expansion about the path will give a PES that is bounded for motion in Q_i even when the frequency $\omega_i(s)$ is very small. Switching functions [1, 56] or similar means of imposing the correct asymptotic behavior are necessary to replace a parabola in Q_i with a flat PES at large Q_i. Classical trajectory studies at any energy and angular momentum can be evaluated, in principle, on such surfaces.

An important practical consideration in the development of PES for dynamical calculations is to ask how we know which parts of the PES are important and how we know that our PES is physically well behaved. In this context, classical trajectory studies provide an excellent testing ground. The trajectories explore only the dynamically relevant part of the surface, and (a practical advantage) the numerical solution of Eq. (5.6) will often break down if the PES is not smooth or suffers a similar, perhaps unexpected, imperfection.

C.   Reactions in Solution

The *ab initio* energy profile and associated reaction path for a chemical reaction in the gas phase might not seem especially relevant to a discussion of mechanism or energetics for the same reaction occurring in a solvent. However, a number of features of the reaction path concept have led to applications of such *ab initio* calculations to reactions in solution.

As Miller and Schwartz pointed out [78, 98], the reaction path Hamiltonian view leads naturally to a separation of the polyatomic vibrations into those strongly coupled to the reaction coordinate and those which are only weakly coupled. Motion along the reaction coordinate, weakly coupled to a bath of vibrational modes, might naturally be treated via a Langevin equation; the bath modes supply both fluctuating forces and dissipation. This is clearly relevant to part of the effect we expect that solvent molecules will have on motion along a reaction coordinate for solute molecules undergoing chemical reaction [82]. Substantial advances have been made in the theory of chemical reaction dynamics in solution, including an understanding of how solvent rearrangement takes place as solute reactants move along a reaction coordinate [99, 113, 182–185]. However, this subject lies outside our focus.

Nevertheless, it is perhaps relevant to point out briefly how *ab initio* calculations of gas-phase reaction paths are being applied directly to understanding the mechanism and thermodynamics of reactions in solution. This work is closely associated with methodology used in biodynamics. For example, in an early study a progression of molecular geometries on an *ab initio* reaction path (a distinguished coordinate path) for $OH^- + H_2CO$ were solvated using Monte Carlo calculations to equilibrate each geometry in a solvent of many water molecules [186]. The solute–solvent and solvent–solvent interactions in such calculations are usually empirical functions but may be based on *ab initio* calculations [186, 187]. This approach provides a means of estimating the progressive change in the free energy for reactions in solution (though based on the gas-phase mechanism). More recently, *ab initio* reaction paths have been used in conjunction with free-energy perturbation calculations of the type that have been applied to conformational changes in biomolecules [60, 62, 188, 189].

VI.   LIMITATIONS OF SIMPLE PATH-BASED SURFACES

At this stage we have at least skimmed through most of the recent applications of electronic structure theory to dynamics via reaction path methods. Although further improvements will no doubt occur, we can already efficiently find a reaction path, construct a simple PES based on the path, and use that PES in dynamical theories, particularly for the reaction rate constant. Somewhat more sophisticated PES have been derived: The simple low-order Taylor expansion method has been extended for some degrees of freedom (although only in a separable mode approximation), and global surfaces have been developed for a few transitional degrees of freedom in cases with "loose" transition states. However, chemical reactions are often more complicated than we might imagine from the picture of motion along the valley of a single reaction path. There may be multiple reaction paths and multiple products.

A. Symmetry Considerations

Multiple reaction paths can arise from symmetry. Let's take some time to consider one example in detail. Consider again the energy profile of Fig. 7 along a reaction path for the abstraction reaction [52, 53] (with D_2 rather than H_2)

$$NH_3^+ + D_2 \rightarrow NH_3D^+ + D . \qquad (6.1)$$

Although the energy profile looks classically simple, the path structure is not. Notice that the molecular geometry has C_{3v} symmetry at the saddle point. However, the geometry at the minimum in the entrance valley has C_{2v} symmetry. As we discussed in Section III.D, this symmetry breaking implies that the reaction path must bifurcate or branch as it falls from the saddle point to the reactant minimum [43]. The location of this branching point is indicated in the figure. We can see from Fig. 7 that either of the three hydrogens in the NH_3 group could rotate toward the H_2 group on the path from the saddle point to the reactant minimum. Thus this IRP trifurcates. In practice, numerical error in following the IRP for reaction (5.1) leads to "falling off" the ridge crest below the valley ridge inflection point (where the valley splits into three) so that one of the equivalent paths was followed. From the reactant minimum shown in Fig. 7 it is also clear that either hydrogen in the H_2 group could rotate toward the NH_3 group as one progresses up the reaction path; in one case we reach the saddle point shown, in the other we reach an equivalent saddle point where H_γ is closer to the NH_3 group. Thus we have yet another set of three equivalent reaction paths.

To pass from one equivalent C_{2v} minimum to another involves internal relative rotation of the NH_3 and H_2 moieties, which we can think of as motions partway along the reaction paths (to near the regions where the trifurcations occur). These are symmetry operations, not operations belonging to the C_{2v} point group, but operations of the relevant *molecular symmetry group* [190]. The molecular symmetry group is the group of all *feasible* symmetry operations: that is, all operations that convert a molecule from one equivalent configuration to another for which the conversion process can be carried out at the energy of interest. Since we are interested in chemical reactions that take place at energies near or above the saddle point energy, converting between these C_{2v} minima is clearly *feasible*. A PES that treats all six C_{2v} minima equivalently must be symmetric with respect to the operations of the molecular symmetry group. The PES is said to be an *invariant* of the group [52, 191]. The same can be said for a PES which contains all six equivalent reactions paths. Clearly, a realistic PES for reaction (6.1) should contain all six paths. However, the situation is actually a bit more complicated still. Since the reaction takes place at energies above the saddle point, the product configurations, $NH_3D^+ + D$, are energetically accessible. Hence the reverse reactions

$$NH_3D^+ + D \rightarrow NH_3^+ + D_2$$
$$\rightarrow NH_2D^+ + HD \tag{6.2}$$

are also energetically feasible. Within the Born–Oppenheimer approximation, the PES cannot distinguish between H and D atoms. Hence, according to (6.2), the symmetry operation that converts $NH_3^+ + D_2$ to $NH_2D^+ + HD$ is a feasible operation

$$NH_3^+ + D_2 \rightarrow NH_2D^+ + HD . \tag{6.3}$$

In this way we can see that for this $NH_3D_2^+$ system, all permutations of the five indistinguishable particles are feasible. The effective symmetry group that governs the reaction (6.1) is, in fact, the fundamental *complete nuclear permutation and inversion* (CNPI) group [190]. There are $5! = 120$ different permutations of the five indistinguishable particles. Any chiral geometry can be converted to its enantiomer by *inversion* of the whole molecule. Thus there are up to $2 \times 120 = 240$ reaction paths on the global PES, which all have the energy profile of Fig. 7. A single reaction path potential is clearly not enough: Not only does it not have the correct symmetry (totally symmetric with respect to the CNPI group), but that failing has a practical consequence—a single path-based PES does not allow both abstraction (6.1) and exchange (6.3) reactions to compete. Indeed, both reactions are observed in competition [192–196].

Another well-studied example of symmetry-breaking bifurcation of the reaction path was provided by Ruedenberg and co-workers. They examine the ring opening of cyclopropylidene to allene, in which the reaction path bifurcations not far from the maximum on the path and two mirror-image products result [67, 69, 70, 181].

It is useful to pose this shortcoming of a single reaction path PES as an opportunity: If we can modify a single path-based PES to have the correct symmetry, that modified PES would be more globally valid and would open up other channels for reaction, such as (6.3). If we take the trouble to do an *ab initio* quantum chemistry calculation at some geometry, the information acquired is also valid at all symmetrically equivalent geometries (up to 240 geometries in total for $NH_3D_2^+$, for example). If we can usefully employ the data at all these equivalent geometries, we might have "got something for nothing."

B. Different Mechanisms

Multiple reaction paths can coexist on a PES because there is more than one mechanism for the same reaction, or because there is more than one possible product [197]. As an example of the former, Fig. 8 presented two reaction paths for the dissociation of $HCOH^+$, as given by Eq. (3.15). One pathway involves an out-of-plane rotation from the most stable *trans* structure to a *cis* structure, followed by direct cleavage of the oxygen–hydrogen bond. The other pathway is a rearrangement–fragmentation involving the intermediacy of the formaldehyde radical cation (H_2CO^+). Experiments with mixed labeled species ($HCOD^+$, $DCOH^+$) indicate that dissociation of both hydrogens contribute to the reaction [54, 55]. We note that since there are two indistinguishable atoms in this molecule, each of these inequivalent pathways is one of four symmetrically equivalent paths. As an aside, we note that Fig. 8 provides a simple example of the doubling of paths due to inversion symmetry. The saddle point geometry (**2a** of Fig. 8) lies on a path where the hydroxyl hydrogen is above the plane of the CHO group; clearly, there is an equivalent path with the hydroxyl hydrogen below the plane. There are other similar examples of reactions with multiple mechanisms that have been investigated using *ab initio* quantum chemistry [198, 199].

C. Coordinates

In developing an *ab initio* PES for a chemical reaction in terms of a single reaction path potential, we first located the stationary points on the PES, then followed the reaction path by the steepest descent or distinguished coordinate methods, then chose coordinates, natural collision coordinates or internals, and finally expressed the PES as a series expansion about the reaction path (using interpolation over a discrete set of data). When multiple paths are present, we can surely (with patience) find all the stationary points and map out the paths. However, choosing suitable coordinates and using truncated series expansions for the PES pose problems.

Natural collision coordinates are clearly hopeless. They are based firmly on one path (which one to choose?) and will probably be undefined, due to path curvature, at large distances from the defining path, where other important paths may reside. Moreover, a Taylor expansion in Cartesian (or MWC) displacements from one path will not give us a PES that is properly symmetric with respect to the CNPI group or even symmetric under rotation and inversion of the molecule.

We can ensure that a PES is invariant to rotation and inversion of a molecule by using internal or valence coordinates. However, a path-based PES is not exempt from the general difficulties experienced in using internal coordinates. When $N > 4$ we have to face the problem of choosing coordinates from a set of redundant bond lengths or other, potentially ill-defined, valence coordinates [191, 200].

This difficulty has two faces. First, valence coordinates such as bond angles and dihedral angles can be undefined or numerically illbehaved for some molecular geometries. For example, a set of coordinates for $HCOH^+$ which contains the torsion angle τ_{HCOH} may be complete near configuration **1** of Fig. 8 but will not be complete near configuration **5** since τ_{HCOH} is undefined when the HCO^+ group is linear. Moreover, τ_{HCOH} will vary alarmingly with the atomic positions when the HCO^+ group is *nearly* linear. Also, the Cartesian derivatives of a bond angle $\angle ABC$ become undefined numerically when $\angle ABC = 180°$. Decius [201] showed how one can always find at least one set of $3N-6$ valence coordinates that are well defined at some geometry. However, using valence coordinates means changing our choice of coordinates from place to place in the configuration space. This is not necessary if one is just interested in molecular vibrations about some relatively stable equilibrium geometry but is almost invariably necessary when one studies chemical reactions where very large changes in the molecular geometry are unavoidable. So unless we are willing, and able, to go through a continual process of changing variables during a dynamical calculation, bond angle and torsion (dihedral) angle valence coordinates should be abandoned in a general approach. Second, we could try using only bond lengths (interatomic distances) as coordinates. Here we run up against the problem of redundant coordinates. There are $N(N-1)/2$ interatomic distances and only $3N-6$ can be algebraically independent. There are too many bond lengths when $N > 4$. Moreover, any one set of $3N-6$ bond

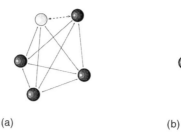

(a)   (b)

Figure 13. Two distinct geometries of a five-atom molecule, which have the same values for nine bond lengths. A redundant coordinate is the length of the tenth bond, denoted by the dashed line (see Ref. 200).

lengths does not globally determine the geometry [200]. By way of illustration, Fig. 13 depicts two configurations of a molecule containing five atoms. The nine bond lengths represented by solid lines in Fig. 13 give a locally complete description of the molecular geometry at both these configurations but fail to distinguish between the two. Hence any PES expressed in terms of these nine bond lengths would have the same value at these two obviously inequivalent configurations. The tenth bond length (dashed line) differentiates the two configurations. However, using this bond length in place of one of the nine clearly would only move the difficulty to another configuration. Generally, then, all but $3N - 6$ atom–atom distances (or linear combinations of same) are redundant as coordinates, but the choice of which $3N - 6$ depends on the region of configuration space considered. Again, one would be faced with the task of switching between different choices of coordinates. Similarly, one can show that a potential which is an *analytic* function of the Cartesian atomic coordinates must in general be a function of all the bond lengths [200]. The "redundant" coordinates are not redundant at all, in the sense that each of the bond lengths must be employed to describe the PES in some region of configuration space. These problems have also been recognized in the context of evaluating cubic and higher-order expansions of the PES in internal coordinates about an equilibrium geometry [84].

These problems with internal (valence) coordinates are generic. For path-based descriptions of the PES, we have in addition to consider how the PES can be expanded (say, in a power series) about one reaction path and still give an accurate description at all relevant reaction paths; or, as seems more sensible, if the PES is expanded locally about all reaction paths on the PES, how those expansions can be matched up to give a single well-defined PES, suitable for dynamics.

VII. GLOBAL SURFACES FROM REACTION PATH POTENTIALS

A. Surfaces from the Theory of Invariants

This group and others have reported some success in overcoming some of the difficulties associated with CNPI and rotation symmetry for PES which are based on reaction paths, using the theory of symmetry invariants [52, 57, 200, 202]. This work has been reviewed only recently [191], so we do not cover that ground again here. We simply state that the group theory of invariants can be applied practically to construct symmetry-invariant PES which describe all the symmetry equivalent paths for a single mechanism such as route A of Fig. 8. Such surfaces can be constructed from the type of *ab initio* data already used in simple reaction path potentials (energies, gradients, and force constant matrices at discrete points on the reaction path), plus the same type of data on related paths (defined by symmetry considerations).

However, the theory of symmetry invariants also strikes the redundant coordinate problem when $N > 4$. As an example of the problems encountered, the reaction of Eq. (3.14) requires all 15 atom–atom distances to form a representation of the CNPI group; no subset of $3N - 6 = 12$ forms a set of irreducible representations. The theory of invariants, as applied to the symmetry of the PES, begins with coordinates that form a set of irreducible representations [191]. Thus we cannot even begin to discuss the symmetry of the PES in terms of as few as $3N - 6$ atom–atom distances. In addition there is no known way to improve such a PES to arbitrary accuracy, and this approach cannot deal with the existence of multiple reaction paths which are not related by symmetry. A more easily applied method, which can deal with all these difficulties, has recently been developed by our group, and we concentrate here on this interpolation approach [203, 204].

B. Surfaces by Interpolation

1. Interpolation Method

Let's go back and consider again the picture of a reaction path, as shown in Fig. 14. Again we represent the molecular geometry in terms of a vector $\mathbf{R}$ of $3N - 6$ internal coordinates, and let $\mathbf{R}(n)$ represent the value of the coordinates at the nth point along the reaction path. Let's assume for the moment that the internal coordinates are, in fact, bond lengths

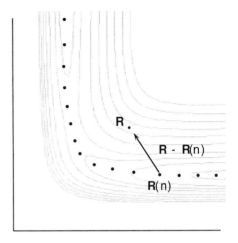

Figure 14. Discrete points on a reaction path, shown on a contour diagram of the PES of Fig. 1. The figure indicates how the PES at any point off the path, **R**, can be given by a Taylor expansion about any point on the path, **R**(n), as in Eq. (7.1).

(this is acceptable for $N \leq 4$). Near any of the points shown on the reaction path, we can expand the PES as a Taylor series:

$$V(\mathbf{R}; n) = V[\mathbf{R}(n)] + [\mathbf{R} - \mathbf{R}(n)]^T \frac{\partial V}{\partial \mathbf{R}} \bigg|_{\mathbf{R} = \mathbf{R}(n)}$$

$$+ \frac{1}{2} [\mathbf{R} - \mathbf{R}(n)]^T \mathbf{F}|_{\mathbf{R} = \mathbf{R}(n)} [\mathbf{R} - \mathbf{R}(n)] + \cdots. \tag{7.1}$$

Natural collision coordinates were introduced so that we would know on which point on the path to base the Taylor expansion. Why? Because we expect that the point on the path closest to **R** will probably give the most accurate truncated Taylor expansion, while points farther away would result in less accuracy. But we do not need to base the expansion solely on the closest point. We could use some or all of the points if we weight their contributions to favor the closest points. For example, if we have *ab initio* calculations at N_{data} points on the reaction path, we can put [203]

$$V(\mathbf{R}) = \sum_{n=1}^{N_{\text{data}}} w_n(\mathbf{R}) V(\mathbf{R}; n), \tag{7.2}$$

where $w_n(\mathbf{R})$ is a weight that approaches 1 as $\mathbf{R} \rightarrow \mathbf{R}(n)$. We take [203]

$$w_n(R) = \frac{v_n(R)}{\sum_{j=1}^{N_{\text{data}}} v_j(R)}, \tag{7.3}$$

$$v_n(R) = \left[\sum_{k=1}^{3N-6} [R_k - R_k(n)]^2 \right]^{-p} . \qquad (7.4)$$

Equation (7.3) ensures that the weights are normalized [so Eq. (7.2) is an average], and Eq. (7.4) ensures that $w_n(\mathbf{R}) \to 1$ as $|\mathbf{R} - \mathbf{R}(n)| \to 0$, and $w_n(\mathbf{R}) \to 0$ as $|\mathbf{R} - \mathbf{R}(n)| \to \infty$. Suppose that the Taylor expansions in Eq. (7.1) are truncated at order n_{Tay}. One can show that if the power p in Eq. (7.4) exceeds $n_{Tay}/2$, Eq. (7.2) is an *interpolation* formula [205] for the Taylor expansions in Eq. (7.1) up to order n_{Tay}. That is, as $|\mathbf{R} - \mathbf{R}(n)| \to 0$, $V(\mathbf{R})$ and all derivatives of $V(\mathbf{R})$ up to order n_{Tay} approach the values given at $\mathbf{R}(n)$. We note that Suhm has successfully applied a zeroth-order version of this interpolation scheme to quantum Monte Carlo calculations [206].

So Eq. (7.2) is a new type of reaction path potential, with some significant advantages. Natural collision coordinates are not involved, so the PES is well-defined independent of the curvature of the path. For "distinguished coordinate" paths, the PES of Eq. (4.14) depends critically on that coordinate; this is sometimes unreasonable since that may be just one of a number of coordinates that change along the path. Equation (7.4) places equal weight on all coordinates in determining which region of the reaction path dominates the PES for a particular geometry off the path. Most important, Eq. (7.2) opens the door to extending reaction path potentials to exact global PES.

2. Multiple Paths

First, we can easily incorporate the correct CNPI symmetry into this PES; all we have to do is include all the symmetry-equivalent geometries of the data points in the sum in Eq. (7.2). If g represents an element of the CNPI group for the molecule, we write

$$V(\mathbf{R}) = \sum_{g \in CNPI} \sum_{n=1}^{N_{data}} w_n[\mathbf{R} - g \circ \mathbf{R}(n)] V[\mathbf{R} - g \circ \mathbf{R}(n); n] , \qquad (7.5)$$

where the symmetry operators take effect on the data points [permuting the elements of $\mathbf{R}(n)$, $\partial V/\partial \mathbf{R}(n)$, and $\mathbf{F}(n)$]. Equation (7.5) is a reaction path potential based equally on all the symmetry-equivalent paths. A little thought makes it clear that swapping the coordinates of indistinguishable particles in $\mathbf{R}$ has no effect on $V(\mathbf{R})$ in Eq. (7.5). This means that the value and shape of the PES are the same at all three geometries in Fig. 15, for example. This is a valuable advance. Suppose that geometry **1** of Fig. 15 lies on the MEP for the reaction (5.4). Geometries

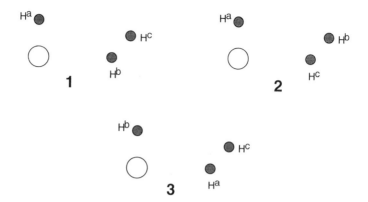

Figure 15. Three (of six possible) labeled versions of a geometry relevant to the $OH + H_2 \rightarrow H_2O + H$ reaction.

1 and **2** are both contained in the data set determining $V(\mathbf{R})$ in Eq. (7.5). A simple reaction path PES such as that of Eq. (4.9), near geometry **1**, would describe harmonic angle bending of the H_2 moiety relative to OH. However, the PES of Eq. (7.5), with geometries **1** and **2** both included in the data set, will describe a hindered rotational potential for the H_2 orientational motion No extra *ab initio* calculations were required to achieve this. Moreover, because geometries such as **3** are also contained in the data set, exchange reaction such as

$$OH + D_2 \rightarrow OD + HD \qquad (7.6)$$

may be included in the PES.

Second, the PES of Eqs. (7.2) and (7.5) can be extended to include reaction paths for other mechanisms and other reactions. For this example, the fragmentation reaction

$$OH + H_2 \rightarrow OH + H + H \qquad (7.7)$$

could be included if *ab initio* data for the associated reaction path were evaluated. Similarly, both the reaction mechanisms of Fig. 8 (and the permutationally equivalent paths) could be included in a reaction path potential like that of Eq. (7.5).

Thus we have a means of constructing a very general reaction path potential, one based on all the reaction paths on the PES. Before we explore the accuracy of such a PES, it is important to note that the PES of Eq. (7.5) can be made exact. It has been shown that if the power p in the weight function of Eq. (7.4) obeys $p > (3N - 6)/2$, Eq. (7.5) for $V(\mathbf{R})$ becomes exact as the density of data points increases [203, 205]. That is, the value of $V(\mathbf{R})$ given by Eq. (7.5) will be the same as that given by an *ab initio* calculation at the level of theory used for the calculations at the data points. This is important because in the spirit of *ab initio* quantum chemistry, one wishes to be able to improve the PES progressively to convergence, in the same way that one wishes to improve the accuracy of the *ab initio* calculations themselves. The practical question is: How many data points are required for the PES to be accurate enough? Or, how can we minimize the number of data points required?

3. Convergence

Let's look once again at Eq. (7.5). We introduced this form of the PES as a new type of reaction path potential, since we chose the data points to lie on the reaction path. As we will see below, choosing points on the reaction path is not only intuitively appealing but practically useful. However, there is no restriction in Eq. (7.5) as to where the data points lie. We could perform *ab initio* calculations of $V(\mathbf{R})$, $\partial V/\partial \mathbf{R}$, and $\mathbf{F}$ at any $\mathbf{R}$ and add this point to the data set [add another weighted Taylor expansion to Eq. (7.5)]. If we add enough data points to the set, Eq. (7.5) will converge to the exact result at this level of *ab initio* theory. How do we judge convergence, and at which $\mathbf{R}$ should we perform new calculations to achieve convergence for the least effort?

We take the view that a PES should be measured by the values of the *observables* which are determined by the PES. Since Eq. (7.5) must converge as $N_{data} \to \infty$, any calculated observable must also converge. So the PES is converged for some observable when the value of that observable stops changing as N_{data} increases. This means that the PES obtained is guaranteed to be "exact" only for that observable; other observables may require a larger N_{data} value to converge. However, the PES has clear physical significance. A convergence measure such as a root-mean-square deviation of $V(\mathbf{R})$ from energies calculated at selected points has no clear physical significance.

Where should additional data points be located? We have developed a selection procedure based on the following simple ideas: Data are needed only in the regions that are sampled by the dynamics; additional data points are needed primarily in regions where data points are absent or sparse; additional data points are needed in regions where the PES is in serious error. To put these simple ideas into effect, we construct the PES iteratively using classical trajectories [203, 204].

We start with a reaction path PES of the form of Eq. (7.5).

Trajectories are initiated in the reactant region, collisions are evaluated classically on the interpolated surface, and the molecular configuration is recorded periodically during the simulation. The N_t configurations recorded indicate the regions of the surface most often sampled by the trajectories, presumably regions of chemical importance. Each of the sampled trajectory points, $\mathbf{R}(j)$, $j = 1, \ldots, N_t$, are then allocated a weight, $h[\mathbf{R}(j)]$:

$$h[\mathbf{R}(j)] = \frac{1}{N_t - 1} \sum_{\substack{n=1 \\ n \neq j}}^{N_t} \frac{v_j[\mathbf{R}(n)]}{\sum_{k=1}^{N_{\text{data}}} v_k[\mathbf{R}(j)]} . \qquad (7.8)$$

Points $\mathbf{R}(j)$ which are close to large numbers of other trajectory points (i.e., points in a dynamically important region of the PES) and trajectory points far away from already existing data points (i.e., in regions of the surface where there are very few data) received the highest weights. An additional *ab initio* calculation of the energy and first and second derivatives are carried out at the point of highest weight, and the point is added to the data set. The procedure is iterated: New trajectories are calculated on the new interpolated surface and their configurations sampled periodically, and so on. This process is particularly efficient in correcting spurious "walls" in the PES, since classical trajectories preferentially sample classical turning points. Unfortunately this process does not correct spurious wells quickly because trajectories moving over a well experience a large increase in kinetic energy and move rapidly out of its vicinity (and are rarely sampled). Such wells represent a serious shortcoming in the iteration process and require prompt correction. The solution used is as follows [204]:

1. The original data set is analyzed and the lowest-energy configuration determined.

2. As the trajectory configurations are recorded, their potential energies on the interpolated surface are also calculated.

3. The trajectory configuration with the lowest potential energy (provided that this energy is lower than the previous minimum) is then added to the data set automatically.

4. The exact energies of the new data are then checked against the previous minimum, and if necessary a new minimum energy recorded.

So at each iteration, the trajectory point of highest weight, h, is added to the data set along with one (if any) trajectory point whose energy was unphysically low on the previous PES. After adding 20 or so data points, we calculate some hundreds or thousands of classical trajectories and evaluate the probability of reaction (an obvious observable to choose), then add another 20, and so on. The entire iteration process stops when the probability of reaction stops changing (to within some tolerance).

An important practical point needs to be made here. As the iteration procedure continues, the total number of data points in Eq. (7.5) may become very large, and the computational effort in evaluating the PES could grow substantially. However, because of the weight, w, there will only be a relatively small proportion of the data set influencing the potential in any given region of the PES, and the computational effort should only be proportional to this small number of data points. This is achieved by introducing a cutoff for the normalized weight function, Eq. (7.4) [203, 204]. Thus data points whose weights are below a given tolerance are assumed to have a negligible effect on the potential. This approach employs the molecular dynamics concept of a *neighbor list* (see, e.g., Ref. 207). The tolerance in the weight function is used to identify a neighbor list for each trajectory configuration at time zero. Trajectories are then allowed to evolve for a fixed time, during which the only data points contributing to the potential energy along each trajectory's path are those on its initial neighbor list. Each trajectory's neighbor list is recalculated from the entire data set after a fixed number of time steps and the trajectory then allowed to continue. Such a process necessarily introduces discontinuities to the PES as the identities of the neighbors change. These discontinuities are small, however, if the original tolerance used to define the neighbor list was small and the identity of the neighbors changes slowly. The tolerance may be chosen so that, independent of the size of the data set, only about 10–20 data points contribute to $V(\mathbf{R})$, at reasonable computational expense.

4. Test Case: $OH + H_2 \rightarrow H_2O + H$

Schatz and Elgersma [161] have published an analytic surface for the reaction of Eq. (5.4). This surface has a relatively simple reaction path, somewhat like that of Figs. 4 and 5. To test the interpolation procedure thoroughly at reasonable cost, we pretend that this analytic surface supplies us with the *ab initio* data, and we require that our converged interpolated surface will give us the same classical trajectory result for the probability of reaction as the analytic surface [204].

The initial conditions for the $OH + H_2$ trajectories were chosen from a microcanonical distribution for a range of appropriate impact parameters, with approximately five quanta of energy in the H_2 vibration, and a translational energy high enough to surmount the saddle point on the PES (and no rotational energy in H_2 or OH). These initial conditions ensure a conveniently high probability of reaction (about 0.4 on the Schatz and Elgersma surface). In a slight variation from Eq. (7.1), the Taylor series is actually evaluated in reciprocal bond lengths; in practice, this appears to extend the radius of accuracy of the series.

Figure 16 shows how the probability of reaction converges as we increase the number of points in the data set of Eq. (7.5). It is encouraging to note from Fig. 16 that the reaction probability calculated from the initial data set of 30 points along the MEP is only a factor of 2 smaller than the result from the Schatz and Elgersma surface, despite the fact that the initial conditions selected involve an energy well above the MEP. Moreover, initial test calculations showed that the obvious convergence in Fig. 16 could not be obtained so easily unless the initial PES contained an adequate number of points on the MEP. This lends further weight to the importance of the MEP potential and suggests that an expansion solely about the MEP may be useful when the total energies considered ensure that trajectories lie close to this path.

However, at these energies the trajectories do not lie close to the path. Figure 17a indicates the locations of 500 points that have been added to the data set by the iterative method. By comparison, Fig. 17b depicts the location of 500 geometries chosen at random in a reasonably confined region of the configuration space containing the reaction path [204]. Figure 16 also shows the probability of reaction for the interpolation

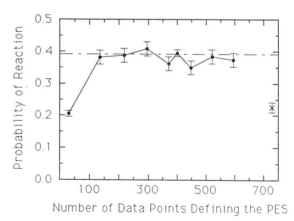

Figure 16. Reaction probability versus the number of data points defining the PES via Eq. (7.5) and the iteration procedure described in the text (filled circles), and via a data set composed of 729 randomly generated, energetically allowed configurations (cross). In all cases the value $p = 9$ was used in Eq. (7.4). The iteration procedure started with 30 points along the MEP. The reaction probability obtained from the surface of Schatz and Elgersma is shown as a dashed line and error bars represent standard deviations.

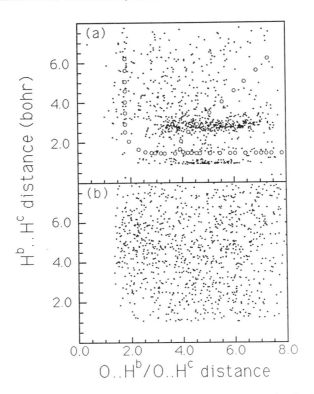

Figure 17. Two-dimensional projection of the configuration space for the OH$_3$ system onto a plane formed by two bond lengths. One bond length is the $H^b \cdots H^c$ distance, with labeling as in Fig. 15(1), the other bond is either of the $O \cdots H^b$ or $O \cdots H^c$ distances [the PES is invariant to permutation of H^b and H^c as labeled in Fig. 15(1)]. (*a*) The original data points along the MEP are represented in each frame by open circles; the small dots represent the first 500 configurations of Fig. 16 added to the data point set. (*b*) 500 randomly generated, energetically allowed configurations (in both labeled versions).

potential of Eq. (7.5) with data given by $3^6 = 729$ such randomly chosen points. It is clear that the selective, iterative choice of data points in Eq. (7.5), based initially on the reaction path, results in a very efficient and accurate potential.

Generally, we expect that a classical trajectory or quantum dynamical calculation (say, by discrete variable methods) of the reaction rate for a tetraatomic system would require at least 10^5–10^6 evaluations of the energy (or energy and gradient). This selective iterative interpolation procedure appears to produce an accurate PES with about *three orders of magnitude* less *ab initio* computation than that of the best direct method [208].

An important aspect of this automated "growing" of a PES is that the final surface is independent of our intuition. Formally, the surface is exact if we only iterate long enough. The PES is physically significant for a modest number of data because we start from data on the reaction path (definitely a use of intuition), but also because data are added where the dynamics demands (independent of intuition). This means that if other processes can occur at the energy considered, other reactions or mechanisms will automatically be discovered and included in the PES. For example, for the highest energy at which the $OH + H_2$ reaction was studied, the Schatz and Elgersma surface allows the fragmentation reaction of Eq. (7.7) to occur as a minor product. Although no reaction path for this process was initially included in the interpolated PES, the final PES does indeed give this reaction in an approximately correct yield.

It is also important to note that while the iterative interpolation procedure gives an intuition-free PES, it also gives a property-dependent PES. That is, the PES is converged for the energy range and observables studied in the iteration process. One would have to investigate whether the surface is converged for any other property of interest at some other energy. This is a simple enough matter. If, for example, the iteration process was terminated at $N_{data} = 400$, one can calculate other properties for a range of N_{data} values up to 400 to see if the property is converged (no further *ab initio* calculations are required). Figure 18 shows the final distribution of the angular momentum in reactive and inelastic collisions for the $OH + H_2$ PES. These properties have converged along with the total reaction probability.

However, not all observables will be converged. Figure 19 compares the probability of reaction for the $OH + H_2$ reaction calculated using the Schatz and Elgersma surface and the interpolated PES for various initial vibrational excitations of the H_2. We see that the reaction probability is less accurate at energies well away from that at which the PES was grown. Again, no effort has been lost here. To improve the PES, one would simply start with the already grown surface and begin the iteration process again, this time using trajectories with a range of initial H_2 vibrational energies. This suggests a general strategy: Choose the ensemble of trajectories with which the PES is grown to reflect the measurement that you wish to make.

5. Higher-Order Interpolated Surfaces

Jordan et al. [209] have investigated the convergence behavior of Eq. (7.5) when the order (n_{Tay}) at which the Taylor expansions are truncated is varied from 1 to 4. At the Hartree–Fock level of theory, analytic

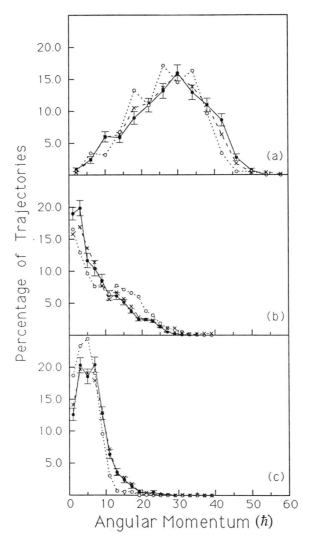

Figure 18. Angular momentum distributions of the (*a*) H_2O reaction product and the (*b*) OH and (*c*) H_2 inelastic collision products as the number of data points defining the interpolated surface is iteratively increased from 30 (open circles) to 400 (crosses) as in Fig. 16. The angular momentum distributions obtained from the surface of Schatz and Elgersma are also shown (filled circles) where the error bars represent standard deviations. The H_2O distributions were obtained using bin sizes of $4\hbar$, whereas the OH and H_2 distributions used bin sizes of $2\hbar$.

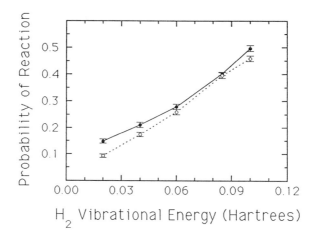

Figure 19. Reaction probability versus initial vibrational excitation of the H_2 molecule for the surface of Schatz and Elgersma (filled circles) and for the interpolated PES (open circles) generated from the first 400 data points of Fig. 16. Error bars represent standard deviations. The interpolated PES was generated for an H_2 initial vibrational excitation of about 0.085 hartree.

derivatives of the energy are available up to third and fourth order. Again, this analysis was performed for the $OH + H_2$ test case using the Schatz and Elgersma surface with Taylor expansions in reciprocal bond lengths. Not surprisingly, a third-order Taylor expansion results in more rapid convergence than that of Fig. 16, with only about 50–100 data points required. Delaying the truncation to fourth order provides a much less dramatic improvement in convergence. Given the additional computational cost even for analytic derivatives, expansions higher than second order do not appear to be particularly efficient. Truncating the expansion at first order leads to extremely slow convergence, so that second-order Taylor expansions are definitely necessary. The third- and fourth-order Taylor expansions do show much improved accuracy across the energy range of Fig. 19, as one might hope.

6. Surfaces for General Polyatomics

The interpolated PES, Eq. (7.5), is generally application to two-, three-, and four-atom systems only. This PES is a function of $3N - 6$ internal coordinates (here interatomic distances). When $N > 4$ we have to face the problem of choosing coordinates from a set of redundant bond lengths or other, potentially ill-defined valence coordinates, as discussed above.

So if internal coordinates are too much trouble, how do we progress to a general PES for many atoms? We cut this particular Gordian knot by using Cartesian coordinates. But a PES in Cartesian coordinates will not have the right symmetry unless we are very careful [200]. For a PES must be unchanged if we rotate or invert a molecule. Moreover, a molecular PES in any coordinates must be an *invariant* of the complete-nuclear-permutation group.

Luckily, our interpolation procedure makes it just as easy to include rotation symmetry as it does to include CNPI symmetry. In Eq. (7.5) we imposed the correct CNPI symmetry by summing over a data set that includes all the symmetry equivalent versions of the data point geometry. The same can be done for rotational symmetry [191, 210]:

$$V(\mathbf{X}) = \int_{SO(3)} d\Omega \sum_{g \in CNPI} \sum_{n=1}^{N_{data}} w_n[\mathbf{X} - \Omega \circ g \circ \mathbf{X}(n)] V[\mathbf{X} - \Omega \circ g \circ \mathbf{X}(n); n] .$$

(7.9)

In Eq. (7.9) the coordinates, $\mathbf{X}$, and all derivatives in the Taylor expansions refer to the $3N$ atomic Cartesian coordinates. The integral over $SO(3)$ (the formal name for the group of all proper rotations in three-dimensional space) is an integral over all possible orientations of the data geometries (with rotation parameterized by the Euler angles, say). We do no coordinate transformations, just substitute the Cartesian coordinates and derivatives and derivatives straight from the *ab initio* programs into Eq. (7.9).

Suppose that we perform an *ab initio* calculation for a particular geometry with the molecule in the standard Z-matrix orientation. The energy, Cartesian geometry, gradients, and force constant matrix go into the data set in Eq. (7.9), and because of the integral over $SO(3)$, so do all the geometries that are just reorientations of the molecule, with reoriented gradients and reoriented force constant matrices. The weight function is that of Eqs. (7.3) and (7.4) with $\mathbf{X}$ in place of $\mathbf{R}$. The data point $\Omega \circ g \circ \mathbf{X}(n)$ with highest weight, w, for some $\mathbf{X}$ is the permuted, inverted, rotated geometry which is "closest" to $\mathbf{X}$ in the $3N$-dimensional configuration space. Although we evaluate the PES of Eq. (7.9) by supplying the Cartesian coordinates $\mathbf{X}$, the result is independent of the orientation of the molecule as well as being independent of which enantiomer or which labeled version we consider.

It is an important practical point to note that we can reorder the sums and integrals in Eq. (7.9) to

$$V(\mathbf{X}) = \sum_{g \in CNPI} \sum_{n=1}^{N_{data}} \int_{SO(3)} d\Omega \, w_n[\mathbf{X} - \Omega \circ g \circ \mathbf{X}(n)] V[\mathbf{X} - \Omega \circ g \circ \mathbf{X}(n); n] .$$

(7.10)

The integration over SO(3) can actually be performed analytically for the inverse power type of weight function. Hence Eq. (7.10) reduces to a computational task of comparable (although larger) magnitude to that required by Eq. (7.5). Implementation of this general Cartesian PES is currently in progress [210].

VIII. CONCLUDING REMARKS

At present, reaction path methods represent the best approach for utilizing *ab initio* electronic structure theory *directly* in chemical reaction dynamics. To study reaction dynamics we need to evaluate accurately the Born–Oppenheimer molecular potential energy surface. Our experience suggests that chemical reaction may take place within in a restricted range of molecular configurations (i.e., there is a defined mechanism for the reaction). Hence we may not need to know the PES everywhere. Reaction path methods provide a means of evaluating the PES for the most relevant molecular geometries and in a form that we can use directly in dynamical calculations.

Analytic gradient and second derivative methods in quantum chemistry provide the type of information about the PES which we need to implement a reaction path approach. Here the major shortcoming is the lack of analytic (relatively inexpensive) gradient and second derivative formulas for highly correlated levels of theory.

Reaction path methods have great promise for future progress because they can be made direct or automated. Direct *ab initio* dynamics is taken to mean the *ab initio* evaluation of the molecular energy whenever it is required in a dynamical calculation. At present this is prohibitively expensive. A more traditional approach uses electronic structure methods to provide information about a PES which is then fitted to a functional form. The dynamics employed thereafter is restricted only by the limitations of current dynamical theories. However, the process of fitting functional forms to a molecular PES is difficult, unsystematic, and extremely time consuming. The reaction path approaches that use interpolation of the reaction path data are affordable methods aimed toward direct dynamics, as they avoid the process of fitting a functional form for the PES. Such methods have been automated (programmed) and are therefore readily usable. Thus reaction path methods have been applied with various levels of *ab initio* theory to statistical theories of the reaction rate, to approximate quantum dynamics, and to classical trajectory studies of reactions.

We have seen that the simple reaction path approach has serious limitations; basically, we cannot apply it reliably to any but the simplest mechanisms for chemical reaction. However, the recent interpolation method, which arose from the reaction path approach, promises to overcome these limitations and provide further advances in direct dynamics. Most important, this procedure, which starts from a reaction path potential, automatically improves the accuracy of the PES to convergence. This certainly interfaces the *ab initio* spirit of systematic convergence with the task of evaluating a usable molecular energy surface. So far, the method has only been used in classical trajectory simulations. The method is reasonably general, however, and should be application to quantum calculations and to statistical theories for the chemical reaction rate.

ACKNOWLEDGMENTS

I would like to thank Peter Taylor, and particularly Meredith Jordan and Keiran Thompson, for very helpful discussions during the preparation of this manuscript.

REFERENCES

1. J. N. Murrell, S. Carter, S. C. Farantos, P. Huxley, and A. J. C. Varandas, *Molecular Potential Energy Functions*, Wiley, Chichester, West Sussex, England, 1984.

2. D. G. Truhlar, Ed., *Potential Energy Surfaces and Dynamics Calculations*, Plenum Press, New York, 1981.

3. P. Jørgensen and J. Simons, Eds., *Geometrical Derivatives of Energy Surfaces and Molecular Properties*, D. Reidel, Dordecht, The Netherlands, 1986.

4. J. F. Gaw, Y. Yamaguchi, and H. F. Schaefer III, *J. Chem. Phys.* **81**, 6395 (1984).

5. J. F. Gaw, Y. Yamaguchi, H. F. Schaefer III, and N. C. Handy, *J. Chem. Phys.* **85**, 5132 (1986).

6. J. F. Gaw, Y. Yamaguchi, R. B. Remington, Y. Osamura, and H. F. Schaefer III, *Chem. Phys.* **109**, 237 (1986).

7. P. Pulay, in *Ab Initio Methods in Quantum Chemistry II*, K. P. Lawley, Ed., Wiley, New York, 1987, p. 241.

8. P. E. Maslen, N. C. Handy, R. D. Amos, and D. Jayatilaka, *J. Chem. Phys.* **97**, 4233 (1992).

9. P. R. Taylor, in *Handbook of Computational Science and Engineering*, CRC Press, Boca Raton, Fla., in preparation.

10. E. Kraka and J. T. H. Dunning, *Adv. Mol. Electron. Struct. Theory* **1**, 129 (1990).

11. D. G. Truhlar and M. S. Gordon, *Science* **249**, 491 (1990).

12. M. Page, *Comput. Phys. Commun.* **84**, 115 (1994).

13. G. C. Schatz, A. F. Wagner, S. P. Walch, and J. M. Bowman, *J. Chem. Phys.* **74**, 4984 (1981).

14. R. J. Duchovic, W. L. Hase, and H. B. Schlegel, *J. Phys. Chem.* **88**, 1339 (1984).

15. D. W. Schwenke, S. C. Tucker, R. Steckler, F. B. Brown, G. C. Lynch, D. G. Truhlar, and B. C. Garrett, *J. Chem. Phys.* **90**, 3110 (1989).

16. L. B. Harding, *Adv. Mol. Electron. Struct. Theory* **1**, 45 (1990).

17. G. C. Schatz, *Adv. Mol. Electron. Struct. Theory*, **1**, 85 (1990).

18. A. D. Isaacson, *J. Phys. Chem.* **96**, 531 (1992).

19. S. Glasstone, K. J. Laidler, and H. Eyring, *The Theory of Rate Processes*, McGraw-Hill, New York, 1941, Chap. 3.

20. J. C. Polanyi and W. H. Wong, *J. Chem. Phys.* **51**, 1439 (1969).

21. M. H. Mok and J. C. Polanyi, *J. Chem. Phys.* **51**, 1451 (1969).

22. J. C. Polanyi, *Acc. Chem. Res.* **5**, 161 (1972).

23. C. J. Cerjan and W. H. Miller, *J. Chem. Phys.* **75**, 2800 (1981).

24. H. B. Schlegel, *Adv. Chem. Phys.* **67**, 249 (1987).

25. H. B. Schlegel, in *Modern Electronic Structure Theory*, D. R. Yarkony, Ed., World Scientific, Singapore, 1994.

26. J. Nichols, H. Taylor, P. Schmidt, and J. Simons, *J. Chem. Phys.* **92**, 340 (1990).

27. P. Pulay and G. Fogarasi, *J. Chem. Phys.* **96**, 2856 (1992).

28. K. Ishida, K. Morokuma, and A. Komornicki, *J. Chem. Phys.* **66**, 2153 (1977).

29. P. Russegger and J. R. Huber, *Chem. Phys.* **89**, 33 (1984).

30. C. Gonzalez and H. B. Schlegel, *J. Chem. Phys.* **90**, 2154 (1989).

31. C. Gonzales and H. B. Schlegel, *J. Phys. Chem.* **94**, 5523 (1990).

32. C. Gonzalez and H. B. Schlegel, *J. Chem. Phys.* **95**, 5853 (1991).

33. I. Shavitt, *J. Chem. Phys.* **49**, 4048 (1968).

34. G. L. Hofacker, *Z. Naturforsch. A* **18**, 607 (1963).

35. R. A. Marcus, *J. Chem. Phys.* **45**, 4493 (1966).

36. R. A. Marcus, *J. Chem. Phys.* **49**, 2610 (1968).

37. D. G. Truhlar and A. Kupperman, *J. Am. Chem. Soc.* **93**, 1840 (1971).

38. K. Fukui, in *The World of Quantum Chemistry*, R. Daudel and B. Pullman, Eds., D. Reidel, Dordrecht, The Netherlands, 1974, p. 113.

39. K. Fukui, *Acc. Chem. Res.* **14**, 363 (1981).

40. B. A. Ruf and W. H. Miller, *J. Chem. Soc. Faraday Trans. 2* **84**, 1523 (1988).

41. J. Ischtwan and M. A. Collins, *J. Chem. Phys.* **89**, 2881 (1988).

42. B. C. Garrett, M. J. Redmon, R. Steckler, D. G. Truhlar, K. K. Baldridge, D. Bartol, M. W. Schmidt, and M. S. Gordon, *J. Phys. Chem.* **92**, 1476 (1988).

43. H. B. Schlegel, *J. Chem. Soc. Faraday Trans.* **90**, 1569 (1994).

44. M. Page and J. W. McIver, *J. Chem. Phys.* **88**, 922 (1988).

45. M. Page, C. Doubleday, and J. W. McIver, *J. Chem. Phys.* **93**, 5634 (1990).

46. General discussion, *J. Chem. Soc. Faraday Trans.* **90**, 1607 (1994).

47. R. Elber and M. Karplus, *Chem. Phys. Lett.* **139**, 375 (1987).

48. S. S.-L. Chiu, J. J. W. McDoull, and I. H. Hiller, *J. Chem. Soc. Faraday Trans.* **90**, 1575 (1994).

49. V. S. Melissas, D. G. Truhlar, and B. C. Garrett, *J. Chem. Phys.* **96**, 5758 (1992).

50. J.-Q. Sun and K. Ruedenberg, *J. Chem. Phys.* **99**, 5257; 5269; 5276 (1993).

51. I. Ischtwan, P. Schwerdtfeger, S. D. Peyerimhoff, M. A. Collins, T. Helgakar, P. Jørgensen, and H. J. A. Jensen, *Theor. Chim. Acta* **89**, 157 (1994).

52. J. Ischtwan and M. A. Collins, *J. Chem. Phys.* **94**, 7084 (1991).

53. J. Ischtwan, B. J. Smith, M. A. Collins, and L. Radom, *J. Chem. Phys.* **97**, 1191 (1992).

54. W. J. Bouma, P. C. Burgers, J. L. Holmes, and L. Radom, *J. Am. Chem. Soc.* **108**, 1767 (1986).

55. P. C. Burgers, A. A. Mommers, and J. L. Holmes, *J. Am. Chem. Soc.* **105**, 5976 (1983).

56. N. L. Ma and M. A. Collins, *J. Chem. Phys.* **97**, 4913 (1992).

57. N. L. Ma, L. Radom, and M. A. Collins, *J. Chem. Phys.* **96**, 1093; **97**, 1612 (1992).

58. A. K. Chandra and V. S. Rao, *Intern. J. Quantum Chem.* **47**, 437 (1993).

59. S. J. Klippenstein and Y.-W. Kim, *J. Chem. Phys.* **99**, 5790 (1993).

60. Z. Peng and K. M. Merz, Jr., *J. Am. Chem. Soc.* **115**, 9640 (1993).

61. B. D. Wladkowski, A. L. L. East, J. E. Mihalick, W. D. Allen, and J. I. Brauman, *J. Chem. Phys.* **100**, 2058 (1993).

62. N. A. Burton, S. S.-L. Chiu, M. M. Davidson, D. V. S. Green, I. A. Hiller, J. J. W. McDouall, and M. A. Vincent, *J. Chem. Soc. Faraday Trans.* **89**, 2631 (1993).

63. W. H. Miller, B. A. Ruf, and Y.-T. Chang, *J. Chem. Phys.* **89**, 6298 (1988).

64. E. Bosch, M. Moreno, J. M. Lluch, and J. Betran, *J. Chem. Phys.* **93**, 5685 (1990).

65. D. K. Hoffman, R. S. Nord, and K. Ruedenberg, *Theor. Chim. Acta* **69**, 265 (1986).

66. H. B. Schlegel, *Theor. Chim. Acta* **83**, 15 (1992).

67. P. Valtazanos and K. Ruedenberg, *Theor. Chim. Acta* **69**, 281 (1986).

68. W. Quapp, *Theor. Chim. Acta* **75**, 447 (1989).

69. P. Valtazanos, S. T. Elbert, and K. Ruedenberg, *J. Am. Chem. Soc.* **108**, 3147 (1986).

70. P. Valtazanos, S. T. Elbert, S. Xantheas, and K. Ruedenberg, *Theor. Chim. Acta* **78**, 287 (1991).

71. J. Baker and P. M. W. Gill, *J. Comput. Chem.* **9**, 465 (1988).

72. S. F. Fischer, G. L. Hofacker, and R. Seiler, *J. Chem. Phys.* **51**, 3951 (1969).

73. N. H. Hijazi and K. J. Laidler, *J. Chem. Phys.* **58**, 349 (1972).

74. R. A. Marcus, *J. Chem. Phys.* **45**, 4500 (1966).

75. W. H. Miller, N. C. Handy, and J. E. Adams, *J. Chem. Phys.* **72**, 99 (1980).

76. S. K. Gray, W. H. Miller, Y. Yamaguchi, and H. F. Schaefer III, *J. Chem. Phys.* **73**, 2733 (1980).

77. W. H. Miller and S. Shi, *J. Chem. Phys.* **75**, 2258 (1981).

78. W. H. Miller and S. Schwartz, *J. Chem. Phys.* **77**, 2378 (1982).

79. N. Makri, *Chem. Phys. Lett.* **169**, 541 (1990).

80. N. Rom, V. Ryaboy, and N. Moiseyev, *Chem. Phys. Lett.* **204**, 175 (1993).

81. J. T. Hougen, P. R. Bunker, and J. W. C. Johns, *J. Mol. Spectrosc.* **34**, 136 (1970).

82. W. H. Miller, *J. Phys. Chem.* **87**, 3811 (1983).

83. W. H. Miller, in *Potential Energy Surfaces and Dynamics Calculations*, D. Truhlar, Ed., Plenum Press, New York, 1981, p. 265.

84. A. R. Hoy, I. M. Mills, and G. Strey, *Mol. Phys.* **24**, 1265 (1972).

85. G. A. Natanson, *Chem. Phys. Lett.* **178**, 49 (1991).

86. M. V. Basilevsky, *J. Mol. Struct.* **103**, 139 (1983).

87. T. Carrington, Jr., and W. H. Miller, *J. Chem. Phys.* **81**, 3942 (1984).

88. T. Carrington, Jr., and W. H. Miller, *J. Chem. Phys.* **84**, 4364 (1986).

89. G. D. Billing, *Chem. Phys.* **146**, 63 (1990).

90. N. Shida, P. F. Barbara, and J. Almlöf, *J. Chem. Phys.* **94**, 3633 (1991).

91. D. G. Truhlar, *Comput. Phys. Commun.* **84**, 78 (1994).

92. D. C. Clary, *J. Phys. Chem.* **98**, 10678 (1994).

93. G. D. Billing, *Chem. Phys.* **135**, 423 (1989).

94. W. H. Miller and F. T. Smith, *Phys. Rev. A* **17**, 939 (1978).

95. C. J. Cerjan, S. Shi, and W. H. Miller, *J. Phys. Chem.* **86**, 2244 (1982).

96. T. Carrington Jr., L. M. Hubbard, H. F. Schaefer III, and W. H. Miller, *J. Chem. Phys.* **80**, 4347 (1984).

97. M. Quack and J. Troe, *Ber. Bunsenges, Phys. Chem.* **78**, 240 (1974).

98. S. D. Schwartz and W. H. Miller, *J. Chem. Phys.* **79**, 3759 (1983).

99. D. G. Truhlar, Y.-P. Liu, G. K. Schenter, and B. C. Garrett, *J. Phys. Chem.* **98**, 8396 (1994).

100. D. G. Truhlar and A. D. Isaacson, *J. Chem. Phys.* **77**, 3516 (1982).

101. T. N. Truong, *J. Chem. Phys.* **102**, 5335 (1994).

102. S. M. Colwell and N. C. Handy, *J. Chem. Phys.* **82**, 1281 (1985).

103. C. Zuhrt, L. Zulicke, and X. Chapuisat, *Chem. Phys.* **166**, 1 (1992).

104. J. F. Blake, S. G. Wierschke, and W. L. Jorgensen, *J. Am. Chem. Soc.* **111**, 1919 (1989).

105. J. A. Boatz and M. S. Gordon, *J. Phys. Chem.* **93**, 5774 (1989)

106. J. C. Corchado, J. Espinosa-Garcia, W.-P. Hu, I. Rossi, and D. G. Truhlar, *J. Phys. Chem.* **99**, 687 (1994).

107. M. R. Soso, M. Page, and M. L. McKee, *Chem. Phys.* **153**, 415 (1991).

108. M. R. Soto and M. Page, *J. Chem. Phys.* **97**, 7287 (1992).

109. B. Hartke and J. Manz, *J. Am. Chem. Soc.* **110**, 3063 (1988).

110. Y. Liu, D. Lu, A. G. Lafont, D. G. Truhlar, and B. C. Garrett, *J. Am. Chem. Soc.* **115**, 7806 (1993).

111. R. A. Marcus and M. E. Coltrin, *J. Chem. Phys.* **67**, 2609 (1977).

112. A. Gonzales-Lafont, T. N. Truong, and D. G. Truhlar, *J. Chem. Phys.* **95**, 8875 (1991).

113. G. K. Schenter, R. P. McRae, and B. C. Garrett, *J. Chem. Phys.* **97**, 9116 (1992).

114. R. T. Skodje, D. G. Truhlar, and B. C. Garrett, *J. Phys. Chem.* **85**, 3019 (1981).

115. R. T. Skodje, D. G. Truhlar, and B. C. Garrett, *J. Chem. Phys.* **77**, 5955 (1982).

116. D. G. Truhlar and B. C. Garrett, *Acc. Chem. Res.* **13**, 440 (1980).

117. D. G. Truhlar and B. C. Garrett, *Ann. Rev. Phys. Chem.* **35**, 159 (1984).

118. D. G. Truhlar, A. D. Isaacson, and B. C. Garrett, in *Theory of Chemical Reaction Dynamics*, M. Baer, Ed., CRC Press, Boca Raton, Fla., 1985, p. 65.

119. D. G. Truhlar, F. B. Brown, D. W. Schwenke, R. Steckler, and B. C. Garrett, in *Comparison of Ab Initio Quantum Chemistry with Experiment for Small Molecules*, R. J. Bartlett, Ed., D. Reidel, Dordrecht, The Netherlands, 1985, p. 95.

120. S. C. Tucker and D. G. Truhlar, in *New Theoretical Methods for Understanding Organic Reactions*, J. Bertran and I. G. Csizmadia, Eds., Kluwer, Dordrecht, The Netherlands, 1989, p. 291.

121. D. M. Wardlaw and R. A. Marcus, *Adv. Chem. Phys.* **70**, 231 (1991).

122. P. Pechukas and F. J. McLafferty, *J. Chem. Phys.* **58**, 1622 (1973).

123. J. C. Keck, *Adv. Chem. Phys.* **13**, 85 (1967), and references therein.

124. D. Lu, T. N. Truong, V. S. Melissas, G. C. Lynch, Y.-P. Liu, B. C. Garrett, R. Steckler, A. D. Isaacson, S. N. Rai, G. C. Hancock, J. G. Lauderdale, T. Joseph, and D. G. Truhlar, *Comput. Phys. Commun.* **71**, 235 (1992).

125. B. C. Garrett and D. G. Truhlar, *J. Phys. Chem.* **83**, 1052, 3058E (1979).

126. R. G. Gilbert and S. C. Smith, in *Theory of Unimolecular and Recombination Reactions*, J. P. Simons, Ed., Blackwell, Oxford, 1990, p. 149.

127. C. Doubleday, Jr., R. N. Camp, H. F. King, J. W. McIver, D. Mullally, and M. Page, *J. Am. Chem. Soc.* **106**, 447 (1984).

128. B. C. Garrett and D. G. Truhlar, *J. Phys. Chem.* **84**, 805 (1980).

129. B. C. Garrett and D. G. Truhlar, *Proc. Natl. Acad. Sci. USA* **76**, 4755 (1979).

130. A. D. Isaacson and D. G. Truhlar, *J. Chem. Phys.* **76**, 1380 (1982).

131. D. G. Truhlar, R. S. Grev, and B. C. Garrett, *J. Phys. Chem.* **87**, 3415 (1983).

132. B. C. Garrett, D. G. Truhlar, R. S. Grev, and A. W. Magnuson, *J. Phys. Chem.* **84**, 1730 (1980).

133. B. C. Garrett and D. G. Truhlar, *J. Chem. Phys.* **72**, 3460 (1980).

134. S. R. Rai and D. G. Truhlar, *J. Chem. Phys.* **79**, 6046 (1983).

135. C. Doublday, J. McIver, M. Page, and T. Zielinski, *J. Am. Chem. Soc.* **107**, 5800 (1985).

136. B. C. Garrett and D. G. Truhlar, *Intern. J. Quantum Chem.* **29**, 1463 (1986).

137. C. Doubleday, Jr., J. W. McIver, and M. Page, *J. Phys. Chem.* **92**, 4367 (1988).

138. N. J. Caldwell, J. K. Rice, H. H. Nelson, G. F. Adams, and M. Page, *J. Chem. Phys.* **93**, 479 (1990).

139. V. S. Melissas and D. G. Truhlar, *J. Chem. Phys.* **99**, 1013 (1993).

140. M. Page and M. R. Soto, *J. Chem. Phys.* **99**, 7709 (1993).

141. V. S. Melissas and D. G. Truhlar, *J. Chem. Phys.* **99**, 3542 (1993).

142. V. S. Melissas and D. G. Truhlar, *J. Chem. Phys.* **98**, 875 (1994).

143. S. K. Gray, W. H. Miller, Y. Yamaguchi, and H. F. Schaefer III, *J. Am. Chem. Soc.* **103**, 1900 (1981).

144. B. C. Garrett, D. G. Truhlar, A. F. Wagner, and T. H. Dunning, Jr., *J. Chem. Phys.* **78**, 4400 (1983).

145. B. C. Garrett, D. G. Truhlar, and G. C. Schatz, *J. Am. Chem. Soc.* **108**, 2876 (1986).

146. D. G. Truhlar and B. C. Garrett, *J. Am. Chem. Soc.* **111**, 1232 (1989).

147. K. K. Baldridge, M. S. Gordon, R. Steckler, and D. G. Truhlar, *J. Phys. Chem.* **93**, 5107 (1989).

148. B. C. Garrett, M. L. Koszykowski, C. F. Melius, and M. Page, *J. Phys. Chem.* **94**, 7096 (1990).

149. T. N. Truong and D. G. Truhlar, *J. Chem. Phys.* **93**, 1761 (1990).

150. D. G. Truhlar, D. Lu, S. C. Tucker, X. G. Zhao, A. Gonzalez-Lafont, T. N. Truong, D. Maurice, Y.-P. Liu, and G. C. Lynch, *ACS Symp. Ser.* **502**, 16 (1992).

151. A. D. Isaacson, L. Wang, and S. Scheiner, *J. Chem. Phys.* **97**, 1765 (1993).

152. R. L. Bell and T. N. Truong, *J. Chem. Phys.* **101**, 10441 (1994).

153. S. M. Colwell, *Theor. Chim. Acta* **74**, 123 (1988).

154. Y.-P. Liu, G. C. Lynch, T. N. Truong, D. Lu, D. G. Truhlar, and B. C. Garrett, *J. Am. Chem. Soc.* **115**, 2408 (1993).

155. T. N. Truong, D.-H. Lu, G. C. Lynch, Y.-P. Liu, V. Melissas, J. J. P. Stewart, R. Steckler, B. C. Garrett, A. D. Isaacson, A. Gonzalez-Lafont, S. N. Rai, G. C. Hancock, J. G. Lauderdale, T. Joseph, and D. G. Truhlar, *Comput. Phys. Commun.* **75**, 143 (1993).

156. G. C. Schatz, *Rev. Mod. Phys.* **61**, 669 (1989).

157. D. G. Truhlar, R. Steckler, and M. S. Gordon, *Chem. Rev.* **87**, 217 (1987).

158. A. J. C. Varandas, *Adv. Chem. Phys.* **74**, 255 (1988).

159. X. Hu and W. L. Hase, *J. Phys. Chem.* **93**, 6029 (1989).

160. X. Hu and W. L. Hase, *J. Chem. Phys.* **96**, 7535 (1992).

161. G. C. Schatz and H. Elgersma, *Chem. Phys. Lett.* **73**, 21 (1980).

162. S. P. Walch and T. H. Dunning Jr., *J. Chem. Phys.* **72**, 1303 (1980).

163. D. G. Truhlar, N. J. Kilpatrick, and B. C. Garrett, *J. Chem. Phys.* **78**, 2438 (1983).

164. W.-P. Hu, Y.-P. Lin, and D. G. Truhlar, *J. Chem. Soc. Faraday Trans.* **90**, 1715 (1994).

165. R. Liu, S. Ma, and Z. Li, *Chem. Phys. Lett.* **219**, 143 (1994).

166. M. A. Collins and J. Ischtwan, *J. Chem. Phys.* **93**, 4938 (1990).

167. P. J. Jasien and R. Shepard, *Intern. J. Quantum Chem. Quantum Chem. Symp.* **22**, 183 (1988).

168. G. A. Natanson, *Mol. Phys.* **46**, 481 (1982).

169. G. A. Natanson, B. C. Garrett, T. N. Troung, T. Joseph, and D. G. Truhlar, *J. Chem. Phys.* **94**, 7875 (1991).

170. R. G. Gilbert and S. C. Smith, in *Theory of Unimolecular and Recombination Reactions*, J. P. Simons, Ed., Blackwell, Oxford, 1990.

171. D. M. Wardlaw and R. A. Marcus, *Chem. Phys. Lett.* **110**, 230 (1984).

172. D. M. Wardlaw and R. A. Marcus, *J. Chem. Phys.* **83**, 3462 (1985).

173. D. M. Wardlaw and R. A. Marcus, *J. Phys. Chem.* **90**, 5383 (1986).

174. S. J. Klippenstein and R. A. Marcus, *J. Chem. Phys.* **87**, 3410 (1987).

175. S. J. Klippenstein and R. A. Marcus, *J. Phys. Chem.* **92**, 3105 (1988).

176. S. J. Klippenstein, *J. Phys. Chem.* **98**, 11459 (1994).

177. J. Yu and S. J. Klippenstein, *J. Phys. Chem.* **95**, 9882 (1991).

178. S. J. Klippenstein and T. Radivoyevitch, *J. Chem. Phys.* **99**, 3644 (1993).

179. S. J. Klippenstein, A. L. L. East, and W. D. Allen, *J. Chem. Phys.* **101**, 9198 (1994).

180. R. N. Porter and L. M. Raff, in *Dynamics of Molecular Collisions*, W. H. Miller, Ed., Plenum Press, New York, 1976, p. 1.

181. W. A. Kraus and A. E. DePristo, *Theor. Chim. Acta* **69**, 309 (1986).

182. W. Hu and D. G. Truhlar, preprint.

183. D. G. Truhlar, G. K. Schenter, and B. C. Garrett, *J. Chem. Phys.* **98**, 5756 (1993).

184. B. J. Gertner, K. R. Wilson, and J. T. Hynes, *J. Chem. Phys.* **90**, 3537 (1989).

185. W. P. Keirstead, K. R. Wilson, and J. T. Hynes, *J. Chem. Phys.* **95**, 5256 (1991).

186. J. D. Madura and W. L. Jorgensen, *J. Am. Chem. Soc.* **108**, 2517 (1986).

187. J. Chandrasekhar, S. F. Smith, and W. L. Jorgensen, *J. Am. Chem. Soc.* **106**, 3049 (1984).

188. D. L. Severance and W. L. Jorgensen, *J. Am. Chem. Soc.* **114**, 10966 (1992).

189. Z. Peng and K. M. Merz, Jr., *J. Am. Chem. Soc.* **114**, 2733 (1992).

190. P. R. Bunker, *Molecular Symmetry and Spectroscopy*, Academic Press, New York, 1979.

191. M. A. Collins and K. C. Thompson, in *Mathematical Chemistry, Vol. 4, Chemical Group Theory*, Gordon and Breach, London, 1995, p. 191.

192. G. Eisele, A. Henglein, P. Botschwina, and W. Meyer, *Ber. Bunsenges. Phys. Chem.* **78**, 1090 (1974).

193. H. Bohringer, *Chem. Phys. Lett.* **122**, 185 (1985).

194. R. J. S. Morrison, W. E. Conaway, T. Ebata, and R. N. Zare, *J. Chem. Phys.* **84**, 5527 (1986).

195. P. R. Kemper and M. T. Bowers, *J. Phys. Chem.* **90**, 477 (1986).

196. J. W. Winniczek, A. L. Braveman, M. H. Shen, S. G. Kelley, and J. M. Farrar, *J. Chem. Phys.* **86**, 2818 (1987).

197. R. Polak, I. Paidarova, and P. J. Kuntz, *Chem. Phys.* **178**, 245 (1993).

198. C. Gonzales, H. B. Schlegel, and J. S. Franscisco, *Mol. Phys.* **66**, 849 (1989).

199. R. M. Minyaev, *J. Mol. Struct.* (*Theochem.*) **262**, 79 (1992).

200. M. A. Collins and D. F. Parsons, *J. Chem. Phys.* **99**, 6756 (1993).

201. J. C. Decius, *J. Chem. Phys.* **17**, 1315 (1949).

202. J. Ischtwan and S. D. Peyerimhoff, *Intern. J. Quantum Chem.* **45**, 471 (1993).

203. J. Ischtwan and M. A. Collins, *J. Chem. Phys.* **100**, 8080 (1994).

204. M. J. T. Jordan, K. C. Thompson, and M. A. Collins, *J. Chem. Phys.* **102**, 5647 (1995).

205. R. Farwig, in *Algorithms for Approximation*, J. C. Mason and M. G. Cox, Eds., Clarendon Press, Oxford, 1987, p. 194.

206. M. A. Suhm, *Chem. Phys. Lett.* **214**, 373 (1993).

207. M. P. Allen and D. J. Tildesley, *Computer Simulations of Liquids*, Clarendon Press, Oxford, 1987.

208. W. Chen, W. L. Hase, and H. B. Schlegel, *Chem. Phys. Lett.* **228**, 436 (1994).

209. M. J. T. Jordan, K. C. Thompson, and M. A. Collins, *J. Chem. Phys.* **103**, 9669 (1995).

210. K. C. Thompson, M. J. T. Jordan, and M. A. Collins, in preparation.

ALGEBRAIC MODELS IN MOLECULAR SPECTROSCOPY

STEFANO OSS

Dipartimento di Fisica, Università di Trento and Istituto Nazionale di Fisica della Materia, Unità di Trento, Povo (TN), Italy

CONTENTS

Advances in Chemical Physics, Volume XCIII, Edited by I. Prigogine and Stuart A. Rice.
ISBN 0-471-14321-9 © 1996 John Wiley & Sons, Inc.

ABSTRACT

In the last few years, algebraic models have been introduced as a computational tool for the analysis and interpretation of experimental rovibrational spectral of small and medium-sized molecules. These models are based on the idea of dynamic symmetry, which, in turn, is expressed through the language of unitary Lie algebras. By applying algebraic techniques, one obtains an effective Hamiltonian operator that conveniently describes the rovibrational degrees of freedom of the physical system. Within this framework, any specific mechanism relevant for the correct characterization of the molecular dynamics and spectroscopy can be accounted. Algebraic models are formulated such that they contain the same physical information for both *ab initio* theories (based on the solution of the Schrödinger equation) and semiempirical approaches (making use of phenomenological expansions in powers of appropriate quantum numbers). However, by employing the powerful method of group theory, the results can be obtained in a more rapid and straightforward way. A comprehensive and up-to-date review of mathematical concepts, physical aspects, practical applications, and numerical implementation of algebraic models in molecular spectroscopy are presented in this paper.

I. INTRODUCTION

A. Molecular Spectroscopy and Theoretical Models

Molecular spectroscopy is an area of active interest from many standpoints. Due to its numerous connections with other scientific areas, this branch of modern physics is playing an essential role in both experimental and theoretical approaches to understanding a huge number of important problems. From the very onset, the interplay between quantum theory and experiments has revealed the importance of molecular spectroscopy. With the development of more powerful experimental techniques, such significance is attracting a wider scientific community. Presently, molecular spectroscopy is going through an exciting time of renewed interest, which, once again, is being fueled by the rapid development of sophisticated experimental approaches. Recent developments include the following: Tunable, stable, and powerful lasers are available to create complex excitation schemes, thus allowing for the study of highly excited levels with unprecedented resolution. Second, new detection techniques are constantly being developed with sensitivities far exceeding the limits of detectors used just a few years ago. Finally, a variety of molecules can be chosen among a very large number of extremely pure samples with careful initial state preparation.

To keep up with the vast production of experimental observations, theoretical physics is constantly being tested to provide a collection of satisfactory models that can account for the observations. However, it should be clear that molecular spectroscopy is undergoing a radical change not simply as a result of technical advancements. One should also realize that as a consequence of new (and quite often unexpected) experimental results, an unprecedented effort toward constructing alternative theoretical models and formulating novel ideas has taken place in recent years. A particular striking example is time-dependent analysis of molecular dynamics. This approach is now being applied in both long and short time ranges, by virtue of a corresponding interest in high-resolution stationary-state spectroscopy and in low-resolution spectroscopy, inclusive of resonant anharmonic couplings. At the same time, molecular beam techniques, color center lasers, and pump/probe methods allow for a better arrangement of situations involving not only the spectroscopic characterization of fundamental modes, but also highly excited levels, intramolecular vibrational energy redistribution (IVR) processes, and atomic clusters. Such experimental interest, along with a more-or-less persevering concern for traditional aspects of molecular spectroscopy, is the starting point for understanding the actual panorama of theoretical approaches in this field of activity.

A comprehensive theoretical treatment for most aspects of molecular spectroscopy necessarily has to rely on a Hamiltonian formulation. As a matter of fact, the typical theoretical procedure used to study a given molecule consists of (1) separating the electronic and nuclear motions (assuming the Born–Oppenheimer approximation) and (2) solving the Schrödinger equation in the potential surface for the rovibrating nuclei. If the molecule is larger than a diatomic, the potential energy surface is a very complex function, composed of a discouragingly large number of coordinates. A standard approach to this problem involves approximating the potential energy surface by convenient analytical functions. A widely used procedure of this type is the force field method, in which one considers small displacements of coordinates from their equilibrium values, thus obtaining a harmonic limit for the potential surface. This kind of approach, of course, encounters difficulties as soon as one considers highly excited levels, and once more, a large number of parameters are needed to achieve meaningful results. The need for practical methods for computing molecular spectra is, in part, satisfied by introducing semiempirical expansions in powers and products of rotational and vibrational quantum numbers. The preeminent expansion is provided by the Dunham series [1]. The most serious drawbacks of this approach are that (1) no Hamiltonian operator is available (not directly, at least), and (2) for large polyatomic molecules, one needs a correspondingly large number of parameters. These parameters, in turn, have to be adjusted by a fitting procedure over a conveniently large experimental database, which is not always available.

In general, despite the numerous theoretical and experimental advances made in molecular spectroscopy, many fundamental questions remain unanswered. While certain answers could be attained by extending current theoretical models, new ones must be developed to gain a deeper perspective on these problems.

B. Algebraic Models in Modern Physics: From Nuclear to Molecular Physics

Symmetry is an extremely important concept in the development of scientific knowledge. The use of symmetry in physics, chemistry, and related areas is widely recognized from both purely theoretical and computational viewpoints. As is well known, the word *symmetry* does not necessarily have a geometric meaning when applied in modern physics; in fact, it was after the introduction of quantum mechanics that symmetries beyond geometric ones appeared and started to demonstrate their usefulness. For example, spin, isospin, permutational, and gauge symmetries gained a large popularity very quickly because of the ease with which one could attack otherwise inaccessible situations. The beauty of symmetry rests in its connection to a possible invariance in a physical system. Such invariance leads directly to conserved quantities, which in a quantum mechanical framework allow one to observe specific degeneracies in the energy spectrum and to introduce a meaningful labeling scheme for the corresponding eigenstates.

Group theory is the natural tool for dealing with symmetry arguments. For addressing quantum mechanical problems embedded in a group theoretical framework, a particularly powerful mathematical technique is that of continuous Lie groups and algebras. As a direct consequence, the matrix or algebraic formulation of quantum mechanics started to show its greater suitability compared with the differential or wave formulation, at least in regard to matters inherent in symmetry problems. The systematic use of Lie algebras dates back to the 1930s, with the pioneering work by Weyl, Wigner, Racah, and others [2–7]. In particular, point and translation group theory played a fundamental role in solid-state physics, while continuous groups, especially unitary groups, were shown to be invaluable in studies related to nuclear and particle physics.

The highest expression of group theoretical methods in nuclear physics is definitely offered by the *interacting boson model*. This approach, introduced for even–even nuclei in 1975 [8] and extended to more complex nuclear systems in the following years [9], is important for its unprecedented power in describing virtually any kind of experimental situation in nuclear physics. More significantly, the interacting boson model is the first example of a comprehensive theoretical model based on a dynamical symmetry environment. In connection with molecular spectroscopy, dynamical symmetries are explored in detail throughout this paper; however, it is important to mention here that dynamical symmetries constitute a big step forward over a conventional use of symmetry arguments, expecially those concerning the description and classification of energy spectra denoting specific degeneracy patterns. One could state, in a rather restrictive parallel, that the amount of information gained in going from a degeneracy symmetry to a dynamical one is similar to that obtained in going from a static to a dynamical study of forces acting in a conventional mechanical system. As a matter of fact, dynamical symmetries contain within themselves both the degeneracy aspects of a physical system and the complete machinery for describing transitions among different states (i.e., the dynamical behavior of the physical object at issue). All these tasks can be carried out in the extremely compact and convenient framework of Lie groups and algebras. It is not just a matter of language, however: The use of group theoretical tools very often allows one to address situations inaccessible by means of conventional methods of quantum mechanics.

The interacting boson model is a beautiful example of how to specialize, from the aforementioned viewpoint, the algebraic, second-quantized formulation of quantum mechanics. As will become clearer in the following description, such specialization involves recognizing, within the commutation relations of the creation/annihilation operators, certain Lie algebraic structures and to limit the choice of such operators within specific (invariant) bilinear forms. The existing mathematical apparatus concerning Lie algebras, tensor calculus, and related topics (such as the Wigner–Eckart theorem) is typically applied, in the case of nuclear physics and the interacting boson model, to the building block of the problem at issue, namely an object with five internal degrees of freedom (i.e., an electric quadrupole). Observable quantities such as excitation spectra and transition probabilities are then obtained in terms of boson operators, whose transformation properties are intimately connected with specific abstract symmetries. Such boson operators are the algebraic version of both the monopole and quadrupole pairing character of the strong interaction between identical nuclear particles.

This kind of strategy was soon exported to the world of molecules. As a matter of fact, the only difference between nuclei and molecules from the viewpoint of the building blocks used is that when dealing with molecules, one has to start with a diatomic unit. This is equivalent to considering boson operators related to the dipole character of the diatom. Consequently, a different dynamical symmetry (of smaller dimension than that used in nuclear physics) is adopted in the description of molecular systems. The bulk of this review is devoted to a complete description of these arguments. It is perhaps worthwhile to comment here that algebraic models, after an initial introductory period during which they were considered to be elegant, if formal, capitulations of already known physical facts, are now used in practical problems of current interest. As in the case of nuclear systems, it was realized that the dynamical symmetry framework was far from simply being just a different language used to address a problem. For molecular spectroscopy as well, algebraic techniques are demonstrating their suitability to address successfully even quite difficult situations. Generally speaking, they seem to offer a concrete and complementary technique to conventional approaches such as those mentioned in Section I.A.

C. Outline

Sections I.A and I.B were, of course, intended only to give the general flavor of many varied and active topics of interest. The topics of concern here, molecular spectroscopy and algebraic methods, themselves cover thousands of papers and specific books. Moreover, several complementary studies in molecular physics, such as electronic spectroscopy, intramolecular redistribution of energy, time evolution, and dynamics, make this field a very involved one. At the same time, group theoretical approaches extend their possibilities basically to every field of physical science. Consequently, it would be impossible to include all these topics in sufficient detail within a single book or paper. However, due to the increasing interest by the scientific community in algebraic approaches to the rovibrational spectroscopy of molecules, it seems judicious to provide a comprehensive, up-to-date review at least on this specific subset of group theoretical methods in modern physics. As a matter of fact, one of the most worrisome drawbacks of any algebraic formulation of quantum mechanical models is found in the formalism employed to express it. Indeed, group theory is an elegant mathematical way of describing situations ranging from subnuclear to solid-state physics, but still, many scientists find it a difficult language. This seems to be especially true in the field of molecular spectroscopy, in which, aside from the use of conventional point groups, the wave formulation of quantum mechanics still dominates. However, an algebraic formulation of quantum mechanics would be a powerful addition to the field of molecular spectroscopy. Two books are now available on algebraic methods in molecular physics. However, these books have been regarded as a general introduction to either abstract algebraic questions [10] or general algebraic-molecular problems [11]. Currently, no practical, brief handbooks exist for scientists willing to apply algebraic models to their problems. It is the hope of this review to fill this gap, as least partially; besides providing a state-of-the-art presentation of algebraic-molecular methods, it should serve to guide both newcomers and experienced scientists through a panorama of possible alternative approaches to a well-consolidated branch of science.

The structure of this paper is as follows. Section II introduces the basic aspects of dynamical symmetries. This topic is addressed by means of an accessible, yet rigorous language. The accent is placed on current practical applications, such as dynamical symmetries leading to algebraic models for the rovibrational spectroscopy of diatomic molecules. All the specific aspects of Lie algebras necessary for a proper formulation of algebraic models are also discussed. Section III is devoted to the study of the one-dimensional vibron model, and in Section IV we present the three-dimensional approach to this model. The ordering of this presentation is based on a conceptual rather than a historical approach. Although the three-dimensional algebraic model was introduced well before than the one-dimensional approach, the mathematical simplicity of the one-dimensional version will be of essential importance for introducing the concepts underlying the far more complex situations typically arising in three-dimensional problems. The reader will find many similarities in the structure of these two approaches. Going from the one-dimensional model to the three-dimensional model is, in fact, a matter of different mathematical techniques rather than strictly introducing new physics. This allows one to construct a logical scheme for algebraic models quite unrelated to the specific dimensionality of the physical problem at issue. To achieve a deeper understanding of how these approaches are employed, we suggest a sequential reading of the paper without omitting any part. Whenever possible, each algebraic statement, theorem, and result is illustrated with practical (i.e., numerical) examples.

At present, it would be an exaggeration to claim that algebraic techniques constitute a completely formalized way of solving problems in molecular spectroscopy. Much work has been done, but many possibilities remain to be explored in detail. However, we believe that more topics in this field will be addressed successfully within this algebraic framework in the near future, and with this in mind, we present preliminary algebraic treatments of certain problems in molecular spectroscopy. Despite the preliminary nature of these studies, they are quite significant because they apply to rather large molecules. Ultimately, the algebraic method could even be extended to biological systems of interest to spectroscopists. In Section V we present essential aspects concerning the numerical implementation of algebraic methods, with special emphasis on the use of simple computer routines, and the study of a simple geometric interpretation of the algebraic model.

II. DYNAMICAL SYMMETRIES

A. Introduction

We start our review by addressing, in a semirigorous fashion, some purely mathematical arguments leading to the definition of the *dynamical symmetry* of a physical system. A proper discussion of this subject would require a lengthy digression into many technical aspects of continuous group theory. However, it is possible to obtain a satisfactory intuitive description of the general essence of a dynamical symmetry by means of following some rather simple considerations. The reader interested in more precise definitions concerning conventional Lie groups and algebras will find some useful information in the following section. It should be noted that in this introductory section, we do not distinguish between *group* and *algebra*. This is rather questionable from a strictly mathematical viewpoint but will not interfere with the general meaning of our discussion. A more formal distinction is presented in Section II.B.

As pointed out in Section I, it is well known that a systematic study of symmetry properties of a given physical system leads to very important consequences for the description of its energy spectrum, the most relevant being related to the classification of specific degeneracies and of selection rules for transitions among different levels. For example, by starting from a Hamiltonian operator rotationally invariant in a three-dimensional space, it is expected that a complete set of quantum states will necessarily be labeled at least by quantum numbers j and m_j, which are related to the angular momentum operator and its projection along a given axis, respectively [12]. It is also well known that for those levels with a given number j there will be $(2j + 1)$-fold degeneracy, each sublevel being labeled by a different value of the quantum number $m_j = -j, -j + 1, \ldots, j - 1, j$. The absence of a privileged direction in the three-dimensional space is such that for a given total angular momentum j, $2j + 1$ orthogonal states share the same energy and transform among themselves in a very precise way. It is then possible to convert this result in the language of group theory, simply by recalling the fact that the spherical symmetric Hamiltonian operator, the angular momentum operator, and its projection operator constitute a set of commuting operators. This means that these operators can simultaneously be diagonalized in the same basis of states, which we denote as $|\alpha; jm_j\rangle$. In this ket, α is a collective quantum number related to the behavior of the state with respect to the specific functional dependence on other coordinates of the potential energy operator. The most important result of this procedure is that one obtains a meaningful classification of quantum states related to the three-dimensional rotational invariance of the Hamiltonian operator (i.e., of the physical system itself). The next step is to realize that such rotational invariance can be expressed in terms of invariance with respect to rotations now intended as elements of the continuous (Lie) group of special, orthogonal transformations in three dimensions, denoted by SO(3). The link between quantum mechanics and rotational invariance is found in the aforementioned commutation relations. In fact, it can easily be shown that a symmetry operator in Hilbert space does not change the Hamiltonian operator (i.e., it is left invariant) if and only if these two operators commute [13]. As a consequence, it is also possible to use the eigenstates to generate representations of the real-space geometric rotations in which the matrices are in block-diagonal form, each block corresponding to a specific irreducible representation of the group elements (i.e., related to invariant subspaces). The situation is quite typical for the situation where the Hamiltonian operator is spherically symmetric. In similar cases, the $(2j + 1)$-fold degeneracy provides unquestionable evidence of the fact that SO(3) is a symmetry group for the physical system.

The important point is that in the spectrum observed, a manifold of levels with a given number j can happen to lie at the same energy as another manifold of levels with a different j. This could be due to accidental reasons, such as the choice of a numerical parameter in the Hamiltonian operator for which certain sets of levels cross in the energy spectrum. It can also happen, however, that such a degeneracy is by no means an accidental one; rather, it appears on a systematic, regular basis. In the language of group theory, a situation of unexpectedly large symmetry (larger than that predicted by the symmetry group) manifests itself through the existence of several irreducible representations (generated by the corresponding degenerate levels) brought together within a single, larger irreducible representation. If this is the case, one has necessarily to invoke the structure of a correspondingly larger symmetry group including within it the original SO(3) group. The reason why one has to consider a symmetry larger than the evident rotational one is to be found in the specific coordinate dependence of the potential energy function. Two case of noticeable import and extended implication are to be cited here: Whenever it happens that (1) $V(r) \propto r^2$ or (2) $V(r) \propto 1/r$, a systematic degeneracy pattern higher than the $(2j + 1)$-fold degeneracy of the SO(3) group is readily recognized. These two cases correspond to (1) the harmonic oscillator in three dimensions and (2) the Coulomb potential. It can be shown that the degeneracy of the corresponding energy spectra can be related in a very precise way to irreducible representations (and invariant operators, see below) of (1) the unitary group in three dimensions, U(3), and (2) the orthogonal group in four dimensions, O(4). Many technical aspects and detailed discussion on this specific subject can be found elsewhere [14, 15]. However, it is important to stress here that the idea of accounting for a global symmetry group, often referred to as a degeneracy group, constitutes the starting point for the realization of far-reaching studies in a group theoretical framework. To be a bit more precise, let us examine the above-mentioned case of the three-dimensional harmonic oscillator. In Fig. 1 we show a portion of the energy spectrum based on the well-known expression [12]

$$E(n_x, n_y, n_z) = \hbar\omega(n_x + n_y + n_z + \tfrac{3}{2}), \qquad n_x, n_y, n_z = 0, 1, 2, \ldots . \quad (2.1)$$

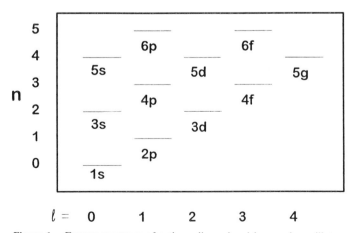

Figure 1. Energy spectrum of a three-dimensional harmonic oscillator.

It is important to notice that each level corresponds at least to $2j + 1$ degenerate states, in view of the rotational invariance of the Hamiltonian operator. In this figure it is also evident that the (aforementioned) systematic degeneracy corresponds to a symmetric group larger than SO(3). Such enhanced degeneracy is in fact obtained by looking explicitly at the eigenvalues in (2.1), which for different choices of the integer numbers n_x, n_y, n_z such that $n \equiv n_x + n_y + n_z$, one obtains the same energy value $\hbar\omega(n + \frac{3}{2})$. It can be shown that the degeneracy pattern characterizing a three-dimensional harmonic oscillator is an $(n + 1)(n + 2)/2$-fold one. A very simple, yet convincing explanation of this fact can be obtained by observing that the Hamiltonian operator

$$\hat{H} = \frac{p^2 + q^2}{2} \tag{2.2}$$

can be written as

$$\hat{H} = |\hat{z}|^2 , \tag{2.3}$$

where we have introduced the square of the modulus of $\hat{z}$, the complex three-dimensional vector operator given by

$$\hat{z} = \frac{p + iq}{\sqrt{2}} . \tag{2.4}$$

By means of the notation (2.3), one can easily see that the Hamiltonian operator must be an invariant quantity under *unitary* transformations in a three-dimensional space. In this scheme and by virtue of the r^2 functional form of the potential energy operator, it can be shown that the irreducible representations of U(3) can be used to label, along with irreducible representations of SO(3) and SO(2), the eigenspectrum of the harmonic oscillator (actually, a *truncated* one; see the discussion below). Even though we address this subject more rigorously below, it is interesting to underscore the role of the underlying SO(3) (rotational) symmetry in the present situation. Following well-established group theoretical arguments, it is possible to show that one can easily find all those irreducible representations (i.e., states transforming among themselves within an invariance subspace) of SO(3) within a given irreducible representation of U(3). This corresponds to describing how the degeneracy group U(3) contains within itself the (still existing) normal group of symmetry SO(3). It can be shown that for a given $n \equiv n_x + n_y + n_z$ of Eq. (2.1), one has

$$j = n, n - 2, \ldots , 1 \text{ or } 0 \quad (\text{for } n \text{ odd or even}) . \tag{2.5}$$

Once again, through standard group theoretical "recipes," one can obtain the branching rule (2.5) for the symmetry reduction $U(3) \supset SO(3)$. This is exactly the same strategy as that adopted for determining those states with different m_j numbers (related to a specific orientation axis) belonging to a representation (state) of $SO(3)$ with a given angular momentum j. In this case, one considers the symmetry reduction $SO(3) \supset SO(2)$.

At this point one can introduce an algebraic "bookkeeping" of states for the three-dimensional (truncated) harmonic oscillator by taking into account the degeneracy chain of transformation groups given by

$$U(3) \supset SO(3) \supset SO(2) . \tag{2.6}$$

To summarize, we have seen that it is possible to explain the degeneracy pattern characteristic of a three-dimensional harmonic oscillator by introducing, as a proper symmetry group of the Hamiltonian operator (here referred to as the *degeneracy group*), a group of unitary transformations in a three-dimensional complex space.

We complete these introductory remarks by sketching the basic rules for delineating the central argument of our discussion, the *dynamical* symmetry for a physical system. The point is that we are looking for an algebraic (i.e., group theoretical) formulation of both degeneracies *and* dynamics related to the Hamiltonian operator. What we mean here is that we need a procedure allowing one to (1) account for the complete set of quantum states and (2) describe transitions among them. If one considers the simple situation of defined angular momentum states $|jm_j\rangle$ corresponding to the group reduction scheme $SO(3) \supset SO(2)$, it is possible to construct transition operators between states belonging to the same irreducible representation of $SO(3)$ (i.e., with the same quantum number j). This is done by introducing ladder operators (or step-up, step-down, raising, lowering operators) of $SO(3)$, which eventually change (by unitary steps) the quantum number m_j. The degeneracy group, $U(3)$, can then be considered to justify the introduction of ladder operators (which will be now second-rank tensor or quadrupole operators) capable of describing transitions between states belonging to the same irreducible representation of $U(3)$ (i.e., with the same n quantum number but with different angular momentum j).

The final step is to obtain a global algebraic picture of states which includes transitions between levels belonging to different irreducible representations of $U(3)$. This can be achieved, in principle, by construction of proper combinations of raising and lowering operators. In fact, such operators exist in Heisenberg or $N(3)$ (non-Lie) algebra and couple states belong to various irreducible representations of $U(3)$ [16]. The problem is that Heisenberg algebra does not include the degeneracy algebra. A detailed study of mathematical properties of these objects and the use of a procedure referred to as *group contraction* leads to the important result that group U(4) is instead suited for this purpose. Thus the unitary group in four dimensions, U(4), is called the dynamical group for the three-dimensional harmonic oscillator. The corresponding chain of groups will thus be given by

$$U(4) \supset U(3) \supset SO(3) \supset SO(2) \ . \qquad (2.7)$$

Figure 2 is a schematic diagram illustrating how one includes transition operators associated with the rotational, degeneracy and dynamical group of the three-dimensional harmonic oscillator.

   So far, we have reviewed, in a semirigorous fashion, the basic aspects of the construction of a dynamical group for a specific Hamiltonian operator in three-dimensional space. One could consider other Hamiltonian operators as well; the case of the Coulomb problem for a single particle falls within the same logical scheme, the only difference bing that now the degeneracy group turns out to be O(4), which contains orthogonal transformations in a four-dimensional space. What is really important from our viewpoint, however, is that we are laying the basis for a systematic algebraic treatment of the quantum problem, including exactly solvable potential energy operators (i.e., in analytical form). This topic is addressed explicitly in the following sections. Here we simply hint that formulation of the harmonic oscillator problem in a dynamical algebra framework is to be viewed as the zeroth-order step for the construction of building blocks for any molecular problem of interest. Ultimately, this method is applied to the rovibrating three-dimensional anharmonic

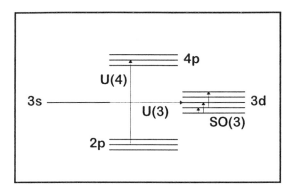

Figure 2. Effects of ladder operators in the U(4) framework.

oscillator. Starting at chain (2.7), we are now ready to define an algebraic Hamiltonian operator. We see that in its more complete form, such an operator corresponds to the usual second-quantized expression in terms of annihilation and creation operators acting on a proper (Fock) model space. The use of dynamical symmetry, a very powerful technique related to the dynamical group discussed above, leads to a conveniently simple form of the second-quantized Hamiltonian operator. This constitutes the starting point for the realization of algebraic molecular Hamiltonian operators.

B. Lie Algebras and Dynamical Symmetries

In this section we consider in more detail the important points leading to a precise algebraic formulation of a quantum problem. We make use of the general ideas sketched in Section II.A, but we also try to give a mathematically rigorous exposition of the algebraic formulation desired. To accomplish this, we need to introduce specific concepts arising from Lie algebra. However, we make physical rather than mathematical use of these powerful tools. Consequently, the reader should not expect to find a comprehensive treatise on groups and algebras. We do not address the problem of how many and which algebras exist, nor do we address how to characterize and classify them [17]. Rather, we focus our interest on those (unitary) algebras that directly apply to a proper understanding of molecular models.

We define a Lie algebra G of order r as the collection of operators $\{\hat{X}_1, \hat{X}_2, \ldots, \hat{X}_r\}$ satisfying the commutation relations

$$[\hat{X}_i, \hat{X}_j] = \sum_k c_{ij}^k \hat{X}_k , \qquad (2.8)$$

where the *structure constants* c_{ij}^k are such that

$$c_{ij}^k = -c_{ji}^k . \qquad (2.9)$$

The following relation, often referred to as the *Jacobi identity*, also holds:

$$[[\hat{X}_i, \hat{X}_j], \hat{X}_k] + [[\hat{X}_j, \hat{X}_k], \hat{X}_i] + [[\hat{X}_k, \hat{X}_i], \hat{X}_j] = 0 . \qquad (2.10)$$

For a given Lie algebra, one can find a corresponding continuous Lie group of elements. The link between groups and Lie algebraic structures is expressed in terms of the commutation relations (2.8), which can be thought as the algebraic counterpart of the underlying multiplication (or combination) law between group elements. To be more precise, one should consider the opposite procedure, for which the starting point is a group of transformations (actually, a continuous set of infinite elements) that can be obtained in terms of a finite number of generating operators. In the simplest case of the group of rotations around a given axis, SO(2), the situation is such that in a two-dimensional vector space, the matrix form of the rotation, in terms of the single parameter θ (here the angle of rotation), is given by

$$R(\theta) = \begin{pmatrix} \cos\theta & -\sin\theta \\ \sin\theta & \cos\theta \end{pmatrix}. \tag{2.11}$$

It is then possible to generate any rotation about this same axis by considering the infinitesimal operator $\hat{X}$, whose matrix form is

$$\hat{X} = \begin{pmatrix} 0 & -1 \\ 1 & 0 \end{pmatrix}. \tag{2.12}$$

By using matrix exponentiation, one obtains

$$\hat{R}(\theta) = \exp(\theta\hat{X}). \tag{2.13}$$

This is equivalent to constructing an operative link between an algebraic environment ($\hat{X}$) and a geometric space [i.e., the group of transformations, $\hat{R}(\theta)$]. The connection between group multiplication laws and algebraic commutation relations (2.8) can be better understood by taking into account a situation less trivial than the single-parameter group SO(2). The case of three-dimensional rotations corresponds to considering the three-parameter continuous group SO(3). It is not difficult to show that by starting from generalization of the matrix (2.11) to the three-dimensional case, one can construct the known rule for the composition law of rotations about an arbitrary space direction by introducing three generating operators $\hat{X}_x$, $\hat{X}_y$, $\hat{X}_z$ whose matrix form (in a Cartesian basis) is

$$\hat{X}_x = \begin{pmatrix} 0 & 0 & 0 \\ 0 & 0 & -1 \\ 0 & 1 & 0 \end{pmatrix}, \qquad \hat{X}_y = \begin{pmatrix} 0 & 0 & 1 \\ 0 & 0 & 0 \\ 1 & 0 & 0 \end{pmatrix},$$

$$\hat{X}_z = \begin{pmatrix} 0 & -1 & 0 \\ 1 & 0 & 0 \\ 0 & 0 & 0 \end{pmatrix}. \tag{2.14}$$

These operators satisfy the commutation relations

$$[\hat{X}_x, \hat{X}_y] = \hat{X}_z, \qquad [\hat{X}_y, \hat{X}_z] = \hat{X}_x, \qquad [\hat{X}_z, \hat{X}_x] = \hat{X}_y \tag{2.15}$$

[i.e., they constitute a Lie algebra whose structure constants include the essential "geometric" nature of the associated SO(3) Lie group]. It is worthwhile noting that a decidedly more convenient way of describing the effect of operators such as (2.12) or (2.14) can be obtained by examining a differential *realization* of the algebraic elements; this is particularly useful in quantum mechanical problems, in which the natural environment is constituted by the Hilbert space of (wave) functions. These operators transform among themselves according to the specific form of group elements or, more precisely, of their differential realization. Going back to the SO(2) case, for example, it is easy to show (by using a Taylor series describing a finite angle rotation in terms of infinitesimal rotations) that in a proper function space, the matrix form (2.12) is replaced by the differential operator

$$\hat{X} = -\frac{\partial}{\partial \phi}.$$
(2.16)

Equivalently, the three-dimensional case of SO(3) is described by using in place of the three matrices (2.14), the (Cartesian) components of the angular momentum operator. The precise way in which this connection is realized requires that

$$\hat{J}_q = i\hat{X}_q, \qquad q = x, y, z.$$
(2.17)

One can show that this definition implies that the $\hat{J}_q$ are Hermitian and that they satisfy the usual commutation relations

$$[\hat{J}_x, \hat{J}_y] = i\hat{J}_z, \qquad [\hat{J}_y, \hat{J}_z] = i\hat{J}_x, \qquad [\hat{J}_z, \hat{J}_x] = i\hat{J}_y.$$
(2.18)

Another definition of interest here involves the subalgebra G' of a given Lie algebra G. One has to consider a subset G' ⊂ G that is closed with respect to the same commutation laws of G,

$$[\hat{X}'_i, \hat{X}'_j] = \sum_k c_{ij}^k \hat{X}'_k.$$
(2.19)

As a simple example, the algebraic structure corresponding to SO(2) is a subalgebra of SO(3). This is true because SO(2) is generated by $\hat{J}_z$, which obviously commutes with itself. As a result, we can write that

$$SO(2) \subset SO(3) .\tag{2.20}$$

This is the exact meaning of relations such as (2.6) and (2.7) from the viewpoint of the definition of a Lie subalgebra.

At this stage we need to address the problem concerning the construction of the *representations* of elements of a given Lie algebra. This turns out to be of particular relevance to our discussion because of the manifest link with the basis states in which the operators act. Once again it is beyond our immediate concern to give both a rigorous and general answer to this problem. One can make extended use of conventional Lie group theory (with particular emphasis for tensor calculus, which in the present framework establishes a fascinating connection between permutational and unitary symmetries [15]) to attain a systematic rule for constructing irreducible representations (or basis states) of any continuous Lie group. The most relevant results of concern here are the following. (1) An irreducible representation of $U(n)$, the unitary Lie group in n dimensions, is labeled by a set of n numbers denoted here as $[\alpha_1 \alpha_2 \cdots \alpha_n]$. It is possible to show that this fact is equivalent to introducing n rows each containing α_i $(i = 1, \ldots, n)$ boxes. This set of $\Sigma_i \, \alpha_i$ boxes is called a *Young tableau*. It is then a matter of algebraic manipulation to understand how to fill this pattern of boxes with nonnegative integer numbers specifying the particular irreducible representation at issue. (2) As will become clear in the next section, "symmetric" irreducible representations of unitary groups play a special role in this framework. Such symmetry is somehow related to the boson character of the algebraic realizations of concern here. To be more precise, it is possible to demonstrate that a symmetric irreducible representation of $U(n)$ is specified in terms of a single row (in place of n of them) containing a given number of boxes. As a result, one needs just a single number (corresponding to the number of boxes appearing in the single row) to label a symmetric irreducible representation of $U(n)$, any n. (3) One then typically has to tackle the problem of finding, for a given irreducible representation of $U(n)$, the allowed irreducible representations of subalgebras contained in $U(n)$ itself. This is important, as it points to the construction of a complete basis set for the starting algebraic structure. It is probably much easier to perceive the actual meaning of this procedure by taking into account the case of rotations in a three-dimensional space described by $SO(3)$. Here the irreducible representations are labeled by a single number j, while the irreducible representations of the subalgebra $SO(2)$ are labeled by the single number m_j. Even though we already know which is an exact form of the complete basis set for the SO(3) algebra [i.e., the spherical harmonics $Y_{jm_j}(\theta, \phi)$], we can now make use of a typical algebraic notation by introducing, for the complete basis set, the following ket:

$$\left| \begin{matrix} SO(3) \supset SO(2) \\ j \qquad\quad m_j \end{matrix} \right\rangle . \qquad (2.21)$$

As an example, the chain of subalgebras (2.7) based on U(4) has, as a complete basis set associated to symmetric irreducible representations, the ket

$$\left| \begin{matrix} U(4) \supset U(3) \supset SO(3) \supset SO(2) \\ N \qquad k \qquad\quad j \qquad\quad m_j \end{matrix} \right\rangle , \qquad (2.22)$$

in which N and k label symmetric irreducible representations of U(4) and U(3), respectively. The aforementioned explicit problem of finding, for a given N, the permitted values of k (and for a given k, those of j) are discussed briefly below. At the moment we simply state that this is a straightforward problem [14] for algebraic structures that have been adopted in the study of molecular situations.

We now go to the crucial point of our scheme, the practical realization of a Hamiltonian operator in the algebraic framework. We are now in a position to establish the basic mathematical foundation for using this alternative procedure in place of the typical wave formulation. Once again, we do not attempt to provide a deep explanation for this quite involved topic. Rather, we give a simple outline establishing the construction of algebraic Hamiltonian operators and, eventually, molecular Hamiltonian operators. As discussed in Section II.A, the quantum problem of the harmonic oscillator (in a space with arbitrary dimensions) can be taken as a good starting point for justifying the introduction of abstract (degeneracy and dynamical) group arguments. Even though we soon move our attention to anharmonic situations, it is nonetheless important to complete the debate about harmonic oscillators by looking at their Hamiltonian in an algebraic framework. We then generalize our results to anharmonic potential functions to properly address the general recipes for a dynamical symmetry formulation of quantum mechanical problems.

The algebraic version of an n-dimensional harmonic oscillator is a completely standard exercise in modern quantum mechanics textbooks [12]. Here we simply recall those aspects of some import to our pursuit. First, we recall that one has to replace usual x_i and p_i space coordinates with quantum (differential) operators x_i, $-i\hbar\partial/\partial x_i$ $(i = 1, \ldots, n)$. This corresponds to constructing a *first quantization* of the problem at issue. The algebraic realization is then obtained in terms of a *second quantization* procedure, by means of replacing the differential space-momentum operators with creation and annihilation operators, the typical replacement rule being (for a harmonic oscillator)

$$\hat{a}_i = \frac{x_i + \partial/\partial x_i}{\sqrt{2}}, \qquad \hat{a}_i^\dagger = \frac{x_i - \partial/\partial x_i}{\sqrt{2}}. \qquad (2.23)$$

The essential point is that by virtue of the quantum nature of the problem, the operators x_i and p_i satisfy certain commutation relations (which in turn come, through proper use of the correspondence principle, from the Poisson brackets). These relations, which contain within themselves the specific aspects of the physical interaction between particles (nuclei, nucleons, etc.), lead to a set of precise commutation relations of the operators $\hat{a}_i$ and $\hat{a}_i^\dagger$, defined in (2.23). The case of a three-dimensional harmonic oscillator is such that starting from

$$\hat{a}_x^\dagger = (x - ip_x)/\sqrt{2}, \qquad \hat{a}_y^\dagger = (y - ip_y)/\sqrt{2},$$

$$\hat{a}_z^\dagger = (z - ip_z)/\sqrt{2}, \qquad (2.24)$$

and knowing that

$$[x, p_x] = [y, p_y] = [z, p_z] = 1, \qquad (2.25)$$

and all the remaining commutators being equal to zero, one easily obtains

$$[\hat{a}_x, \hat{a}_x^\dagger] = [\hat{a}_y, \hat{a}_y^\dagger] = [\hat{a}_z, \hat{a}_z^\dagger] = 1,$$

$$[\hat{a}_x, \hat{a}_x] = [\hat{a}_y, \hat{a}_y] = [\hat{a}_z, \hat{a}_z] = 0, \qquad (2.26)$$

$$[\hat{a}_x^\dagger, \hat{a}_x^\dagger] = [\hat{a}_y^\dagger, \hat{a}_y^\dagger] = [\hat{a}_z^\dagger, \hat{a}_z^\dagger] = 0.$$

The Hamiltonian operator (2.2) can now be written in terms of the operators $\hat{a}_i$ and $\hat{a}_i^\dagger$: One has, in units of $\hbar\omega$,

$$\hat{H} = \hat{N} + \frac{n}{2}, \qquad (2.27)$$

in which we use the "number" operator

$$\hat{N} = \sum_{i=1}^{n} \hat{a}_i^\dagger \hat{a}_i. \qquad (2.28)$$

The essential feature of this realization is that the states of the system can be obtained by consecutive action of creation and annihilation operators on a vacuum model (algebraic) space. This is usually written as

$$|v_1 v_2 \cdots v_n\rangle = \frac{1}{A}(\hat{a}_1^\dagger)^{v_1}(\hat{a}_2^\dagger)^{v_2} \cdots (\hat{a}_n^\dagger)^{v_n}|00\cdots0\rangle , \qquad (2.29)$$

in which A is a normalization constant. This space of states is usually referred to as the *Fock space* [18]. As a consequence, the Hamiltonian operator (2.27) acts on the Fock space of states in such a way that the eigenspectrum (2.1) is readily recovered (for arbitrary dimension n):

$$E(v_1, v_2, \ldots, v_n) = \hbar\omega\left(\frac{n}{2} + v_1 + v_2 + \cdots + v_n\right) . \qquad (2.30)$$

To say it in a slightly different fashion, in the algebraic formulation one introduces a model space of states (the Fock space) that is spanned entirely by the second-quantized version of the space-momentum operators. The most appealing features of this procedure are that (1) the Hamiltonian operator is now written in terms of a "counter" of harmonic quanta of excitation, and (2) eigenstates and eigenvalues are obtained in a purely algebraic fashion (i.e., one does not have to perform integral or differential calculations as required by the wave formulation of this same problem). If needed, one can, of course, go back to the conventional geometric environment, in which wavefunctions are written in terms of the usual space coordinates.

We are now interested in including larger groups (the degeneracy and dynamical groups) mentioned earlier. This is easily done by observing that the algebraic Hamiltonian operator (2.27) is given in terms of n^2 (3^2 in the three-dimensional case) annihilation–creation operators $\hat{a}_i$ and $\hat{a}_j^\dagger$ ($i, j = 1, \ldots, n$). The important point is that one can easily show that

$$[\hat{H}, a_i^\dagger a_j] = 0 \qquad \text{(any } i, j) . \qquad (2.31)$$

These commutation relations are an unequivocal sign of symmetry for the Hamiltonian operator, $\hat{H}$. Such symmetry is made clear through a careful study of group theoretical properties of the bilinear forms $\hat{a}_i^\dagger \hat{a}_j$. As in similar cases, one can carry out certain transformations of these bilinear forms to convert them into irreducible tensors with respect to rotations [7, 19] (see also a much more detailed discussion in Section II.C.2). The final result is that the n^2 bilinear forms $\hat{a}_i^\dagger \hat{a}_j$ can be viewed as generators of the n-dimensional unitary group $U(n)$. The underlying rotational structure of the harmonic oscillator is then recovered by means of a typical group theoretical procedure. In fact, it is possible to acknowledge the vectorial character of boson operators $\hat{a}_i^\dagger$ and $\hat{a}_j$ by taking them as an operator (irreducible tensor) basis for the representation of SO(3) labeled by $j = 1$, which we denote as $D^{(j=1)}$. This means that the bilinear forms $\hat{a}_i^\dagger \hat{a}_j$ will transform according to the direct (group) product of representations $D^{(j=1)} \otimes D^{(j=1)}$. Once again, by using standard arguments of group theory, one can show that this direct product representation can be decomposed in terms of irreducible representations of SO(3). Equivalently, such a product of representations contains within itself the irreducible representations of SO(3) associated with angular momenta $j = 0, 1, 2$. This is usually written as

$$D^{(j=1)} \otimes D^{(j=1)} = D^{(j=0)} \oplus D^{(j=1)} \oplus D^{(j=2)} . \tag{2.32}$$

The actual meaning of this decomposition is that the set of n^2 operators generating U(n) can be grouped in families of operators behaving in a very precise way with regard to certain symmetries included in U(n). For example, in the three-dimensional case, one can see that those operators related to the irreducible representation $D^{(j=0)}$ are in the form of the invariant operator (2.28). This, in turn, is basically the Hamiltonian operator (2.27). Equivalently, this Hamiltonian is itself a scalar operator (an invariant quantity). One is then left with $3^2 - 1 = 8$ operators transforming according to irreducible representations with $j = 1$ and $j = 2$ of SO(3), corresponding to vector and second-rank tensor operators. It can be demonstrated that the scalar part (the Hamiltonian) and the remaining eight vector and tensor operators define, respectively, the unitary algebras U(1) and SU(3). In the latter, S denotes the special character of these unitary transformations [14]. This is usually written as

$$U(3) = U(1) \otimes SU(3) . \tag{2.33}$$

What is really important to note is that the Lie algebra SU(3) contains within itself ladder operators capable of describing transitions between states differing in their SO(3) and SO(2) quantum numbers.

In brief, what we are looking for is an algebraic device for including transitions between states belonging to different irreducible representations of U(3) as well. This cannot be achieved in the aforementioned framework, as the Hamiltonian operator is an invariant quantity under the symmetry U(3) (i.e., the total number of harmonic quanta cannot be changed). This number, k, is the label of the irreducible representation of U(3) at issue, which is in turn the eigenvalue of the scalar operator (2.28). The solution to this problem is to extend the algebraic treatment of this situation to account for a larger (dynamic) group whose irreducible representations include several irreducible representations of U(3). Moreover, such a procedure will make available new ladder operators suited to describing transitions between states differing in their numbers of vibrational excitation. The group theoretical answer to this problem is the $(n + 1)$-dimensional unitary group $U(n + 1)$ (U(4) in the three-dimensional case). The chain of subalgebras (2.7) can now be completed to construct the algebraic ket (2.22), in which the branching laws for the labels k, j, and m_j are

$$k = N, N - 1, \ldots, 0,$$

$$j = k, k - 2, \ldots, 1 \text{ or } 0, \qquad (2.34)$$

$$m_j = -j, -j + 1, \ldots, j - 1, j.$$

The corresponding Hamiltonian operator will still be given in terms of proper expansions over bilinear forms of (boson) creation and annihilation operators. (The more complex situations including half-spin particles can be addressed as well by using fermion operators [20].) The general rule is that one introduces a set of $(n + 1)^2$ boson operators $\hat{b}_i$ and $\hat{b}_j^{\dagger}$ $(i, j = 1, \ldots, n + 1)$ satisfying the commutation relations

$$[\hat{b}_i, \hat{b}_j^{\dagger}] = \delta_{ij}, \qquad [\hat{b}_i, \hat{b}_j] = [\hat{b}_i^{\dagger}, \hat{b}_j^{\dagger}] = 0, \qquad i, j = 1, \ldots, n + 1.$$

$$(2.35)$$

The next step is to construct the algebraic (second-quantized) version of the operator of interest. The case of the Hamiltonian operator is such that one can write

$$\hat{H} = E_0 + \sum_{i,j} e_{ij} \hat{b}_i^{\dagger} \hat{b}_j + \sum_{i,j,h,k} f_{ijhk} \hat{b}_i^{\dagger} \hat{b}_j^{\dagger} \hat{b}_h \hat{b}_k + \cdots. \qquad (2.36)$$

This expression includes terms up to two-body interactions. The algebraic Hamiltonian (2.27) of the (n-dimensional) harmonic oscillator is, of course, a special case of Eq. (2.36). One then observes that it is possible to arrange the Hamiltonian (2.36), in the framework of a dynamical algebra, by explicitly introducing the bilinear products

$$\hat{G}_{ij} \equiv \hat{b}_i^{\dagger} \hat{b}_j, \qquad i, j = 1, \ldots, n + 1; \qquad (2.37)$$

the $(n + 1)^2$ operators $\hat{G}_{ij}$ satisfy [along with the Jacobi identity (2.10)] the commutation relations

$$[\hat{G}_{ij}, \hat{G}_{hk}] = \hat{G}_{ik}\delta_{jh} - \hat{G}_{hj}\delta_{ki} . \qquad (2.38)$$

This is the equivalent of stating that the $\hat{G}_{ij}$'s give the unitary algebra $U(n + 1)$. Moreover, it is possible to write the Hamiltonian operator (2.36) directly in terms of the generators (2.37),

$$\hat{H} = E_0 + \sum_{i,j} e_{ij}\hat{G}_{ij} + \sum_{i,j,h,k} f_{ijhk}\hat{G}_{ih}\hat{G}_{jk} + \cdots . \qquad (2.39)$$

In this expression, the coefficients e_{ij}, f_{ijhk}, ... are slightly different from those used in expansion (2.36). At this stage it is worthwhile noticing that the algebraic Hamiltonian (2.39), expressed in terms of elements of $U(n + 1)$, is completely general and holds for any n-dimensional problem of quantum mechanics. This means that the dynamical group for any three-dimensional problem is $U(4)$, while for any one-dimensional situation the dynamical group is $U(2)$. We postpone until Sections II.C.1 and II.C.2 the explicit realization of anharmonic Hamiltonian operators within this framework. In the following sections we shall also understand why $U(n + 1)$ is called the *spectrum-generating algebra* (SGA) for the single n-dimensional degree of freedom.

The last point to be discussed here concerns an extremely convenient procedure, often referred to as *dynamical symmetry*, which allows one to greatly simplify the outcomes of a second-quantization realization. The basic idea is to choose the parameters e_{ij}, f_{ijhk}, ... of the general expansion (2.39) in such a way that only certain operators of the subalgebras of the dynamical algebras are taken into account. As a matter of fact, if one includes in this expansion only the *invariant* or *Casimir* operators of the subalgebras, the Hamiltonian operator can be written as

$$\hat{H} = E_0 + A\hat{C} + A'\hat{C}' + A''\hat{C}'' + \cdots , \qquad (2.40)$$

in which the $\hat{C}$'s are invariant operators of the subalgebras G', G'', ... of the dynamical algebra G:

$$G \supset G' \supset G'' \supset \cdots . \qquad (2.41)$$

In this case it is customary to say that the Hamiltonian (2.40) is in the *enveloping* algebra of G.

To appreciate the actual meaning of this procedure, we need to clarify briefly the concept of invariant or Casimir operator. Returning to the rotation group SO(3), we know that the square of the angular momentum operator, $\hat{J}^2$, is such that

$$[\hat{J}^2, \hat{J}_x] = [\hat{J}^2, \hat{J}_y] = [\hat{J}^2, \hat{J}_z] = 0 . \tag{2.42}$$

In other words, $\hat{J}^2$ commutes with the generators of SO(3) and $\hat{J}^2$ must be invariant within a given irreducible representation $D^{(j)}$ of SO(3). Consequently, this operator must be a multiple of the unit operator. This can be seen from the known eigenvalue equation

$$\hat{J}^2 Y_{jm_j} = j(j + 1)Y_{jm_j} . \tag{2.43}$$

The considerations above have general validity, as it can be demonstrated that for every Lie group, one can obtain scalar (invariant) quantities which can be linear, quadratic, cubic, and so on, in the elements of the algebra. These objects are called *Casimir operators* [21]. From this viewpoint, one can also define, as the *rank* of a given Lie algebra, the total number of its independent Casimir operators. The most important result here is that Casimir operators are diagonal in the basis corresponding to an irreducible representation of the given algebra. We also mention that the algebraic problem of finding the eigenvalues of any Casimir operator for any Lie algebra has been already solved [14]. We do, however, provide explicit formulas only for molecular problems of direct interest.

Having defined the Casimir operators, it is now possible to understand the actual significance of the Hamiltonian expansion (2.39) converted to the dynamical symmetry form (2.40). Since the Casimir operators $\hat{C}$'s are diagonal in the basis of the dynamical algebra, their expectation values can be obtained in closed (analytical) form. This is particularly useful whenever one needs to construct an effective, algebraic Hamiltonian operator for the purpose of studying an experimental spectrum. To be more precise, let us examine, in some detail, the case of the three-dimensional harmonic oscillator. Starting from the algebra U(4) and adopting as an algebraic complete basis the ket (2.22), we write the dynamical symmetric Hamiltonian operator

$$\hat{H} = E_0 + A_1\hat{C}_1[\text{U}(3)] + A_2\hat{C}_2[\text{U}(3)] + A_2'\hat{C}_2[\text{SO}(3)] , \tag{2.44}$$

in which we introduce the invariant operators of U(3) ($\hat{C}_1$ is linear while $\hat{C}_2$ is quadratic in terms of elements of the algebra) and of SO(3), which possesses only the quadratic invariant operator $\hat{J}^2$. This choice corresponds (1) to including up to quadratic terms in the elements of the algebra (i.e., up to two-body interactions in the Hamiltonian), and (2) to disregarding the SO(2) symmetry, associated with an SO(3) symmetry breaking. (Such a situation is obtained with the application of an external magnetic field.) It is a matter of standard Lie algebraic theory to compute the eigenvalues of $\hat{H}$ in the basis (2.22):

$$E = E(N, k, j) = E_0 + A_1 k + A_2 k(k + 3) + A_2' j(j + 1) . \qquad (2.45)$$

In this equation, by letting $A_2 = A_2' = 0$, one immediately recovers the energy spectrum of the three-dimensional harmonic oscillator (truncated up to $N + 1$ vibrational levels),

$$E = E_0 + A_1 k , \qquad k = 0, 1, \ldots, N - 1, N . \qquad (2.46)$$

Instead, by letting $A_2 \neq 0$ and $A_2' \neq 0$, one can describe more complex (nonharmonic) situations, such as the nonrigid rotator in three dimensions. Another important aspect is that by having embedded the degeneracy symmetry U(3) in the larger dynamical structure U(4), one has the possibility of representing, for each irreducible representation of U(4), a "physically" different harmonic oscillator. It should be clear from Eq. (2.46) that the label N establishes the total number of bound states supported by the harmonic (truncated, indeed) potential. Such a procedure provides us with an "extra" quantum mechanical number to adjust when describing an entire set of apparently different problems with one technique. This advantage will become clearer subsequent to the discussion of molecular algebraic models.

To summarize, the most important steps leading to the formulation of a dynamical symmetry have been presented. This formulation should be thought as a very effective, specialized version of the usual second-quantized realization of a quantum problem. In such a realization (1) the wave equation is replaced by an algebraic equation, (2) the wave functions are replaced with a Fock space, and (3) the most general algebraic expansion, in terms of (boson) creation–annihilation operators, is restricted to invariant or Casimir operators of subalgebras of the dynamical algebra. Such "ultimate" algebraic structure turns out to be, for n-dimensional problems, the Lie algebra U($n + 1$). These three steps constitute the basic components for the definition of the dynamical symmetry realization of the Hamiltonian operator. We are now ready to address, in a more explicit fashion, the question of how to construct algebraic models for anharmonic degrees of freedom. This question is of obvious central interest in molecular studies.

C. Dynamical Symmetries for Molecular Spectroscopy: The Vibron Model

The main purpose of this section is to illustrate some specific dynamical symmetries suitable for molecular Hamiltonian operators. In this section we bridge groups, algebras, and related abstract arguments on one side and rovibrational modes, infrared, Raman spectra, and related physical subjects on the other.

The most remarkable outcome of the previous sections is as follows: The dynamical group $U(n + 1)$ can be used to realize a compact, effective, algebraic Hamiltonian operator (as well as other observable quantities) for any exactly solvable physical problem in n dimensions for which the potential energy operator can be written in an analytical form of its own coordinates. It should be clear that if properly interpreted and put into practice, such a technique can have extremely far-reaching consequences in a huge number of applications. This is indeed the case, as dynamical symmetries play a significant role in the general understanding and interpretation of experimental facts ranging from hadron physics to biological systems. We must now apply the dynamical symmetry approach to the world of molecules. Fortunately, the basic idea is quite simple. However, anharmonicity is an important factor in the application of dynamical symmetry to molecules. In the Born–Oppenheimer approximation, the molecule is regarded as a collection of point masses whose interactions have an anharmonic character. As a consequence, our goal will be to provide both (1) a dynamical symmetry description of single anharmonic degrees of freedom, and (2) a method for unifying these objects in an algebraic sense.

The first step is to understand how a dynamical symmetry approach can lead to practical results regarding uncoupled anharmonic oscillators. This will be accomplished in two distinct subsections, addressing the one- and three-dimensional problems, respectively. (Two-dimensional questions are presently under study and are not considered in this paper. We do, however, provide some information for those two-dimensional situations of direct interest to our immediate goals whenever the opportunity will arise.)

Before proceeding, it is necessary to specify more clearly what *kind* of anharmonic behavior will be dealt with in the framework of rovibrational modes. Consequently, a brief but important digression concerning the conventional treatment of interatomic potential functions in molecular studies is presented. We are interested in having at our disposal a potential energy function (either in analytical form or expressed as a series in powers of proper analytical terms) for reproducing the sequence

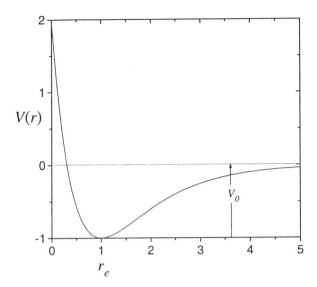

Figure 3. Potential energy function of a diatomic molecule in its electronic ground state. V_0 is the dissociation energy, D_e, in units of hc, and r_e is the equilibrium position.

of rovibrational levels observed experimentally for a single diatomic bond. The typical situation is shown in Fig. 3. Here an asymmetric potential curve is shown with a deep minimum for $r = r_e$, the equilibrium distance for the scalar interatomic coordinate r. As one can clearly see, this potential curve attains its dissociation energy value (V_0) for $r \rightarrow \infty$, while for $r \rightarrow 0$ the function is characterized by a steep repulsive behavior. A convenient way of approximating this type of a curve is by the Morse potential [22],

$$V_M(r) = V_0\{1 - \exp[-\beta(r - r_e)]\}^2,\qquad(2.47)$$

where β is a parameter related to the flex point of the curve. This potential is especially well suited to describe vibrational levels up to, say, 90% of the dissociation energy in realistic situations. At higher energy one needs to consider more sophisticated models. The most appealing feature of the Morse potential is that one can solve the associated Schrödinger equation in an exact way for one-dimensional problems or in a quasi-exact way for two- or three-dimensional problems, the error often being smaller than 1 part in 10^8–10^{10} (for zero angular momentum). Such an approximation comes from the fact that analytical solutions can be achieved only under the constraint that $V(r) \rightarrow \infty$ for $r \rightarrow 0$. Such a condition is only approximately fulfilled by the Morse potential. The bound-state part of the rovibrational energy spectrum obtained is given by

$$E(v, j) = \omega_e(v + \tfrac{1}{2}) - \omega_e x_e(v + \tfrac{1}{2})^2 + B_e j(j + 1) , \qquad (2.48)$$

where

$$\omega_e = \hbar\beta\sqrt{\frac{2V_0}{\mu}} , \qquad \omega_e x_e = \frac{\hbar^2\beta^2}{2\mu} , \qquad B_e = \frac{\hbar^2}{2\mu r_e} \qquad (2.49)$$

and μ denotes the reduced mass of the two-body system. For nonzero values of the angular momentum, one has to face in Eq. (2.48) a further source of error, related to the centrifugal coefficient B_e. To solve the Schrödinger equation in closed form, one is obliged to use the equilibrium value (B_e) in place of the actual expression $\hbar^2/2\mu r$. As a consequence, one should expect nonnegligible numerical errors when considering rovibrational states involving large deviations from the radial coordinate from its equilibrium value. The Morse spectrum is a special case of the general expansion

$$E(v, j) = \sum_{h,k} y_{hk} v^h [j(j + 1)]^k , \qquad (2.50)$$

which, in the form of Eq. (2.48), is often referred to as *Dunham expansion* [1]. The aforementioned problems concerning the approximate character of the solutions of the Schrödinger equation cease to exist in the one-dimensional version of this situation. In this case one can write, as an *exact* solution of the one-dimensional Schrödinger equation, the eigenspectrum

$$E(v) = \omega_e(v + \tfrac{1}{2}) - \omega_e x_e(v + \tfrac{1}{2})^2 , \qquad (2.51)$$

in which the constants ω_e and $\omega_e x_e$ are the same as the ones defined in Eq. (2.49). Wavefunctions may be obtained as well in closed form [23].

As will soon become clear, the one-dimensional version of the Morse potential problem is a useful heuristic device, despite the three-dimensional nature of the molecules. In fact, certain vibrational modes in medium-sized to large molecules can be treated as uncoupled diatomic molecules, at least in the absence of strong interactions with other, nearby energy levels. The specific problem concerning the geometric nature of vibrational modes (i.e., whether stretching or bending degrees of freedom) is addressed later in distinct sections. At this stage, bending vibrations must be treated separately because one needs to consider potential energy curves completely different from the Morse functional form. For example, in case of nondegenerate bending motions, a possibility is to adopt the Pöschl–Teller potential [24]:

$$V_{PT}(\theta) = \frac{W_0}{\cosh^2 \alpha\theta}.$$ (2.52)

Two points of interest in our algebraic analysis are that (1) this function is symmetrical under $\theta \to -\theta$ (as expected in a bending motion), and (2) it can be shown that for one-dimensional problems, the Pöschl–Teller and Morse potentials are *isospectral* (i.e., they have the same bound-state spectrum) [25]. Figure 4 shows the typical behavior of the Pöschl–Teller potential function.

To reiterate, we prefer to describe the one-dimensional model first because of its mathematical simplicity in comparison to the three-dimensional model. From a strictly historical point of view, the situation is slightly more involved. The *vibron model* was officially introduced in 1981 by Iachello [26]. In his work one can find the fundamental idea of the dynamical symmetry, based on U(4), for realizing an algebraic version of the three-dimensional Hamiltonian operator of a single diatomic molecule. After this work, many other realizations followed (see specific

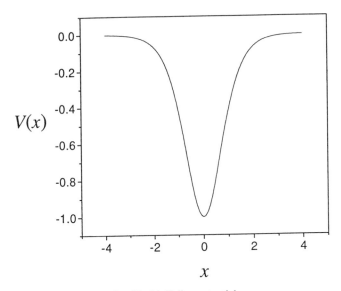

Figure 4. Pöschl–Teller potential curve.

references throughout this article). However, the procedures required to apply the one-dimensional algebraic approach in a systematic fashion to the study of complex molecular systems are quite recent [27]. It should also be noted that several preliminary (and very important) works on simple one-dimensional problems predate the official introduction of the three-dimensional vibron model [28, 29].

1. U(2) Model

The main purpose of this section is to illustrate the explicit algebraic realization of a one-dimensional uncoupled anharmonic oscillator, which will be then used as a building block for the construction of an algebraic molecular model in Section III. As specified in Sections II.B and II.C, we want to use a U(2) dynamical algebra to reproduce, by means of a proper dynamical symmetry procedure, the energy spectrum of either a one-dimensional Morse potential (for stretching modes) or a one-dimensional Pöschl–Teller potential (for nondegenerate bending modes). The fact that we have to start from the algebra U(2) leads to a very simple mathematical situation on one side and to a severe physical limitation on the other side. Using a one-dimensional approach is equivalent to disregarding completely the rotational degrees of freedom. Such a drawback will be overcome with the introduction of the three-dimensional approach. However, we recall that in many situations of practical interest, it is more important to provide a global vibrational picture of a certain molecule rather than to give a detailed description of the rotational part. Whenever the experimental resolution is not very high (permitting simultaneous observation of a very broad portion of the spectrum), a Hamiltonian model of vibrational polyads, anharmonic couplings, and infrared and Raman intensities is the preferred answer in quantitative terms.

Let us start by considering the algebraic study of the U(2) model. First, we need to realize this algebra by means of four (n^2) operators in terms of two creation–annihilation boson operators $\hat{t}^\dagger$ and $\hat{s}^\dagger$ (and $\hat{t}$ and $\hat{s}$):

$$\hat{S}_+ = \hat{t}^\dagger \hat{s} , \qquad \hat{S}_- = \hat{s}^\dagger \hat{t} , \qquad \hat{S}_z = (\hat{t}^\dagger \hat{t} - \hat{s}^\dagger \hat{s})/2 ,$$

$$\hat{N} = (\hat{t}^\dagger \hat{t} + \hat{s}^\dagger \hat{s}) , \tag{2.53}$$

which is often referred to as the Schwinger realization of U(2) [30]. The operators $\hat{t}^\dagger$ and $\hat{s}^\dagger$ (and $\hat{t}$ and $\hat{s}$) satisfy the commutation relations

$$[\hat{s}, \hat{t}] = [\hat{s}^\dagger, \hat{t}^\dagger] = 0 , \qquad [\hat{s}, \hat{s}^\dagger] = [\hat{t}, \hat{t}^\dagger] = 1 . \tag{2.54}$$

This procedure can readily be related to the general equations (2.36), (2.38), and (2.39). By introducing the linear combinations

$$\hat{S}_x = (\hat{S}_+ + \hat{S}_-)/2 , \qquad \hat{S}_y = (\hat{S}_+ - \hat{S}_-)/2i , \qquad (2.55)$$

one recognizes the same algebraic structure underlying the rotational group SO(3). The operators $\hat{S}_x$, $\hat{S}_y$, and $\hat{S}_z$ satisfy the same commutation relations as the angular momentum operators $\hat{J}_x$, $\hat{J}_y$, and $\hat{J}_z$ [Eq. (2.18)]. This corresponds to realizing in an explicit way the decomposition of the U(2) algebra into direct factors [as already done for the algebra U(3), Eq. (2.33)]:

$$U(2) = U(1) \otimes SU(2) . \qquad (2.56)$$

One can also state that SU(2) is *isomorphic* to SO(3) itself. The remaining number $\hat{N}$ operator of Eq. (2.53) is associated with U(1). For this reason, whenever the eigenvalue N (number of boson states) is fixed, it is not necessary to distinguish between U(2) and SU(2). At the same time, it is important to keep in mind that because of (2.56), U(2) and SU(2) have different algebraic structures from a general point of view. For orthogonal groups, on the other hand, $O(n)$ and $SO(n)$ have the same algebraic structure and we shall make use of the same notation convention in both cases.

The next step is to construct the second-quantized version of the Hamiltonian operator in terms of (2.53) and powers thereof. By following the procedure outlined in Eq. (2.39), we obtain

$$\hat{H} = E_0 + e_z \hat{S}_z + e'_x \hat{S}_x^2 + e'_y \hat{S}_y^2 + e'_z \hat{S}_z^2 + \cdots . \qquad (2.57)$$

Such a Hamiltonian corresponds to the dynamical algebraic study of a generic one-dimensional potential $V(r)$. In principle, one can find the spectrum of (2.57) by means of a numerical diagonalization of the Hamiltonian operator in the $(N + 1)$-dimensional basis corresponding to the irreducible representation of U(2) labeled by N [see below for a comment on the labeling scheme holding for U(2)]. The advantage of the present procedure is, however, that one can consider certain dynamical symmetries by including in (2.57) only invariant operators of the subalgebras U(1) and O(2) of U(2). These, in turn, are isomorphic algebraic structures [i.e., the dynamical symmetry procedure based on U(2) is unique from a strictly topological (algebraic) viewpoint]. Nonetheless, it is useful to treat these two cases separately, as they will lead to some interesting consequences when extended to higher-dimensional models.

Starting from U(2), we introduce two dynamical symmetries, (a) and (b), corresponding to the chains

$$\text{(a)} \ \ U(2) \supset U(1) \,,$$
$$\text{(b)} \ \ U(2) \supset O(2) \,. \tag{2.58}$$

Chain (a) is characterized by the following algebraic ket:

$$\left| \begin{matrix} U(2) & \supset & U(1) \\ N & & n \end{matrix} \right\rangle, \qquad n = N, N-1, \ldots, 0 \,. \tag{2.59}$$

To begin with, we observe that the irreducible representation of U(2) is labeled by the single number N. A generic irreducible representation of U(2) would require two independent numbers for a complete labeling. However, as pointed out in Section II.B, by virtue of the boson character of the algebraic realization of U(2), one just has to use symmetric irreducible representations of the algebra (corresponding to a Young tableau with a single row of N boxes). Chain (b) is characterized by the following algebraic ket:

$$\left| \begin{matrix} U(2) & \supset & O(2) \\ N & & m \end{matrix} \right\rangle, \qquad m = \pm N, \pm(N-2), \ldots, \pm 1 \ \text{or} \ 0 \,. \tag{2.60}$$

Second, within a given irreducible representation N of U(2) (i.e., for fixed N), one can use U(2) and SU(2) as equivalent algebraic structures. We also know that SU(2) is isomorphic to SO(3), which in turn is intimately connected to the action of the angular momentum operator. Consequently, in view of the further isomorphism between U(1) and O(2), one could suspect that both kets (2.59) and (2.60) recall to mind the usual branching law for the quantum numbers j and m_j of the rotational group chain $SO(3) \supset SO(2)$. This is indeed the case, and it can be shown by means of a simple algebraic manipulation that the branching laws for chains (a) and (b) are essentially the same as that of the angular momentum.

We now go to the problem of construction of a dynamical symmetric Hamiltonian operator based on chains (a) and (b). Starting from chain (a), one can write

$$\hat{H}^{(a)} = E_0 + e_1 \hat{C}^{(1)}_{U(1)} + e_2 \hat{C}^{(2)}_{U(1)} \,, \tag{2.61}$$

in which are introduced linear and quadratic invariant (Casimir) operators of U(1). By recalling the decomposition (2.56) of U(2) in terms of U(1) and SU(2), it should be clear that the invariant operator of U(1) corresponds to the number operator $\hat{n}$, whose eigenvalue is n ($n = N, N - 1, \ldots, 0$). In place of (2.61) we can write

$$\hat{H}^{(a)} = E_0 + e_1 \hat{n} + e_2 \hat{n}^2 , \tag{2.62}$$

which is trivially diagonal in the basis (2.59) with eigenvalues

$$E^{(a)}(n) = E_0 + e_1 n + e_2 n^2 , \qquad n = N, N - 1, \ldots, 0 . \tag{2.63}$$

By considering, for example, $e_2 = 0$, one recovers the energy spectrum of the one-dimensional *harmonic* oscillator truncated to the maximum vibrational quantum number $n = N$. Instead, by letting $e_2 \neq 0$, one introduces a certain amount of anharmonicity in the energy spectrum. We postpone until Section V.C the general discussion concerning the geometric (or semiclassical) interpretation of this simple result. Let us go now to chain (b). The Hamiltonian operator with this dynamical symmetry has the following form:

$$\hat{H}^{(b)} = E_0 + A_1 \hat{C}^{(1)}_{O(2)} + A_2 \hat{C}^{(2)}_{O(2)} , \tag{2.64}$$

in which, again, we introduce the invariant operators (linear and quadratic) of O(2). These are obviously in close connection with $\hat{J}_z$, since we are basically writing, in a different formulation, the operator structure of the symmetry scheme $SO(3) \supset SO(2)$. The eigenvalues of this Hamiltonian operator, in the basis (2.60), are

$$E^{(b)}(m) = E_0 + A_1 m + A_2 m^2 ,$$
$$m = \pm N, \pm(N - 2), \ldots, 1 \text{ or } 0 . \tag{2.65}$$

This corresponds to an anharmonic sequence of levels labeled by m (the total number of states being once more equal to $N + 1$). This quantum number, m, should not be confused with the quantum number m_j associated with the projection of the angular momentum operator. The most interesting situations occurs with the particular choice $A_1 = 0$, $A \equiv A_2 \neq 0$ in (2.64) and (2.65). In this case it is possible to put the spectrum (2.65) in a one-to-one correspondence with the bound-state spectrum of the one-dimensional Morse potential. This can be done by choosing in Eq. (2.65) only the nonnegative branch of the quantum number m. Correspondingly, we obtain

$$E^{(b)}(m) = E_0 + Am^2, \qquad m = N, N - 2, \ldots, 1 \text{ or } 0. \qquad (2.66)$$

It can be shown that due to special characteristics of the one-dimensional scheme, the originally double sequence of positive and negative m numbers is somehow related to a mirror image of the anharmonic potential around $x = 0$. This rather artificial feature disappears in two- and three-dimensional approaches. We can easily recognize in (2.66) the Morse spectrum (2.51) by introducing the usual vibrational quantum number

$$v = \frac{N - m}{2} = 0, 1, \ldots, \frac{N}{2} \text{ or } \frac{N-1}{2} \quad (N \text{ even or odd}). \qquad (2.67)$$

Written in function of v, Eq. (2.66) becomes

$$E^{(b)}(v) = E_0 + A(N - 2v)^2 = e_0 - 4Av(N - v), \qquad (2.68)$$

where $e_0 = E_0 + AN^2$. By comparing (2.68) directly with the stopped Dunham expansion (2.51), we readily obtain

$$e_0 = \frac{\omega_e}{2}\left(1 - \frac{x_e}{2}\right), \qquad A = -\frac{\omega_e x_e}{4}, \qquad N = \frac{1}{x_e} - 1. \qquad (2.69)$$

We can write these same equations in terms of the Morse potential parameters V_0, β, and μ by using Eq. (2.49):

$$e_0 = \hbar\beta\left(\sqrt{\frac{V_0}{2\mu}} - \frac{\hbar\beta}{16\mu}\right), \qquad A = -\frac{\hbar^2\beta^2}{8\mu}, \qquad N = \frac{\sqrt{8\mu V_0}}{\hbar\beta} - 1.$$

$$(2.70)$$

Let us consider more carefully the real meaning of the algebraic outcome of the dynamical symmetry based on chain (b). By looking at the eigenvalues (2.68), we understand that the dynamical Hamiltonian (2.64) [with $A_1 = 0$, that is, expressed in terms of the quadratic invariant operator of O(2)] is in *exact* correspondence with the Schrödinger equation of the one-dimensional Morse potential. Such a potential depends, basically, on two critical parameters: its depth V_0 (i.e., the dissociation energy D_e) and its distortion constant β. The equilibrium coordinate, r_e, does not play any physical role in this one-dimensional (rotationless) framework. By considering Eq. (2.70), we notice that β and V_0 are uniquely determined (for given μ) in terms of the *algebraic parameters* A and N. We can consider this same situation in terms of the Dunham parameters ω_e and $\omega_e x_e$, as shown in Eq. (2.69). We have to recall that N corresponds, on the one hand, to the particular irreducible representation of U(2) and, on the other hand, that it gives direct information on the total number of vibrational bound states, which is in fact $(N/2) + 1$ or $(N + 1)/2$ (for N even or odd; we are working only with the positive branch of the quantum number m). At this point, from Eq. (2.69), we can write

$$x_e = \frac{1}{N + 1}, \qquad (2.71)$$

which establishes a significant link between the algebraic representation and an important physical aspect of this situation: The number N [labeling the irreducible representation of U(2)] is directly related to the anharmonicity of the Morse potential. This fact suggests a possible strategy for computing the vibrational spectrum of a diatomic molecules in the present framework. The first step is to obtain reasonable numerical estimates of ω_e and $\omega_e x_e$ (or x_e) by means of either a direct computation or a numerical fitting procedure over a convenient portion of the observed spectrum. Having obtained these numbers, one gets N [thus fixing a particular irreducible representation of U(2)] and A. Finally, these values can be inserted in Eq. (2.68) to extend the calculation eventually to hitherto unknown levels, up to the dissociation energy. The first, quite obvious comment is that this procedure is precisely equivalent to a Dunham analysis based on the two-parameter expansion (2.51), the only difference being in the language adopted. As a matter of fact, in such circumstances, the algebraic model is merely a formal reiteration of traditional wave mechanics. However, one can also consider higher powers of Casimir operators in the Hamiltonian (2.64):

$$\hat{H} = E_0 + \sum_{k=2} A_k [\hat{C}^{(1)}_{O(2)}]^k . \qquad (2.72)$$

From a traditional point of view, such a Hamiltonian corresponds to a potential function given as a sum over integer powers of one-dimensional Morse potentials,

$$V(r) = \sum_k V_k \{1 - \exp[-\beta(x - x_e)]\}^k . \qquad (2.73)$$

A remarkable difference between the set of eigenvalues of (2.72) and the corresponding Dunham series is that the latter has to be seen as an expansion around a *harmonic* limit, while the former is an expansion

around a (Morse) *anharmonic* limit. Two relevant consequences stem-
ming for this fact are that (1) the Hamiltonian (2.72) should give reliable
eigenvalues, typically within few terms, and (2) a direct comparison
between the Dunham and the Morse expansions demonstrates the
existence of certain relations among the respective parameters. As a
simple example, let us consider a Morse expansion stopped within two
terms such as the one associated with the algebraic Hamiltonian,

$$\hat{H} = E_0 + A^{(1)}\hat{C}^{(2)}_{O(2)} + A^{(2)}[\hat{C}^{(2)}_{O(2)}]^2 . \tag{2.74}$$

The corresponding eigenspectrum is given by

$$E(v) = E_0 + A^{(1)}(N - 2v)^2 + A^{(2)}(N - 2v)^4 . \tag{2.75}$$

The Dunham expansion to the same order in v must be written as

$$E_D(v) = \omega_e(v + \tfrac{1}{2}) - \omega_e x_e(v + \tfrac{1}{2})^2 + y_{30}(v + \tfrac{1}{2})^3 + y_{40}(v + \tfrac{1}{2})^4 . \tag{2.76}$$

This expression depends on the four parameters ω_e, $\omega_e x_e$, y_{30}, and y_{40},
while the Morse expansion (2.75) depends on the three parameters N,
$A^{(1)}$, and $A^{(2)}$. One then concludes that the Dunham parameters cannot
be independent. We discuss this specific problem more generally in
Section II.C.2.

Let us now consider a simple numerical example to highlight the
physical meaning of the algebraic parameters involved. The diatomic
molecule HBr possesses a vibrational spectrum whose levels can be
reproduced fairly well by means of the Dunham expansion (2.51) with
the following parameters: $\omega_e = 2649.67 \text{ cm}^{-1}$ and $\omega_e x_e = 45.2 \text{ cm}^{-1}$. By
using Eqs. (2.69), we obtain $e_0 = 1313.57 \text{ cm}^{-1}$, $A = -11.30 \text{ cm}^{-1}$, and
$N = 57.82$. In light of its algebraic nature, we round off N to the nearest
integer, $N = 58$. The corresponding Morse potential will thus support
$(N/2) + 1 = 30$ bound vibrational states (including the zero-energy level).
The resulting spectrum is shown schematically in Fig. 5. By computing
the highest excited level ($v_{MAX} = 29$), in the present approximation one
obtains $E(v_{MAX}) = E_{MAX} = 38{,}821 \text{ cm}^{-1}$, which is not in agreement with
the experimental value of the dissociation energy, $D_e = 31{,}590 \text{ cm}^{-1}$. This
can be explained in terms of an increasing discrepancy that occurs
between the Morse curve and the true potential energy at higher and
higher vibrational energies. It is, however, possible to improve the
numerical description of this situation by either adding higher powers of v
(in the Dunham expansion) or including higher powers of Casimir

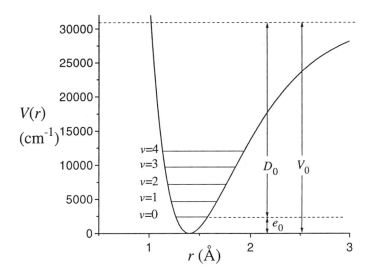

Figure 5.   Vibrational levels of HBr in the Morse potential approximation.

operators (in the algebraic model). Such higher-order terms need to be calibrated numerically over a sufficiently extended experimental database. However, it should be clear that the algebraic expansion (2.72) (equivalent to the Morse expansion) should give better results than the corresponding Dunham series with the same number of independent parameters.

We now show that the algebraic realization of the one-dimensional Morse potential can be adopted as a starting point for recovering this same problem in a conventional wave-mechanics formulation. This will be useful for several reasons: (1) The connection between algebraic and conventional coordinate spaces is a rigorous one, which can be depicted explicitly, however, only in very simple cases, such as in the present one-dimensional situation; (2) for traditional spectroscopy it can be useful to know that boson operators have a well-defined differential operator counterpart, which will be appreciated particularly in the study of transition operators and related quantities; and (3) the one-dimensional Morse potential is not the unique outcome of the dynamical symmetry based on U(2). As already mentioned, the Pöschl–Teller potential, being isospectral with the Morse potential in the bound-state portion of the spectrum, can be also described in an algebraic fashion. This is particularly apparent after a detailed study of the differential version of these two anharmonic potential models. Here we limit ourselves to a brief description. A more complete analysis can be found elsewhere [25]. As a first step, we introduce the following realization of U(2) boson operators, in terms of differential operators, in a two-dimensional harmonic oscillator space:

$$\hat{s} = (x_1 + \partial/\partial x_1)/\sqrt{2}, \qquad \hat{s}^\dagger = (x_1 - \partial/\partial x_1)/\sqrt{2},$$
$$\hat{t} = (x_2 + \partial/\partial x_2)/\sqrt{2}, \qquad \hat{t}^\dagger = (x_2 - \partial/\partial x_2)/\sqrt{2}. \tag{2.77}$$

The next step consists of the translation of the algebraic eigenequations

$$\hat{N}|N, m\rangle = N|N, m\rangle,$$
$$\hat{C}^{(1)}_{O(2)}|N, m\rangle = m|N, m\rangle \tag{2.78}$$

by explicitly using the differential realization (2.77). This can be carried out conveniently by introducing, in place of coordinates (x_1, x_2),

$$x_1 = \rho \cos\varphi,$$
$$x_2 = \rho \sin\varphi, \qquad 0 \leq \rho < \infty, \quad 0 \leq \varphi < 2\pi. \tag{2.79}$$

It is now possible to show that the eigenequations (2.78) are solved by

$$\Phi_{Nm}(\rho, \varphi) = R_{Nm}(\rho) \exp(2im\varphi), \tag{2.80}$$

where $2m$ = integer, while $R_{Nm}(\rho)$ satisfies the following eigenvalue differential equation:

$$\left(-\frac{1}{2}\frac{\partial}{\partial\rho}\rho\frac{\partial}{\partial\rho} + \frac{4m^2}{\rho^2} + \rho^2 \right) R_{Nm}(\rho) = 2(N+1)R_{Nm}(\rho). \tag{2.81}$$

This equation can be put in a more familiar form by transforming the variable r according to

$$\rho^2 = (N+1)\exp(-r). \tag{2.82}$$

As a consequence, Eq. (2.81) becomes (in convenient units)

$$\left[-\frac{\hbar^2}{2\mu}\frac{d^2}{dr^2} + V(r) \right] R_{Nm}(\rho) = E(m)R_{Nm}(\rho), \tag{2.83}$$

in which $V(r)$ is the one-dimensional Morse potential and $E(m)$ has the form of Eq. (2.51). A similar procedure can also be used to obtain a different potential functional form, the one-dimensional Pöschl–Teller potential. Knowing that the Morse and Pöschl–Teller potential curves have the same bound-state spectrum (2.68) is, by itself, sufficient to assure that the U(2) dynamical algebra will provide an adequate description of both cases. For the bending potential, an explicit differential realization is achieved by considering, in place of Eqs. (2.78), the well-known spherical realization of SO(3) [or SU(2)], leading to the differential eigenvalue equation

$$\left[-\frac{1}{\sin \vartheta} \frac{\partial}{\partial \vartheta} \left(\sin \vartheta \frac{\partial}{\partial \vartheta} \right) + \frac{m^2}{\sin^2 \vartheta} \right] F_{jm_j}(\vartheta) = j(j+1)F_{jm_j}(\vartheta) , \quad (2.84)$$

in which j and m_j have the usual definitions. Finally, in place of the change of variable (2.82), one replaces ϑ with

$$\cos \vartheta = \tanh \xi , \qquad -\infty < \xi < \infty , \qquad (2.85)$$

which leads to

$$\left[-\frac{d^2}{d\xi^2} - \frac{j(j+1)}{\cosh^2 \xi} \right] F_{jm_j}(\xi) = -m^2 F_{jm_j}(\xi) \qquad (2.86)$$

[i.e., the one-dimensional Schrödinger equation for the Pöschl–Teller potential (2.52).] Further aspects concerning bending degrees of freedom are considered in other parts of this article.

A final comment about wavefunctions. We already have analytical expressions for wavefunctions of the harmonic, Morse, and Pöschl–Teller potentials in the one-dimensional case. They can be obtained in terms of the single coordinate in a conventional differential approach. Instead, in the algebraic framework (or in a second-quantization scheme, more generally speaking), wavefunctions are expressed in terms of boson annihilation–creation operators acting on the Fock "vacuum" state, as shown in Eq. (2.29). In the specific case of U(2), for example, the basis state $|N, m\rangle$ can be written (created) as

$$|N, m\rangle = \frac{1}{\sqrt{m! \, (N-m)!}} (\hat{t}^\dagger)^m (\hat{s}^\dagger)^{N-m} |0, 0\rangle . \qquad (2.87)$$

Starting from the vacuum state, one can directly obtain any excited state.

To summarize, we have seen how the simple Lie algebra U(2) embeds within itself the analytical solutions of a potentially infinite class of one-dimensional quantum problems. Furthermore, the harmonic, Morse, and Pöschl–Teller oscillators are solved instantaneously in terms of the trivial expansions (2.61) and (2.64). However, the most important aspect of this study rests in the construction of a zeroth-order Hamiltonian which describes a single, uncoupled *anharmonic* degree of freedom. A similar result will be obtained in the three-dimensional case, with the important difference that rotational motions are explicitly included.

2. U(4) Model

The full treatment of molecular rovibrational degrees of freedom can be achieved only within a three-dimensional framework. As alluded to in Section II.B, the Schrödinger equation for the three-dimensional spherically symmetric potential $V(r)$ has an algebraic counterpart based on the Lie group U(4) and its operators. We will now describe the basic mathematical recipes leading to the proper formulation of such an alternative approach. The following scheme closely resembles the one adopted in the one-dimensional case. We will see, however, how the larger group U(4) implies a correspondingly richer algebraic structure, thus allowing for a more sophisticated description of physical situations. Unfortunately, we also expect to face more exacting mathematical problems. However, most problems have already been solved in detail and, quite often, converted into computer routines. So it is our intention to focus on the physical meaning of the algebraic questions at issue rather than on a detailed mathematical treatment. At the same time, some attention to abstract group theoretical arguments helps in achieving a real understanding of the advantages, limitations, and possible extensions of this method. Again the focus will be on those elements of the algebraic model needed to describe a single diatomic unit. A larger molecular system is treated fully in Section IV.

To begin with, it is important to specify that we are dealing with a three-dimensional diatomic molecule which can be either a rigid rotor (i.e., the rotational states are not affected by vibrational states) or a nonrigid rotor. Its three degrees of freedom can be represented by the coordinates (r, θ, φ) of Fig. 6. These coordinates define the vector character of the diatomic unit and, as a consequence, the use of U $(n + 1) = U(4)$ as a dynamical algebra. The rovibrational spectrum should be obtained in terms of an anharmonic (spherically symmetric) interaction. A convenient functional form for such a three-dimensional anharmonic potential energy is the Morse potential (2.47). If we accept the aforementioned small numerical errors (Section II.C), the Morse potential leads to the rovibrational spectrum (2.48). The most striking difference with the one-dimensional case is the existence of the rotational degree of freedom, labeled by the quantum number j. In the simplest case of the rigid rotor, it is possible to factor out completely the radial and angular parts of the overall wavefunction. In the nonrigid case, one finds floppy molecules characterized by a spacing of rotational levels compar-

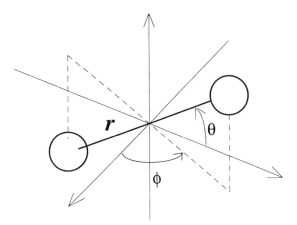

Figure 6.   Space coordinates for a diatomic molecule.

able to those of vibrational levels. Real molecules occur between two such limiting cases. We will see here how U(4) contains a reasonable description of both (rigid and nonrigid) situations.

The first step is to provide the minimum amount of algebraic information concerning U(4) and its dynamical arrangement. So we introduce a set of four boson creation (and four annihilation) operators $\hat{b}^{\dagger}_{k}(\hat{b}_{k})$, $k = 1, 2, 3, 4$. We then need to construct 16 $(=4^2)$ bilinear forms to generate U(4). By close analogy to Section II.B, we introduce the operators (2.37) satisfying the commutation relations (2.38). A key point of concern here is that we already realize that our dynamical symmetry scheme will necessarily lead to a rotationally invariant Hamiltonian operator [i.e., one respecting the physical SO(3) symmetry of the potential energy operator]. We thus expect to conserve the angular momentum throughout the several steps required by the second-quantization process. As a consequence, it is useful to emphasize the rotational features of the involved operators; one can exploit the large arbitrariness characterizing the boson operators [or, equivalently, the corresponding bilinear forms (2.37)] of the algebra of interest. In practice, these operators are replaced by boson operators behaving in a very precise way under rotations in a three-dimensional space. In the specific case of U(4), the four boson operators $\hat{b}^{\dagger}_{k}(\hat{b}_{k})$ are superseded by a scalar operator $\hat{s}^{\dagger}(\hat{s})$ and a vector operator $\hat{p}^{\dagger}_{\mu}(\hat{p}_{\mu})$, $\mu = 0, \pm 1$. The scalar and vector boson operators have defined rotational and parity characters 0^{+} and 1^{-}, respectively. In other words, the rotational invariance of the physical system strongly suggests the adoption of spherical "algebraic coordinates" (boson operators) in place of Cartesian coordinates. A deeper analysis of

tensor calculus and rotational symmetries leads to the important definition of the spherical tensor operator in the Racah sense [19]: the operator
$\hat{T}^{(\lambda)}_{\mu}$ (a rank-λ tensor, with $2\lambda + 1$ components, $\mu = -\lambda,\ -\lambda + 1, \ldots,$
$\lambda - 1,\ \lambda$) is said to constitute a spherical tensor [with respect to SO(3)
operations] if it satisfies the commutation relations given by

$$[\hat{J}_z, \hat{T}^{(\lambda)}_{\mu}] = \mu \hat{T}^{(\lambda)}_{\mu} ,$$

$$[\hat{J}_{\pm}, \hat{T}^{(\lambda)}_{\mu}] = \sqrt{(\lambda \mp \mu)(\lambda \pm \mu + 1)}\,\hat{T}^{(\lambda)}_{\mu \pm 1} .$$

(2.88)

(A more general form exists for operators acting under algebras larger
than SO(3). This aspect can be further addressed by extension of the
Wigner–Eckart theorem to larger symmetries [14].) Following the defini-
tion above, one can show that while $\hat{s}^{\dagger}$ and $\hat{p}^{\dagger}_{\mu}$ transform as spherical
tensor operators, the corresponding annihilation operators $\hat{s}$ and $\hat{p}_{\mu}$ do
not transform as requested. This problem is usually solved by introducing, in place of $\hat{s}$ and $\hat{p}_{\mu}$, the operators

$$\tilde{s} = \hat{s} , \qquad \tilde{p}_{\mu} = (-)^{\mu} \hat{p}_{-\mu} ,$$

(2.89)

in which we prefer to transform both the scalar and the vector operators,
although this should not be necessary for the scalar part. The four boson
operators (2.89) constitute a set of spherical tensor operators which can
now be coupled, in bilinear products, to give U(4). It is also worthwhile
to note here that the definition of spherical tensor operator could lead in
the specific case of the angular momentum operator to some confusion
between raising or lowering ($\hat{J}_{\pm}$) and spherical ($\hat{J}_{\mu}$) components. The
former, defined as $\hat{J}_{\pm} = \hat{J}_x \pm i\hat{J}_y$, are *not* spherical tensor operators, as
they do not satisfy (2.88). Instead, the latter are defined as

$$\hat{J}_{\pm 1} = \mp \frac{1}{\sqrt{2}}(\hat{J}_x \pm i\hat{J}_y) , \qquad \hat{J}_0 = \hat{J}_z .$$

(2.90)

These are indeed spherical tensor operators; they are obviously related to
step operators by means of $\hat{J}_{\pm 1} = \mp \hat{J}_{\pm}/\sqrt{2}$. The typical group theoretical
route for constructing bilinear forms respecting the SO(3) symmetry,
contained in U(4), is to make use of tensor products of boson operators.

  Before providing more detail on this procedure, we prefer to outline
the strategy specifically adopted in the U(4) problem. By analogy with
the U(2) case, we need to perform two distinct steps: (1) to identify the
subalgebra chains of U(4) closing in SO(3), and (2) to construct the
Hamiltonian operator in a dynamical symmetry sense. The first step is a
strictly Lie algebraic question, which can be solved by making explicit use

of commutation relations and of tensor product methods. The second step requires a precise knowledge of invariant operators (i.e., the algebraic structure) of subalgebra chains of U(4). More generally speaking, though, the starting point of this scheme is, again, the second-quantized expansion of the Hamiltonian operator given by Eq. (2.36). In its most general form, such expansion depends on a certain set of arbitrary parameters. By explicitly taking into account the transformation properties of the boson operators, the total number of arbitrary parameters can be reduced dramatically. In practice this is done by using, in place of the original expansion (2.36) [often referred to an an "uncoupled form" of the U(4) realization], a different expansion based on boson operators coupled with respect to SO(3) group elements (i.e., rotations in a three-dimensional space).

We define the tensor product [31] [with respect to SO(3)] between two (spherical) tensor operators $\hat{T}^{(\lambda_1)}_{\mu_1}$, $\hat{S}^{(\lambda_2)}_{\mu_2}$, $\mu_i = -\lambda_i, \ldots, \lambda_i$ ($i = 1, 2$) as the quantity

$$\hat{W}^{(\lambda)}_{\mu} \equiv [\hat{T}^{(\lambda_1)} \times \hat{S}^{(\lambda_2)}]^{(\lambda)}_{\mu} = \sum_{\mu_1, \mu_2} \langle \lambda_1 \mu_1 \lambda_2 \mu_2 | \lambda \mu \rangle \hat{T}^{(\lambda_1)}_{\mu_1} \hat{S}^{(\lambda_2)}_{\mu_2} , \quad (2.91)$$

where $\langle \lambda_1 \mu_1 \lambda_2 \mu_2 | \lambda \mu \rangle$ denotes a Clebsch–Gordan coefficient [32]. With such a definition we obtain a rank-λ spherical tensor operator (with $2\lambda + 1$ components labeled by μ) starting from the two rank-λ_i spherical tensor operators $\hat{T}$ and $\hat{S}$. Let us examine the actual meaning of this definition by taking a simple example. By considering the angular momentum operator $\hat{J} = (\hat{J}_x, \hat{J}_y, \hat{J}_z)$, we can construct the spherical components of such rank 1 tensor operator by using Eq. (2.90). It is then possible to obtain several tensor products of $\hat{J}$ with itself. By making explicit use of some elementary properties of Clebsh–Gordon coefficients and of the well-known commutation relations of angular momentum operators, we compute the following products:

$$\hat{B}_{00} \equiv [\hat{J} \times \hat{J}]^{(0)}_0 = \frac{1}{\sqrt{3}} (\hat{J}_{+1}\hat{J}_{-1} + \hat{J}_{-1}\hat{J}_{+1} - \hat{J}_0\hat{J}_0) = -\frac{1}{\sqrt{3}} \hat{J}^2 ,$$

$$\hat{B}_{10} \equiv [\hat{J} \times \hat{J}]^{(1)}_0 = \frac{1}{\sqrt{2}} (\hat{J}_{+1}\hat{J}_{-1} - \hat{J}_{-1}\hat{J}_{+1}) = -\frac{1}{\sqrt{2}} \hat{J}_0 ,$$

$$\hat{B}_{1-1} \equiv [\hat{J} \times \hat{J}]^{(1)}_{-1} = \frac{1}{\sqrt{2}} (\hat{J}_0\hat{J}_{-1} - \hat{J}_{-1}\hat{J}_0) = -\frac{1}{\sqrt{2}} \hat{J}_{-1} ,$$

$$\hat{B}_{11} \equiv [\hat{J} \times \hat{J}]^{(1)}_1 = \frac{1}{\sqrt{2}} (\hat{J}_{+1}\hat{J}_0 - \hat{J}_0\hat{J}_{+1}) = -\frac{1}{\sqrt{2}} \hat{J}_{+1} .$$

$$(2.92)$$

These relations show that by starting from the tensor product of $\hat{J}$ (rank 1 tensor or vector operator) with itself, we can construct a scalar quantity ($\hat{B}_{00}$), a vector quantity ($\hat{B}_{1\mu}$, $\mu = 0, \pm 1$), and a quadrupole ($\hat{B}_{2\mu}$, $\mu = 0, \pm 1, \pm 2$, not shown here). It is important to understand that we use the terms *scalar*, *vector*, and so on, with respect to spherical [SO(3)] transformation laws. This is equivalent to saying that the $\hat{B}_{\lambda\mu}$'s must satisfy (2.88). We see, for example, that

$$[\hat{J}_z, \hat{B}_{10}] = -\frac{1}{\sqrt{2}}[\hat{J}_z, \hat{J}_z] = 0 \, ,$$

$$[\hat{J}_\pm, \hat{B}_{10}] = -\frac{1}{\sqrt{2}}[\hat{J}_\pm, \hat{J}_z] = \pm\frac{1}{\sqrt{2}}\hat{J}_\pm = -\hat{J}_{\pm 1} = \sqrt{2}\hat{B}_{1\pm 1} \, ,$$

(2.93)

is in manifest agreement with Eq. (2.88). As a particular case of tensor product, we can also introduce the scalar product defined as

$$(\hat{T}^{(\lambda)} \cdot \hat{S}^{(\lambda)}) = (-)^\lambda \sqrt{2\lambda + 1} \, [\hat{T}^{(\lambda)} \times \hat{S}^{(\lambda)}]_0^{(0)} = \sum_\mu (-)^\mu \hat{T}_\mu^{(\lambda)} \hat{S}_\mu^{(\lambda)} \, .$$

(2.94)

As an alternative set of U(4) generators, it is now possible to consider the following *coupled* spherical tensor operators:

$$\left.\begin{array}{l} \hat{n}_s \equiv [\hat{s}^\dagger \times \tilde{s}]_0^{(0)} \\[4pt] \hat{n}_p \equiv -\sqrt{3}[\hat{p}^\dagger \times \tilde{p}]_0^{(1)} \end{array}\right\} \qquad \text{scalar (one component)} \, ,$$

$$\hat{J}_\mu \equiv \sqrt{2}[\hat{p}^\dagger \times \tilde{p}]_\mu^{(1)} \qquad \text{pseudovector (three components)} \, ,$$

$$\left.\begin{array}{l} \hat{D}_\mu \equiv [\hat{p}^\dagger \times \tilde{s} + \hat{s}^\dagger \times \tilde{p}]_\mu^{(1)} \\[4pt] \hat{D}_\mu' \equiv i[\hat{p}^\dagger \times \tilde{s} - \hat{s}^\dagger \times \tilde{p}]_\mu^{(1)} \end{array}\right\} \qquad \text{vector (three components)} \, ;$$

$$\hat{Q}_\mu \equiv [\hat{p}^\dagger \times \tilde{p}]_\mu^{(2)} \qquad \text{rank 2 tensor (five components)} \, .$$

(2.95)

These operators must be seen as a replacement for the uncoupled bilinear operators $\hat{G}_{ij}$ [Eq. (2.37)]. For example, the operator $\hat{Q}$ is given explicitly by

$$\hat{Q}_\mu = \sum_{\mu_1, \mu_2} \langle 1\mu_1 1\mu_2 | 2\mu \rangle \hat{p}_{\mu_1}^\dagger \tilde{p}_{\mu_2} \, , \qquad \mu = 0, \pm 1, \pm 2 \, . \qquad (2.96)$$

The single components $\hat{Q}_0$, $\hat{Q}_{\pm 1}$, $\hat{Q}_{\pm 2}$ can be obtained by computing the

Clebsch–Gordan coefficients in

$$\hat{Q}_0 = \sum_{\mu_1,\mu_2} \langle 1\mu_1 1\mu_2 | 20 \rangle \hat{p}^{\dagger}_{\mu_1} \tilde{p}_{\mu_2} \,, \qquad \mu_1 + \mu_2 = 0 \,, \qquad (2.97)$$

which leads to

$$\hat{Q}_0 = \langle 1-111|20 \rangle \hat{p}^{\dagger}_{-1}\tilde{p}_1 + \langle 1010|20 \rangle \hat{p}^{\dagger}_0\tilde{p}_0 + \langle 111-1|20 \rangle \hat{p}^{\dagger}_1\tilde{p}_{-1}$$

$$= \frac{1}{\sqrt{6}} (\hat{p}^{\dagger}_{-1}\tilde{p}_1 + 2\hat{p}^{\dagger}_0\tilde{p}_0 + \hat{p}^{\dagger}_1\tilde{p}_{-1}) \,, \qquad (2.98)$$

with similar relations for the remaining components. In Eq. (2.98), $\hat{p}^{\dagger}_{\mu}(\tilde{p}_{\mu})$ denote the spherical components of the vector operator $\hat{p}^{\dagger}(\tilde{p})$, that is,

$$\hat{p}^{\dagger}_0 = \hat{p}^{\dagger}_z \,, \qquad \hat{p}^{\dagger}_{\pm 1} = \mp \frac{1}{\sqrt{2}} (\hat{p}^{\dagger}_x \pm i\hat{p}^{\dagger}_y) \,. \qquad (2.99)$$

For reasons that will become clear in Section V, the operators $\hat{D}_{\mu}$ and $\hat{D}'_{\mu}$ in (2.95) have basically the same physical meaning as the coordinate and momentum operators. The final result of this replacement is that one can write the second-quantized form of the spherical invariant Hamiltonian operator in terms of U(4) coupled boson operators (up to two-body interactions and with the compact notation $\hat{b}^{\dagger}_1 = \hat{s}^{\dagger}$, $\hat{b}^{\dagger}_{\kappa} = \hat{p}^{\dagger}_{\kappa-3}$, $\kappa = 2, 3, 4$)

$$\hat{H} = E_0 + \sum_{\kappa} e_{\kappa}[\hat{b}^{\dagger}_{\kappa} \times \tilde{b}_{\kappa}]^{(0)}_0$$

$$+ \sum_{\lambda=0,1,2} \sum_{\kappa\iota\kappa'\iota'} f^{(\lambda)}_{\kappa\iota\kappa'\iota'} [[\hat{b}^{\dagger}_{\kappa} \times \hat{b}_{\iota}]^{(\lambda)} \times [\tilde{b}^{\dagger}_{\kappa'} \times \tilde{b}_{\iota'}]^{(\lambda)}] \,. \qquad (2.100)$$

Equation (2.100) should be compared with Eq. (2.36). Several terms of Eq. (2.100) can be related by virtue of the fact that the Hamiltonian is a Hermitian operator, and by virtue of the conservation of the total boson number, as stated by the invariance of the U(4) number operator, within a given irreducible representation.

The next step is to identify what subalgebra chains, closing in SO(3), can be obtained starting from U(4). This is a standard group theoretical problem [14]. As discussed in Section II.B, a subalgebra of U(4) closes under the same commutation relations as U(4) itself. Moreover, we are looking only for those subalgebras of U(4) containing the generators of

SO(3), which, in coupled notation, can be written as

$$[\hat{p}^\dagger \times \tilde{p}]^{(1)}_\mu , \qquad \mu = 0, \pm 1 . \tag{2.101}$$

By making explicit use of the commutation relations defining the U(4) structure, one can show that two subalgebra chains can be identified, namely those generated by

$$\hat{n}_p , \quad \hat{J}_\mu , \quad \hat{Q}_\mu \tag{2.102}$$

and

$$\hat{J}_\mu , \quad \hat{D}_\mu \quad (\text{or } \hat{D}'_\mu) . \tag{2.103}$$

Operators (2.102) give the U(3) algebra because they are defined only in terms of p boson operators, which results in them being closed under the same commutation relations for U(4). Operators (2.103) give the SO(4) algebra for similar, if slightly more subtle reasons. To be more precise, the double choice between operators $\hat{D}_\mu$ and $\hat{D}'_\mu$ is such that one actually obtains *two* distinct subalgebra chains of U(4) based on SO(4). These two chains are, however, closely related through a canonical transformation among p boson operators. In common practice, it will not be necessary to use both chains, as their properties are extremely similar.

As a consequence of the aforementioned discussion, U(4) is considered the starting point for the two chains given by

$$
\begin{array}{cl}
\text{(a)} & \left| \begin{array}{cccc} U(4) \supset U(3) \supset O(3) \supset O(2) \\ N \qquad n_p \qquad j \qquad m_j \end{array} \right\rangle , \\[2ex]
\text{(b)} & \left| \begin{array}{cccc} U(4) \supset O(4) \supset O(3) \supset O(2) \\ N \qquad \omega \qquad j \qquad m_j \end{array} \right\rangle .
\end{array}
\tag{2.104}
$$

In these expressions, we use O(n) in place of SO(n) and we introduce certain labels for the pertinent irreducible representations of subalgebras of U(4). The algebraic rules for obtaining these quantum numbers are as follows:

Chain (a): $\quad n_p = N, N - 1, \ldots, 1, 0 ,$

$\qquad\qquad j = n_p, n_p - 2, \ldots, 1 \text{ or } 0 \ (n_p \text{ odd or even}) ,$

$\qquad\qquad m_j = -j, -j + 1, \ldots, j - 1, j .$

Chain (b): $\quad \omega = N, N - 2, \ldots, 1 \text{ or } 0 \ (N \text{ odd or even}) ,$

$$\tag{2.105}$$

$$j = \omega, \omega - 1, \ldots, 1, 0,$$

$$m_j = -j, -j + 1, \ldots, j - 1, j.$$

In both chains (a) and (b), N denotes a given, totally symmetric irreducible representation of U(4) (corresponding to a Young tableau with N boxes arranged in a single row). In chain (a) n_p denotes the irreducible representations of U(3) contained in N. It can be shown that starting from a symmetric irreducible representation of U(4) [or U(n) with arbitrary n] only symmetric irreducible representations of U(3) [or U($n - 1$)] can be obtained. This fact explains why a single number n_p (corresponding to a Young tableau with n_p boxes in a single row) is needed here. In chain (b), ω denotes the irreducible representations of O(4) contained in N. Once again, we consider here only symmetric irreducible representations of O(4). [Otherwise, we should use *two* numbers to label a generic irreducible representation of O(4).] In both chains, j and m_j are defined as in Section II.A for the reduction O(3) $\supset$ O(2).

For a given irreducible representation of U(4) we are left with three quantum numbers to label the physical states. This is the natural outcome of the quantum mechanical treatment of a three-dimensional system, in which, besides j and m_j, one has to deal with a radial quantum number [n_p in chain (a) and ω in chain (b)]. The advantage of the dynamical symmetry approach is found in the *fourth* quantum number N. Such an "extra" quantum number has the important role of allowing access to entire families of distinct physical situations. In the specific three-dimensional case, we will see how the number N spans situations characterized by different anharmonicities or, equivalently, by a different number of bound states.

In regard to constructing Hamiltonian operators based on algebraic chains arising from U(4), we have to consider the two possibilities, corresponding to chains (a) and (b) of Eq. (2.104), in terms of Casimir operators of subalgebras of U(4). One realizes special cases of the general Hamiltonian (2.100), which can be referred to as dynamical symmetries U(3) [chain (a)] and O(4) [chain (b)]. Chain (a) leads to the following general Hamiltonian operator:

$$\hat{H}^{(a)} = E_0 + \alpha^{(1)} \hat{C}_{U(3)}^{(1)} + \alpha^{(2)} \hat{C}_{U(3)}^{(2)} + \beta \hat{C}_{O(3)}^{(2)}, \qquad (2.106)$$

in which, as usual, we disregard the O(2) term related to the symmetry breaking caused by external fields. Standard rules of Lie algebra allow one to write the previous expression, in terms of coupled U(4) tensor operators, as

$$\hat{H}^{(a)} = E_0 + \alpha^{(1)}\hat{n}_p + \alpha^{(2)}\hat{n}_p(\hat{n}_p + 3) + \beta\hat{J}^2 , \qquad (2.107)$$

which is a definitely more convenient form. In fact, this operator is trivially diagonal, with eigenvalues given by

$$E^{(a)} = E^{(a)}(N, n_p, j) = E_0 + \alpha^{(1)}n_p + \alpha^{(2)}n_p(n_p + 3) + \beta j(j + 1) , \quad (2.108)$$

where the branching laws (2.105) for the quantum numbers n_p, j must be used to span the complete spectrum of eigenstates. Chain (b) leads to the algebraic Hamiltonian

$$\hat{H}^{(b)} = E_0 + A\hat{C}^{(2)}_{O(4)} + B\hat{C}^{(2)}_{O(3)} , \qquad (2.109)$$

which, in view of well-established results of algebraic theory, can also be written as

$$\hat{H}^{(b)} = E_0 + A(\hat{D}^2 + \hat{J}^2) + B\hat{J}^2 , \qquad (2.110)$$

where we have introduced the coupled tensor operators of U(4) [Eq. (2.95)]. The eigenvalues of this Hamiltonian can now be obtained in the basis (b) of Eq. (2.104), which is, again, in diagonal form:

$$E^{(b)} = E^{(b)}(N, \omega, j) = E_0 + A\omega(\omega + 2) + Bj(j + 1) . \qquad (2.111)$$

These eigenvalues have to be calculated in terms of the branching laws (2.105) for the quantum numbers ω and j.

Let us now consider the physical meaning of the Hamiltonian operators introduced above. To start with, Fig. 7 reproduces the spectrum obtained from Eq. (2.108) with the numerical values $N = 10$, $\alpha^{(1)} = 10.0$, $\alpha^{(2)} = 0.5$, $\beta = 1.0$. Such a sequence of levels can be better understood by noticing that the particular choice $\alpha^{(2)} = \beta = 0.0$ leads to the purely isotropic harmonic oscillator in three dimensions. Its correct degeneracy is recovered by means of Eqs. (2.105) and (2.108), which provide for a given n_p, $(n_p + 1)(n_p + 2)/2$ degenerate levels. It is important to stress that the present physical picture of a three-dimensional oscillator is such that the minimum of potential energy is attained at the origin of the radial coordinate, $r = 0$. As a consequence, the rotational structure is somewhat "embedded" in the vibrational one, as is shown clearly in the typical degeneracy pattern for states of well-defined angular momentum. By assuming that the minimum potential energy is attained for $r \neq 0$ or that the potential energy departs from a purely parabolic

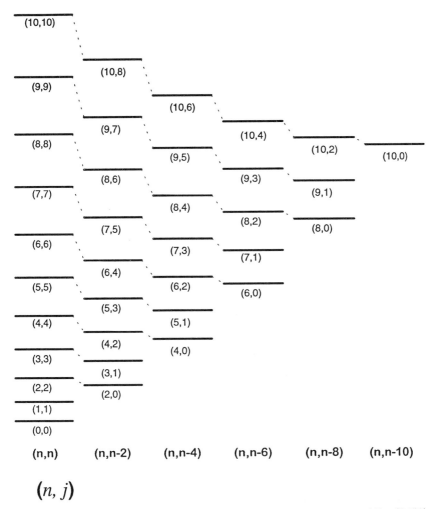

Figure 7.   Energy spectrum corresponding to the Hamiltonian operator of Eq. (2.108).

dependence, the physical picture corresponds to that of a nonrigid, three-dimensional rovibrator. In the algebraic framework, its energy spectrum is achieved by letting $\alpha^{(2)} \neq 0$ and $\beta \neq 0$ in Eq. (2.107). It is also important to notice how $\hat{J}^2$, the invariant operator of SO(3), acts on the spectrum, by removing the original degeneracy for certain states. This can also be realized by comparing Figs. 7 and 1. By virtue of its peculiar characteristics, the Hamiltonian operator, based on chain (a), is well suited for describing physical situations in which very soft (floppy) rovibrational degrees of freedom occur [33]. An example of such a situation might be a single atom enclosed in the fullerene cage or, more generally, those situations in which one can adopt, as a zeroth-order model, the three-dimensional isotropic oscillator. It is also possible to show that this dynamical symmetry provides an excellent approximation for the computation of the bound-state spectrum of a three-dimensional Pöschl–Teller potential.

From the molecular point of view, the dynamical symmetry based on chain (b) of Eq. (2.104) definitely deserves more attention. If we start from eigenvalues (2.111) and then introduce the vibrational quantum number v defined as in Eq. (2.67), we can write the eigenvalues of the Hamiltonian operator (2.110) in the following form:

$$E^{(b)} = E^{(b)}(n, v, j) = E_0' - 4Av(N + 1 - v) + Bj(j + 1) , \quad (2.112)$$

where $E_0' = E_0 + AN(N + 2)$. By analogy with the one-dimensional case, Eq. (2.112) represents an anharmonic expansion in the quantum number v. We insist, of course, on the essential difference between the one- and three-dimensional case, as in the latter case, we also obtain the rotational contribution to the energy spectrum. As a matter of fact, the final outcome of the dynamical symmetry O(4) is the (well-approximated) rovibrational spectrum of a three-dimensional Morse oscillator. Its physical parameters can easily be related to the algebraic quantities A, N, E_0', and B by means of the following relations:

$$A = -\frac{\hbar^2 \beta^2}{8\mu} , \qquad N = \frac{\sqrt{8V_0 \mu}}{\hbar \beta} - 2 ,$$

$$E_0' = \hbar \beta \left(\sqrt{\frac{V_0}{2\mu}} + \frac{3\hbar \beta}{8\mu} \right) , \qquad B = \frac{\hbar^2}{2\mu r_e^2} , \tag{2.113}$$

which can be obtained by direct comparison of (2.112) with the spectrum (2.48). We also note that the number N [total number of bosons, label of the irreducible representation of U(4)] is related to the total number of bound states supported by the potential well. Equivalently, it can be put in a one-to-one correspondence with the anharmonicity parameters x_e by means of

$$x_e = \frac{1}{N + 2} , \tag{2.114}$$

by analogy with the one-dimensional result (2.71). Accordingly, the same conclusions can be drawn for the three-dimensional model, namely that the relation (2.114) shows the actual meaning of the additional quantum number N. One can, in fact, study an infinite class of (Morse) potentials differing in terms of their anharmonic behavior by changing the starting irreducible representation of U(4). As explained for the U(2) model, a possible way of using the U(4) approach in practical applications consists of (1) obtaining an estimate of N and (2) fitting the algebraic parameters A and B over a convenient set of experimental rovibrational levels. However, it is more interesting to discuss, in some detail, the possibility of extending this method to higher-order terms. This is equivalent to considering a Dunham-like expansion, in which one adds powers and products of Casimir operators,

$$\hat{H} = E_0 + \sum_{h,k} Y_{hk} [\hat{C}^{(2)}_{O(4)}]^h [\hat{C}^{(2)}_{O(3)}]^k . \qquad (2.115)$$

The eigenvalues of this Hamiltonian operator can be related directly to the conventional Dunham expansion in rovibrational quantum numbers given by

$$E(v, j) = E_0 + \sum_{h,k} y_{hk} (v + 1/2)^h [j(j+1)]^k . \qquad (2.116)$$

It should be clear that the use of Hamiltonian (2.115) is equivalent to an expansion in powers of Morse potential functions. As a consequence, one is expected to find certain relations between Dunham (y_{hk}) and algebraic (Y_{hk}) parameters. To be more precise, let us consider the Dunham expansion written in the form

$$E^D(v, j) = \sum_{h,k} x_{hk} v^h [j(j+1)]^k , \qquad (2.117)$$

where we basically change the zero of the energy scale for simpler notation. The algebraic expansion (2.115) leads to the following series of eigenvalues:

$$E^A(v, j) = \sum_{h,k} A_{hk} [-4v(N+1-v)]^h [j(j+1)]^k , \qquad (2.118)$$

in which the A_{hk}'s are trivially related to the Y_{hk}'s. We notice that the "vibrational unit" of Eq. (2.118) is anharmonic by virtue of the v^2 term. As a consequence, by stopping both the Dunham and the algebraic expansion at the same number of terms, we obtain higher powers of v in the algebraic expansion. This can be shown explicitly by reorganizing Eq.

(2.118) in such a way that equal powers of v are collected together:

$$E^A(v, j) = \sum_{h,k} Q_{hk} v^h [j(j + 1)]^k , \qquad (2.119)$$

where, as is easily obtained, we introduce the quantities

$$Q_{hk} = (-1)^h \sum_{m=q_h}^{h} \frac{4^m m!}{(h - m)!(2m - h)!} A_{mk} M^{2m-h} . \qquad (2.120)$$

In this expression we use, as a shorthand, $M = N + 1$. The lower limit in the sum, q_h, can be written as

$$q_h = (h + p_h)/2 , \qquad (2.121)$$

where $p_h = 0$ or 1 for h even or odd. We now obtain for the first few terms of the algebraic expansion written in the form of Eq. (2.119):

$$Q_{10} = -4A_{10}M , \qquad\qquad Q_{11} = -4A_{11}M , \qquad\qquad \ldots .$$

$$Q_{20} = 4(A_{10} + 4A_{20}M^2) , \qquad Q_{21} = 4(A_{11} + 4A_{21}M^2) , \qquad \ldots$$

$$Q_{30} = -32M(A_{20} + 2A_{30}M^2) , \quad Q_{31} = -32M(A_{21} + 2A_{31}M^2) , \quad \ldots$$

$$Q_{40} = 16(A_{20} + 12A_{30}M^2 \qquad\qquad Q_{41} = 16(A_{21} + 12A_{31}M^2$$

$$\qquad + 16A_{40}M^4) , \qquad\qquad\qquad + 16A_{41}M^4) , \qquad\qquad \ldots .$$

$$\qquad\qquad\qquad\qquad\qquad\qquad\qquad\qquad\qquad\qquad\qquad (2.122)$$

As both the Dunham and the algebraic expansions are now written in the same form, the respective parameters must be the same, $y_{hk} = Q_{hk}$. By limiting ourselves to the first four terms above (disregarding rotational contributions, $k = 0$), we can write the following relations:

$$\frac{x_{20}}{x_{10}} = \frac{Q_{20}}{Q_{10}} = -\frac{A_{10} + 4A_{20}M^2}{A_{10}M} ,$$

$$\frac{x_{30}}{x_{10}} = \frac{Q_{30}}{Q_{10}} = \frac{8(A_{20} + 2A_{30}M^2)}{A_{10}} , \qquad (2.123)$$

$$\frac{x_{40}}{x_{10}} = \frac{Q_{40}}{Q_{10}} = -\frac{4(A_{20} + 12A_{30}M^2 + 16A_{40}M^4)}{A_{10}M} .$$

These relations have general validity in that they hold for expansion of arbitrary order. Moreover, they provide a direct comparison between traditional and algebraic coefficients. Let us now consider now a Dunham expansion stopped at the v^2 term, that is, based on the two independent parameters x_{10} and x_{20}, where we ignore the rotational part. The corresponding algebraic expansion is obtained in terms of the parameters N and A_{10}, the remaining A_{20}, A_{30}, ... being zero. By putting $A_{20} = 0$ in the first of equations (2.123), we have

$$x_{20} = -\frac{x_{10}}{M} = -\frac{x_{10}}{N+1}.$$

(2.124)

This is equivalent to expression (2.114). Here we obtain $N+1$ in place of $N+2$ because of the slightly different forms adopted for the general Dunham expansion [Eqs. (2.116) and (2.117)]. If we now go to the Dunham expansion stopped at the v^3 term, we have to extend the algebraic series by adding an independent parameter, A_{20}. We can now use (2.123) with $A_{30} = A_{40} = 0$ to obtain

$$\frac{x_{20}}{x_{10}} = -\frac{1}{M} - \frac{4A_{20}M}{A_{10}}, \qquad \frac{x_{30}}{x_{10}} = \frac{8A_{20}}{A_{10}},$$

(2.125)

leading to

$$x_{30} = -\frac{2}{M}\left(\frac{x_{10}}{M} + x_{20}\right).$$

(2.126)

This equation shows a possible way of expressing x_{30} in terms of x_{10}, x_{20}, and M. This does not mean that the three Dunham parameters are dependent, as M must be seen as an arbitrary (algebraic) parameter. However, it is sometimes convenient to consider M (i.e., $N+1$) as a predetermined quantity (e.g., by using tabulated values of $\omega_e x_e$ and x_e) and to attempt a fitting procedure only of the parameters A_{hk}. As soon as one goes to expansions including terms higher than the third, the number of independent parameters starts to be effectively smaller in the algebraic than in the traditional approach. For example, by considering the v^4 expansion in the algebraic method we do not need to add further parameters as in the v^3 case, since v^4 is already included in the first two terms, A_{10} and A_{20}. This is seen explicitly by writing the fourth relation for the expansion coefficients [besides (2.125) and (2.126)] given by

$$\frac{x_{40}}{x_{10}} = -\frac{4A_{20}}{A_{10}M},$$

(2.127)

leading to

$$x_{40} = \frac{1}{M}\left(\frac{x_{10}}{M} + x_{20}\right).$$

(2.128)

These expressions can be read as a possible way of expressing x_{30} and x_{40} in terms of x_{10}, x_{20}, and M (predetermined). More generally, in the algebraic expansion including up to v^4, one needs only *three* independent parameters, A_{10}, A_{20}, and M (i.e., $N + 1$). It is possible, in fact, to find an explicit relation among the x_{hk}'s, which in this specific case is given by

$$x_{20} = \frac{2x_{10}x_{40}}{x_{30}} + \frac{x_{30}^2}{4x_{40}}. \qquad (2.129)$$

The final result is that, as suggested clearly by Eqs. (2.119) and (2.121), a Dunham expansion including terms up to v^h, based on h independent parameters, can be replaced by an algebraic expansion with q_h parameters, where q_h is the same as in Eq. (2.121). This result is the direct consequence of having considered a power series of Morse potential functions instead of harmonic terms. It is very important to understand the actual meaning of this procedure. To use fewer independent parameters implies that a different (and hopefully better) sequence of levels is now the basis of our model. The experimental evidence indicates that most (rigid or quasirigid) diatomic units behave, in first approximation, according to a Morse potential description. Consequently, the algebraic model is naturally suited to provide a good set of eigenvalues but with a smaller number of terms than those need in the traditional method.

This procedure can easily be extended to include rovibrational terms as well, for which one has to consider the general expansion

$$E(v, j) = E(v) = B_v j(j + 1) + D_v[j(j + 1)]^2 + \cdots. \qquad (2.130)$$

In this expression, the purely vibrational terms have been included in $E(v)$, while B_v depends on v according to

$$\begin{aligned} B_v &= y_{01} + y_{11}(v + \tfrac{1}{2}) + y_{12}(v + \tfrac{1}{2})^2 + \cdots \\ &= B_e - \alpha_e(v + \tfrac{1}{2}) - \gamma_e(v + \tfrac{1}{2})^2 + \cdots. \end{aligned} \qquad (2.131)$$

We can compare this expression with the algebraic one by writing

$$\begin{aligned} E(v, j) &= (Q_{00} + Q_{10}v + Q_{20}v^2 + \cdots) \\ &\quad + (Q_{01} + Q_{11}v + Q_{21}v^2 + \cdots)j(j + 1) + \cdots, \end{aligned} \qquad (2.132)$$

where we collect in the first set of parentheses those terms giving $E(v)$

and, in the second set, B_v. This means that in terms of the Q_{hk}'s, one has

$$B_v = \sum_{h=0} Q_{h1} v^h .$$ (2.133)

As an example of a possible manipulation of algebraic rovibrational series, we are now interested in obtaining an algebraic Dunham-like expansion of the experimentally significant quantity:

$$\Delta B_v \equiv B_{v+1} - B_v .$$ (2.134)

In terms of (2.133), we easily obtain

$$\Delta B_v = \sum_{h=0} \sum_{k=h+1} Q_{k1} \binom{k}{h} v^h .$$ (2.135)

This means that ΔB_v will contain higher and higher powers in v with the following coefficients:

$$v^0 \mapsto Q_{11} + Q_{21} + Q_{31} + \cdots \quad = \sum_{p=1} Q_{p1} ;$$

$$v^1 \mapsto 2Q_{21} + 3Q_{31} + 4Q_{41} + \cdots = \sum_{p=2} p Q_{p1} ;$$ (2.136)

$$v^2 \mapsto 3Q_{31} + 6Q_{41} + 10Q_{51} + \cdots = \sum_{p=2} \frac{p(p+1)}{2} Q_{p1} ;$$

$$\vdots$$

If we limit ourselves to terms up to the second order, one now accounts for up to the fourth power of v. Let us now consider explicit contributions up to v^2. By substituting (2.122) in (2.136) and by letting $A_{31} = A_{41} = 0$, we obtain

$$\Delta B_v = [-4A_{11}N + 16A_{21}(N^2 - 1)]$$
$$+ [8A_{11} + 16A_{21}(N^2 - N - 1)]v - 32A_{21}(3N + 1)v^2 .$$ (2.137)

This expression allows one to carry out a three-parameter (A_{11}, A_{21}, and N) fit of the experimentally obtained ΔB_v's. The physical meaning of the A_{hk}'s can be inferred by observing that Eq. (2.137) can also be written as

$$\Delta B_v = -4A_{11}N(1 - 2v/N) + 16A_{21}(N^2 - 1)$$
$$\times \left[1 + \frac{N^2 - 4N - 1}{N^2 - 1} v - \frac{2(3N + 1)}{N^2 - 1} v^2 \right] .$$ (2.138)

By letting $A_{21} = 0$ (algebraic first-order model), we have

$$\Delta B_v = -4A_{11}N(1 - 2v/N) , \qquad (2.139)$$

leading to the conclusion that $4A_{11}N$ is (within orders of $1/N$) the algebraic counterpart of the Dunham parameter y_{11}. This, in turn, is related directly to the lowest order nonrigidity parameter α_e, usually defined as

$$B_v = B_e - \alpha_e(v + \tfrac{1}{2}) , \qquad (2.140)$$

giving $\Delta B_v = -\alpha_e$. Expression (2.139) is somewhat more general, as it includes at its lowest order anharmonic corrections (v/N, i.e., linear in v). For even more precise fitting procedures, one can use the second-order expansion (2.138), which can be written (for $N \gg 1$) as

$$\Delta B_v \simeq -4A_{11}N(1 - 2v/N) + 16A_{21}N^2(1 + v - 6v^2/N) . \quad (2.141)$$

One can also develop algebraic expansion for other quantities of interest. A typical example is the complete analysis of rotational bands (both in infrared and Raman regimes), for which one introduces the usual R- and P-branch difference transitions $v = 1 \leftarrow v = 0$, given by

$$v_R(j + 1 \leftarrow j) - v_P(j \leftarrow j - 1) = 2B_1(2j + 1) ,$$
$$v_R(j + 1 \leftarrow j) - v_P(j + 2 \leftarrow j + 1) = 2B_0(2j + 3) . \qquad (2.142)$$

These can be written in terms of proper algebraic parameters by using the general expansion (2.133) and the definition (2.120) to compute R- and P-branch transitions [34]. Once again, the interesting aspect of this process is found in the anharmonicity of the power development in rovibrational quantum numbers. This same idea can also be applied to include higher powers of the $j(j + 1)$ term, thus taking into account the centrifugal distortion and its anharmonic dependence on v. Other essential aspects, concerning transition rules and intensities obtained in the algebraic framework, are addressed in Sections III and IV.

To summarize, we have seen how the U(4) algebraic structure is a convenient starting point for the construction of model Hamiltonian operators, widely applicable to a three-dimensional space. Depending on the specific physical situation, one has to choose between two dynamical symmetries: the first, U(3), is suited for describing very shallow potential energy functions, floppy vibrational modes, and nonrigid rovibrators; the second, O(4), is capable of providing a convenient framework for the study of rigid or quasirigid rovibrators (i.e., the typical diatomic rotational/vibrational motions). By properly adjusting the algebraic parameters, one can fit the Hamiltonian operator to give a good representation of even more complex spectra. A particularly significant feature of this approach is its anharmonic character, which, for higher-order models, requires fewer arbitrary parameters than does the traditional Dunham expansion.

Finally, in physical situations characterized by potential energy functions intermediate between purely rigid and nonrigid rovibrators, one should consider more complex algebraic treatments in which both U(3) and O(4) invariant operators are included. Consequently, the Hamiltonian operator can no longer be diagonal in the chosen algebraic basis (related to either one or the other of the two dynamical symmetries). However, matrix elements for any operator of interest have already been explicitly computed in analytical form [35].

III. ONE-DIMENSIONAL ALGEBRAIC MODELS FOR POLYATOMIC MOLECULES

A. Introduction

In Section II we explored several important aspects of the algebraic formulation for both one- and three-dimensional (exactly solvable) quantum problems by means of proper dynamical symmetries. As a result, one obtains effective Hamiltonian operators suited to describe rovibrational spectra of diatomic molecules. We have seen how these models can readily be converted to Dunham-like expansions, in product and powers of rovibrational quantum numbers, based on anharmonic terms, thereby allowing for quickly convergent fitting procedures. However, the real advantage of using Lie algebras over traditional approaches is not so obvious in the case of diatomic molecules; this becomes manifest only in the treatment of polyatomic molecules. In this section we discuss how to extend the Lie algebra techniques to polyatomic systems. However, as stated previously, this model does not take into account rotational motions. Nonetheless, it can be used to obtain a complete picture of the vibrational behavior of complex situations, falling even beyond the possibilities of a three-dimensional approach. We will see how this simple model can account for anharmonic couplings between local modes (both stretching *and* nondegenerate bending vibrations), anharmonic (Fermi) resonances, symmetry adaption of wavefunctions, and some other important aspects of molecular spectroscopy.

To build a polyatomic molecule in an algebraic sense, one has to replace each one-dimensional internal degree of freedom with a U(2) Lie algebra. It is best to start with the simplest case of two coupled stretching modes, as those occurring in a triatomic molecule, XYZ. The three-body system is characterized by $3N - 3 = 6$ rovibrational degrees of freedom. For a bent geometry there are three vibrational modes, while for a linear configuration, there are, instead, four (i.e., $3N - 5 = 4$ versus $3N - 6 = 3$, where N = number of atoms in molecule). We focus our attention on the two stretching motions involving changes in the X-Y and Y-Z interatomic coordinates. As such, these coordinates have to be seen as one-dimensional variables to be studied within the U(2) model. As will soon become clear, the XYZ bending angle can also be described as a one-dimensional coordinate in the particular case of nondegenerate bending motions. From a strictly physical standpoint, as soon as one considers two vibrating masses interacting, by means of certain force fields, with a third common mass, it is important to establish the most convenient interparticle coordinates. A typical procedure is to use *local* X-Y, Y-Z coordinates, which is the natural choice whenever there is weak coupling between X and Z atoms. Such coordinates, often referred to as *bond coordinates*, are particularly well suited for use in an algebraic language. Bond coordinates can be regarded as the second quantized version of the algebraic blocks used to construct a polyatomic molecule. For a three-atomic molecule, with its two chemical bonds, we consider, as a starting (i.e., spectrum generating) algebra, the composition of Lie algebras

$$U_1(2) \otimes U_2(2) , \qquad (3.1)$$

in which the indexes 1 and 2 refer to bonds X-Y and Y-Z, respectively. Our first goal is to understand the real meaning of the algebraic product (3.1). However, before doing so, two issues must be addressed: (1) product operations for (3.1) and (2) the assumption of weak coupling between atoms. With regard to the first issue, the algebraic structures should be added rather than multiplied among themselves, and this multiplication should be applied to the corresponding group elements. This somewhat unorthodox convention is used to emphasize the underlying product of basis states, which is the desired objective of this manipulation. Second, in the case where the three atoms have similar mass (e.g., SO_2 or CO_2), strong coupling should occur between the vibrational modes. Consequently, bond coordinates are no longer the best choice of coordinates. However, our one-dimensional algebraic model can incorporate other type of coordinate systems which are better suited to model strong coupling between atoms. So the algebraic method can accommodate both local and strongly coupled vibrational motions.

Finally, we anticipate that such an algebraic analysis will produce a starting spectrum generating algebra that contains two subalgebra chains. The first one describes two basically uncoupled, independent Morse oscillators, while the second one includes a nondiagonal, anharmonic coupling between the oscillators. As a consequence, within the product (3.1) one will recognize both the *local* and *normal* limits as regards the overall molecular behavior. Depending on the relative magnitude of the Hamiltonian parameters, it will be possible to obtain a fair good and simple description of the vibrational spectrum (for stretching modes) of any kind of triatomic molecule. It is possible to attack this algebraic problem from a very general, systematic, and detailed point of view [10]. We prefer here to adopt a milder approach, in which all the mathematical statements are slightly softened in favor of a direct physical (less algebraic, at least) insight in this subject.

B. Anharmonic Coupling Between Two Oscillators

The algebraic structure of the product (3.1) can be understood in terms of the separate commutation laws for the two (independent) $U_i(2)$ $(i = 1, 2)$ Lie algebras. These are obtained directly from the commutation laws (2.54), which are here extended to include two families of boson operators:

$$[\hat{s}_\alpha, \hat{s}_\beta^\dagger] = [\hat{t}_\alpha, \hat{t}_\beta^\dagger] = \delta_{\alpha\beta} , \qquad \alpha, \beta = 1, 2 , \qquad (3.2)$$

with all the remaining commutators being zero. We also introduce the bilinear generators [corresponding to the single algebra generators (2.37)] given by

$$\hat{G}_{11}^{(\alpha)} = \hat{s}_\alpha^\dagger \hat{s}_\alpha , \qquad \hat{G}_{12}^{(\alpha)} = \hat{s}_\alpha^\dagger \hat{t}_\alpha , \qquad \hat{G}_{21}^{(\alpha)} = \hat{t}_\alpha^\dagger \hat{s}_\alpha ,$$
$$\hat{G}_{22}^{(\alpha)} = \hat{t}_\alpha^\dagger \hat{t}_\alpha , \qquad \alpha = 1, 2 \qquad\qquad (3.3)$$

and satisfying the commutation relations [corresponding to the relations (2.38)]

$$[\hat{G}_{ij}^{(\alpha)}, \hat{G}_{hk}^{(\beta)}] = (\hat{G}_{ik}^{(\alpha)} \delta_{jh} - \hat{G}_{hj}^{(\beta)} \delta_{ki}) \delta_{\alpha\beta} , \qquad i, j, h, k, \alpha, \beta = 1, 2 . \quad (3.4)$$

As can easily be verified, the relations above imply that the $\hat{G}_{ij}^{(\alpha)}$'s give the direct product (3.1). It is then possible to recover the decomposition $U(2) = SU(2) \otimes U(1)$ for the two algebras by considering the Schwinger realization (2.53) for the corresponding boson operators. This is equivalent to saying that the algebraic structure, supporting two nonequivalent one-dimensional systems, is closely related to the well-known problem of the coupling of states with different angular momenta. This problem is expressed through the group composition $O_1(3) \otimes O_2(3)$ (for two subsystems), which is, in turn, isomorphic to the above-mentioned $SU_1(2) \otimes SU_2(2)$ product. Thus we are lead to face a well-established treatise of traditional quantum mechanics, the only difference being that the mathematical outcome of this analysis will have to be adapted here to match a different situation from a physical standpoint.

As for the single degree of freedom, we need to construct an algebraic Hamiltonian operator based on boson operators of the spectrum-generating algebra (3.1). A possible procedure is simply to extend the single oscillator Hamiltonian (2.39) [and (2.57) as well] to include both degrees of freedom, 1 and 2. The most general Hamiltonian operator will, however, contain products between boson operators 1 and 2, thus incorporating some type of interaction. Such a Hamiltonian operator must be given by

$$\hat{H} = \hat{H}_1 + \hat{H}_2 + \hat{H}_{12} , \tag{3.5}$$

where $\hat{H}_1$ and $\hat{H}_2$ are in the form of Eq. (2.57) with indexes 1 and 2 and

$$\begin{aligned}
\hat{H}_{12} &= f_1 \hat{s}_1^\dagger \hat{s}_1 \hat{s}_2^\dagger \hat{s}_2 + f_2 \hat{t}_1^\dagger \hat{t}_1 \hat{t}_2^\dagger \hat{t}_2 + f_3 \hat{s}_1^\dagger \hat{s}_1 \hat{t}_2^\dagger \hat{t}_2 + f_4 \hat{t}_1^\dagger \hat{t}_1 \hat{s}_2^\dagger \hat{s}_2 \\
&\quad + f_5 (\hat{s}_1^\dagger \hat{t}_1 \hat{s}_2^\dagger \hat{t}_2 + \hat{t}_1^\dagger \hat{s}_1 \hat{t}_2^\dagger \hat{s}_2) + f_6 (\hat{s}_1^\dagger \hat{t}_1 \hat{t}_2^\dagger \hat{s}_2 + \hat{t}_1^\dagger \hat{s}_1 \hat{s}_2^\dagger \hat{t}_2) .
\end{aligned} \tag{3.6}$$

This Hamiltonian is Hermitian and parity invariant and inclusive up to two-body terms. This operator has the role of mixing states 1 and 2 among themselves. The precise way in which such mixing is carried out can be seen explicitly by introducing a convenient set of complete algebraic kets. To do that, we need to specify to some extent the algebraic structure of the product (3.1). A very simple method is to exploit the aforementioned similarity between the product (3.1) and the coupling between two systems of well-defined angular momentum. As is well known one can consider two different schemes. We start with two subsystems (1 and 2) in a state of well-defined total angular momentum $|j_{12}\mu_{12}\rangle$. This ket can be considered as the product of states $|j_1\mu_1\rangle$, $|j_2\mu_2\rangle$ associated with the separated subsystems. This is usually expressed as

$$|j_{12}\mu_{12}\rangle = \sum_{\mu_1\mu_2} \langle j_1\mu_1 j_2\mu_2 | j_{12}\mu_{12}\rangle (|j_1\mu_1 j_2\mu_2\rangle) , \tag{3.7}$$

in which $\langle j_1\mu_1 j_2\mu_2 | j_{12}\mu_{12}\rangle$ is a Clebsh–Gordan coefficient. Here the

Clebsh–Gordan coefficient plays the role of matrix elements of a unitary transformation between the coupled and uncoupled schemes. In view of the isomorphism $SU_1(2) \otimes SU_2(2) \simeq SO_1(3) \otimes SO_2(3)$, it is possible to recognize, starting from the product (3.1), two different coupling schemes,

$$U_1(2) \otimes U_2(2) \supset O_1(2) \otimes O_2(2) \supset O_{12}(2) \qquad (3.8)$$

and

$$U_1(2) \otimes U_2(2) \supset U_{12}(2) \supset O_{12}(2) , \qquad (3.9)$$

the first corresponding to product states $|j_1 \mu_1 j_2 \mu_2\rangle$, the second to coupled states $|j_{12} \mu_{12}\rangle$. The subalgebra chain (3.9) can be obtained directly by considering, in place of the $\hat{G}_{ij}^{(\alpha)}$'s as generators of the direct product (3.1), the following bilinear operators:

$$\hat{G}_{ij} = \hat{G}_{ij}^{(1)} + \hat{G}_{ij}^{(2)} , \qquad i, j = 1, 2 . \qquad (3.10)$$

The correspondence between rotations and unitary subalgebra chains can be made more precise by taking into account the explicit branching laws for the labels of the involved irreducible representations. Within the usual angular momentum framework, one has

$$|j_1 - j_2| \leq j_{12} \leq j_1 + j_2 , \qquad -j_{12} \leq \mu_{12} \leq j_{12} , \qquad \mu_{12} = \mu_1 + \mu_2 . \quad (3.11)$$

Algebraic chains (3.8) and (3.9) are labeled similarly, the only differences being in the meaning of labels used for unitary groups. As already hinted (see Section II.B), the decomposition $U(2) = SU(2) \otimes U(1)$ automatically introduces a corresponding decomposition of the labels used to denote irreducible representations of $U(2)$ and $SU(2)$. It can be shown that the number N of $U(2)$ is related to j of $SU(2)$ by means of $j = N/2$. Consequently, the label m of $O(2)$ [intended as a subalgebra of $U(2)$] is related to the label μ of $O(2)$ [intended as a subalgebra of $O(3)$] by means of $\mu = m/2$. Moreover, while μ changes by unitary steps, m is related to N through $m = \pm N, \pm(N-2), \ldots$ [see also Eq. (2.60)]. A further distinction between algebraic rules involving $U(2)$ and $O(3)$ groups should be made when considering irreducible representations of $U(2)$, intended as a subalgebra of the direct product $U_1(2) \otimes U_2(2)$. As already pointed out, we are interested in symmetric irreducible representations of unitary groups [i.e., labeled by a Young tableau with a single row of N boxes (a single number)]. However, the subalgebra $U_{12}(2)$ of $U_1(2) \otimes U_2(2)$ can also lead to nonsymmetric irreducible representations, labeled by two quantum numbers and denoted by $[f_1, f_2]$ [14]. It is possible to relate these labels to j_{12} of Eq. (3.11) and $N_{12} \equiv N_1 + N_2$ by means of

$$f_1 - f_2 = 2j_{12}, \qquad f_1 + f_2 = N_{12}. \tag{3.12}$$

In this equation, the N_i's are the labels for the irreducible representations of $U_i(2)$ $(i = 1, 2)$.

Let us now consider a simple example in which we couple two systems of equal angular momentum, $j_1 = j_2 = 1$. Following rules (3.11), the overall system $(1 + 2)$ is seen to be compatible with states of total angular momentum $|j_{12} \mu_{12}\rangle$ given by

$$|0, 0\rangle; \qquad |1, 0\rangle, |1, \pm 1\rangle; \qquad |2, 0\rangle, |2, \pm 1\rangle, |2, \pm 2\rangle. \tag{3.13}$$

These states can be expressed in terms of (direct) products of uncoupled basis states $|j_1 \mu_1 j_2 \mu_2\rangle$ by direct application of (3.7), giving, for example,

$$|0, 0\rangle = \frac{1}{\sqrt{3}} \{|1, 1; 1, -1\rangle - |1, 0; 1, 0\rangle + |1, -1; 1, 1\rangle\},$$

$$|1, 0\rangle = \frac{1}{\sqrt{2}} \{|1, 1; 1, -1\rangle - |1, -1; 1, 1\rangle\}, \tag{3.14}$$

$$|1, 1\rangle = \frac{1}{\sqrt{2}} \{|1, 1; 1, 0\rangle - |1, 0; 1, 1\rangle\},$$

and so on, for the remaining states [notice that both schemes lead to a total number of states given by $(2j_1 + 1)(2j_2 + 1) = 9$]. We now go to the original problem of finding the irreducible representations contained in chains (3.8) and (3.9), starting from the irreducible representation of $U_1(2) \otimes U_2(2)$. We denote the irreducible representation of $U_1(2) \otimes U_2(2)$ as $[N_1, 0] \otimes [N_2, 0]$, since $f_2 = 0$ for symmetric representations. We assume, in view of the relation $j_i = N_i/2$ $(i = 1, 2)$, that $N_1 = N_2 = 2$. We are thus left with the algebraic problem of reducing the representation $[2, 0] \otimes [2, 0]$ in irreducible representations of $O_1(2) \otimes O_2(2)$ [chain (3.8)] and of $U_{12}(2)$ [chain (3.9)]. Chain (3.8) is solved directly by applying the single algebra branching law (2.60), which now becomes

$$m_1 = 0, \pm 2; \qquad m_2 = 0, \pm 2. \tag{3.15}$$

The final coupling $O_1(2) \otimes O_2(2) \supset O_{12}(2)$ is trivially additive (as the algebras are by themselves additive), thus giving

$$m_{12} = m_1 + m_2 = -4, -2, 0; \quad -2, 0, 2; \quad 0, 2, 4. \tag{3.16}$$

By limiting ourselves to the positive branch of m_1 and m_2 we are finally left with the labels

$$m_{12} = 0, 2; \quad 2, 4. \tag{3.17}$$

This is quite an obvious result, as it corresponds to the classification in terms of direct product states of $O_1(2)$ and $O_2(2)$, in which we have

$$\mu_1, \mu_2 = 0, \pm 1, \quad \mu = \mu_1 + \mu_2. \tag{3.18}$$

These states will transform among themselves to give coupled states $|j_{12}\mu_{12}\rangle$, as shown in Eqs. (3.7) and (3.14). Let us consider the explicit branching laws for chain (3.9). By making explicit use of Young tableaux, it is possible to show that one can reduce the direct product representation $[2, 0] \otimes [2, 0]$ in the following way:

$$[2, 0] \otimes [2, 0] = [4, 0] \oplus [3, 1] \oplus [2, 2], \tag{3.19}$$

in which the right side contains (nonsymmetric) irreducible representations of U(2). These in turn, are trivially related to the underlying SU(2) (rotational) structure by means of (3.12), thus leading to

$$[2, 0] \otimes [2, 0] \mapsto (j_{12} = 2) \oplus (j_{12} = 1) \oplus (j_{12} = 0), \tag{3.20}$$

corresponding to the usual coupling $(j_1 = 1) \otimes (j_2 = 1)$ in terms of angular momenta. The problem of obtaining the irreducible representations of the final coupling $U_{12}(2) \supset O_{12}(2)$ of chain (3.9) is solved either in terms of irreducible representations $[f_1, f_2]$ of $U_{12}(2)$ contained in (3.19) or simply by applying the relation $\mu_{12} = m_{12}/2$; the result in both cases is already given by Eq. (3.16) or (3.17). To summarize, we can write for the chains (3.8) and (3.9) the complete algebraic kets as follows:

$$\left| \begin{array}{cccc} U_1(2) \otimes U_2(2) \supset & O_1(2) \otimes O_2(2) & \supset & O_{12}(2) \\ [2, 0] \; [2, 0] & m_1 = 0, \pm 2 \; m_2 = 0, \pm 2 & m_{12} = -4, -2, 0; -2, 0, 2; 0, 2, 4 \end{array} \right\rangle, \tag{3.21}$$

$$\left| \begin{array}{cccc} U_1(2) \otimes U_2(2) \supset & U_{12}(2) & \supset & O_{12}(2) \\ [2, 0] \; [2, 0] & [4, 0][3, 1][2, 2] & m_{12} = -4, -2, 0; -2, 0, 2; 0, 2, 4 \end{array} \right\rangle. \tag{3.22}$$

It is possible, of course, to write the general result for two irreducible representations $[N_1, 0]$ and $[N_2, 0]$ of U(2) by means of quite simple formulas related to angular momentum coupling schemes.

The next step is the construction of Hamiltonian operators in the dynamical symmetry framework. The general procedure is to restrict the expansion (3.6) to invariant operators of the subalgebra chains, thus leading to two distinct models:

$$\hat{H}^{(a)} = E_0 + A_1 \hat{C}^{(2)}_{O_1(2)} + A_2 \hat{C}^{(2)}_{O_2(2)} + A_{12} \hat{C}^{(2)}_{O_{12}(2)} , \qquad (3.23)$$

$$\hat{H}^{(b)} = E_0 + \alpha \hat{C}^{(2)}_{U_{12}(2)} + \beta \hat{C}^{(2)}_{O_{12}(2)} . \qquad (3.24)$$

Written in terms of O(3) and O(2) quantum numbers, the eigenvalues of these operators are readily computed:

$$E^{(a)} = E^{(a)}(\mu_1, \mu_2) = E_0 + A_1 \mu_1^2 + A_2 \mu_2^2 + A_{12}(\mu_1 + \mu_2)^2 , \qquad (3.25)$$

$$E^{(b)} = E^{(b)}(j_{12}, \mu_{12}) = E_0 + \alpha j_{12}(j_{12} + 1) + \beta \mu_{12}^2 . \qquad (3.26)$$

We notice, in Eq. (3.26), the simple meaning of the quadratic invariant operator of U(2) as intended from the rotation group viewpoint. It is also possible to write these same eigenvalues in terms of U(2) and O(2) quantum numbers. By adopting (for matters of convenience) the same parameters as in the relations above, we find that

$$E^{(a)} = E^{(a)}(N_1, N_2; m_1, m_2; m_{12})$$
$$= E_0 + A_1 m_1^2 + A_2 m_2^2 + A_{12}(m_1 + m_2)^2 , \qquad (3.27)$$

$$E^{(b)} = E^{(b)}(N_1, N_2; f_1, f_2; \mu_{12})$$
$$= E_0 + \alpha[f_1(f_1 + 1) + f_2(f_2 - 1)] + \beta \mu_{12}^2 , \qquad (3.28)$$

in which

$$m_i = N_i, N_i - 2, \ldots, 1 \text{ or } 0 ; \quad m_{12} = m_1 + m_2 ,$$
$$f_1 = N_1 + N_2, N_1 + N_2 - 1, \ldots, \max(N_1, N_2) ,$$
$$f_2 = 0, 1, \ldots, \min(N_1, N_2) , \qquad (3.29)$$
$$\mu_{12} = f_2 - f_1, f_2 - f_1 + 2, \ldots, f_1 - f_2 \quad (f_1 > f_2) .$$

We can now consider a preliminary, intuitive physical description of these mathematical results. A *very* important point is that the use of rotational invariance and angular momentum composition laws has to be thought as a purely *algebraic* tool, without any real connection to the three-dimensional world. Our models are one-dimensional; the use of rotationlike algebraic structures is nothing more than a very convenient way of exploiting certain mathematical facts. The first chain of subalgebras (3.8) and the associated Hamiltonian operator (3.23) with eigenvalues (3.27) can be seen as the "superposition" of two (not necessarily equivalent) one-dimensional anharmonic (Morse-like) oscillators, each described by the single-term Hamiltonian operator discussed in Section II.C.1. This is completely equivalent to considering a *local* approximation, in which the natural choice for describing molecular vibrations is the replacement of bond coordinates by algebraic realizations of anharmonic oscillators. A convenient form of the spectrum (3.27) is obtained by introducing the usual local quantum numbers v_a and v_b related to bonds 1 and 2, respectively (we shall make use of v_1, v_2, v_3 quantum numbers to denote normal or symmetrized modes; see below). Local vibrations are thus labeled according to

$$v_a = (N_1 - m_1)/2, \qquad\qquad v_b = (N_2 - m_2),$$
$$v_a = 0, 1, \ldots, N_1/2 \text{ or } (N_1 - 1)/2, \qquad v_b = 0, 1, \ldots, N_2/2 \text{ or } (N_2 - 1)/2, \qquad (3.30)$$

by means of which we obtain, in place of Eq. (3.27),

$$E^{(a)}(v_a, v_b) = E_0' + A_1[-4v_a(N_1 - v_a)] + A_2[-4v_b(N_2 - v_b)]$$
$$+ A_{12}[-4(v_a + v_b)(N_{12} - v_a - v_b)], \qquad (3.31)$$

where $E_0' = A_1 N_1^2 + A_2 N_2^2 + A_{12} N_{12}^2$. We recognize in Eq. (3.31) the anharmonic sequences (in the quantum numbers v_a and v_b) of the independent Morse oscillators 1 and 2. We also have a (diagonal) term associated with the invariant operator of the coupled algebra $O_{12}(2)$ whose eigenvalues suggest a simple coupling between (local) vibrational modes. We discuss this particular aspect in the following sections. At this stage we note simply that such a coupling provides equal eigenvalues for these states satisfying the condition $v_a + v_b = $ constant. Later we see how the contribution of this coupling term to the energy spectrum will account for a "global" anharmonicity of polyads (i.e., families) of interacting levels. Such anharmonic terms are of decided importance in the correct description of combination bands [e.g., states in which two (or more) vibrational modes are simultaneously excited]. Let us consider the simplest example, in which v_a, $v_b = 0, 1$. We are thus lead to include in our scheme the following modes:

$$|v_a, v_b\rangle = |0, 0\rangle, \text{ ground state:} \qquad E(0, 0) = 0,$$

$$|v_a, v_b\rangle = |1, 0\rangle, v_a \text{ fundamental:} \qquad E(1, 0) = -4A_1(N_1 - 1) - 4A_{12}(N_1 + N_2 - 1),$$

$$(3.32)$$

$|v_a, v_b\rangle = |0, 1\rangle$, v_b fundamental: $E(0, 1) = -4A_2(N_2 - 1) - 4A_{12}(N_1 + N_2 - 1)$,

$|v_a, v_b\rangle = |1, 1\rangle$, combination band: $E(1, 1) = -4A_1(N_1 - 1) - 4A_2(N_2 - 1)$

$$- 8A_{12}(N_1 + N_2 - 2),$$

in which we choose $E_0 = -(A_1 N_1^2 + A_2 N_2^2 + A_{12} N_{12}^2)$, thus giving $E_0' = 0$. We observe in Eqs. (3.32) that $E(1, 1) \neq E(1, 0) + E(0, 1)$, being in fact

$$E(1, 1) - [E(1, 0) + E(0, 1)] = 8A_{12} . \qquad (3.33)$$

This simple result clearly shows the role played by the invariant operator of $O_{12}(2)$: Its parameter, A_{12}, gives a direct measure of the deviation from the harmonic sum of different local modes in a combination band.

Before presenting a more realistic example, let us consider the second model Hamiltonian (3.24) based on the subalgebra chain (3.9). A simple glance at this chain hints at the intrinsically coupled nature of the corresponding Hamiltonian operator. In a rather naive fashion, we can emphasize the fact that the symmetry reduction scheme $U(2) \supset O(2)$ (with the corresponding operators) is still present in the coupled model; as a consequence, once again, we expect to deal with some kind of anharmonic sequence of levels which are somehow related to a "collective" or *normal* vibrational behavior of the physical system. We have seen in fact, that the uncoupled picture based on the first chain (3.8) is in a one-to-one correspondence with a rotationlike representation of the direct product states $|j_1 \mu_1 j_2 \mu_2\rangle$. These states, in turn, can be transformed by means of Eq. (3.7) to give basis states in the coupled form, $|j_{12} \mu_{12}\rangle$, which can be put in direct correspondence with coupled *vibrational* states. Such correspondence between the basis set and the coupled states will now be discussed, once again, by means of a simple example. Let us go back to the previous case where $j_1 = j_2 = 1$ ($N_1 = N_2 = 2$) and consider the fundamental modes $|v_a, v_b\rangle = |1, 0\rangle$, $|0, 1\rangle$, which correspond now to direct product states $|j_1 \mu_1 j_2 \mu_2\rangle = |1011\rangle$, $|1110\rangle$, as can trivially be seen by applying Eq. (3.30) and the fact that $j_i = N_i/2$, $\mu_i = m_i/2$ ($i = 1, 2$). Let us now use Eq. (3.7) to transform uncoupled (product) states in the coupled representation. We obtain

$$\frac{1}{\sqrt{2}}(|1011\rangle + |1110\rangle) = |21\rangle ,$$

$$\frac{1}{\sqrt{2}}(-|1011\rangle + |1110\rangle) = |11\rangle . \qquad (3.34)$$

This shows that symmetric and antisymmetric combinations of the local modes $|v_a, v_b\rangle = |1, 0\rangle$, $|0, 1\rangle$ lead to the two states $|j_{12}\mu_{12}\rangle = |21\rangle$, $|11\rangle$ of the algebraic ket associated with the chain (3.9). To recover a convenient physical counterpart for this simple result, we recall that in the case of *identical* coupled oscillators, we expect to obtain exactly symmetric and antisymmetric normal oscillations, as shown in Fig. 8. The natural solution to our problem is to attribute a normal vibration character to the coupled states $|21\rangle$, $|11\rangle$ through a convenient transformation of quantum numbers. A more general study shows that such a transformation is given by

$$v_A = N - j_{12},$$

$$v_S = j_{12} - \mu_{12}, \tag{3.35}$$

where $N \equiv N_1 = N_2$ and v_A and v_S denote vibrational quantum numbers for antisymmetric and symmetric combinations of local modes states, respectively. In traditional molecular spectroscopy problems, the symbols v_1 and v_3 are used in preference of v_S and v_A. v_1 and v_3 can be used for nonsymmetric triatomic molecules, in which one expects some amount of coupling between local modes but without any symmetry character to bond exchange (v_2 is left to denote bending modes). In our specific example, starting from Eq. (3.34) and by means of Eq. (3.35), we readily

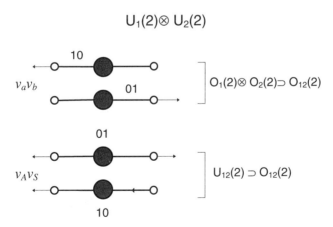

Figure 8. Local and normal algebraic pictures of two identical coupled one-dimensional anharmonic oscillators.

obtain

$$
|v_A, v_S\rangle = \begin{cases} |0, 1\rangle = \dfrac{1}{\sqrt{2}} \left(v_a = 1, v_b = 0\rangle + |v_a = 0, v_b = 1\rangle \right) \\[2mm] |1, 0\rangle = \dfrac{1}{\sqrt{2}} \left(-|v_a = 1, v_b = 0\rangle + |v_a = 0, v_b = 1\rangle \right). \end{cases} \tag{3.36}
$$

To gain a deeper physical insight, it is possible to consider this problem from a different viewpoint. The key point arising from the previous discussion is the *exact* equivalence of the two oscillators. Our triatomic molecule, XYX, is characterized by a permutational symmetry with respect to the exchange of the X atoms. This symmetry corresponds to a reflection through the plane containing the Y atom and perpendicular to the line connecting the X atoms. The exchange of the X atom is carried out algebraically simply by interchanging the N_1, N_2 (or j_1, j_2) quantum numbers. It is not difficult to show by exploiting the symmetry properties of the Clebsh–Gordan transformation brackets that the $j_1 \leftrightarrow j_2$ exchange leads to a transformation of the coupled basis $|j_{12}\mu_{12}\rangle$ given by

$$
|j_{12}\mu_{12}\rangle \to (-)^{N-j_{12}} |j_{12}\mu_{12}\rangle. \tag{3.37}
$$

This simple result, compared with the transformation (3.35), justifies the introduction of symmetric and antisymmetric combinations of local modes with respect to the aforementioned reflection operation. It is, of course, possible to write, in place of Eq. (3.26) or (3.28), the energy spectrum of the normal limit expressed as a function of the v_A and v_S vibrational quantum numbers (for identical oscillators, $N_1 = N_2 \equiv N$):

$$
E^{(b)}(v_A, v_S) = E_0' + \alpha v_A (v_A - 2N - 1) + \beta (v_A + v_S)(v_A + v_S - 2N), \tag{3.38}
$$

where $E_0' = E_0 + \alpha N(N + 1) + \beta N^2$. Figures 9 and 10 show the energy spectra obtained in the local and normal limits, respectively.

Having studied the local and normal limits, we can now return to the original problem of constructing a Hamiltonian model for the realistic case of two stretching modes in a triatomic molecule. We clearly expect that a real molecule will behave in some intermediate way between the local and normal limits considered previously. At this stage, it is important to point out that a typical algebraic strategy will be more advantageous to a local picture than for a normal one, for several reasons: (1) it is possible to account for normal situations within the local scheme by means of a convenient breaking of the local dynamical

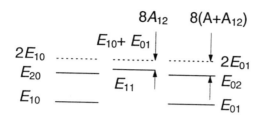

Figure 9. Low-lying levels in the uncoupled picture for two identical anharmonic oscillators.

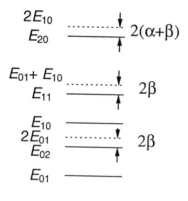

Figure 10. Low-lying levels in the coupled picture for two identical anharmonic oscillators.

symmetry (3.8); (2) symmetric molecules YX_2 find in the local (i.e., based on bond coordinates) realization a natural way of accounting for the invariance of the Hamiltonian operator under bond exchange; (3) the local Hamiltonian (3.23) can be modified for describing more or less strong normal behaviors in such a way that its diagonalization requires very little algebraic manipulation; and (4) transition intensities, relying on proper bond electric dipole or quadrupole operators, can be obtained in a more straightforward fashion for the local formulation than for the normal formulation.

The basic idea is very simple. Both the local model [Eq. (3.23)] and the normal model [Eq. (3.24)] are such that $v_a + v_b = v_A + v_S$ is a conserved quantity. In the local model, v_a and v_b are separately good quantum numbers, as are v_A and v_S in the normal model. This means that in the normal limit v_a and v_b are mixed, as is clearly shown by Eq. (3.36). Such mixing is caused by the invariant operator of $U_{12}(2)$: we must remember that from a rotationlike viewpoint, this operator is related to the *total* angular momentum, in which the single components somehow have lost their own identity. In other words, in the normal picture, some amount of energy flows between the oscillators, the strength of such coupling being related to the effective action of the operator $\hat{C}^{(2)}_{U_{12}(2)}$ compared with the single-oscillator Hamiltonian terms. From a physical viewpoint we can describe the realistic system of two oscillators by letting such oscillators interact between themselves, either for purely mechanical reasons (atoms of comparable masses) or by means of chemical (electrostatic) forces. The most obvious algebraic realization of this situation consists of introducing the $\hat{C}^{(2)}_{U_{12}(2)}$ operator in the local model. As a matter of fact, its effect is one of breaking the original dynamical symmetry, in which v_a and v_b were good quantum numbers. As stated earlier, the interbond coupling is such that these local modes are, to some extent, mixed, and as a consequence, $v_a + v_b$ (and $v_A + v_S$ as well) is the only conserved quantity. This procedure is formally expressed by the *algebraic lattice*, defined as

$$U_1(2) \otimes U_2(2) \supset \begin{Bmatrix} O_1(2) \otimes O_2(2) \\ U_{12}(2) \end{Bmatrix} \supset O_{12}(2) . \tag{3.39}$$

The corresponding Hamiltonian operator can thus be written as

$$\hat{H} = E_0 + A_1 \hat{C}^{(2)}_{O_1(2)} + A_2 \hat{C}^{(2)}_{O_2(2)} + A_{12} \hat{C}^{(2)}_{O_{12}(2)} + \lambda_{12} \hat{C}^{(2)}_{U_{12}(2)} . \tag{3.40}$$

Let us now consider the problem of obtaining the spectrum of this Hamiltonian, working in the local basis $|v_a, v_b\rangle$. Matrix elements of local operators in Eq. (3.40) have already been obtained. Thus we are left with the computation of matrix elements for the operator $\hat{C}^{(2)}_{U_{12}(2)}$. To this end we observe that this operator can be seen in a rotationlike framework, as

$$\hat{C}^{(2)}_{U_{12}(2)} \propto \hat{J}^2 = (\hat{J}_1 + \hat{J}_2)^2 = \hat{J}_1^2 + \hat{J}_2^2 + 2\hat{J}_1 \cdot \hat{J}_2 . \tag{3.41}$$

The operators $\hat{J}_1^2$ and $\hat{J}_2^2$ are constant within a given irreducible representation of the spectrum-generating algebra; they are not of interest here. The effective coupling is given by the scalar product $\hat{J}_1 \cdot \hat{J}_2$, which can be written as [see Eq. (2.94)]

$$\hat{J}_1 \cdot \hat{J}_2 = \sum_\mu (-)^\mu \hat{J}_{1\mu} \hat{J}_{2\mu} , \tag{3.42}$$

where we make use of the spherical components $\hat{J}_{i\mu}$ ($\mu = 0, \pm 1$) of the angular momentum operators. Thus we need the following matrix elements:

$$\sum_{\mu} (-)^{\mu} \langle j_1 \mu_1' j_2 \mu_2' | \hat{J}_{1\mu} \hat{J}_{2\mu} | j_1 \mu_1 j_2 \mu_2 \rangle = \sum_{\mu} (-)^{\mu} \langle j_1 \mu_1' | \hat{J}_{1\mu} | j_1 \mu_1 \rangle \langle j_2 \mu_2' | \hat{J}_{2\mu} | j_2 \mu_2 \rangle$$

$$(3.43)$$

written in the angular momentum basis rather than in the vibrational (local) basis. The matrix elements above can easily be obtained by recalling the well-known result

$$\langle j\mu \pm 1 | \hat{J} | j\mu \rangle = \sqrt{(j \mp \mu)(j \pm \mu + 1)}, \qquad (3.44)$$

leading to

$$\langle j_1 \mu_1' j_2 \mu_2' | \hat{J}_1 \cdot \hat{J}_2 | j_1 \mu_1 j_2 \mu_2 \rangle = \mu_1 \mu_2 \delta_{\mu_1' \mu_1} \delta_{\mu_2' \mu_2}$$

$$+ \tfrac{1}{2} \sqrt{(j_1 \mp \mu_1)(j_1 \pm \mu_1 + 1)(j_2 \pm \mu_2)(j_2 \mp \mu_2 + 1)} \, \delta_{\mu_1' \mu_1 \pm 1} \delta_{\mu_2' \mu_2 \mp 1} . \quad (3.45)$$

These matrix elements can also be written in the local basis $|v_a v_b \rangle$. For this purpose it is convenient to introduce a slightly different form of the interaction term, often referred to as the *Majorana operator*, $\hat{M}_{12}$, which is related to $\hat{C}^{(2)}_{U_{12}(2)}$ or $\hat{J}_1 \cdot \hat{J}_2$ by means of several conventions. We choose to define $\hat{M}_{12}$ such that in place of (2.45), we obtain the following matrix elements:

$$\langle v_a' v_b' | \hat{M}_{12} | v_a v_b \rangle = (v_a N_2 + v_b N_1 - 2 v_a v_b) \delta_{v_a' v_a} \delta_{v_b' v_b}$$

$$- \sqrt{(v_a + 1)(N_1 - v_a) v_b (N_2 - v_b + 1)} \, \delta_{v_a' - 1, v_a} \delta_{v_b' + 1, v_b}$$

$$- \sqrt{(v_b + 1)(N_2 - v_b) v_a (N_1 - v_a + 1)} \, \delta_{v_a' + 1, v_a} \delta_{v_b' - 1, v_b} .$$

$$(3.46)$$

The physical meaning of this apparently complex result can be understood by recalling the previous example of two identical oscillators. Let us start from the local basis states $|v_a v_b \rangle = |1, 0 \rangle$, $|0, 1 \rangle$. In manifest agreement with the previous discussion, the Majorana operator $\hat{M}_{12}$ has the effect of mixing these states, so we obtain

$$\langle 1, 0 | \hat{M}_{12} | 0, 1 \rangle = \langle 0, 1 | \hat{M}_{12} | 1, 0 \rangle = -\sqrt{N_1 N_2} . \qquad (3.47)$$

If we start from the Hamiltonian operator (3.40), in which we formally replace $\hat{C}^{(2)}_{U_{12}(2)}$ with $\hat{M}_{12}$, the eigenvalue problem for the first two local modes is expressed in terms of the Hamiltonian matrix,

$$
\begin{pmatrix}
-4A(N-1) - 4A_{12}(2N-1) + \lambda_{12}N & -\lambda_{12}N \\
-\lambda_{12}N & -4A(N-1) - 4A_{12}(2N-1) + \lambda_{12}N
\end{pmatrix}, \quad (3.48)
$$

where we put $A_1 = A_2 \equiv A$, $N_1 = N_2 \equiv N$. The corresponding secular determinant shows that the two local modes $|1, 0\rangle$ and $|0, 1\rangle$, initially with the same energy $\varepsilon_0 \equiv -4A(N-1) - 4A_{12}(2N-1)$, are now mixed, shifted, and split under the action of $\hat{M}_{12}$. We obtain an antisymmetric combination (with energy $\varepsilon_0 + 2\lambda_{12}N$) and a symmetric combination (with energy ε_0). The sign of λ_{12} determines which of these combinations has the highest energy. This simple example can readily be generalized to an entire manifold of levels founded on the local basis of two anharmonic modes. Matrix elements of $\hat{M}_{12}$ [Eq. (3.46)] are such that the Hamiltonian matrix is in tridiagonal form, thus leading to the block structure shown in Fig. 11. Specifically, we observe that each block contains states with the same value of $v_a + v_b$. This result is in obvious agreement with the fact that $v_a + v_b$ is a conserved quantity for both chains (3.8) and (3.9). The inclusion of $\hat{M}_{12}$ in the local Hamiltonian operator cannot affect the $v_a + v_b$ conservation rule. We will see how this specific feature

v_a+v_b

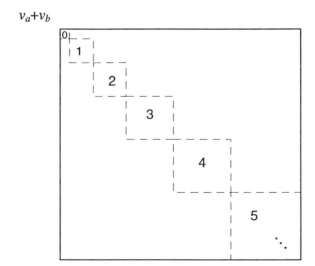

Figure 11. Block structure of the Hamiltonian matrix caused by the Majorana operator.

has far-reaching consequences for both computational and physical aspects for these kinds of problems.

Finally, we mention the possibility of assigning a measure of locality (or, equivalently, of normality) to a given situation by means of the dimensionless parameter

$$\xi \equiv \frac{2}{\pi} \tan^{-1} \frac{8\lambda_{12}}{A + A_{12}} , \tag{3.49}$$

which is obviously a very small quantity in local or quasilocal molecules, while for normal cases ξ is close to 1 [36].

As discussed in Section II.C.2 for the single three-dimensional diatomic molecule, the algebraic formulation is particularly well suited for the inclusion of higher-order terms, given the intrinsically anharmonic nature of the zeroth-order contributions to the rovibrational spectrum. These same considerations hold in the algebraic description of two (or more) interacting oscillators. We expect to construct expansions requiring a smaller number of arbitrary parameters than those needed in the corresponding Dunham series, stopped to the same order. However, in this case, the explicit relations between algebraic and Dunham parameters are more difficult to obtain for two quite distinct reasons. First, one typically has many parameters to deal with, thus leading to difficult algebraic relations. Second, the *local* algebraic expansion must be diagonalized to account for even small intermode couplings (as well as for exactly local situations in which the molecule is symmetric). Instead, the corresponding Dunham expansion is written in terms of normal parameters, thus making direct comparison very difficult, if not impossible, at least from an analytical point of view. However, it is worthwhile to sketch the procedure of the two-oscillator system in the purely local limit. The Dunham expansion can be written as

$$E^D(v_a, v_b) = \sum_{i,j} x_{ij} v_a^i v_b^j , \tag{3.50}$$

while the algebraic eigenvalues can be expressed as

$$E^A(v_a, v_b) = \sum_{i,j,k} y_{ijk} [\langle \hat{C}_1 \rangle]^i [\langle \hat{C}_2 \rangle]^j [\langle \hat{C}_{12} \rangle]^k , \tag{3.51}$$

in which $\langle \hat{C}_i \rangle$ denotes the expectation value in the local basis of the invariant $\hat{C}_{O_i(2)}^{(2)}$, $i = 1, 2$, and 12. By limiting ourselves to algebraic terms with $k = 0$, we recover the "product" version of the single-oscillator case discussed in its three-dimensional version in Section II.C.2. The diagonal coupling term $\hat{C}_{12}$ has, however an independent role in the expansion

(2.51), thus leading to further independent relations among the Dunham parameters. We will see how, in some practical cases, these higher-order terms can affect a fitting procedure.

Before concluding this section, we want to discuss a simple but realistic situation. We must, however, reiterate that the real advantages of the one-dimensional model will become evident only when dealing with more complex problems, such as those including several coupled bonds. So the following example is presented for purely pedagogical purposes. Using algebraic formulation, let us construct the Hamiltonian operator that describes the stretching modes of the SO_2 molecule. This molecule is symmetric and has a bent equilibrium geometry. So SO_2 is a good example of a rather strongly coupled system, because of the comparable masses of the three atoms. This fact is easily recognized by looking at the relative splitting between the fundamental stretching modes, located at 1151.0 and 1361.2 cm^{-1}. In comparison, H_2O, which is a local system, has the fundamental local modes located 3657.1 and 3755.9 cm^{-1}. We now want to fit, over a convenient set of experimental data, the four parameters $A_1 = A_2 \equiv A$, A_{12}, λ_{12}, and $N_1 = N_2 \equiv N$ for the Hamiltonian operator (3.40). First, we obtain an independent estimate for N by means of Eq. (2.71). This can be given in terms of tabulated values of ω_e and $\omega_e x_e$ for the diatom $^{16}O^{32}S$, thus leading to

$$N = \frac{1}{x_e} - 1 = \frac{\omega_e}{\omega_e x_e} - 1 = \frac{1149.2 \text{ cm}^{-1}}{5.6 \text{ cm}^{-1}} - 1 \simeq 204 . \qquad (3.52)$$

This numerical value must be seen as an initial guess; depending on the specific molecular structure, one can expect changes in such an estimate, which, however, should not be larger than $\pm 20\%$ of the original value (3.52). The second step is to obtain a starting guess for the parameter A. As such, we make explicit use of Eq. (2.68), which expresses the single-oscillator fundamental mode as

$$E(v = 1) = -4A(N - 1) . \qquad (3.53)$$

In the present case we have two different energies, corresponding to symmetric and antisymmetric combinations of the two local modes. A possible strategy is to use the center of gravity of these modes, so the guess for $\bar{A}$ is given by

$$\bar{A} = \frac{\bar{E}}{4(1 - N)} = \frac{(1151.0 + 1361.2) \text{ cm}^{-1}}{2} \frac{1}{4(1 - N)} \simeq -1.5 \text{ cm}^{-1} . \qquad (3.54)$$

The third step is to obtain an initial guess for λ_{12}. Its role is to split the initially degenerate local modes, placed here at the common value $\bar{E}$ used in Eq. (3.54). Such an estimate is obtained by considering the simple matrix structure (3.48). We easily find that

$$\lambda_{12} \simeq \frac{|E_a - E_b|}{2N} \simeq 0.5 \text{ cm}^{-1}. \tag{3.55}$$

We are now ready to let the computer carry out a numerical fitting procedure to adjust (in a least-squares sense, for example) the parameters A and λ_{12}, starting from values (3.54) and (3.55), and A_{12} (whose initial guess can be zero). One can also try to change (within, say, no more than $\pm 20\%$) the value of N to get better results. This is equivalent to changing the single-bond anharmonicity according to the specific molecular environment, in which it can be slightly different. The final result of the fitting procedure for this simple example is reported in Table I. First, we notice the relatively large value of the parameter λ_{12} (0.6220 cm^{-1}) in comparison with A (-1.6290 cm^{-1}). Such values correspond to the Majorana operator hardly being involved in the splitting of the local modes. The small value of A_{12} (-0.0470 cm^{-1}) explains the correspondingly small contribution of the diagonal term, which in a conventional (local) coordinate space could be written as $r_1 r_2$. This can be

TABLE I
One-Dimensional Algebraic Model Fit to SO_2 Stretching Modes[a]

$\nu_1 \nu_2 \nu_3$	Expt.	Calc.	Expt.–Calc.
100	1151.0	1151.1	−0.1
001	1361.2	1360.1	1.1
200	2295.9	2295.3	0.6
101	2499.1	2497.9	1.2
002	2714.0	2712.2	1.7
300	3435.4	3432.5	2.9
201	3629.6	3628.5	1.1
102	3837.0	3836.7	0.3
003	4055.0	4056.3	−1.3
400	4560.0	4562.7	−2.7
301	4751.2	4751.9	−0.7
202	—	4954.0	—
103	5167.0	5167.7	−0.7
004	—	5392.2	—

[a] $N = 168$, $A = -1.6290$, $A_{12} = -0.0470$, $\lambda_{12} = 0.6220$. All values in cm^{-1}, except N, which is dimensionless.

attributed partially to the contribution coming from the diagonal part of the Majorana operator [see Eq. (3.46)], which contains a $v_a v_b$ term.

The second comment concerns the comparison with the traditional (Dunham) analysis of this situation. To include the same powers and products of (normal) vibrational quantum numbers, the Dunham expansion should be written as

$$E^D(v_A, v_S) = x_{10} v_A + x_{01} v_S + x_{20} v_A^2 + x_{02} v_S^2 + x_{11} v_A v_S . \quad (3.56)$$

Thus we are left with the problem of fitting five independent parameters, while the algebraic expansion depends only on three parameters (four, including N). The root-mean-square (rms) error of the algebraic fit of Table I is $1.5 \, \text{cm}^{-1}$, while the error achieved with the Dunham expansion (3.56) is $1.1 \, \text{cm}^{-1}$. As pointed out earlier, it is difficult to highlight the advantages of the algebraic method in this simple case. However, we will soon see how the economy of parameters achieved here becomes substantial in more complex situations.

As a final comment, the diagonalization of the Hamiltonian operator (3.40), in the local basis $|v_a v_b\rangle$, leads to eigenvectors whose usefulness can be realized in the assignment of meaningful labels to the vibrational states. For example, we obtain

$$\phi_1^{(1)} = \frac{1}{\sqrt{2}} (|1,0\rangle + |0,1\rangle) ,$$

$$\phi_2^{(1)} = \frac{1}{\sqrt{2}} (-|1,0\rangle + |0,1\rangle) ,$$

$$\phi_1^{(2)} = a(|2,0\rangle + |0,2\rangle) + b|1,1\rangle , \quad (3.57)$$

$$\phi_2^{(2)} = \frac{1}{\sqrt{2}} (|2,0\rangle - |0,2\rangle) ,$$

$$\phi_3^{(2)} = a'(|2,0\rangle + |0,2\rangle) - b'|1,1\rangle ,$$

where $a = 0.507$, $b = 0.697$, $a' = 0.493$, and $b' = 0.717$; the $\phi_k^{(h)}$ denote wavefunctions of the block (polyad) corresponding to $v_a + v_b = h$, while k labels different states of increasing energy within the same block. The situation for the first block, $h = 1$, has already been discussed, [see Eq. (3.36)]. The second block contains eigenstates that are linear combinations of the local modes $|2,0\rangle$, $|0,2\rangle$, and $|1,1\rangle$. It is clear that $\phi_1^{(2)}$ and $\phi_3^{(2)}$ are symmetric under bond exchange, while $\phi_2^{(2)}$ is antisymmetric. The molecular point symmetry affects the specific form of wavefunctions, which, in turn, is obtained correctly by the action of the Majorana operator. It is customary to denote these well-behaved states with the symmetrized notation

$$|v_a, v_b^{\pm}\rangle \equiv \tfrac{1}{\sqrt{2}}(|v_a, v_b\rangle \pm |v_b, v_a\rangle)\,, \qquad v_a \neq v_b$$

$$|v_a, v_b^{+}\rangle \equiv |v_a, v_b\rangle\,, \qquad\qquad\qquad v_a = v_b\,.$$

(3.58)

Consequently, states $\phi_1^{(1)}$ and $\phi_2^{(1)}$ can be denoted as $|10^{\pm}\rangle$ and $\phi_2^{(2)}$ as $|20^{-}\rangle$. States $\phi_1^{(2)}$ and $\phi_3^{(2)}$ contain a nonnegligible contribution coming from the local mode $|1, 1\rangle$. Nonetheless, it is a bad habit to denote such states, following Eq. (3.58), as $|20^{+}\rangle$ and $|11^{+}\rangle$, respectively. This notation should be employed only for local or quasilocal situations. In the present case it is clearly more appropriate to give the detailed information on the wavefunctions as expressed in Eq. (3.57). The final outcome of this procedure is that we can rely on a set of wavefunctions (expressed in the local basis) that lead to an unambiguous assignment of vibrational modes. In a normal or quasinormal situation we obviously expect to deal with strong couplings of local modes. This fact basically renders the local labeling scheme useless. However, situations between the local and normal limits can be addressed properly by using, as good effective quantum numbers, (1) the index $v_a + v_b$ of the polyad, and (2) the position of a given level within the polyad itself. We consider, in a more systematic fashion the same problem in the following section, where we discuss the extension of the one-dimensional model to polyatomic molecules. Further information on interesting applications of this simple model can be found in several papers [37–39].

C.   Anharmonic Coupling of Many Oscillators

In this section we extend the one-dimensional algebraic model to multicoupled, anharmonic oscillators. Moreover, any real polyatomic molecule can be seen as an ensemble of *three*-dimensional oscillators. Consequently, we have to delineate the actual meaning of the one-dimensional approach to a polyatomic molecule; it goes beyond neglecting rotational degrees of freedom. Let us consider the ethylene molecule, shown schematically in Fig. 12. Each CH bond can be excited in terms of stretching motions (in which the bond length r changes), in-plane bending motions (where the variable is the angle θ), and out-of-plane bending motions (in which the variable is the angle ϕ). For CH vibrational modes the overall molecular behavior can be approximated as these three types of one-dimensional motions attached to the four hydrogen atoms. However, at this stage, several points need to be clarified. First, we notice how both stretching *and* bending modes have been included equally in the present scheme. This is not accidental, as we can recover

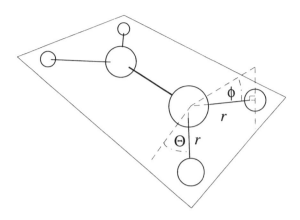

Figure 12. Stretching and bending coordinates for CH bonds in C_2H_2.

the algebraic description for the one-dimensional Morse potential (for stretching modes) and of the Pöschl–Teller potential (for bending modes) basically through the *same* dynamical symmetry, $U(2) \supset O(2)$, as discussed in Section II.C.1. This is in agreement with the statement made in Section III.A, where we were considering the possibility of replacing internal (anharmonic) degrees of freedom with $U(2)$ algebras. In Section III.B we have studied in some detail the procedure for algebraically expressing the coupling of two oscillators. Thus we must face the problem of extending such procedure to many oscillators. We will see that the inclusion of several oscillators in the algebraic framework has interesting consequences for both abstract and physical discussions related to this problem. Among such consequences, the two most important are that (1) it is possible to account for the point molecular symmetry in a systematic fashion, and (2) spurious modes (i.e., nonvibrational modes naturally arising from a description based on internal molecular coordinates) can be treated properly and selectively removed from the manifold of true vibrational levels. In the present section we make extended use of previous arguments because, to some extent, the polyatomic problem is quite a straightforward extension of the two-oscillator system.

To begin with, we generalize the spectrum-generating algebra (3.1), introduced for two interacting oscillators, by considering, for N interacting oscillators, the product

$$U_1(2) \otimes U_2(2) \otimes \cdots \otimes U_N(2) . \tag{3.59}$$

Consequently, the algebraic Hamiltonian for N uncoupled anharmonic oscillators, based on the $U(2) \supset O(2)$ dynamic symmetry, will be given by

$$\hat{H}_{\text{uncoupled}} = \sum_{i=1}^{N} A_i \hat{C}_{O_i(2)}^{(2)} .$$ (3.60)

We also introduce the algebraic local basis

$$|m_1 m_2 \cdots m_N\rangle , \qquad m_i = N_i, N_i - 2, \ldots , \quad i = 1, \ldots , N , \quad (3.61)$$

or, equivalently, the local vibrational basis

$$|v_1 v_2 \cdots v_N\rangle , \qquad v_i = 0, 1, 2, \ldots , \quad i = 1, \ldots , N . \quad (3.62)$$

In the basis shown, the eigenvalues can be computed according to Eq. (2.68):

$$E_{\text{uncoupled}}(v_1, \ldots , v_N) = -4 \sum_{i=1}^{N} A_i v_i (N_i - v_i) .$$ (3.63)

The local basis (3.62) can be arranged in polyads characterized by a well-defined total vibrational number, $\Sigma_{i=1}^{N} v_i \equiv p$ [once again for the trivially additive properties of the $O_i(2)$ algebras]. This means that within the same polyad (i.e., for given p), single basis states are expressed in terms of integer partitions of p in N parts. For example, if we consider three oscillators ($N = 3$), the first three polyads will be given by the following states:

$$p = 1 \mapsto |100\rangle, |010\rangle, |001\rangle ,$$

$$p = 2 \mapsto \begin{cases} |200\rangle, |110\rangle, |020\rangle , \\ |101\rangle, |011\rangle, |002\rangle , \end{cases}$$ (3.64)

$$p = 3 \mapsto \begin{cases} |300\rangle, |210\rangle, |120\rangle, |030\rangle, |201\rangle , \\ |102\rangle, |021\rangle, |012\rangle, |003\rangle, |111\rangle . \end{cases}$$

We now have to account for some type of interaction among local modes. This is done by operators that are related strictly to the invariant operators of the coupled algebras $U_{12}(2)$ and $O_{12}(2)$, introduced in the two-oscillator case. In the N-oscillator case we expect to deal with coupling terms involving pairs of oscillators; this is equivalent to considering algebraic lattices, starting from the product (3.59), of the following types:

$$O_1(2) \otimes O_2(2) \otimes \cdots \otimes O_N(2) \supset \begin{cases} O_{12}(2) \otimes O_3(2) \otimes \cdots \otimes O_N(2) \\ O_{13}(2) \otimes O_2(2) \otimes \cdots \otimes O_N(2) \\ \vdots \\ O_{1N}(2) \otimes O_2(2) \otimes \cdots \otimes O_{N-1}(2) \end{cases}$$

(3.65)

and

$$U_1(2) \otimes U_2(2) \otimes \cdots \otimes U_N(2) \supset \begin{cases} U_{12}(2) \otimes U_3(2) \otimes \cdots \otimes U_N(2) \\ U_{13}(2) \otimes U_2(2) \otimes \cdots \otimes U_N(2) \\ \vdots \\ U_{1N}(2) \otimes U_2(2) \otimes \cdots \otimes U_{N-1}(2) \,. \end{cases}$$

(3.66)

Thus we are lead to the following Hamiltonian operator for N interacting bonds:

$$\hat{H} = E_0 + \sum_{i=1}^{N} A_i \hat{C}^{(2)}_{O_i(2)} + \sum_{i<j=1}^{N} A_{ij} \hat{C}^{(2)}_{O_{ij}(2)} + \sum_{i<j=1}^{N} \lambda_{ij} \hat{M}_{ij} \,, \quad (3.67)$$

which is the straightforward generalization of the Hamiltonian (3.40). Before doing a careful analysis of these operators, it is convenient to adopt a simplified notation, in which we denote the invariant operator $\hat{C}^{(2)}_{O_i(2)}$ as $\hat{C}_i$. Moreover, it is convenient, for matters of physical significance of the algebraic parameters, to use in place of the invariant $\hat{C}^{(2)}_{O_{ij}(2)}$ the slightly different operator, defined as

$$\hat{C}_{ij} \equiv \hat{C}^{(2)}_{O_{ij}(2)} - N_{ij} \left(\frac{\hat{C}_i}{N_i} + \frac{\hat{C}_j}{N_j} \right) , \quad (3.68)$$

where, as usual, $N_{ij} \equiv N_i + N_j$. The eigenvalues of the Hamiltonian operator (3.67) can be obtained by employing Eqs. (3.31) and (3.46). In particular, the Majorana operator has eigenvalues (in the local basis) in the form of Eq. (3.46), where one replaces the indices a and b with i and

j. According to the definition (3.68), one has

$$\langle v_i v_j | \hat{C}_{ij} | v_i v_j \rangle = -4(v_i + v_j)(N_{ij} - v_i - v_j)$$

$$+ 4N_{ij} \left[\frac{v_i(v_i - N_i)}{N_i} + \frac{v_j(v_j - N_j)}{N_j} \right]. \tag{3.69}$$

We are now ready to address the problem of the physical characterization of these mathematical terms. The first point to note is that the Majorana operators are still responsible for intermode coupling; in our description, by letting $\lambda_{ij} = 0$, we recover the purely local limit of N oscillators. Nonetheless, these oscillators are somehow correlated with each other through the $\hat{C}_{ij}$ operators, which account for (diagonal) cross-anharmonicities. Furthermore, it should be noted that by following Eq. (3.68), one basically subtracts from $\hat{C}^{(2)}_{O_{ij}(2)}$ those terms arising from uncoupled single-oscillator contributions. In the special case of a pair of equivalent oscillators i and j ($N_i = N_j$), we obtain, in place of Eq. (3.69), the following matrix elements:

$$\langle v_i v_j | \hat{C}_{ij} | v_i v_j \rangle = -4(v_i - v_j)^2 \tag{3.70}$$

[i.e., the matrix elements do not depend on N_i (N_j)]. As a result, $\hat{C}_{ij}$ will account for different contributions throughout different polyads *and* within the same polyad; the most important aspect of $\hat{C}_{ij}$ is the dependence of its matrix elements on the product $v_i v_j$.

The role of the Majorana operators $\hat{M}_{ij}$ is to introduce nondiagonal couplings between pairs of local modes. We have seen how in the simplest case of equivalent interacting bonds, the Majorana operator naturally leads to a solution for symmetrized coupled modes, in which the invariance of the Hamiltonian operator, under bond exchange, is explicitly taken into account [see Eq. (3.48)]. A rather appealing feature of this algebraic model is that such a "symmetrizing" property of the Majorana operator, actually quite a trivial one for two equal bonds, can readily be extended to any molecular geometry, even a very complex one. The key point is that the basic information characterizing the specific molecular geometry can easily be incorporated by introducing proper linear combinations of Majorana operators. We discuss this particular problem in the following section.

In the one-dimensional algebraic model, the study of molecular vibrations requires proper treatment of bending modes as well. As alluded to earlier, the one-dimensional Hamiltonian operator is equally well suited for the description of both stretches and bends, by virtue of the *isospectrality* of the one-dimensional Morse and Pöschl–Teller potential curves (for the bound-state part of the spectrum). This means that the operators introduced in the dynamical symmetry $U(2) \supset O(2)$ can be adapted directly (through a proper choice of parameters) to reproduce the energy levels of a single bending degree of freedom and its interactions with other bending or stretching modes. We postpone the complete analysis of such a general situation until Section III.C.2.

1. *Majorana Operators as Symmetry Adapters*

Let us start with a simple example. We consider a triatomic planar molecule XY_3, with three equal XY bonds, in which the mass of the X atom is much larger than that of the Y atoms. The Hamiltonian operator (3.67) can easily be adapted to describe the three stretching modes involving the XY bonds. This is ideally accomplished in two distinct steps. In the first step, one accounts for (equivalent) local modes $|v_1 v_2 v_3\rangle$ by means of the Hamiltonian operator

$$\hat{H}_{\text{local}} = -4A \left(\sum_{i=1}^{3} \hat{C}_i \right) - 4A' \left(\sum_{i<j=1}^{3} \hat{C}_{ij} \right), \tag{3.71}$$

whose eigenvalues are given by

$$E_{\text{local}}(v_1, v_2, v_3) = -4A \left[\sum_{i=1}^{3} v_i(N - v_i) \right] - 4A' \sum_{i<j=1}^{3} (v_i - v_j)^2. \tag{3.72}$$

In the equations above, we put $N_1 = N_2 = N_3 \equiv N$, $A_1 = A_2 = A_3 \equiv A$, and $A_{12} = A_{23} = A_{13} \equiv A'$ because of the equivalence of the three oscillators. Figure 13 shows a portion of this local energy spectrum, in which we adopt, as a convenient way of labeling stretching modes, $(v_I v_{II} v_{III})$, the notation of Child and Halonen [36]. Such notation allows one to use a unique label for a degenerate polyad. We have v_I quanta in one of the three local modes, v_{II} in a second, and v_{III} in the remaining mode. For example, (310) is shorthand denoting the six degenerate levels $|310\rangle$, $|301\rangle$, $|130\rangle$, $|103\rangle$, $|013\rangle$, and $|031\rangle$. We can see by inspection that the algebraic eigenvalues (3.72) are, in fact, grouped in $(p+1)(p+2)/2$-fold degenerate polyads, where as usual p is the polyad index, $p = v_1 + v_2 + v_3 = v_I + v_{II} + v_{III}$.

   In the second step we account for the essential physical effect of intermode couplings, as explained in the two-oscillator case. With three equal oscillators, such interactions will result in (1) the splitting of degeneracies associated with the purely local behavior and (2) the correspondingly specific symmetry of wavefunctions, under bond permu-

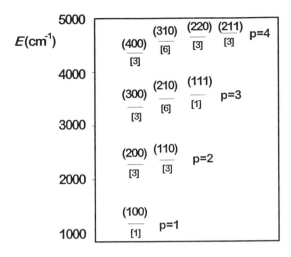

Figure 13. Spectrum of a planar triatomic molecule in the local limit. $N = 100$, $A = -3\,\text{cm}^{-1}$, and $A' = 3\,\text{cm}^{-1}$. (v_1, v_2, v_3) is the degeneracy local mode notation (see the text). The degeneracy associated with the true local mode basis appears in brackets. p is the total number of stretching vibrational quanta in a given polyad.

tation. This can be seen directly by adding to the local Hamiltonian (3.71) the nondiagonal operator given by

$$\hat{S} = \lambda(\hat{M}_{12} + \hat{M}_{13} + \hat{M}_{23}),\tag{3.73}$$

which corresponds to the nondiagonal contribution of the Hamiltonian (3.67), where we put $\lambda_{12} = \lambda_{13} = \lambda_{23} = \lambda$, as a consequence of the supposed physical equivalence of the three interatomic bonds and of their interactions in pairs. Thus the complete Hamiltonian operator is given by

$$\hat{H} = \hat{H}_{\text{local}} + \hat{S}.\tag{3.74}$$

The secular problem is solved easily by using the local eigenvalues [Eq. (3.72)] and the matrix elements of Majorana operators [Eq. (3.46)]. Let us consider the explicit form of the secular determinant for the first polyad, $p = 1$. We obtain

$$\hat{H} = \begin{pmatrix} \varepsilon + 2\lambda N & -\lambda N & -\lambda N \\ -\lambda N & \varepsilon + 2\lambda N & -\lambda N \\ -\lambda N & -\lambda N & \varepsilon + 2\lambda N \end{pmatrix},\tag{3.75}$$

where $\varepsilon = -4A(N-1) - 8A'$ denotes the single local-mode energy as obtained from Eq. (3.72). The eigenvalues of (3.75) can be computed exactly, the result being

$$E_{(A)} = \varepsilon \, ,$$

$$E_{(E)} = \varepsilon + 3\lambda N \, . \tag{3.76}$$

We notice that the three initially degenerate local modes (with energy ε) split into a nondegenerate level, $E_{(A)}$ (with energy ε), and into a twofold-degenerate level, $E_{(E)}$ (with energy $\varepsilon + 3\lambda N$). We use the labels "A" and "E" because they refer to irreducible representations of the symmetry group of the equilateral triangle or, more specifically, to any point group based on a threefold principal axis, such as C_3, C_{3v}, C_{3h}, D_3, D_{3d}, D_{3h}. The corresponding wavefunctions, expressed in the local basis, are found to be in exact agreement with symmetry coordinates [40] (for stretching modes) associated with the point group:

$$\Psi_{(A)} = \frac{1}{\sqrt{3}} \left(r_1 + r_2 + r_3 \right) \, ,$$

$$\Psi_{(E)_1} = \frac{1}{\sqrt{6}} \left(2r_1 - r_2 - r_3 \right) \, , \tag{3.77}$$

$$\Psi_{(E)_2} = \frac{1}{\sqrt{2}} \left(r_2 - r_3 \right) \, .$$

This result shows that the operator $\hat{S}$ [Eq. (3.73)] acts as a "symmetrizer" in regard to the construction of (vibrational) wavefunctions, which transform according to irreducible representations of the molecular point group. As explained later through additional examples, this is a completely general (and important) result of the algebraic approach. It can be shown [10, 27, 41] that Majorana operators have the appealing property of generating symmetry-adapted wavefunctions, according to the discrete symmetry rules imposed on the operator $\hat{S}$. In our example, the C_3 symmetry requires that $\lambda_{12} = \lambda_{13} = \lambda_{23}$. This is the only requirement to be satisfied to generate (1) the expected degeneracy pattern in the energy spectrum and (2) wavefunctions carrying irreducible representations of species A and E of the discrete group. Moreover, as will become clear, the specific outcome of this procedure is of great convenience for strictly computational purposes. At this stage we limit ourselves to saying that the operator $\hat{S}$ produces consistent results for overtone and combination bands in a completely automatic way (as shown in Fig. 14). Analytical formulas for excited levels can be obtained easily but are not shown here. A further useful aspect of the operator $\hat{S}$ is that it can be "specialized" to act on just a specific species of symmetry, allowing for a precise tuning of

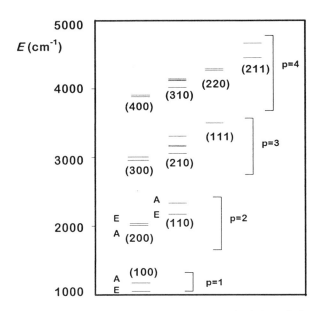

Figure 14. Spectrum of a planar triatomic molecule obtained through the symmetrizer operator $\hat{S}$ (see the text for explanations). The same notation as in Fig. 13: $N = 100$, $A = -3\,\text{cm}^{-1}$, $A' = 3\,\text{cm}^{-1}$, $\lambda = -0.5\,\text{cm}^{-1}$.

energy levels in a realistic fitting procedure. Although this feature will be used more in complex situations, the eigenvalues (3.76) already show that the $\hat{S}$ operator, for the tetratomic symmetric molecule, has the property of acting as a shifter (or projection operator) for the E mode, without affecting the totally symmetric mode A. This result suggests the following possible strategy: One first calibrates the parameter A to reproduce the experimental energy $E_{(A)}$; as a second step, the experimental energy $E_{(E)}$ is fitted by varying the parameter λ.

Before continuing with other examples, we want to emphasize that the aforementioned property of Majorana operators and their linear combinations can be explained on intuitive grounds in the following way. We recall that Majorana operators act basically as permutation operators on the bond indices. In the previous case of three equal bonds, the states obtained by diagonalizing $\hat{S}$ are shown to carry irreducible representations of S_3, the (symmetric) group of permutations for three objects. The Cayley theorem [17] states that every point group is isomorphic to a subgroup of S_n (the permutation group of n objects). Thus we can see the intuitive link between irreducible representations of discrete groups and the specific effect of linear combinations of Majorana operators. As a matter of fact, Majorana operators are invariant operators for unitary algebraic structures; it is also a well-known result of group theory that unitary and permutational structures are closely related, mainly through their respective irreducible representations and basis states. It is possible, of course by starting from the wavefunctions obtained within the present formulation to compute their characters. This can be of great help in nontrivial situations, where it can be very difficult to assign symmetry labels because this procedure requires the explicitly use of involved linear combinations of internal coordinates. We finally notice that $\hat{S}$ (like any Majorana operator or, more generally, an algebraic operator in the present model) induces intrinsically *anharmonic* couplings between local modes. Such anharmonic behavior is an important characteristic in any realistic situations, where nonnegligible deviations from the harmonic limit can strongly affect the overall vibrational picture of the molecule.

Let us now consider further some algebraic treatments of stretching motions and of molecular symmetry adaptation. We start with XF_6 molecules, $X = S, W, U$. These molecules belong to the octahedral group O_h, and the typical arrangement of bonds and labels is shown in Fig. 15. For XF stretching motions, the algebraic scheme provides the following sixfold local Hamiltonian operator:

$$\hat{H}_{\text{local}} = A \sum_{i=1}^{6} \hat{C}_i + A' \sum_{i<j=1}^{6} \hat{C}_{ij}, \qquad (3.78)$$

which is diagonal in the local basis $|v_1 v_2 v_3 v_4 v_5 v_6\rangle$ or, equivalently, in the "degeneracy" basis introduced in the previous example. In this basis one expects to obtain six degenerate fundamental modes labeled as (100000), 20 levels in the first excited polyad, containing two groups of degenerate levels, labeled as (200000) (six modes) and (110000) (14 modes), and so on. The specific molecular geometry is incorporated simply by observing

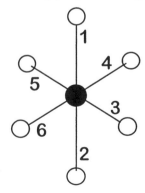

Figure 15. Bond numbering (XF stretches) of an octahedral molecule.

that the XF_6 molecule can be characterized by two different types of intermode couplings, relying on either a purely physical description or a topological one. We have adjacent (or perpendicular) pairs of bonds, such as 1–3, 3–4, ... and opposite (or parallel) pairs, such as 1–2, 3–5, ... We then account for these bond relations by imposing on the symmetrizer operator $\hat{S}$, defined as

$$\hat{S} = \sum_{i<j=1}^{6} \lambda_{ij} \hat{M}_{ij} , \qquad (3.79)$$

the following relations:

$$\lambda_{12} = \lambda_{35} = \lambda_{46} \equiv \lambda_{(opp)} ,$$

$$\lambda_{13} = \lambda_{14} = \lambda_{15} = \lambda_{16} = \lambda_{23} = \lambda_{24} = \lambda_{25} = \lambda_{26} \qquad (3.80)$$

$$= \lambda_{45} = \lambda_{56} = \lambda_{36} = \lambda_{34} \equiv \lambda_{(adj)} .$$

Thus we reduce the number of arbitrary parameters in $\hat{S}$ from 15 to only 2. Let us go now to the secular determinant, where once again we are limiting ourselves to the fundamental polyad of levels, $p = 1$. We easily obtain the following Hamiltonian matrix:

$$
\hat{H} = \varepsilon \hat{I} + \lambda_{(adj)} N
\begin{pmatrix}
4 & 0 & -1 & -1 & -1 & -1 \\
0 & 4 & -1 & -1 & -1 & -1 \\
-1 & -1 & 4 & -1 & 0 & -1 \\
-1 & -1 & -1 & 4 & -1 & 0 \\
-1 & -1 & 0 & -1 & 4 & -1 \\
-1 & -1 & -1 & 0 & -1 & 4
\end{pmatrix}
$$

$$
+ \lambda_{(opp)} N
\begin{pmatrix}
1 & -1 & 0 & 0 & 0 & 0 \\
-1 & 1 & 0 & 0 & 0 & 0 \\
0 & 0 & 1 & 0 & -1 & 0 \\
0 & 0 & 0 & 1 & 0 & -1 \\
0 & 0 & -1 & 0 & 1 & 0 \\
0 & 0 & 0 & -1 & 0 & 1
\end{pmatrix} , \qquad (3.81)
$$

where $\hat{I}$ denotes a 6×6 identity matrix, ε is the local-mode energy [Eq. (3.72)], and $N \equiv N_1 = N_2 = N_3 = N_4 = N_5 = N_6$. By explicit diagonalization of (3.81), the originally sixfold-degenerate mode (100000) is split in three levels and, within an O_h point group, these levels belong to the following symmetry species: A_{1g} (nondegenerate), E_g (twofold degenerate), and F_{1u} (threefold degenerate). As hinted previously, $\hat{S}$ can be written conveniently in terms of specialized shifter or projection operators. Either explicitly or by inspection, the following equations hold:

$$\hat{S} = \lambda_{(\text{opp})}\hat{M}_{(\text{opp})} + \lambda_{(\text{adj})}\hat{M}_{(\text{adj})} = \lambda_{(E_g)}\hat{M}_{(E_g)} + \lambda_{(F_{1u})}\hat{M}_{(F_{1u})} \; ; \quad (3.82)$$

here $\hat{M}_{(\text{opp})} = \hat{M}_{12} + \hat{M}_{35} + \hat{M}_{46}$ and $\hat{M}_{(\text{adj})} = \hat{M}_{13} + \hat{M}_{14} + \cdots$, according to (3.80). On the right side, we also introduce the symmetry mover or projection operators $\hat{M}_{(E_g)}$ and $\hat{M}_{(F_{1u})}$ (with their respective parameters), which can be expressed through the following linear combinations of Majorana operators:

$$\hat{M}_{(E_g)} = \hat{M}_{(\text{opp})} - 2\hat{M}_{(\text{adj})} \, , \qquad \hat{M}_{(F_{1u})} = \hat{M}_{(\text{adj})} \, . \quad (3.83)$$

It is then possible to discover (once again, either explicitly or by direct inspection of matrix elements) that $\hat{M}_{(E_g)}$ moves *only* the E_g mode by the amount $6\lambda_{(E_g)}N$, while $\hat{M}_{(F_{1u})}$ acts *only* on the F_{1u} mode, which is displaced by $2\lambda_{(F_{1u})}N$. This is the equivalent to stating that for the fundamental stretching modes of a XF_6 molecule, the following *exact* relation holds:

$$E_{(A_{1g})} = \varepsilon = -4A(N-1) - 8A' \, ,$$

$$E_{(E_g)} = \varepsilon + 6\lambda_{(E_g)}N \, , \quad (3.84)$$

$$E_{(F_{1u})} = \varepsilon + 2\lambda_{(F_{1u})}N \, .$$

These relations are quite useful because they allow one to obtain a precise estimate of the algebraic parameters A, $\lambda_{(E_g)}$, and $\lambda_{(F_{1u})}$. The Hamiltonian is still in the form of Eq. (3.74), regardless of what expression of $\hat{S}$ in Eq. (3.82) is used. Moreover, this same equation also provides a trivial connection between projection and general Majorana parameters, which can be written as

$$\lambda_{(E_g)} = \lambda_{(\text{opp})} \, , \qquad \lambda_{(F_{1u})} = \lambda_{(\text{adj})} - 2\lambda_{(\text{opp})} \, . \quad (3.85)$$

To appreciate the facility of using this procedure, let us consider some numerical cases for the "real" molecules SF_6, WF_6, and UF_6 [27].

We start with the experimental values [36] for the fundamental XF stretching modes of these three molecules (in cm^{-1}):

	SF_6	WF_6	UF_6	
E_g	643.95	678.00	534.10	
A_{1g}	774.54	772.14	667.10	(3.86)
F_{1u}	948.10	712.60	625.50	

Then we obtain, from spectroscopic tables of ω_e and $\omega_e x_3$, the following (dimensionless) values of N:

$$
\begin{array}{cccc}
 & \mathrm{SF}_6 & \mathrm{WF}_6 & \mathrm{UF}_6 \\
N & 180 & 200 & 250
\end{array}
\tag{3.87}
$$

We finally use the equation for the local mode energy, ε, to calculate, for A_{1g} levels, the estimates (in cm^{-1}) of the parameter A:

$$
\begin{array}{cccc}
 & \mathrm{SF}_6 & \mathrm{WF}_6 & \mathrm{UF}_6 \\
A = -\dfrac{E_{(A_{1g})}}{4(N-1)} & -1.08 & -0.97 & -0.67
\end{array}
\tag{3.88}
$$

We can now split the local modes by introducing the Majorana (projection) operators. Their parameters can be obtained directly by application of Eq. (3.84) to the experimental energies for levels E_g and F_{1u}. As a result, one obtains

$$
\begin{array}{cccc}
 & \mathrm{SF}_6 & \mathrm{WF}_6 & \mathrm{UF}_6 \\[4pt]
\lambda_{(F_{1u})} = \dfrac{E_{(F_{1u})} - E_{(A_g)}}{2N} & 0.482 & -0.150 & -0.028 \\[12pt]
\lambda_{(E_g)} = \dfrac{E_{(E_g)} - E_{(A_g)}}{6N} & -0.121 & -0.078 & -0.089
\end{array}
\tag{3.89}
$$

With these values for N, A, $\lambda_{(F_{1u})}$, and $\lambda_{(E_g)}$, we can compute the vibrational spectra (for XF stretches) for the three molecules SF_6, WF_6, and UF_6. The complete procedure obviously requires further refinement of the Hamiltonian operator. Typically, this is achieved by adding cross-anharmonicities (i.e., operators $\hat{C}_{ij}$) and/or higher-order terms, which are calibrated against experimental values of levels belonging to exited polyads [27]. Figure 16 shows the results of such a complete computation for the first two polyads of stretching modes. We notice, in particular, the total amount of splitting characterizing the fundamental bands, which ranges from $94\ \mathrm{cm}^{-1}$ for WF_6 to $305\ \mathrm{cm}^{-1}$ for SF_6. This corresponds to considering increasing amounts of normal molecular behavior.

A second, nontrivial example is benzene, $\mathrm{C}_6\mathrm{H}_6$. We address its complete vibrational spectroscopy in Section III.C.2. However, we can study the manifold of CH stretching modes in exactly the same way as done for the XF_6 molecule. The benzene molecule possesses six equivalent CH oscillators attached to the carbon ring, which will be neglected at

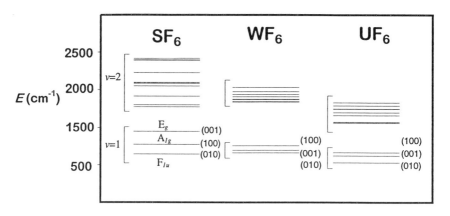

Figure 16. Calculated vibrational spectra (XF stretches) of SF_6, WF_6, and UF_6.

this stage of the analysis. Thus we are left with a physical situation characterized by six anharmonic local oscillators. The specific geometric configuration of this molecule (shown in Fig. 17) can now be inserted in the algebraic framework by using the symmetrizer Majorana operator, which is expressed as

$$\hat{S} = \lambda_{I}\hat{M}_{I} + \lambda_{II}\hat{M}_{II} + \lambda_{III}\hat{M}_{III} , \qquad (3.90)$$

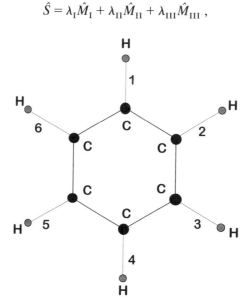

Figure 17. Bond numbering (CH stretches) of benzene.

where

$$\hat{M}_I = \hat{M}_{12} + \hat{M}_{23} + \hat{M}_{34} + \hat{M}_{45} + \hat{M}_{56} + \hat{M}_{16},$$

$$\hat{M}_{II} = \hat{M}_{13} + \hat{M}_{24} + \hat{M}_{35} + \hat{M}_{46} + \hat{M}_{15} + \hat{M}_{26},$$

$$\hat{M}_{III} = \hat{M}_{14} + \hat{M}_{25} + \hat{M}_{36}.$$

We introduce here the $\hat{M}_I$, $\hat{M}_{II}$, and $\hat{M}_{III}$ operators, which represent the first, second, and opposite neighbor interactions, respectively. This is requested by the hexagonal geometry and by the equivalence among the six CH bonds. The operator (3.90) produces a splitting of the degenerate polyads of local modes in agreement with the symmetry of the D_{6h} point group; the six modes, belonging to the first polyad (fundamental levels), give two nondegenerate levels (A_{1g}, B_{1u}) and two double degenerate levels (E_{1u}, E_{2g}), as shown in Fig. 18. Again, we can introduce projection operators for the symmetry species of in-plane stretching vibrations of C_6H_6. These operators can be written as

$$\hat{M}_{(E_{2g})} = \hat{M}_I + \hat{M}_{II} - 2\hat{M}_{III},$$

$$\hat{M}_{(B_{1u})} = \hat{M}_I - \hat{M}_{II} + \hat{M}_{III}, \qquad (3.91)$$

$$\hat{M}_{(E_{1u})} = -\hat{M}_I + \hat{M}_{II} + 2\hat{M}_{III}.$$

The totally symmetric mode, A_{1g}, does not move under the action of $\hat{S}$.

As a final example, we consider CH stretches of the ethylene molecule, C_2H_4. In this molecule, one starts from four equivalent CH modes (see Fig. 19). The splitting of the fundamental polyad of levels

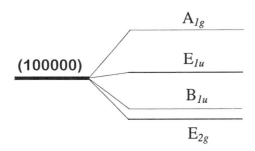

Figure 18. Splitting of the degenerate CH stretches of C_6H_6 by the Majorana operators.

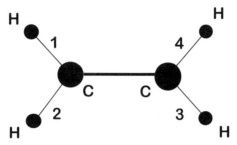

Figure 19. Bond numbering (CH stretches) of ethylene.

(1000) is then obtained through the symmetrizer operator given by

$$\hat{S} = \lambda_a \hat{M}_a + \lambda_c \hat{M}_c + \lambda_t \hat{M}_t ,$$

$$\hat{M}_a = \hat{M}_{12} + \hat{M}_{34} , \qquad \hat{M}_c = \hat{M}_{14} + \hat{M}_{23} , \qquad \hat{M}_t = \hat{M}_{13} + \hat{M}_{24} ,$$

$$(3.92)$$

where the operators $\hat{M}_a$, $\hat{M}_c$, and $\hat{M}_t$ account for adjacent, *cis-*, and *trans*-neighbor interactions, respectively. The diagonalization of the complete Hamiltonian operator splits the (1000) polyad in four non-degenerate levels. These levels correspond to the A_g, B_{1g}, B_{2u}, and B_{3u} irreducible representations of the point group, D_{2h}. Again, projection operators can be applied to specific symmetry species. In the present case we find

$$\hat{M}_{(B_{1g})} = \hat{M}_a + \hat{M}_c - \hat{M}_t ,$$

$$\hat{M}_{(B_{2u})} = \hat{M}_a - \hat{M}_c + \hat{M}_t ,$$

$$(3.93)$$

$$\hat{M}_{(B_{3u})} = -\hat{M}_a + \hat{M}_c + \hat{M}_t ;$$

thus the operator $\hat{S}$ is written as

$$\hat{S} = \lambda_{(B_{1g})} \hat{M}_{(B_{1g})} + \lambda_{(B_{2u})} \hat{M}_{(B_{2u})} + \lambda_{(B_{3u})} \hat{M}_{(B_{3u})} . \qquad (3.94)$$

The action of this operator is shown schematically in Fig. 20. As already noted, for these examples, the common denominator is a simple procedure that provides coupled wavefunctions carrying irreducible representations of the molecular point group, which leads to initial guesses for the algebraic parameters in a straightforward way. To better reproduce more excited levels, the subsequent fitting procedure will require a increased number of higher-order terms.

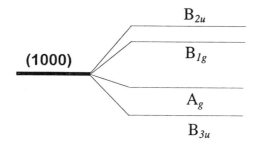

Figure 20. Splitting of the degenerate CH stretches in C_2H_4 by the Majorana operators.

2. Vibrational Spectroscopy of Polyatomic Molecules

We are now ready to address the problem of obtaining a complete algebraic description of a polyatomic molecule (i.e., including both stretching and bending modes). As stated in Section III.A, the basic idea is to replace every one-dimensional internal degree of freedom with a U(2) algebra. To understand the basic features of this approach, we could apply it to some very simple systems; in fact, this has already been done for some triatomic molecules [39], where the one-dimensional model has been used to describe their complete vibrational spectra. The strategy is simple. The spectrum-generating algebra consists of the product $U_1(2) \otimes U_2(2) \otimes U_B(2)$, in which $U_1(2)$ and $U_2(2)$ replace the two interatomic bond coordinates r_1 and r_2, while $U_B(2)$ describes the bending angle $\widehat{r_1 r_2}$. This is possible by virtue of the one-to-one correspondence between the representations of the algebra chain $U(2) \supset O(2)$ and the energy spectrum of the bending model potential given by the Pöschl–Teller function. It is then possible to account for stretch–stretch and stretch–bend coupling terms through the operators $\hat{C}_{ij}$ and $\hat{M}_{ij}$ $(i, j = 1, 2, B)$. However, this use of a one-dimensional model does not lead to real advantages in comparison to traditional approaches, with the exception that fewer arbitrary parameters are needed and it is somewhat easier to obtain infrared intensities, as discussed later. Consequently, a less trivial use of this model will be oriented toward describing the complete vibrational spectroscopy of a complex molecule, such as benzene. Benzene is an excellent example of a highly symmetric, many-body system in which the most interesting spectroscopic features are undoubtedly related to CH stretching modes. At the same time, other internal degrees of freedom affect the overall vibrational behavior in a complicated way. As a result, proper understanding of this molecule requires a reliable Hamiltonian operator accounting for all the vibrational modes and their anharmonic interaction. Benzene has 12 atoms and thus $3N - 6 = 30$ internal degrees of freedom. One could introduce 30 unitary algebras $U(2)$ to construct a one-dimensional algebraic model of this molecule. However, it is definitely more convenient to attach three $U(2)$ algebras to each atom such that the complete set of 36 coordinates is "algebrized." Nonetheless, this procedure still generates some "spurious" modes (i.e., nonvibrational ones), which have to be removed from the eigenspectrum.

The problem of spurious modes is not negligible, as it typically affects any theoretical model (either algebraic or not) based on atomic coordinates. However, it can be solved easily within the algebraic framework. Spurious modes can be understood easily by considering the simple case of a (symmetric) triatomic molecule, such as H_2O. We see in Fig. 21 that two $U(2)$ algebras are attached to each hydrogen atom. As already explained, the $U_1(2)$ and $U_2(2)$ algebras describe stretching modes. Moreover, we introduce the "perpendicular" algebras $U_3(2)$ and $U_4(2)$, which can be seen as generators of transverse anharmonic motions for the OH bonds. In light of their equivalence, we expect to find symmetric (ϕ_S) and antisymmetric (ϕ_A) combinations of local transversal modes. Therefore, it should be clear that ϕ_A corresponds to a true (bending) vibrational motion, while ϕ_S correlates to a rotation of the whole molecule about an axis perpendicular to the atomic plane. The latter is a spurious mode, which results from the introduction of two independent "coordinates" for describing an effective one-dimensional degree of freedom. Although we consider the explicit solution of this problem for benzene, the general technique for removing spurious modes can be understood by realizing that for a given molecular geometry, translation-

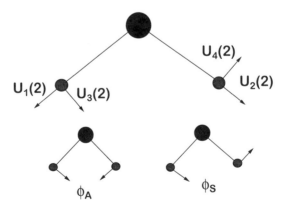

Figure 21. Origin of spurious modes and redundant (algebraic) coordinates.

al, rotational, and vibrational degrees of freedom transform according to specific irreducible representations of the molecular point group. So, by using Majorana projection operators, it is possible selectively to remove the nonvibrational (either translational or rotational) modes from the true vibrations. For H_2O, the z rotation transforms according to the A_2 species of C_{2v}. So we will use the A_2 shifter operator to remove this rotational, spurious mode.

Let us now return to our original problem. The choice of local coordinates for hydrogen atoms in benzene is shown in Fig. 22. A similar choice can be done for the carbon atoms of the molecular ring. In fact, this problem has already been studied from both a traditional viewpoint [40, 42–45] and in the algebraic framework [46–49], so it is possible to compare the correspondingly different limits and advantages of each formulation. The spectrum-generating algebra introduced here is based on 36 U(2) algebras to which we associate a 36-dimensional Hamiltonian matrix for describing the fundamental modes. Unfortunately, excited levels generate prohibitively large matrices. We simplify our problem by considering a blocking procedure of the original matrix, where we separate the local basis into six completely independent families of vibrational modes and of a different nature. The idea is to neglect interactions between CH and CC modes, stretching and bending modes, in-plane and out-of-plane modes. This type of an approximation can be fairly well justified because of (1) the small mechanical coupling resulting from large mass differences, and (2) different vibrational energies of bending and stretching modes (as well as in- and out-of-plane modes).

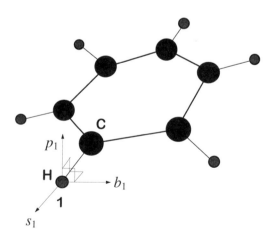

Figure 22. Local coordinates of the H atom 1 in benzene, s denotes a stretching motion; b and p denote in- and out-of-plane bending motions, respectively.

However, this approximation is inadequate if we want to account for the redistribution of vibrational energy among molecular modes. In these processes, anharmonic quasiresonant couplings involving different families of modes play a significant role. We account for these phenomena in Section III.E.

The blocking procedure leads us to the uncoupled families of (fundamental) vibrational modes given by

$$\text{CH stretches} \rightarrow E_{2g} + B_{1u} + E_{1u} + A_{1g},$$

$$\text{CH in-plane bends} \rightarrow E_{2g} + B_{2u} + E_{1u} + A_{2g},$$

$$\text{CH out-of-plane bends} \rightarrow E_{2u} + B_{2g} + E_{1g} + A_{2u},$$

$$\text{CC stretches} \rightarrow E_{2g} + B_{2u} + E_{1u} + A_{1g}, \tag{3.95}$$

$$\text{CC in-plane bends} \rightarrow E_{2g} + B_{1u},$$

$$\text{CC out-of-plane bends} \rightarrow E_{2u} + B_{2g}.$$

We notice that the first four families are divided into four levels, two of which are twofold degenerate (species E). This is in agreement with our previous analysis (carried out by means of the algebraic projection operators) of the six equivalent CH bonds of C_6H_6. Each of the last two families (in- and out-of-plane CC bending modes) contains two levels. This result is indicative of spurious modes, which must associate with three rotational and three translational degrees of freedom of the "heavy" part of the molecule (i.e., the carbon ring). Through standard procedures of spurious modes analysis [40], it is found that these six redundant coordinates combine among themselves to give the following symmetry species:

$$\text{CC in-plane bends} \rightarrow A_{1g} + E_{1u},$$

$$\text{CC out-of-plane bends} \rightarrow A_{1u} + E_{1g}. \tag{3.96}$$

Keeping this result in mind, we now proceed to construct the algebraic Hamiltonian in its blocked form. For those families of levels excluding spurious modes, we apply the technique discussed in Section III.C.1. Thus we must fit the algebraic parameters A, A', and λ_{ij} (plus N, separately fixed) over a convenient database of experimental levels. A complete list of parameters can be found in Ref. 47. In Table II we report only the fundamental vibrational energies of benzene.

In regard to the families of modes (3.96) containing spurious species, we can apply the Majorana projection operators to selectively act either

TABLE II
Fundamental Vibrations of Benzene[a]

Wilson Label	Symmetry Label	Expt.	Calc.
	CH Stretches		
7	E_{2g}	3056.6	3056.99
13	B_{1u}	3057	3057.19
20	E_{1u}	3064.367	3064.74
2	A_{1g}	3073.94	3074.11
	CC Stretches		
1	A_{1g}	993.1	993.10
14	B_{2u}	1309.8	1309.80
19	E_{1u}	1484	1484.00
8	E_{2g}	1601	1601.00
	CH Bends in Plane		
18	E_{1u}	1038.3	1037.73
15	B_{2u}	1148.5	1149.07
9	E_{2g}	1177.8	1177.89
3	A_{2g}	1350	1349.91
	CH Bends out of Plane		
11	A_{2u}	674	674.62
4	B_{2g}	707	707.68
10	E_{1g}	848.9	848.20
17	E_{2u}	967	966.36
	CC Bends in Plane		
6	E_{2g}	609	609.00
12	B_{1u}	1010	1010.00
	CC Bends out of Plane		
16	E_{2u}	398.8	398.80
5	B_{2g}	990	990.01

[a]All values in cm^{-1}. The mode ν_{20} (E_{1u}) is a deperturbed value (see the text).

on species A_{1g}, E_{1u} (for in-plane CC bends) or A_{1u}, E_{1g} (for out-of-plane CC bends). In practice, such unwanted modes are moved at energies 10 (or even more) times than those of the true vibrational levels. Removal of the spurious mode could be accomplished by giving the parameter λ a large value. However, we proceed in a slightly different way, which involves giving a value to N (the vibron number) that is very large,

$N \approx 10^6$. Such a choice implies that the behavior of the corresponding oscillator is essential harmonic and the removal is carried out in an *exact* way. Otherwise, spurious modes could interact anharmonically with nonspurious ones, with the result that the true vibrational modes would be affected seriously by unwanted contributions from the Hamiltonian operator.

3. Mass Scaling Laws and Lowering of Symmetry

Another interesting feature of the one-dimensional algebraic model is the possibility of accounting for atomic substitutions in a given molecule. This problem can be addressed by adopting two distinct strategies. In the first one we use the available experimental information to extract the algebraic parameters associated with those oscillators modified because of the atomic substitution, isotopic or otherwise. In the second strategy we use certain approximate algebraic scaling laws for obtaining, in terms of the already known parameters for the nonsubstituted bonds, the values associated with substituted bonds. It should be clear that these methods account only for isotopic substitutions. Substitutions for one atom for another type create new chemical bonds, with the consequence that the algebraic parameters can be substantially changed. In some cases the fitting procedure must be entirely revised. Regardless of the type of substitution, it modifies the symmetry of the molecule. For example, the replacement of a single hydrogen atom in benzene reduces the symmetry from D_{6h} to C_{2v}. As discussed in Sections III.C.1 and III.C.2, the algebraic arrangement of point group symmetries is implemented through the linear combinations of Majorana operators. Consequently, we expect to obtain meaningful results from the analysis of substituted species for a given molecule. In fact, an algebraic analysis of deuterated benzene has already been done [50]. We illustrate this analysis briefly to highlight the essential features of this model.

First, the experimentally obtained levels of C_6H_6 and C_6D_6 are used to derive the algebraic parameters A, A', λ_{ij}, and N for the two molecular species. Both C_6H_6 and C_6D_6 have an equivalent geometry; therefore, we can employ exactly the same symmetrizer operators as discussed in Section III.C.2. In regard to the algebraic parameters A, A', and N, the isotopic substitution leads to very simple scaling laws which are derived by direct inspection of Eq. (2.70). For the isotopic substitutions $D \rightarrow H$ in a CH bond, we find

$$N_D = \rho N_H, \qquad A_D = A_H/\rho^2, \qquad A'_D = A'_H/\rho^2, \qquad (3.97)$$

where the proportionality factor, ρ, is given by

$$\rho = \sqrt{\frac{\mu_{CD}}{\mu_{CH}}} = \sqrt{\frac{m_D}{m_H} \frac{m_C + m_H}{m_C + m_D}} = 1.363 \ . \tag{3.98}$$

In Eq. (3.97), the subscript H labels algebraic parameters for nonsubstituted bonds (CH), while D represents substituted bonds (CD). The applicability of these scaling formulas can be checked by performing distinct fitting procedures for both molecules. As a result we find that the scaled parameters are in good agreement (within 1%) with fitted values. Unfortunately, for benzene, it is not possible to write down a generalized, simple scaling law for the Majorana operator parameters, λ_{ij}. This arises from the increased coupling between the carbon vibrational modes and the heavier deuterium atoms. The correspondingly more normal behavior obviously reflects this increased intermode coupling of CX (X = H, D, ...) (i.e., the numerical values of the λ_{ij} change in an unpredictable way). Nonetheless, since we have at our disposal a convenient experimental database of vibrational levels for C_6D_6, it is possible to obtain independently (fitted) numerical values for the λ_{ij}. This is useful, as it allows one to predict in a reliable way the vibrational spectrum of partially substituted benzenes, $C_6H_xD_{6-x}$, $x = 0, \ldots, 6$, in any geometric configuration. For example, let us consider the molecule 1,4-$C_6H_4D_2$, shown in Fig. 23. We expect to split the sixfold-degenerate local mode (100000) according to the symmetry reduction $D_{2h} \subset D_{6h}$. Correspondingly, the symmetrized operator must be different from the operator (3.90) of benzene (either C_6H_6 or C_6D_6). From a physical standpoint, we note that in the 1,4-$C_6H_4D_2$ molecule, the interactions for the first, second, and opposite atoms acquire a richer structure because of the presence of both H and D atoms. By starting from Eq. (3.90) and keeping Fig. 23 in mind, we can write

$$\lambda_I \hat{M}_I \rightarrow \lambda_I^{HH}(\hat{M}_{23} + \hat{M}_{56}) + \lambda_I^{HD}(\hat{M}_{12} + \hat{M}_{34} + \hat{M}_{45} + \hat{M}_{16}) \ ,$$

$$\lambda_{II} \hat{M}_{II} \rightarrow \lambda_{II}^{HH}(\hat{M}_{26} + \hat{M}_{35}) + \lambda_{II}^{HD}(\hat{M}_{13} + \hat{M}_{24} + \hat{M}_{46} + \hat{M}_{15}) \ , \tag{3.99}$$

$$\lambda_{III} \hat{M}_{III} \rightarrow \lambda_{III}^{HH}(\hat{M}_{25} + \hat{M}_{36}) + \lambda_{III}^{DD} \hat{M}_{14} \ .$$

This corresponds to considering six independent terms (in place of three) contributing to the symmetrizer $\hat{S}$. In the present case, experimental information is lacking. Therefore, we have to adopt a different strategy, where we assume that the completely unknown "mixed" parameters λ^{HD} of Eq. (3.99) (i.e., involving couplings between CH and CD bonds) can

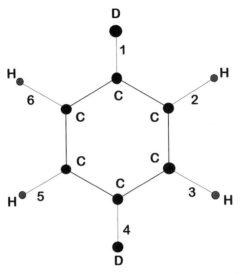

Figure 23. Bond numbering (CX stretches) of the 1,4-$C_6H_4D_2$ substituted benzene.

be calculated in terms of the known "diagonal" parameters λ^{HH} and λ^{DD} (i.e., involving only CH–CH or CD–CD interactions) through the following equation:

$$\lambda^{HD} = (\lambda^{HH}\lambda^{DD})^{1/2}. \tag{3.100}$$

This relation can be and, in fact, has been checked by its direct application to a different, more symmetric, and better known substituted benzene, the 2,4,6-$C_6H_3D_3$ species. In fact, Eq. (3.100) leads to computed vibrational energies in very good agreement with existing experimental data. Thus the algebraically obtained Hamiltonian model can then be used to (1) extend the calculation to (hitherto unknown) excited vibrational modes, and (2) obtain the corresponding vibrational wavefunctions in the local basis. In the present case of a low-symmetry species, it is important to have reliable results for the vibrational wavefunctions because they allow one to calculate the exact amount of mode "scrambling" of substituted species in comparison to the higher-symmetry parent molecule (here C_6H_6). A more complete discussion of this case can be found in Ref. 50.

   To summarize, we have seen how the one-dimensional model can easily be modified to describe molecules of increasing complexity, where the isotopic substitution of one or more atoms leads to lowered symmetry and, consequently, to a larger number of arbitrary parameters. In general, such symmetry lowering is an interesting implication for molecular spectroscopy. In Section V.A we discuss how the one-dimensional model can describe the vibrational spectrum of complex molecular systems, such as the benzene dimer. At this stage we limit ourselves to noting that the construction of a molecular dimer relates to previous arguments involving the reduction of symmetry in a molecular environment. In the benzene dimer we will see that the two participating sites are separately subject to small rearrangements of their respective bond lengths, force constants, and so on. Having at our disposal an algebraic Hamiltonian operator in which the symmetry information is naturally inserted, we can quickly obtain how the vibrational spectrum is modified by this single-site rearrangement. Although we return to this argument later, Fig. 24 shows the energy curves representing the effect by small variations of the algebraic parameter A_1 on the stretching vibrational levels in benzene. This is equivalent to changing the force constant of the CH bond "1" slightly. Thus we obtain an example of correlation curve for

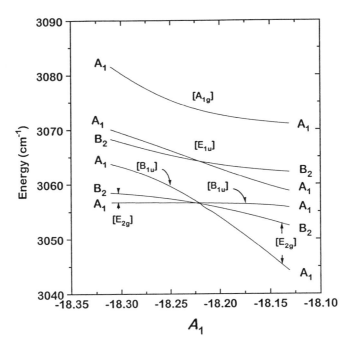

Figure 24. Behavior of vibrational energies of CH stretches in benzene under lowering of symmetry $D_{6h} \rightarrow C_{2v}$.

the "continuous" symmetry lowering in going from D_{6h} to C_{2v} point groups.

D. Electromagnetic Transition Intensities

One of the most important aspects in both experimental and theoretical studies in molecular spectroscopy is, undoubtedly, the characterization of intensities induced by electromagnetic radiation. We are, of course, interested in obtaining information concerning infrared and Raman transitions which are driven by electric dipole and quadrupole operators, respectively. These transitions can be represented as

$$\mathfrak{I}_{vj \to v'j'} = |\langle vj \| \hat{D}^{(1)} \| v'j' \rangle|^2 \qquad (3.101)$$

for infrared transitions and

$$\mathfrak{R}_{vj \to v'j'} = |\langle vj \| \hat{Q}^{(2)} \| v'j' \rangle|^2 \qquad (3.102)$$

for Raman transitions. The key point here is that we are interested in the computation of (reduced) matrix elements for transition operators of rank 1 and 2 between different rovibrational states. In principle, the algebraic approach to this problem is very straightforward, since it is based on the application of operators whose tensor character is well defined and easily adapted to the algebraic approach. In fact, the coupled form of U(4) boson operators (2.95) already contains $\hat{D}^{(1)}$ and $\hat{Q}^{(2)}$. Consequently, we should be able to employ these operators within the algebraic framework in a rather natural way; however, this turns out to be only partially true. To compute matrix elements such as (3.101) and (3.102), we need to know the exact form of the involved operators in terms of their interatomic coordinates, charge distribution, and so on. The algebraic approach can give only partial answers to such questions. By analogy with the calculation of energy levels, we have to face the problem of fixing a certain number of arbitrary parameters in order to reproduce the observed intensities. Within the algebraic model we can provide a very accommodating Hamiltonian formulation of the rovibrational problem. In this formulation, the vibrational wavefunctions are typically expressed in a local (bond) basis and they correspond to an intrinsically anharmonic description of the physical system. If the electric dipole and quadrupole operators can be formulated in the same way, we obtain a theoretical model which, despite its semiempirical nature, will contain several appealing features. In particular, we should be able to construct expansions in higher-order terms of electric operators while retaining the ease of application characteristic of an algebraic scheme.

Obviously, a comprehensive study of transition intensities in molecular spectroscopy requires a full three-dimensional treatment. Nonetheless, it is worthwhile to take a simplified one-dimensional approach to these arguments. Rotations will be excluded from this treatment. As a result, the obtained information will be incomplete since many spectroscopic studies focus on making correct assignments of rotational bands observed at high resolution. Rather, the one-dimensional model allows for a global picture of the rovibrating molecule in which the rotational detail is (from the point of view of the computation of intensities) averaged in its effects. In the last few years the one-dimensional model has been applied to the study of vibrational transitions in a number of cases, particularly complex ones involving several interacting atoms. In fact, medium-sized and large polyatomic molecules tend to defy traditional approaches via difficulty in computation of matrix elements of transition operators. Moreover, those conventional methods, based on harmonic models, are even less efficient in describing highly excited states, which are currently of particular experimental interest. Here we briefly describe the most important aspects of the one-dimensional algebraic treatment of transition intensities. The three-dimensional formulation is discussed in Section IV.E.

Any theoretical model for the computation of transition intensities should account for two distinct aspects of matrix elements (3.101) and (3.102). One needs a reasonable model for describing (1) the dependence of the operators on interatomic coordinates and (2) the corresponding rovibrational selection rules. As necessitated by the one-dimensional picture, we now focus our attention on purely vibrational transitions. Thus we are looking for certain algebraic operators capable of inducing transitions between states differing precisely in their vibrational quantum numbers. This is an easy task, as U(2) ladder operators [defined in Section II.C.1, Eq. (2.53)] act on the single-oscillator local basis in the following way:

$$\hat{S}_{\pm}|N, m\rangle = \tfrac{1}{2}\sqrt{(N \mp m)(n \mp m + 2)}|N, m \pm 2\rangle . \qquad (3.103)$$

We can also write, in the vibrational local basis $|N, v\rangle$,

$$\hat{S}_{+}|N, v\rangle = \sqrt{v(N - v + 1)}|N, v - 1\rangle ,$$
$$\hat{S}_{-}|N, v\rangle = \sqrt{(v + 1)(N - v)}|N, v + 1\rangle . \qquad (3.104)$$

These $\hat{S}_{\pm}$ operators are the algebraic counterpart of the single-bond electric dipole operator in the rigid-rotor, harmonic approximation, which corresponds only to $\Delta v = \pm 1$ infrared transitions. In this elementary model we cannot account for the symmetry character of the vibrating bond in terms of exchange of its terminal atoms. Such a limitation is severe because it excludes diatomic homonuclear molecules which are not infrared active because of their permanently zero electric dipole moment in any vibrational state. In a more realistic formulation, we should also be able to describe overtone transitions, whose infrared intensities decrease according to an approximate exponential law in the vibrational quantum number. Thus we need transition operators more sophisticated than $\hat{S}_+$. A possible solution is to use

$$\hat{T} = t \exp[-\beta(\hat{S}_+ + \hat{S}_-)] \, . \tag{3.105}$$

The matrix elements of this operator can be obtained explicitly in the local basis. This computation employs the close similarity between the $\hat{S}_\pm$ operators and the angular momentum ladder operators, $\hat{J}_\pm$. Thus the operator $\hat{T}$ can be seen as a rotation about a finite angle related to the parameter β. Consequently, $\hat{T}$ acts on the single state of the local basis to generate an expansion (in terms of Jacobi polynomials) over the complete set of vibrational states. This is equivalent to saying that the harmonic selection rules are relaxed and transitions $\Delta v = 0, \pm 1, \pm 2, \ldots$ now have nonzero matrix elements, in accordance with the electric anharmonicity of the dipole operator. A good approximation for these matrix elements is given by the simple formula

$$\langle N, v | \hat{T} | N, v' \rangle \simeq t \exp(-\beta |v - v'|) \, , \tag{3.106}$$

which reproduces fairly well the exponential decay of experimentally obtained infrared intensities. In Eq. (3.106), the parameters t and β have to be fitted numerically. However, such a simple model is not particularly useful in and of itself. Rather, its significance rests in being a heuristic device for constructing, once again, a molecular picture based on a local bond coordinate system. Explicit use of the Hamiltonian operator discussed previously leads to a set of symmetry-adapted wavefunctions in this same local basis. As a consequence, the simple operator (3.105), applied to each interatomic bond, will give the information requested on the infrared activity of the complete molecule. Let us now consider the simple example of the water molecule. As shown in Fig. 25, the total electric operator can be written in terms of the Cartesian components

$$\hat{T}_x = \gamma(\hat{T}_1 - \hat{T}_2) \sin \frac{\theta_{eq}}{2} \, ,$$

$$\hat{T}_y = \gamma(\hat{T}_1 + \hat{T}_2) \cos \frac{\theta_{eq}}{2} \, , \tag{3.107}$$

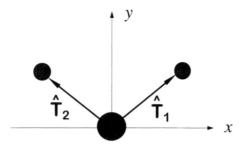

Figure 25. Bond electric dipole operators of H_2O.

in which θ_{eq} is the equilibrium angle of the bent molecular geometry and γ is an overall scaling factor. The infrared intensity of the generic transition $v \to v'$ is then expressed as

$$\Im_{v \to v'} = |\langle v_1 v_2 | \hat{T}_x | v_1' v_2' \rangle|^2 + |\langle v_1 v_2 | \hat{T}_y | v_1' v_2' \rangle|^2 . \tag{3.108}$$

Here $|v_1 v_2\rangle$ denotes the *local* stretching basis state, and as a straightforward generalization of Eq. (3.106), we have

$$\langle v_1 v_2 | \hat{T}_1 | v_1' v_2' \rangle \simeq t_1 \exp(-\beta_1 |v_1 - v_1'|)\delta_{v_2 v_2'} ,$$

$$\langle v_1 v_2 | \hat{T}_2 \| v_1' v_2' \rangle \simeq t_2 \exp(-\beta_2 |v_2 - v_2'|)\delta_{v_1 v_1'} , \tag{3.109}$$

where for quite obvious reasons of molecular symmetry, $t_1 = t_2 = t$ and $\beta_1 = \beta_2 = \beta$. It is very important to understand the local meaning and consequences of this model from a practical, computational point of view. Since water belongs to the C_{2v} point group, both its symmetric (A_1) and antisymmetric (B_2) stretching modes will be infrared active. In our (algebraic) local formulation for transitions from ground to fundamental stretching levels v_S and v_A, we obtain the following relations:

$$\langle v_S | \hat{T}_x | 0 \rangle = \frac{1}{\sqrt{2}} \{ \langle 10 | \hat{T}_x | 00 \rangle + \langle 01 | \hat{T}_x | 00 \rangle \} = 0 ,$$

$$\langle v_A | \hat{T}_x | 0 \rangle = \frac{1}{\sqrt{2}} \{ \langle 10 | \hat{T}_x | 00 \rangle - \langle 01 | \hat{T}_x | 00 \rangle \} = \sqrt{2}\gamma \sin\frac{\theta_{eq}}{2} t \exp(-\beta) ,$$

$$\tag{3.110}$$

$$\langle v_S | \hat{T}_y | 0 \rangle = \frac{1}{\sqrt{2}} \{ \langle 10 | \hat{T}_y | 00 \rangle + \langle 01 | \hat{T}_y | 00 \rangle \} = \sqrt{2} \gamma \cos \frac{\theta_{eq}}{2} t \exp(-\beta) ,$$

$$\langle v_A | \hat{T}_y | 0 \rangle = \frac{1}{\sqrt{2}} \{ \langle 10 | \hat{T}_y | 00 \rangle - \langle 01 | \hat{T}_y | 00 \rangle \} = 0 ,$$

in which we use Eqs. (3.109) to write $\langle 10 | \hat{T}_1 | 00 \rangle = \langle 01 | \hat{T}_2 | 00 \rangle = t \exp(-\beta)$ and $\langle 10 | \hat{T}_2 | 00 \rangle = \langle 01 | \hat{T}_1 | 00 \rangle = 0$. We thus obtain the following transition intensities:

$$\Im_{0 \to v_S} = |\langle v_S | \hat{T}_x | 0 \rangle|^2 + |\langle v_S | \hat{T}_y | 0 \rangle|^2 = 2\gamma^2 \cos^2 \frac{\theta_{eq}}{2} t^2 \exp(-2\beta) ,$$

$$\Im_{0 \to v_A} = |\langle v_A | \hat{T}_x | 0 \rangle|^2 + |\langle v_A | \hat{T}_y | 0 \rangle|^2 = 2\gamma^2 \sin^2 \frac{\theta_{eq}}{2} t^2 \exp(-2\beta) .$$

$$\text{(3.111)}$$

The definition (3.107) of dipole operator can be viewed as the outcome of two distinct but related procedures. On one hand, we can use a purely geometric construction, based on Fig. 25. This is a convenient way of proceeding in cases, such as the present one, in which the molecular geometry is simple and characterized by an highly symmetric environment. On the other hand, we can construct the correct form of the dipole operator by realizing that (1) this operator must transform as a vector, and (2) Majorana projection operators give transformations specific to the symmetry of a given point group. In Eq. (3.107) we see that $\hat{T}_x$ and $\hat{T}_y$ are antisymmetric and symmetric combinations of the bond operators $\hat{T}_1$ and $\hat{T}_2$. This same result can be obtained by the operator $\hat{M}_{12}$. In other words, we can use Majorana projection operators for constructing not only wavefunctions of adapted symmetry, but also for any object of physical interest that satisfies specific symmetry requirements. For water, $\hat{T}_x$ transforms like x (species B_2) and $\hat{T}_y$ like y (species A_1). As shown clearly in Eqs. (3.110), $\hat{T}_x$ corresponds to transitions from the ground vibrational state (species A_1) to antisymmetric stretching levels, while $\hat{T}_y$ acts only on symmetric combinations of local states. This is in obvious agreement with the selection rules; the transition $\phi_i \to \phi_f$ is allowed whenever the product of the irreducible representations (associated with ϕ_i, ϕ_f, $\hat{T}$) contains the totally symmetric irreducible representation (e.g., A_1). This simple result (3.111) allows one to find a relationship between the equilibrium angle and the ratio $\Im_{0 \to v_A} / \Im_{0 \to v_S} \equiv \sigma$:

$$\cos \theta_{eq} = \frac{1 - \sigma}{1 + \sigma} . \qquad \text{(3.112)}$$

The experimental value is $\sigma \sim 15$ ($\theta_{eq} \simeq 105°$), in contrast to the afore-mentioned simple model, which gives $\theta_{eq} \simeq 150°$, quite a poor approximation. Nonetheless, it is possible (and important) to improve this model from several standpoints. To begin with, the operator (3.107) does not account for transitions involving combination bands. These can properly be included by using more sophisticated operators. A possible solution is to introduce products of dipole operators, whose matrix elements can be approximately modeled as

$$\langle v_1 v_2 | \hat{T}_1 \hat{T}_2 | v_1' v_2' \rangle \simeq t_1 t_2 \exp(-\beta_1 |v_1 - v_1'|) \exp(-\beta_2 |v_2 - v_2'|) . \quad (3.113)$$

From this same point of view, one can also consider more general expansions, including product and/or powers of dipole operators with other elements of the enveloping algebra. In other words, one can use Casimir and Majorana operators associated with the various interatomic bonds. We notice finally that the expectation values of the bond dipole operators suggest that in going from a given vibrational level to the next level, the intensity should decrease by a factor of $\sim\exp(-2\beta)$. This provides a simple estimate of the parameter β. As is typical for infrared sequences of XH overtones, the intensity falls by a factor of ~ 10 in going from a given overtone or combination band to the following one. Consequently, $\beta \sim 1 \div 1.5$.

To employ the one-dimensional treatment of infrared transitions efficiently, it is worthwhile to consider some more complex situations. The general procedure, however, remains the same as the one outlined previously for two interacting bonds. We now attach a bond (dipole) operator to each one-dimensional internal degree of freedom of the molecule. Such operators will change the local vibrational quantum number by 0 ± 1, $\pm 2, \ldots$, units leading to exponentially decreasing expectation values of the infrared transition $0 \to v$. The total dipole moment operator for n oscillator is given by

$$\hat{T} = \sum_{i=1}^{n} a_i \hat{T}_i , \quad (3.114)$$

with matrix elements in the local basis

$$\langle v | \hat{T}_i | v' \rangle = t_i \exp(-\beta_i |v_i - v_i'|) , \qquad i = 1, \ldots, n . \quad (3.115)$$

In the equations above, a_i and β_i depend specifically on the molecular geometry, As an example, let us consider benzene, C_6H_6. By recalling

Fig. 17, we obtain

$$\hat{T}_x = \gamma \frac{\sqrt{3}}{2}[(\hat{T}_2 + \hat{T}_3) - (\hat{T}_5 + \hat{T}_6)],$$

$$\hat{T}_y = \gamma[(\hat{T}_1 - \hat{T}_4) + \tfrac{1}{2}(\hat{T}_2 - \hat{T}_3 - \hat{T}_5 + \hat{T}_6)].$$

(3.116)

As stated earlier, we can derive this same result by considering linear combinations of local bond quantities transforming as a vector, under D_{6h} point group operations [i.e., in the same way as the pair of coordinates (x, y) associated with the degenerate species E_{1u} (for in-plane vibrations)]. Thus we are left to determine two arbitrary parameters γ, β. The first one is typically fixed by a normalization procedure, while β is obtained by explicitly accounting for the observed slope of the infrared intensities.

The extension of this model to other vibrational families, such as out-of-plane and bending modes, can be accomplished in a straightforward way. For example, in benzene, out-of-plane bends of CH bonds can be described as shown in Fig. 26, where we use the labels 1', 2', ... to distinguish them from CH stretches, which are denoted by 1, 2, ... Obviously, we expect electric dipole activity along the z axis associated with in-phase out-of-plane oscillations of the hydrogen atoms. Consequently, we can write the following expression for the dipole operator:

$$\hat{T}' = \gamma'(\hat{T}_{1'} + \hat{T}_{2'} + \hat{T}_{3'} + \hat{T}_{4'} + \hat{T}_{5'} + \hat{T}_{6'}).$$

(3.117)

This operator transforms according to the irreducible representation A_{2u}

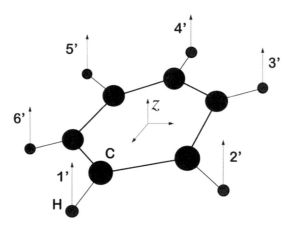

Figure 26. Bond numbering of out-of-plane CH bending modes of benzene.

of D_{6h} (notice the $2u$ label: $\hat{T}'$ is odd for C_2' and i operations for this point group). The single-bond dipole operators $\hat{T}_{i'}$ will obviously depend on β according to Eq. (3.106) or (3.115). However, we will use a different numerical value of β for different families of modes.

As a final illustration of infrared calculations in the one-dimensional model, we demonstrate the possibility of accounting for lower-symmetry (or substituted) molecular species. For example, let us consider the substituted benzene $1,3,5\text{-}C_6H_3D_3$, which was already presented in Section III.C.3. In regard to infrared intensities, CH and CD stretches can be accommodated by means of the following operator (see also Fig. 27):

$$\hat{T}_x = \frac{\sqrt{3}}{2}\left[\gamma_{CH}(\hat{T}_2^{(H)} - \hat{T}_6^{(H)}) + \gamma_{CD}(\hat{T}_3^{(D)} - \hat{T}_5^{(D)})\right],$$

$$\hat{T}_y = \gamma_{CH}\left[\tfrac{1}{2}(\hat{T}_2^{(H)} + \hat{T}_6^{(H)}) - \hat{T}_4^{(H)}\right] - \gamma_{CD}\left[\tfrac{1}{2}(\hat{T}_3^{(D)} + \hat{T}_5^{(D)}) - \hat{T}_1^{(D)}\right],$$

$$(3.118)$$

in which

$$\langle v_i^{(X)}|\hat{T}_i^{(X)}|v_i'^{(X)}\rangle \simeq t^{(X)}\exp(-\beta^{(X)}|v_i^{(X)} - v_i'^{(X)}|), \qquad X = H, D.$$

$$(3.119)$$

Through use of expressions (3.118) and (3.119), we can attempt a fitting

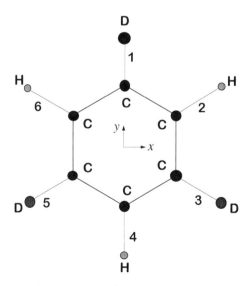

Figure 27. Bond numbering of CH and CD stretches in $1,3,5\text{-}C_6H_3D_3$.

procedure for the unknown parameters $\beta^{(H)}/\beta^{(D)}$ and γ_{CH}/γ_{CD}. The initial guesses for the β's can be made by examining overtone intensities or, more simply, by taking the values directly from the parent molecules, C_6H_6 and C_6D_6. γ_{CH} and γ_{CD} can basically be determined through normalization with respect to experimental data. A detailed analysis of infrared spectra in the one-dimensional model can be found in Refs. 28 and 51.

What about Raman transitions? As alluded to in Eq. (3.102), it is possible to construct an algebraic model where the transition operator is related to molecular polarizability. However, characterization of the transition operator will be different for Raman versus infrared transitions because of the differing selection rules. Infrared (IR) activity corresponds to having a permanent electric dipole, while Raman activity results from induced polarizability that behaves as a symmetric tensor of rank 2. The operator for Raman transitions is typically divided in a scalar part and a traceless symmetric tensor. Thus the algebraic scheme strictly follows the strategy adopted for the electric dipole operator. However, in contrast to IR, far fewer systems and molecules have been modeled with an algebraic approach. Consequently, the algebraic models are less well tested for Raman spectroscopy, and at present, only preliminary studies are available. Nonetheless, we sketch the basic steps leading to a consistent algebraic formulation of Raman transitions. This can be accomplished through two different approaches. In the first, we construct the five components of the polarizability tensor operator by direct inspection of those objects that transform according to certain irreducible representations of the molecular point group. Depending on the specific type of vibrational modes, the effective number of independent, nonzero tensor components can be smaller than five. In the simple case of C_{2v} symmetry, by limiting ourselves to account for vibrational modes occurring in the molecular plane, three components will suffice to characterize the polarizability tensor completely. Let us choose xy as the molecular plane; the Cartesian components, which transform according to a rank 2 tensor operator in C_{2v}, are thus given by $\{x^2, y^2, z^2\}$ (A_1 symmetry) and $\{xy, xz, yz\}$ (A_2, B_1, and B_2 symmetry, respectively). If $z = 0$, we need three independent components of the operator $\hat{Q}$: $\hat{Q}_{x^2}$, $\hat{Q}_{y^2}$, and $\hat{Q}_{xy}$. For these components we can write the following relations:

$$\hat{Q}_{x^2} = h_{x^2}(\hat{Q}_1 + \hat{Q}_2),$$

$$\hat{Q}_{y^2} = h_{y^2}(\hat{Q}_1 + \hat{Q}_2), \qquad (3.120)$$

$$\hat{Q}_{xy} = h_{xy}(\hat{Q}_1 - \hat{Q}_2).$$

These operators transform as required under the action of C_{2v} operations, with $\hat{Q}_{x^2}$ and $\hat{Q}_{y^2}$ carrying the irreducible representation A_1 while $\hat{Q}_{xy}$ carries B_2. In regard to evaluating the matrix elements of the Raman operators, we can expect a dependence on the vibrational quantum number similar to the one adopted for dipole operators,

$$\langle v_i | \hat{Q} | v_i' \rangle = q_i \exp(-\alpha_i |v_i - v_i'|) , \qquad (3.121)$$

where α_i will now account for the (steeper) decrease of Raman intensities along a sequence of vibrational overtones. It should be clear that in the equations above, the parameters h_i, q_i, and α_i are unknown, and they have to be determined, once again, by means of a fitting procedure.

The second approach involves constructing the Raman operators as products of Cartesian components of the dipole operator. By recalling the dipole operator components (3.107) and by introducing the following products, the C_{2v} case is then described:

$$\hat{Q}_{x^2} \equiv \hat{T}_x \cdot \hat{T}_x = \gamma^2 \sin^2 \frac{\theta_{eq}}{2} (\hat{T}_1 - \hat{T}_2)^2 ,$$

$$\hat{Q}_{y^2} \equiv \hat{T}_y \cdot \hat{T}_y = \gamma^2 \cos^2 \frac{\theta_{eq}}{2} (\hat{T}_1 + \hat{T}_2)^2 , \qquad (3.122)$$

$$\hat{Q}_{xy} \equiv \hat{T}_x \cdot \hat{T}_y = \frac{\gamma^2}{2} \sin \theta_{eq} \cos \theta_{eq} (\hat{T}_1 + \hat{T}_2) \cdot (\hat{T}_1 - \hat{T}_2) .$$

Since the Raman operators are derived from the electric dipole operators, the Raman intensity can be described in terms of the slope, β, of the IR intensities. As a result, if the infrared intensity decreases by a factor of g for an overtone transition of $+1$, the corresponding Raman intensity (if active) will decrease by a factor of g^2. Although such a result is merely an approximate rule of thumb, it does seem to correlate well with the few available experimental data for Raman overtone transitions. A similar analysis can be carried out for any molecular point group. For example, an analysis of the CH stretching modes of benzene would involve using the Raman operators $\hat{Q}_{xy}$, and $\hat{Q}_{x^2+y^2}$ and $\hat{Q}_{x^2-y^2}$ to construct the polarization trace.

As is evident from the previous discussion, application of the algebraic framework to the calculation of Raman intensities is far from being comprehensive. First, one cannot use the simple expectation values of dipole operators (3.106) directly when dealing with higher-order terms in $\hat{T}$. A proper treatment of these objects would require (at least in certain cases) using the exact form of matrix elements of $\hat{T}$. Second, the one-dimensional approach does not include rotations, so it lacks information concerning wavefunction parity. Such limitation is significant, expecially in regard to transition operators and selection rules. Nonetheless, a considerable amount of numerical information and predictive power can be (and has been) achieved through one-dimensional strategies. Moreover, these models have wide applicability, due to their facility of use and quick adaptability to any molecular geometry. Such characteristics make algebraic models a good starting point for further, more sophisticated theoretical analysis. Infrared transitions are well described by this algebraic approach, while the description of Raman transitions requires further work.

E. Anharmonic Couplings

Benzene has already been used as a nontrivial, challenging test for the one-dimensional algebraic model. Its highly symmetric geometry and a wealth of experimental data make this molecule an attractive candidate of theoretical studies. In the one-dimensional formulation of C_6H_6 we essentially assumed that the different families of vibrational modes (i.e., stretches and bends, in- and out-of-plane bends, CH and CC modes) can be modeled with noninteracting degrees of freedom. Such a hypothesis can be valid only if one does not compare the theoretical results with high or even intermediate resolution spectra. For example, the actual vibrational spectrum of benzene is much more rich and complicated than is predicted by the theoretical model, which assumes that CH stretching and bending modes are uncoupled. Generally speaking, the assumption of no coupling between modes is a very weak one and, in certain conditions, one is *not* allowed to disregard anharmonic couplings between (local) modes, even if they are of very different nature. In C_6H_6, for example, the CH stretching fundamental vibrations can be excited at energies on the order of $3000\,cm^{-1}$. These CH stretches are the unique fundamental bands at this energy. Consequently, we expect to observe a single peak in the infrared spectrum (because only the E_{1u} mode is infrared active). However, bands for CC ring stretches and CH in-plane bends fundamental bands occur at about $1500\,cm^{-1}$ (i.e., at one-half of the CH stretch fundamental energy). Therefore, it is possible that some of the overtone–combination bands of CH bends/CC stretches are nearly degenerate with the fundamental CH stretching level. Overtone–combination bands have, of course, a much lower intensities than those of the corresponding fundamental transitions. However, by virtue of their closeness to the "bright" CH fundamental mode, the "dark" modes can interact with it and borrow some infrared (or Raman) activity as a result of the mixing among the respective wavefunctions. This simple explanation can readily be generalized to any kind of molecule containing one or more CH chromophores which can interact anharmonically with other molecular modes. Typically, the nth overtone of a CH stretch is close to a certain number of the $(n + 1)$th overtone–combination levels of a CH bending mode. When two of such levels have the same symmetry, they can interact through anharmonic coupling. As a result of such interactions, part of the electric dipole or quadrupole activity is shared among the participating modes. Although this is not the only mechanism that can account for the intensities observed, Fermi resonances is one of the most significant contributions to a high-resolution spectrum of a large molecule. The theoretical formulation and experimental validation of Fermi resonance date back to the 1930s, when Fermi [52] was able to assign anomalous band in the IR spectrum of CO_2. Other important types of anharmonic coupling mechanisms include the Darling–Dennison interaction, which are discussed briefly in Section IV.C.

In this section we present the essential features of the one-dimensional algebraic approach to anharmonic intermode couplings and their effects on the vibrational spectra of polyatomic molecules. A satisfying treatment of these arguments would require a detailed review of a large number of works carried out within more traditional schemes [53–57]. The wide interest in this subject can be attributed partially to the fact that anharmonic Fermi resonances are frequently used to interpret complex experimental data. As an example, the understanding of IVR (intramolecular vibrational energy redistribution) processes (closely related to time-dependent analysis of molecular dynamics) firmly includes Fermi resonances in its theoretical description [58]. Moreover, the algebraic formulation of anharmonic couplings has advanced significantly in the last few years. In fact, from a semiclassical standpoint alone, it is almost impossible to keep track of all the important contributions made by an algebraic treatment of anharmonic couplings in molecular physics [59–65]. So in the following paragraphs, we offer a rather brief introduction to the problem with the accent placed on the most recent computational aspects. A slightly more specialized approach is discussed for the three-dimensional model in Section IV.C.

To understand the theoretical aspects of Fermi resonances, it is worthwhile to recall that in the conventional force-field method, Fermi resonance is described in terms of cubic or higher-order operators in the space of internal coordinates; these internal coordinates can be either rectilinear or curvilinear, local or normal. In the previous example of interaction between $v = n$ CH stretches and $v = n + 1$ CH bends, the corresponding Fermi operator can be written as

$$\hat{F}_{SB} = \phi_{SBB} Q_S Q_B^2 \,, \tag{3.123}$$

where Q_S and Q_B denote stretching and bending coordinates, respectively. This operator effects a nondiagonal Hamiltonian contribution, whose matrix elements are given by

$$\langle v_S, v_B | \hat{F}_{SB} | v_S - 1, v_B + 2 \rangle = k_{SBB} \sqrt{v_S(v_B + 1)(v_B + 2)} . \quad (3.124)$$

It is important to note that $\hat{F}_{SB}$ couples states whose total vibrational quantum numbers $v_T = v_S + v_B$ differ by one. This is, of course, required by the Fermi resonances itself, which can be seen as destroying one vibrational quantum in v_S and creating two quanta in v_B. As a result, the combination level $|v_S v_B\rangle = |10\rangle$ is almost degenerate with $|02\rangle$. Thus we are led to a simple conclusion: A (one-dimensional) algebraic version of the operator (3.123) must have the following characteristics: (1) the total number of vibrational quanta of a polyad of interacting levels is allowed to change, and (2) a local picture of the molecule should be somewhat favored and employed. We already have at our disposal an algebraic operator whose characteristics are almost perfectly suited for constructing the Fermi operator. By recalling Eq. (3.42), the Majorana operator for a stretch–bend interaction can be written as

$$\hat{M}_{SB} \propto \hat{S}_{(S+)}\hat{S}_{(B-)} + \hat{S}_{(S-)}\hat{S}_{(B+)} , \quad (3.125)$$

where $\hat{S}_{(X\pm)}$ denotes the (pseudo) angular momentum ladder operators creating $(+)$ or destroying $(-)$ one quantum in stretch (S) or bend (B) modes. The notation for $\hat{M}$ points to the possibility of extending the action of Majorana operators to change $v_S + v_B$ in an arbitrary way. If we introduce the operator

$$\hat{F}_{SB} = \hat{S}_{(S+)}\hat{S}^2_{(B-)} + \hat{S}_{(S-)}\hat{S}^2_{(B-)} , \quad (3.126)$$

we obtain for its nonzero matrix elements the following result:

$$\langle v_S, v_B | \hat{F}_{SB} | v_S - 1, v_B + 2 \rangle = [v_S(N_S - v_S + 1)(v_B + 1)(v_B + 2)$$

$$\times (N_B - v_B)(N_B - v_B - 1)]^{\frac{1}{2}} . \quad (3.127)$$

These matrix elements are equivalent to those of Eq. (3.124), apart from anharmonic contributions of the order of v/N. So we see that the extended Majorana operator has the required effect on the states involved in the resonance mechanism. At the same time, $\hat{F}_{SB}$ does not preserve the coupled $O_{SB}(2)$ symmetry: in other words, $v_S + v_B$ is not conserved anymore. Consequently, the block-diagonal structure of the Hamiltonian operator is destroyed and the numerical diagonalization of the corresponding matrix could become prohibitively large for complex situations. We can, however, include only those coupling terms that, for various reasons, have a definitely nonnegligible role in the manifolds of resonating levels while maintaining, at a reasonable dimension, the largest block to be diagonalized.

In the more general case of two different modes v_a and v_b involved in a $n:m$ resonance (the previous one was a 1:2 resonance), the corresponding algebraic Fermi operator reads

$$\hat{F}_{a,b}^{n,m} = \hat{S}_{a+}^n \hat{S}_{b-}^m + \hat{S}_{a-}^n \hat{S}_{b+}^m , \qquad (3.128)$$

with matrix elements in the basis $|v_a v_b\rangle$ given by

$$\langle v_a', v_b' | \hat{F}_{a,b}^{n,m} | v_a, v_b \rangle = \left[\frac{v_a!(N_a - v_a + n)!(v_b + m)!(N_b - v_b)!}{(v_a - n)!(N_a - v_a)!v_b!(N_b - v_b - m)!} \right]^{1/2} \delta_{v_a', v_a - n} \delta_{v_b', v_b + m}$$
$$+ \left[\frac{v_b!(N_b - v_b + m)!(v_a + n)!(N_a - v_a)!}{(v_b - m)!(N_b - v_b)!v_a!(N_a - v_a - m)!} \right]^{1/2} \delta_{v_a', v_a + n} \delta_{v_b', v_b - m} . \qquad (3.129)$$

In this case the total number of vibrational quanta will change by $|n - m|$. It is customary to introduce, as a convenient way of grouping together all the levels belonging to a resonating manifold, the Fermi polyad label, v_F, given by

$$v_F \equiv n v_a + m v_b . \qquad (3.130)$$

A simple practical example of use of the algebraic Fermi operator is shown in Fig. 28, where we report a portion of the vibrational spectrum obtained with the Hamiltonian operator

$$\hat{H} = A_S \hat{C}_S + A_B \hat{C}_B + \lambda_{SB} \hat{F}_{SB}^{1,2} , \qquad (3.131)$$

which describes two one-dimensional anharmonic oscillators interacting through a 1:2 Fermi coupling. Figure 29 shows the initial $v_T = v_S + v_B$ (1:1) Majorana blocking and the $v_F = 2v_S + v_B$ (1:2) Fermi blocking. We observe that, generally speaking, the label v_F is *not* a good quantum number because of the coupling induced by the Majorana operator; in other words, $\hat{M}$ and $\hat{F}$ do not commute. However, in the present case, the 1:1 Majorana interaction is negligible, and as a consequence, v_F is an (approximately) good quantum number.

What about practical uses of algebraic Fermi operators? Several interesting molecules are affected by anharmonic interactions, involving several families of vibrational modes. It is possible to account for such complex situations with a straightforward extension of operator (3.128),

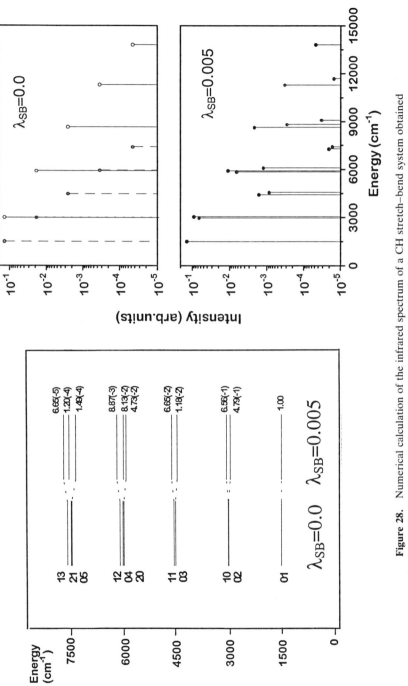

Figure 28. Numerical calculation of the infrared spectrum of a CH stretch–bend system obtained in the algebraic scheme. *Left side*: energy levels labeled by $(v_S v_A)$ and infrared intensities relative to the CH bending fundamental level (01). *Right side*: Stick spectra with and without Fermi interaction; continuous sticks denote stretching mode transitions, dashed sticks refer to bending mode transitions.

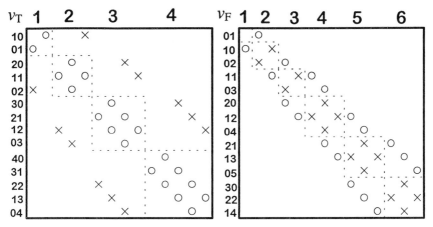

Figure 29. Different blocking forms of the Hamiltonian matrix corresponding to the Majorana (*left side*) and Fermi 2:1 interaction (right side). Off-diagonal matrix elements of the Majorana operator are denoted by circles, crosses refer to the Fermi interaction.

where we include ladder operators (to the appropriate power) for each interacting bond. Once again, the molecular symmetry plays an essential role in determining the most convenient form of the Fermi operator. For the purpose of illustrating this method, we return to benzene as an example [47, 48]. In the energy region of the fundamental CH stretching modes, three rather intense peaks are observed, located at ~ 3048, 3084, and 3102 cm^{-1}. An independent analysis of the rovibrational spectrum of this molecule [66] shows that the $v = 1$ CH stretch E$_{1u}$ should be located at 3065 cm^{-1}. The three observed peaks can be explained through anharmonic resonances between the $v = 1$ unperturbed CH mode and two (or more) combination–overtone bands (or E$_{1u}$ symmetry) of other molecular modes in the same energy region. The block-diagonal approximation of the algebraic Hamiltonian is used to compute E$_{1u}$ modes close to 3065 cm^{-1}. A combination band involving two quanta of CC stretch is located 3083 cm^{-1}; a ternary combination, with two quanta of CC stretch and one quantum of CC in-plane bend, is located at 3085 cm^{-1}; a further ternary combination band, with one quantum CH bend in-plane and two of CC stretch, occurs at 3106 cm^{-1}. We are now ready to introduce the Fermi operator(s) acting on these states. Since these levels are overtone–combination bands with two or more vibrational quanta, they have a much smaller infrared intensities than the $v = 1$ stretching mode. The construction of Fermi operators is closely related to the specific physical mechanism at issue and to molecular symmetry arguments. The anharmonic interaction between CH/CC stretches is shown schematically in

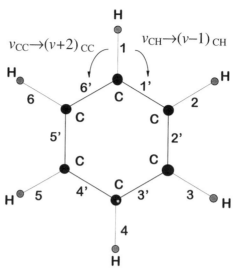

Figure 30. Construction of the 1:2 CH stretch/CC bend anharmonic resonance in benzene.

Fig. 30. The simple 1:2 Fermi operator (3.126) can be modified to describe the present case by considering the following expression:

$$
\begin{aligned}
\hat{F}^{1,2}_{CH_S/CH_B} = {} & \hat{S}_{1-}(\hat{S}_{6'+} + \hat{S}_{1'+})^2 + \hat{S}_{2-}(\hat{S}_{1'+} + \hat{S}_{2'+})^2 \\
& + \hat{S}_{3-}(\hat{S}_{2'+} + \hat{S}_{3'+})^2 + \hat{S}_{4-}(\hat{S}_{3'+} + \hat{S}_{4'+})^2 \\
& + \hat{S}_{5-}(\hat{S}_{4'+} + \hat{S}_{5'+})^2 + \hat{S}_{6-}(\hat{S}_{5'+} + \hat{S}_{6'+})^2 ,
\end{aligned} \tag{3.132}
$$

plus the Hermitian conjugate operator. This formulation expresses the fact that (1) CH/CC stretching modes are coupled (as expected), and (2) only those levels with equal D_{6h} symmetry labels can interact, because the Fermi operator is invariant under bond permutation (see Fig. 30). In other words, the coupling (3.132) plays the role of exchanging two CC stretches with one CH stretch of equal symmetry. We can account for other Fermi interactions as well, by constructing appropriate combinations of ladder operators associated with the pertinent degrees of freedom. As a final step, we need to adjust the strength of Fermi operators through a fitting procedure, in which both energies *and* infrared intensities must agree with observed data. The results of this calculation are shown in Fig. 31 for the CH stretch transitions $0 \rightarrow 1$ and $0 \rightarrow 3$ of benzene. For overtone transitions, one needs to make explicit use of the "tier" model, in which several families of vibrational levels interact through the same resonance mechanism, as discussed elsewhere [43]. By

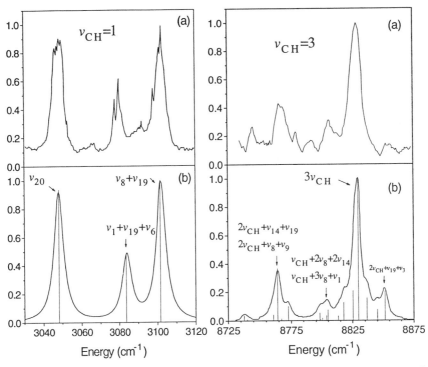

Figure 31. Comparison between observed (panels a) and algebraically computed (panels b) infrared spectra for the $v = 1$ and $v = 3$ CH stretch transitions in benzene. The most important peaks are labeled by Wilson numbers.

starting from a single "bright" level (an excited CH stretch), Fermi resonances couple this mode with other molecular states. These states, in turn, can interact with higher overtone–combination bands of levels of lower and lower energy. The initial single peak is thus involved in a complex sequence of interactions whose complete explanation requires a detailed knowledge of both dynamic and spectroscopic molecular properties. In low- or intermediate-resolution experiments, the transition to the CH overtone is seen as a broad, unstructured band in the infrared spectrum, while at higher resolution, as shown in Fig. 31, several, distinct peaks appear. Therefore, it is important to grasp whether the coupling mechanism can be explained in terms of a detailed calculation of molecular states or rather through purely statistical considerations. So the algebraic realization of Fermi interactions is well suited to describe the tier model, as it provides complete systematic bookkeeping of numerous vibrational levels up to very high energies. Moreover, Fermi operators are easily arranged to act in such a way that the essential information concerning the molecular symmetry is almost automatically inserted in the Hamiltonian operator.

Another appealing outcome of the algebraic model is the possibility of carrying out a simple but reliable time-dependent analysis of Fermi resonances. This is of particular relevance to understanding IVR processes, in which the vibrational energy of a selectively excited levels flows into the bath of molecular states predicted by the tier model. IVR phenomena occur on a few-nanosecond timescale, but more precise measurements in the energy domain could reveal interesting effects occurring on longer times. The time-dependent algebraic analysis can be accomplished because it can derive both eigenvalues and eigenvectors from the Hamiltonian operator. If we denote $|\phi_0\rangle$ as the initial bright state, and $|\phi_k\rangle$ and E_k as the eigenvectors and eigenvalues of our Hamiltonian, we can easily compute the autocorrelation amplitude (often referred to as *survival amplitude*) of $|\phi_0\rangle$ in the following way:

$$A(t) = \langle \phi_0 | e^{-i\hat{H}t} | \phi_0 \rangle = \sum_k \langle \phi_0 | \phi_k \rangle \, e^{-iE_k t} . \qquad (3.133)$$

The survival probability of $|\phi_0\rangle$ is then defined as

$$S(t) = |A(t)|^2 . \qquad (3.134)$$

A plot of $S(t)$ shows the depopulation of $|\phi_0\rangle$ in favor of other molecular vibrational modes. The autocorrelation function is also important because it can be viewed as the Fourier transform of the energy spectrum, thus allowing for a useful and direct comparison between theory and experiment.

Further examples and details involving an algebraic formulation of Fermi resonances can be found in Refs. 57, 67, and 68. Although algebraic solutions are far from being *the* solution for these very exacting problems, they can still be seen as a quick, alternative route capable of providing a global, quantitative picture for very complex situations. In addition, more sophisticated methods exist [44, 45] for dealing with these problems. However, the computational effort required by these approaches is orders of magnitude larger than that of the one-dimensional algebraic model. While the determination of arbitrary parameters for hitherto unclassified molecules can be quite tedious, a judicious analysis of similar molecular complexes can provide algebraic parameters useful for an acceptable first-order approximation Hamiltonian model for an otherwise untenable physical situation.

IV. THREE-DIMENSIONAL ALGEBRAIC MODELS FOR POLYATOMIC MOLECULES

A. Introduction

In this section we employ the mathematical techniques introduced in Section II.C.2 to realize a three-dimensional algebraic model that describes the rovibrational spectrum of a polyatomic molecule. By keeping in mind the general strategy adopted in the construction of the one-dimensional model, our task will be relatively straightforward. The three-dimensional model will lead to a more complex (and realistic) picture of a molecular system as well as producing a more exacting algebraic treatment. The algebraic procedures presented in Section III.C.2 for the multiple interacting oscillators are most applicable to three-dimensional problems. However, the three-dimensional model is definitely more difficult to manipulate than the one-dimensional one, purely for algebraic aspects. Consequently, a comprehensive and rigorous exposition of the technical points characterizing this model is beyond the original intent of this presentation. Nonetheless, it is possible to utilize the essential strengths of the algebraic model by consolidating a detailed study of the single three-dimensional oscillator (Section II.C.2) with the coupling strategies developed in Sections III.B and III.C. The most striking difference between one- and three-dimensional approaches is found in the physical interpretation of the algebraic outcomes. By coupling two rovibrators together, we, in fact, obtain a "real" triatomic molecule in which the complete set of internal degrees of freedom is accounted for from the very beginning (i.e., without any further, artificial assumption regarding the algebraic structures used). We expect to obtain six degrees of freedom (including stretching and bending vibrations) as well as rotational modes. The discussion in Section II.C.2, concerning the existence of two different dynamical symmetries suited for rigid and soft bonds [Eq. (2.104)] maintains its validity as well for several interacting bonds. An algebraic description of floppy vibrational modes has not yet been investigated with any detail for polyatomic molecules. Therefore, we focus on the rigid dynamical symmetry for the single three-dimensional bond, based on the subalgebra chain

$$U(4) \supset O(4) \supset O(3) \supset O(2) . \qquad (4.1)$$

We proceed as follows; after a brief explanation of the basic algebraic aspects required to couple two structures of the type (4.1), we consider the general problem of triatomic molecules with both linear and bent equilibrium geometry. We then address the construction of more sophisticated Hamiltonian operators, including Fermi resonances and rovibrational interactions. This will also constitute the starting point for the construction of new algebraic objects and procedures for four-atomic molecules. At this point we have reached the limits of our algebraic approach. Although not impossible, the full algebraic model has not been applied to molecules with more than four atoms. For such systems, the convenience of the algebraic approach is obscured by the complex dynamical symmetry required by the molecule. However, this size limitation can be circumvented in certain cases involving nuclear problems. For example, the interacting boson model is capable of providing complete and reliable answers to problems involving heavy nuclei (i.e., *real* many-body systems) and it is constructed from dynamical symmetries based on subtle algebraic structures. However, for polyatomic molecules, the problem is a different one: In going from very light to heavy systems, atomic nuclei can be characterized by rather homogeneous behavior as regards their dynamic and spectroscopic features. In other words, nuclear matter is naturally prone to a dynamical symmetry treatment, while molecules are less so. Each molecule reacts differently to our attempts at capturing its spectrum in the dynamic symmetry cage. Despite that, the algebraic method shows high precision and competitive ease of use in comparison to traditional approaches. In the following sections we review the algebraic theory necessary for the description of triatomic molecules [69].

B. Coupling of Two Rovibrators: Triatomic Molecules

We start our discussion with the construction of a proper, spectrum-generating algebra for two interacting oscillators by attaching a unitary algebraic structure to each of them. To be more precise, we need two unitary algebras in four dimensions, $U_1(4)$ and $U_2(4)$, whose boson operators are a replacement of the *vector* coordinates, $\mathbf{r}_1$ and $\mathbf{r}_2$ (as explained in Section II.C.2). Thus we have

$$\text{spectrum-generating algebra} \equiv U_1(4) \otimes U_2(4) . \qquad (4.2)$$

In the one-dimensional case, the essential physical features of the two-body system were derived through two different dynamical symmetries. Once again, we repeat these steps, with some important exceptions due to the richer structure of the starting algebra (4.2). Each U(4) algebra is given by 16 generating operators, expressed in terms of bilinear combinations of coupled tensor operators, similar to those presented in Section II.C.2 [Eq. (2.95)]. The next step is the expansion of the general

Hamiltonian,

$$\hat{H} = \hat{H}_1 + \hat{H}_2 + \hat{H}_{12} ,$$ (4.3)

in terms of the coupled tensor operators [as done in Eqs. (3.5) and (3.6)]. Finally, starting from the SGA (4.2), we examine the existing dynamic symmetries for the Hamiltonian operator (4.3). We proceed by recalling that in the one-dimensional framework, one obtains two chains, depending basically on the priority given to the symmetry reduction $U(2) \supset O(2)$ in comparison with the "overall" algebraic coupling $U_1(2) \otimes U_2(2)$ [see Eqs. (3.8) and (3.9)]. In the three-dimensional case we find ourselves considering the following subalgebra chains:

$$U_1(4) \otimes U_2(4) \supset O_1(4) \otimes O_2(4) \supset O_{12}(4) \supset O_{12})3) \supset O_{12}(2)$$ (4.4)

and

$$U_1(4) \otimes U_2(4) \supset U_{12}(4) \supset O_{12}(4) \supset O_{12}(3) \supset O_{12}(2) .$$ (4.5)

Once again, we achieve a local scheme with the first chain and with the second chain, a normal picture of the physical system. The final $O_{12}(2)$ algebras are of no direct interest here, as they refer to the action of external fields breaking the overall spherical symmetry. Moreover, a completely different algebraic lattice can be obtained by adopting the $U(4) \supset U(3) \supset O(3)$ reduction pattern for one or both oscillators, typically leading to a description of floppy motions. Although perfectly clear from a purely mathematical point of view, such a procedure has yet to be applied to realistic situations such as the vibrational spectrum of a van der Waals bond. By close analogy to the one-dimensional model, we will see how the coupled algebras $O_{12}(4)$ and $U_{12}(4)$ justify the introduction of diagonal and nondiagonal coupling operators in a local basis. However, in the three-dimensional approach, the $O_{12}(4)$ algebra gives something beyond the diagonal cross anharmonic terms because it directly describes true, anharmonic bending motions.

By starting from chains (4.4) and (4.5), we can write the corresponding algebraic kets;

$$\left| \begin{array}{cccccc} U_1(4) & \otimes\ U_2(4) & \supset\ U_1(4)\ \otimes\ O_2(4) & \supset\ O_{12}(4) & \supset\ O_{12}(3) & \supset\ O_{12}(2) \\ N_1 & N_2 & \omega_1 \qquad\quad \omega_2 & (\tau_1, \tau_2) & j & m_j \end{array} \right\rangle ,$$ (4.6)

$$\left| \begin{array}{cccccc} U_1(4) & \otimes\ U_2(4) & \supset\ U_{12}(4) & \supset\ O_{12}(4) & \supset\ O_{12}(3) & \supset\ O_{12}(2) \\ N_1 & N_2 & [n, m]^* & (\sigma_1, \sigma_2) & j & m_j \end{array} \right\rangle .$$ (4.7)

In these expressions the number N_i labels symmetric irreducible representations of $U_i(4)$, $i = 1, 2$. This number is related to the total number of bound states supported by the anharmonic oscillator i, for the Morse potential realization. The number ω_i labels symmetric irreducible representations of $O_i(4)$ [it should be denoted as $(\omega_i, 0)$], while (τ_1, τ_2) and (σ_1, σ_2) denote (nonsymmetric) irreducible representations of $O_{12}(4)$, which are obtained from the reduction of either $O_1(4) \otimes O_2(4)$ or $U_{12}(4)$. Finally, j and m_j represent the usual angular momentum quantum numbers. Before commenting on the second chain [Eq. (4.7)], it is worthwhile to clarify some interesting and important algebraic aspects of the "local" chain (4.6). They will be needed for the purpose of understanding the branching rules of the algebraic quantum numbers. On account of the isomorphism between the algebras $U(2)$ and $O(3)$, the local limit of the one-dimensional model was discussed easily in Section III.B. The reduction $U_1(2) \otimes U_2(2) \supset O_1(2) \otimes O_2(2)$ was realized through the direct products of uncoupled angular momentum states. Although similar, the corresponding three-dimensional chain deserves definitely more attention because it involves more complex, less familiar algebraic structures and their couplings. One can address this problem from a completely general and abstract point of view. In fact, it is possible to obtain the final result in terms of explicit tensor calculus for four-dimensional Lie algebras [70]. It is also possible, however, to sketch, through isomorphism, the essential points of this procedure in a very simple way,

$$O(4) \approx O(3) \otimes O(3) \approx U(2) \otimes U(2) . \qquad (4.8)$$

In this expression we find, on the right side, the previously used relation $O(3) \approx U(2)$. On the left side we observe an intimate connection between orthogonal transformation in four and three dimensions. The decomposition of $O(4)$ into the direct product of two $O(3)$'s is a trivial consequence of the commutation relations characterizing these algebraic structures. Because of this connection, the irreducible representations of $O(4)$ must be labeled by two numbers, which are in correspondence with the labels of the two groups $O(3)$ contained in $O(4)$. The product $O_1(4) \otimes O_2(4)$, appearing in the chain (4.6), can also be studied through the isomorphism (4.8). In the $O(3)$ case, this is the equivalent of considering the Clebsh–Gordan series (i.e., the rule for the addition of angular momenta), which is usually written as

$$(j_1) \otimes (j_2) = (j_1 + j_2) \oplus (j_1 + j_2 - 1) \oplus \cdots \oplus (|j_1 - j_2|) . \qquad (4.9)$$

The corresponding result for the $O(4)$ problem is obtained with a similar procedure. We give the result for the direct product of two symmetric irreducible representations of O(4):

$$(\omega_1, 0) \otimes (\omega_2, 0) = \bigoplus_{\alpha, \beta = 0}^{\min(\omega_1, \omega_2)} (|\omega_1 - \omega_2| + \alpha + \beta, \alpha - \beta), \quad (4.10)$$

where the right side contains irreducible representations of the coupled algebra $O_{12}(4)$. The next step is to find the irreducible representations (j) of O(3) contained in a given irreducible representation (τ_1, τ_2) of O(4). Once again, in view of the isomorphism (4.8) it is not difficult to show that

$$(\tau_1, \tau_2) = \bigoplus_{k=0}^{\tau_1 - |\tau_2|} (|\tau_2| + k). \quad (4.11)$$

The final result is that the algebraic branching rules for chain (4.6) can be written as

$$\omega_i = N_i, N_i - 2, \ldots, 1 \text{ or } 0 \ (i = 1, 2),$$

$$\tau_1 = |\omega_1 - \omega_2| + \alpha + \beta, \quad \tau_2 = \alpha - \beta, \quad \alpha, \beta = 0, 1, \ldots, \min(\omega_1, \omega_2),$$

$$j = |\tau_2|, |\tau_2| + 1, \ldots, \tau_1 - 1, \tau_1. \quad (4.12)$$

We can clarify these seemingly cumbersome laws with the help of a practical example. Let us consider the irreducible representation of $U_1(4) \otimes U_2(4)$, given by $[N_1] \otimes [N_2] = [3] \otimes [2]$. From the first of Eqs. (4.12), we have that $\omega_1 = 3, 1$ and $\omega_2 = 2, 0$. Correspondingly, we find that the irreducible representation $[3] \otimes [2]$ of $U_1(4) \otimes U_2(4)$ contains the following irreducible (symmetric, direct product) representations of $O_1(4) \otimes O_2(4)$:

$$(3, 0) \otimes (2, 0); (3, 0) \otimes (0, 0); (1, 0) \otimes (2, 0); (1, 0) \otimes (0, 0). \quad (4.13)$$

Each of these representations can now be decomposed into irreducible representations of the coupled $O_{12}(4)$ algebra, according to Eq. (4.10). Starting from $(3, 0) \otimes (2, 0)$, we have $|\omega_1 - \omega_2| = 1$, $\min(\omega_1, \omega_2) = 2$ and $\alpha, \beta = 0, 1, 2$. Equation (4.10) tells us that

$$(3, 0) \otimes (2, 0) = (1, 0) \oplus (2, 1) \oplus (3, 2) \oplus (2, -1) \oplus (3, 0)$$
$$\oplus (4, 1) \oplus (3, -2) \oplus (4, -1) \oplus (5, 0), \quad (4.14)$$

in which the right side contains irreducible representations (τ_1, τ_2) of the coupled $O_{12}(4)$ algebra. Each of these representations can, in turn, be decomposed into irreducible representations of $O_{12}(3)$, according to Eq. (4.11). We thus obtain

$$(1,0) = (0) \oplus (1) ; \quad (2,1) = (1) \oplus (2) ; \quad \ldots , \qquad (4.15)$$

where (0), (1), ... denote different values of the total angular momentum label j. A more detailed study of this problem would allow attainment of the correct parities for the angular momentum states contained in the final reduction to $O_{12}(3)$. It is possible to attach an odd (even)-parity label to j states coming from irreducible representations of $O(4)$ of the type $(\tau_1, 0)$ with τ_1 odd (even), while for $\tau_2 \neq 0$, j states have double $(\pm)$ parity. In the previous example, Eq. (4.15) should be written as

$$(1,0) = (0^+) \oplus (1^-) ; \quad (2,1) = (1^\pm) \oplus (2^\pm) ; \quad \ldots \qquad (4.16)$$

As discussed for the one-dimensional model, we can expand the coupled basis in terms of uncoupled states [see Eq. (3.7)]. In the $O(4)$ case, we obtain a similar expansion, in which, besides the usual $O(3) \supset O(2)$ Clebsch–Gordan coefficients, we need the $O(4) \supset O(3)$ transformation brackets, often referred to as $O(4)$ *isoscalar factors*. This is typically written as

$$|[N_1], [N_2]; (\omega_1, 0), (\omega_2, 0); (\tau_1, \tau_1); j, m_j\rangle$$

$$= \sum_{j_1 j_2 m_1 m_2} \left\langle \begin{matrix} (\omega_1, 0) & (\omega_2, 0) \\ j_1 & j_2 \end{matrix} \middle| \begin{matrix} (\tau_1, \tau_1) \\ j \end{matrix} \right\rangle \left\langle \begin{matrix} j_1 & j_2 \\ m_1 & m_2 \end{matrix} \middle| \begin{matrix} j \\ m \end{matrix} \right\rangle \qquad (4.17)$$

$$\times |[N_1]; (\omega_1, 0); j_1, m_1\rangle |[N_2]; (\omega_2, 0); j_2, m_2\rangle ,$$

where

$$\left\langle \begin{matrix} j_1 & j_2 \\ m_1 & m_2 \end{matrix} \middle| \begin{matrix} j \\ m \end{matrix} \right\rangle = \langle j_1 m_1 j_2 m_2 | jm \rangle \qquad (4.18)$$

is the usual Clebsh–Gordan coefficient (or $3j$ symbol) and

$$
\left\langle \begin{matrix} (\omega_1, 0) & (\omega_2, 0) \\ j_1 & j_2 \end{matrix} \middle| \begin{matrix} (\tau_1, \tau_2) \\ j \end{matrix} \right\rangle
$$

$$
= \sqrt{(\tau_1 + \tau_2 + 1)(\tau_1 - \tau_2 + 1)(2j_1 + 1)(2j_2 + 1)} \begin{Bmatrix} \dfrac{\omega_1}{2} & \dfrac{\omega_2}{2} & \dfrac{\tau_1 + \tau_2}{2} \\[1mm] \dfrac{\omega_1}{2} & \dfrac{\omega_2}{2} & \dfrac{\tau_1 - \tau_2}{2} \\[1mm] j_1 & j_2 & j \end{Bmatrix} \quad (4.19)
$$

is the isoscalar factor of $O(4)$, here expressed in terms of a $9j$ symbol. All these objects can be computed with standard numerical routines or, for simple cases, in analytical form.

Before converting these purely abstract quantities into vibrational and rotational quantum numbers, we return to the "normal" chain, Eq. (4.7). Its meaning is closely related to the corresponding normal limit of the one-dimensional approach. Once again, the most important difference is in the increased algebraic complexity. Now we must comment briefly on the label $[n, m]^*$. Here one faces the not so unusual problem of the "missing label," which relates to the multiple appearance of a single irreducible representation of $O_{12}(4)$ in the irreducible representation of $U_{12}(4)$. However, we ignore this problem because it does not really have pertinence to the local picture. As explained earlier for the one-dimensional case, the normal picture is an important limit because it leads to intermode coupling (Majorana) operators. We will see, in the following paragraphs, how to obtain and use the Majorana interaction for the three-dimensional model.

Before utilizing the algebraic quantum numbers previously obtained in a realistic model for triatomic molecules, it is worthwhile to clarify some essential aspects of the physical problem. In the algebraic local basis (4.6), we have, for given N_1, N_2, six independent quantum numbers: ω_1, ω_2, τ_1, τ_2, j, m_j. This is a good omen, since a triatomic molecule possesses six rovibrational degrees of freedom. To construct a one-to-one correspondence between algebraic and rovibrational quantum numbers, it is, however, of absolute importance to distinguish between the two allowed equilibrium shapes of a triatomic molecule: linear or bent. Apart from its particular geometry, any triatomic molecule XYZ has two stretching (XY, YX) degrees of freedom; the total angular momentum can also be discussed, irrespective of the equilibrium geometry, in terms of the usual j and m_j quantum numbers. The remaining two degrees of freedom refer to bending and rotational motions, according to the equilibrium molecular shape. In a bent molecule, we have a nondegenerate bending mode; moreover, we have to introduce a figure projection axis, along which the total angular momentum is quantized (for a symmetric top), according to the rule

$$j = |k|, |k| + 1, |k| + 2 \ldots . \tag{4.20}$$

In a linear molecule, the bending motion is twofold degenerate because it can be decomposed according to two equivalent directions orthogonal to the internuclear axis (Fig. 32). This means that a bending vibration can also be seen in terms of rotational excitation about the same figure axis. As is customary, we introduce a rotational quantum number l of (bending) vibrational origin, which is related to j by means of

$$j = |l|, |l| + 1, |l| + 2, \ldots . \tag{4.21}$$

In other words, a vibrational quantum number, l, for the linear case, is converted in a purely rotational quantum number, k, for the bent case. As amply discussed elsewhere [71, 72], this is a particular example of the general problem concerning the correlation between rovibrational spectra of the same molecule in different equilibrium configurations. Figure 33 shows the bent-to-linear correlation diagram or a triatomic molecule. In light of Eqs. (4.20) and (4.21), l and k states correlate trivially, $l = k$.

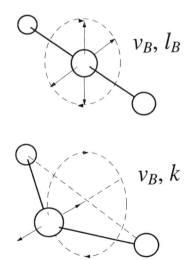

Figure 32. Bending motion in linear and bent triatomic molecules.

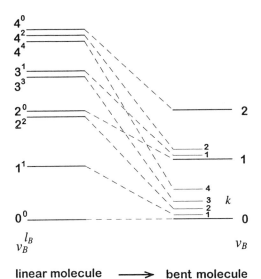

Figure 33. Correlation diagram for linear to bent triatomic molecules.

Instead, the correlation between bending vibrational modes is given by

$$v_B^{(\text{bent molecule})} = \frac{v_B^{(\text{linear molecule})} - |l|}{2}. \tag{4.22}$$

Such a correlation law can find a rigorous physical counterpart in a Hamiltonian operator containing one or more parameters whose continuous variation transforms the molecule (i.e., its spectrum) from one geometric shape to another.

We are ready to account for these important differences within the algebraic model. In light of the previous discussion, we must anticipate the problem of converting the algebraic quantum numbers in rovibrational ones according to the correct molecular equilibrium geometry. To begin with, we introduce the dynamical symmetry Hamiltonian operator, based on the lattice of subalgebra chains (4.6), (4.7), expressed as

$$\hat{H} = E_0 + A_1 \hat{C}_1 + A_2 \hat{C}_2 + A_{12} \hat{C}_{12} + A'_{12} |\hat{C}'_{12}| + \lambda_{12} \hat{M}_{12} + B \hat{J}^2. \tag{4.23}$$

In this expression, $\hat{C}_1$ and $\hat{C}_2$ are Casimir operators of the $O_1(4)$ and $O_2(4)$ algebras. Their matrix elements in the local basis (4.6) have been already obtained in Section II.C.2 [see Eq. (2.111)]. We also recall that $O(4)$ possesses *two* invariant operators, $\hat{C}$ and $\hat{C}'$, which can be written, in terms of dipole $(\hat{D})$ and angular momentum $(\hat{J})$ operators, as

$$\hat{C} = \hat{D}^2 + \hat{J}^2, \qquad \hat{C}' = \hat{D} \cdot \hat{J}. \tag{4.24}$$

Their matrix elements in the local basis are given by

$$\langle \hat{C} \rangle = \tau_1(\tau_1 + 2) + \tau_2^2, \tag{4.25}$$

$$\langle \hat{C}' \rangle = \tau_2(\tau_1 + 1). \tag{4.26}$$

We notice that $\hat{C}'$ has nonzero matrix elements only in a nonsymmetric irreducible representation of O(4) ($\tau_2 \neq 0$). This also explains why the use of $\hat{C}'$ was neglected in the single-oscillator case, where only symmetric irreducible representations of O(4) were considered. The operator $\hat{C}'$ is a pseudoscalar quantity since it is given basically by the scalar product of a polar vector operator with an axial one. Since we want a scalar Hamiltonian operator, we must take either the absolute value or the square of $\hat{C}'$. In practice, in Eq. (4.23), $\hat{C}_{12}$ and $\hat{C}'_{12}$ are both invariant operators (with nonzero matrix elements) of the coupled algebra $O_{12}(4)$. We discuss their exact physical meaning later; for the moment it suffices to say that the operator $\hat{C}'$ has the essential role of realizing the above-mentioned correlation between linear and bent triatomic molecules. The Hamiltonian operator (4.23) contains the Majorana operator $\hat{M}_{12}$, which is closely related (excluding constant terms) to the invariant operator of the coupled algebra $U_{12}(4)$ appearing in the "normal" chain (4.8). Its physical role is to introduce intermode coupling terms in the local basis. By making explicit use of group theoretical techniques [35], it is possible to show that $\hat{M}_{12}$, acting within a given irreducible representation (τ_1, τ_2) of $O_{12}(4)$, couples states satisfying the selection rules $\Delta\omega_1, \Delta\omega_2 = 0, \pm 2$. The nonzero matrix elements can be written in the following way:

Nondiagonal contributions:

$$\langle N_1 N_2 \omega_1' \omega_2'(\tau_1, \tau_2) | \hat{M}_{12} | N_1 N_2 \omega_1 \omega_2(\tau_1, \tau_2) \rangle$$

$$= (-)^{\tau_1 + 1}(\omega_1' + 1)(\omega_2' + 1) \begin{Bmatrix} \dfrac{\omega_1}{2} & \dfrac{\omega_2}{2} & \dfrac{\tau_1 - \tau_2}{2} \\[2mm] \dfrac{\omega_2'}{2} & \dfrac{\omega_1'}{2} & 1 \end{Bmatrix} \begin{Bmatrix} \dfrac{\omega_1}{2} & \dfrac{\omega_2}{2} & \dfrac{\tau_1 + \tau_2}{2} \\[2mm] \dfrac{\omega_2'}{2} & \dfrac{\omega_1'}{2} & 1 \end{Bmatrix}$$

$$\times \langle N_1 \omega_1' \| \hat{D}_1 \| N_1 \omega_1 \rangle \langle N_2 \omega_2' \| \hat{D}_2 \| N_2 \omega_2 \rangle \delta_{\omega_1', \omega_1 \pm 2} \delta_{\omega_2', \omega_2 \mp 2}, \tag{4.27}$$

Diagonal contributions:

$$\langle N_1 N_2 \omega_1 \omega_2 (\tau_1, \tau_2) | \hat{M}_{12} | N_1 N_2 \omega_1 \omega_2 (\tau_1, \tau_2) \rangle$$

$$= -\frac{1}{4} [\tau_1(\tau_1 + 2) + \tau_2^2 - \omega_1(\omega_1 + 2) - \omega_2(\omega_2 + 2)] + \frac{N_1 N_2}{2} \tag{4.28}$$

$$-\frac{1}{4} (-)^{N_1 + N_2 + 1} (N_1 + 1)(N_2 + 1) \begin{Bmatrix} \dfrac{N_2}{2} & \dfrac{N_1}{2} & \dfrac{N_1 + N_2}{2} \\ \dfrac{N_2}{2} & \dfrac{N_1}{2} & 1 \end{Bmatrix},$$

where the nonzero reduced matrix elements of the O(4) dipole operator are given by

$$\langle N\omega \| \hat{D} \| N\omega' \rangle =$$

$$\begin{cases} \dfrac{N + 2}{2}, & \omega' = \omega, \\[2ex] \dfrac{1}{2} \sqrt{\dfrac{(N - \omega + 2)(N + \omega + 2)(\omega + 1)}{\omega - 1}}, & \omega' = \omega - 2, \\[2ex] \dfrac{1}{2} \sqrt{\dfrac{(N - \omega)(N + \omega + 4)(\omega + 1)}{\omega + 3}}, & \omega' = \omega + 2. \end{cases} \tag{4.29}$$

Despite their formidable appearances, these expressions can be computed numerically with minimum effort. We discuss the essential role of this operator not only for molecules close to the normal limit, but also for symmetric molecules, even if very local ones. Finally, the $\hat{J}^2$ or $\hat{C}^2_{O(3)}$ operator gives the eigenvalue of the square of the total angular momentum operator.

We are now almost ready to convert into a rovibrational language both the algebraic branching rules discussed previously and the dynamical symmetry operator (4.23), complete with its matrix elements in the local basis (4.6). The first step is a straightforward one since it does not depend on the specific equilibrium shape of the molecule. We can consider uncoupled anharmonic bond stretching vibrations in a fashion similar to that in the single-oscillator problem. By recalling Eq. (2.67), we obtain

$$v_a = \frac{N_1 - \omega_1}{2} = 0, 1, \ldots, \frac{N_1}{2} \text{ or } \frac{N_1 - 1}{2} \ (N_1 \text{ even or odd}) ;$$

$$v_b = \frac{N_2 - \omega_2}{2} = 0, 1, \ldots, \frac{N_2}{2} \text{ or } \frac{N_2 - 1}{2} \ (N_2 \text{ even or odd}) . \tag{4.30}$$

Therefore, these quantum numbers label local stretching vibrations of the two interatomic bonds 1 and 2 according to a Morse potential model. However, the nontrivial aspect of this approach appears in the algebraic coupling procedure. This part *does* depend on the molecular geometry, whether linear or bent. Consequently, we divide our discussion into two subsections, where the focus is on the vibrational energy spectrum of bent and linear triatomic molecules, respectively. We postpone until Section IV.D any discussion on the rotational part of these spectra.

1. Bent Triatomic Molecules

In this section we study in detail the Hamiltonian operator (4.23) with the aim of describing the rovibrational spectrum of an either symmetric or nonsymmetric bent triatomic molecule. Therefore, we need to complete the conversion from algebraic to rovibrational quantum numbers [Eq. (4.30)]. It is possible to show [69] that the desired relations for the bending–rotation part are given by

$$v_B = \frac{\omega_1 + \omega_2 - \tau_1 - \tau_2}{2}, \qquad k = \tau_2 . \tag{4.31}$$

We now have a complete one-to-one correspondence between algebraic $(\omega_1, \omega_2, \tau_1, \tau_2, j, m_j)$ and rovibrational local quantum numbers $(v_a, v_b, v_B, k, j, m_j)$. Thus the expectation value of the Hamiltonian operator (4.23) can be written in the following form:

$$\begin{aligned}
E = E(\omega_1, \omega_2, \tau_1, \tau_2, j) = &\ E_0 + A_1\omega_1(\omega_1 + 2) + A_2\omega_2(\omega_2 + 2) \\
&+ A_{12}[\tau_1(\tau_1 + 2) + \tau_2^2] + A'_{12}|\tau_2(\tau_1 + 1)| + Bj(j + 1) ,
\end{aligned} \tag{4.32}$$

where, for the moment, we disregard the Majorana operator $\hat{M}_{12}$. By combining the conversion laws (4.30) and (4.31) and inserting them into the eigenvalues (4.32), we readily obtain

$$\begin{aligned}
E = E(v_a, v_b, v_B, k, j) = &\ E'_0 - 4A_1 v_a(N_1 + 1 - v_a) - 4A_2 v_b(N_2 + 1 - v_b) \\
&- 4A_{12}(v_a + v_b + v_B)(N_{12} + 1 - v_a - v_b - v_B) \\
&- 2A_{12}k[N_{12} + 1 - 2(v_a + v_b + v_B) - k] \\
&+ A'_{12}|k(N_{12} + 1 - v_a - v_b - v_B)| + Bj(j + 1) ,
\end{aligned} \tag{4.33}$$

where $N_{12} \equiv N_1 + N_2$. To understand the meaning of this expression, we focus our attention on the vibrational part of the spectrum (i.e., we put $B = 0$). Since we are not interested here in rotational levels, the contributions to Eq. (4.33) from the quantum number k must collapse on the corresponding vibrational level. We see from the equation above that this is achieved simply by letting $A'_{12} = 2A_{12}$. The resulting spectrum is a purely vibrational one and represents a triatomic molecule in the exact local limit. We insist on the fact that rotational bands are still present (labeled by k), but they are intentionally disregarded by letting $A'_{12} = 2A_{12}$.

However, the bending vibration is accounted for through Eq. (4.31) and the $\hat{C}_{12}$ operator. Let us now consider a simple example. By starting from two different diatoms XY and YZ with $N_1 = 100$ and $N_2 = 60$, we generate the sequences of irreducible representations of $O_1(4)$ and $O_2(4)$,

$$\omega_1 = 100, 98, 96, \ldots, 2, 0 \; ; \qquad \omega_2 = 60, 58, 56, \ldots, 2, 0 \, , \quad (4.34)$$

corresponding to the local stretching modes

$$v_a = 0, 1, \ldots, 49, 50 \; ; \qquad v_b = 0, 1, \ldots, 29, 30 \, . \qquad (4.35)$$

We then consider the direct product states giving irreducible representations of the coupled $O_{12}(4)$ through Eq. (4.10). As we are not interested here in the rotational part, we include in our branching law only those states where $(\tau_1, \tau_2) = (\tau_1, 0)$. This is the equivalent of using the following relation (for $\omega_1 > \omega_2$) in place of Eq. (4.10):

$$\tau_1 = \omega_1 + \omega_2, \omega_1 + \omega_2 - 2, \ldots, \omega_1 - \omega_2 \, . \qquad (4.36)$$

Consequently, the conversion law for v_B [Eq. (4.31)] tells us that for each pair of stretching vibrations v_a and v_b (or, equivalently, of quantum numbers ω_1 and ω_2) we obtain a (finite) sequence of bending vibrations, $v_B = 0, 1, 2, \ldots$. Their contribution to the energy spectrum is regulated in Eq. (4.33) by the algebraic parameter A_{12}. We observe a sequence of bending levels whose anharmonicity is basically given by $\sim 1/N_{12}$.

We are now ready to study the realistic case of a triatomic molecule in which local modes (either stretching or bending ones) are allowed to interact among themselves. We have seen that in the algebraic framework, this kind of interaction is properly accounted for by the Majorana operator. We found in the one-dimensional model that $\hat{M}_{12}$ originates a block-diagonal secular matrix, with each block being labeled by $v_T = v_a + v_b = v_1 + v_2$. This results from having preserved the coupled $O_{12}(2)$ symmetry. In the three-dimensional model [see Eqs. (4.7) and (4.8)] the preserved symmetry is $O_{12}(4)$ since it appears in both local and normal subalgebra chains. In the present case we then expect to have a block-diagonal structure of the Hamiltonian matrix. The difference is in the conserved quantum numbers, which are now (τ_1, τ_2) because they label the irreducible representations of $O_{12}(4)$. Consequently, the introduction of intermode coupling terms in the local picture will lead to a symmetry breaking, where only those states with the same (τ_1, τ_2) quantum numbers can interact among themselves. For a bent triatomic molecule, this is equivalent to saying that the Majorana operator acts on polyads of states with the same total vibrational quantum number:

$$v_T \equiv v_a + v_b + v_B .\tag{4.37}$$

Let us consider the explicit effect of the Majorana operator on a *symmetric* triatomic molecule. In light of the symmetry under bond exchange, the Hamiltonian operator (4.23) can be written as

$$\hat{H} = A(\hat{C}_1 + \hat{C}_2) + A_{12}(\hat{C}_{12} + 2\hat{C}'_{12}) + \lambda_{12}\hat{M}_{12} ,\tag{4.38}$$

where $A \equiv A_1 = A_2$, $A'_{12} = 2A_{12}$, and $B = 0$. The purely vibrational spectrum can be obtained by explicit diagonalization of this operator in the local basis $|v_a v_b v_B\rangle$. So let us consider the first polyad of vibrational levels, characterized by the quantum number (4.37)

$$v_a + v_b + v_B = 1 .\tag{4.39}$$

In view of the conversion (4.30) and of the selection rules $\Delta\omega = 0, \pm 2$ for the Majorana operator, we expect that the vibrational levels of the polyad (4.39) (i.e., $|v_a v_b v_B\rangle = |100\rangle, |010\rangle, |001\rangle$) will be coupled as follows:

$$
\begin{array}{c|ccc}
 & 100 & 010 & 001 \\
\hline
100 & E_S & x & y \\
010 & x & E_S & y \\
001 & y & y & E_B .
\end{array}
\tag{4.40}
$$

In this matrix, E_S denotes the energy of the fundamental stretching local mode (which is the same for bonds 1 and 2 in a XY_2 molecule), while E_B is the energy of the fundamental bending mode. In terms of algebraic parameters, we can write

$$E_S = E(100) = E(010) = -4N(A + 2A_{12}) ,$$
$$E_B = E(001) = -8NA_{12} ,\tag{4.41}$$

where $N \equiv N_1 = N_2$. The off-diagonal terms x and y are obtained explicitly by using the exact mathematical definition of the Majorana operator [Eqs. (4.27)–(4.29)]. Although we will not give the general solution of this problem, it is very simple to obtain analytical expressions for the first low-lying states. For example, let us take the matrix element x connecting $|100\rangle$ and $|010\rangle$. Since we have $\omega_i = N - 2v_i$ and $\tau_1 = 2N - 2v_T = 2N - 2$, $\tau_2 = 0$, we need to consider the following matrix elements in terms of algebraic basis states:

$$\langle \omega_1, \omega_2; (\tau_1, \tau_2)|\hat{M}_{12}|\omega_1', \omega_2'; (\tau_1, \tau_2)\rangle$$

$$= \langle N - 2, N; (2N - 2, 0)|\hat{M}_{12}|N, N - 2; (2N - 2, 0)\rangle . \quad (4.42)$$

We are thus lead to the evaluation of $6j$ symbols of the type

$$\begin{Bmatrix} \dfrac{\omega_1}{2} & \dfrac{\omega_2}{2} & \dfrac{\tau_1 \pm \tau_2}{2} \\ \dfrac{\omega_2'}{2} & \dfrac{\omega_1'}{2} & 1 \end{Bmatrix} = \begin{Bmatrix} \dfrac{N}{2} - 1 & \dfrac{N}{2} & N - 1 \\ \dfrac{N}{2} - 1 & \dfrac{N}{2} & 1 \end{Bmatrix} = \sqrt{\dfrac{N - 2}{(N^2 - 1)(N + 2)}} .$$

$$(4.43)$$

We also need the reduced matrix elements (4.29), which in the present example are given by

$$\langle \omega_1'\|\hat{D}_1\|\omega_1\rangle = \langle N - 2\|\hat{D}\|N\rangle = \frac{N + 1}{\sqrt{N - 1}}, \quad (4.44)$$

with similar expressions for the remaining terms. The final result for the matrix representation of the Hamiltonian operator in the first vibrational polyad is

$$\begin{bmatrix} E_S + \lambda_{12}N & -\lambda_{12}N & -2\lambda_{12}\sqrt{N} \\ -\lambda_{12}N & E_S + \lambda_{12}N & -2\lambda_{12}\sqrt{N} \\ -2\lambda_{12}\sqrt{N} & -2\lambda_{12}\sqrt{N} & E_B + 4\lambda_{12}N \end{bmatrix} . \quad (4.45)$$

It is, of course, possible to extend this calculation to obtain, in closed analytical form, the first excited polyad, $v_T = 2$. The result is shown schematically in Fig. 34. In particular, we notice the direct coupling between pairs of stretching modes in light of the selection rule (for $v_B = \text{const}$) $\Delta(v_a + v_b) = 0$ and $\Delta v_a, \Delta v_b = 0, \pm 1$. This means that the (initially degenerate) stretches $|100\rangle$, $|010\rangle$ now mix and split under the effect of $\hat{M}_{12}$. Due to the symmetry under bond exchange, we obtain either symmetric or antisymmetric wavefunctions, as discussed for the one-dimensional case. The difference here is the presence of the bending mode, which is also involved in the coupling scheme induced by the Majorana operator. We can see in both Fig. 34 and Eq. (4.45) that

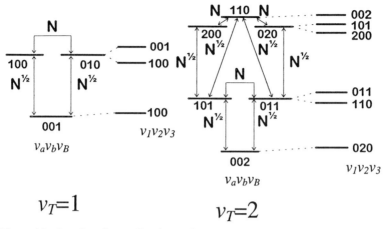

$$v_T=1 \qquad\qquad v_T=2$$

Figure 34. Local-mode coupling (according to the three-dimensional algebraic model) in a bent triatomic molecule for the first two vibrational polyads.

bending modes interact with stretching modes. Although this fact seems quite obvious, we emphasize how such an interaction is included automatically in the algebraic model through an intrinsically *anharmonic* description involving states of different nature (i.e., bends and stretches).

As already discussed for the single-oscillator case in Section II.C.2, it is possible to obtain Dunham-like form for the local-mode eigenspectrum (4.33). Given the anharmonic character of the algebraic formulation, we expect to find a certain number of relations between the Dunham coefficients similar to those of the x–K relations of Lehmann [73, 74]. In fact, we can compare the traditional Dunham series for a triatomic molecule

$$G(v) = \sum_{i=1,2,3} \omega_{e_i}(v_i + \tfrac{1}{2}) + \sum_{i,j=1,2,3} x_{ij}(v_i + \tfrac{1}{2})(v_j + \tfrac{1}{2}), \qquad (4.46)$$

with the algebraic local expansion to find the relations

$$x_{11} = x_{33} = 4(A + A_{12}),$$

$$x_{12} = x_{23} = x_{13} = 2x_{22} = 8A_{12},$$

$$\omega_{e_1} = \omega_{e_3} = -4(2A_{12} + A)(N + 2), \qquad (4.47)$$

$$\omega_{e_2} = -8A_{12}(N + 2).$$

The algebraic model discussed here can describe overtone–combination levels of any triatomic molecule within the first five or six polyads with an rms accuracy of $10 \, \text{cm}^{-1}$. Reference 75 contains the detailed analysis of several molecules. The situation for nonsymmetric bent molecules is very similar to that for symmetric ones if one excludes the absence of symmetric–antisymmetric combinations of local modes (see also Section IV.B.2 for further comments). The general procedure for obtaining a good fit for the vibrational spectrum of the water molecule is presented. By starting from the matrix form (4.45) of the fundamental polyad, we easily obtain, to leading order in N, the following relations:

$$A = \frac{E_S - 2E_A + E_B}{4N},$$

$$A_{12} = \frac{2(E_A - E_S) - E_B}{8N}, \qquad (4.48)$$

$$\lambda_{12} = \frac{E_A - E_S}{2N},$$

where E_A and E_S denote the energies of the antisymmetric and symmetric combinations of stretching local modes, respectively. We know from experiments that $E_A = 3756 \, \text{cm}^{-1}$, $E_S = 3657 \, \text{cm}^{-1}$, and $E_B = 1595 \, \text{cm}^{-1}$. The number N can be estimated with Eq. (2.114) to be $N \simeq 40$. Thus we obtain from the equations above,

$$A = -14.1 \, \text{cm}^{-1}, \qquad A_{12} = -4.4 \, \text{cm}^{-1}, \qquad \lambda_{12} = 1.2 \, \text{cm}^{-1}. \quad (4.49)$$

We notice that the sign of λ_{12} is established by the relative position in the energy spectrum of E_A and E_S. With the values of Eq. (4.49), we can start a fitting procedure over a larger set of experimental levels. In Ref. 75, this model is extended to include higher-order terms (either products or powers of Casimir–Majorana operators). The general aspects of such an extended procedure were discussed in Section II.C.2. We expect to improve the overall quality of the algebraic fit by about one order of magnitude through inclusion of second-order algebraic operators. The corresponding Hamiltonian operator, stopped at quadratic terms, can be written as

$$\hat{H} = A^{(1)}(\hat{C}_1 + \hat{C}_2) + A_{12}^{(1)}\hat{C}_{12} + \lambda_{12}^{(1)}\hat{M}_{12} + A^{(2)}(\hat{C}_1 + \hat{C}_2)^2 + A_{12}^{(2)}\hat{C}_{12}^2$$

$$+ A^{(1,1)}(\hat{C}_1 + \hat{C}_2)\hat{C}_{12} + \lambda_{12}^{(1,1)}[(\hat{C}_1 + \hat{C}_2)\hat{M}_{12} + \hat{M}_{12}(\hat{C}_1 + \hat{C}_2)]$$

$$+ \lambda_{12}^{(2)}\hat{M}_{12}^2, \qquad (4.50)$$

where we introduce the symmetrized product of Casimir and Majorana operators, $\hat{C}\hat{M} + \hat{M}\hat{C}$, for the purpose of obtaining a Hermitian combination.

We finally mention that it is possible to construct simple scaling laws for the algebraic parameters relating to isotopic substitution in a triatomic molecule. In terms of the parameter, ρ, given by

$$\rho = \sqrt{\frac{m_X m_Y}{m_X + m_Y}}, \qquad (4.51)$$

it is found that the following scaling laws can be used for the algebraic parameters:

$$N \propto \rho, \qquad A \propto 1/\rho^2, \qquad A_{12} \propto 1/\rho^2, \qquad \lambda_{12} \propto 1/\rho. \qquad (4.52)$$

These rules lead to predictions of vibrational spectra of isotopic species with an rms accuracy of about 20–40 cm^{-1} [75].

2. *Linear Triatomic Molecules*

The analysis of the algebraic Hamiltonian for a bent triatomic molecule carried out in Section IV.B.1 can readily be adapted to a linear geometry. As stated previously, the most striking difference between bent and linear shapes is found in the rotational structure of the energy spectrum. By comparing Eq. (4.21) with the algebraic branching rule $j = |\tau_2|, |\tau_2| + 1, \ldots$ of Eqs. (4.11) and (4.12), in a linear molecule, the τ_1 and τ_2 quantum numbers must be converted in vibrational quantum numbers according to

$$v_B = \omega_1 + \omega_2 - \tau_1, \qquad l_B = \tau_2; \qquad (4.53)$$

the degenerate bending vibrational mode is denoted as $v_B^{l_B}$, while the labels of the stretching local modes, v_a and v_b, are still obtained by using Eq. (4.30). If we compare Eq. (4.53) with the corresponding relation for bent molecules [Eq. (4.31)], we return to the same correlation law (4.22). So our laws for the algebraic conversion of rovibrational quantum numbers are in agreement with the scheme leading to a unified treatment of linear and bent triatomic molecules.

By restricting our interest to pure vibrational modes, the dynamical symmetry Hamiltonian operator (4.23) (excluding the Majorana operator for the moment) has eigenvalues given by

$$
E = E(v_a, v_b, v_B^{l_B}) = -4A_1 v_a (N_1 + 1 - v_a) - 4A_2 v_b (N_2 + 1 - v_b)
$$
$$
- A_{12}[2(N_{12} + 1)(2v_a + 2v_b + v_B) - (2v_a + 2v_b + v_B)^2 - l_B^2] \qquad (4.54)
$$
$$
+ A'_{12}|l_B[N_{12} + 1 - (2v_a + 2v_b + v_B)]| \, .
$$

The most important difference with the local eigenvalues obtained for the bent case [Eq. (4.33)] is found in the double dependence on the vibrational angular momentum quantum number l_B, which appears in the expectation values of both $\hat{C}_{12}$ and $\hat{C}'_{12}$ operators. In the bent-to-linear correlation pattern for rovibrational energy levels (Fig. 33) we achieve the exact linear limit for $A'_{12} = 0$ (we recall that the bent limit is obtained with $A'_{12} = 2A_{12}$). This means that in the eigenvalues (4.54), the dominant term in l_B is derived from the $\hat{C}_{12}$ operator. However, it is possible to account for minor adjustments of energy terms explicitly dependent on l_B, by adding (small) contributions related to the operator C'_{12}. In the linear case, it is convenient to use, in place of the absolute value, the square of this operator, in such a way that the vibrational spectrum recalls the usual Dunham series (written in normal quantum numbers)

$$
G(v_1, v_2^{l_2}, v_3) = \sum_{i=1}^{3} \omega_i \left(v_i + \frac{d_i}{2} \right)
$$
$$
+ \sum_{i \leq j = 1}^{3} x_{ij} \left(v_i + \frac{d_i}{2} \right) \left(v_j + \frac{d_j}{2} \right) + g_{22} l_2^2 , \qquad (4.55)
$$

where $d_i = 1$ or 2, depending on whether i refers to a nondegenerate or doubly degenerate vibrational mode. In particular, we notice the contribution $g_{22} l_2^2$, which accounts for l–splitting effects in the observed spectrum. Such effects correspond to the different energies of vibrational bands with different quantum number $l_B = l_2$; an example is the Σ–Δ splitting between the bending modes 002^0 and 002^2. In th local expansion (4.54) we recognize the algebraic equivalent of the Dunham parameter g_{22} (or g_{BB}) in the combination of terms $-A_{12} + A'_{12}[N_{12} + 1 - (2v_a + 2v_b + v_B)]^2$.

As discussed previously for bent molecules, the local model (4.54) is a poor approximation when intermode coupling occurs, so we now need to introduce the Majorana operator. The explicit analysis of this problem is perfectly analogous to the previous one, apart from the different conversion law between algebraic and vibrational quantum numbers. Moreover, in a linear molecule we expect to obtain vibrational wavefunctions carrying irreducible representations of the $C_{\infty v}$ and $D_{\infty h}$ point groups [i.e., for a symmetric linear molecule species labeled by Σ_g, Σ_u ($l_B = 0$), Π_g, Π_u ($l_B = \pm 1$), Δ_g, Δ_u ($l_B = \pm 2$), etc.]. A particularly noteworthy difference is also found in the structure of vibrational polyads. By analogy to the bent case, the coupled $O_{12}(4)$ algebra is preserved (i.e., τ_1 and τ_2 remain good quantum numbers). For linear triatomic molecules, on account of Eq. (4.53), this is equivalent to grouping vibrational levels within blocks or polyads characterized by the quantum numbers

$$v_T \equiv v_a + v_b + \frac{v_B - l_B}{2}, \quad l_B. \tag{4.56}$$

This is quite a different classification of states than that in Eq. (4.37). Consequently, the block-diagonal Hamiltonian matrix, introduced by the action of the Majorana operator, will contain the following polyads of levels:

$$
\begin{aligned}
(v_T, l_B) &\quad \text{local modes} \\
(1, 0) &\rightarrow |100^0\rangle, |010^0\rangle, |002^0\rangle \\
(0, 1) &\rightarrow \quad |001^1\rangle \\
(1, 1) &\rightarrow |101^1\rangle, |011^1\rangle, (003^1) \\
(2, 0) &\rightarrow \begin{cases} |200^0\rangle, |110^0\rangle, |020^0\rangle, \\ |102^0\rangle, |012^0\rangle, |004^0\rangle \end{cases} \\
(0, 2) &\rightarrow \quad |002^2\rangle \\
&\quad \cdots
\end{aligned}
\tag{4.57}
$$

We notice, in particular, that states with different vibrational angular momentum (i.e., Σ, Π, Δ, ... bands) are never mixed by $\hat{M}_{12}$. By adopting the technique outlined earlier, we obtain the matrix representation of the Hamiltonian operator inclusive of the Majorana operator. The first polyad (local states 002^0, 100^0, 010^0) is given by

$$
\begin{bmatrix}
-2N_{12}(2A_{12} - \lambda_{12}) & -\lambda_{12}\dfrac{N_{12}}{\sqrt{N_1}} & -\lambda_{12}\dfrac{N_{12}}{\sqrt{N_1}} \\[2ex]
-\lambda_{12}\dfrac{N_{12}}{\sqrt{N_1}} & -4(A_1 N_1 + A_{12} N_{12}) + \lambda_{12} N_2 & -\lambda_{12}\sqrt{N_1 N_2} \\[2ex]
-\lambda_{12}\dfrac{N_{12}}{\sqrt{N_1}} & -\lambda_{12}\sqrt{N_1 N_2} & -4(A_2 N_2 + A_{12} N_{12}) + \lambda_{12} N_1
\end{bmatrix}.
$$

$$\tag{4.58}$$

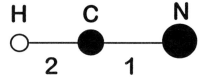

Figure 35. Bond numbering of HCN.

Let us consider a practical example. We want to use the algebraic model for describing the vibrational spectrum of HCN, a linear nonsymmetric molecule (Fig. 35). As per custom, we first determine the vibron numbers N_1 and N_2 of the CN and HC bonds, respectively, by using Eq. (4.114). We obtain $N_1 = 156$ and $N_2 = 43$. As in Section IV.B.1 [Eq. (4.41)] we then recover from the purely local model, the initial guesses of the algebraic parameters

$$A_1 = -\frac{E(100^0)}{4N_1}, \qquad A_2 = -\frac{E(010^0)}{4N_2},$$

$$A_{12} = -\frac{E(002^0)}{4N_{12}} = -\frac{E(001^1)}{2N_{12}}.$$

(4.59)

We observe that the equation for A_{12} is a particular case of $x\text{–}K$ relations: the algebraic result for A_{12} is, in fact, such that $E(002^0) = 2E(001^1)$. By imposing this approximate relation to the energies calculated through an equivalent Dunham expansion, we obtain certain relations among the parameters ω_i and x_{ij}. By starting from the appropriate experimental information, we obtain the zeroth-order (local, anharmonic) Hamiltonian model for the HCN molecule. Although HCN is, indeed, a good example of a local molecule, we can expect that especially at higher excitation energies, local vibrations start to mix among themselves to some extent. Such coupling can only be accommodated by the Majorana operator. Consequently, we have to add nondiagonal interactions to our initial local model. In brief, for symmetric molecules, local-mode splitting gives a direct measure of the action of the Majorana operator. This does not happen in asymmetric molecules, where the splitting between local modes is due to their different chemical or structural nature. An alternative way for determining the strength of the Majorana operator involves explicitly computing intensities for infrared (or Raman) transitions because these intensities depend quite critically on the detailed vibrational wavefunctions. So the idea is to calibrate the Majorana operator indirectly by comparing computed and observed transition intensities.

Returning to Eq. (4.59), we obtain the following numbers:

$$A_1 = -3.4 \, \text{cm}^{-1}, \qquad A_2 = -19.2 \, \text{cm}^{-1}, \qquad A_{12} = -1.8 \, \text{cm}^{-1}. \quad (4.60)$$

These numbers are used in Eq. (4.54) to compute further vibrational levels. In Table III the results of this computation are reported for 30 vibrational bands of HCN. In particular, we see that the trivial guesses of

TABLE III
Three-Dimensional Algebraic Model Fits to HCN[a]

$\nu_1 \nu_2^{l_2} \nu_3$	Expt.	Fit I	Expt.-Fit I	Fit II	Expt.-Fit II
$02^0 0$	1,411.43	1,413.93	−2.50	1,405.20	6.23
$10^0 0$	2,096.85	2,096.85	0.00	2,104.99	8.14
$00^0 1$	3,311.48	3,311.47	0.01	3,302.44	9.04
$04^0 0$	2,802.85	2,813.66	−10.81	2,805.78	−2.93
$12^0 0$	3,501.13	3,496.58	4.55	3,494.76	6.37
$20^0 0$	4,173.07	4,170.74	2.33	4,187.43	−14.36
$02^0 1$	4,684.32	4,711.20	−26.88	4,685.71	−1.39
$10^0 1$	5,393.70	5,394.12	−0.42	5,391.00	2.70
$00^0 2$	6,519.61	6,520.48	−0.87	6,509.26	10.35
$22^0 0$	5,571.89	5,556.25	15.64	5,561.75	10.14
$12^0 1$	6,761.33	6,779.63	−18.30	6,758.94	2.39
$10^0 2$	8,585.57	8,588.91	−3.34	8,581.27	4.30
$00^0 3$	9,627.02	9,627.02	0.00	9,620.52	6.50
$12^0 2$	9,914.41	9,960.22	−45.81	9,926.86	−12.45
$10^0 3$	11,674.46	11,681.24	−6.78	11,675.84	−1.84
$00^0 4$	12,635.90	12,631.09	4.81	12,636.33	−0.43
$00^0 5$	15,551.94	15,532.69	19.25	15,556.80	−4.86
$01^1 0$	711.98	706.97	5.01	707.89	4.09
$03^1 0$	2,113.46	2,113.80	−0.34	2,110.64	2.82
$11^1 0$	2,805.58	2,796.71	8.87	2,805.01	0.57
$01^1 1$	4,004.17	4,011.34	−7.17	3,999.06	5.11
$13^1 0$	4,201.29	4,189.33	11.96	4,192.32	8.97
$21^1 0$	4,878.27	4,863.49	14.78	4,879.58	−1.31
$03^1 1$	5,366.86	5,403.96	−37.10	5,380.20	−13.34
$11^1 1$	6,083.35	6,086.87	−3.52	6,079.81	3.54
$01^1 2$	7,194.75	7,213.24	−18.49	7,194.34	0.41
$02^2 0$	1,426.53	1,406.83	19.70	1,423.76	2.77
$04^2 0$	2,818.16	2,806.55	11.61	2,823.75	−5.59
$12^2 0$	3,516.88	3,489.47	27.41	3,512.71	4.17
$02^2 1$	4,699.21	4,704.09	−4.88	4,703.40	−4.19

[a]All values in cm^{-1}, except N_1 and N_2, which are dimensionless. Algebraic parameters: $N_1 = 156$, $N_2 = 43$

	A_1	A_2	A_{12}	A'_{12}	λ_{12}
Fit I	−3.357	−19.106	−1.776	—	—
Fit II	−3.314	−18.212	−1.759	1.93×10^{-4}	0.919

Eq. (4.60) lead to a satisfactory description of the spectrum, with an rms error of only 16 cm^{-1}. In Table III we also report the result of a fitting procedure in which the three parameters of Eq. (4.60) and λ_{12} and A'_{12} have been adjusted against the same 30 observed levels. As a result, the rms error goes down to 6.7 cm^{-1}. This is due partially to the action of $\hat{M}_{12}$, although its effect will be more pronounced at higher energy. We also notice the relatively important contributions of the $\hat{C}'_{12}$ operator. The levels 002^0 and 002^2 have experimental energies of 1411.4 and 1426.5 cm^{-1}, respectively. Their observed splitting ($+15.1$ cm^{-1}) is poorly reproduced in the first calculation (-7.1 cm^{-1}). The inclusion of $\hat{C}'_{12}$ brings down the overall rms error and leads to a more correct value of the Σ–Δ splitting, $+18.6$ cm^{-1}.

As emphasized, one of the advantages of this model is that it provides explicit wavefunctions which can be used in the computation of expectation values for various operators of interest. Due to limitations of space, we cannot reproduce here the complete set of vibrational wavefunctions obtained in the HCN calculation [76]. However, the typical outcome of the algebraic procedure can be outlined. We obtain a polyad of levels labeled by the numbers v_T and l_B of Eq. (4.56). Each polyad contains a number of local states, such as those listed in Eq. (4.57). The numerical diagonalization of the Hamiltonian matrix is performed separately for each polyad. Thus the eigenvectors derived represent the vibrational wavefunctions in the local basis. A possible outcome of the analysis of the HCN molecule could therefore be given by the following sequence of numbers:

$$\text{Polyad: } (\tau_1, \tau_2) = (193, 0) \Leftrightarrow (v_T, l_B) = (3, 0) \,,$$

$$\text{Local basis: } \{006^0, 014^0, 104^0, \ldots, 210^0, 300^0\} \,,$$

$$\text{Energy 5 (say) } 6247.3 \text{ cm}^{-1} \,, \tag{4.61}$$

$$\text{Wave function 5: } 0.030 \times \phi_6 - 0.011 \times \phi_9 + 0.980 \times \phi_{10} \,.$$

Let us clarify the meaning of these numbers. First, the algebraic multiplet, labeled by the $O_{12}(4)$ irreducible representation $(\tau_1, \tau_2) = (193, 0)$, corresponds to the vibrational polyad $v_T = (N_{12} - \tau_1)/2 = 3$, $l_B = \tau_2 = 0$. Second, the 10 local states, belonging to this polyad, are obtained by using Eq. (4.56). Third, the 10×10 Hamiltonian matrix is diagonalized. In order of increasing energy, the fifth level is located (computed) at 6247.3 cm^{-1}. Finally, its conventional spectroscopic labels can be assigned by looking at the corresponding eigenvector in the local basis. The label assignment can be done very precisely due to the nearly perfect local behavior. The last line of Eq. (4.61) simply tells us that level 5 is basically a local 300^0 mode (ϕ_{10}) with nonnegligible contributions coming from 202^0 (ϕ_6) and 210^0 (ϕ_9).

The aforementioned discussion is also valid for bent triatomic molecules, the only difference being in the absence of the vibrational angular momentum. For a detailed numerical study of these problems, see Ref. 76.

C. Anharmonic (Fermi) Interactions

As discussed in Section III.E, one of the most interesting features of the vibrational spectra of polyatomic molecules is the existence of intense anharmonic resonances involving XH stretches and other molecular modes. We have seen how the one-dimensional approach can address this problem successfully through the use of generalized Fermi–Majorana operators. In principle, we could adopt this same strategy for the three-dimensional scheme. The problem here is a purely mathematical one; we have seen [Eqs. (4.27)–(4.29)] that the $U_{12}(4)$ invariant operator can be somewhat cumbersome to use. Consequently, the idea of constructing Fermi–Majorana operators suitable for the creation–annihilation of an arbitrary number of vibrational quanta is a bit discouraging. As stated previously, the first acknowledged experimental evidence (and theoretical explanation) of a Fermi resonance was provided by the study of CO_2, a rather modest molecule. So it is very important to see whether the prototype Fermi interaction found in CO_2 can be described by algebraic methods. First, we recall that in CO_2 the stretching fundamental 100^0 (Σ_g^+) is nearly degenerate with the bending overtone 002^0 (Σ_g^+); in the Raman spectrum one observes two intense lines approximately located at energies for these levels. A simple theoretical treatment leads to an incorrect result—that only the 100^0 band should give an intense transition. Correct assignment of these bands is done by considering the existence of strong coupling between the 100^0 and 002^0 levels, which is allowed because these modes have the same symmetry [52].

From the onset, the three-dimensional algebraic model presents a consistent treatment of this problem. Let us recall the Hamiltonian matrix for the first polyad of a linear triatomic molecules [Eq. (4.58)], written here in the symmetrical case, $A \equiv A_1 = A_2$ and $N \equiv N_1 = N_2$:

$$\begin{bmatrix} -4N(2A_{12} - \lambda_{12}) & -2\lambda_{12}\sqrt{N} & -2\lambda_{12}\sqrt{N} \\ -2\lambda_{12}\sqrt{N} & -4N(A + 2A_{12}) + \lambda_{12}N & -\lambda_{12}N \\ -2\lambda_{12}\sqrt{N} & -\lambda_{12}N & -4N(A + 2A_{12}) + \lambda_{12}N \end{bmatrix}.$$

$$(4.62)$$

In this expression we notice that $\hat{M}_{12}$ does two things simultaneously; it splits the degenerate local stretches (entry $-\lambda_{12}N$), but it also couples these levels with the bending overtone (entry $-2\lambda_{12}\sqrt{N}$). By adjusting λ_{12} in such a way that the stretching modes are split by the experimental value (about $1200\ \text{cm}^{-1}$, *very* normal behavior, which corresponds to a large value of λ_{12}), the 002^0 bending mode will interact with the $100^0/010^0$ levels. In the resulting normal picture, (1) local stretches $100^0/010^0$ become the g/u combination modes usually denoted by $(v_1 v_2^{l_2} v_3) = (10^0 0)\Sigma_g^+$, $(00^0 1)\Sigma_u^+$; and (2) the bending overtone is strongly coupled only to the symmetric stretch. Figure 36 shows a schematic representation of the effects of $\hat{M}_{12}$. In short, we let the Majorana operator remove the local stretch degeneracy until the symmetric combination is basically degenerate with the bending overtone. Although this description is quite satisfactory, we cannot expect to obtain perfect agreement with the experimental results using only the Majorana operator. In light of its anomalous behavior, the CO_2 molecule requires a more sophisticated treatment, where the exact amount of Fermi mixing can be adjusted independent of the Majorana operator. For the purpose of obtaining a widely applicable operator, we introduce the Fermi operator, $\hat{F}$, whose matrix elements, in the first vibrational polyad, are given by

$$\begin{bmatrix} 0 & F & F \\ F & 0 & 0 \\ F & 0 & 0 \end{bmatrix}, \tag{4.63}$$

where $F = 2(1 - f_{12})\lambda_{12}\sqrt{N}$. The total Hamiltonian operator is then

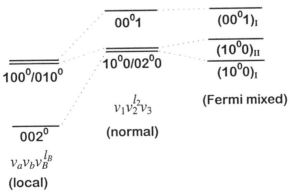

Figure 36. Fermi coupling in CO_2 between $100^0/010^0$ local modes and 002^0. Energy axis not on scale.

written as

$$\hat{H} = \hat{H}_{\text{local}} + \lambda_{12}\hat{M}_{12} + \hat{F} . \tag{4.64}$$

In the first polyad of levels, its matrix form is

$$\begin{bmatrix} -4N(2A_{12} - \lambda_{12}) & -2f_{12}\lambda_{12}\sqrt{N} & -2f_{12}\lambda_{12}\sqrt{N} \\ -2f_{12}\lambda_{12}\sqrt{N} & -4N(A + 2A_{12}) + \lambda_{12}N & -\lambda_{12}N \\ -2f_{12}\lambda_{12}\sqrt{N} & -\lambda_{12}N & -4N(A + 2A_{12}) + \lambda_{12}N \end{bmatrix} . \tag{4.65}$$

It is then possible, by proper adjustments of the Fermi parameter f_{12}, to calibrate both the energy position and the amount of wavefunction mixing against the experimental values. The Fermi operator introduced here is a special case of Majorana interaction, which can readily be generalized to higher polyads of vibrational levels.

Besides Fermi interactions, another class of anharmonic couplings among vibrational levels exist and they are often referred to as *Darling–Dennison interaction* [77]. They typically arise in conjunction with a normal-mode molecular picture, since the associated matrix elements can be written as

$$\langle v_1 v_2^{l_2} v_3 | \hat{V}_{DD} | v_1 \mp 2, v_2^{l_2}, v_3 \pm 2 \rangle . \tag{4.66}$$

Although we will not discuss in detail this particular aspect of anharmonic resonances, it is important to note that Darling–Dennison couplings are automatically included by the action of the Majorana operator. A practical way to convince ourselves of this inclusion is to diagonalize (either numerically or in closed form) the Hamiltonian matrix explicitly for the first two polyads of levels and then to convert, in normal-mode notation, the vibrational states obtained. As discussed in Ref. 11, the Hamiltonian (4.38) can also be written (neglecting $\hat{C}_{12}$ and $\hat{C}'_{12}$ interactions) as

$$\hat{H} = \hat{H}_{\text{normal}} + A(\hat{C}_1 + \hat{C}_2) . \tag{4.67}$$

Here $\hat{H}_{\text{normal}} = \lambda_{12}\hat{M}_{12}$. In a *normal* basis $\hat{H}_{\text{normal}}$ is diagonal while $\hat{C}_1$ and $\hat{C}_2$ are not, so they break the dynamic symmetry. The parameter A is directly related to the local-mode anharmonicity. Consequently, we expect to obtain strong Darling–Dennison normal-mode couplings in the presence of very anharmonic local modes.

As a final comment, it should be clear that the local-to-normal transition parameter introduced in Section III.B [Eq. (3.49)] can be applied equally well to triatomic molecules in the three-dimensional model. Here we need to introduce two parameters, one for each bond, according to

$$\xi_k = \frac{2}{\pi} \tan^{-1} \frac{8\lambda_{12}}{A_k + A_{12}}, \qquad k = 1, 2 \qquad (4.68)$$

and to define the overall parameter of locality–normality for triatomic molecules according to

$$\xi = \sqrt{|\xi_1 \xi_2|} . \qquad (4.69)$$

As an example, we obtain $\xi = 0.12$ for the HCN molecule and $\xi = 0.93$ for CO_2, in agreement with the local behavior of HCN and the more normal one of CO_2.

D. Rotational Spectroscopy

It would be excessive to include in this article a comprehensive treatise on rotational molecular spectroscopy. It is, however, worthwhile to address this important subject for the purpose of demonstrating what kind of theoretical constructions can be handled within the algebraic framework. So far, most applications of algebraic models have been dealt with vibrational rather than rotational spectroscopy. However, it is only a matter of time before the algebraic treatment is applied to rotational spectroscopy since it has a unique association with the three-dimensional model.

To begin with, we recall some basic information about the rotational Hamiltonian operator, related to the rotational motions of a quantum rigid body [78, 79]. Its basis states are completely determined by three quantum numbers $|jkm_j\rangle$, where k refers to the projection along the figure axis of the angular momentum operator and m_j to the projection along the fixed frame z axis. For the rigid-rotor Hamiltonian operator, we can write the expression

$$\hat{H}_{\rm rot} = \frac{1}{2} \left(\frac{\hat{J}_x^2}{I_{xx}^0} + \frac{\hat{J}_y^2}{I_{yy}^0} + \frac{\hat{J}_z^2}{I_{zz}^0} \right) \equiv \frac{1}{2} \left(\frac{\hat{J}_A^2}{I_A} + \frac{\hat{J}_B^2}{I_B} + \frac{\hat{J}_C^2}{I_C} \right), \qquad (4.70)$$

where we adopt the usual spectroscopic notation for molecular momenta of inertia, $I_A < I_B < I_C$. The eigenvalue problem $\hat{H}_{\rm rot}|jkm_j\rangle = E_{\rm rot}|jkm_j\rangle$ is better addressed by considering distinct cases corresponding to various molecular symmetries.

1. Spherical Top, $I_A = I_B = I_C$. This is the simplest case. Spherical top molecules such as CH_4 and SF_6 have an highly degenerate spectrum whose levels are given by

$$E_{rot}(j, k, m_j) = Bj(j + 1) .\tag{4.71}$$

Thus the overall rotational degeneracy is $(2j + 1)^2$ [i.e., $(2j + 1)$ fold-degenerate in both k and m_j quantum numbers].

2. Symmetric Top, $I_A < I_B = I_C$ (prolate top) or $I_A = I_B < I_C$ (oblate top). The rotational spectrum is still degenerate in the m_j quantum number; however, part of the k degeneracy is lift. Only those levels with $k \neq 0$ are now doubly degenerate (in $\pm k$). The rotational spectrum of symmetric top molecules (such as NH_3, C_6H_6) can be written as

$$E_{rot}(j, k, m_j) = Bj(j + 1) + (A - B)k^2 \quad \text{(prolate)} ,$$
$$E_{rot}(j, k, m_j) = Bj(j + 1) + (C - B)k^2 \quad \text{(oblate)} ,\tag{4.72}$$

where $A = h/(8\pi^2 c I_A)$, $C = h/(8\pi^2 c I_C)$, and $B = h/(8\pi^2 c I_B)$.

3. Linear Top, $I_A = 0$, $I_B = I_C$. As already discussed, this is a special case since it requires a specific treatment associated with the existence of an angular momentum originating from vibrational motions. The final result [80] is that k is replaced by l_B and the rotational spectrum can be written as

$$E_{rot}(j, l_B) = B[j(j + 1) - l_B^2] .\tag{4.73}$$

4. Asymmetric top, $I_A < I_B < I_C$. This is the most difficult case. Asymmetric top molecules (e.g., H_2O, NH_2D) lead to off-diagonal elements in the quantum number k. Consequently, it is necessary to perform a numerical diagonalization of the rotational Hamiltonian operator (4.70), although slightly asymmetric molecules can be studied properly through perturbation methods.

The aforementioned simple rotational spectra are readily reproduced, in their essential features, within the three-dimensional algebraic approach. If we limit ourselves to distinguishing between linear and nonlinear rigid rotors, it is evident from Section IV.B [Eq. (4.23)] that the rotational part of the algebraic Hamiltonian operator is given by

$$\hat{H}_{rot} = A'_{12}(\hat{C}'_{12})^2 + B\hat{j}^2 .\tag{4.74}$$

The eigenvalues of $\hat{H}_{rot}$ can be written in different ways, according to the conversion laws for rovibrational quantum numbers for linear or bent molecules [see Eqs. (4.33) and (4.54)]:

Bent molecule:

$$E_{rot}(j, k) = A'_{12}(N_{12} + 1 - v_a - v_b - v_B)^2 k^2 + Bj(j + 1) ; \quad (4.75)$$

Linear molecule:

$$E_{rot}(j, k) = [A'_{12}(N_{12} + 1 - 2v_a - 2v_b - v_B)^2 - A_{12}]l_B^2 + Bj(j + 1) . \quad (4.76)$$

We can also write these equations as

Bent molecule:

$$E_{rot}(j, k) = B'k^2 + Bj(j + 1) ; \quad (4.77)$$

Linear molecule:

$$E_{rot}(j, l_B) = g_{BB}l_B^2 + Bj(j + 1) , \quad (4.78)$$

where

$$B' = A'_{12}(N_{12} + 1 - v_a - v_b - v_B)^2 , \quad (4.79)$$

$$g_{BB} = A'_{12}(N_{12} + 1 - 2v_a - 2v_b - v_B)^2 - A_{12} . \quad (4.80)$$

If we compare Eq. (4.78) with Eq. (4.73), it is clear that the algebraic three-dimensional model provides the correct rotational spectrum of a rigid linear rotor, where the (vibrational) angular momentum coefficient, g_{BB}, is described by the algebraic parameters A_{12} and A'_{12}. The j-rotational band is obtained by recalling in Eq. (4.12), the branching law

$$\begin{aligned} j &= 0^+, 1^-, 2^+, \ldots & (l_B = 0) , \\ j &= l_B^\pm, (l_B + 1)^\mp, (l_B + 2)^\pm, \ldots & (l_B \neq 0) . \end{aligned} \quad (4.81)$$

The corresponding spectrum is shown schematically in Fig. 37. If we now compare Eq. (4.77) with Eqs. (4.71) and (4.72), we recognize the rotational structure of either a spherical top ($B' = 0$) or a symmetric top ($B' = A - B$, prolate; $B' = C - B$, oblate). This result justifies, a posteriori, use of the square of the pseudoscalar operator $\hat{C}' = \hat{D} \cdot \hat{J}$ in the dynamic symmetry Hamiltonian (4.23). The rotational spectrum of a nonsymmetric rotor requires the introduction of a more complex operator

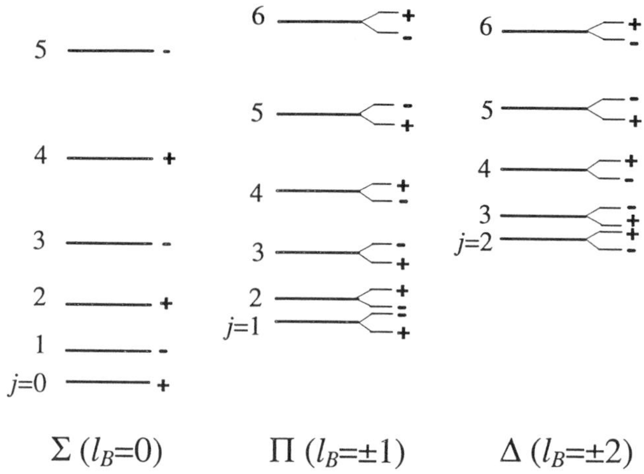

Figure 37. Rotational spectrum of a linear top.

capable of lifting the twofold degeneracy in the quantum number k. The j-rotational band is, again, obtained via the branching law

$$j = 0^+, 1^-, 2^+, \ldots \quad (k = 0),$$

$$j = 1^\mp, 2^\pm, 3^\mp, \ldots \quad (k = 1), \qquad (4.82)$$

$$j = 2^\pm, 3^\mp, 4^\pm, \ldots \quad (k = 2),$$

$$\vdots$$

The corresponding spectrum is shown schematically in Fig. 38. For further details and references on an algebraic approach to rotational spectroscopy of triatomic molecules, see Refs. 81–83.

The rigid-rotor approximation is a very poor one for a detailed study of high-resolution experimental spectra with complex rotational structure. Although a complete analysis of these arguments is not feasible within the scope of this treatise, once again we emphasize the most relevant aspects of rotational dynamics for nonrigid rotors. (For a more detailed review, see, for example, Ref. 84.) We can summarize the situation by stating that the separation of vibrational and rotational degrees of freedom can no longer be satisfied exactly. As a result, we expect to write a Dunham-like expression for the rovibrational energies given by

$$E(v, j) = E_{\text{vib}}(v) + E_{\text{rot}}(j, v), \qquad (4.83)$$

where E_{vib} denotes the usual Dunham expansion for purely vibrational

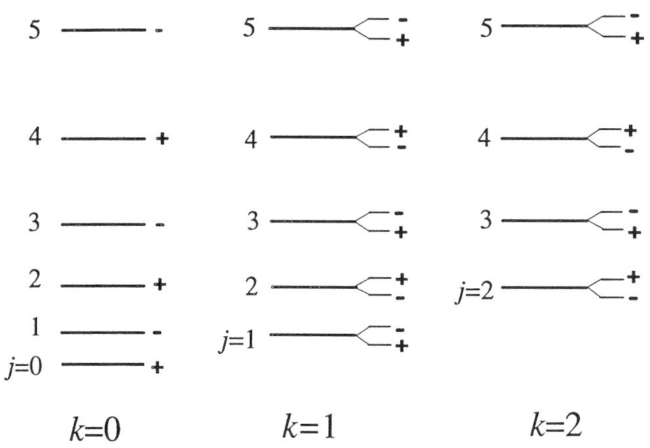

Figure 38. Rotational spectrum of a symmetric top.

terms, while E_{rot} corresponds to the rotational part, which is, to some extent, affected by the specific vibrational level v (here corresponding to the generic set of vibrational quantum numbers $v_1, v_2, \ldots, v_n$). In Section II.C.2 we discussed a simple case of rovibrational interaction associated with diatomic molecules. The triatomic (or polyatomic) case falls exactly within this same scheme but presents far more exacting technical problems.

To be more specific, let us consider the explicit form of E_{rot} in the linear rotor case (neglecting l-doubling effects, which we discuss later):

$$E_{\text{rot}}(j, v) = B_v[j(j+1) - l_B^2] - D_{j,v}[j(j+1)]^2$$

$$- D_{j,v,l_B}j(j+1)l_B^2 - D_{l_B,v}l_B^4 + \cdots, \qquad (4.84)$$

where B_v is the usual rotational "constant" and the various D_j, D_{jl_B}, and D_{l_B} are the centrifugal distortion constants, whose typical numerical values are on the order of $10^{-4} \times B_v$. The explicit dependence of B_v on the vibrational (normal) quantum numbers is usually written as

$$B_v = B_e - \sum_{r=1}^{3N-5} \alpha_r^B\left(v_r + \frac{d_r}{2}\right) + \sum_{r' \geq r} \gamma_{rr'}^B\left(v_r + \frac{d_r}{2}\right)\left(v_{r'} + \frac{d_{r'}}{2}\right) + \cdots.$$

$$(4.85)$$

The typical experimental procedure consists in the measurement of $\Delta B_v \equiv B_v - B_0$. A consistent theoretical model, supporting the aforementioned expansions, can be obtained by an approximate separation of vibrations and rotations, where the complete Hamiltonian operator is averaged over all the vibrational coordinates. In practice, the rovibrational Hamiltonian matrix is constructed in the basis of rigid, harmonic oscillator eigenstates; a perturbation treatment is applied to this matrix to remove off-diagonal elements in the vibrational quantum numbers ("contact" transformation); the transformed matrix elements provide an effective rotational Hamiltonian operator in a (convergent) series of partial operators. The partial operators correspond to expansion of the effective tensor of inertia and of the potential energy surface in terms of vibrational coordinates about their equilibrium values. Finally, the Hamiltonian operator can be written as

$$\hat{H} = \sum_{m,n} \hat{h}_{mn} \,, \qquad (4.86)$$

where m and n denote powers of vibrational and rotational operators, respectively. This means, for example, that $\hat{h}_{m0}$ refer to purely vibrational terms; that is, $\hat{h}_{20}$ gives the usual anharmonic approximation, $\hat{h}_{40}$ a quartic anharmonicity, and so on; $\hat{h}_{02}$ corresponds to the rigid-rotor approximation, while $\hat{h}_{12}$ and $\hat{h}_{22}$ give centrifugal distortion terms. So we are lead to a general scheme of classification [85] where the vibrational dependence of the rotational constants is given by the diagonal matrix elements of $\hat{h}_{22}$. Among such effects, we also mention the Coriolis interaction, whose algebraic description has not yet been considered in sufficient detail.

We must note that we have been disregarding off-diagonal elements in rotational l_B (or k) quantum numbers of the transformed Hamiltonian matrix. Such off-diagonal elements can be of noticeable importance in the study of vibrational l splitting and l doubling, as well as rotational l-doubling (or l-resonance) effects in polyatomic molecules. Vibrational l splitting (Σ/Δ separation) and l doubling (Σ^+/Σ^- separation) are not of direct interest here because they can be accommodated by a purely vibrational scheme, where the predominant effect is given by the quartic anharmonicity operator $\hat{h}_{40}$. Instead, the rotational l-doubling effect is related to off-diagonal elements of the operator $\hat{h}_{22}$, whose diagonal elements are related to the vibrational dependence of B_v. In the simpler case of a linear top, the off-diagonal matrix elements are

$$\langle v_B^{l_B+1} j | \hat{h}_{22} | v_B^{l_B-1} j \rangle = q\sqrt{(v_B+1)^2 - l_B^2}$$

$$\times \sqrt{[j(j+1) - l_B(l_B+1)][j(j+1) - l_B(l_B-1)]} \,,$$

$$(4.87)$$

in manifest agreement with removal of the degeneracy between the Π bending states 1^{+1} and 1^{-1}. The interaction above is governed by the so called l-doubling constant q. Higher-order effects can also be included through operators $\hat{h}_{42}$, $\hat{h}_{24}$, ..., leading to experimentally accessible interactions associated with $\Delta l_B > 2$ selection rules.

The three-dimensional algebraic model can reproduce, in detail, the aforementioned classification of the rovibrational Hamiltonian operator in "partials" $\hat{h}_{22}$. To achieve this goal, we start by writing a compact form of this operator in its usual (normal coordinate) notation:

$$\hat{h}_{22} = \hat{h}_{22}^{D} + \hat{h}_{22}^{ND} , \tag{4.88}$$

where we separate the diagonal part,

$$\hat{h}_{22}^{D} \sim \left[\sum_{r,s} \frac{\partial^2 \mu}{\partial Q_r \partial Q_s} Q_r Q_s \right] \hat{j}^2 , \tag{4.89}$$

and the nondiagonal part,

$$\hat{h}_{22}^{ND} \sim \sum_{r,s} (Q_r \pm Q_s)^2 (\hat{J}_x \pm i\hat{J}_y)^2 . \tag{4.90}$$

The diagonal part can be written in terms of algebraic operators as

$$\hat{h}_{22}^{D} = \left(\sum_i A_i^{RV} \hat{C}_i \right) \hat{j}^2 , \tag{4.91}$$

whose eigenvalues, to lowest order and for linear molecules in the local limit, are given by

$$E_{22}^{D}(v_a, v_b, v_B^{l_B}) = \{ -4A_1^{RV} v_a (N_1 + 1 - v_a) - 4A_2^{RV} v_b (N_2 + 1 - v_b)$$
$$- A_{12}^{RV} [(2v_a + 2v_b + v_B)(2N_{12} + 2 - 2v_a - 2v_b - v_B)$$
$$- l_B^2] \} j(j+1) . \tag{4.92}$$

If one adds this expression to the rotational energy (3.78), it is possible to recover the algebraic counterpart of ΔB_v:

$$\Delta B_v = \frac{E_{22}^{D}(v_a, v_b, v_B^{l_B})}{j(j+1)} . \tag{4.93}$$

For example, we have

$$\Delta B_{(100^0)} = -4A_1^{RV}N_1 - 4A_{12}^{RV}N_{12} \, ,$$
$$\Delta B_{(001^1)} = -2A_{12}^{RV} \, . \tag{4.94}$$

The expressions above allow one to perform purely rotational fitting procedures, where the parameters A_i^{RV} can be determined by the experimental values obtained for the ΔB_v's. Examples of this method can be found in Ref. 86. A similar treatment can, of course, be applied to bent molecules, where the quantum number l_B is replaced by k and the conversion laws for vibrational quantum numbers are changed according to our previous prescription.

For an algebraic point of view, the nondiagonal part of $\hat{h}_{22}$ deserves more attention. The coupled $O_{12}(4)$ symmetry must now be broken to account for l-resonance terms, by means of which τ_2 [i.e., l (or k)] is no longer conserved. As a consequence, we expect to introduce rather complex algebraic operators which do not respect the initial $O(4)$ dynamic symmetry. This can be done in a straightforward way by requiring a specific tensor character for these operators. Unfortunately, detailed study of such operators requires an extended foray in the realm of Racah algebra and tensor calculus. We prefer to omit all the mathematical details. The general idea, however, is to calculate matrix elements of operators of the following form:

$$\hat{V}_{\alpha\beta}^{(L)} \equiv [[\hat{D}_\alpha \times \hat{D}_\beta]^{(L)} \times [\hat{J} \times \hat{J}]^{(L)}]^{(0)} \, , \qquad \alpha, \beta = 1, 2 \, , \tag{4.95}$$

where $\hat{D}_\alpha$ are the usual bond dipole operators of $O_\alpha(4)$ and the products are in the tensor format discussed in Section II.C.2. The operators $\hat{V}_{\alpha\beta}^{(L)}$ are coupled to zero total angular momentum to give scalar terms in the Hamiltonian operator, while L can be even only because the operators are Hermitian. It is possible to show that the operators $\hat{V}_{\alpha\beta}^{(0)}$ correspond basically to the diagonal part of $\hat{h}_{22}$. We are left with the explicit computation of the matrix elements of $\hat{V}_{\alpha\beta}^{(L)}$. The final result, expressed in the algebraic basis for $\alpha\beta = 12$, is

$$\langle \omega_1' \omega_2'(\tau_1'\tau_2')jm_j | \hat{V}_{12}^{(L)} | \omega_1 \omega_2(\tau_1\tau_2)jm_j \rangle$$

$$= \tfrac{1}{4} j(j+1)(2j+1)\sqrt{2L+1}\sqrt{\omega_1(\omega_1+2)\omega_2(\omega_2+2)}(-)^L \begin{Bmatrix} L & j & j \\ j & 1 & 1 \end{Bmatrix}$$

$$\times \Sigma_{l_1 l_2 l_1' l_2'}(-)^{l_1+l_2+l_1'+l_2'}[1-(-)^{l_1+l_1'}][1-(-)^{l_2+l_2'}]\sqrt{(2l_1'+1)(2l_2'+1)}$$

$$\times \begin{Bmatrix} l_1 & l_2 & j \\ l_1' & l_2' & j \\ 1 & 1 & L \end{Bmatrix} \left\langle \begin{matrix} \omega_1 & \omega_2 \\ l_1 & l_2 \end{matrix} \middle| \begin{matrix} (\tau_1 \tau_2) \\ j \end{matrix} \right\rangle \left\langle \begin{matrix} \omega_1 & \omega_2 \\ l_1' & l_2' \end{matrix} \middle| \begin{matrix} (\tau_1' \tau_2') \\ j \end{matrix} \right\rangle \tag{4.96}$$

$$\times \left\langle \begin{matrix} \omega_1 & (1,1) \\ l_1 & 1 \end{matrix} \middle| \begin{matrix} \omega_1 \\ l_1' \end{matrix} \right\rangle \left\langle \begin{matrix} \omega_2 & (1,1) \\ l_2 & 1 \end{matrix} \middle| \begin{matrix} \omega_2 \\ l_2' \end{matrix} \right\rangle ,$$

with similar expressions for $\hat{V}_{11}^{(2)}$ and $\hat{V}_{22}^{(L)}$. In the formula above, we make use of the O(4) isoscalar factors defined in Section IV.B [Eq. (4.19)]. In brief, it is possible to use the operators $\hat{V}_{\alpha\beta}^{(L)}$ to describe, in detail, the rotational l-resonance effect for a linear molecule.

To begin with, the formidable aspect of Eq. (4.96) is, after all, a minor problem, because these matrix elements can be directly obtained with appropriate computer routines. Now we must examine the structure of rovibrational couplings associated with l-doubling interactions. In place of states $|v_B^{l_B}, j\rangle$, it is customary to introduce a rovibrational basis of levels with well-defined parity

$$\hat{p}|v_B^{\pm l_B}, j\rangle \equiv (-)^{j \mp l_B}|v_B^{\mp l_B}, j\rangle . \tag{4.97}$$

The transformed basis, often referred to as *Wang's basis* [87], is defined as

$$|v_B^0, j\rangle_e = |v_B^0, j\rangle , \quad l_B = 0 ,$$

$$|v_B^{l_B}, j\rangle_e = \frac{1}{\sqrt{2}}[|v_B^{l_B}, j\rangle + (-)^{l_B}|v_B^{-l_B}, j\rangle] , \tag{4.98}$$

$$|v_B^{l_B}, j\rangle_f = \frac{1}{\sqrt{2}}[|v_B^{l_B}, j\rangle - (-)^{l_B}|v_B^{-l_B}, j\rangle] , \quad l_B \neq 0 .$$

According to this definition, states e and f are eigenstates of the parity operator $\hat{p}$ with eigenvalues $(-)^j$ and $(-)^{j+1}$, respectively. As a consequence of the goodness of the parity quantum number, the Hamiltonian matrix must be in block-diagonal form, each block being labeled by e or f parities. Let us consider first the interaction Π/Π, leading to a direct l-coupling effect. In this case the rovibrational Hamiltonian operator is represented by a 2×2 matrix with off-diagonal elements coupling the bending sublevels $\Pi_e = |1^{+1}\rangle$ and $\Pi_f = |1^{-1}\rangle$. By using Eq. (4.96) we obtain

$$E(\Pi_e) = E_v(\Pi) + \Delta B_e j(j+1) ,$$

$$E(\Pi_f) = E_v(\Pi) + \Delta B_j j(j+1) , \tag{4.99}$$

where $E_v(\Pi)$ is the energy of the Π_e / Π_f doublet in the absence of l doubling and

$$\Delta B_e = \Delta B_v + q/2 \,,$$
$$\Delta B_f = \Delta B_v - q/2 \qquad\qquad (4.100)$$

represent the rovibrational displacements of the e/f states, expressed in terms of the displacement ΔB_v of the band center and of the l-doubling strength q. If we now go to the next bending level, $v_B = 2$, we obtain (for $j \geq 2$) a 3×3 matrix, since we now include Σ (2^0) and Δ $(2^{\pm 2})$ bending sublevels. The original form of the $\hat{V}^{(2)}$ matrix is of the type

$$\begin{bmatrix} E_\Sigma & q & q \\ q & E_\Delta & 0 \\ q & 0 & E_\Delta \end{bmatrix}, \qquad\qquad (4.101)$$

because this operator connects only states with $\Delta l_B = \pm 2$. By applying the transformation (4.98), we obtain states of well-defined parity Σ_e, Δ_e, and Δ_f whose energies, for $E_\Delta > E_\Sigma$ and $q \ll |E_\Delta - E_\Sigma|$, are given by

$$E_{\Delta_f} = E_\Delta \,, \qquad E_{\Delta_e} \simeq E_\Delta - \frac{2q^2}{E_\Delta - E_\Sigma} \,, \qquad E_{\Sigma_e} \simeq E_\Sigma - \frac{2q^2}{E_\Delta - E_\Sigma}$$

$$(4.102)$$

(i.e., the initially degenerate Δ sublevels are split into the components Δ_e and Δ_f). Figure 39 shows an example of rovibrational coupling pattern for purely bending-mode excitation obtained with the algebraic operator (4.96).

A detailed study of further properties of the $\hat{V}^{(2)}$ operator shows that it behaves fairly well compared to realistic situations. In this brief review we have not discussed the dependence of the l-doubling operator on the stretching v_a and v_b quantum numbers or on the angular momentum quantum number j. A discussion on this dependence can be found in Refs. 86 and 88. Nonetheless, the algebraic realization can lead to results superior to those obtained with standard approaches. Moreover, the algebraic formulation includes vibrational anharmonicity from the very onset of formulating the model for a given system.

E. Electromagnetic Transition Intensities

As discussed in Section III.D, the algebraic model is well suited for providing a straightforward analysis of infrared and Raman intensities. This is emphasized particularly in the three-dimensional model, where,

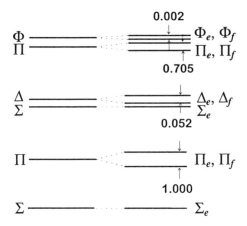

Figure 39. l-doubling effect in states with $j = 3$ and $v_B = 0, 1, 2, 3$; v_a, $v_b = 0$. Energy axis not on scale.

from the beginning, both vibrations and rotations are present. Moreover, a proper treatment of matrix elements for dipole or quadrupole operators must account for the rotational degrees of freedom. If one excludes these important aspects, the strategy for the calculation of infrared and Raman intensities strictly follows the steps outlined in Section III.D. However, the inclusion of rotational degrees of freedom leads to somewhat more involved expressions, which are discussed in subsequent paragraphs. A detailed study of infrared operators can be found in Ref. 89, while Raman transitions are discussed in Ref. 90.

For electric dipole (E1) transitions, the most obvious choice is the O(4) dipole operator introduced in Section II.C.2. We thus consider the operator

$$\hat{T} = t_0 \hat{D} , \tag{4.103}$$

whose tensor character (and parity) is $j^P = 1^-$. It leads to the rovibrational selection rules $\Delta v = 0$ and $\Delta j = \pm 1$. This means that the operator (4.103) accounts for transitions only within the same vibrational band. As explained previously, this result does not correspond to realistic situations; mechanical anharmonicity effects originate transitions $\Delta v = \pm 1$, $\pm 2, \ldots$. Similar to Eq. (3.105), we then introduce more complex forms of the dipole operator, such as

$$\hat{T} = \tfrac{1}{2}[(t_0 + t_1 \hat{n}_p + t_2 \hat{n}_p^2 + \cdots)\hat{D} + \hat{D}(t_0 + t_1 \hat{n}_p + t_2 \hat{n}_p^2 + \cdots)] \tag{4.104}$$

or

$$\hat{T} = t_0 \hat{D} + t_1 [\hat{D} \exp(-\alpha \hat{n}_p) + \exp(-\alpha \hat{n}_p) \hat{D}] \,, \tag{4.105}$$

where $\hat{n}_p$ is the number operator for p bosons and we use symmetrized products (i.e., $\hat{n}_p \hat{D} + \hat{D} \hat{n}_p$) because the factors do not commute. It is also possible to account for more subtle effects in the dipole moment function, such as two (or more) different slopes in the exponential dependence:

$$\hat{T} = t_0 \hat{D} + t'[\hat{D}\hat{t}' + \hat{t}'\hat{D}] \,,$$

$$\hat{t}' = [t_1 \exp(-\alpha_1 \hat{n}_p) + t_2 \exp(-\alpha_2 \hat{n}_p)] \,. \tag{4.106}$$

The aforementioned matrix elements of electric dipole operators cannot be calculated analytically. Fortunately, the realistic limit for large N leads to closed forms for the matrix elements of $\exp(-\alpha \hat{n}_p)$. It is customary to introduce the standard Herman–Wallis notation for denoting the *line strength*:

$$S_{vj \to v'j'} \equiv |\langle v'j' \| \hat{T}^{(1)} \| vj \rangle|^2 = |\mu| |R_{v'v}|^2 F_{v'v}(\mu) \,, \tag{4.107}$$

where

$$\mu = \tfrac{1}{2}[j'(j'+1) - j(j+1)] = \begin{cases} j+1 & \text{(R-branch, } j' = j+1) \\ -j & \text{(P-branch, } j' = j-1) \,. \end{cases} \tag{4.108}$$

It is possible to show that if one starts from the simple operator (4.103), the line strength (4.107) is given in terms of the following quantities:

$$R_{v'v} = t_0(N - 2v + 1) \,, \qquad F_{v'v}(\mu) = 1 - \frac{\mu^2}{(N - 2v + 1)^2} \,. \tag{4.109}$$

By starting from more complex operators, such as (4.104), (4.105), or (4.106), one obtains approximate solutions expressed as power series in the rovibrational quantum numbers. (For a complete discussion, see Ref. 11.) The rotational factor, $F_{v'v}(\mu)$, corresponds to a rigid-rotor model. We can improve its description by allowing the inclusion of rovibrational interactions in the dipole operator. The simplest form of dipole operator, coupled with the angular momentum $\hat{J}$, is given by

$$\hat{D} + \zeta[\hat{J} \times \hat{D}' + \hat{D}' \times \hat{J}]^{(1)} \,, \tag{4.110}$$

where $\hat{D}'$ is the second O(4) dipole operator [see Eq. (2.95)].

Raman intensities are obtained in a similar way, with the important differences discussed in Section III.D. So we have to consider two separate contributions. These are the monopole (scalar) operator

$$\hat{R}^{(0)} = \alpha^{(0)} , \tag{4.111}$$

with selection rules $\Delta v = 0$ and $\Delta j = 0$ and the (traceless) rank 2 tensor operator

$$\hat{R}^{(2)} = \alpha^{(2)}\hat{Q}^{(2)} , \tag{4.112}$$

where, once again, $\hat{Q}^{(2)}$ is the O(4) quadrupole operator of Eq. (2.95). Its selection rules are $\Delta v = 0, \pm 1$ and $\Delta j = 0, \pm 2$, and its matrix elements can be found in Ref. 90. An important result is that in the realistic limit of large N, these matrix elements can be written in compact form. In terms of the associated line strength we have

$$S_{vj \to v'j'} = G_j |R^{(2)}_{v'v}|^2 F^{(2)}_{v'v}(j) , \tag{4.113}$$

where

$$
\begin{aligned}
G_j &= \frac{3(j+1)(j+2)}{2(2j+3)} , & (\Delta j = +2, \text{ S branch}) , \\
G_j &= \frac{3(j-1)j}{2(2j-3)} , & (\Delta j = -2, \text{ O branch}) , \qquad (4.114) \\
G_j &= \frac{\alpha^{(0)}}{\alpha^{(2)}}(2j+1) + \frac{j(j+1)(2j+1)}{(2j-1)(2j+3)} , & (\Delta j = 0, \text{ Q branch}) ,
\end{aligned}
$$

and

$$R^{(2)}_{v'v}(v) = \alpha^{(2)}\sqrt{\frac{N}{6}}(\sqrt{N}\delta_{vv'} + \sqrt{v}\,\delta_{v-1,v'} + \sqrt{v+1}\,\delta_{v+1,v'}) , \qquad F^{(2)}_{vv'} = 1 . \tag{4.115}$$

Electric Raman anharmonicities can also be introduced by considering, for example, operators of the form

$$\hat{R}^{(2)} = \sum q_i^{(2)} \exp(-\alpha_i \hat{n}_p)\hat{Q}^{(2)} , \tag{4.116}$$

whose matrix elements can be obtained either numerically or in closed form, in the limit of large N.

Regarding the computation of transition intensities for polyatomic molecules, the three-dimensional algebraic model has only been used for infrared transitions in some triatomic molecules (H_2O, CO_2, HCN) [75, 91]. Although it is possible to extend the aforementioned strategy from diatomic molecules to larger molecules in order to obtain dipole (or quadrupole) operators, from a strictly computational point of view, the situation is rather complex. For HCN we use the total dipole operator

$$\hat{T} = \hat{T}_1 + \hat{T}_2 , \qquad (4.117)$$

where $\hat{T}_1$ and $\hat{T}_2$ are the dipole operators for bonds HC and CN. Both a good and thorough parametrization of these operators (including electric anharmonicities) is given by

$$\hat{T}_i = t_i^{(0)} \hat{D}_i + [\hat{D}_i \hat{t}_i + \hat{t}_i \hat{D}_i] , \qquad i = 1, 2 ,$$

$$\hat{t}_i = [t_i^{(1)} \exp(-\alpha_{i1} \hat{n}_{p1}) \exp(-\alpha_{i2} \hat{n}_{p2}) + t_i^{(2)} \exp(-\beta_{i1} \hat{n}_{p1}) \exp(-\beta_{i2} \hat{n}_{p2})] ,$$

$$(4.118)$$

where the parameters α_{ij}, β_{ij}, and $t_i^{(j)}$ must be determined through a fitting procedure over the observed intensities. Via these operators, one obtains an acceptable, if not excellent description of the experimental infrared spectrum. Further improvements are achieved by considering more complex dipole operators, where the bond Casimir operators are explicitly taken into account. For example, in lieu of the bond dipole operators (4.118) we can use the expansion

$$\hat{T}_i = t_i^{(0)} \hat{D}_i + \hat{Y}[\hat{D}_i \hat{t}_i + \hat{t}_i \hat{D}_i] + [\hat{D}_i \hat{t}_i + \hat{t}_i \hat{D}_i] \hat{Y} , \qquad (4.119)$$

where

$$\hat{Y} \equiv \sum_{k,h} \rho_{kh} (\hat{C}_k)^h \qquad (4.120)$$

is the operator explicitly accounting for smaller (anharmonic) effects in the computation of infrared intensities for certain "difficult" overtone–combination bands. As is clear from the relations above, these computations require determining a large number of arbitrary parameters. Moreover, the computed intensities are extremely sensitive to even very small changes in the parameters. At the present time it is not clear whether such difficulties are related to numerical problems in the computation of the (approximate) matrix elements of the exponential operators or whether the aforementioned dipole operators should be improved further. Nonetheless, with dipole operators much simpler than the operator (4.119) it is possible to reproduce, with high precision, the infrared spectrum of CO_2 (Table IV). We recall that this molecule is affected by strong Fermi resonances, giving rise to nonnegligible mixing of the vibrational wavefunctions. The good agreement between the algebraic calculation and the experimental result is encouraging in that besides providing a reliable model for the electric dipole operator of this molecule, the algebraic model also accounts for the specific Fermi effect of mixing of the wavefunctions.

F. Tetratomic Molecules

The passage from diatomic to triatomic molecules leads to the double-sided world of linear and bent molecular shapes. Although strictly correlated, the respective rovibrational spectra require distinct treat-

TABLE IV

Algebraic Computed Infrared Intensities $(00^00) \rightarrow (\nu_1 \nu_2^{l_2} \nu_3)$ (Σ Bands) in the CO_2 Molecule

| | Energy (cm^{-1}) | Intensity[a] | |
$\nu_1 \nu_2^{l_2} \nu_3$		Expt. IR	Calc.
00^01	2349.2	9550	9550^b
02^01	3612.8	104	170.8
10^01	3714.8	158	219.3
04^01	4853.6	0.778	1.173
12^01	4977.8	3.520	5.302
20^01	5099.6	1.090	1.487
06^01	6075.9	5.23(−3)	5.10(−3)
14^01	6227.9	4.61(−2)	4.99(−2)
22^01	6347.8	4.58(−2)	4.90(−2)
30^01	6503.1	5.93(−3)	4.47(−3)
08^01	7284.0	1.97(−5)	1.87(−5)
16^01	7460.5	3.78(−4)	2.63(−4)
24^01	7594.0	1.02(−3)	6.51(−4)
32^01	7734.0	2.79(−4)	1.75(−4)
40^01	7920.8	1.66(−5)	7.45(−6)
00^03	6972.6	0.146	0.212
02^03	8192.5	4.31(−3)	5.45(−3)
10^03	8294.0	6.13(−3)	3.01(−3)
04^03	9389.0	3.85(−5)	5.64(−5)
12^03	9517.0	2.31(−4)	1.01(−4)
20^03	9631.4	9.04(−5)	1.17(−5)

[a] 10^{-20} cm^{-1}/molec cm^{-2} at 296 K.
[b] Normalized value.

ments, due primarily to their different rotational dynamics. Such a difference is very important in correct formulation of the algebraic model; as a matter of fact, the spectrum-generating algebra $U_1(4) \otimes U_2(4)$ and its dynamic symmetry chains are the same for both linear and bent cases. It is only when we convert the algebraic quantum numbers into rovibrational ones that the molecular equilibrium shape manifests its fundamental role. For three-body systems, the correlation between linear and bent is a very simple and well-established one [71]. What if we consider four-body systems from this same viewpoint of the molecular geometry? Looking at Fig. 40 we see that the situation is becoming more involved; we can have either chain or nonchain molecules (HCCF, H_2CO); linear or half-linear molecules (HCCH, HCNO); nonlinear planar, either chain or nonchain molecules; and nonplanar, either chain or nonchain molecules (HOOH, NH_3). Moreover, discrete symmetries require using numerous point group tables for a proper classification of molecular modes. Consequently, a complete algebraic treatment of tetratomic molecules will find its grounds in the corresponding problem of correlation between rovibrational states in the various configurations shown in Fig. 40. This is, in fact, a fundamental step in obtaining a reliable set of conversion formulas of quantum numbers, going from the purely algebraic world to the rovibrational one. As discussed briefly here, this is a relatively easy task in the simplest case of linear tetratomic molecules;

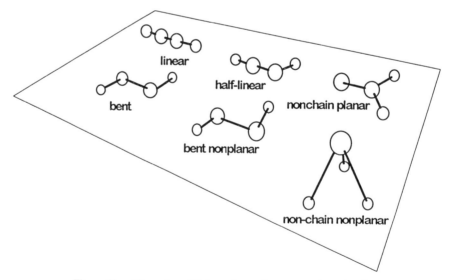

Figure 40.   Allowed equilibrium shapes for tetratomic molecules.

here the algebraic model has been studied carefully and adapted to deal with even very fine structures of the vibrational spectrum [68, 92, 93]. The main point is that the conversion formulas holding for linear triatomic molecules are still valid for linear tetratomic molecules. After all, only one kind of linear molecule exists, which is triatomic, tetratomic, pentatomic, and so on. However, the conversion from linear to nonlinear four-body systems is not well established. Thus we expect to deal with a different algebraic-vibrational model for every nonlinear equilibrium shape. In and of itself, this would not be a major problem. However, from general principles, linear ↔ nonlinear correlation laws are not clearly understood. Without a complete theoretical explanation for the correlation laws connecting different equilibrium shapes, it is perfectly useless to attempt an algebraic study of the rovibrational problem. At the same time, it is important to observed that in a few cases [94], algebraic model did suggest a correct solution to the correlation problem. [An alternative (nonalgebraic) study of this situation is given in Ref. 72.]

These introductory remarks have two distinct purposes. First, the algebraic model *can* be extended to molecules of increasing complexity, even if this can be a cumbersome procedure because of the ambiguous correlation laws or because of the presence of anomalous effects in the observed spectra. Second, the algebraic model *has not* been extended to molecules more complex than a linear tetratomic one. The reason for this is the preference of the one-dimensional model over the three-dimensional model due to the former's simplicity in making calculations. Due to its ease and practicability, the U(2) model is definitely more attractive than the U(4) model, at least from a spectroscopy viewpoint. Nonetheless, we find it both interesting and important to conclude our work with a brief digression into the algebraic analysis of a linear tetratomic molecule. A very detailed account of this subject can be found in Refs. 88 and 95.

The algebraic model of a four-body system is based on the spectrum-generating algebra

$$U_1(4) \otimes U_2(4) \otimes U_3(4) . \tag{4.121}$$

After the assignment of the three vibron numbers N_i, $i = 1, 2, 3$, the Hilbert model space of the physical problem is given by the basic states of the symmetric irreducible representations $[N_1] \otimes [N_2] \otimes [N_3]$ of the SGA (4.121). To specify these basic states unambiguously, we choose a complete subalgebra chain of (4.121). The first step is the usual local (bond) assignment of Morse rovibrating units, $U_i(4) \supset O_i(4)$ ($i = 1, 2, 3$), leading to

$$U_1(4) \otimes U_2(4) \otimes U_3(4) \supset O_1(4) \otimes O_2(4) \otimes O_3(4) . \tag{4.122}$$

The second step is the realization of algebraic couplings. Here we encounter a completely new problem: whether we first want to couple bonds 1 and 2, leaving bond 3 as a bystander, or whether we first couple 1 and 3, or 2 and 3. After having made a decision the rest is easy, since we couple the already linked system $(1 + 2)$ to the third bond, to achieve the overall coupled picture $(1 + 2 + 3)$ or, better, $(12)3$. What if we first couple $1 + 3$? Due to the associative property of the bond combination law, the final result *must be the same*. This can be written symbolically as

$$(1 + 2) + 3 = (1 + 3) + 2 = (2 + 3) + 1 = (1 + 2 + 3) . \qquad (4.123)$$

These relations constitute a typical example of different couplings schemes; even if it is true that the final result $(1 + 2 + 3)$ is the same for any coupling scheme, it is important to consider whether certain couplings are physically more significant than others. The answer to this question can be only given by the specific molecule under study. For example, in the HCCF molecule we expect that the CC + CF coupling is somewhat more important than either HC + CF or HC + CC. In HCCH, for rather obvious reasons of symmetry, it is convenient first to couple (HC + CH). It is very important to understand the actual significance of this initial coupling. In brief, it corresponds to the (algebraic) construction of a triatomic subsection, where the bending vibration, originating from $1 + 2$, must be seen relative to bond 3. If we choose to denote 1 and 2 as the directly coupled bonds, the complete chain of subalgebras obtained from (4.122) can be written as

$$\begin{vmatrix} U_1(4) \otimes U_2(4) \otimes U_3(4) \supset O_1(4) \otimes O_2(4) \otimes O_3(4) \\ N_1 \qquad N_2 \qquad N_3 \qquad \omega_1 \qquad \omega_2 \qquad \omega_3 \end{vmatrix}$$

$$(4.124)$$

$$\left. \begin{array}{c} \supset O_{12}(4) \otimes O_3(4) \supset O_{123}(4) \supset O_{123}(3) \\ (\tau_1, \tau_2) \qquad\qquad (\sigma_1, \sigma_2) \quad j , \end{array} \right\rangle$$

where the overall coupled algebra $O_{123}(4)$ eventually refers to the collective molecular bending vibration, while $O_{123}(3)$ provides the total angular momentum j.

The algebraic branching laws for the quantum numbers of the above ket are obtained readily by using similar equations to those discussed in triatomic problems (Section IV.B). In light of this discussion, the rule for the ω_i's, τ_1, and τ_2 are exactly the same as in the triatomic case [Eqs. (4.10) and (4.12)]; j is obtained by replacing τ_1 and τ_2 by σ_1 and σ_2 in

Eq. (4.11). So we obtain

$$\sigma_1 = \tfrac{1}{2}[|\tau_1 - \omega_3 + \tau_2| + |\tau_1 - \omega_3 - \tau_2|] + \gamma + \delta ,$$

$$\sigma_2 = \tfrac{1}{2}[|\tau_1 - \omega_3 + \tau_2| - |\tau_1 - \omega_3 - \tau_2|] + \gamma - \delta , \qquad (4.125)$$

$$\gamma = 0, 1, \ldots, \min(\tau_1 + \tau_2, \omega_3) , \qquad \delta = 0, 1, \ldots, \min(\tau_1 - \tau_2, \omega_3) .$$

With the local basis, we are ready to construct a triatomic-like Hamiltonian operator where most of the physically relevant interactions should be either diagonal or in the nondiagonal form of the Majorana operator. This is a direct consequence of our choice for the coupling scheme $(1 + 2) + 3$, which is, in fact, done to favor interactions of the type $(1 + 2)$. So for a linear tetratomic molecule we write the following Hamiltonian operator:

$$\hat{H} = A_1 \hat{C}_1 + A_2 \hat{C}_2 + A_3 \hat{C}_3 + A_{12} \hat{C}_{12} + A_{123} \hat{C}_{123} + \lambda_{12} \hat{M}_{12} . \quad (4.126)$$

In this operator the Majorana term originates the off-diagonal elements in the basis (4.124). Such elements are very important for a linear symmetric molecule such as HCCH or DCCD. The purely local part of the Hamiltonian (4.126) has eigenvalues given by

$$E_{\text{local}} = E_{\text{local}}(\omega_1, \omega_2, \omega_3, \tau_1, \tau_2, \sigma_1, \sigma_2) = A_1 \omega_1 (\omega_1 + 2)$$

$$+ A_2 \omega_2 (\omega_2 + 2) + A_3 \omega_3 (\omega_3 + 2) + A_{12}[\tau_1(\tau_1 + 2) + \tau_2^2]$$

$$+ A_{123}[\sigma_1(\sigma_1 + 2) + \sigma_2^2] . \qquad (4.127)$$

We can now introduce local vibrational labels by means of the relations

$$\omega_1 = N_1 - 2v_a , \qquad \omega_2 = N_2 - 2v_b , \qquad \omega_3 = N_3 - 2v_c ,$$

$$\tau_1 = N_{12} - 2v_a - 2v_c - v_e , \qquad \tau_2 = l_e , \qquad (4.128)$$

$$\sigma_1 = N_{123} - 2v_a - 2v_b - 2v_c - v_d - v_e , \qquad \sigma_2 = l_d + l_e ,$$

where, as usual, $N_{12} = N_1 + N_2$ and $N_{123} = N_1 + N_2 + N_3$. Moreover, v_a, v_b, and v_c denote local stretching quantum numbers and v_d and v_e (l_d and l_e for the rotational part) bending quantum numbers. We notice that the expressions above are very similar to those used in the linear triatomic case. This agrees with the previously mentioned universality of the linear geometry. Written as vibrational quantum numbers, the local spectrum

(4.127) becomes

$$E_{local}(v_a, v_b, v_c, v_d^{l_d}, v_e^{l_e})$$

$$= -4A_1 v_a(N_1 + 1 - v_a) - 4A_2 v_b(N_2 + 1 - v_b) - 4A_3 v_c(N_3 + 1 - v_c)$$

$$- A_{12}[(2v_a + 2v_c + v_e)(2N_{12} + 2 - 2v_a - 2v_c - v_e)$$

$$- l_e^2] - A_{123}[(2v_a + 2v_b + 2v_c + v_d + v_e)$$

$$\times (2N_{123} + 2 - 2v_a - 2v_b - 2v_c - v_d - v_e) - (l_d + l_e)^2] . \qquad (4.129)$$

Equation (4.129) can describe, to a first approximation, the spectrum of molecules such as HCCD or HCCF (see also Fig. 41). The same techniques, described in Sections IV.B.1 and IV.B.2, can be used for the explicit calculation of vibrational levels. The experimental and algebraic values of the fundamental bands for HCCF can be related through the following:

$$E_{1000^00} = -4A_3 N_3 - 4A_{123}N_{123} \qquad \text{(CH stretch, 3357.0 cm}^{-1}),$$

$$E_{0100^00} = -4A_2 N_2 - 4A_{12}N_{12} - 4A_{123}N_{123} \quad \text{(CH stretch, 2239.2 cm}^{-1}),$$

$$E_{0010^00} = -4A_1 N_1 - 4A_{12}N_{12} - 4A_{123}N_{123} \quad \text{(CF stretch, 1061.4 cm}^{-1}), \qquad (4.130)$$

$$E_{0001^10} = -2A_{123}N_{123} \qquad \text{(\overset{\frown}{CCH} bend, 583.7 cm}^{-1}),$$

$$E_{0000^011} = -2A_{12}N_{12} - 2A_{123}N_{123} \qquad \text{(\overset{\frown}{CCF} bend, 366.6 cm}^{-1}),$$

which allows for estimates of the algebraic parameters A_1, A_2, A_3, A_{12}, and A_{123}. In a symmetric molecule (such as HCCH), the procedure is slightly different; by necessity $N_1 = N_2$, $A_1 = A_2$, and the local (stretching) modes are now split by the Majorana operator, whose parameter λ_{12}

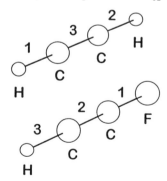

Figure 41. Linear four-atom molecules: notice the different bond numbering adopted according to the $(1 + 2) + 3$ coupling scheme.

in Eq. (4.126) can be determined by the method described in Section IV.B.1 [Eq. (4.48)] for H_2O.

So far, extension of the algebraic technique to tetratomic molecule does not require any substantial change compared to the triatomic case. Nevertheless, let us consider the eventuality that we want to include some kind of direct coupling between bonds 1 and 3 (or 2 and 3). This is, of course, a sensible request, since our original model gives total priority to the coupling $1 + 2$ but the role of the third bond is banished in the overall coupled algebra $O_{123}(4)$, which is dedicated basically to the description of one of the two bending modes. However, it could very well be that the direct physical interactions $(1 + 3)$ or $(1 + 2)$ are of nonnegligible relevance for the correct description of vibrational energies and wavefunctions. Consequently, we are left with the problem of calculating the matrix elements of operators such as $\hat{C}_{13}$, $\hat{C}_{23}$, $\hat{M}_{13}$, and $\hat{M}_{23}$. Although it is, in principle, straightforward, the solution of this problem deserves a thorough examination of our initial choice for the coupling scheme, $(1 + 2) + 3$. In short, any operator of the form $\hat{C}_{ij}$, diagonal in the coupling scheme $(i + j) + h$, is not necessarily diagonal in the coupling schemes $(i + h) + j$ and $(j + h) + i$. This is a well-known result of operator (tensor) calculus in quantum mechanics; for example to couple three angular momenta, starting from

$$\hat{J} = \hat{J}_1 + \hat{J}_2 + \hat{J}_3 , \qquad (4.131)$$

we can construct the wavefunctions in three different ways, depending on whether we first couple $\hat{J}_1$ and $\hat{J}_2$, $\hat{J}_2$ and $\hat{J}_3$, or $\hat{J}_1$ and $\hat{J}_3$. The angular momentum states corresponding to the first coupling are usually denoted by $|(j_1 j_2)j_{12}, j_3; jm_j\rangle$, those associated with the second scheme by $|(j_2 j_3)j_{23}, j_1; jm_j\rangle$, and $|(j_1 j_2)j_{13}, j_2; jm_j\rangle$ for the third coupling scheme. Since each scheme gives the same total angular momentum states jm_j, the three sets of wavefunctions must be related by a transformation, expressed through a recoupling coefficient,

$$|(j_2 j_3)j_{23}, j_1; jm_j\rangle$$

$$= \Sigma_{j12}\langle(j_1 j_2)j_{12}, j_3; j|(j_2 j_3)j_{23}, j_1; j\rangle|(j_1 j_2)j_{12}, j_3; jm_j\rangle , \qquad (4.132)$$

where the coefficient

$$\langle(j_1 j_2)j_{12}, j_3; j|(j_2 j_3)j_{23}, j_1; j\rangle$$

$$= (-)^{j_1 + j_2 + j_3 + j}\sqrt{(2j_{12} + 1)(2j_{23} + 1)}\begin{Bmatrix} j_1 & j_2 & j_{12} \\ j_3 & j & j_{23} \end{Bmatrix} \qquad (4.133)$$

is given in terms of a $6j$ symbol. The situation of the algebraic molecular model is very similar, because we now have to carry out a recoupling procedure explicitly with $O(4)$ basis states in place of the $O(3)$ angular momentum wavefunctions. This is necessary to obtain, in a straightforward way, the matrix elements of operators such as $\hat{C}_{13}$ in the basis $(1+2)+3$ [Eq. (4.124)]. The operator $\hat{C}_{13}$ is diagonal in the basis $(1+3)+2$, since we can write, by recalling Eq. (4.25) for the expectation value of $\hat{C}_{ij}$ in the "right" basis,

$$\langle \omega_1, \omega_3(\rho_1, \rho_2)\omega_2(\sigma_1, \sigma_2)|\hat{C}_{13}|\omega_1, \omega_3(\rho_1, \rho_2)\omega_2(\sigma_1, \sigma_2)\rangle$$

$$= \rho_1(\rho_1 + 2) + \rho_2^2, \tag{4.134}$$

that is, we introduce the new (algebraic) coupling scheme

$$\left|
\begin{array}{cccccc}
U_1(4) \otimes & U_2(4) \otimes & U_3(4) \supset & O_1(4) \otimes & O_2(4) \otimes & O_3(4) \\
N_1 & N_2 & N_3 & \omega_1 & \omega_2 & \omega_3
\end{array}
\right.$$

$$\left.
\begin{array}{ccc}
\supset O_{13}(4) \otimes & O_2(4) \supset O_{123}(4) \supset O_{123}(3) \\
(\rho_1, \rho_2) & (\sigma_1, \sigma_2) \quad j \,,
\end{array}
\right\rangle \tag{4.135}$$

whose basis states are denoted by

$$|\omega_1, \omega_3(\rho_1, \rho_2)\omega_2(\sigma_1, \sigma_2)\rangle . \tag{4.136}$$

Let us introduce the following shorthand notation for basis states, corresponding to different coupling schemes:

$$|\omega_1, \omega_2(\tau_1, \tau_2)\omega_3(\sigma_1, \sigma_2)\rangle \equiv |(12)3\rangle ,$$

$$|\omega_1, \omega_3(\rho_1, \rho_2)\omega_2(\sigma_1, \sigma_2)\rangle \equiv |(13)2\rangle . \tag{4.137}$$

In light of our previous discussion, a recoupling transformation coefficient $\mathscr{C}$ must exist such that

$$|(12)3\rangle = \sum \mathscr{C}_{(13)2 \to (12)3}|(13)2\rangle . \tag{4.138}$$

If this is the case, we must also have

$$\langle (12)3'|\hat{C}_{13}|(12)3\rangle$$

$$= \sum \mathscr{C}_{(13)2 \to (12)3} \mathscr{C}_{(13)2' \to (12)3'} \langle (13)2'|\hat{C}_{13}|(13)2\rangle . \tag{4.139}$$

A detailed study of $O(4)$ transformation properties leads to an exact formula which is not explicitly shown. However, the important points are that (1) in the exact formula, the recoupling coefficient is written as

$$\mathcal{C}_{(13)2\to(12)3} = \langle \omega_1, \omega_2(\tau_1, \tau_2)\omega_3(\sigma_1, \sigma_2)|\omega_1, \omega_3(\rho_1, \rho_2)\omega_2(\sigma_1, \sigma_2)\rangle$$

$$= \sqrt{(\tau_1 + \tau_2 + 1)(\tau_1 - \tau_2 + 1)(\rho_1 + \rho_2 + 1)(\rho_1 - \rho_2 + 1)}$$

$$\times \left\{ \begin{matrix} \dfrac{\omega_2}{2} & \dfrac{\omega_1}{2} & \dfrac{\tau_1 + \tau_2}{2} \\ \dfrac{\omega_3}{2} & \dfrac{\sigma_1 + \sigma_2}{2} & \dfrac{\rho_1 + \rho_2}{2} \end{matrix} \right\} \left\{ \begin{matrix} \dfrac{\omega_2}{2} & \dfrac{\omega_1}{2} & \dfrac{\tau_1 - \tau_2}{2} \\ \dfrac{\omega_3}{2} & \dfrac{\sigma_1 - \sigma_2}{2} & \dfrac{\rho_1 - \rho_2}{2} \end{matrix} \right\},$$

(4.140)

and (2) in Eq. (4.138) the matrix elements of $\hat{C}_{13}$, in the *wrong* basis $|(12)3\rangle$, are calculated in the *right* basis $|(13)2\rangle$, where $\hat{C}_{13}$ is diagonal. This same method can obviously be applied to calculation of the Majorana operators $\hat{M}_{13}$ and $\hat{M}_{23}$; their matrix elements are easily obtained in the *right* bases $|(13)2\rangle$ and $|(23)1\rangle$ by using Eqs. (4.27)–(4.29) and then recoupled by means of (4.138) in the *wrong* basis $|(12)3\rangle$.

Another interesting effect, arising from the analysis of tetratomic molecules and addressed successfully within the algebraic framework, is the vibrational l doubling of bending modes. The simplest case is found in the combination band of the two fundamental bending Π modes $v_4^{l_4} = v_5^{l_5} = 1^1$. These modes originate three sublevels of different angular momentum, according to the following scheme:

$$\Pi \times \Pi: |0001^1 1^1\rangle \mapsto \begin{cases} \Delta \ (l = l_4 + l_5 = +1 + 1, -1 - 1 = \pm 2) ; \\ \Sigma \ (l = l_4 + l_5 = +1 - 1, -1 + 1 = 0) . \end{cases}$$

(4.141)

In the conventional force-field treatment, these sublevels are split by the interaction (often referred to as Amat–Nielsen coupling [96])

$$\langle v_4^{l_4} v_5^{l_5} | \hat{V}_{AN} | v_4^{l_4 \pm 2} v_5^{l_5 \mp 2} \rangle$$

$$= \mathscr{A} \sqrt{(v_4 \pm l_4 + 2)(v_4 \mp l_4)(v_5 \mp l_5 + 2)(v_5 \pm l_5)} .$$

(4.142)

The splitting produced by the Amat–Nielsen interaction, shown schematically in Fig. 42 for the HCCH molecule, can also be described in an algebraic fashion by operators of the type $\hat{C}'_{ij}$. As a mater of fact, $\hat{C}'_{12}$ was used in linear triatomic molecules to account for Σ/Δ splitting (see Section IV.B.2). In linear tetratomic molecules, the operators $\hat{C}'_{12}$ and

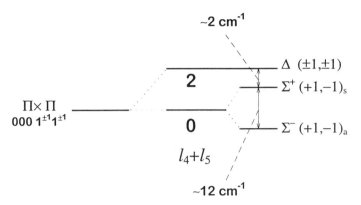

Figure 42. Amat–Nielsen splitting induced on the 0001^11^1 combination band of the HCCH molecule. Energy axis not on scale.

$\hat{C}'_{123}$ are diagonal in the local basis (4.124) and they lead to the separation of levels with different total vibrational angular momentum, labeled by $l_4 + l_5$ [i.e., the combination bands Σ $(+1 - 1, -1 + 1)$ and $\Delta(+1 + 1, -1 - 1)$]. The operators $\hat{C}'_{23}$ and $\hat{C}'_{13}$ are *not* diagonal in the scheme $(1 + 2) + 3$, but their matrix elements can readily be obtained with the recoupling method explained above. The final result is that these operators have the correct behavior, in that they mix symmetric–antisymmetric linear combinations of the Σ sublevels $1^{+1}1^{-1}$ and $1^{-1}1^{+1}$, which are also split into states Σ^+ and Σ^-, as observed experimentally. In Table V we report direct comparison between some matrix elements of $\hat{C}'^{2}_{13}$ and those of the Amat–Nielsen operator [Eq., (4.142)]. However, we should comment on the fact that HCCH, besides being a symmetric molecule, has a particularly well-behaved nature from the dynamic symmetry viewpoint. A similar analysis carried out on the HCCF molecule leads to several disparate problems. Just to mention some, (1) the $\hat{C}'_{13}$ and $\hat{C}'_{23}$ operators work as expected, but after the removal of their diagonal matrix elements; (2) quite a large set of Fermi resonances affect the rovibrational spectrum; and (3) the stronger normal behavior of HCCF in comparison to the more local acetylene molecule requires careful inclusion of several Majorana terms and/or higher-order operators in the Hamiltonian. However, all three of these problems have been solved successfully [68, 93].

This basically concludes the presentation of the current state of three-dimensional models in molecular spectroscopy. Needless to say, from an algebraic standpoint many aspects still must be considered to compete this topic. First, molecular equilibrium shapes other than the linear

TABLE V
Selected Matrix elements of the Amat–Nielsen and Algebraic l-Doubling Operators

ϕ_a	ϕ_b	$\langle \phi_a |$ Amat–Nielsen $| \phi_b \rangle$	$\langle \phi_a |$ Algebraic $| \phi_b \rangle$
		Σ Bands	
$000\ 1^{+1}1^{-1}$	$000\ 1^{-1}1^{+1}$	1.000	1.000^a
$000\ 3^{+1}1^{-1}$	$000\ 3^{-1}1^{+1}$	2.000	1.953
		Π Bands	
$000\ 2^{+2}1^{-1}$	$000\ 2^{0}1^{+1}$	1.414	1.398
$000\ 4^{+2}1^{-1}$	$000\ 4^{0}1^{+1}$	2.449	2.364
		Δ Bands	
$000\ 2^{+3}1^{-1}$	$000\ 3^{+1}1^{+1}$	1.732	1.691
$000\ 5^{+3}1^{-1}$	$000\ 5^{+1}1^{+1}$	2.828	2.697
		Φ Bands	
$000\ 4^{+2}1^{+1}$	$000\ 4^{+4}1^{-1}$	2.000	1.930

^{a}Normalized value.

geometry have to be studied in detail. A preliminary attempt can be found in Ref. 94 for the analysis of a slightly nonlinear geometry. The problem of real nonlinear shapes, nonchain molecules, or even worse, nonplanar molecules is yet to be addressed within the algebraic model. Second, the rotational spectrum is also far from being clearly described in algebraic terms. Although the one-dimensional model can be of great help in providing a unified and simple picture for overall vibrational behavior, it is only through the full three-dimensional model that we can have complete access to the rotational part of the spectrum. For these reasons we still believe that despite its sometimes worrisome mathematical aspects, the three-dimensional algebraic model is a very powerful tool for solving very complex spectroscopic problems of current interest.

V. FURTHER ASPECTS

In this part we discuss briefly some additional aspects of the algebraic approach to problems in molecular spectroscopy. However, given the intentionally introductory nature of this article, it is impossible to address all areas in this field; in fact, this article is only intended to interest spectrocopists in an algebraic formulation of the rovibrational spectroscopy of small and medium-sized molecules. Moreover, this paper should signal that new routes are available for understanding the most difficult features of the spectroscopic landscape. Further references on the algebraic approach can be found in Iachello and Levine [11]. However, it is worthwhile to conclude with a brief discussion of some arguments that can clearly highlight the power of the algebraic approach. As an example of its powerful capabilities in describing a complex system, we outline in Section V.A, the algebraic study of the vibrational spectrum of the benzene dimer.

A question frequently asked by traditional spectroscopists is: Where is the potential surface? Indeed, in many cases we could live equally well *without* knowing the exact potential surface of a given molecule arranged in an algebraic framework. As amply described, the potential energy is embedded in the algebraic formulation without being an explicitly required ingredient for the calculation of spectra and transition intensities. Yet, either for comparison or for more specifically technical reasons, one may want to know the potential surface associated with the spectroscopic problem at hand. Since the algebraic model can provide the complete rovibrational spectrum, one could try to set out the classical inverse problem for determination of the potential surface starting from the known spectrum. Actually, we should include the continuum part of the spectrum, which can be obtained, in principle, by using noncompact groups [25]. However, an alternative efficient method can be formulated through the semiclassical limit of algebraic dynamic theory, by means of which a considerable amount of information on the underlying classical interactions can be extracted by starting from algebraic parameters. We discuss some basic aspects of this method in Section V.B.

The purported ease of calculating various quantities via the algebraic method has been cited throughout this article. As such, this ease of computation deserves further comment and explanation. Therefore, we outline in Section V.C the computer routines used most commonly in the algebraic model.

A. Case Study: The Benzene Dimer

As stated earlier on several occasions, the algebraic method should not be viewed on a mere mimicking of other well-established approaches to solving molecular spectroscopy problems. However, one could have just such an impression if the problems are limited to very simple cases that can be addressed equally well by traditional methods and do not carry the embarrassing burden of Lie algebras or Racah's tensor calculus. Nonetheless, every introductory article must start with simple examples and only then proceed to more complex ones. Sections III.C.2 and IV.B reveal the algebraic approach as capable of providing reliable and alternative solutions to nontrivial questions. Here *alternative* means "in a faster way and with fewer arbitrary parameters." In this section we basically consolidate the essential features of an algebraic model to describe a simple portion of the vibrational spectrum of $(C_6H_6)_2$ (i.e., the dimer of the benzene molecule).

Weakly bound aromatic complexes are currently under intensive spectroscopic investigation. A detailed understanding of weakly interacting carbon rings such as benzene or pyrrole is important both for organic chemistry problems and for proper characterization of intersite (van der Waals) rovibrational modes. From this standpoint the dimer has been one of the most studied compounds [97–99]. However, it presents quite a formidable problem, as the dimer has 24 interacting atoms, leading to 60 fundamental vibrations and six van der Waals modes. There are three aspects to consider in this study: (1) *site shifts* (i.e., changes in the force constants of the individual monomer due to the influence of the other), (2) *excitation exchange* (i.e., the interaction between modes of the different sites), and (3) the existence of van der Waals rovibrational modes. Although we address only the first two aspects, the third could also be discussed within an algebraic formulation through the "floppy" three-dimensional model introduced in Section II.C.2. We now apply the one-dimensional model to discuss the most important features of site shifts and excitation exchange in the benzene dimer. A comprehensive examination of this problem can be found in Ref. 100.

First, one must consider the accepted model of the geometric configuration of the dimer, shown in Fig. 43 and referred to as the *T-shaped dimer*. The geometric picture is of crucial importance for a correct arrangement of the algebraic model and for comparison with the (few

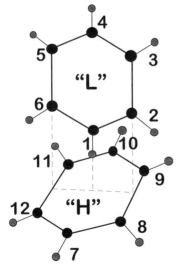

Figure 43. T-shaped model of the benzene dimer.

existing) experimental data on the dimer. The T-shaped configuration is accepted for gas-phase dimers, while different models (such as the *stacked dimer*; see below) are used for studying solid samples. Figure 43 shows that the two monomers are distinguishable by virtue of their different arrangement in the overall molecular structure. For site L its CH bond 1 (pointing downwards) is affected differently by the presence of the other site than are bonds 2–6. The corresponding force constants must then change such that in place of six equal constants, k, we have

$$k_1, \quad k_2 = k_6, \quad k_3 = k_5, \quad k_4. \tag{5.1}$$

The same conclusions apply to the force constants of CH bends as well as CC modes. By virtue of its immersion in the environment of the other site, the point symmetry of L changes from D_{6h} to C_{2v}. If we now consider site H, it is simple to realize that the CH bond force constants must satisfy

$$k_7 = k_9 = k_{10} = k_{12}, \quad k_8 = k_{11}. \tag{5.2}$$

This means that the point symmetry of H changes from D_{6h} to D_{2h} ("L" and "H" stand for "low-" and "high"-symmetry site, respectively, as should now be evident). The first step in the construction of the benzene dimer is to modify force constants of the two sites according to Eqs. (5.1) and (5.2). A similar study can be easily and systematically implemented within the one-dimensional algebraic model by taking the Hamiltonian operator for CH stretching modes of the benzene molecule (Section III.C.2),

$$\hat{H} = A \sum_{i=1}^{6} \hat{C}_i + A' \sum_{i<j=1}^{6} \hat{C}_{ij} + \sum_{i<j=1}^{6} \lambda_{ij} \hat{M}_{ij}, \tag{5.3}$$

and by computing its eigenvalues and eigenvectors, as a function of different values of the various algebraic parameters. To be more precise, let us consider the effects of the reduced symmetry environment on the L site. According to Eq. (5.1), the Hamiltonian operator (5.3) should now be written as

$$\hat{H}_{\text{``L''}} = A_1 \hat{C}_1 + A_{2=6}(\hat{C}_2 + \hat{C}_6) + A_{3=5}(\hat{C}_3 + \hat{C}_5) + A_4 \hat{C}_4$$

$$+ A' \sum_{i<j=1}^{6} \hat{C}_{ij} + \sum_{i<j=1}^{6} \lambda_{ij} \hat{M}_{ij}, \tag{5.4}$$

where, to the first approximation, we neglect possible modifications in the coupling operators $\hat{C}_{ij}$ and $\hat{M}_{ij}$. The diagonalization of this Hamiltonian leads automatically to states carrying irreducible representations of the point group C_{2v}. Figure 44 shows the effect on energy positions and on infrared and Raman intensities resulting from varying the parameter A_1 (starting from the unperturbed value, corresponding to the individual benzene molecule). A similar study can be done for variations of the Majorana parameters λ_{ij}. Regarding infrared and Raman transitions intensities, the corresponding dipole and quadrupole operators are also modified by symmetry breaking, but this is a second-order effect. Rather, we focus on the changes in intensities due to modifications of the vibrational wavefunctions. We can see in Fig. 44 that in addition to explaining the splitting of energy levels in a very definite way, symmetry breaking results in some of the infrared inactive modes of the monomer (i.e., A_{1g}, E_{2g}, and B_{1u} modes) now becoming active. This effect is seen more dramatically for Raman transition intensities. A completely similar study can also be carried out for the H site. In this case we obtain the behavior of energies and transition intensities under lowering of symmetry $D_{6h} \rightarrow D_{2h}$. As a result of this preliminary analysis of the two moieties, the suitability of the algebraic model in emphasizing the "fingerprint" left by the lowering of symmetry on energy spectra and intensities is clearly discovered.

The next step is to introduce a specific interaction between the two sites. This can be done very easily within the algebraic framework. If examining Fig. 44, we expect that bond 1 of site L will be coupled with equal strength to the six bonds of site H; bonds 2–6 will be only very weakly coupled with site H. For the fundamental polyad of CH stretches of both sites the final result is given by the 12×12 Hamiltonian matrix shown in Fig. 45. In this figure we see that the blocks along the main diagonal are 6×6 matrix representations of the single-site Hamiltonian operators of the type (5.4), while the off-diagonal blocks contain the excitation exchange information in algebraic form. We can also study the effect on vibrational energies and intensities of the Hamiltonian of Fig. 45 as a function of variations of the intersite coupling x. For $x = 0$ (no excitation exchange) we simply double the spectra of the two sites; however, for even very small values of x, we expect to introduce nontrivial features in the computed spectrum. As an example, let us consider the CH "breathing" mode $v_2(A_g)$ of the single site. From experimental studies two slightly different peaks are found to be separate by about $2.5\,\mathrm{cm}^{-1}$ (Fig. 46). If the dimer had two equivalent sites, we would just observe a single transition instead of two because the excitation exchange would lead to a symmetric (Raman active) mode and an antisymmetric (Raman inactive) combination of the A_{1g} single-site

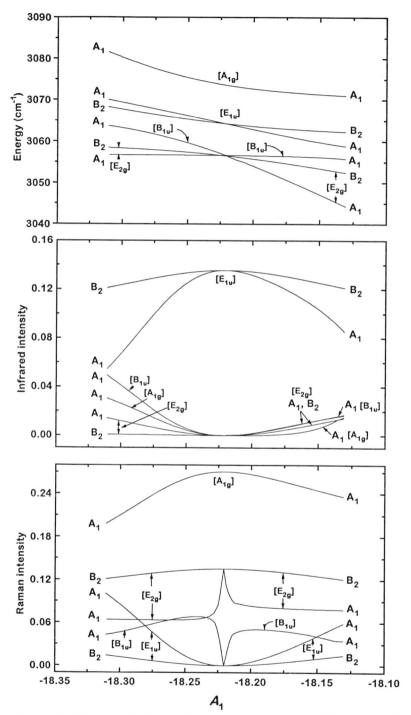

Figure 44. Behavior of vibrational energies and infrared and Raman transition intensities as a function of A_1, under lowering of symmetry $D_{6h} \rightarrow C_{2v}$ in the benzene dimer, site L. States are labeled by D_{6h} species (in brackets) and C_{2v} species. $A_1 = -18.2238$ in benzene.

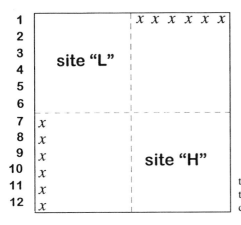

Figure 45. Hamiltonian matrix of the fundamental CH stretching modes of the T-shaped model of the benzene dimer.

modes. Since this is not the case, we have to support a T-shaped model of inequivalent sites. The algebraic model can now reproduce precisely, in the 12-dimensional local basis of two sets of CH benzene oscillators, energy positions, splitting, and intensities of the doubled breathing mode of Fig. 46. To accomplish this, one needs to introduce an excitation exchange of about $0.11 \, \text{cm}^{-1}$ and to modify the algebraic parameters of site L by 2.4% (bond 1), 1.1% (bonds 2 and 6), and 0.2% (bonds 3–5). The algebraic parameters of site H are modified only to lower its symmetry from D_{6h} to D_{2h}. An overall decrease of about 1% for all the algebraic force constants has also been applied to account for the small (but expected) chemical shift effect of the modified molecular environment for both sites. A similar analysis has also been carried out for the infrared spectrum, where the E_{1u} active mode of the benzene molecule is double due to the excitation exchange, in addition to being affected, in a complicated way, by Fermi resonances (see also Section III.E).

B. Geometric Interpretation of Algebraic Models

In the dynamic symmetry formulation of one- and three-dimensional problems considered so far, kinetic and potential energy functions do not play an active role because they simply do not appear in the algebraic Hamiltonian operator. However, as stated previously, it is possible to return to the potential energy surface by a direct study of the inverse problem, where one uses the (algebraically obtained) rovibrational energy spectrum to compute the corresponding potential function in a convenient geometric space of coordinates [101, 102]. Nonetheless, in practice, the calculation can be very difficult. In this section we discuss a different way to obtain potential surfaces of molecules described within the algebraic model. This technique, however, constitutes a very long and

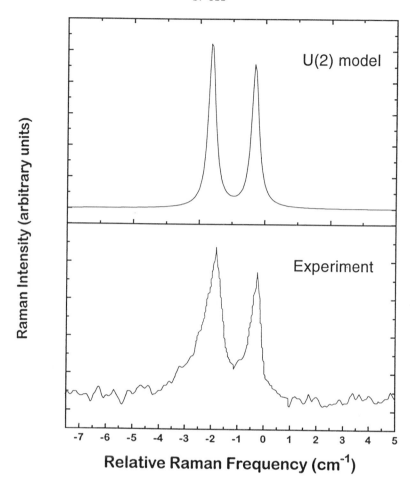

Figure 46. Comparison of calculated (*top*) and observed (*bottom*) Raman intensities of the benzene dimer in the region of the breathing mode v_2 (A_{1g}) of the benzene molecule.

involved chapter of mathematical physics. Consequently, this section is limited to a quick introduction to this problem.

To begin with, we recall that in certain cases, the algebraic model has been already put in a one-to-one correspondence with a specific potential function for the usual space coordinates. We have already studied dynamic symmetries providing exact solutions for the one-, two-, and three-dimensional truncated harmonic oscillators, the Morse and Pöschl–Teller potential functions. When we consider more complicated algebraic expansions in terms of Casimir operators, or when we deal with coupled oscillators, the corresponding potential function must be calculated in an appropriate way. An existing mathematical technique allows one to associate, in a systematic fashion, a geometric space to a given algebraic structure. The properties of this geometric space are suitable for construction of a representative space of (eventually) conventional (either action-angle or space-momentum) coordinates. This "coset" space [103] is intimately related to the starting algebraic structure and leads to classical limits of boson operators. In one such limit, this different realization of the system's quantum states is called a *coherent state realization* [35, 104]. It is possible to show that the expectation values of the algebraic Hamiltonian operator, in the coherent state basis, provide the classical Hamiltonian, expressed as a function of the coset space variables. The calculation of such expectation values can be carried out explicitly for any operator of interest, but this can be a rather difficult procedure, for purely algebraic technical reasons. Here we discuss a simple way for obtaining, to order $1/N$, the classical Hamiltonian for even complex algebraic situations. This method, introduced by Gilmore [105] and nicely reviewed in Refs. 11 and 106, makes use of classically commuting quantities, referred to as *intensive boson operators*. Their construction is based on the exact commutation relations of quantum boson operators $\hat{s}$ and $\hat{t}$ ($\hat{s}^{\dagger}$ and $\hat{t}^{\dagger}$) considered in Section II.C.1 (for one-dimensional problems) and boson operators $\tilde{s}$ and $\tilde{p}_{\mu}$ ($\hat{s}^{\dagger}$ and $\hat{p}^{\dagger}_{\mu}$) of Section II.C.2 (for three-dimensional problems). If we divide the one-dimensional operators by $\sqrt{N}$, we obtain

$$\left[\frac{\hat{s}}{\sqrt{N}}, \frac{\hat{s}^{\dagger}}{\sqrt{N}}\right] = \frac{1}{N}, \qquad \left[\frac{\hat{t}}{\sqrt{N}}, \frac{\hat{t}^{\dagger}}{\sqrt{N}}\right] = \frac{1}{N}. \tag{5.5}$$

We thus recover commuting operators in the limit $N \to \infty$. Let us now introduce the two complex quantities α and β, defined as

$$\alpha = \frac{\hat{s}}{\sqrt{N}}, \qquad \beta = \frac{\hat{t}}{\sqrt{N}} \tag{5.6}$$

with the complex conjugate

$$\alpha^{*} = \frac{\hat{s}^{\dagger}}{\sqrt{N}}, \qquad \beta^{*} = \frac{\hat{t}^{\dagger}}{\sqrt{N}}. \tag{5.7}$$

It is possible to show that replacement of the operators $\hat{s}$ and $\hat{t}$ ($\hat{s}^{\dagger}$ and $\hat{t}^{\dagger}$) according to Eqs. (5.6) and (5.7) in the algebraic Hamiltonian is equivalent to averaging $\hat{H}$ itself in a classical sense (i.e., in the coherent state basis). The replacement of $\hat{s}$ and $\hat{t}$ ($\hat{s}^\dagger$ and $\hat{t}^\dagger$) operators with α and β (and α^* and β^*) actually depends on only two quantities, since the conservation of the total boson number leads to the relations

$$\hat{t}^\dagger \hat{t} + \hat{s}^\dagger \hat{s} = |\alpha|^2 + |\beta|^2 = 1 , \tag{5.8}$$

moreover, in which the overall phase cannot be determined. Consequently,

$$\alpha = \sqrt{1 - |\beta|^2} . \tag{5.9}$$

Let us consider the simple example where the algebraic Hamiltonian operator is of the dynamic symmetry type $U(2) \supset U(1)$. Thus we have

$$\hat{H} = E_0 + A_1 \hat{C}^{(1)}_{U(1)} + A_2 \hat{C}^{(2)}_{U(1)} , \tag{5.10}$$

where $\hat{C}^{(1)}_{U(1)} = \hat{t}^\dagger \hat{t}$. By applying the replacement of Eqs. (5.6) and (5.7), we obtain

$$H_{cl} = H_{cl}(\beta, \beta^*) = E_0 + A_1 N |\beta|^2 + A_2 N^2 |\beta|^4 . \tag{5.11}$$

We can introduce a more familiar form of this Hamiltonian by considering the canonical transformation

$$\beta = \frac{q + ip}{\sqrt{2}} , \qquad \beta^* = \frac{q - ip}{\sqrt{2}} , \tag{5.12}$$

where (q, p) are the usual phase space (one-dimensional) coordinates. With this transformation the Hamiltonian (5.11) reads

$$H_{cl} = H_{cl}(q, p) = E_0 + \frac{A_1 N}{2}(q^2 + p^2) + \frac{A_2 N^2}{4}(q^2 + p^2)^2 . \tag{5.13}$$

Let us take, for a moment, $E_0 = A_2 = 0$. The resulting Hamiltonian is

$$H_{cl}(q, p) = k(q^2 + p^2) , \qquad k = \frac{A_1 N}{2} , \tag{5.14}$$

which corresponds to the classical Hamiltonian of the one-dimensional harmonic oscillator. From this simple case, we can trivially extract, from the Hamiltonian function, the kinetic and potential energy contributions. However, if we return to the case $A_2 \neq 0$ [Eq. (5.13)], such a direct separation is not possible anymore. This means that the Hamiltonian can only be written in the general form

$$H_{cl}(q, p) = \frac{p^2}{2\mu} + V(q) , \tag{5.15}$$

where the reduced mass μ depends, in an effective way, on q and p, and the potential function $V(q)$ is defined as

$$V(q) \equiv H_{cl}(q, p = 0) . \tag{5.16}$$

In the previous example, we therefore obtain

$$V(q) = \frac{q^2 N}{2} \left(A_1 + \frac{A_2 N}{2} q^2 \right) ;$$

$$\mu = \mu(q, p) = \frac{2}{N[2A_1 + A_2 N(p^2 + 2q^2)]} . \tag{5.17}$$

The second example is a $U(2) \supset O(2)$ dynamic symmetry Hamiltonian operator,

$$\hat{H} = E_0 + A\hat{C}_{O(2)}^{(2)} , \tag{5.18}$$

where $\hat{C}_{O(2)}^{(2)} = (\hat{t}^\dagger \hat{s} + \hat{s}^\dagger \hat{t})^2$. So we obtain, by applying Eqs. (5.6)–(5.9),

$$H_{cl} = H_{cl}(\beta, \beta^*) = E_0 + AN^2(1 - |\beta|^2)(\beta + \beta^*)^2 . \tag{5.19}$$

This can be written in the (q, p) space as

$$H_{cl} = H_{cl}(q, p) = E_0 + AN^2 q^2 (2 - q^2 - p^2) , \tag{5.20}$$

which corresponds to the potential function

$$V(q) = H_{cl}(q, p = 0) = E_0 + AN^2 q^2 (2 - q^2) . \tag{5.21}$$

The Morse potential function can be recovered by means of the further canonical transformation

$$q^2 = \exp[-\beta(r - r_e)] , \tag{5.22}$$

where r is the one-dimensional bond coordinate. By transforming q^2 in Eq. (5.21), according to Eq. (5.22), we obtain

$$V(r) = E_0 + AN^2 \{2 \exp[-\beta(r - r_e)] - \exp[-2\beta(r - r_e)]\} . \tag{5.23}$$

So we see that this method allows one to relate the strength of the Morse potential with the algebraic parameters A, N (and E_0). We also note that β and r_e cannot be determined directly using this technique. The equilibrium coordinate r_e is simply not relevant in one-dimensional problems, while β can be obtained by considering excited states of the Hamiltonian operator [11].

As a further example of the classical limit, let us now consider the algebraic description of two coupled oscillators. Here we study the simplest problem of two one-dimensional oscillators, given in terms of the Hamiltonian operator

$$\hat{H} = E_0 + A_1 \hat{C}_1 + A_2 \hat{C}_2 + A_{12} \hat{C}_{12} + \lambda_{12} \hat{M}_{12} . \tag{5.24}$$

The geometrical interpretation is constructed through a two-dimensional coset space spanned by two complex variables

$$\beta_k = \frac{q_k + ip_k}{\sqrt{2}}, \qquad k = 1, 2 . \tag{5.25}$$

In the language of intensive boson operators, we simply have to double the definitions (5.6) and (5.7) to account for two families of boson operators $\hat{s}_k$ and $\hat{t}_k$ ($\hat{s}_k^\dagger$ and $\hat{t}_k^\dagger$), $k = 1, 2$. This method, when applied to Eq. (5.24), leads to the following potential surface:

$$V(q_1, q_2) = E_0 + (A_1 + A_{12})N_1^2 q_1^2 (2 - q_1^2) + (A_2 + A_{12})N_2^2 q_2^2 (2 - q_2^2)$$
$$+ 2A_{12}N_1N_2 q_1 q_2 \sqrt{(2 - q_1^2)(2 - q_2^2)} + 2\lambda_{12}N_1N_2 q_1^2 q_2^2 , \tag{5.26}$$

which, by using the transformation of bond coordinates $q_k^2 = \exp[-\beta_k(r_k - r_{e_k})]$, $k = 1, 2$, can also be written as

$$V(r_1, r_2) = E_0 + a_1 V_M(r_1) + a_2 V_M(r_2) + a_{12}\sqrt{V_M(r_1)V_M(r_2)} + b_{12}V_{12}(r_1, r_2) ,$$

$$a_1 = (A_1 + A_{12})N_1^2 , \qquad a_2 = (A_2 + A_{12})N_2^2 , \tag{5.27}$$

$$a_{12} = 2A_{12}N_1N_2 , \qquad b_{12} = 2\lambda_{12}N_1N_2 ,$$

where $V_M(r)$ denotes the "unit" Morse potential function,

$$V_M(r) = \{2 \exp[-\beta(r - r_e)] - \exp[-2\beta(r - r_e)]\} \tag{5.28}$$

and $V_{12}(r_1, r_2)$ correspond to the classical limit of the Majorana operator.

It is given by

$$V_{12}(r_1, r_2) = \exp[-\beta_1(r_1 - r_{e_1}) - \beta_2(r_2 - r_{e_2})] \, . \tag{5.29}$$

This expression can be obtained by applying the intensive boson method to the invariant operator of the coupled algebra $U_{12}(2)$.

As a final example of application of the intensive boson operator technique to the one-dimensional algebraic model, we consider the case of the $n{:}m$ Fermi operator introduced in Section III.E [Eq. (3.128)]. A straightforward use of the aforementioned method leads to the classical potential surface

$$V(q_a, q_b) = \mathscr{F}[q_a^2(2 - q_a^2)]^{n/2}[q_b^2(2 - q_b^2)]^{m/2} \, , \tag{5.30}$$

where q_a and q_b are space coordinates, representative of the two Fermi-interacting degrees of freedom. Once again, in terms of conventional bond coordinates r_a and r_b, we can also write Eq. (5.30) as

$$V(r_a, r_b) = \mathscr{F}\sqrt{V_M^n(r_a)V_M^m(r_b)} \, , \tag{5.31}$$

where V_M is the Morse function of Eq. (5.28). The classical potential surface for Fermi couplings, involving stretch–bend interactions, can be obtained through a similar procedure, the only difference being in the canonical transformation (5.22). By starting from the potential for a single oscillator [Eq. (5.21)], we observe that the transformation

$$q^2 = 1 + \tanh[\chi_e(\theta - \theta_e)] \tag{5.32}$$

leads to the potential function

$$V_{PT}(\theta) = \frac{N}{\cosh^2[\chi_e(\theta - \theta_e)]} \, . \tag{5.33}$$

This is the precise analytical form of the one-dimensional Pöschl–Teller potential, suited for bending vibrations. The corresponding $n{:}m$ Fermi interactions is thus given by

$$V(r, \theta) = \mathscr{F}\sqrt{V_M^n(r)V_{PT}^m(\theta)} \, . \tag{5.34}$$

Three-dimensional problems can also be addressed within the framework of the intensive boson operators method, by simply introducing, in place of the scalar quantities α and β of Eq. (5.6), two (complex) scalars and two vectors associated with the scalar and vector boson operators of $U(4)$, respectively. In this way it is possible to obtain the classical limit of three-dimensional algebraic Hamiltonian operators, as discussed in Ref. 106. Further work on more sophisticated approaches to the classical limit, such as those done with the mean-field technique, can be found in Ref. 107.

C. Algebraic Methods and Computer Routines

Throughout this paper, we have seen that algebraic techniques often provide extremely simple numerical results with small computational effort. This is particularly true in the preliminary phases of one-dimensional calculations, where almost trivial relations can be found for the initial guesses for the algebraic parameters, as shown in Sections II.C.1 and III.C.2. However, it is also true that as soon as real calculations of more complex vibrational spectra are requested, the problem of adapting the various algebraic Hamiltonian and transition operators to suitable computer routines must be resolved. The construction of a computer interface between theoretical models and numerical results is absolutely necessary. Nonetheless, it is rather atypical to discuss these problems explicitly in a theoretical paper such as this one. However, the novelty of these methods itself justifies further explanation and comment on the computational procedures required in practical applications. In this section we present only a brief outline of the development of algebraic software in the last few years, as well as the most peculiar situations one expects to encounter.

First, *the* computer program for algebraic molecular calculations does not exist. However, it is possible to collect a list of specific routines for numerically addressing the most typical steps characteristic of an algebraic analysis of a rovibrational spectra. Also, given the distinct nature of any given molecule, it is impossible to have one, generalized computer program. Despite that, VIBR3AT, a computer program based on the three-dimensional model, has been developed for triatomic molecules. The ultimate aim is to extend this program to do calculations on any kind of molecule regardless of its symmetry or dynamic behavior. The documentation of this program, as well as the most commonly used strategies, can be found in Ref. 76. It is also possible to obtain this program by accessing the anonymous internet node `itncpl.science.unitn.it` (`130.186.34.12`), directory `/pub/physics/vibron`. VIBR3AT is a typical example of an algebraic programming technique. Within its code, all the standard procedures and routines required to implement numerically the equations discussed mainly in Sections IV.B.1 and IV.B.2 can be found. The most important steps are the following:

1. Construction of the algebraic local basis and its conversion in terms of vibrational quantum numbers
2. Construction of the Hamiltonian operator, including both diagonal and nondiagonal contributions
3. Fitting procedure of the algebraic parameters over the experimental database of vibrational levels
4. Calculation of the vibrational spectrum extended to hitherto unknown levels

The first step can be addressed from a purely group theoretical viewpoint [i.e., using formulas such as (4.12)]. In lieu of the simple conversion laws for vibrational quantum numbers [Eqs. (4.31) and (4.53)], it is easier to directly reconstruct local vibrational basis states rather than purely algebraic ones. The routine for generating a convenient vibrational basis will thus be given by a simple code for the calculation of integer partitions of the total vibrational quantum number v_T.

In the second step we are left with the problem of evaluating, in the local basis, matrix elements of the various Casimir operators and functions thereof. This is basically a trivial task for diagonal operators. However, the Majorana operator deserves more computational skill, for obvious reasons. To accomplish this task, one needs standard routines for the evaluation of $6j$ symbols. It is then necessary to arrange the basis states in such a way that the block diagonal form of the Majorana operator is conveniently emphasized.

The third step is rather critical because the final reliability of the fitted algebraic parameters is based on its efficiency. Several minimization procedures have been tested in the VIBR3AT program. A useful property of the algebraic Hamiltonian operator is that it is a linear function of its own parameters. This means that a simple matrix inversion is needed to minimize, in a least-squares sense, the rms error. However, due to the nondiagonal contributions coming from the Majorana operator, the minimum searching routine requires an iterative procedure. It is also possible to adopt simplex minimization strategies for the fitting procedure. They have the advantage of being easily constrained within the range of variability of the parameters, but they also have the drawback of consuming large amounts of CPU time.

In the final step, it is important to produce a highly readable output file. The strategy adopted by VIBR3AT is to present both the local basis and the eigenvectors of the Hamiltonian matrix for the purpose of obtaining a quick and convenient way of labeling the vibrational wavefunctions within a given polyad of levels.

Although the program is not available in a public domain format, the three-dimensional model has also been implemented in the tetratomic version. As pointed out in Section IV.F, the extension from three to four atomic molecules is not that difficult. The real problem occurs in the presence of nonlinear tetratomic configurations. Even by limiting ourselves to linear molecules, the tetratomic program is definitely more complex than the triatomic program. The complexity arises for two reasons: on one hand, the bookkeeping, either algebraic or vibrational, of states involves rather cumbersome relations defining the polyads of vibrational levels. This is the direct, unavoidable consequence of having coupled three $U(4)$ algebras in place of two. The routine for computing the local basis states for tetratomic molecules is therefore a rather complex piece of software, at least if one desires an efficient, nonredundant code. On the other hand, the general treatment of tetratomic molecules often requires explicit realization of the recoupling scheme discussed in Section IV.F. Once again, despite its relative theoretical simplicity, the practical implementation of an algebraic recoupling scheme is a nontrivial task.

As a final note on numerical aspects in the three-dimensional model, we mention the availability of a complete set of computer routines for the calculation of rotational spectra and rovibrational interactions for triatomic molecules. These routines are quite complex because they must deal with matrix elements of the operators $\hat{V}^{(L)}$ introduced in Section IV.D. In short, the practical calculations of rovibrational spectra of triatomic molecules can be rather demanding from a purely computational point of view, because of the large, complex matrices necessary to account for highly excited rotational states.

Fortunately, computer programs in the one-dimensional algebraic model are much simpler. To begin with there is very little actual group theory involved in the one-dimensional approach to molecular spectroscopy problems. The $U(2) \supset O(2)$ chain is trivial, and couplings of many $O(2)$ algebras are simply additive. Consequently, it would be somewhat dishonest to claim that Lie algebraic structures and related group properties are deeply exploited in the construction of an algebraic polyatomic molecule. It is more accurate to say that the initial correspondence between one-dimensional anharmonic potential functions and certain algebraic rules constitutes the groundwork for the realization of a very handy numerical and symbolic approach to spectroscopic problems. So the essential flowchart of a one-dimensional computer code is an extremely simple one. The most important steps are the same as those listed in the three-dimensional approach, but the bookkeeping of the vibrational levels is definitely much easier in this model, since it is systematically realized through a partition scheme of a given number, in integer parts. The Majorana operator is also easily computed in terms of square roots of integer numbers (see Section III.D). As the one-dimensional model can be used to study even quite complex molecules, it is, however, important to have easy access to the large amount of information concerning the computed vibrational spectrum. Needless to say, a general program for computing spectra of polyatomic molecules with the one-dimensional model does not exist. It is preferable to deal with a number of exportable pieces of code and to combine them in different ways, according to a specific molecule. As a practical example, let us consider the basic structure of the program for C_6H_6.

1. We start by constructing local vibrational basis states for, say, CH stretching modes. We have the total vibrational (polyad) quantum number, $n = 0, 1, \ldots, nmax$, where $nmax$ is the number of the highest excited polyad desired in our computation. Consequently, we need an initial routine giving the vibrational states distributed over six available local modes (one for each bond):

$$n = 0: v_1v_2v_3v_4v_5v_6 = 000000$$

$$n = 1: v_1v_2v_3v_4v_5v_6 = 100000$$

$$010000$$

$$\ldots$$

$$000001$$

$$n = 2: v_1v_2v_3v_4v_5v_6 = 200000 \tag{5.35}$$

$$110000$$

$$020000$$

$$\ldots$$

$$000002$$

$$\ldots$$

2. The matrix elements of the Hamiltonian operator are then obtained in the local basis (5.35). If we consider the local basis state, given by the array $v(i,j)$, $i = 1, \ldots, 6$; $j = 1, \ldots, ndim$, where $ndim$ is the dimension of the given polyad, the matrix elements of $\hat{C}_k$ ($k = 1, \ldots, 6$) are simply calculated as

$$C(k) = 4*v(k,j)*(v(k,j)-N), \tag{5.36}$$

where N is the vibron number (the same for all bonds). A similar sequence of instructions can be written for the calculation of matrix elements of the Majorana operators. We recall that in the C_6H_6 case, the Majorana operators are grouped according to first, second, and third neighbor interactions in order to construct the symmetrizer or projection operator $\hat{S}$. If we use such a parametrization and then diagonalize the Hamiltonian operator, we obtain symmetry-adapted wavefunctions.

3. The next step is to obtain the explicit symmetry label for each eigenvector according to its molecular point group. For CH stretching modes, in benzene, we must obtain species A_{1g}, B_{1u}, E_{1u}, and E_{2g}. To attach these labels to each vibrational state, we proceed as follows. The generic state will be given as an array of numbers, evec(i, j), where the first index refers to the eigenvalue and the second one to the local basis component. The symmetry label is obtained by looking at the character of the eigenvector under point group operations. In the algebraic scheme, such operations (i.e., rotations, reflections, etc.) are easily reproduced in terms of permutations of bonds. Figure 47 demonstrates a C_6 rotation of the group D_{6h}. It can be represented by the

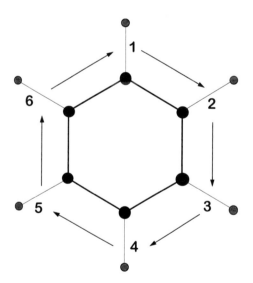

Figure 47.   Permutation of bonds in C_6H_6 equivalent to the C_6 operation of the D_{6h} point group.

permutation matrix

$$
\hat{P}_{C_6} =
\begin{bmatrix}
0 & 1 & 0 & 0 & 0 & 0 \\
0 & 0 & 1 & 0 & 0 & 0 \\
0 & 0 & 0 & 1 & 0 & 0 \\
0 & 0 & 0 & 0 & 1 & 0 \\
0 & 0 & 0 & 0 & 0 & 1 \\
1 & 0 & 0 & 0 & 0 & 0
\end{bmatrix}. \tag{5.37}
$$

It is now possible to apply $\hat{P}_{C_6}$ to any (local) basis state:

$$
\hat{P}_{C_6} |v_1 v_2 v_3 v_4 v_5 v_6\rangle = |v_6 v_1 v_2 v_3 v_4 v_5\rangle . \tag{5.38}
$$

The transformed (permuted) state is necessarily in the same polyad as the nontransformed one. This means that permutation operators act by changing the order of the basis states in a given polyad. This is very important because it allows one to obtain the transformed eigenvector explicitly (i.e., the linear combination of local modes) in terms of a simple linear (permutation) transformation. By applying the well-known formula of discrete group theory for the calculation of the character, we can schematically write

$$
\texttt{char(i)} = \texttt{sum}_{j,k} (\texttt{evec(i,j)} * \texttt{Perm(evec(i,k)))} , \tag{5.39}
$$

where Perm denotes the permutation operation of the type (5.37) acting on the sequence of local basis states. A direct comparison with the point group table allows one to attach the correct symmetry label to the complete set of eigenstates.

4. The algebraic parameters are fitted over an experimental database of levels. Within the one-dimensional scheme, it is possible to adopt minimization routines based on the simplex method. This results from the typically faster and easier calculations characterizing the one-dimensional model in comparison with the three-dimensional one.

5. The one-dimensional model is also more suitable for the direct computation of infrared and Raman transition intensities. They are obtained through known vibrational wavefunctions and through simple calculations involving linear combinations of exponential operators. If we adopt the approximate formulas discussed in Section III.D, these calculations can be carried out within the same program used for the vibrational energies.

All of these programs have been developed and written in FORTRAN language. However, it is possible to obtain very important and useful results in a symbolic language environment, offered by software products such as Mathematica, Maple, Mathcad, Macsyma, Derive, and so on. This is, of course, particularly true in the one-dimensional approach, which is often characterized by straightforward but tedious polynomial expansions of the vibrational quantum numbers.

VI.   CONCLUSIONS

In this article we have attempted to provide a complete review of algebraic techniques in the field of molecular spectroscopy. Since the typical concern of newcomers to the algebraic world is one of the language used, we have tried throughout this paper to adopt a milder mathematical approach to these algebraic arguments. Although this could be very dangerous in certain circumstances, where the rigorous and sophisticated mathematical apparatus of Lie groups and algebras requires precise, unambiguous statements and equations, we have, nonetheless, tried to satisfy the requirements of both simplicity and precision. Of course, this article cannot contain the complete information on spectroscopic studies within the algebraic framework. As explained in the introductory sections, the basic goal of this paper is to orient both the expert and the neophyte toward an alternative route for doing molecular spectroscopy while placing a constant accent on the applied aspects of these methods. The general approach has been to provide a detailed study of a simple example before moving to a more complex one. Hopefully, a reader, determined to carry out a thorough examination of algebraic techniques to apply them to actual problems, will soon become convinced that the initial investment in the acquisition of a new language can be repaid quickly in terms of productivity. In fact, the hope is for this article to be seen as an official invitation to enter this exciting area of research for the purpose of actively contributing new ideas, comments, and criticism. Nonetheless, in many respects the meticulous reader could judge this paper an incomplete work. However, we hope that the large number of cited references will be of some help in obtaining a deeper insight into several technical aspects either intentionally disregarded or only hinted at briefly in this article.

Molecular spectroscopy problems constitute just a small portion of the accessible applications of algebraic methods. Nonetheless, molecular spectroscopy provides a fertile for the algebraic approach to take root even though it can be applied to virtually any discipline involving molecules. For example, electronic spectroscopy time-dependent dynamics, and scattering are currently playing a remarkable role in physical–

chemical studies of molecules. Electronic spectroscopy, particularly, is of noticeable importance for its connection with coupled rovibrational–electronic degrees of freedom for relatively small, highly excited molecules [108]. Several studies of these problems have been accomplished in using an algebraic framework, where the Hamiltonian operator is expressed by generators for both electronic and nuclear degrees of freedom [109–112]. We also want to mention several algebraic studies of scattering problems, where the interaction potential is described in terms of (noncompact) algebraic structures [25, 113–117]. As discussed briefly on several occasions throughout this paper, the time-dependent analysis of inter- and intramolecular dynamics is naturally allied with the algebraic model by virtue of its intrinsic Hamiltonian, anharmonic formulation. The example of the benzene dimer presented in Section V.A is a first step in pushing the algebraic models toward very complex problems; such a developments would eventually attract people working with clusters, polymers, or molecules of biochemical interest. Moreover, it has been shown [118] that the algebraic model is also suited to describing absolute thermodynamic properties of complex molecules. As a particularly significant outcome of this study, the partition function and the density of vibrational states for the C_2H_2 molecule have been calculated very accurately.

In other words, the algebraic studies discussed in this article are only a small sampling of the huge number of still unsolved and fascinating questions concerning molecular physics. With this paper we would like symbolically to close the era of preliminary test studies of algebraic techniques and to unveil a new period of discoveries and nontrivial applications of these methods to this vigorous area of modern science.

ACKNOWLEDGMENTS

My thanks goes to F. Iachello, whose constant encouragement and inspiring guidance have been extremely important throughout our years of collaboration. I also want to thank G. Scoles, K. K. Lehmann, M. Herman, W. Klemperer, M. Quack, and D. Bassi for their continuous interest and support of my work. I am very grateful to Sondra E. Vitols, who carefully and patiently revised the manuscript despite her busy schedule. I would like to thank my parents and, most of all, my wife, Saveria, for her love and participation in my work. Without her, this article would probably never have been finished. Without my children, Lorenzo and Elena, this article would have appeared several years ago, but life would not have been so delightful.

REFERENCES

1. J. L. Dunham, *Phys. Rev.* **41**, 721 (1932).
2. H. Weyl, *Gruppentheorie und Quantenmechanik*, Hirzel, Leipzig, Germany, 1931.
3. E. Wigner, *Gruppentheorie*, Vieweg, Brunswick, Germany, 1931.
4. E. Wigner, *Phys. Rev.* **51**, 106 (1937).
5. G. Racah, *Phys. Rev.* **62**, 438 (1942).
6. G. Racah, *Phys. Rev.* **76**, 1352 (1949).
7. G. Racah, *Lincei Rend. Sci. Fis. Mat. Nat.* **8**, 108 (1950).
8. A. Arima and F. Iachello, *Phys. Rev. Lett.* **35**, 1069 (1975); A. Arima and F. Iachello, *Ann. Phys. (NY)* **99**, 253 (1976); F. Iachello and A. Arima, *The Interacting Boson Model*, Cambridge University Press, Cambridge, 1987.
9. A. Arima and F. Iachello, *Ann. Phys. (NY)* **111**, 201 (1978); **123**, 468 (1979); F. Iachello and P. van Isacker, *The Interacting Boson–Fermion Model*, Cambridge University Press, Cambridge, 1991.
10. A. Frank and P. van Isacker, *Algebraic Methods in Molecular and Nuclear Structure Physics*, Wiley, New York, 1994.
11. F. Iachello and R. D. Levine, *Algebraic Theory of Molecules*, Oxford University Press, Oxford, 1994.
12. A. Messiah, *Quantum Mechanics*, North-Holland, Amsterdam, 1975.
13. J. P. Elliot and P. G. Dawber, *Symmetry in Physics*, Oxford University Press, Oxford, 1979.
14. B. G. Wybourne, *Classical Groups for Physicists*, Wiley, New York, 1974.
15. C. D. H. Chisholm, *Group Theoretical Techniques in Quantum Chemistry*, Academic Press, London, 1976.
16. R. F. Streater, *Commun. Math. Phys.* **4**, 217 (1967).
17. M. Hamermesh, *Group Theory and Its Applications to Physical Problems*, Addison-Wesley, Reading, Mass., 1964; J. F. Cornwell, *Group Theory in Physics*, Academic Press, London, 1984.
18. V. Fock, *Z. Phys.* **98**, 145 (1935).
19. U. Fano and G. Racah, *Irreducible Tensorial Sets*, Academic Press, New York, 1959.
20. J. Hinze, ed., *The Unitary Group for the Evaluation of Electronic Matrix Elements*, Springer-Verlag, Berlin, 1981.
21. H. Casimir, *Proc. Roy. Acad. Amsterdam* **34**, 844 (1931).
22. P. M. Morse, *Phys. Rev.* **34**, 57 (1929).
23. P. M. Morse and H. Feshbach, *Methods of Theoretical Physics*, McGraw-Hill, New York, 1953.
24. G. Pöschl and S. Teller, *Z. Phys.* **83**, 143 (1933).
25. Y. Alhassid, F. Gürsey, and F. Iachello, *Ann. Phys.* **148**, 346 (1983).
26. F. Iachello, *Chem. Phys. Lett.* **78**, 581 (1981).
27. F. Iachello and S. Oss, *Phys. Rev. Lett.* **66**, 2776 (1991).
28. R. D. Levine and C. E. Wulfman, *Chem. Phys. Lett.* **60**, 372 (1979).
29. M. Berrondo and A. Palma, *J. Phys. A* **13**, 773 (1980).

30. J. Schwinger, in *Quantum Theory of Angular Momentum*, L. C. Biedenharn and H. Van Dam, Eds., Academic Press, New York, 1965.

31. A. de Shalit and I. Talmi, *Nucleal Shell Theory*, Academic Press, London, 1963.

32. M. Rotenberg, R. Bivins, N. Metropolis, and J. K. Wooten, Jr., *The 3 − j and 6 − j Symbols*, Technology Press, Cambridge, Mass., 1959.

33. R. S. Berry, in *Quantum Dynamics of Molecules*, R. G. Woolley, Ed., Plenum Press, New York, 1980.

34. J. D. Graybeal, *Molecular Spectroscopy*, McGraw-Hill, New York, 1988.

35. O. S. van Roosmalen, Ph.D. thesis, University of Groningen, The Netherlands, 1982.

36. M. S. Child and L. O. Halonen, *Adv. Chem. Phys.* **57**, 1 (1984).

37. O. S. van Roosmalen, I. Benjamin, and R. D. Levine, *J. Chem. Phys.* **81**, 5986 (1984).

38. I. Benjamin and R. D. Levine, *J. Mol. Spectrosc.* **126**, 486 (1987).

39. A. Mengoni and T. Shirai, *J. Mol. Spectrosc.* **162**, 246 (1993).

40. E. B. Wilson, Jr., J. C. Decius, and P. C. Cross, *Molecular Vibrations*, Dover, New York, 1955.

41. A. Frank and R. Lemus, *Phys. Rev. Lett.* **68**, 413 (1992).

42. R. H. Page, Y. R. Shen, and Y. T. Lee, *J. Chem. Phys.* **88**, 4621 (1988); **88**, 5362 (1988).

43. E. L. Siebert III, W. P. Reinhardt, and J. Y. Huyes, *J. Chem. Phys.* **81**, 1115 (1984); **81**, 1135 (1984).

44. Y. Zhang, S. J. Klippenstein, and R. A. Marcus, *J. Chem. Phys.* **94**, 7319 (1991); Y. Zhang and R. A. Marcus, *J. Chem. Phys.* **96**, 6065 (1992); **97**, 5283 (1992).

45. R. E. Wyatt, C. Yung, and C. Leforestier, *J. Chem. Phys.* **97**, 3458 (1992).

46. F. Iachello and S. Oss, *Chem. Phys. Lett.* **187**, 500 (1991).

47. F. Iachello and S. Oss, *J. Chem. Phys.* **99**, 7337 (1993).

48. D. Bassi, L. Menegotti, S. Oss, M. Scotoni, and F. Iachello, *Chem. Phys. Lett.* **207**, 167 (1993).

49. F. Iachello and S. Oss, *Chem. Phys. Lett.* **205**, 285 (1993).

50. F. Iachello and S. Oss, *J. Mol. Spectrosc.* **153**, 225 (1992).

51. I. Benjamin, I. L. Cooper, and R. D. Levine, *Chem. Phys. Lett.* **139**, 285 (1987).

52. E. Fermi, *Z. Phys.* **71**, 250 (1931).

53. H. R. Dübak and M. Quack, *J. Chem. Phys.* **81**, 3779 (1984).

54. G. J. Scherer, K. K. Lehmann, and W. Klemperer, *J. Chem. Phys.* **81**, 5319 (1984).

55. G. A. Voth, R. A. Marcus, and A. H. Zeway, *J. Chem. Phys.* **81**, 5494 (1984).

56. A. Campargue, F. Stoeckel, M. Cherevier, and H. B. Kraiem, *J. Chem. Phys.* **87**, 5598 (1987).

57. S. Oss, *Z. Phys. D*, in press (1995).

58. M. Quack, *Ann. Rev. Phys. Chem.* **41**, 839 (1990).

59. M. E. Kellman, *J. Chem. Phys.* **83**, 3843 (1985).

60. M. E. Kellman and E. D. Lynch, *J. Chem. Phys.* **85**, 7216 (1986); **88**, 2205 (1988).

61. L. Xiao and M. E. Kellman, *J. Chem. Phys.* **93**, 5805 (1990).

62. M. E. Kellman and L. Xiao, *J. Chem. Phys.* **93**, 5821 (1990).

63. G. Wu, *Chem. Phys. Lett.* **195**, 115 (1992); *Chem. Phys.* **173**, 1 (1993).

64. R. D. Levine, *Chem. Phys. Lett.* **95**, 87 (1993).

65. I. Benjamin, K. H. Bisseling, R. Kosloff, R. D. Levine, J. Manz, and H. H. Schor, *Chem. Phys. Lett.* **116**, 225 (1985).

66. J. Pliva and A. S. Pine, *J. Mol. Spectrosc.* **126**, 82 (1987).

67. M. Scotoni, S. Oss, L. Lubich, S. Furlani, and D. Bassi, *J. Chem. Phys.* **103**, 897 (1995).

68. F. Iachello, S. Oss, and L. Viola, *Mol. Phys.* **78**, 561 (1993).

69. O. S. van Roosmalen, F. Iachello, R. D. Levine, and A. E. L. Dieperink, *J. Chem. Phys.* **79**, 2515 (1983).

70. R. Lemus Casillas, Ph.D. thesis, Universidad Nacional Autonoma de Mexico, 1988.

71. G. Herzberg, *Molecular Spectra and Molecular Structure. Vol. 2. Infrared and Raman Spectra of Polyatomic Molecules*, Van Nostrand Reinhold, New York, 1945.

72. N. Manini and S. Oss, *Z. Phys. D* **32**, 85 (1994).

73. K. K. Lehmann, *J. Chem. Phys.* **79**, 1098 (1983).

74. I. M. Mills and A. G. Robiette, *Mol. Phys.* **56**, 743 (1985).

75. F. Iachello and S. Oss, *J. Mol. Spectrosc.* **142**, 85 (1990).

76. S. Oss, N. Manini, and R. Lemus, *Comput. Phys. Comnmun.* **74**, 164 (1993).

77. B. T. Darling and D. M. Dennison, *Phys. Rev.* **15**, 128 (1940).

78. D. Papousèk and M. R. Aliev, *Molecular Vibrational-Rotational Spectra*, Elsevier, Moscow, 1982.

79. H. W. Kroto, *Molecular Rotation Spectra*, Wiley, London, 1975.

80. J. K. Watson, *Mol. Phys.* **15**, 479 (1968); **19**, 465 (1970).

81. R. L. Anderson, S. Kumei, and C. E. Wulfman, *J. Math. Phys.* **14**, 11 (1973).

82. R. Gilmore and J. P. Draayer, *J. Math. Phys.* **26**, 3053 (1985).

83. C. C. Martens, *J. Chem. Phys.* **96**, 8971 (1992).

84. I. M. Mills, in *Molecular Spectroscopy: Modern Research*, Vol. I, K. N. Rao and C. W. Mathews, Eds., Academic Press, New York, 1972.

85. T. Oka, *J. Chem. Phys.* **47**, 5410 (1967).

86. F. Iachello, S. Oss, and L. Viola, *J. Chem. Phys.* **101**, 3531 (1994).

87. S. C. Wang, *Phys. Rev.* **34**, 243 (1929).

88. L. Viola, M.S. thesis, Università di Trento, Italy, 1991.

89. F. Iachello, A. Leviatan, and A. Mengoni, *J. Chem. Phys.* **95**, 1449 (1991).

90. A. Leviatan, in preparation.

91. F. Iachello, S. Oss, and R. Lemus, *J. Mol. Spectrosc.* **146**, 56 (1991).

92. F. Iachello, S. Oss, and R. Lemus, *J. Mol. Spectrosc.* **149**, 132 (1991).

93. F. Iachello, S. Oss, and L. Viola, *Mol. Phys.* **78**, 545 (1993).

94. F. Iachello, N. Manini, and S. Oss, *J. Mol. Spectrosc.* **156**, 190 (1992).

95. N. Manini, M.S. thesis, Università di Trento, Italy, 1991.

96. G. Amat, M. Goldsmith, and H. H. Nielsen, *J. Chem. Phys.* **24**, 44 (1956); **25**, 800 (1956).

97. B. F. Henson, G. V. Hartland, V. A. Venturo, and P. M. Felker, *J. Chem. Phys.* **91**, 2751 (1989); **97**, 2189 (1992).

98. B. F. Henson, G. V. Hartland, V. A. Venturo, R. A. Hertz, and P. M. Felker, *Chem. Phys. Lett.* **176**, 91 (1991).

99. H. Krause, B. Ernstberger, and H. J. Neusser, *Chem. Phys. Lett.* **184**, 411 (1991).

100. F. Iachello and S. Oss, *J. Chem. Phys.* **102**, 1141 (1995).

101. A. R. Hoy, I. M. Mills, and G. Strey, *Mol. Phys.* **24**, 1265 (1972).

102. A. B. McCoy and E. L. Siebert III, *J. Chem. Phys.* **97**, 2938 (1992).

103. R. Gilmore, *Lie Groups, Lie Algebras and Some of Their Applications*, Wiley, New York, 1974.

104. O. S. van Roosmalen and A. E. L. Dieperink, *Ann. Phys. (NY)* **139**, 198 (1982).

105. R. Gilmore, *Catastrophe Theory for Scientists and Engineers*, Wiley, New York, 1981.

106. I. L. Cooper and R. D. Levine, *J. Mol. Spectrosc.* **148**, 391 (1991).

107. B. Shao, N. Walet, and R. D. Amado, *Phys. Rev. A* **46**, 4037 (1992); **47**, 2064 (1993).

108. A. Delon, R. Jost, and M. Lombardi, *J. Chem. Phys.* **95**, 5701 (1991).

109. A. Frank, F. Iachello, and R. Lemus, *Chem. Phys. Lett.* **131**, 380 (1986).

110. A. Frank, R. Lemus, and F. Iachello, *J. Chem. Phys.* **91**, 29 (1989).

111. R. Lemus, A. Leviatan, and A. Frank, *Chem. Phys. Lett.* **194**, 327 (1992).

112. R. Lemus and A. Frank, *Ann. Phys.* **206**, 122 (1991).

113. C. E. Wulfman and R. D. Levine, *Chem. Phys. Lett.* **97**, 361 (1983).

114. A. Frank and K. B. Wolf, *Phys. Rev. Lett.* **52**, 1737 (1984).

115. Y. Alhassid, F. Iachello, and J. Wu, *Phys. Rev. Lett.* **56**, 271 (1986).

116. J. Wu, F. Iachello, and Y. Alhassid, *Ann. Phys.* **173**, 68 (1987).

117. J. Wu, Y. Alhassid, and F. Gürsey, *Ann. Phys.* **196**, 163 (1989).

118. D. Kusnesov, *J. Chem. Phys.* **101**, 2289 (1994).

TIGHT-BINDING MOLECULAR DYNAMICS STUDIES OF COVALENT SYSTEMS

C. Z. WANG AND K. M. HO

Ames Laboratory and Department of Physics, Iowa State University, Ames, Iowa

CONTENTS

ABSTRACT

Tight-binding molecular dynamics has recently emerged as a useful method for studying the structural, dynamical, and electronic properties

Advances in Chemical Physics, Volume XCIII, Edited by I. Prigogine and Stuart A. Rice.
ISBN 0-471-14321-9 © 1996 John Wiley & Sons, Inc.

of covalent systems. The method incorporates electronic structure calculation into molecular dynamics through an empirical tight-binding Hamiltonian and bridges the gap between *ab initio* molecular dynamics and simulations using empirical classical potentials. In this article we review some recent achievements of the tight-binding molecular dynamics method and discuss some opportunities for future development.

I. INTRODUCTION

The modeling and simulation of covalent materials have long been the prime concern of many scientists in the fields of physics, chemistry, and materials science. First-principles techniques such as local density functional formalism are well developed and have already produced a wealth of significant results [1]. However, the computational cost is prohibitive for large systems. Even though recent breakthroughs such as the Car–Parrinello method [2] (or similar algorithms) have made molecular dynamics (MD) possible within the local density functional regime, simulations are still limited to a small number of atoms and short time scales. On the other hand, quite a number of empirical interatomic potentials have been proposed for covalent materials such as silicon and carbon [3–10]. These potentials include three-body terms which are usually fitted to energy–volume phase diagrams and other properties of the crystalline structures obtained by first-principles calculations or from experiments. Although simulations with classical empirical potentials are fast, these empirical potentials do not always give correct descriptions for properties that are not explicitly included in the fitting database [11, 12]. Electronic structure information cannot be obtained, nor can we expect these classical potentials to give accurate descriptions of phenomena where quantum mechanical interference effects are essential (e.g., Jahn–Teller distortions around vacancies, conjugated π-state effects in carbon systems). There is a large class of problems that requires more atoms than first-principles techniques can handle and demands more accuracy than classical potentials can provide. Thus, it is imperative to have a scheme that is powerful enough to treat several hundred atoms (several thousand atoms with massively parallel computers) while being accurate enough so that we can trust the results.

Recently, a simplified quantum mechanical molecular dynamics scheme, [i.e., tight-binding molecular dynamics (TBMD)] has been developed [13–16] which bridges the gap between classical-potential simulations and the Car–Parrinello scheme. In the same spirit as the Car–Parrinello scheme, TBMD incorporates electronic structure effects into molecular dynamics through an empirical tight-binding Hamiltonian

H_{TB}. The quantum mechanical many-body nature of the interatomic forces is taken into account naturally through the Hellmann–Feynman theorem. Since the scheme usually uses a minimal basis set for the electronic structure calculation and the Hamiltonian matrix elements are parametrized, large numbers of atoms can be tackled within the present computer capabilities. One of the distinctive features of this scheme in comparison with other empirical schemes is that all the parameters in the model can be obtained theoretically. It is therefore very useful for studying novel materials where experimental data are not readily available. The scheme has been demonstrated to be a powerful method for studying various structural, dynamical, and electronic properties of covalent systems.

Several other computational advantages unique to the tight-binding scheme are worth noting. First, the computational cost is about the same for different elements, such as silicon and carbon, in the tight-binding scheme, while the memory and computer time requirements for the Car–Parrinello method rise sharply if the element under consideration has valence states that require a large number of plane waves for proper convergence (e.g., carbon). Moreover, plane-wave-based schemes such as the Car–Parrinello method are designed for three-dimensional periodic systems. For problems that do not fit into this category, such as clusters and surface problems, a large amount of empty space has to be created artificially by embedding the system inside a huge unit cell (supercell), creating a heavy burden on computational resources. In contrast, three-dimensional periodicity is not required for tight-binding methods. Tight-binding methods also have extra flexibility since the electronic structure problem can be handled either in k space or in r space. The latter means that the complexity of the problem in principle scales only linearly with the number of atoms under consideration, and parallel computing algorithm can be implemented easily.

As an empirical scheme, TBMD certainly has disadvantages and limitations. A very difficult and crucial step in the TBMD approach is to determine the tight-binding parameters for the systems of interest. It usually takes a lot of work to generate a tight-binding model with good accuracy and transferability. Although tight-binding parametrization of the electronic band structures of crystalline lattices within the framework of two-center approximation for the hopping integrals was formulated by Slater and Koster 40 years ago [17], tight binding models for total energy calculations was not developed until the late 1970s following the work of Harrison [18] and Chadi [19]. While it was shown that the tight-binding models developed by Harrison and by Chadi are very reliable for describing the properties of silicon and carbon in a crystalline tetrahedral bonding environment [13, 18–20], more transferable tight-binding models that can offer a satisfactory description for more general atomic coordinations and disordered systems are needed to perform tight-binding molecular dynamics simulations. Recent work has shown that it is indeed possible to construct transferable tight-binding models describing accurately several phases of silicon and carbon with drastically different coordination and environments [21–27].

In this article we review some achievements of TBMD in the last several years and discuss some opportunities for future development. The paper is organized as follows: In Section II we review several tight-binding models for silicon and carbon. TBMD algorithms and formalism for the force calculations are described in Section III. In Section IV we summarize some achievements of TBMD for silicon and carbon systems. Recent developments in algorithms for TBMD simulations and in tight-binding modeling are discussed in Section V. Finally, in Section VI, we look at the future of TBMD.

II. TIGHT-BINDING TOTAL ENERGY MODELS

In this section we review several tight-binding total energy models for silicon and carbon that have been widely used in TBMD simulations. Some other TB models are discussed in the later sections when we address recent developments in tight-binding modeling.

The tight-binding expression for the binding energy of a system with N atoms is given by

$$E_{\text{binding}} = E_{\text{bs}} + E_{\text{rep}} - E_0 . \tag{1}$$

The first term on the right-hand side is the band structure energy from the tight-binding Hamiltonian H_{TB}. The second term is a repulsive energy representing the repulsion due to the overlap between the orbitals, corrections to the exchange-correlation energy, and to the double counting of the electron–electron interaction in the first term. The third term is a constant that can be adjusted to give the correct cohesive energy for the condensed system.

A. Early Versions of Tight-Binding Models

In the early versions of tight-binding total energy models for silicon and carbon developed by Harrison [18] and Chadi [19], a minimal orthogonal sp^3 basis set is used to construct the tight-binding Hamiltonian matrix. The diagonal elements of H_{TB} are set equal to the atomic energy levels E_s and E_p, and the off-diagonal elements are constructed, following Slater and Koster [17], from two-center hopping parameters that scale as the inverse square of the interatomic distance r:

$$h_\alpha(r) = h_\alpha(r_0)\left(\frac{r_0}{r}\right)^2,$$ (2)

where r_0 is the equilibrium nearest-neighbor distance of the diamond structure. The subscript α denotes the four possible types of interatomic hopping: $ss\sigma$, $sp\sigma$, $pp\sigma$, and $pp\pi$. These tight-binding parameter are determined by fitting to the band structure of the diamond structure.

The repulsive energy is calculated from a sum of pairwise potentials $\phi(r)$. Harrison argued that the repulsive interaction originates from the overlap of the atomic orbitals and therefore, should scale as [18]

$$\phi(r) = \phi_0\left(\frac{r_0}{r}\right)^4,$$ (3)

while Chadi expressed the pairwise repulsion as [19]

$$\phi(r) = v_0 + v_1\left(\frac{r}{r_0} - 1\right) + v_2\left(\frac{r}{r_0} - 1\right)^2.$$ (4)

The parameters ϕ_0 or v_0, v_1, and v_2 are determined by fitting to the lattice constant and bulk modulus of the diamond structure. In later work, Wang and co-workers modified the expression for the pairwise repulsive potential from the Chadi model by fitting to the binding energy of diamond structure as a function of volume obtained from first-principles density functional (LDA) calculations. In addition to being accurate for the lattice constant, bulk modulus, and phonon frequencies of the diamond structure, the resulting potential gives an excellent description of the anharmonic properties such as Gruneisen parameters, thermal expansion, and thermal broadening and shifts of the phonons [13, 20, 28].

B. Transferable TB Model for Silicon

While the early versions of tight-binding energy models were quite successful in describing the properties of the solid near the diamond structure, they were not accurate when extrapolated to other crystalline or disordered structures. Tomanek and Schluter [31] tried to overcome this discrepancy by introducing a term dependent on the total number of bonds and the total number of atoms into the binding energy expression. However, this bond-counting term in their model is not suitable for molecular dynamics simulations. In 1989, Goodwin et al. (GSP) [21] and Sawada [22] independently showed that by multiplying an attenuation function to the simple power-law scaling of the tight-binding parameters and the pairwise repulsion, it is possible to obtain a tight-binding model that describes fairly well the energy–volume behavior of silicon in crystalline phases with different atomic coordinations. Subsequent models developed by Kwon et al. [24], Kohyama [27], and Mercer and Chou [25] also give accurate energy–volume phase diagrams for higher-coordinated metallic structures as well as the diamond structure. These developments correct the deficiency in earlier tight-binding models.

Goodwin et al. noted that the Harrison tight-binding model gives reasonably correct equilibrium energies for the crystalline structures of silicon with different coordination numbers, but failed to predict correctly the equilibrium volumes for structures other than the diamond structure. They also noted that in the Harrison model, the binding energy of silicon in a given crystalline lattice is a function of one free parameter: the nearest-neighbor distance r or, equivalently, r_0/r:

$$E_{binding}^{lattice}\left(\frac{r_0}{r}\right) = E_{bond}^{lattice}\left(\frac{r_0}{r}\right) + E_{rep}^{lattice}\left(\frac{r_0}{r}\right) \tag{5}$$

One can rescale the energy–volume curves for silicon crystalline structures by replacing r_0/r with some function of it,

$$E_{rescaled}^{lattice}\left(\frac{r_0}{r}\right) = E_{binding}^{lattice}\left[\frac{f(r_0/r)}{f(1)}\right] \tag{6}$$

Dividing by $f(1)$ fixes the diamond equilibrium volume. This rescaling of the energy expression defines new functional forms for the TB parameters and the pairwise repulsive interaction.

$$h_\alpha^{rescaled}\left(\frac{r_0}{r}\right) = h_\alpha(1)\left[\frac{f(r_0/r)}{f(1)}\right]^n \tag{7}$$

$$\phi^{rescaled}\left(\frac{r_0}{r}\right) = \phi(1)\left[\frac{f(r_0/r)}{f(1)}\right]^m \tag{8}$$

Because the equilibrium energies of all the lattices and the diamond equilibrium volume are invariant under rescaling, the function f can be used to improve the calculated values of equilibrium volumes of the remaining lattices. By choosing the function f as

$$f\left(\frac{r_0}{r}\right) = \frac{r_0}{r}\exp\left[-\left(\frac{r}{r_c}\right)^{n_c}\right] \tag{9}$$

the hopping parameters and pairwise potential are scaled as a power-law function multiplied by an exponential function:

$$h_\alpha(r) = h_\alpha(r_0)\left(\frac{r_0}{r}\right)^n \exp\left\{n\left[-\left(\frac{r}{r_c}\right)^{n_c} + \left(\frac{r_0}{r_c}\right)^{n_c}\right]\right\} \tag{10}$$

and

$$\phi(r) = \phi(r_0)\left(\frac{r_0}{r}\right)^m \exp\left\{m\left[-\left(\frac{r}{r_c}\right)^{n_c} + \left(\frac{r_0}{r_c}\right)^{n_c}\right]\right\}. \tag{11}$$

The parameters in the GSP model are listed in Table I.

The GSP model describes reasonably well the energy–volume behavior of silicon in crystalline phases with different atomic coordination (Fig. 1) as well as the structures of small clusters ($n \leq 10$). The model has been applied to study complex systems such as point defects in crystalline silicon, as well as amorphous and liquid silicon [15, 32–35].

The model developed by Kwon et al. is a modified version of the GSP model. It follows the same scaling form for the TB parameters and the pairwise potential as in the GSP model but uses different scaling parameters for different hopping integrals as well as the pairwise potential. It also expresses the repulsive energy in a functional form of pairwise potential following the TB model for carbon developed by Xu et al. [23] (see Section II.C). The model thus has extra degrees of freedom for adjusting not only the equilibrium volumes (as did in GSP model) but also the energy differences for different coordinated structures. The models developed by Sawada, by Kohyama, and by Mercer and Chou use a different family of radial functions (of the form $[1 + \exp(r - r_c)/\nu]^{-1}$) instead of the exponential function as the attenuation function. In addition, the environment dependence of the interactions was also taken

TABLE I
Parameters for the GSP Model for Si

Parameter	Value
$E_s - E_p$ (eV)	8.295
$h_{ss\sigma}(r_0)$ (eV)	−1.82
$h_{sp\sigma}(r_0)$ (eV)	1.96
$h_{pp\sigma}(r_0)$ (eV)	3.06
$h_{pp\pi}(r_0)$ (eV)	−0.87
$\phi(r_0)$ (eV)	3.4581
n	2
m	4.54
r_c	3.67
n_c	6.48

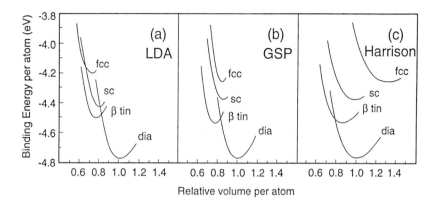

Figure 1. Binding energies of silicon versus volume for four crystalline structures: (*a*) results from first-principles LDA calculations; (*b*) results from the GSP TB model; (*c*) results from the Harrison TB model. The results are quoted from Ref. 21.

into account in the expression for the repulsive energy in order to improve the transferability of the models. For more details of these models, we refer to the original publications [22, 24, 25, 27].

C. Transferable TB Model for Carbon

Following the work of GSP, Xu and co-workers (XWCH) developed a similar tight-binding model for carbon [23] based on the scaling form given by GSP [21] for the dependence of the TB hopping parameters and the pairwise potential on the interatomic separation:

$$h_\alpha(r) = h_\alpha(r_0)\left(\frac{r_0}{r}\right)^n \exp\left\{n\left[-\left(\frac{r}{r_c}\right)^{n_c} + \left(\frac{r_0}{r_c}\right)^{n_c}\right]\right\} \qquad (12)$$

and

$$\phi(r) = \phi_0\left(\frac{d_0}{r}\right)^m \exp\left\{m\left[-\left(\frac{r}{d_c}\right)^{m_c} + \left(\frac{d_0}{d_c}\right)^{m_c}\right]\right\}, \qquad (13)$$

where r_0 denotes the nearest-neighbor atomic separations in diamond and n, n_c, r_c, ϕ_0, m, d_c, and m_c are fitting parameters.

Unlike the GSP model, the scaling parameters r_c and n_c for $h_\alpha(r)$ are not necessarily the same as the corresponding d_c and m_c for $\phi(r)$ in this model. The expression for the repulsive energy is also different from the

GSP model. It is given by

$$E_{rep} = \sum_i f\left(\sum_j \phi(r_{ij})\right),$$ (14)

where $\phi(r_{ij})$ is a pairwise potential between atoms i and j, and f is a functional expressed as a fourth-order polynomial with argument $\Sigma_j \phi(r_{ij})$.

The potential has a relatively short cutoff distance of 2.6 Å, which allows the Hellmann–Feynman forces to be calculated fairly quickly. The parameters in the model are chosen primarily by fitting first-principles LDA results of energy–volume curves of different carbon polytypes: diamond, graphite, linear chain, simple cubic, and face-centered cubic structures, with special emphasis on the diamond, graphite, and linear chain structures. Additional checks were made to ensure that the model gives reasonable results for the electronic band structure, elastic moduli, and phonon frequencies in the diamond and graphite structures, although these properties do not enter explicitly into the fitting procedure.

The resulting sp^3 tight-binding parameters are $E_s = -2.99$ eV, $E_p = 3.71$ eV, $h_{ss\sigma}(r_0) = -5.0$ eV, $h_{sp\sigma}(r_0) = 4.7$ eV, $h_{pp\sigma}(r_0) = 5.5$ eV, and $h_{pp\pi}(r_0) = -1.55$ eV. The scaling functions and the polynomial function $f(x) = \Sigma_{n=0}^4 c_n x^n$, with $x = \Sigma_j \phi(r_{ij})$, are shown in Fig. 2. The parameters can be found in Ref. 23. To handle the effects of charge transfer in disordered systems correctly, particularly in the presence of dangling bonds, a Hubbard-like term,

$$H_u = \tfrac{1}{2} u(q_i - q_i^0)^2,$$ (15)

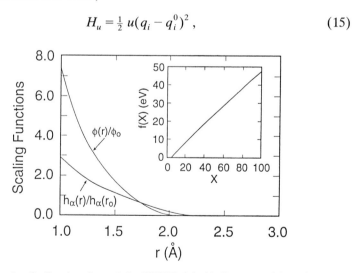

Figure 2. Scaling functions of the XWCH tight-binding potential model.

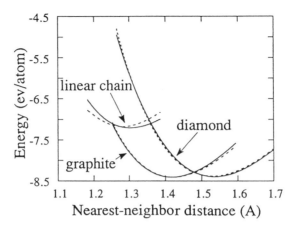

Figure 3. Binding energy versus nearest-neighbor atomic distance for carbon in the diamond, graphite, and linear chain structures. Solid and dashed lines show the results from the XWCH tight-binding model and LDA calculations, respectively. (From Refs. 23 and 53.)

can be added to the tight-binding Hamiltonian H_{TB}, where q_i is the Mulliken population at atomic site i and q_i^0 is the number of valence electrons of atom i. The parameter u is taken to be 4 eV for carbon.

The model has been shown to have good transferability when applied to a variety of crystal structures. This can be seen from Fig. 3 and Tables II and III, where the energies, vibrational and elastic properties for different coordinated crystalline structures obtained from this model are compared with first-principles LDA calculations and experiments. Applications in molecular-dynamics study of the liquid and amorphous phases of carbon as well as the structures of carbon clusters indicate that the potential does a good job of describing carbon systems over a wide range of bonding environments. These applications are reviewed in Section IV.

III. TBMD ALGORITHMS AND FORCE CALCULATIONS

In this section we describe two common algorithms for TBMD simulations: TBMD using the standard matrix diagonalization method and TBMD using the Car–Parrinello algorithm. We also discuss a method to include the effects of electronic states at finite temperature in the TBMD simulations. Recent developments on order N algorithms for TBMD simulations are discussed in Section V.

TABLE II

Elastic Constants, Phonon Frequencies, and Grünneisen Parameters of Diamond Calculated from the XWCH Tight-Binding Model Compared with Experimental Results from Ref. 29

	Model	Expt.
$c_{11} - c_{12}$ (10^{12} dyn/cm^2)	6.22	9.51
c_{44}^{0} [a] (10^{12} dyn/cm^2)	5.42	—
c_{44}^{b} (10^{12} dyn/cm^2)	4.75	5.76
$\nu_{LTO(\Gamma)}$ (THz)	37.80	39.90
$\nu_{TA(X)}$ (THz)	22.42	24.20
$\nu_{TO(X)}$ (THz)	33.75	32.0
$\nu_{LA(X)}$ (THz)	34.75	35.5
$\gamma_{LTO(\Gamma)}$	1.03	0.96
$\gamma_{TA(X)}$	−0.16	—
$\gamma_{TO(X)}$	1.10	—
$\gamma_{LA(X)}$	0.62	—

Source: Ref. 23.

[a] C_{44}^{0} denotes the shear constant that would appear in the absence of internal displacements.

[b] C_{44} is the shear constant in the absence of internal stress.

TABLE III

Elastic Constants, Phonon Frequencies, and Grüneisen Parameters of Graphite Calculated from the XWCH Tight-Binding Model Compared with Experimental results from Ref. 30.

	Model	Expt.
$c_{11} - c_{12}$ (10^{12} dyn/cm^2)	8.40	8.80
E_{2g_2} (THz)	49.92	47.46
A_{2u} (THz)	29.19	26.04
$\gamma(E_{2g_2})$	2.00	—
$\gamma(A_{2u})$	0.10	—

Source: Ref. 23.

A. TBMD Using Standard Matrix Diagonalization

In the tight-binding molecular-dynamics scheme [13], the atomic motion in the system is governed by the following Hamiltonian:

$$H(\{\mathbf{r}_i\}) = \sum_i \frac{P_i^2}{2m} + 2 \sum_n <\psi_n|H_{TB}(\{\mathbf{r}_i\})|\psi_n> + E_{rep}(\{\mathbf{r}_i\}), \quad (16)$$

where $\{\mathbf{r}_i\}$ denotes the positions of the atoms $(i = 1, 2, \ldots, N)$, and $\mathbf{P}_i$ denotes the momentum of the ith atom. The first term in (16) is the kinetic energy of the ions. The second and third terms are from the tight-binding energy model, as discussed in Section II. The summation in the second term runs over all occupied electronic states, and the factor 2 accounts for the spin degeneracy.

The Hamiltonian modeled as (16) is easy to insert into the standard molecular dynamics routine, as illustrated in Fig. 4. The interatomic forces necessary for performing the molecular dynamics run are expressed as a sum of the repulsive forces arising from $E_{rep}(\{\mathbf{r}_i\})$ and a contribution coming from the band-structure energy. In the empirical tight-binding scheme, such electronic forces or Hellmann–Feynman forces are easy to calculate, especially when the two-center approximation is used for the tight-binding hopping parameters. The force acting on atom l arising from the band-structure energy E_{bs} is given by

$$\mathbf{F}_l = -2 \sum_n \left\langle \psi_n \left| \frac{\partial H}{\partial \mathbf{R}_l} \right| \psi_n \right\rangle, \quad (17)$$

where H is the tight-binding Hamiltonian matrix and $\mathbf{R}_l$ is the position vector of atom l. The sum index n is over occupied levels ε_n with eigenvectors ψ_n. For a periodic system, the sum over n can be replaced by sums over the band index and wave vector.

By writing the eigenfunction in terms of the basis function $\varphi_{l\alpha}(\mathbf{r})$,

$$\psi_n(\mathbf{r}) = \sum_{l,\alpha} c_{nl\alpha} \varphi_{l\alpha}(\mathbf{r}), \quad (18)$$

where l runs over all atoms and α over all orbital types, it is easy to find

$$\mathbf{F}_l = -2 \sum_{l'} \sum_n \sum_\alpha \sum_{\alpha'} c_{nl\alpha}^* c_{nl'\alpha'} \frac{\partial H_{l\alpha,l'\alpha'}}{\partial \mathbf{R}_l} = \sum_{l'} \mathbf{F}_{ll'}, \quad (19)$$

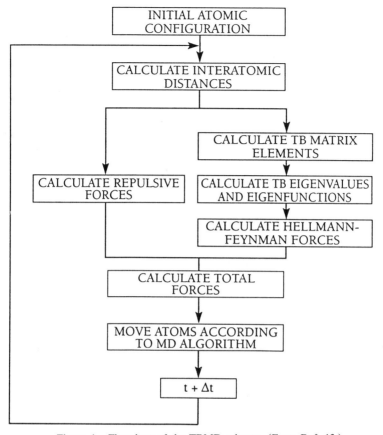

Figure 4. Flowchart of the TBMD scheme. (From Ref. 13.)

with

$$\mathbf{F}_{ll'} = -2 \sum_n \sum_\alpha \sum_{\alpha'} c_{nl\alpha}^* c_{nl'\alpha'} \frac{\partial H_{l\alpha,l'\alpha'}}{\partial \mathbf{R}_l}. \tag{20}$$

The $\mathbf{F}_{ll'}$ in Eq. (20) is a many-body force because the electronic states sample the environment of the interacting atom l and l' in the system. The formula above for the Hellmann–Feynman force calculation is also applicable to the tight-binding Hamiltonians when the Hubbard-like u term is present. Inclusion of the overlap matrix S in a nonorthogonal tight-binding scheme can also be handled by replacing the Hamiltonian matrix H by $H - \varepsilon_n S$ in the equations above.

B.   TBMD Using Car–Parrinello Algorithm

An algorithm for TBMD similar to the method of Car and Parrinello has been developed by Khan and Broughton [14]. In this procedure the trajectories of both the ionic and electronic coordinates are predicted via molecular dynamics, with the forces on the electrons and ions being calculated from a "fictitious Lagrangian" L,

$$L = \frac{1}{2} \mu \sum_{n,l,\alpha} (\dot{c}_{nl\alpha})^2 + \frac{1}{2} M \sum_{l} \dot{\mathbf{R}}_1^2 - E_{\text{tot}}[\{\mathbf{R}_l\}, \{c_{nl\alpha}\}], \qquad (21)$$

and the constraining equations imposed by the orthonormality of the occupied states,

$$\sum_{l,\alpha} c_{nl\alpha} c_{n'l\alpha} - \delta_{nn'} = 0. \qquad (22)$$

In the fictitious Lagrangian above, M is the mass of the ion and the electronic degrees of freedom $c_{nl\alpha}$ are treated as "position" variables of classical particles of fictitious mass μ. The equations of motion of the system are

$$M\ddot{\mathbf{R}}_l = - \frac{\partial E_{\text{tot}}[\{\mathbf{R}_l\}, \{c_{nl\alpha}\}]}{\partial \mathbf{R}_l} \qquad (23)$$

for the ionic coordinates and

$$\mu \ddot{c}_{nl\alpha} = - \frac{\partial E_{\text{tot}}[\{\mathbf{R}_l\}, \{c_{nl\alpha}\}]}{\partial c_{nl\alpha}} + \sum_{n'} \Lambda_{nn'} c_{n'l\alpha} \qquad (24)$$

for the electronic coordinates. Here n and n' run over the occupied states. The matrix $\Lambda_{nn'}$ is a symmetric matrix whose elements are the Lagrange multipliers introduced to satisfy the orthonormality constraints of the electronic coordinates given by Eq. (22). The Lagrange multipliers are easily determined by differentiating Eq. (22) twice with respect to time, giving

$$\Lambda_{nn'} = \sum_{l,\alpha} \left(\frac{1}{2} c_{n'l\alpha} \frac{\partial E_{\text{tot}}}{\partial c_{nl\alpha}} + \frac{1}{2} c_{nl\alpha} \frac{\partial E_{\text{tot}}}{\partial c_{n'l\alpha}} - \mu \dot{c}_{nl\alpha} \dot{c}_{n'l\alpha} \right). \qquad (25)$$

The inclusion of the electronic degrees of freedom in the molecular dynamics step is an important innovation by Car and Parrinello. Since this scheme operates only on the electronic coordinates belonging to the occupied subspace of the electronic eigenvectors, it speeds up the calculation considerably for the cases where wavefunctions are expanded using a plane-wave basis. In the case of TBMD, the occupied subspace is about half of the total space spanned by the expansion coefficients. The Car–Parrinello scheme does not provide significant savings in this respect. Furthermore, the fictitious dynamics of the electronic degrees of freedom greatly reduces the MD time step compared to the direct diagonalization method. However, because of the sparseness of the Hamiltonian matrix, the computational effort in this scheme should grow quadratically with the number of basis orbitals instead of the cubic growth for the direct diagonalization method in the limit of large number of atoms.

C.   TBMD Including Electronic Temperature

To include the effects of electronic states at finite temperature, the Fermi–Dirac (F–D) distribution is used to describe the occupation of the electronic states in the energy and force calculations in the TBMD simulations:

$$E_{TB} = 2 \sum_i \varepsilon_i f_i \,, \tag{26}$$

$$\mathbf{F}_l = -2 \sum_i \left\langle \psi_i \left| \frac{\partial H_{TB}}{\partial \mathbf{R}_l} \right| \psi_i \right\rangle f_i - 2 \sum_i \varepsilon_i \frac{\partial f_i}{\partial(\varepsilon_i - \mu)} \frac{\partial(\varepsilon_i - \mu)}{\partial \mathbf{R}_l} \,, \tag{27}$$

where

$$f_i = \frac{1}{e^{(\varepsilon_i - \mu)/k_B T_{el}} + 1} \,. \tag{28}$$

μ, the chemical potential, is adjusted every time step to guarantee the conservation of the total number of electrons:

$$2 \sum_i f_i = N_{el} \,. \tag{29}$$

It was pointed out by Pederson and Jackson [36] that it is very difficult to calculate the second term in the equation (27) in first-principles MD simulations. However, Wentzcovitch et al. [37] introduced the Mermin free energy [38]:

$$G = E_{total} + K_I - T_{el} S_{el} \,, \tag{30}$$

$$S_{el} = -2k_B \sum_i [f_i \ln f_i + (1 - f_i) \ln(1 - f_i)] \tag{31}$$

and showed numerically that the first-principles MD simulation conserves the free energy G if one drops the second term in Eq. (27). It can also be shown analytically that only the first term in Eq. (27) is required if the Hellmann–Feynman forces are calculated using the electronic free energy instead of electronic energy E_{TB}. The second term in Eq. (27) is canceled by the derivative of the electronic entropy:

$$\frac{\partial(T_{el}S_{el})}{\partial \mathbf{R}_l} = T_{el} \sum_i \frac{\partial S_{el}}{\partial f_i} \frac{\partial f_i}{\partial(\varepsilon_i - \mu)} \frac{\partial(\varepsilon_i - \mu)}{\partial \mathbf{R}_l}. \tag{32}$$

Equation (32) can be rewritten as

$$\frac{\partial(T_{el}S_{el})}{\partial \mathbf{R}_l} = -2k_B T_{el} \sum_i \frac{\partial[f_i \ln f_i + (1 - f_i)\ln(1 - f_i)]}{\partial f_i} \frac{\partial f_i}{\partial(\varepsilon_i - \mu)} \frac{\partial(\varepsilon_i - \mu)}{\partial \mathbf{R}_l}, \tag{33}$$

and after some simple algebra, Eq. (32) becomes

$$\frac{\partial(T_{el}S_{el})}{\partial \mathbf{R}_l} = 2 \sum_i \varepsilon_i \frac{\partial f_i}{\partial(\varepsilon_i - \mu)} \frac{\partial(\varepsilon_i - \mu)}{\partial \mathbf{R}_l}. \tag{34}$$

Here conservation of the total number of electrons is assumed:

$$\sum_i \frac{\partial f_i}{\partial(\varepsilon_i - \mu)} \frac{\partial(\varepsilon_i - \mu)}{\partial \mathbf{R}_l} = \frac{\partial}{\partial \mathbf{R}_l} \sum_i f_i = \frac{1}{2} \frac{\partial}{\partial \mathbf{R}_l} N_{el} = 0. \tag{35}$$

Thus the second term in Eq. (27) is canceled by the derivative of electronic entropy and the first term is $-\partial(E_{TB} - T_{el}S_{el})/\partial \mathbf{R}_l$. The inclusion of electronic temperature effects not only avoids the instability caused by the change of occupancies of states near the Fermi level in metallic systems, but also includes the effects of electronic entropy into the calculation in a very convenient manner.

IV. APPLICATIONS

In the last six years, TBMD has been applied to the study of structural, dynamical, and electronic properties of various covalent systems, including crystalline structures, disordered liquid and amorphous phases, clusters, surfaces, defects, and even more complex hydrogenated systems. In this section we highlight some of these applications. We focus primarily on the results of our work; work from other groups will be covered only briefly.

A. Study of Anharmonic Effects in Solids

One of the first successes of TBMD has been the study of the anharmonic properties of crystalline silicon and diamond [13, 20]. Since the system stays close to the diamond structure throughout the simulation, the tight-binding parameters proposed by Chadi provide a good description. Wang et al. [13] modified the original Chadi model by changing the pairwise repulsive potential to reproduce accurately the binding energy curve of the perfect diamond lattice obtained from first-principles LDA calculations. Listed in Table IV are some properties calculated from the resulting tight-binding model in comparison with *ab initio* and classical potential calculation results [11, 39, 40], and with experimental data [29]. The model gives a good description of the Grüneisen parameters, although these properties are not included in the fitting database to determine the tight-binding parameters. At low temperatures, the tight-binding model also predicts correctly the negative thermal expansion in silicon and positive thermal expansion in diamond, respectively [28], as one can see from Fig. 5, where the tight-binding calculation results are

TABLE IV

Some Properties of Silicon Obtained by the Tight-Binding (TB) Model of Ref. 20 Compared with the Results Obtained by First-Principles (LDA), the Stillinger-Weber (S-W) Potential, and Experiments

	TB	LDA	S-W	Expt.[a]
B (10^{11} erg/cm^3)	9.20	9.20[b]	10.14[c]	9.78
$C_{11} - C_{12}$ (10^{11} erg/cm^3)	7.25	9.80[d]	7.50[c]	10.12
C_{44}^{0} [e] (10^{11} erg/cm^3)	10.26	11.10[d]	—	—
C_{44} [f] (10^{11} erg/cm^3)	6.16	8.50[d]	5.64[c]	7.96
$\nu_{\text{LTO}(\Gamma)}$ (THz)	16.95	15.16[b]	17.83[c]	15.53
$\nu_{\text{TA}(X)}$ (THz)	4.96	4.45[b]	5.96[c]	4.49
$\nu_{\text{TO}(X)}$ (THz)	14.71	13.48[b]	—	13.90
$\nu_{\text{LOA}(X)}$ (THz)	12.37	12.16[b]	—	12.32
$\gamma_{\text{LTO}(\Gamma)}$	0.98	0.92[b]	0.80[g]	0.98
$\gamma_{\text{TA}(X)}$	−1.12	−1.50[b]	−0.04[g]	−1.40
$\gamma_{\text{TO}(X)}$	1.37	1.34[b]	0.89[g]	1.50
$\gamma_{\text{LO}(X)}$	1.02	0.92[b]	0.83[g]	0.90

[a] From Ref. 29.
[b] From Ref. 39.
[c] From Ref. 11.
[d] From Ref. 40.
[e] C_{44}^{0} denotes the shear constant that would appear in the absence of internal displacements.
[f] C_{44} is the shear constant in the absence of internal stress.
[g] From Ref. 20.

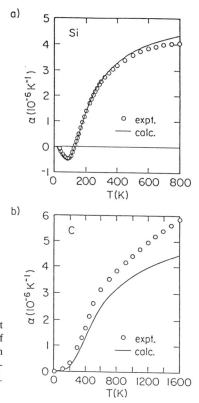

Figure 5. Linear thermal-expansion coefficient of (*a*) silicon and (*b*) diamond as a function of temperature. The solid lines are the results from TB calculations and the open circles are experimental data from Refs. 29 and 41. (From Ref. 28.)

compared with experimental data. By performing TBMD simulations at various temperatures, Wang et al. obtained the anharmonic phonon frequency shifts and phonon line widths in silicon and diamond as a function of temperature [20]. The results, shown in Figs. 6 and 7 are in excellent agreement with experimental data obtained from Raman scattering experiments [42, 43]. TBMD simulations have also been used to study the sharp two-phonon Raman peak in diamond observed by experiment more than 25 years ago [44, 45]. These successes indicate that the empirical tight-binding approach, although simple, is able to capture the essential interactions in the system and can describe systems having directional bonding with an accuracy comparable to *ab initio* approaches. By comparison, many classical potentials failed to describe the anharmonic behavior of covalent materials correctly. For example, the Stillinger–Weber (SW) potential, used in numerous studies of silicon, fails to

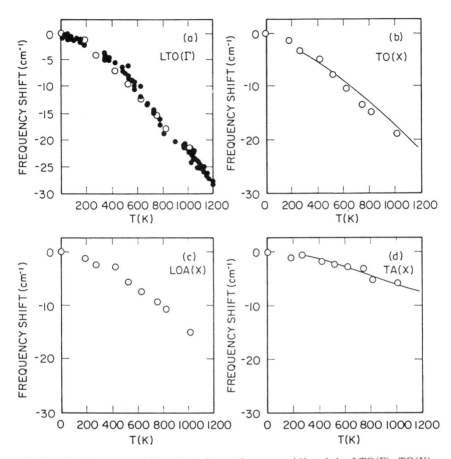

Figure 6. Temperature-dependent phonon frequency shifts of the LTO(Γ), TO(X), LOA(X), and TA(X) modes of silicon. The open circles are the TBMD results. The solid dots in (a) are Raman scattering data quoted from Ref. 42. The lines in (b) and (d) are experimental data quoted from Ref. 43. (From Ref. 20.)

reproduce the strong negative Grüneisen parameter of the TA(X) phonon mode in silicon (see Table IV).

B. Simulation of Liquids

The structure and properties of liquid silicon have been studied by several groups [15, 24, 32, 33, 35] using the GSP tight-binding model and the model developed by Kwon et al. Results from these studies demonstrated that the structure and properties of liquid silicon can be well reproduced

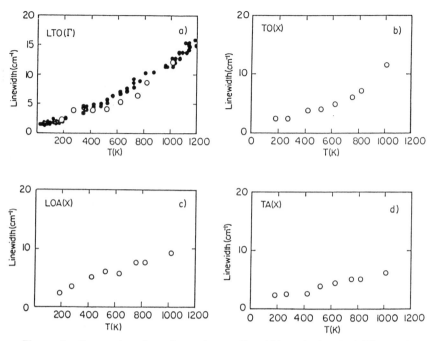

Figure 7. Temperature-dependent phonon linewidths of the LTO(Γ), TO(X), LOA(X), and TA(X) modes of silicon. The open circles are the TBMD results. The solid dots in (a) are Raman scattering data quoted from Ref. 42 (From Ref. 20.)

by TBMD simulation. The average coordination number obtained from the TBMD simulations is found to be close to the experimental value and *ab initio* molecular dynamics simulation result of 6.4 [46, 47]. The radial and angular distribution functions and the electronic structures obtained from the TBMD simulations are in quite close agreement with *ab initio* molecular dynamics results using the Car–Parrinello method [47], as one can see from Figs. 8–10. In particular, the bond-angle distribution function (with a bond-length cutoff of 3.22 Å) exhibits two peaks, at about 100 and 60°, respectively. This double peak structure agrees with the Car–Parrinello simulation result but differs qualitatively from the simulation result using the classical Stillinger–Weber (SW) potential, which fails to predict the second peak at 60° besides the main peak near the tetrahedral bond angle [48]. The latent heat of fusion for silicon obtained from simulation of Ref. 33 is about 0.43 eV, which is in good agreement with the experimental value of 0.47 eV [49]. It has been noted [33] that the GSP model tends to overestimate the average nearest-neighbor distance (2.54 Å, compared to the experimental value of 2.50 Å

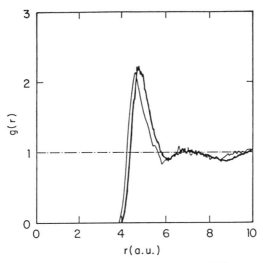

Figure 8. Radial distribution function of liquid silicon. Thick line represents the TBMD result. Thin line is the result obtained by the Car–Parrinello method [47]. (From Ref. 33.)

and Car–Parrinello result of 2.46 Å; the SW potential gives 2.56 Å). The model may also overestimate the pressure and melting temperature. The diffusion constant deduced from the TBMD simulation [33] is 1.25×10^{-4} cm^2/s at 1780 K, which is smaller than the *ab initio* MD simulation results of 2.26×10^{-4} cm^2/s [47] and 1.90×10^{-4} cm^2/s [50] at 1800 K, but larger than the values of 0.694×10^{-4} cm^2/s at 1691 K [51] and 0.98×10^{-4} cm^2/s at 2010 K [52] obtained by the SW potential.

The structural, dynamical, and electronic properties of liquid carbon over a wide range of densities have been studied extensively using the XWCH tight-binding model [53]. The results showed that liquid carbon at low densities (about the graphitic density) is metallic and dominated by twofold- and threefold-coordinated atoms, while the high-density liquid (about diamond density or higher) favors tetrahedral arrangements. The radial distribution functions of liquid carbon at low and high densities are shown in Figs. 11 and 12. These results are consistent with Car–Parrinello simulation results, which are available at densities of 2.0 and 4.4 g/cm^3 [55–57], although the TBMD simulations tend to underestimate the coordination slightly in comparison with the Car–Parrinello simulation results. The diffusion constant for the liquid at density of 2.0 g/cm^3 and temperature of 5000 K is found to be 2.05×10^{-4} cm^2/s [53], which is slightly smaller than the value of 2.40×10^{-4} cm^2/s estimated from Car–Parrinello simulations at the same density and temperature [55, 57]. Very

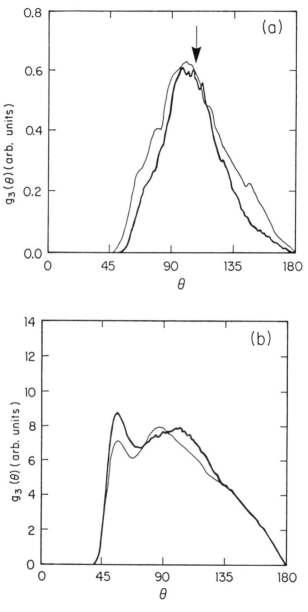

Figure 9. Bond-angle distribution functions of liquid silicon for (*a*) bond lengths less than 2.51 Å (2.40 Å in the Car–Parrinello simulation) and (*b*) bond lengths less than 3.22 Å (3.12 Å in the Car–Parrinello simulation). The thick line represents the TBMD result and the thin line is the result obtained by the Car–Parrinello method [47]). The arrow indicates the position of the tetrahedral angle. (From Ref. 33.)

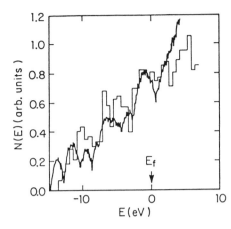

Figure 10. Electronic density of states of liquid silicon. The thick line represents the TBMD result and the thin line is the result obtained by the Car–Parrinello method [47]. (From Ref. 33.)

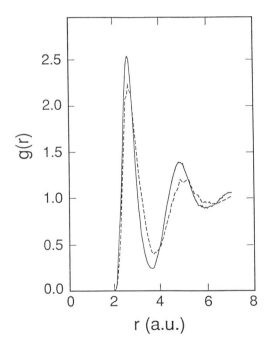

Figure 11. Pair correlation function $g(r)$ of liquid carbon obtained from TBMD (solid line) compared with the *ab initio* MD result of Ref. 55 (dashed line). Both simulations are for a 54-atom unit cell with a density of $2.0\,g/cm^3$ (From Ref. 53.)

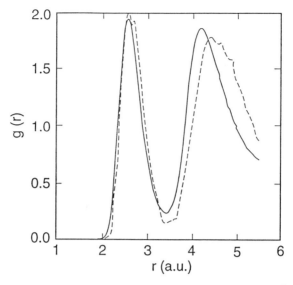

Figure 12. Pair correlation function $g(r)$ of high-density $(4.4 \, g/cm^3)$ liquid carbon obtained from TBMD (solid line) compared with the *ab initio* MD result of Ref. 56 (dashed line). (From Ref. 53.)

recently, Morris et al. have also performed TBMD to study the electrical conductivity in liquid carbon at various pressures [54]. The simulation results show that in the low- to intermediate-pressure regime, the conductivity of liquid carbon increases with pressure. This pressure dependence is in agreement with experimental data reported by Bundy more than 30 years ago [59]. The simulations indicated that the change in conductivity is due to the increase in the number of threefold atoms over twofold atoms as the pressure increases. At the present, there are no reliable experimental data on the structure of liquid carbon available for comparison with the TBMD or Car–Parrinello simulation results.

C. Simulation of Amorphous Structures

Wang and Ho have performed extensive TBMD simulations to study the structure and properties of amorphous carbon (a-C) over a wide range of densities [60–63]. The simulations showed that a-C samples produced under different densities (i.e., under different compressive stresses) have very different structures. The general trend is that the sp^3 bonding concentration increases when the a-C sample is generated under higher densities. Near the graphitic density $(2.27 \, g/cm^3)$, a-C is dominated by threefold atoms. Diamond-like sp^3-dominated a-C structures can be

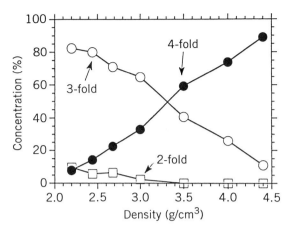

Figure 13. Concentration of various coordinated atoms in the a-C samples as a function of quenching density. (From Ref. 62.)

obtained by quenching carbon liquid under high pressure. Figure 13 and Table V show the concentration of three- and fourfold atoms as a function of the density of the sample. The simulations also show that the shape and the position of the first peak of the radial distribution function in a-C are correlated strongly with the relative concentration of sp-, sp^2-, and sp^3-bonded atoms in the sample. The peak position shifts toward larger distances with increasing density as the percentage of fourfold sites increases, as one can see from Fig. 14. The position of the nearest-neighbor peak gives a good measurement of the relative population of three- to fourfold sites in the system. The TBMD simulation results are

TABLE V
Concentration of Various Coordinated Atoms in the Amorphous Carbon Networks.[a]

Sample	Density (g/cm^3)	Two-fold (%)	Three-fold (%)	Four-fold (%)	Energy (eV/atom)
A	2.20 (2.20)	9.7	82.4	7.9	+0.6579
B	2.44 (2.44)	5.8	80.0	14.2	+0.6527
C	2.68 (2.68)	6.5	71.0	22.5	+0.6551
D	3.00 (3.00)	2.3	64.7	33.0	+0.5788
E	3.50 (3.30)	0.0	40.7	59.3	+0.4943
F	4.00 (3.35)	0.0	26.0	74.0	+0.5404
G	4.40 (3.40)	0.0	11.0	89.0	+0.5328

Source: Ref. 62.

[a] Both quenching density and final equilibrium density (in parentheses) are shown for each sample. The table also shows the binding energies of the a-C samples (relative to that of diamond).

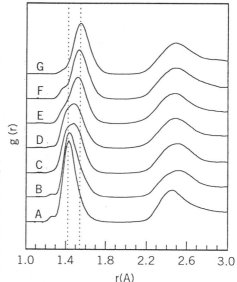

Figure 14. Pair-correlation functions of a-C samples generated under various densities as specified in Table V. Note that the first peak position shifts systematically toward larger bond lengths as the density increases. The two dashed vertical lines indicate the bond lengths of graphite (1.42 Å) and diamond (1.54 Å), respectively. (From Ref. 62.)

consistent with experimental observations. In Figs. 15 and 16 the radial distribution functions of the low-density graphitic and high-density diamond-like a-C obtained from TBMD simulations are compared with available experimental neutron scattering data [64, 65]. The agreement between the TBMD theory and experiments suggests that the TBMD results are accurate and reliable for these complex systems. The TBMD

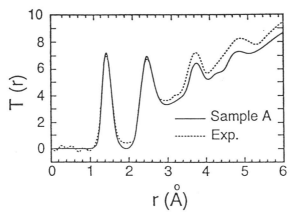

Figure 15. Radial distribution function $T(r)$ of the low-density a-C ($\rho = 2.20 \, \mathrm{g/cm^3}$) obtained from TBMD simulation (solid curve) compared with the neutron scattering data of Ref. 64 (dotted curve). The simulation result has been broadened by A Gaussian function with a width of 0.085 Å. (From Ref. 62.)

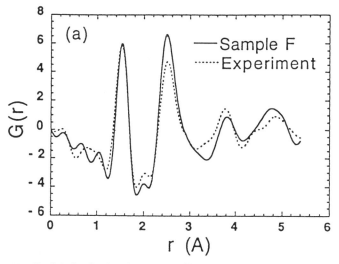

Figure 16. Radial distribution functions $G(r)$ of the diamondlike amorphous carbon samples generated by tight-binding molecular dynamics (solid curve) compared with the neutron scattering data of Ref. 65 (dotted curve). The theoretical results have been convoluted with the experimental resolution corresponding to the termination of the Fourier transform at the experimental maximum scattering vector $Q_{max} = 16 \, Å^{-1}$. (From Ref. 62.)

simulations can also give microscopic pictures about the atomic arrangement and electronic structure of this disordered system. There is some tendency for segregation of three- and four-fold atoms in the a-C sample (Fig. 17).

Besides a-C, TBMD has also been applied to study amorphous silicon [34, 66, 67] and amorphous GaAs [68]. For details of these studies, we refer the reader to the original publications.

D. Study of Clusters and Buckyballs

Tight-binding calculations and tight-binding molecular dynamics simulations have been applied very successfully to the study of carbon clusters, buckyballs, and buckytubes [69–85, 91–93]. The accuracy and efficiency of the TBMD scheme enabled us to predict the ground-state structures of every even-numbered carbon cluster ranging from C_{20} to C_{102} [71–76]. In Fig. 18 the formation energies of these fullerenes are plotted as a function of cluster size. It is found that the C_{60}, C_{70}, and C_{84} clusters are especially stable compared with their neighbors. Indeed, these magic-number clusters correspond to the most abundant fullerenes observed in experiments. For the case of larger fullerenes, such as C_{78}, C_{82}, and C_{84}, TBMD theoretical results preceded the experimental determination of the structures of these molecules. In particular, for C_{84}, experimental determination of the structure is hindered by the presence of two isomers

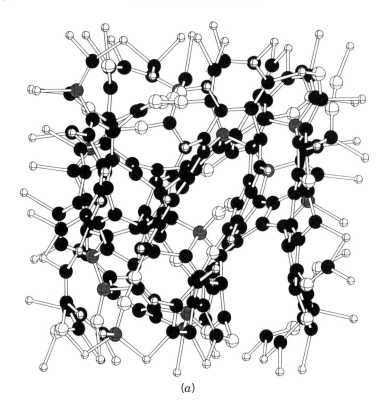

(a)

Figure 17. Microscopic structures of the TBMD-generated (*a*) amorphous carbon and (*b*) diamondlike amorphous carbon networks. The densities of the samples are 2.2 and 3.35 g/cm^3 respectively. Various coordinated atoms are shaded differently (two-fold: white balls; three-fold: black balls; four-fold: shaded balls). Smaller white balls represent carbon atoms outside the central cell (periodic boundary conditions are used). (From Refs. 60 and 61.)

in the experimental extract. Our TBMD prediction [71, 72] of the identities of the two ground-state isomers as shown in Fig. 19 has been verified by subsequent experiments [86–88]. In general, the energy ordering of the different fullerene isomers for a given cluster size obtained from the TBMD optimization are in good agreement with first-principles calculations [89, 90], even though the energy differences among the isomers are very small. In Table VI, the binding energies of some C_{78}, C_{82}, and C_{84} isomers obtained from the tight-binding calculations are compared with the results from the first-principles LDA calculations using the DMol package from Biosym.

Tight-binding molecular dynamics with the XWCH potential has also

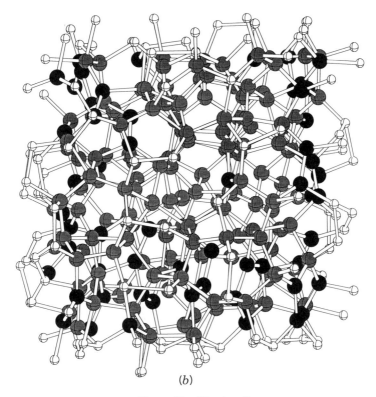

(b)

Figure 17. (*Continued*)

been used to study the dynamics of the carbon fullerenes, including fullerene formation, fragmentation, disintegration, and collision processes [78, 80–82]. Vibrational properties of fullerenes and buckytubes have also been studied extensively with TBMD [77, 79, 91]. By incorporating a Lennard-Jones potential plus a bond charge model for the intermolecular interactions and the XWCH model for intra molecular interactions, Yu et al. [92, 93] have also performed TBMD simulations to study the effects of pressure on the structural and vibrational properties of solid C_{60} (Fig. 20). The results obtained from these simulations are in good agreement with available experimental data. In contrast to the case of carbon clusters, TBMD studies of silicon clusters have been limited. Only results on small silicon clusters (mostly $n \leq 10$) have been reported [21, 24, 31, 95, 96].

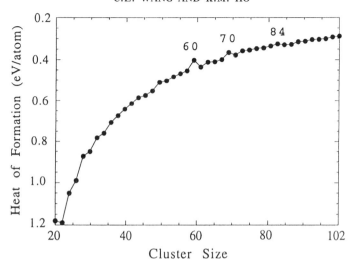

Figure 18. Heat of formation of carbon fullerenes relative to that of graphite as a function of fullerene size obtained from the TBMD calculations. (From Ref. 76.)

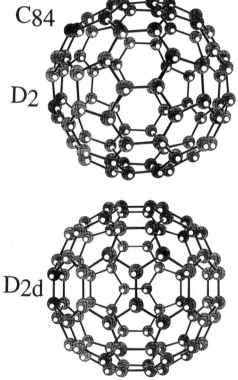

Figure 19. Lowest-energy isomers of C_{84} fullerene predicted by the TBMD calculations. (From Ref. 72.)

TABLE VI
Energy Ordering (eV/molecule) and HOMO-LUMO Gaps (eV) of Some C_{78}, C_{82}, and C_{84}
Isomers Obtained from Tight-Binding and LDA Calculations

Isomers	ΔE_{TB}	ΔE_{LDA}	HOMO-LUMO (TB)	HOMO-LUMO (LDA)
$C_{78}-C_{2v'}$	0.000	0.000	0.493	0.746
$C_{78}-D_{3h}$	0.101	0.160	0.353	0.562
$C_{78}-C_{2v}$	0.302	0.210	0.545	1.123
$C_{78}-D_3$	0.344	0.361	0.443	0.775
$C_{78}-D_{3h'}$	0.983	1.200	1.373	1.668
$C_{82}-C_2$	0.000	0.000	0.649	0.767
$C_{82}-C_s$	0.076	0.123	0.568	0.703
$C_{82}-C_{2'}$	0.166	0.312	0.362	0.703
$C_{82}-C_{2''}$	0.191	0.418	0.477	0.595
$C_{82}-C_{s'}$	0.226	0.385	0.773	0.793
$C_{82}-C_{s''}$	0.229	0.467	0.234	0.292
$C_{82}-C_{2v}(C_2)$	0.285	0.656	0.081	0.084
$C_{82}-C_{3v}$	0.566	1.287	0.120	0.003
$C_{82}-C_{3v'}(C_s)$	0.731	1.025	0.088	0.082
$C_{84}-D_2(22)$	0.000	0.000	0.82	1.06
$C_{84}-D_{2d}(23)$	0.033	0.034	0.84	1.12
$C_{84}-C_2(11)$	0.273	0.412	0.66	0.76
$C_{84}-D_2(21)$	0.429	0.630	0.40	0.76
$C_{84}-D_2(5)$	0.677	0.655	0.61	1.02
$C_{84}-D_2(1)$	1.908	2.260	1.28	1.47

Source: Refs. 76, 89, and 90.

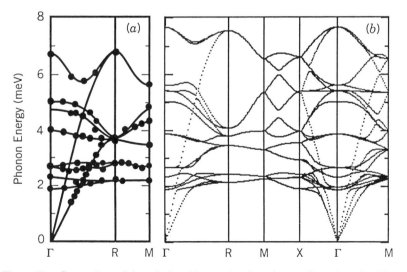

Figure 20. Comparison of the calculated intermolecular–phonon dispersion of solid C_{60} with experiment measurement: (*a*) experiment (data from Ref. 94); (*b*) TBMD calculation. (From Ref. 93.)

E.   Study of Defects, Surfaces, and Hydrogenated Systems

Tight-binding calculations and TBMD have also been performed to study the properties of more complex systems, such as defects, surfaces, and hydrogenated silicon and carbon systems. Using the tight-binding model developed by Goodwin et al., Wang et al. have studied the properties of point defects in crystalline silicon [32]. The formation energies of vacancy and self-interstitials in silicon obtained from TBMD are listed in Table VII. These results are in good agreement with first-principles calculation results [97–100]. The Jahn–Teller distortion around the silicon vacancy is also well described by the TBMD approach. Calculations using the TB model developed by Kwon et al. give similar results [24]. The XWCH model for carbon has also been applied to study defects in graphite by Xu et al. [101]. The results are also found to be in good agreement with first-principles LDA calculations and experimental data. The XWCH model is not suited for studying the defects in three-dimensional bulk graphite due to the short-ranged cutoff distance. Several TB models have also been applied to study the structures, energies, and electronic properties of grain boundaries in silicon and diamond by Paxton and Sutton [102] and Kohyama et al. [103–105].

Empirical tight-binding calculation of silicon and carbon surfaces can be traced back to the early work of Chadi [19] and Mele et al. [106, 107]. Recently, several newly developed models have also been applied to the studies of silicon, germanium, and diamond surfaces [22, 108–110, 112]. Almost all of these studies were zero-temperature static calculations except for the work of Stokbro et al., who applied the newly developed effective-medium TB model to study surface melting and defect-induced premelting behavior on the Si(100) surface [112].

TABLE VII

Comparison of TBMD Calculated Point-Defect Formation Energies with the First Principles (LDA) Calculation Results from Refs. 97–100[a]

Defect	64 atoms	216 atoms	512 atoms	LDA
Vacancy	3.67 (5.02)	3.96 (5.82)	4.12 (6.03)	3.6–5.0
T-inter	4.39 (6.42)	4.40 (6.54)	4.41 (6.55)	4.3–6.2
H-inter	5.78 (10.08)	5.90 (10.32)	5.93 (10.35)	5.0–6.0

Source: Ref. 32.

[a] The TBMD calculations have been performed with 64-, 216-, and 512-atom supercells, respectively. The values in parentheses are the formation energies of the unrelaxed defects. All energies are in units of eV.

Recently, several Si–H and C–H tight-binding models have been developed to simulate the behavior of hydrogen–silicon and hydrogen–carbon systems [113–118]. Most of these models describe fairly well the structural and vibrational properties of small hydrogenated silicon and carbon molecules and hydrogenated silicon and diamond surfaces. Dynamical simulations of hydrogen diffusion in crystalline silicon have been performed by Colombo et al. [115] and by Boucher and DeLeo [114]. The hydrogen diffusion constants as a function of temperature obtained from the TBMD simulation of Colombo et al. are in good agreement with those from *ab initio* MD simulations using the Car–Parrinello method [119]. TBMD simulations of defects in hydrogenated amorphous silicon have also been carried out recently by Li and Biswas [67].

V. RECENT DEVELOPMENTS

Recent developments in computers indicate great promise for a rapid increase in computing capability through the use of massively parallel computers that contain a large number of processors working in parallel on a single problem. However, the architecture of these powerful computers is quite different from that of traditional computers and requires the development of entirely new algorithms to take advantage of the parallel processing capability through the minimization of interprocessor communication overhead. The area of TBMD is no exception. Recently, there have been two interesting developments toward adapting the method to treat big systems and long time scales. The development of "order N" methods enables solution of the electronic structure problem with a computational effort that scales linearly with the system size. Compared with traditional techniques that scale as the cube of the system size, this makes possible the study of larger systems. The other is the development of a genetic optimization algorithm which is much more efficient than traditional simulated-annealing methods. Both of the above-mentioned algorithms are well suited for use on parallel computers. In addition to algorithm developments, there are also efforts to go beyond the traditional two-center approximation in attempts to generate more accurate and transferable tight-binding models. In the following subsections we discuss these new developments in more detail.

A. O(N) Algorithms for Electronic Structure Calculations

Recently, many schemes were proposed for solving the electronic structure problem with a computational effort that scales linearly with system

size [called *order N* or O(N) scaling] [120–128]. The development of
order N methods is motivated by the fact that standard matrix diagonalization algorithms have a complexity that grows as cube of the system size
and thus the use of first-principles or tight-binding calculations are limited
to small systems. With the development of computer capabilities and
computational methods, the sizes of the systems studied are becoming
larger and larger and we are now at a point where linear scaling methods
can compete effectively against traditional algorithms. An order N
algorithm, if it is efficient and accurate, will evidently broaden the
application of first-principles and tight-binding calculations to a wide
variety of new problems involving larger and more complicated systems.
Order N algorithms also tend to be more naturally adaptable to parallel
computers. With an order N scheme implemented on parallel machines, it
is expected that many simulations that were too expensive to be carried
out previously can now be performed fairly quickly.

   The density-matrix (DM) algorithm proposed by Li et al. [120] and
Daw [121] is one of the promising order N algorithms. In their approach
they introduce a variational method for solving the electron density
matrix. The method takes advantage of the locality of the density matrix
in real space to achieve linear scaling. The approximation is made by
truncating the off-diagonal elements of the density matrix beyond a cutoff
radius R_c. The method becomes exact as $R_c \to \infty$.

   Instead of calculating E_{TB} through direct-diagonalization (DD) of
H_{TB}, which scales as the cube of the system size, the DM method
performs the energy calculation through the following variational approach. The grand potential, defined as

$$\Omega = E_{\mathrm{TB}} - \mu N_e = \mathrm{tr}[\tilde{\rho}(H_{\mathrm{TB}} - \mu)] \,, \qquad (36)$$

is minimized with respect to the density matrix $\tilde{\rho}$, where N_e is the total
number of valence electrons in the system and μ is the chemical
potential. The density matrix $\tilde{\rho}$ is related to a variational matrix ρ
through

$$\tilde{\rho} = 3\rho^2 - 2\rho^3 \,, \qquad (37)$$

where off-diagonal elements of ρ beyond a cutoff range are set to zero by
assuming that ρ is well localized. In normal situations, the density matrix
is required to be idempotent (i.e., $\rho^2 = \rho$). However, the use of
expression (37) for $\tilde{\rho}$ greatly simplifies the problem: When ρ is idempotent, $\tilde{\rho} = \rho$. Moreover, it can be shown that an unconstrained optimization over ρ of the form above yields the same answer as a constrained

search restricted to idempotent matrices. The minimization of Ω can be carried out either with conjugate gradient or steepest descent algorithms.

The Hellmann–Feynman force can be calculated through the derivative of the grand potential Ω with respect to atomic coordinates at fixed μ,

$$\frac{d\Omega}{d\xi} = \frac{\partial\Omega}{\partial\rho}\frac{d\rho}{d\xi} + \frac{\partial\Omega}{\partial H_{TB}}\frac{dH_{TB}}{d\xi}, \tag{38}$$

but the first term vanishes at the variational solution, so that the force is given by

$$\frac{d\Omega}{d\xi} = \text{tr}\left[\tilde{\rho}\,\frac{dH_{TB}}{d\xi}\right]. \tag{39}$$

More details about the DM method can be found in Refs. 120 and 121.

In addition to the density-matrix scheme, $O(N)$ methods based on an orbital formulation of the electronic problem [122–125] also work quite well. A key idea of $O(N)$ orbital schemes is to use wavefunctions forced to be localized in given regions of space. The solution of the eigenstates directly is therefore abandoned in favor of a search for a set of localized wavefunctions spanning the same subspace as that spanned by the occupied eigenstates. In this way the total number of expansion coefficients used to represent the localized electronic orbitals depends linearly on the size of the system, and the number of operations needed for the evaluation of $H\phi$ can be reduced to $O(N)$. A general discussion of the density matrix scheme in comparison with the orbital schemes has been given by Qiu et al. [129] and Mauri and Galli [130].

Implementations of order N algorithms into TBMD have been reported by several groups [128–130]. Benchmark results on crystalline, amorphous and liquid carbon show that the computational efforts of these new $O(N)$ schemes do scale linearly with the number of atoms in the system. A comparison of the CPU time used in the standard diagonalization method and the density matrix approach as the function of system size for the diamond structure is shown in Fig. 21. The linear scaling of the variational density matrix method can clearly be seen. The crossover of CPU time between the two methods is about 60 atoms. For amorphous and liquid simulations, the crossovers are estimated to be 120 and 230 atoms, respectively [129]. The quality of the results obtained by the $O(N)$ method are also very good in comparison with the exact diagonalization method. These can be seen from Figs. 22 and 23, where the evolution of the potential energy ($E_{TB} + E_{rep}$), and the total energy of the system,

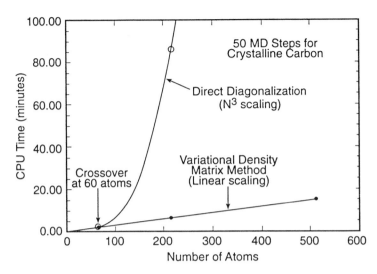

Figure 21. Measure of crossover point between the DM-TBMD and DD-TBMD method. The data are collected from the simulation of crystalline carbon over 50 MD time steps for system sizes of 64, 216, and 512 carbon atoms per unit cell. (From Ref. 129.)

versus simulation time are presented for the crystalline and amorphous simulations. For the crystalline simulation, the evolution of potential energy calculated from the DM-TBMD method agrees very well with that from the DD-TBMD. Although the system total energy calculated from DM-TBMD is not a constant, it fluctuates around a constant value, which is acceptable for MD simulations. The DM-TBMD scheme also works quite well for amorphous carbon. Although the potential energy versus simulation time does not match the exact solution as well as the crystalline case, it can be improved by increasing the cutoff range, improving the minimization tolerance, and reducing the MD time step. The total energy of the system fluctuates around a constant value that is conserved quite well in the entire simulation. For liquid carbon the evolution of the system potential energy calculated from DM-TBMD does not match what is obtained from the DD-TBMD, as can be seen in Fig. 24. The trajectories in phase space begin to diverge after about 50 MD time steps. The origin of this divergence can be traced to the chaotic nature of the dynamics in the liquid. Any two schemes with a slight difference will lead to a divergence over a long enough time period. Although the trajectories are different, the statistically averaged results of liquid carbon are very similar from the two simulations. The results of pair-correlation functions and atomic distributions obtained from the

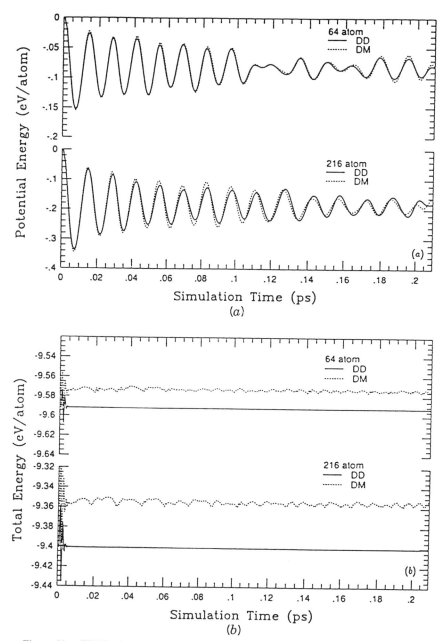

Figure 22. TBMD simulation for crystalline carbon for the period of 0.21 ps (300 MD time steps) with conjugate gradient tolerance $\tau = 10^{-5}$ for system sizes of 64 and 216 carbon atoms: (*a*) potential energy; (*b*) system total energy. (From Ref. 129.)

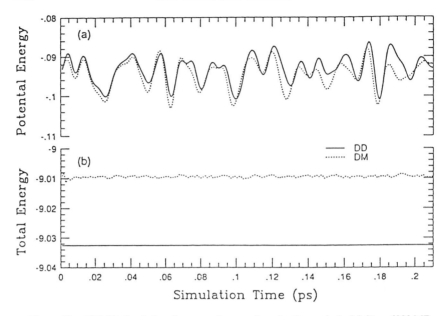

Figure 23. TBMD simulation for amorphous carbon for the period of 0.21 ps (200 MD time steps) with conjugate gradient tolerance $\tau = 10^{-5}$ for system sizes of 216 carbon atoms: (*a*) potential energy (eV/atom); (*b*) system total energy (eV/atom). (From Ref. 129.)

DM-TBMD and DD-TBMD schemes are compared in Fig. 25 and Table VIII. These results show that the DM-TBMD scheme reproduces the results of DD-TBMD quite well. Similar benchmark results have been reported by Mauri et al. using the O(N) orbital scheme [130].

With these newly developed O(N) TBMD algorithms, simulations with more than 1000 atoms can be performed on sequential computers such as the IBM RISC-6000 workstation or vector computers such as the Cray. Qiu et al. have applied the O(N) method to study the structure and energetics of giant fullerenes. Shown in Fig. 26 is the optimized geometry of a 1620-atom fullerene obtained using an IBM RISC-6000 workstation. Mauri and Galli have applied the O(N) TBMD to study the structure and dynamic of C_{60} striking a diamond surface [131]; 1140 carbon atoms have been used in their simulation.

The density-matrix O(N) TBMD scheme has recently been implemented successfully on Intel Paragon parallel computers by our group at Ames Laboratory, Iowa State University. Although optimization is still ongoing, extrapolation from current work indicates that practical simulations of 10,000-atom systems should be easily achievable on the 512-node

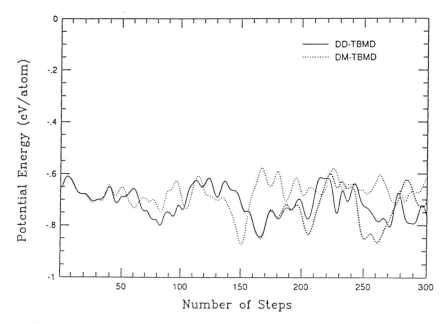

Figure 24. Evolution of potential energy versus simulation time for liquid carbon, where the heavy dashed line represents the DM-TBMD result starting from the configuration of DD-TBMD at the 150th time step. (From Ref. 129.)

XP/S-35, taking an estimated 5–20 hours for 1000 MD steps, depending on whether the system is crystalline or liquid. This development should enable us to perform TBMD simulations for a wide variety of new problems requiring larger number of atoms and longer simulation time.

B. Genetic Algorithm for Structural Optimization

While recent developments in parallel computer design and algorithms have made considerable progress in enlarging the system size that can be accessed using atomistic simulations, methods for effectively shortening the simulation time remain relatively unexplored. For example, determination of the lowest-energy configurations of a collection of atoms using molecular dynamics or Monte Carlo computer simulations is a hard problem. Because the number of candidate local energy minima grows exponentially with the number of atoms, the computational effort also increases exponentially [133]. In practice, realistic potentials describing covalently bonded materials possess significantly more rugged energy landscapes than those of the simple two-body potentials, further increasing the difficulty. Traditionally, simulated annealing technique has been

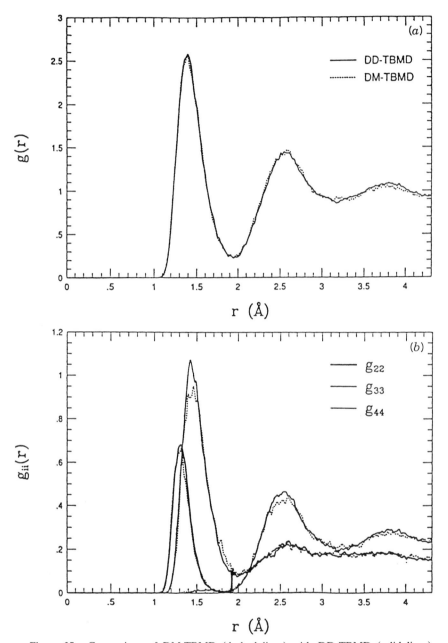

Figure 25. Comparison of DM-TBMD (dashed lines) with DD-TBMD (solid lines) simulation for liquid carbon in (*a*) pair correlation, (*b*, *c*) partial radial distribution, and (*d*) angular distribution function. (From Ref. 129.)

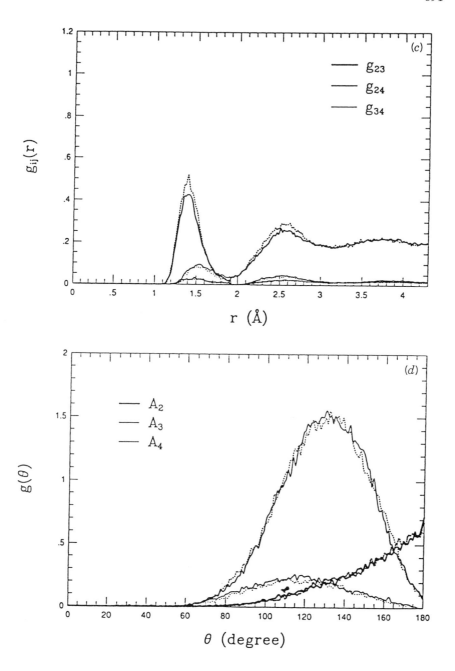

Figure 25. (*Continued*)

TABLE VIII
Ratios of Various Coordinated Atoms of Liquid Carbon with a Density of $2.0\,\mathrm{g/cm^3}$
Calculated from DD-TBMD and DM-TBMD

Scheme	Onefold (%)	Twofold (%)	Threefold (%)	Fourfold (%)	Average Coordination Number, n_c
DD-TBMD	3.85	40.85	50.75	4.54	2.56
DM-TBMD	3.62	42.63	49.59	4.15	2.54

Source: Ref. 129.

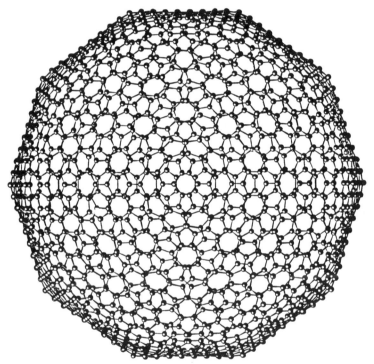

Figure 26. Geometry of a giant carbon fullerene (C_{1620}) optimized using the $O(N)$ TBMD scheme.

used to search for the ground-state structures. However, attempts to use simulated annealing to find the global energy minimum in these systems are frustrated by high-energy barriers that trap the simulation in one of the numerous metastable configurations. Thus an algorithm is needed that can "hop" from one minimum to another and permit an efficient sampling of phase space.

Very recently, an approach based on the genetic algorithm (GA), an optimization strategy inspired by the Darwinian evolution process [136], has been developed. Starting with a population of candidate structures, the method relaxes these candidates to the nearest local minimum. Using the relaxed energies as the criterion of fitness, a fraction of the population is selected as "parents." The next generation of candidate structures is produced by "mating" these parents. The process is repeated until the ground-state structure is located.

Deaven and Ho have applied this algorithm to optimize the geometry of carbon clusters up to C_{60} [132]. In all cases of study, the algorithm finds the ground-state structures efficiently, starting from an unbiased population of random atomic coordinates. This performance is very impressive since carbon clusters are bound by strong directional bonds, which result in large energy barriers between different isomers. Although there have been many previous attempts to generate the C_{60} buckyball structure from simulated annealing, none has yielded the ground-state structure [84, 134, 135]. The genetic approach dramatically outperforms simulated annealing and can arrive at the lowest-energy structure of C_{60} (the icosahedral buckminsterfullerene cage) in a relatively short simulation time, as one can see from Fig. 27.

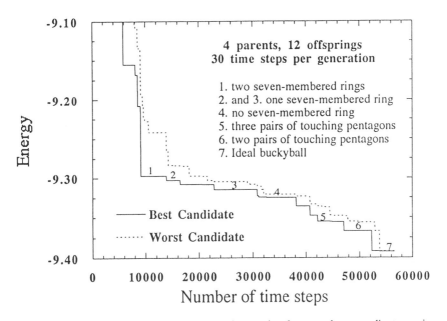

Figure 27. Generation of the C_{60} molecule, starting from random coordinates, using the genetic algorithm with four candidates ($p = 4$). The energy per atom is plotted for the lowest-energy (solid line) and highest-energy (dashed line) candidate structure as a function of MD time steps. Mating operations among the four candidates are performed every 30 MD time steps. (From Ref. 132.)

While the artificial dynamics of the genetic algorithm cannot be expected to mimic the natural annealing process in which atomic clusters are formed, Deaven and Ho [132] found that the intermediate structures located by the genetic algorithm on its way to the ground-state structure are very similar to the results of simulated annealing. Thus it appears that the same kinetic factors which influence the annealing process also affect the ease with which a particular candidate is generated by the genetic algorithm. It would be very interesting to compare the results from the genetic algorithm with those of extended conventional simulated annealing runs.

C. Tight-Binding Models Beyond Two-Center Approximation

Developments in new algorithms and optimization techniques together with progress in computer capabilities have made TBMD calculations quite efficient and fast. At the present stage, it seems that the main time-consuming bottleneck in the TBMD approach is the construction of an accurate and reliable tight-binding model to describe the interatomic interactions in a given material. Most previous TB models basically followed the two-center approximation for the hopping integrals formulated by Slater and Koster [17] four decades ago. The accuracy and transferability of these models are limited by the approximation, particularly for systems where metallic effects are significant. Recently, efforts have tried to go beyond the traditional two-center approximation with tight-binding models that include environment-dependent interactions.

Tight-binding models including explicit three-center interactions in the expression for the repulsive energy were first suggested by Sawada [22] and subsequently by Kohyama [27] and Mercer and Chou [25]. They showed that the inclusion of three-body interactions in the repulsive energy gives a better description of the energy–volume curves for silicon and germanium than that of the GSP model. Tight-binding models that allow the diagonal matrix elements to be dependent on the environment of the atoms have also been developed for silicon by Mercer and Chou [26] and for transition metals by Cohen et al. [137]. They showed that allowing the diagonal matrix elements of the TB Hamiltonian to be environment dependent can simplify the expression for the repulsive energy term in the tight-binding models. Recently, neighbor-dependent hopping integrals between silicon and hydrogen atoms has also been considered by Li and Biswas [116] in their Si–H tight-binding model. They found that an environment-dependent Si–H interaction is essential for a correct description of the properties of the interstitial hydrogen atom in the silicon lattice.

A tight-binding model that considers the environment dependence for off-diagonal as well as diagonal matrix elements is the effective-medium tight-binding model recently developed by Stokbro et al. [110]. In this model, a LMTO-TB Hamiltonian [111] is used to describe the one-electron energy in the total energy model. Although the LMTO-TB Hamiltonian appears in a two-center form, multicenter effects are taken into account through the structure factor S and through the environmental dependence of the diagonal elements. Application of this model to the study of Si surfaces has produced some interesting results [110, 112].

In another model recently developed by Tang et al. [138], the strength of the TB hopping parameters and the repulsive interaction between atoms i and j are modeled to be dependent on the environments of atoms i and j through two scaling factors. The first factor scales the distance between the two atoms according to their effective coordination numbers. Longer effective bond lengths are assumed for higher coordinated atoms. Another is a screening factor that is designed such that the interaction between two atoms is weaker due to screening effects if there are intervening atoms between the interacting atoms. In addition, the diagonal matrix elements are also dependent on the environment, Preliminary results on carbon show that this new approach improves significantly the transferability of the model in describing the band structures, binding energies, and other properties of higher-energy hypothetical metallic structures. Applications of the new approach to silicon and transition metal elements are in progress.

VI. CONCLUDING REMARKS

In this article we have shown that tight-binding molecular dynamics is a useful and powerful method for studying covalent systems. The scheme is much more accurate and transferable than simulations using classical potentials and is substantially faster than *ab initio* methods. Recent developments in the $O(N)$ algorithm for tight-binding electronic calculation and parallel computing have greatly enhanced the capability of the TBMD scheme.

The transferability of the tight-binding model is a key issue. Among the tight-binding models proposed so far, the model for carbon developed by Xu et al. [23] is the most successful. The model has been applied to various carbon systems, ranging from clusters to crystalline structures and to the liquid and amorphous structures of carbon. The results from TBMD simulations not only agree well with available *ab initio* calculations and experimental data, but also predict correctly the structures and properties of fullerenes and provide useful insights into the structure and properties of liquid and amorphous carbon.

In comparison, the accuracy and transferability of TB models for silicon are not as good as that for carbon. Although several tight-binding models for silicon have been developed, each model seems to work well only for certain structures and properties of silicon. In particular, none of these tight-binding models can predict correctly the geometry of Si_{13} cluster in comparison with *ab initio* calculations. A more transferable tight-binding model for silicon is needed. Tight-binding models for binary systems (e.g., Si–H, C–H) are still not well established. Although several models have been proposed, the transferability of these models need to be demonstrated.

We note that most tight-binding models developed previously are based on the two-center approximation. While the two-center approximation greatly simplifies the TB parametrization, neglecting multicenter interactions is justified only when the bonding in the system is very strongly covalent, so that electrons are well localized in the covalent-bond region (e.g., the lower-coordinated structures of carbon). The approximation is inadequate for systems where metallic effects are significant. To improve the accuracy and transferability of the tight-binding model, the effects of multicenter interactions need to be taken into account. In several recent developments, some steps in this direction have been taken.

It is also known that the use of a minimal basis set is not adequate for describing the electronic conduction band and the optical properties of tetrahedral silicon. A larger basis set, including the effects of higher-energy orbitals, may be needed to improve the accuracy of the model further. It is still not clear whether a nonorthogonal basis is better than an orthogonal basis.

Before closing the paper, we would like to list some areas where we believe TBMD will be very useful in future studies. We anticipate that TBMD will continue to play an important role in simulation study of carbon systems. For example, further applications of this scheme to examine the pressure–temperature phase diagram of carbon will be very intriguing. This type of simulation has the following advantages: (1) it will lead to better understanding of the experimental data available; (2) regions of the phase diagram at very high temperature and very high pressure can be accessed; and (3) any "unexpected" new phase from the computer experiment may stimulate further experimental investigations. Carbon clusters will continue to be a topic of interest throughout the 1990s. Although fullerene products are dominated by C_{60} and C_{70}, larger fullerenes and metallofullerenes have been fabricated in experiments. Characterization of the structural, dynamical, and electronic properties of these fullerenes and understanding the formation mechanism of the fullerenes remain a great challenge.

Other semiconductor clusters, such as silicon, silicon–oxygen, silicon–carbon, and boron–nitrogen clusters, not only are fascinating from the scientific point of view but have a variety of important technological applications. The properties of these clusters are not yet well understood and may be suitable for TBMD studies.

Disordered covalent materials such as amorphous silicon, hydrogenated amorphous silicon, amorphous carbon, and diamondlike carbon are of current interest because they are important in technological applications. The TBMD scheme will be very useful in the study of the microscopic structural, dynamical, and electronic properties of these complex systems.

Besides semiconductor elements (e.g., silicon, carbon), there have been considerable efforts in the last several years devoted to the tight-binding modeling of metals. Since the strong directional d-band bonding in many body-centered cubic transition metals resembles the covalent bonding in semiconductors to some extent, they may provide good opportunities for TBMD approaches.

The structures and properties of polymers and biomolecules provide a very exciting research area for potential applications with the TBMD method. Knowledge of the properties of polymers and biomolecules at the atomic and electronic levels is very important for a fundamental understanding of the properties and performance of these materials. In recent years, computer simulations have been emerging as a very useful approach to gaining insights into the microscopic nature of realistic materials. Nevertheless, polymers and biomolecules are still too complex for *ab initio* calculations, and most current computational studies are being performed using classical empirical potentials. Since these systems consist primarily of strongly covalent-bonding elements such as carbon, hydrogen, oxygen, nitrogen, and phosphorus, we anticipate that the interatomic interactions should be fairly well modeled within the tight-binding framework. It would be interesting to see what can be achieved by application of the TBMD approach in this area.

ACKNOWLEDGMENTS

We would very much like to thank C. T. Chan for his close collaboration in this TBMD project and for his critical reading of this manuscript. We are also grateful to B. L. Zhang, C. H. Xu, B. J. Min, S. Y. Qiu, J. R.

Morris, D. Deaven, D. Turner, J. Corkill, Y. F. Li, M. S. Tang, B. C. Pan, and H. Haas for their contributions to the TBMD project, and to X. Q. Wang, R. Biswas, Z. Zhu, Y. H. Lee, I. Kwon, G. Kopidakis, and C. M. Soukoulis for their collaboration. We also would like to thank B. N. Harmon for his encouragement and many useful discussions throughout the course of these studies. This work was supported by the Director for Energy Research, Office of Basic Energy Science, the High Performance Computing and Communications Initiative, including a grant of computer time at the NERC at Livermore and by grants from the U.S. AFOSR and the National Science Foundation. Ames Laboratory is operated for the U.S. Department of Energy by Iowa State University under contract W-7405-Eng-82.

REFERENCES

1. G. P. Srivastava and D. Weaire, *Adv. Phys.* **36**, 463 (1987).

2. R. Car and M. Parrinello, *Phys. Rev. Lett.* **55**, 2471 (1985).

3. F. H. Stillinger and T. A. Weber, *Phys Rev. B* **31**, 5262 (1985).

4. J. Tersoff, *Phys. Rev. Lett.* **56**, 632 (1986); *Phys. Rev. B* **37**, 6991 (1988).

5. J. Tersoff, *Phys. Rev. Lett.* **61**, 2879 (1988).

6. R. Biswas and D. R. Hamann, *Phys. Rev. Lett.* **55**, 2001 (1985).

7. B. W. Dodson, *Phys. Rev. B* **35**, 2795 (1987).

8. J. R. Chelikowsky, J. C. Phillips, M. Kamal, and M. Strauss, *Phys. Rev. Lett.* **62**, 292 (1989).

9. M. I. Baskes, *Phys. Rev. Lett.* **59**, 2666 (1987).

10. K. E. Khor and Das Sarma, *Phys. Rev. B* **38**, 3318 (1988).

11. E. R. Cowley, *Phys. Rev. Lett.* **60**, 2379 (1988).

12. P. Ballone and P. Milani, *Phys. Rev. B* **42**, 3201 (1990).

13. C. Z. Wang, C. T. Chan, and K. M. Ho, *Phys. Rev. B* **39**, 8592 (1989).

14. F. S. Khan and J. Q. Broughton, *Phys. Rev. B* **39**, 3688 (1989).

15. K. Laasonen and R. M. Nieminen, *J. Phys. Condensed Matter* **2**, 1509 (1991); R. Virkkunen, K. Laasonen, and R. M. Nieminen, *J. Phys. Condensed Matter* **3**, 7455 (1991).

16. M. Menon and R. E. Allen, *Phys. Rev. B* **33**, 7099 (1986).

17. J. C. Slater and G. F. Koster, *Phys. Rev.* **94**, 1498 (1954).

18. W. A. Harrison, *Electronic Structure and the Properties of Solids*, W. H. Freeman, San Francisco, 1980.

19. D. J. Chadi, *Phys. Rev. Lett.* **41**, 1062 (1978); *Phys. Rev. B* **29**, 785 (1984).

20. C. Z. Wang, C. T. Chan, and K. M. Ho, *Phys. Rev. B* **42**, 11276 (1990).

21. L. Goodwin, A. J. Skinner, and D. G. Pettifor, *Europhys. Lett.* **9**, 701 (1989).

22. S. Sawada, *Vacuum* **41**, 612 (1990).

23. C. H. Xu, C. Z. Wang, C. T. Chan, and K. M. Ho, *J. Phys. Condensed Matter* **4**, 6047 (1992).

24. I. Kwon, R. Biswas, C. Z. Wang, K. M Ho, and C. M. Soukoulis, *Phys. Rev. B* **49**, 7242 (1994).

25. J. L. Mercer, Jr., and M. Y. Chou, *Phys. Rev. B* **47**, 9366 (1993).

26. J. L. Mercer, Jr., and M. Y. Chou, *Phys. Rev. B* **49**, 8506 (1994).

27. M. Kohyama, *J. Phys. Condensed Matter* **3**, 2193 (1991).

28. C. H. Xu, C. Z. Wang, C. T. Chan, and K. M. Ho, *Phys. Rev. B* **43**, 5024 (1991).

29. O. Madelung, M. Schulz, and H. Weiss, Eds., *Semiconductors: Physics of Group IV Elements and III-V Compounds*, Landolt–Börnstein New Series III/17a, Springer-Verlag, New York, 1982; O. Madelung and M. Schulz, Eds., *Semiconductors: Intrinsic Properties of Group IV Elements and III-V, II-VI and I-VII Compounds*, Landolt–Börnstein New Series III/22a, Springer-Verlag, New York, 1987.

30. M. S. Dresselhaus and G. Dresselhaus, in *Light Scattering in Solids III*, M. Cardona and G. Guntherodt, Eds., Springer-Verlag, Berlin, 1982, p. 8.

31. D. Tomanek and M. A. Schluter, *Phys. Rev. Lett.* **56**, 1055 (1986); *Phys. Rev. B* **36**, 1208 (1987).

32. C. Z. Wang, C. T. Chan, and K. M. Ho, *Phys. Rev. Lett.* **66**, 189 (1991).

33. C. Z. Wang, C. T. Chan, and K. M. Ho, *Phys. Rev. B* **45**, 12227 (1992).

34. G. Servalli and L. Colombo, *Europhys. Lett.* **22**, 107 (1993).

35. A. P. Horsfield and P. Clancy, *Modelling Simul. Mater. Sci. Eng.* **2**, 277 (1994).

36. M. Pederson and K. Jackson, *Phys. Rev. B* **43**, 7312 (1991).

37. R. M. Wentzcovitch, J. L. Martin, and P. B. Allen, *Phys. Rev. B* **45**, 11372 (1992).

38. N. D. Mermin, *Phys. Rev.* **137** A1441 (1965).

39. M. T. Yin and M. L. Cohen, *Phys. Rev. B* **26**, 3259 (1982).

40. O. H. Nielsen and R. M. Martin, *Phys. Rev. B* **32**, 3798 (1985).

41. H. Ibach, *Phys. Status Solidi* **31**, 625 (1969).

42. M. Balkanski, R. F. Wallis, and E. Haro, *Phys. Rev. B* **28**, 1928 (1983).

43. R. Tsu and G. Hernandez, *Appl. Phys. Lett* **41**, 1016 (1982).

44. C. Z. Wang, C. T. Chan, and K. M. Ho, *Solid State Commun.* **76**, 483 (1990).

45. C. Z. Wang, C. T. Chan, and K. M. Ho, *Solid State COmmun.* **80**, 469 (1990).

46. Y. Waseda and K. Suzuki, *Z. Phys. B* **20**, 339 (1975).

47. I. Stich, R. Car, and M. Parrinello, *Phys. Rev. Lett.* **63**, 2240 (1989); *Phys. Rev. B* **44**, 4262 (1991).

48. W. D. Leudtke and U. Landman, *Phys. Rev. B* **37**, 4656 (1988).

49. A. R. Ubbelohbe, *The Molten State of Matter*, Wiley, New York, 1978, p. 239.

50. J. R. Chelikowsky, N. Troullier, and N. Binggeli, *Phys. Rev. B* **49**, 114 (1994).

51. J. Q. Broughton and X. P. Li, *Phys. Rev. B* **35**, 9120 (1987).

52. P. Allen and J. Q. Broughton, *J. Phys. Chem.* **91**, 4964 (1987).

53. C. Z. Wang, K. M. Ho, and C. T. Chan, *Phys. Rev. B* **47**, 14835 (1993).

54. J. R. Morris, C. Z. Wang, and K. M. Ho *Phys. Rev. B* **52**, 4138 (1995).

55. G. Galli, R. M. Martin, R. Car, and M. Parrinello, *Phys. Rev. Lett.* **62**, 555 (1989).

56. G. Galli, R. M. Martin, R. Car, and M. Parrinello, *Science* **250**, 1547 (1990).

57. G. Galli, R. M. Martin, R. Car, and M. Parrinello, *Phys. Rev. B* **42**, 7470 (1990).

58. G. Galli, R. M. Martin, R. Car, and M. Parrinello, *Phys. Rev. Lett.* **63**, 988 (1989).

59. F. P. Bundy, *J. Chem. Phys.* **38**, 618 (1963).

60. C. Z. Wang, K. M. Ho, and C. T. Chan, *Phys. Rev. Lett.* **70**, 611 (1993).

61. C. Z. Wang and K. M. Ho, *Phys. Rev. Lett.* **71**, 1184 (1993).

62. C. Z. Wang and K. M. Ho, *Phys. Rev. B* **50**, 12429 (1994).

63. C. Z. Wang and K. M. Ho, *J. Phys. Condensed Matter* (letter to the Editor) **6**, L239 (1994).

64. F. Li and J. S. Lannin, *Phys. Rev. Lett.* **65**, 1905 (1990).

65. P. H. Gaskell, A. Saeed, P. Chieux, and D. R. McKenzie, *Phys. Rev. Lett.* **67**, 1286 (1991).

66. R. Biswas, C. Z. Wang, C. T. Chan, K. M. Ho, and C. M. Soukoulis, *Phys. Rev. Lett.* **63**, 1491 (1989).

67. Q. M. Li and R. Biswas, *Phys. Rev. B* **52**, 10705 (1995).

68. C. Molteni, L. Colombo, and L. Miglio, *Europhys. Lett.* **24**, 659 (1993).

69. D. Tomanek and M. Schluter, *Phys. Rev. Lett.* **67**, 2331 (1991).

70. C. H. Xu, C. Z. Wang, C. T. Chan, and K. M. Ho, *Phys. Rev. B* **47**, 9878 (1993).

71. B. L. Zhang, C. Z. Wang, and K. M. Ho, *J. Chem. Phys.* **96**, 7183 (1992).

72. B. L. Zhang, C. Z. Wang, and K. M. Ho, *Chem. Phys. Lett.* **193**, 225 (1992).

73. B. L. Zhang, C. H. Xu, C. Z. Wang, C. T. Chan, and K. M. Ho, *Phys. Rev. B* **46**, 7333 (1992).

74. B. L. Zhang, C. Z. Wang, K. M. Ho, C. H. Xu, and C. T. Chan, *J. Chem. Phys.* **97**, 5007 (1992).

75. B. L. Zhang, C. Z. Wang, K. M. Ho, C. H. Xu, and C. T. Chan, *J. Chem. Phys.* **98**, 3095 (1993).

76. C. Z. Wang, B. L. Zhang, K. M. Ho, and X. Q. Wang, *Intern. J. Mod. Phys. B* **7**, 4305 (1993).

77. C. Z. Wang, C. T. Chan, and K. M. Ho, *Phys. Rev. B* **46**, 9761 (1992).

78. C. Z. Wang, C. H. Xu, C. T. Chan, and K. M. Ho, *J. Phys. Chem. Lett.* **96**, 3563 (1992).

79. B. L. Zhang, C. Z. Wang, and K. M. Ho, *Phys. Rev. B* **47**, 1643 (1993).

80. B. L. Zhang, C. Z. Wang, C. T. Chan, and K. M. Ho, *J. Chem. Phys.* **97**, 3134 (1993).

81. B. L. Zhang, C. Z. Wang, C. T. Chan, and K. M. Ho, *Phys. Rev. B* **48**, 11381 (1993).

82. B. L. Zhang, C. Z. Wang, K. M. Ho, and C. T. Chan, *Europhys. Phys. Lett.* **28**, 219 (1994).

83. C. H. Xu and G. E. Scuseria, *Phys. Rev. Lett.* **72**, 669 (1994); **74**, 274 (1995).

84. C. Z. Wang, C. H. Xu, C. T. Chan, and K. M. Ho, *J. Phys. Chem.* **96**, 3563 (1992).

85. S. G. Kim and D. Tomanek, *Phys. Rev. Lett.* **72**, 2418 (1994).

86. K. Kikuchi, N. Nakahara, T. Wakabayashi, S. Suzuki, H. Shiromaru, Y. Miyake, K. Saito, I. Ikemoto, M. Kainosho, and Y. Achiba, *Nature (London)* **357**, 142 (1992).

87. F. Diederich and R. L. Whetten, *Acc. Chem. Res.* **25**, 119 (1992); C. Thilgen, F. Diederich, and R. L. Whetten, in *The Fullerenes*, W. Billups, Ed., VCH, Deerfield, Fl., 1992 p. 408.

88. D. E. Manolopoulos, P. W. Fowler, R. Taylor, H. W. Kroto, and D. R. M. Walton, *J. Chem. Soc. Faraday Trans.* **88**, 3117 (1992).

89. X. Q. Wang, C. Z. Wang, B. L. Zhang, and K. M. Ho, *Phys. Rev. Lett.* **69**, 69 (1992); *Chem. Phys. Lett.* **207**, 349 (1993).

90. X. Q. Wang, C. Z. Wang, B. L. Zhang, and K. M. Ho, *Chem. Phys. Lett.* **200**, 35 (1992); **217**, 199 (1994).

91. J. Yu, R. K. Kalia, and P. Vashishta, to be published.

92. J. Yu, R. K. Kalia, and P. Vashishta, *Appl. Phys. Lett.* **63**, 3152 (1993).

93. J. Yu, R. K. Kalia, and P. Vashishta, *Phys. Rev. B* **49**, 5008 (1994).

94. L. Pintschovius, S. L. Chaplot, R. Heid, M. Haluska, and H. Kuzmany, *Proceedings of the International Winterschool on Electronic Properties of Fullerenes and Other Novel Materials, J. Fink, Ed.*, Springer Series in Solid State Science, Vol. 117, Springer-Verlag, Berlin, 1993).

95. M. Menon and K. R. Subbaswamy, *Phys. Rev. B* **47**, 12754 (1993).

96. P. Ordejon, D. Lebedenko, and M. Menon, *Phys. Rev. B* **50**, 5645 (1994); M. Menon and R. Subbaswamy, *Phys. Rev. B* **50**, 11577 (1994).

97. Y. Bar-Yam and J. D. Joannopoulos, *Phys. Rev. Lett.* **52**, 1129 (1984); and in *Proceedings of 13th International Conference on Defects in Semiconductors*, L. C. Kimerling and J. M. Parsey, Eds., The Metallurgical Society of AIME, New York, 1984.

98. A. Antonelli and J. Bernholc, *Phys. Rev. B* **40**, 10643 (1989).

99. R. Car, P. Kelly, A. Oshiyama, and S. T. Pantelides, *Phys. Rev. Lett.* **52**, 1814 (1984); and in *Proceedings of 13th International Conference on Defects in Semiconductors*, L. C. Kimerling and J. M. Parsey, Eds., The Metallurgical Society of AIME, New York, 1984.

100. G. A. Baraff and M. Schlüter, *Phys. Rev. B* **30**, 3460 (1984).

101. C. H. Xu, C. L. Fu, and D. F. Pedraza, *Phys. Rev. B* **48**, 13273 (1993).

102. A. T. Paxton and A. P. Sutton, *Acta Metall.* **37**, 1693 (1989).

103. M. Kohyama, R. Yamamoto, Y. Ebata, and M. Kinoshita, *J. Phys. C Solid State Phys.* **21**, 3205 (1988).

104. M. Kohyama and R. Yamamoto, *Phys. Rev. B* **49**, 17102 (1994); **50**, 8502 (1994).

105. M. Kohyama, H. Ichinose, and Y. Ishida, *MRS Symp. Proc.* **339**, C. H. Carter, Jr., et al., EDS., MRS, 1994.

106. D. C. Allan and E. J. Mele, *Phys. Rev. Lett.* **53**, 826 (1984); *Phys. Rev. B* **31**, 5565 (1985).

107. O. L. Alerhand and E. J. Mele, *Phys. Rev. B* **35**, 5533 (1987).

108. J. L. Mercer, Jr., and M. Y. Chou, *Phys. Rev. B* **48**, 5374 (1993).

109. B. N. Davison and W. E. Pickett, *Phys. Rev. B* **49**, 11253 (1994).

110. K. Stokbro, N. Chetty, K. W. Jacobsen, and J. K. Norskov, *Phys. Rev. B* **50**, 10727 (1994); K. Stokbro, Ph.D. thesis, Danish Technical University, August 1994.

111. O. K. Andersen and O. Jepsen, *Phys. Rev. Lett.* **53**, 2571 (1984).

112. K. Stokbro, K. W. Jacobsen, J. K. Norskov, D. Deaven, C. Z. Wang, and K. M. Ho, to be published.

113. B. J. Min, Y. H. Lee, C. Z. Wang, C. T. Chan, and K. M. Ho, *Phys. Rev. B* **45**, 6839 (1992); **46**, 9677 (1992).

114. D. E. Boucher and G. DeLeo, *Phys. Rev. B* **50**, 5247 (1994).

115. G. Panzarini and L. Colombo, *Phys. Rev. Lett.* **73**, 1636 (1994).

116. Q. M. Li and R. Biswas, *Phys. Rev. B* **50**, 18090 (1994).

117. E. Kim, Y. H. Lee, and J. M. Lee, *J. Phys. Condensed Matter* **6**, 9561 (1994).

118. Y. Wang and C. H. Mak, *Chem. Phys. Lett.* **235**, 37 (1995).

119. F. Buda, G. Chiarotti, R. Car, and M. Parrinello, *Phys. Rev. Lett.* **63**, 294 (1989).

120. X.-P. Li, R. W. Nunes, and D. Vanderbilt, *Phys. Rev. B* **47**, 10891 (1993).

121. M. S. Daw, *Phys. Rev. B* **47**, 10895 (1993).

122. F. Mauri, G. Galli, and R. Car, *Phys. Rev. B* **47**, 9973 (1993).

123. P. Ordejon, D. Drabold, M. P. Grumbach, and R. M. Martin, *Phys. Rev. B* **48**, 14646 (1993).

124. L. W. Wang and M. P. Teter, *Phys. Rev. B* **46**, 12798 (1992).

125. W. Yang, *Phys. Rev. Lett.* **66**, 1438 (1991).

126. S. Baroni and P. Giannozzi, *Europhys. Lett.* **17**, 547 (1992).

127. D. A. Drabold and O. F. Sankey, *Phys. Rev. Lett.* **70**, 3631 (1993).

128. S. Goedecker and L. Colombo, *Phys. Rev. Lett.* **73**, 122 (1994).

129. S. Y. Qiu, C. Z. Wang, K. M. Ho, and C. T. Chan, *J. Phys. Condensed Matter* **6**, 9153 (1994).

130. F. Mauri and G. Galli, *Phys. Rev. B* **50**, 4316 (1994).

131. F. Mauri and G. Galli, *Phys. Rev. lett.* **73**, 3471 (1994).

132. D. Deaven and K. M. Ho, *Phys. Rev. Lett.* **75**, 288 (1995).

133. L. T. Wille, and J. Vennik, *J. Phys. A* **18**, L419 (1985).

134. P. Ballone and P. Milani, *Phys. Rev. B* **42**, 3201 (1990).

135. J. Chelikowsky, *Phys. Rev. Lett.* **67**, 2970 (1991).

136. J. H. Holland, *Adaptation in Natural and Artificial Systems*, University of Michigan Press, Ann Arbor Mich., 1975.

137. R. E. Cohen, M. J. Mehl, and D. A. Papaconstantopoulos, *Phys. Rev. B* **50**, 14694 (1994).

138. M. S. Tang, C. Z. Wang, C. T. Chan, and K. M. Ho, *Phys. Rev. B*, Jan. 15, 1996.

PERSPECTIVES ON SEMIEMPIRICAL
MOLECULAR ORBITAL THEORY

WALTER THIEL

Institut für Organische Chemie, Universität Zürich, Zürich, Switzerland

CONTENTS

I. INTRODUCTION

Over the past decades the semiempirical molecular orbital (MO) methods of quantum chemistry have been used widely in computational studies. Self-consistent-field (SCF) π-electron calculations have been carried out since the 1950s and valence-electron calculations since the 1960s. Several books [1–8] and reviews [9–15] describe the underlying theory, the variants of semiempirical methods, and the numerical results in much

Advances in Chemical Physics, Volume XCIII, Edited by I. Prigogine and Stuart A. Rice.
ISBN 0-471-14321-9 © 1996 John Wiley & Sons, Inc.

detail. Therefore the present article will not attempt a comprehensive coverage of the available material. Instead, we shall try to assess the current status of the field, to describe some recent methodological advances, and to outline the perspectives for the future.

Semiempirical methods play a dual role as electronic structure models and and as computational tools. The currently accepted treatments are based on the orbital approximation. In their implementation, they neglect many of the less important integrals that occur in the *ab initio* MO formalism. These severe simplifications call for the need to represent the remaining integrals by suitable parametric expressions and to calibrate them against reliable experimental or accurate theoretical reference data. This strategy can be successful only if the semiempirical model retains the essential physics to describe the properties of interest. Provided that this is the case, the parametrization can account for all other effects in an average sense, and it is then a matter of testing and validation to establish the numerical accuracy of a given approach. To the extent that the semiempirical results represent reality, they can be used to explain and to predict chemical behavior in terms of the underlying concepts (e.g., orbital interactions or electrostatic interactions). In their role as electronic structure models, semiempirical methods may thus provide qualitative insights into chemical behavior, especially when comparing classes of related compounds. In their role as efficient computational tools, they can yield fast quantitative estimates for a number of properties. This may be particularly useful for correlating large sets of experimental and theoretical data, for establishing trends, and for scanning a computational problem before proceeding with more expensive and more accurate theoretical calculations.

This review is organized as follows: Section II describes the current status of semiempirical methodology. The currently accepted treatments are characterized briefly with regard to their basic assumptions, computational speed, and accuracy, and their relation to *ab initio* and density functional methods is discussed in this context. Selected recent applications from fullerene chemistry are included to illustrate the potential and the limitations of the currently available semiempirical approaches. Section III deals with recent and ongoing methodological developments. This includes work on the foundations of semiempirical theory, the improvement of semiempirical methods, the parametrization of such methods, and the progress in computational aspects. Based on these advances, new applications are identified which should soon become more tractable by semiempirical calculations. Section IV offers concluding remarks and an outlook.

II. CURRENT STATUS

A. Available Methods

Quantum chemical semiempirical treatments are defined by the following specifications:

(a) The underlying theoretical approach
(b) The integral approximation and the types of interactions included
(c) The integral evaluation and the associated parametric expressions
(d) The parametrization

In our terminology, the specifications (a) and (b) define a semiempirical model, (a)–(c) an implementation of a given model, and (a)–(d) a particular method. We first give a general overview over possible choices for (a)–(d).

The underlying *theoretical approach* is characterized by the type of the wavefunction and the choice of the basis set. Most current general-purpose semiempirical methods are based on molecular orbital theory and employ a minimal basis set for the valence electrons. Electron correlation is treated explicitly only if this is necessary for an appropriate zero-order description (e.g., in the case of electronically excited states or transition states in chemical reactions). Correlation effects are often included in an average sense by a suitable representation of the two-electron integrals and by the overall parametrization.

Traditionally, there are three levels of *integral approximation* [4, 16]: CNDO (complete neglect of differential overlap), INDO (intermediate neglect of differential overlap), and NDDO (neglect of diatomic differential overlap). NDDO is the best of these approximations since it retains the higher multipoles of charge distributions in the two-center interactions (unlike CNDO and INDO, which truncate after the monopole). It has been shown [17] that these additional terms are responsible for significant improvements of NDDO over INDO models that have otherwise been parametrized analogously (MNDO and MINDO/3, see below). Furthermore, the time-determining step in CNDO, INDO, and NDDO SCF-MO calculations is not the evaluation of the integrals, but always the solution of the secular equations. Hence all the available evidence suggests the use of the NDDO integral approximation. Even at this level, however, the overlap matrix is replaced by the unit matrix in the Hartree–Fock secular equations. Hence, in the CNDO, INDO, and NDDO approximations, the integrals are implicitly assumed to refer to an orthogonal basis. The orthogonalization corrections to the integrals are sometimes neglected and sometimes included to a certain extent in the existing methods. Their best representation is the subject of current work (see Section III.B).

Integral evaluation (at a given level of integral approximation) usually proceeds by one of three approaches: The integrals are either determined directly from experimental data, calculated from the corresponding analytical formulas, or computed from suitable parametric expressions. The first option is generally feasible only for the one-center integrals, which may be derived from atomic spectroscopic data. The choice between the second and third options is influenced by the ease of implementation of the analytical formulas, but mainly by the assessment of which interactions are essential and need to be parametrized in a semiempirical framework. The selection of appropriate parametric expressions is normally guided either by an analysis of the corresponding analytical integrals (e.g., with regard to the distance dependence in two-center terms) or by intuition (e.g., when assuming proportionality between resonance and overlap integrals). The existing semiempirical methods incorporate parametric functions which ensure a realistic balance between attractive and repulsive forces in the molecule. The final choice of these functions is usually made empirically by carrying out analogous parametrizations for various reasonable combinations of functions and then selecting the one with the best performance.

Originally, there have been two basic strategies for *parametrization*. Approximate MO methods aim at reproducing *ab initio* MO calculations with the same minimal basis set (MBS), whereas semiempirical MO methods attempt to reproduce experimental data. Nowadays the limitations of MBS *ab initio* calculations are well known and the predominant feeling is that approximate MO methods would not be useful enough in practice even if they would exactly mimic MBS *ab initio* calculations. Hence with the exception of PRDDO [18], current parametrizations usually adhere to the semiempirical philosophy and employ experimental reference data (or possibly, accurate high-level theoretical predictions as substitutes for experimental data; see Section III.E).

This completes the general overview over choices (a)–(d) in a semiempirical framework. We shall now briefly characterize some current treatments: MNDO [19], MNDOC [20], AM1 [21], PM3 [22], SAM1 [23, 24], SINDO1 [25, 26], and INDO/S [27, 28]. These methods are widely available in distributed software packages and have generally replaced older methods in practical applications (e.g., CNDO/2 [29], CNDO/S [30], INDO [31], and MINDO/3 [32]). Recent developments from our group, such as MNDO/d [33–36] and approaches with orthogonalization corrections [37, 38] are discussed later (see Section III.B).

As a point of reference we first outline the MNDO formalism for closed-shell molecules [19]. MNDO is a valence-electron SCF-MO treatment, which employs a minimal basis of atomic orbitals (AOs, ϕ_μ) and the NDDO integral approximation. The molecular orbitals ψ_i and the corresponding orbital energies ε_i are obtained from the solution of the secular equations ($S_{\mu\nu} = \delta_{\mu\nu}$ for NDDO):

$$\psi_i = \sum_\mu c_{\mu i} \phi_u \, , \tag{1}$$

$$0 = \sum_\nu (F_{\mu\nu} - \delta_{\mu\nu} \varepsilon_i) c_{\nu i} \, . \tag{2}$$

Using superscripts to assign an AO (with index μ, ν, λ, σ) to an atom A or B, the NDDO Fock matrix elements $F_{\mu\nu}$ are given as

$$F_{\mu^A \nu^A} = H_{\mu^A \nu^A} + \sum_{\lambda^A} \sum_{\sigma^A} P_{\lambda^A \sigma^A}[(\mu^A \nu^A, \lambda^A \sigma^A) - \tfrac{1}{2}(\mu^A \lambda^A, \nu^A \sigma^A)]$$
$$+ \sum_B \sum_{\lambda^B} \sum_{\sigma^B} P_{\lambda^B \sigma^B}(\mu^A \nu^A, \lambda^B \sigma^B) \, , \tag{3}$$

$$F_{\mu^A \nu^B} = H_{\mu^A \nu^B} - \tfrac{1}{2} \sum_{\lambda^A} \sum_{\sigma^B} P_{\lambda^A \sigma^B}(\mu^A \lambda^A, \nu^B \sigma^B) \, , \tag{4}$$

where $H_{\mu\nu}$ and $P_{\lambda\sigma}$ are elements of the one-electron core Hamiltonian and the density matrix, respectively, and $(\mu\nu, \lambda\sigma)$ denotes a two-electron integral. In the case of an sp basis, Eq. (3) may be simplified further [19]. The total energy E_{tot} of a molecule is the sum of its electronic energy E_{el} and the repulsions E_{AB}^{core} between the cores of all atoms A and B.

$$E_{el} = \tfrac{1}{2} \sum_\mu \sum_\nu P_{\mu\nu}(H_{\mu\nu} + F_{\mu\nu}) \, , \tag{5}$$

$$E_{tot} = E_{el} + \sum_{A<} \sum_B E_{AB}^{core} \, . \tag{6}$$

The following energy terms and interactions are included in MNDO:

1. One-center one-electron energies $U_{\mu\mu}$ (as part of $H_{\mu\mu}$)
2. One-center two-electron repulsion integrals $(\mu^A \nu^A, \lambda^A \sigma^A)$ [e.g., Coulomb integrals $g_{\mu\nu} = (\mu\mu, \nu\nu)$ and exchange integrals $h_{\mu\nu} = (\mu\nu, \mu\nu)$]
3. Two-center one-electron resonance integrals $\beta_{\mu\nu} = H_{\mu^A \nu^B}$
4. Two-center one-electron integrals $(\mu^A \nu^A, B)$ representing electrostatic core–electron attractions (as part of $H_{\mu^A \nu^A}$)

5. Two-center two-electron repulsion integrals $(\mu^A\nu^A, \lambda^B\sigma^B)$
6. Two-center core–core repulsions E_{AB}^{core} composed of an electrostatic term E_{AB}^{coul} and an additional effective term E_{AB}^{eff} (see below)

These specifications define choices (a) and (b) in the MNDO formalism and thus constitute the MNDO model. The original implementation of the model may be summarized as follows [19]. Conceptually the one-center terms are taken from atomic spectroscopic data, with the refinement that slight adjustments of the $U_{\mu\mu}$ parameters are allowed in the optimization to account for possible differences between free atoms and atoms in a molecule. Any such adjustments should be minor to ensure that the one-center parameters remain close to their spectroscopic values and thus retain their physical significance. The one-center two-electron integrals derived from atomic spectroscopic data are considerably smaller than their analytically calculated values, which is (at least partly) attributed to an average incorporation of electron correlation effects. For reasons of internal consistency, these integrals provide the one-center limit $(R_{AB} = 0)$ of the two-center two-electron integrals $(\mu^A\nu^A, \lambda^B\sigma^B)$, whereas the asymptotic limit of $(\mu^A\nu^A, \lambda^B\sigma^B)$ for $R_{AB} \to \infty$ is determined by classical electrostatics. The semiempirical calculation of $(\mu^A\nu^A, \lambda^B\sigma^B)$ conforms to these limits and evaluates these integrals from semiempirical multipole–multipole interactions [39]: The relevant multipoles are represented by suitable point-charge configurations whose interaction is damped according to the Klopman–Ohno formula [40, 41]. Therefore, at intermediate distances, the semiempirical two-electron integrals are smaller than their analytical counterparts which again reflects some inclusion of electron correlation effects. Aiming for a reasonable balance between electrostatic attractions and repulsions within a molecule, MNDO treats the core–electron attractions $(\mu^A\nu^A, B)$ and the core–core repulsions E_{AB}^{coul} in terms of the two-electron integrals $(\mu^A\nu^A, s^Bs^B)$ and (s^As^A, s^Bs^B), respectively, neglecting, for example, penetration effects. The additional effective terms E_{AB}^{eff} (with an essentially exponential repulsion) attempt to compensate for errors introduced by the above assumptions, but mainly represent the Pauli exchange repulsions. Finally, following semiempirical tradition, the resonance integrals are taken to be proportional to the corresponding overlap integrals.

Hence choices (c) in the MNDO implementation are characterized by five simple guidelines:

1. A realistic description of the atoms in a molecule
2. A balance between electrostatic attractions and repulsions
3. An average incorporation of dynamic electron correlation effects

4. A description of covalent bonding mainly through overlap-dependent resonance integrals

5. An inclusion of Pauli exchange repulsions and other corrections through an effective atom-pair function

The parametrization (d) of MNDO has focused on ground-state properties, mainly heats of formation and geometries, with the use of ionization potentials and dipole moments as additional reference data. Heats of formation at 298 K are derived from the calculated total energies [Eq. (6)], by subtracting the calculated electronic energies of the atoms in the molecule and by adding their experimental heats of formation at 298 K. This traditional choice [3] of heats of formation as reference data implies that the parametrization must account for zero-point vibrational energies and for thermal corrections between 0 K and 298 K in an average sense. This is not satisfactory theoretically, but it has been shown empirically [12, 42] that the overall performance of MNDO is not affected much by this choice. MNDO has been parametrized for many elements, and the MNDO results have been documented in detail (see, e.g., Refs. 8, 11–15, and 43).

AM1 [21] and PM3 [22] are based on exactly the same model (a)–(b) as MNDO and differ from MNDO only in one aspect of the implementation (c): The effective atom-pair term E_{AB}^{eff} in the core–core repulsion function is represented by a more flexible function with several additional adjustable parameters. The additional Gaussian terms in E_{AB}^{eff} are not derived theoretically but justified empirically as providing more opportunities for fine tuning, especially for reducing overestimated nonbonded repulsions in MNDO. The parametrization (d) in AM1 and PM3 follows the same philosophy as in MNDO. However, more effort has been spent on the parametrization of AM1 and PM3, and additional terms have been treated as adjustable parameters (e.g. the one-center Coulomb and exchange integrals) so that the number of optimized parameters per element has typically increased from 5–7 in MNDO to 18 in PM3. Hence AM1 and PM3 may be regarded as methods that attempt to explore the limits of the MNDO model through careful and extensive parametrization.

To our knowledge, the SAM1 [23, 24] formalism and parameters have not yet been published in detail. The available description of the method [23] indicates that SAM1 also employs the basic MNDO model (a)–(b) but differs in two aspects of the implementation (c): In addition to using an AM1-type effective atom-pair function E_{AB}^{eff}, the two-center two-electron integrals are now first calculated analytically and then scaled by an empirical function to allow for electron correlation. Therefore, apart from the actual parametrization, the main distinction between AM1 and SAM1 appears to be a somewhat different distance dependence of the two-center electrostatic interactions.

MNDOC [20] is a correlated version of MNDO. Unlike all methods discussed previously, MNDOC includes electron correlation explicitly and thus differs from MNDO at the level of the underlying quantum-chemical approach (a) while being completely analogous to MNDO in all other aspects (b)–(d) except for the actual values of the parameters. In MNDOC electron correlation is treated conceptually [20] by full configuration (CI), and practically by second-order perturbation theory (BWEN [20]) in simple cases (e.g., closed-shell ground states) and by a variation-perturbation treatment (PERTCI [44]) in more complicated cases (e.g., electronically excited states). The MNDOC parameters have been determined at the correlated level and should thus be appropriate in all MNDO-type applications that require an explicit correlation treatment for a qualitatively suitable zero-order description. In closed-shell ground states of typical organic molecules, the correlation effects are fairly uniform [12, 20] at both the semiempirical and *ab initio* [45, 46] levels, and it should therefore be possible to absorb them by a semiempirical SCF parametrization. As expected, MNDO and MNDOC indeed show similar average errors for such molecules [20]. On the other hand, MNDOC offers significant advantages over MNDO for systems with specific correlation effects, including certain reactive intermediates [47], transition states in chemical reactions [47–49], and electronically excited states [50]. Guidelines have been recommended concerning the correlation treatment in semiempirical studies of reactions (MNDOC vs. MNDO) [49], and there have been a number of successful MNDOC-CI applications to excited states and photochemistry (see, e.g., Refs. 51–53).

SINDO1 [25, 26] is a semiempirical SCF-MO treatment that is based on the INDO integral approximation and thus neglects all electrostatic two-center interactions involving higher multipoles (other than the monopole). In contrast to the methods discussed previously SINDO1 explicitly includes orthogonality corrections and effective core potentials to account for exchange repulsions and core–valence interactions, respectively. Hence, with regard to choices (a)–(b), the SINDO1 model is quite distinct from the MNDO model, and there are also significant differences in the implementation (c): SINDO1 analytically evaluates all one- and two-center Coulomb integrals of type (ss, ss) that survive at the INDO level as well as the core–core repulsions $E_{AB}^{core} = Z_A Z_B / R_{AB}$ (core charges Z_A and Z_B). The SINDO1 resonance integrals are also related to overlap integrals, but in a fairly intricate manner [25]. As in MNDO-type methods, the parametrization (d) of SINDO1 focuses on ground-state properties (employing binding energies rather than heats of formation as reference data). The published SINDO1 results often appear to be of similar accuracy as those from MNDO-type methods, but there are considerably fewer published SINDO1 applications, and a comprehensive and systematic comparison of the relative performance of these methods still seems to be missing (see Ref. 15 for further remarks).

INDO/S [27, 28] has been designed for calculating electronic spectra, particularly vertical excitation energies at given ground-state geometries. It is thus different from the methods considered previously, which have been developed for computing ground-state potential surfaces. INDO/S is a semiempirical CI method parametrized at the CIS level (CI with single excitations), like the older CNDO/S method [30]. It employs the INDO approximation (like SINDO1; see above) and is implemented with the Mataga–Nishimoto approximation [54] for the two-center two-electron integrals and a rather flexible expression for the resonance integrals. Without using an extensive parametrization, INDO/S has been quite successful in calculating electronic spectra of larger molecules [15], especially for excitation energies below $40,000\,\mathrm{cm}^{-1}$. An alternative semiempirical CI approach (LNDO/S [55]) parametrized at a higher CI level (CI with single and double excitations from all relevant main configurations) has shown promising results for vertical excitation energies and ionization potentials [55, 56] but has apparently never been distributed.

Summarizing this section, there are a number of semiempirical valence–electron SCF-MO methods that are currently accepted and widely used. A closer inspection of the underlying theoretical models reveals a much smaller variety: The MNDO model forms the common basis of several popular methods (MNDO, AM1, PM3, etc.) which are often applied in studies of ground-state potential surfaces (with SINDO1 as a possible alternative). Electronic spectra are usually calculated with semiempirical CI methods specifically designed for this purpose (most notably INDO/S).

B. Accuracy and Efficiency

At present, *ab initio* methods, density functional methods, and semiempirical methods serve as the major computational tools of quantum chemistry. There is an obvious trade-off between accuracy and computational effort in these methods. The most accurate results are obtained from high-level correlated *ab initio* calculations (e.g., multireference CI or coupled cluster calculations with large basis sets) which also require the highest computational effort. On the other end of the spectrum, semiempirical MO calculations are very fast, and it is therefore realistic to expect only a limited accuracy from such calculations. Actual applications will usually have to balance the required accuracy against the available computational resources. The combined use of several computational tools may well be the best approach to solve a given problem, for example, by initial explorations at the semiempirical level followed by more accurate density functional or *ab initio* calculations.

In recent years there have been considerable technical improvements in *ab initio* methodology and in density functional theory (DFT) which have made the corresponding computer programs more efficient and applicable to ever larger molecules. These advances include the implementation of direct and semidirect *ab initio* methods, the faster evaluation of two-electron integrals over Gaussian basis functions, and the development of better numerical integration schemes in DFT codes. In view of these advances, it seems appropriate to emphasize that semiempirical calculations are still much faster than standard DFT or *ab initio* SCF calculations on the same molecule. This is illustrated in Table I which lists corresponding computation times for one energy evaluation in tetracene $C_{18}H_{12}$ using three state-of-the-art programs [57] (MNDO93 [58], DGAUSS [59], CADPAC [60]). The MNDO SCF calculation is faster by a factor of more than 1000 compared with the lowest standard DFT or *ab initio* level normally applied (i.e., DFT with a DZVP basis and a local exchange-correlation potential [59], or *ab initio* SCF with the 3-21G basis [61]). Similar observations have been made with other *ab initio* programs, for example, in our semiempirical and *ab initio* studies on fullerenes (see Section II.C), where the *ab initio* SCF computation times with TURBOMOLE [62] on standard workstations were higher

TABLE I
Comparison of Computation Times[a]

Program	Method[b]	T (s)
MNDO93 (4.1)	SCF, MNDO	0.35
DGAUSS (2.0)	DFT, DZVP, local	551
	DFT, DZVP, nonlocal	930
CADPAC (5.1)	SCF, 3-21G	524
	SCF, 6-31G*	4,287
	MP2, 6-31G*	12,599[c]

[a] User CPU times on a single processor of a Cray Y-MP 8 E/8128 (6 ns, sn1802). Test case: tetracene $C_{18}H_{12}$, one energy evaluation. Software: UniChem (2.0.1) of October 1993.

[b] Standard notation; see the text.

[c] Using the large memory algorithm (26 MW in this case).

than the MNDO SCF times by factors of about 1300 (C_{78}, C_{2v}, 3-21G basis [63]) and 7000 (He@C_{84}, C_2, tzp(He)/dz(C) basis [64]). These factors increase further, of course, when using larger basis sets or correlated *ab initio* treatments (see Table I).

The reasons for the computational speed of semiempirical SCF methods are obvious from Eqs. (1)–(6): There are only one- and two-center integrals in the formalism, so that integral evaluation scales as N^2 for N basis functions. For large molecules this becomes negligible compared with the N^3 steps in the solution of the secular equations and the formation of the density matrix. Whenever possible, full diagonalizations of the Fock matrix are avoided in favor of a pseudodiagonalization procedure [65]. As a consequence, most of the work in the N^3 steps occurs in matrix multiplications which are handled very efficiently on current workstations as well as on vector and some parallel machines. The relatively small number of nonzero integrals in semiempirical methods also implies that there are usually no input/output bottlenecks. The calculations can normally be performed completely in memory, with memory requirements that scale as N^2 (e.g., around 20 MB for C_{120} and 80 MB for C_{240}). The semiempirical integral approximations also allow for a rapid gradient evaluation in the case of variational SCF wavefunctions. Here the time for the gradient calculation again scales as N^2 and is therefore usually negligible relative to the SCF time (see above).

In summary, semiempirical and *ab initio* or DFT calculations differ in their computational effort by at least three orders of magnitude. Hence for any given hardware, there will be certain applications that are feasible only with a semiempirical approach, and others where a combined use of semiempirical and *ab initio* or DFT methods would seem attractive. In such cases, the essential question is whether the available semiempirical methods provide the accuracy desired.

There is a vast literature [3–8, 11–15, 20–28, 36, 43, 48] documenting the accuracy and the errors of the existing semiempirical methods. Without going into any details, we present only one example: Table II shows the mean absolute errors for some ground-state properties of typical organic molecules which correspond to the validation set from the original MNDO evaluation [43]. The comparisons cover MNDO, AM1, PM3, and a newly developed method (Section III.B). The mean absolute errors lie between 3.5 and 6.3 kcal/mol for heats of formation, between 0.01 and 0.02 Å for bond lengths, around 2° for bond angles, between 0.3 and 0.5 eV for ionization potentials, and around 0.3 D for dipole moments. Semiempirical heats of formation are essentially equivalent to total atomization energies. Considering the difficulties in *ab initio* calculations to predict accurate absolute atomization energies, the errors for

TABLE II
Mean Absolute Errors for Organic Molecules (C, H, N, O)

Property[a]	N[b]	MNDO	AM1	PM3	Ref. 37[c]
ΔH_f (kcal/mol)	133	6.3	5.5	4.2	3.5
R (Å)	228	0.015	0.017	0.011	0.012
θ (deg)	92	2.69	2.01	2.22	1.88
IP (eV)	51	0.47	0.36	0.43	0.32
μ (D)	57	0.32	0.25	0.27	0.25

Source: Ref. 37.

[a] Heats of formation ΔH_f, bond lengths R, bond angles θ, ionization potentials IP (Koopmans' theorem), dipole moments μ.
[b] Number of comparisons.
[c] New method with orthogonalization corrections; see Section III.B.

heats of formation in Table II would seem surprisingly low (owing to the parametrization, of course). It should be stressed, however, that relative energies are of primary importance in most application. *Ab initio* methods often provide fairly reliable predictions for such relative energies (e.g., in isodesmic reactions) [61], because the individual absolute errors tend to be systematic. Semiempirical methods often show a larger scatter in the errors for individual heats of formation, which can cause larger errors in relative energies. These remarks may be generalized to other properties as well: The available experience indicates that the deviations between theory and experiment are normally more systematic in *ab initio* than in semiempirical calculations.

When selecting the strategy for a computational study, we advocate the use of accurate *ab initio* calculations whenever this is feasible, because reliable theoretical predictions can be obtained at high theoretical levels (ideally through a convergence toward the solution of the time-independent Schrödinger equation). We recommend the use of semiempirical methods in cases where a complete *ab initio* or DFT study is impractical. It is essential in such cases to establish the expected accuracy of the semiempirical results, either by comparisons with the available literature data or by separate calibrations. In this context it is usually helpful to perform some checks against *ab initio* or DFT calculations (if possible). The next section includes examples of such studies.

C. Selected Recent Applications

Since the preparation of macroscopic amounts [66] of buckminsterfullerene [67], there has been extensive experimental and theoretical work on fullerene chemistry [68–70]. Fullerenes C_n are carbon cages with 12 pentagons and $(n/2-10)$ hexagons. The smallest possible fullerene is C_{20}, the most prominent ones are C_{60} and C_{70}, and there are giant fullerenes that may contain hundreds of carbon atoms. The size of the fullerenes makes them appropriate targets for systematic semiempirical studies, which may be supplemented with (computationally rather demanding) *ab initio* and DFT calculations in cases where higher accuracy is desired. Some selected theoretical investigations of this kind are summarized here to illustrate current applications of semiempirical methods.

Fullerenes are strained molecules consisting of pyramidalized sp^2 carbon atoms. Introducing a π-orbital axis vector [71] with equal angles $\theta_{\sigma\pi} > 90°$ to the three σ-bonds at a given carbon atom (POAV1 convention) the pyramidalization angle may be defined as $\theta_p = \theta_{\sigma\pi} - 90°$. These angles are usually quite large (e.g., $\theta_p = 11.6°$ in C_{60}) and imply a large strain energy. For practical purposes it is often sufficient to assess the global strain by the average of the squared pyramidalization angles, $\theta^2 = \Sigma_p \theta_p^2 / n$. Systematic MNDO calculations for fullerenes from C_{20} to C_{540} indeed show a good correlation between θ^2 and thermodynamic stability [72]. Based on these results and on experimental data, the strain per carbon atom in C_{60} is estimated [73] as 8 kcal/mol, so that fullerenes such as C_{60} are among the most strained organic molecules ever isolated. The MNDO calculations predict increasing thermodynamic stability with increasing fullerene size. This is not too surprising, because the proportion of hexagons must grow with increasing fullerene size which is normally accompanied by a decrease in θ^2 and increased stability. In fact, the MNDO optimized structures of large icosahedral fullerenes such as C_{540} and C_{960} (Fig. 1) may be viewed as almost planar graphite segments with 12 pentagonal defects where curvature and strain are localized [72]. Such structures are considerably more stable than alternative spherical I_h structures (with all atoms constrained to lie on the surface of a sphere), and the latter structures collapse to the former ones in unconstrained geometry optimizations [74]. In the case of C_{180} and C_{240}, *ab initio* SCF calculations support these MNDO predictions: For example, the energy differences between the corresponding two C_{180} structures are almost identical at the MNDO and *ab initio* SCF levels (0.77 and 0.81 kcal/mol per carbon atom, respectively) [74].

The correlation between structure (θ^2) and stability is also useful in rationalizing the structural preferences in giant fullerenes which may occur in spheroidal and cylindrical shapes (buckyspheres vs. buckytubes). Since the buckytubes have a higher curvature and hence higher strain, they are expected to be less stable thermodynamically. This is confirmed both by calculations [75] and by the experimental observation [76] that

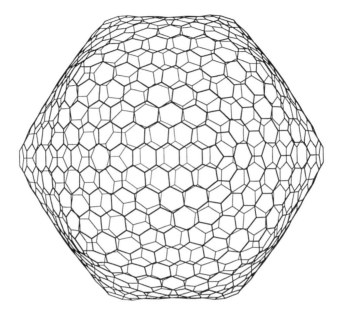

Figure 1. MNDO optimized structure of the icosahedral C_{960} fullerene. The view along a C_3 axis emphasizes the localization of curvature at the pentagons.

buckytubes in the originally formed fullerene soot are converted to buckyspheres upon high-energy electron bombardment.

Many isomers exist for a given fullerene size (e.g., 1812 for C_{60}) [77]. Their relative stability is governed by strain and π conjugation. The most important stability criterion is the isolated pentagon rule (IPR), which states that fused pentagons are energetically unfavorable. In this case, strain and π conjugation work in the same direction, because a pentalene substructure in a fullerene causes a higher local curvature and contains an antiaromatic 8π subsystem. The combined effect amounts to a destabilization of about 20 kcal/mol per pair of fused pentagons, according to MNDO calculations [78]. Generally, the observed fullerenes have isolated pentagons (e.g., C_{60}, C_{70}). For the higher fullerenes (C_n, $n \geq 76$) there are several IPR isomers [79], and usually there is a competition between strain and π conjugation in determining the relative stabilities. Sometimes (e.g., in C_{78} [80]) high-level *ab initio* calculations are required for a reliable assessment of these competing factors, but in most cases semiempirical treatments are sufficient to identify the most stable IPR isomers. It appears that minimization of strain is usually more important in the higher fullerenes than maximization of π conjugation. For example, in C_{84} [81], the isomer with the highest Hückel resonance energy (and a fairly "edgy" shape) is the least stable of the 24 IPR isomers, and the two lowest, almost isoenergetic isomers of D_{2d} and D_2 symmetry (according to semiempirical and *ab initio* SCF calculations) are those with the lowest overall curvature (θ^2). The experimental ^{13}C NMR spectrum [79] of C_{84} is consistent with a $1:2$ mixture of a D_{2d} and a D_2 isomer, in agreement with the theoretical results.

Curvature influences not only the thermodynamic stabilities of fullerenes through differences in global strain (see above), but also the ease of reactions and the regiochemistry through the relief of local strain. It is well established both experimentally and theoretically that additions to fullerenes occur most easily at a 6–6 bond (shared by two hexagons), and distinct patterns of addition to C_{60} are predicted from systematic semiempirical calculations [82–86]. In C_{70} and higher fullerenes there are different types of 6–6 bonds, and one may expect that the most strained 6–6 bond will react most easily because the addition will lead to the largest possible decrease in strain energy. This is indeed found in the η_2 complexation of C_{70} with transition metals [73] and in the osmylation of C_{70} [87] and C_{76} [88]. In view of these findings, recent theoretical work on the 46 IPR isomers of C_{90} and the 187 IPR isomers of C_{96} has marked the atoms with the highest pyramidalization angles in the low-energy species, in order to predict the probable regiochemistry simply from the calculated geometries [89].

The mechanism of spontaneous fullerene formation has been the subject of much debate and speculation [90]. One of the major advances in this area comes from gas-phase ion chromatography, which allows to separate charged carbon clusters of specific mass into families of isomers based on their mobilities through helium-filled drift tubes [91–93]. These families (e.g., monocyclic rings, bi- and tricyclic compounds, fullerenes) are identified by comparing the observed mobilities with those calculated from optimized semiempirical equilibrium geometries [92], which are accurate enough for this purpose. Such studies lead to product distributions of the cations $C_n{}^+$ as a function of n [90–93] and allow the tentative formulation of a detailed mechanism for fullerene synthesis [93].

The mechanisms for fullerene annealing and fragmentation have recently been investigated theoretically in much detail [94–96] using semiempirical geometry optimizations with higher-level single-point energy calculations as well as tight-binding molecular dynamics simulations. Annealing denotes rearrangements that move the pentagons and hexagons on a fullerene surface by changing the interconnections between the carbon atoms. The simplest annealing process is the Stone–Wales [97] pyracylene rearrangement with an "in-plane" rotation of a 6–6 bond by 90°, which may yield an IPR-violating isomer as, for example, in C_{60}, where two pairs of fused pentagons are generated in this manner. The Stone–Wales rearrangement is formally forbidden [97] and requires much activation. An alternative "out-of-plane" pathway for this rearrangement has been found computationally on the MNDO potential surface [94, 96] which involves two successive 1,2 carbon shifts via an intermediate with one sp^3 carbon atom and which needs almost 2 eV less activation than the traditional "in-plane" pathway. Under the conditions of the usual fullerene synthesis ($T > 1000$ K) such a relatively easy annealing mechanism will allow the sampling of many different fullerene isomers and thus increase the chances to find the most stable one.

Experimentally, fragmentation of fullerenes normally occurs by loss of C_2 units, down to C_{32} [98]. The originally proposed mechanism assumes the elimination of C_2 from a 5–5 bond (shared by two pentagons, i.e., from an IPR-violating isomer that may have to be generated by annealing) [98]. Computational studies show [94–96] that for IPR isomers, there is an energetically competitive mechanism on the MNDO potential surface where the leaving C_2 unit comes from a 5–6 bond and which proceeds through the sp^3 intermediate of the out-of-plane annealing pathway. However, when comparing C_2 loss from an IPR and an IPR-violating fullerene, the calculated barrier is appreciably lower in the latter case [96]. Hence under the usual experimental conditions, fullerenes with isolated pentagons such as C_{60} and C_{70} should be kinetically more stable against C_2 loss than their neighbors (C_{58}/C_{62} and C_{68}/C_{72}, respectively), which do have fused pairs of pentagons. C_{60} and C_{70} are thus both thermodynamically and kinetically more stable than their neighbors, which further rationalizes their prominence in fullerene production.

There are many other examples in the literature of how semiempirical calculations have contributed to recent progress in fullerene chemistry (often in combination with other quantum chemical calculations). The cases discussed above should be sufficient to illustrate typical conclusions from such studies which provide not only specific numerical results, but also new qualitative insights, for example, with regard to the correlation between curvature and stability, the relative importance of strain and π conjugation, the connection between structure and reactivity, or the mechanism for fullerene annealing and fragmentation.

III.   CURRENT DEVELOPMENTS

Based on the description of the current status of the field (Section II), this part of the review will address recent and ongoing developments in semiempirical methodology. It will cover a number of topics ranging from theoretical foundations to computational aspects. The selection of these topics is subjective and partly reflects personal interests, but it also represents an attempt to identify developments which are conceptually promising and may lead to improved semiempirical treatments that are useful in practical applications.

A. Theoretical Foundations

The most rigorous theoretical justification for semiempirical methods rests on the theory of effective valence shell Hamiltonians [99–102]. The basic concept of this approach is summarized as follows [100]. Consider an infinite basis set consisting of core orbitals, valence orbitals, and excited orbitals, and distinguish between two types of N-electron determinants: Those in the primary space P with core orbitals filled, excited orbitals empty, and valence electrons distributed among the valence orbitals in any conceivable manner, and those in the secondary space Q with excitations from the core orbitals and/or into the excited orbitals. Now the objective is to derive an effective valence shell Hamiltonian $\mathcal{H}^v$ such that a full CI treatment in the primary valence space P yields the exact experimental energies E for the valence states of a given system. This can be achieved by writing the exact Schrödinger equation in block matrix form (with the use of projection operators P and $Q = 1 - P$):

$$\begin{bmatrix} PHP & PHQ \\ QHP & QHQ \end{bmatrix} \begin{bmatrix} P\psi \\ Q\psi \end{bmatrix} = E \begin{bmatrix} P\psi \\ Q\psi \end{bmatrix}. \tag{7}$$

Algebraic manipulations then yield an effective Hamiltonian $\mathcal{H}^v$ which acts only on the valence space and nevertheless produces the exact energies E_α for all valence states:

$$\mathcal{H}^v P\psi_\alpha = E_\alpha P\psi_\alpha , \tag{8}$$

$$\mathcal{H}^v(E_\alpha) = PHP + PHQ(E_\alpha 1 - QHQ)^{-1}QHP . \tag{9}$$

It is this effective Hamiltonian which semiempirical methods try to approximate more or less successfully. Therefore, an *ab initio* evaluation of $\mathcal{H}^v$ is certainly desirable for an analysis of semiempirical methods. The computational implementation of Eqs. (7)–(9) requires a number of approximations [103]: Use of a large but finite basis set, a Rayleigh–Schrödinger perturbation expansion of the inverse matrix in Eq. (9) around zero-order energies, and a quasidegenerate perturbation treatment through second or third order. This leads to an energy-independent Hamiltonian which includes nonclassical effective many-electron operators (i.e., three-electron terms in second order and four-electron terms in third order). Using these approximations, *ab initio* $\mathcal{H}^v$ calculations have been carried out for many atoms and molecules (see, e.g., Refs. 104–119). These calculations have established the effective Hamiltonian approach as a viable *ab initio* technique because the results are of similar accuracy as in more conventional *ab initio* studies employing basis sets of comparable quality.

By its design the effective valence shell Hamiltonian method provides a completely *ab initio* basis for semiempirical methods. Both approaches employ a minimal basis set of valence orbitals and attempt to generate exact solutions (corresponding to the exact experimental results). Therefore, the matrix elements of $\mathcal{H}^v$ represent *ab initio* analogs of the semiempirical parameters. These matrix elements contain first-order contributions evaluated from the *PHP* term in Eq. (9) which correspond to the analytical one- and two-electron integrals in the usual notation (cf. Section II.A). The second term in Eq. (9) gives rise to higher-order contributions from dynamic electron correlation which may strongly modify the first-order one- and two-electron matrix elements of $\mathcal{H}^v$ and which also introduce nonclassical three- and four-electron matrix elements of $\mathcal{H}^v$ that have no counterparts in traditional semiempirical methods. It is obvious that an analysis of these *ab initio* matrix elements can provide much valuable information concerning the justification of semiempirical model assumptions and guidelines for further developments.

Systematic *ab initio* $\mathcal{H}^v$ calculations with large basis sets have recently been performed [115–118] for several unsaturated hydrocarbons (ethylene, *trans*-butadiene, cyclobutadiene, benzene, and hexatriene) to examine the theoretical basis of semiempirical π-electron theories [1–3] [Pariser–Parr–Pople (PPP)]. The results from these studies are remarkable [115–118]: As expected from formal $\mathcal{H}^v$ theory, the vertical excitation energies of the different excited states are indeed reproduced with good accuracy by the *ab initio* $\mathcal{H}^v$ method in its current implementation when using a molecule-specific valence space. More important, little accuracy is lost for the low-lying valence-like states when using a common set of valence orbitals constrained to be linear combinations of transferable carbon-atom p_π orbitals (analogous to the common minimal basis set in the PPP method). The zero-differential-overlap (ZDO) approximation is satisfied surprisingly well by the *ab initio* $\mathcal{H}^v$ matrix elements in the Löwdin basis, because the higher-order correlation contributions further diminish the corresponding analytical one- and two-electron integrals (which are often already small in the Löwdin basis). The $\mathcal{H}^v$ matrix elements surviving at the ZDO level are normally large and well transferable between different molecules, particularly in the Löwdin basis. The nonclassical three- and four-electron terms turn out to be less important for the relevant excitation energies and ionization potentials, and can therefore be neglected as a leading approximation. In summary, the *ab initio* $\mathcal{H}^v$ calculations [115–118] thus justify several basic assumptions of semiempirical π-electron theory (PPP), including the choice of a common minimal basis set, the ZDO approximation, and the transferability of the parameters.

It is one of the advantages of the $\mathcal{H}^v$ approach that it allows an explicit analysis of the correlation contributions to the matrix elements. The $\mathcal{H}^v$ calculations for atoms [104, 108] and for the π-electron systems [114–118] support the concept of reduced effective two-electron Coulomb interactions in current semiempirical methods [19–22] (see Section II.A). The semiempirical Coulomb integrals are relatively close to the correlated $\mathcal{H}^v$ matrix elements, which are considerably lower than the corresponding analytical two-electron integrals. The two-center two-electron Coulomb interactions from the $\mathcal{H}^v$ approach can be fitted [110] better with a Klopman–Ohno [40, 41] than with a Mataga–Nishimoto [54] formula, which is in line with the preferred choice in current semiempirical treatments [19–22]. On the other hand, the one-center two-electron matrix elements of $\mathcal{H}^v$ show a correlation-induced bond-length dependence [109, 110, 112] that is not included in the present semiempirical schemes. Similarly, novel bond-length dependencies are suggested for several other interactions on the basis of $\mathcal{H}^v$ calculations [109, 110, 112], as well as novel bond-angle dependencies in the case of polyatomic molecules such as CH_2 [113].

Ideally, the *ab initio* $\mathcal{H}^v$ approach would directly produce a useful semiempirical model Hamiltonian and provide the necessary parameters from the nonvanishing *ab initio* $\mathcal{H}^v$ matrix elements. The most recent $\mathcal{H}^v$ study on hexatriene [118] comes close to this goal by proposing two PPP-type models (with parameters derived from the *ab initio* $\mathcal{H}^v$ calculations), one of which ($\mathcal{H}^v$-PPP$_{34B}$) already outperforms the original PPP method [118]. This success is facilitated by the fact that the nonclassical three- and four-electron terms do not affect the π-electron excitation energies of hexatriene much [118]. In the general case, these nonclassical terms may be quite large and must then be averaged suitably into the one- and two-electron integrals [104], which is still an unsolved problem at present [118].

In conclusion, the *ab initio* effective Hamiltonian method has theoretically justified several basic assumptions of semiempirical models, particularly through recent studies on π-electron systems [115–118]. It seems to approach the point where it can provide useful theoretical guidance for further improvements in valence-electron semiempirical methods (e.g., with regard to the inclusion of additional interactions in the model or the use of different parametric functions for integral evaluation). Many of the specific suggestions from the $\mathcal{H}^v$ investigations arise from a detailed analysis of electron correlation effects and have not been considered in traditional semiempirical methods where dynamic correlation is taken into account only in an average sense through a global reduction of the two-electron interactions (see Section II.A). It would therefore seem worthwhile to test the specific suggestions from correlated $\mathcal{H}^v$ theory in the context of the usual semiempirical parametrizations. This would be a logical next step after completing current methodological developments which are motivated by considerations at the SCF-MO level and which are described in the next section.

B. General-Purpose Methods

The discussion in Section II.A has shown that many of the currently accepted semiempirical methods for computing potential surfaces are based on the MNDO model. These methods differ mainly in their actual implementation and parametrization. Given the considerable effort that has gone into their development, we believe that further significant overall improvements in general-purpose semiempirical methods require improvements in the underlying theoretical model. In this spirit we describe two recent developments: The extension of MNDO to d orbitals and the incorporation of orthogonalization corrections and related one-electron terms into MNDO-type methods.

MNDO, MNDOC, AM1, and PM3 employ an sp basis without d orbitals in their original implementation [19–22]. Therefore, they cannot be applied to most transition metal compounds, and difficulties are expected for hypervalent compounds of main-group elements, where the importance of d orbitals for quantitative accuracy is well documented at the *ab initio* level [61]. The inclusion of d orbitals is easier in CNDO and INDO than in NDDO. For example, both CNDO and INDO represent all two-center Coulomb interactions by a single common integral $(s^A s^A, s^B s^B)$, wheres NDDO retains all nonvanishing two-center two-electron integrals $(\mu^A \nu^A, \lambda^B \sigma^B)$ (i.e., 204 unique integrals for an spd basis [33]). Consequently, extensions to d orbitals have long been available for CNDO methods [4], SINDO1 [26], and INDO/S [28], but only more recently for MNDO [33–36].

In the MNDO/d approach, the established MNDO formalism and parameters remain unchanged for hydrogen, helium, and the first-row elements. The inclusion of d orbitals for the heavier elements requires a generalized semiempirical treatment of the two-electron interactions. The *two-center two-electrons integrals* are calculated by an extension [33] of the original point-charge model [39] that is currently used in MNDO, MNDOC, AM1, and PM3. These integrals are expanded in terms of semiempirical multipole–multipole interactions. For an spd basis, there are 45 distinct one-center charge distributions that are associated with multipoles up to hexadecapoles. All monopoles, dipoles, and quadrupoles of these charge distributions are included whereas all higher multipoles are neglected [33]. Using suitable point-charge representations the remaining multipole–multipole interactions are calculated by applying the Klopman–Ohno formula [40, 41], which does not involve any new adjustable parameters. The computational implementation of the extended point-charge model has been specified in detail [33]. All nonzero *one-center two-electron integrals* are retained to ensure rotational invariance [120]. Because of the large number of such unique integrals in an spd basis [33], it is generally not feasible to determine all of them from experimental data or to optimize all of them in the semiempirical parametrization. Therefore, the one-center Coulomb integrals g_{ss}, g_{pp}, and g_{dd} are optimized in the parametrization and used to derive corresponding orbital exponents which analytically reproduce these semiempirical integrals. Using these orbital exponents, all other one-center two-electron integrals are calculated analytically, except for g_{sp} and h_{sp}, which are optimized independently to allow for a finer tuning of the results.

The MNDO/d formalism for the two-electron integrals has several advantages: Conceptually, it employs the same approach as the established MNDO-type methods [19–22], including an effective reduction of the Coulomb interactions to allow for dynamic electron correlation. Moreover, it is rotationally invariant and computationally efficient, and analytical first and second derivatives of the integrals are available [121]. An alternative formalism has been implemented more recently in the SAM1 approach [23, 122], which apparently involves the analytical evaluation of the two-electron integrals in an spd basis and subsequent scaling (details still unpublished).

The implementation and the parametrization of the MNDO/d approach are analogous to MNDO (see Section II.A), with only minor and presently irrelevant variations in certain details [33, 36]. Optimized final parameters are available [34–36] for the second-row elements, the halogens, and the zinc group elements. MNDO/d employs an spd basis for Al, Si, P, S, Cl, Br, and I but only an sp basis for Na, Mg, Zn, Cd, and Hg. For the latter five elements, parametrizations with an sp and an spd basis yield results of similar quality (as expected in a semiempirical framework) which allows us to adopt the simpler sp basis in these cases [36]. Table III reports a statistical evaluation of extensive test calculations

TABLE III

Mean Absolute Errors for Molecules Containing Second-Row and Heavier Elements[a]

Property[b]	N[c]	MNDO	AM1	PM3	MNDO/d
All Compounds					
ΔH_f (kcal/mol)	575	29.2	15.3	10.9	5.4
R (Å)	441	0.072	0.065	0.065	0.056
θ (deg)	243	3.7	3.4	7.4	2.5
IP (eV)	200	0.89	0.53	0.64	0.45
μ (D)	133	0.55	0.50	0.60	0.35
Normalvalent Compounds					
ΔH_f (kcal/mol)	508	11.0	8.0	9.6	5.4
θ (deg)	179	2.9	2.7	7.7	2.7
μ (D)	117	0.48	0.38	0.50	0.27
Hypervalent Compounds					
ΔH_f (kcal/mol)	67	143.2	61.3	19.9	5.4
θ (deg)	64	5.6	5.1	6.5	2.1
μ (D)	16	1.04	1.31	1.29	0.93

Source: Ref. 36.

[a] Elements Na, Mg, Al, Si, P, S, Cl, Br, I, Zn, Cd, and Hg are included.

[b] See footnote *a* in Table II.

[c] Number of comparisons for MNDO/d (slightly lower for the other methods due to missing parameters; see Ref. 36 for details).

covering several properties and more than 600 molecules containing the foregoing elements. MNDO/d provides significant improvements over the established semiempirical methods (MNDO, AM1, PM3), especially for hypervalent compounds. The mean absolute error in MNDO/d heats of formation amounts to 5.4 kcal/mol for the complete validation set of 575 molecules and is identical for the subsets of 508 normalvalent and 67 hypervalent molecules. By contrast, the established semiempirical methods with an *sp* basis yield heats of formation for hypervalent compounds that show much larger errors [34, 36], even in the case of PM3 [22], where a determined effort has been made in the parametrization to remove such shortcomings. In addition, MNDO/d consistently predicts the correct shape (point group) of each hypervalent molecule studied (e.g., C_{2v} for ClF_3), which is not true for the older methods with an *sp* basis [34, 36]. In an overall assessment, the MNDO/d predictions for compounds with heavy elements are mostly of similar accuracy as the MNDO predictions for organic compounds (H, C, N, O) [36, 43].

Apart from a better overall performance (see the statistics in Table

III), the inclusion of d orbitals for the heavier elements leads to qualitative improvements for hypervalent compounds in MNDO/d. Similar observations have been made in SINDO1 [26] and in SAM1/ SAM1d test calculations [36]. Hence the available evidence clearly indicates that d orbitals are essential for a balanced semiempirical description of both normalvalent and hypervalent molecules.

The documented successes of MNDO/d (see above) suggest an extension to transition metals. The MNDO/d parametrization has been completed for several transition metals [123] and is in progress for others, but a systematic assessment of the results is not yet feasible. Elsewhere the published MNDO/d formalism for the two-electron integrals [33] has been implemented independently [124] and combined with PM3 in another parametrization attempt for transition metals [125]. Moreover, the SAM1 approach is also being parametrized for transition metals [122, 126]. Given these complementary activities it seems likely that the performance of MNDO-type methods with an spd basis for transition metal compounds will be established in the near future.

As discussed above, the extension of the MNDO model to d orbitals has deliberately retained as many features from the original MNDO formalism as possible (see Section II.A). From a theoretical point of view, the least satisfactory of these features is the effective atom-pair term E_{AB}^{eff}, which is included in the core–core repulsion E_{AB}^{core} and accounts globally for Pauli exchange repulsions (and other corrections). Efforts to go beyond the MNDO model should thus aim at an explicit representation of Pauli exchange repulsions (and other corrections) in the core Hamiltonian and the Fock matrix which would make the term E_{AB}^{eff} obsolete and allow its removal [12, 37]. In the following we describe such recent developments [37, 38]. The inclusion of explicit orthogonalization corrections is the spirit of, for example, the SINDO1 method [25] or the CNDO-S^2 method [127], but the model assumptions and the implementation at the NDDO level are quite different in the approach described here [37, 38].

As a point of reference, consider the eigenvalue problem solved by *ab initio* SCF methods:

$$^{\lambda}\mathbf{F}^{\lambda}\mathbf{C} = {}^{\lambda}\mathbf{C}\mathbf{E} \tag{10}$$

$$^{\lambda}\mathbf{F} = \mathbf{S}^{-1/2}\mathbf{F}\mathbf{S}^{-1/2} \tag{11}$$

$$^{\lambda}\mathbf{C} = \mathbf{S}^{1/2}\mathbf{C} \tag{12}$$

where $^{\lambda}\mathbf{F}$ and $^{\lambda}\mathbf{C}$ denote the Fock matrix and the eigenvector matrix in the Löwdin orthogonalized basis ($^\Lambda\Phi$) [128, 129], respectively. $\mathbf{F}$ and $\mathbf{C}$ are the corresponding matrices in the basis of the nonorthogonal atomic orbitals (Φ), $\mathbf{S}$ the overlap matrix, and $\mathbf{E}$ the diagonal eigenvalue matrix.

Semiempirical SCF methods solve an eigenvalue problem analogous to Eq. (10). Neglecting correlation effects for the moment (see Section III.A), the semiempirical integrals can thus be associated with theoretical integrals in the $^\Lambda\Phi$ basis. In the case of the two-electron integrals, this provides the traditional justification for the NDDO approximation [130–133]: The three- and four-center as well as certain two-center two-electron integrals are small in the $^\Lambda\Phi$ basis and may therefore be neglected. For the remaining two-center two-electron NDDO integrals, the orthogonalization transformation ($\Phi \to {}^\Lambda\Phi$) leads to a reduction of these integrals at intermediate distances [130–133], which is consistent with the Klopman–Ohno scaling [40, 41] of the semiempirical integrals. Hence current MNDO-type methods account for major orthogonalization effects on the two-electron integrals in that they neglect integrals that are small in the $^\Lambda\Phi$ basis and employ reduced values for the nonzero two-electron integrals at intermediate distances. It is fortunate that both these choices also reflect the behavior of the correlation contributions to the two-electron integrals [116, 117] (see Section III.A).

The orthogonalization transformation ($\Phi \to {}^\Lambda\Phi$) for the one-electron integrals causes substantial changes in the core Hamiltonian matrix:

$$^\Lambda\mathbf{H} = \mathbf{S}^{-1/2}\mathbf{H}\mathbf{S}^{-1/2} . \tag{13}$$

From a computational point of view, $^\Lambda\mathbf{H} - \mathbf{H}$ could be evaluated analytically (*ab initio*) and then be added to the semiempirical core Hamiltonian matrix. This procedure, however, introduces an imbalance between the one- and two-electron parts of the Fock matrix as long as the two-electron integrals are not subjected to the same exact transformation ($\Phi \to {}^\Lambda\Phi$), which would sacrifice the computational efficiency of semiempirical methods and is therefore not feasible. Hence the orthogonalization corrections to the one-electron integrals must instead be represented by suitable parametric functions. Their essential features can be recognized from the analytic expressions for the matrix elements of $^\Lambda\mathbf{H}$ in the simple case of a homonuclear diatomic molecule with two orbitals (ϕ_μ at atom A, ϕ_ν at atom B):

$$^\Lambda H_{\mu\mu} = H_{\mu\mu} - S_{\mu\nu}M_{\mu\nu}/(1 - S_{\mu\nu}^2) \tag{14}$$

$$^\Lambda H_{\mu\nu} = M_{\mu\nu}/(1 - S_{\mu\nu}^2) \tag{15}$$

$$M_{\mu\nu} = H_{\mu\nu} - S_{\mu\nu}(H_{\mu\mu} + H_{\nu\nu})/2 \tag{16}$$

$$^{\lambda}H_{\mu\mu} = H_{\mu\mu} - S_{\mu\nu}\,^{\lambda}H_{\mu\nu}\,. \tag{17}$$

Numerical calculations for the hydrogen molecule with a minimal basis show that $H_{\mu\mu}$ and $^{\lambda}H_{\mu\mu}$ are of the same order of magnitude, whereas $H_{\mu\nu}$ and $^{\lambda}H_{\mu\nu}$ differ strongly [134]. Therefore, $^{\lambda}H_{\mu\mu} - H_{\mu\mu}$ may be regarded as a relatively small correction, and it should be possible to represent this term by a suitable parametric expression. On the other hand, since $|H_{\mu\nu}|$ and $|^{\lambda}H_{\mu\nu} - H_{\mu\nu}|$ are of similar magnitude due to $|^{\lambda}H_{\mu\nu}| \ll |H_{\mu\nu}|$, it would seem more appropriate in this case to parametrize the resonance integrals $\beta_{\mu\nu}$ directly such that they mimic the integrals $|^{\lambda}H_{\mu\nu}|$ in the orthogonalized basis. With this choice, the orthogonalization corrections $^{\lambda}H_{\mu\mu} - H_{\mu\mu}$ are $-S_{\mu\nu}\beta_{\mu\nu}$ in our simple example [see Eq. (17)], which implies a balanced treatment of attractive and repulsive one-electron interactions: The resonance integrals represent the two-center core Hamiltonian matrix elements, which are mainly responsible for bonding, but they also enter the core Hamiltonian matrix elements in an antibonding manner because the orthogonalization corrections are normally repulsive.

The actual implementation [37] of these ideas is guided by power series expansions for $^{\lambda}\mathbf{H}$ according to Eq. (13) and by numerical results from *ab initio* SCF calculations. The orthogonalization corrections are given by a two-parametric function which involves the resonance integrals (and some other terms appearing in the second-order expansion of $^{\lambda}\mathbf{H}$). The resonance integrals contain an empirical parametrized radial part (with a distance dependence similar to that in *ab initio* $^{\lambda}H_{\mu\nu}$ integrals) and the same angular part as the corresponding overlap integrals. The explicit treatment of the orthogonalization corrections causes several other changes in comparison with the standard MNDO model (see Section II.A): First, as motivated in the introductory discussion, the effective atom-pair terms E_{AB}^{eff} in the core–core repulsions are removed. Second, all two-center Coulomb interactions (i.e., two-electron integrals, core–electron attractions, and core–core repulsions) are calculated analytically and then subjected to a uniform Klopman–Ohno scaling, which implies that the penetration integrals [29, 135] are no longer neglected. Third, the core–valence interactions and orthogonality effects are included through an effective core potential, which is again determined analytically with subsequent Klopman–Ohno scaling (see Ref. 37 for further discussion of these choices).

The parametrization of the semiempirical treatment outlined [37] employs ground-state reference data and follows the same strategy as in MNDO (see Section II.A). Parameters are available for the elements H, C, N, O, and F. In a statistical evaluation for ground-state properties

[37], this treatment shows slight but consistent improvements over MNDO, AM1, and PM3 (see Table II). Significant improvements are found for excited states, transition states, and hydrogen bonds. For example, the mean absolute error in vertical excitation energies (33 comparisons) is 0.28 eV compared with 1.20 eV for AM1 and 1.18 eV for PM3. These improvements are readily rationalized from the underlying theoretical model: In the simple two-orbital case considered above [see Eqs. (14)–(17)] the destabilization of the antibonding molecular orbital (relative to the AO energy) is greater than the stabilization of the bonding molecular orbital at the *ab initio* level. In the established MNDO-type methods [19–22], however, the ZDO approximation leads to a symmetric splitting of bonding and antibonding molecular orbitals. Since MNDO-type methods describe atoms realistically and are parametrized for ground-state properties (effectively including the energies of occupied MOs), they should tend to underestimate the energies of the unoccupied MOs due to the ZDO approximation. Hence one should expect a systematic underestimation of excited-state energies. This is indeed observed for MNDO, AM1, and PM3. The explicit inclusion of electron correlation in MNDOC provides a suitable theoretical description for excited states and some improvement in the calculated vertical excitation energies [50] through the reparametrization at a correlated level, but the basic shortcomings of the ZDO approximation remain in effect and again lead to excitation energies that are usually too small. These problems are alleviated by the introduction of orthogonalization corrections into the theoretical model [37]. These repulsive corrections effectively increase the energies of the relevant interacting AOs [e.g., by $-S_{\mu\nu}{}^{\lambda}H_{\mu\nu}$ in Eq. (17)]. The subsequent symmetric splitting starts at this higher energy and must reach a realistic energy for the bonding MO (see above), which automatically places the antibonding MO at a relatively high energy. Therefore, the inclusion of the orthogonalization corrections into the model will generally yield higher excitation energies, compared with the established MNDO-type methods [19–22]. This is indeed confirmed by the numerical results: The proposed treatment [37] predicts reasonable excitation energies (without large systematic errors) even though it has not been parametrized for excited states. The improvements thus seem to be inherent to the underlying theoretical model, and it is not too surprising that transition states in thermal reactions are also described well by this treatment [37], probably for similar reasons.

Conformational properties such as rotational barriers and energy differences between conformers are often poorly reproduced by existing semiempirical methods [19–22, 37]. These problems have been attributed to the neglect of orthogonalization corrections to the resonance integrals, particularly those for nonbonded 1,3 interactions [136, 137]. These corrections include three-center contributions and are not included in the treatment discussed previously [37] where the resonance integrals attempt to mimic the analytical $^\lambda H_{\mu\nu}$ integrals in a diatomic system without accounting for multicenter effects. Recent work [38] has extended the previous approach [37] to include orthogonalization corrections to the resonance integrals, and it has indeed been found in preliminary parametrizations (C, H, O) that this refinement works well: The mean absolute errors for the usual ground-state properties are similar to those before [37] (with some improvements for ionization potentials and dipole moments), and as expected, conformational properties are predicted much better than before (e.g., with regard to the rotational barrier in ethane, the conformational preferences in cyclohexane or cyclopentane and the relative energies of n-pentane and neopentane). Optimization of this new approach is in progress.

In concluding this section it should be emphasized again that in our opinion, improvements in the underlying theoretical models are the best strategy for improving general-purpose semiempirical methods. This belief is supported by the following observations:

1. Advances of MNDO over MINDO/3 can be traced back to the inclusion of higher multipoles in electrostatic two-center interactions [17].

2. MNDOC performs better than MNDO in cases [47–53] where specific correlation effects require the explicit treatment of electron correlation.

3. The inclusion of d orbitals in MNDO/d is responsible for the balanced description of normalvalent and hypervalent main-group compounds [33–36], which has not been achieved in semiempirical treatments with an sp basis.

4. The introduction of orthogonalization corrections for the one-center part of the core Hamiltonian matrix [37] generates improved excitation energies by correcting for deficiencies inherent to the ZDO approximation.

5. As expected theoretically [136, 137], better predictions for conformational properties become possible [38] when including orthogonalization corrections also for the two-center resonance integrals.

A careful and balanced parametrization is clearly important in exploring the limits of a given semiempirical model, but it cannot replace the theoretically guided search for better models, which offer the most promising perspective for general-purpose semiempirical methods with better overall performance.

C. Specialized Methods

In *ab initio* quantum chemistry, it is well established that different classes of compounds and different properties require different levels of theory for a reliable description. By contrast, general-purpose semiempirical methods attempt to describe all classes of compounds and many properties simultaneously and equally well. This is an ambitious goal, and it is obvious that compromises cannot be avoided in such an endeavor (e.g., during the parametrization). It would therefore seem attractive to develop specialized semiempirical methods for certain classes of compounds or for specific properties because such methods ought to be more accurate in their area of applicability than the general-purpose methods. This advantage must be weighed against the additional parametrization work and the danger of parameter proliferation.

Several specialized semiempirical treatments exist. Many of them are based on the MNDO model in one of its standard implementations (e.g., MNDO, AM1, PM3), which implies the assumption that this model captures the essential physics to describe the special class of compounds or the property of interest. As discussed previously (see Section II.A), the MNDO model is characterized by a self-consistent determination of the electronic distribution, a realistic treatment of electrostatic attractions and repulsions, an average incorporation of dynamic electron correlation, a derivation of atomic parameters from experiment, a representation of covalent bonding mainly through empirical resonance integrals, and an approximate inclusion of effective Pauli exchange repulsions and other corrections. The parameters in the general-purpose methods are chosen such that there is a reasonable overall balance between these terms and interactions. For specific classes of compounds or specific properties this balance may be somewhat different, and it may then be justified to adjust the parameters or aspects of the implementation accordingly. To retain the physical significance of the underlying model, however, it seems essential to require that such adjustments remain relatively small.

Examples of specialized semiempirical methods include two MNDO variants for small carbon clusters [138] and for fullerenes [139]. Small carbon clusters (C_2–C_{10}) show many peculiar features and are difficult to handle by semiempirical or low-level *ab initio* calculations (e.g., with regard to the relative stabilities of isomers). A reparametrization [138] of the standard MNDO approach [19] (using experimental and high-level *ab initio* reference data for C_3 and C_4) leads to a much improved MNDO-type treatment that affords reasonable predictions for the clusters not included in the parametrizations (C_5–C_{10}). Many properties of fullerenes are well described by the general-purpose MNDO-type methods (see Section II.C) but the absolute heats of formation are systematically and severely overestimated. A reparametrization [139] of MNDO (using mainly C_{60} reference data) remedies this problem without deteriorating the good results for other fullerene properties (e.g., relative stabilities or geometries). The reparametrized MNDO variant [139] yields realistic heats of formation for fullerenes not included in the parametrization (e.g., C_{70}) and also improved vibrational frequencies. In both cases [138, 139] the new MNDO parameters do not differ too much from the original values [19] (e.g., in the new fullerene parametrization [139] by at most 1%). It may seem surprising that such small changes can lower the calculated heat of formation of C_{60} by about 270 kcal/mol [139], but it should be kept in mind that the corresponding change per CC bond is only 3 kcal/mol.

Hydrogen bonding is poorly described by the standard MNDO method [19], and there are several MNDO versions [140–143] with a special treatment of hydrogen bonds which involve modifications of the core–core repulsion function without any changes in the electronic part of the calculation. These versions involve significant advances over standard MNDO, but this also applies to AM1 [21] and PM3 [22]. The specialized methods [140–143] do not seem to perform so much better than the recent general-purpose methods [21, 22] that they would have replaced the latter in practice. In our opinion, none of the available semiempirical variants is completely satisfactory for weak hydrogen bonds between neutral molecules, whereas strong hydrogen bonds involving cations are well described by recent methods [37]. Given the great importance of hydrogen bonding in biological systems, it would seem worthwhile to continue work on improved specialized treatments in this area.

A promising recent development concerns the use of semiempirical NDDO methods with specific reaction parameters (NDDO-SRP) [144–147] in direct dynamics calculations. In these studies the parameters in the standard AM1 method are carefully adjusted to optimize the potential surface for an individual reaction or a set of related reactions (typically allowing parameter variations up to 10% from the original values). When adjusting with respect to experimental data, NDDO-SRP is required to reproduce the exothermicity and the barrier (or rate constant) of the reaction investigated. Under these circumstances NDDO-SRP then predicts reasonable transition structures and force fields for the reaction which is consistent with previous experience [48, 49]. Direct dynamics calculations on such NDDO-SRP surfaces have provided very encouraging results [144–146] (e.g., for kinetic isotopic effects) which agree well with experimental measurements. The qualitative picture of corner-cutting tunneling processes (in 21 dimensions) emerges from such calculations [146].

The latest NDDO-SRP study [147] emphasizes a different perspective: A reference *ab initio* surface (UMP2/cc-pVDZ) for the reaction $CH_4 + Cl \rightarrow CH_3 + HCl$ is fitted by the NDDO-SRP method (allowing variations up to 10% in a small subset of the usual AM1 parameters). Most remarkably, the description of the surface given by the AM1 Hamiltonian changes from qualitatively incorrect to semiquantitatively accurate when modifying the AM1 parameters in a fit involving only four points from the *ab initio* surface (reactants, products, transition state, and van der Waals cluster). A more extended fit involving 13 points on the reaction path leads to a mean absolute deviation of only 1.08 kcal/mol between the correlated *ab initio* (UMP2) and the semiempirical (UHF) surface, for 23 points on and off the reaction path covering an energy range of 48 kcal/mol. Obviously, the UHF-AM1 approach requires very little information about the *ab initio* surface to achieve a reasonable accuracy. This indicates that the MNDO model in its AM1 implementation includes the essential interactions to describe the reaction above, and it seems superior in this regard to explicit analytic functions that have traditionally been used to fit *ab initio* potential surfaces [148]. The NDDO-SRP scheme may therefore serve as a robust and economical protocol to generate approximate potential surfaces which closely reproduce the corresponding reference *ab initio* surfaces and which may be employed in direct reactive dynamics calculation of small or medium-sized systems [147]. This is technically feasible because the semiempirical calculations are so much faster than the *ab initio* reference calculations (by almost four orders of magnitude in this example [147]; see also Section II.B).

The specialized semiempirical methods discussed up to this point are designed for an improved description of potential surfaces, either for certain classes of compounds or for specific reactions. We now consider other properties.

Electrostatic potentials are important tools in modeling because they indicate the preferred points of nucleophilic and electrophilic attack in large molecules. Their evaluation is straightforward at the *ab initio* level, and RHF/6-31G* potentials are established as the current de facto standard in the field. For semiempirical wavefunctions, there are two traditional procedures for calculating electrostatic potentials: The "quasi ab initio" approach [149] transforms the wavefunction to a deorthogonalized basis and then computes the potential analytically [7, 150–153] whereas the "purely semiempirical" approach [7, 153–156] evaluates the potential from the usual semiempirical wavefunction employing the same formulas and parameters as for the interaction between atoms (e.g., treating the positive test charge as a hydrogen atom [7, 154]). The results from both approaches differ appreciably from the RHF/6-31G* electrostatic potentials, and it is therefore tempting to consider a direct calibration against RHF/6-31G* reference data. This has been done independently by two groups, with different objectives: The first parametrization focuses on the minima of the electrostatic potential [157, 158], while the second one aims at reproducing the *ab initio* electrostatic potential everywhere [159, 160]. The parametric functions are similar in both cases, with one or two parameters per element [157–160]. The resulting "parametrized semiempirical" potentials are significantly closer to their *ab initio* RHF/6-31G* counterparts than those from previous approaches. Hence these special parametrizations provide improved semiempirical electrostatic potentials that are expected to be useful in hybrid methods (see Section III.D) and in modeling studies quite generally (see Refs. 161 and 162 for another promising semiempirical development).

Partial atomic charges have been employed by chemists for many years in qualitative discussions of structure and reactivity. They play an important role in many classical force fields used for molecular simulations and in the quantitative modeling of structure and reactivity [163–165]. They cannot, however, be defined unambiguously because they are nonmeasurable and do not correspond to the expectation value of a quantum mechanical operator. Theoretical studies have proposed many operational definitions for partial atomic charges which mostly fall into two classes [166]: Those that extract charges from an analysis of the quantum mechanical wavefunction (e.g., via population analysis [167, 168] or the partitioning of the three-dimensional electron density into atomic regions [169]) and those that extract them from an analysis of a physical observable predicted from the quantum mechanical wavefunction (e.g., via fitting of predicted interaction energies [170, 171] or electrostatic potentials [172–174]). Both these approaches can be applied with different types of wavefunctions (e.g., *ab initio* or semiempirical). In actual applications, the preferred choice of a charge model will depend partly on the ease of evaluation, but mainly on the context in which the partial charges are being used. For example, classical force fields without an explicit treatment of polarization tend to employ charges that are higher than usual, to account for the formally neglected polarization effects in an average manner [173]. When partial charges in isolated molecules are derived from *ab initio* calculations for modeling purposes, the current de facto standard involves the fitting of RHF/6-31G* electrostatic potentials [172–174] to yield potential-derived (PD) charges. It is therefore of interest to investigate whether these (relatively expensive) *ab initio* charges can be obtained efficiently and with reasonable accuracy from a specialized semiempirical treatment. A scaling of MNDO-PD charges has been recommended for this purpose [173]. Alternatively, it is also possible to invoke the principle of electronegativity equalization proposed by Sanderson [175, 176] which has been justified theoretically [177, 178] and which forms the basis of several charge models [179–181]. According to this principle, the electronegativity χ_A^0 of the free atom changes upon molecule formation and assumes a common value χ for all atoms in the molecule. Simple considerations [159, 160] suggest the relation

$$\chi = \chi_A^0 + \sum_B J_{AB} q_B \qquad \text{for all atoms A}, \qquad (18)$$

where J_{AB} and q_B denote effective one- or two-center Coulomb interactions and partial charges, respectively. A semiempirical reformulation [159, 160] of an *ab initio*–based model [181] considers χ_A^0 and J_{AA} as adjustable atomic parameters (with J_{AB} determined from the Klopman–Ohno formula [40, 41]) and optimizes these parameters such that RHF/6-31G* partial charges are reproduced as closely as possible for a representative set of molecules. Given these optimized parameters, the partial charges for other molecules are easily obtained by solving the linear system implied by Eq. (18). This specialized semiempirical treatment of partial charges is very economical because it does not require quantum chemical calculation. Using two separately optimized parameter sets [159, 160] it reproduces RHF/6-31G* Mulliken charges typically to within $0.05e$, and the corresponding potential-derived charges to within $0.10e$. In the latter case, the largest deviations occur for buried atoms in the molecular skeleton whose *ab initio* PD charges may well be statistically ill defined [182–184], so that the semiempirical charge model may indeed perform better for the PD charges than indicated by the statistical evaluations. In any event, the charges from the suggested parametrized model [159, 160] seem accurate enough to be an attractive choice for hybrid methods (see Section III.D) and for molecular simulations.

Another recent semiempirical charge model [166] is based on a different concept: Here the usual atomic charges from an established semiempirical method (e.g., AM1 or PM3) are taken as the starting point for a subsequent semiempirical mapping which further refines these charges such that physical observables calculated from these modified charges are as accurate as possible [166]. Hence the general-purpose semiempirical methods are assumed to provide a reasonable zero-order description, which is then further improved for a particular application by a special parametrization. The published example [166] involves fitting to dipole moments, which is equivalent to fitting electrostatic potentials at large distances where only the effect of the leading multipole moment survives. The results from this approach are impressive [166]. For a validation set of 186 neutral molecules, the mean absolute errors in the calculated dipole moments are 0.25 D for both proposed parametrized models (AM1-CM1A, PM3-CM1P), much lower than those for AM1 and PM3 itself (0.45 D with this set of molecules). Moreover, the predicted partial charges are reasonably close to state-of-the-art *ab initio* PD charges, and the deviations of the corresponding dipole moments from experiment even tend to be somewhat smaller than in the *ab initio* case [166].

In our opinion, there is no best procedure for determining partial atomic charges, simply because these charges are nonmeasurable and hence arbitrary to some extent. The choice of a particular definition will depend largely on the intended application. The preceding discussion emphasizes that it is possible to develop parametrized semiempirical charge models [159, 160, 166] which are very efficient and compare well with state-of-the-art *ab intio* approaches, both with regard to the actual charges and to physical observables derived therefrom. This is achieved through a special parametrization, of course, which may be considered a disadvantage, but also allows for a careful tuning of the charge distribution. In view of their computational speed, the proposed semiempirical charge models [159, 160, 166] should be particularly useful in modeling studies of large molecules.

Quantum chemical calculations refer to isolated molecules in the gas phase, whereas most of the experimental work in chemistry takes place in liquid solutions. The modeling of solvation is therefore of great practical importance. A detailed discussion of this topic is beyond the scope of this article (see Refs. 185 and 186 for recent reviews), and the following remarks are only intended to provide some brief comments on the use of semiempirical methods in solvation models. There are two complementary theoretical approaches to solvent effects [185]: Explicit solvation models involving a statistical sampling of the solute surrounded by many solvent molecules (e.g., by molecular dynamics simulations) and continuum solvation models treating the solvent as a continuous medium with a bulk dielectric constant. In the first approach, the interactions in solution are usually described by classical force fields, although there are an increasing number of recent studies with hybrid models (see Section

III.D) where the solute is treated quantum mechanically and the solvent molecules classically [186–197]. The second approach [185] may be implemented in an entirely classical framework (e.g., through the solution of the Poisson equation or the introduction of the generalized Born model in molecular mechanics) or in a quantum mechanical framework where the wavefunction of the solute is optimized self-consistently in the presence of the reaction field which represents the mutual polarization of the solute and the bulk solvent. Due to the complexity of solvation phenomena, both approaches contain a number of severe approximations, and if a quantum chemical description is employed at all, it is usually restricted to the solute molecule. When choosing such a quantum chemical description from the usual alternatives (*ab initio*, DFT, or semiempirical methods) it should be kept in mind that *ab initio* or DFT calculations may provide an accuracy that is far beyond the overall accuracy of the underlying solvation model. For a balanced treatment it may be attractive to employ efficient semiempirical methods provided that they capture the essential physics of solvation. The performance and predictive power of such semiempirical solvation models may then be improved further by a specific parametrization.

This strategy has been followed in several studies on continuum solvation models [185]. The reaction field concept accounts for the mutual solute–solvent polarization in a self-consistent manner. Since current semiempirical methods include electrostatic interactions self-consistently in their formalism (see Section II.A) it seems appropriate to implement reaction fields in such methods. This has been done by several groups in recent years, with implementations that involve Born–Kirkwood–Onsager reaction fields for spherical cavities [198–203], reaction fields from higher-order multipole expansions for spherical or ellipsoidal cavities [204–208], and generalized rection fields from surface charge densities evaluated for realistically shaped molecular cavities [209–214]. Conceptually, the generalized reaction fields should be most accurate, and among these the COSMO approach [214] seems to be particularly attractive since it provides analytical gradients. The semiempirical implementations listed above cover electrostatic and polarization interactions between solute and solvent but do not include first-solvation-shell effects [185] such as cavitation energy, dispersion energy, and solvent-structural rearrangements (CDS terms). These effects are treated in the parametrized SMx solvation models [215–220], which are the most elaborate semiempirical solvation models presently available. In the SMx approach, the electrostatic and polarization interactions are described by a generalized Born model (employing especially adjusted semiempirical charges [166] in the newest version), whereas the CDS terms are represented by intuitive parametric expressions (e.g., involving surface tensions and surface areas). The SMx models are parametrized against experimental solvation energies, and they reach an impressive accuracy for those solvents where parameters are available (water, alkanes). Further conceptual developments and parametrizations of the SMx models are in progress [221].

Summarizing this section, we conclude that specialized semiempirical methods can indeed be developed which perform well in their area of applicability. The examples given include special parametrizations for certain classes of compounds (e.g., carbon clusters), for individual reactions (NDDO-SRP), and for specific properties (electrostatic potentials, charges, and solvent effects). The success in these areas suggests that there is some promise in developing specially parametrized semiempirical treatments for other applications as well, provided that the relevant interactions are included in the model. From this perspective, improvements in general-purpose semiempirical models (see Section III.B) may be useful in providing better starting points for special parametrizations.

D. Hybrid Methods

Hybrid methods are characterized by a combination of quantum mechanical (QM) and molecular mechanical (MM) potentials. They treat the electronically important part of a large system by a quantum chemical method (e.g., *ab initio*, DFT, or semiempirical) and the remainder by a classical force field. They can provide an appropriate description for systems which are too large for a purely quantum chemical approach (even at the semiempirical level) and which contain regions that cannot be described classically (e.g., reactive centers with breaking and forming bonds, or chromophores where an electronic excitation takes place).

A comprehensive discussion of hybrid methods and their application is beyond the scope of this article, and we refer the reader to recent reviews in this field [186, 222–224] and to other methodological papers [225–228] for more detailed information. In the following we focus on the use of semiempirical methods in QM/MM approaches and on some conceptual issues. One typical example for the use of hybrid methods is the study of solvent effects through explicit solvation models (see Section III.C). In this case there is an obvious partitioning between the QM part (solute) and the MM part (solvent) of the system, with well-known QM/MM interactions. The effective Hamiltonian is written as the sum of three terms [186, 227]:

$$\hat{H}_{eff} = \hat{H}_{QM} + \hat{H}_{QM/MM} + \hat{H}_{MM} ,\qquad(19)$$

where $\hat{H}_{QM}$ is the usual quantum mechanical Hamiltonian of the solute, $\hat{H}_{MM}$ denotes the classical interaction energy in the solvent, and $\hat{H}_{QM/MM}$ represents the solute–solvent interaction Hamiltonian. The latter is usually approximated as the sum of an electrostatic contribution $\hat{H}^{el}_{QM/MM}$ and a van der Waals contribution $\hat{H}^{vdW}_{QM/MM}$. The electronic wavefunction of the solute is then determined with the use of the operator $\hat{H}_{QM}$ + $\hat{H}^{el}_{QM/MM}$ [i.e., in SCF-MO treatments from a modified Fock matrix which includes additional one-electron terms $(H^{el}_{QM/MM})_{\mu\nu}$ describing the electrostatic solute–solvent interaction]. The total energy E_{tot} is given as the expectation value of the effective Hamiltonian using the normalized wavefunction Ψ of the solute:

$$E_{tot} = \langle \Psi | \hat{H}_{eff} | \Psi \rangle = E_{QM} + E_{QM/MM} + E_{MM} , \qquad (20)$$

where E_{QM}, $E_{QM/MM}$, and E_{MM} denote the quantum mechanical energy of the solute, the solute–solvent interaction energy, and the molecular mechanical energy of the solvent, respectively. Conceptually, E_{QM} can be decomposed into the energy of an isolated solute molecule and the induction energy in the solute molecule due to surrounding solvent, whereas $E_{QM/MM}$ consists of the electrostatic and the van der Waals solute–solvent interaction energies.

Semiempirical implementations of this QM/MM approach [186] first require a choice of the methods to be combined (e.g., AM1/CHARMM [227], AM1/TIP3P [188–190], or AM1/OPLS [188, 195]). In addition, the one-electron coupling terms $(H^{el}_{QM/MM})_{\mu\nu}$ in the Fock matrix and the van der Waals interaction parameters need to be determined. In the available implementations, this is done by reasonable ad hoc assumptions (in the spirit of the methods being combined) and some limited parametrization [186, 188, 195, 227]. Successful applications [186] of these implementations have addressed bimolecular hydrogen-bonded complexes, free energies of solvation, solvent effects on conformational and tautomeric equilibria, solvent effects on chemical reactions (including several types of S_N2 reactions), and solvatochromic shifts of electronic transitions. The established semiempirical methods (e.g., AM1) used in these QM/MM implementations generally appear to be accurate enough to predict the correct qualitative trends upon solvation, and they may also be useful quantitatively (especially with some parametrization). In principle, the accuracy of such QM/MM approaches can be improved by employing *ab initio* or DFT methods, but most of the available QM/MM studies on solvation have used semiempirical wavefunctions [186].

Another and more difficult application of hybrid methods involves the division of a system across covalent bonds. Such a partitioning into QM and MM regions may be unavoidable, for example, in the treatment of enzymatic reactions [229, 230], where the enzyme may participate in bond forming and breaking processes. Cutting covalent bonds at the QM/MM boundary clearly causes a major perturbation of the electronic structure and immediately raises the problem of an appropriate description of the QM region at this boundary. Two approaches are available for this purpose: First, in the spirit of early work [225], it has been suggested [231–233] to represent the bonds being cut by localized hybrid orbitals (built from the AOs of the respective QM atoms) which are determined from suitable model compounds and are assumed to be transferable. The SCF-MO calculation for the QM region then includes the field from the electron density of these hybrid orbitals (which are kept fixed themselves). In the second approach, the free valences of the QM atoms at the boundary are satisfied by *link atoms* [226, 227], typically hydrogen atoms which are placed along the broken QM/MM bonds. Such hydrogen atoms are then treated as usual in the QM calculation. It is possible to employ other types of link atoms [186], and in the context of semiempirical QM methods, it would also seem feasible to introduce a special parametrization for the link atoms such that they mimic the electronic influence of the MM region in an optimum manner. The choice between these approaches and their variants is not clearcut and may depend on the exact location of the QM/MM boundary in a particular application. Semiempirical QM/MM treatments can afford to use relatively large QM regions, with QM/MM boundaries that are quite distant from the electronically important part of the QM region (e.g., the reactive center). Under these circumstances the choice discussed above seems less critical, and we shall therefore only consider the most convenient concept (i.e., the use of hydrogen atoms as link atoms).

Previous semiempirical QM/MM approaches with link atoms are based on intuitively reasonable model assumptions and implementations [186]. Recent work in our group [159, 234] has attempted to define a hierarchy of semiempirical QM/MM models (A, B, C) and to achieve an optimum implementation for combinations of MNDO or AM1 with the MM3 force field [235, 236]. In the following we outline the main ideas of this development: Consider a molecule $X - Y$, which is partitioned into an MM part X and a QM part Y (treated as Y-$\mathscr{L}$ with a link atom $\mathscr{L} = H$). At the simplest level A, the total energy may be written in an obvious notation as

$$E_{\text{tot}}^{\text{A}}(X - Y) = E_{\text{MM}}(X - Y) - E_{\text{MM}}(Y - \mathscr{L}) + E_{\text{QM}}(Y - \mathscr{L}) . \quad (21)$$

The force field energies E_{MM} are composed of contributions referring to a single fragment [e.g., $E_{MM}(X)$] and those representing the interactions between two fragments [e.g., $E_{MM}(X, Y)$]. Inserting these sums into Eq. (21) and exploiting the resulting cancellations yields

$$E_{tot}^{A}(X - Y) = E_{MM}(X) + E_{MM}(X, Y) + E_{QM}(Y - \mathscr{L}) + E_{LINK} \quad (22)$$

$$E_{LINK} = -E_{MM}(\mathscr{L}) - E_{MM}(Y, \mathscr{L}), \quad (23)$$

where E_{LINK} denotes a link atom correction. Model A represents a mechanical embedding of the QM region since all QM/MM interactions are taken from the classical force field and since the MM region does not influence the QM wavefunction. In most force fields, the interaction energy $E_{MM}(X, Y)$ consists of bonded terms $[E_{MM}^{bonded}(X, Y)]$, nonbonded electrostatic terms $[E_{MM}^{elstat}(X, Y)]$, and nonbonded van der Waals terms $[E_{MM}^{vdW}(X, Y)]$. Model A may therefore be refined by introducing a quantum chemical description of the nonbonded electrostatic interactions. The Coulomb interaction $E^{coul}(X, Y)$ can be evaluated from

$$E^{coul}(X, Y) = \sum_{I}^{MM} q_I \langle \Phi^I \rangle, \quad (24)$$

where q_I and $\langle \Phi^I \rangle$ denote the atomic charges of the MM atoms I and the electrostatic potentials generated by the QM region at the positions of the MM atoms I, respectively. In our implementation [159, 160], these quantities are obtained from a special semiempirical parametrization aimed at reproducing the corresponding *ab initio* RHF/6-31G* reference data (see Section III.C). The induction energy $E^{ind}(Y)$ in the QM region is taken into account by optimizing the SCF-MO wavefunction of the QM part in the presence of the atomic charges q_I of the MM atoms. The total energy of the resulting model B is given by

$$E_{tot}^{B}(X - Y) = E_{MM}(X) + E_{MM}^{bonded}(X, Y) + E_{MM}^{vdW}(X, Y) + E^{coul}(X, Y)$$
$$+ E^{ind}(Y) + E_{QM}(Y - \mathscr{L}) + E_{LINK}. \quad (25)$$

Model B represents an electronic embedding of the QM region. In many aspects it is similar to a previous QM/MM model [227] which has been discussed above [see Eqs. (19)–(20)], although there are differences, for example, in the link atom correction and the actual implementation. Model B may be refined further by including the induction energy $E^{ind}(X)$ in the MM region, which can be calculated from the electric field generated by the QM region and the induced dipole moments in the MM region. The latter are obtained from a published dipole interaction model [237] which involves the use of parametrized atomic polarizabilities [237]. The total energy in model C is

$$E_{tot}^{C}(X - Y) = E_{tot}^{B}(X - Y) + E^{ind}(X) . \tag{26}$$

A detailed analysis [159] shows that the link atom correction E_{LINK} can become unphysical for geometries far away from equilibrium (e.g., for distances approaching zero). The potential surface for geometry optimizations is therefore defined as

$$E_{pot}(X - Y) = E_{tot}(X - Y) - E_{LINK} \tag{27}$$

in each model (A, B, C). Gradients on such surfaces can be evaluated efficiently for models A and B but not for model C, where finite-difference calculations are required [159].

These semiempirical QM/MM models have been applied [159, 234] to study proton affinities of alcohols and substituted pyridines, deprotonation enthalpies of alcohols and carbon acids, nucleophilic additions of lithium hydride to carbonyl compounds, hydrid transfer reactions in deprotonated hydroxyketones, and nucleophilic ring openings in oxiranes (using MNDO/MM3 and AM1/MM3). A comparison of the results for models A–C allows to identify the appropriate description of the QM/MM interactions in a given application. Not too surprisingly, the treatment of protonation and deprotonation requires an electronic embedding of the QM region: For example, the trends in the proton affinities and deprotonation enthalpies of alcohols are reproduced qualitatively only at levels B and C, respectively, indicating the general importance of electrostatic interactions in these processes as well as the special importance of polarization for the anions. On the other hand, the hydrid transfer reactions in deprotonated hydroxyketones with rigid carbon skeletons [238–240] can already be described at level A (mechanical embedding). In this case, the bicyclic or polycyclic backbones of the hydroxyketones studied [238–240] impose characteristic transition-state geometries which determine the trends in the barriers for hydride transfer. Transition-state modeling [241] has also been applied [242, 243] to these reactions, but the QM/MM approach describes the available experimental data at least as well. Generally speaking, the semiempirical QM/MM models discussed (A–C) are not always accurate quantitatively, but they often reproduce qualitative trends reasonably.

From an overall perspective, hybrid methods seem very promising, but much work remains to be done. There is considerable flexibility in selecting a suitable QM component (e.g., *ab initio*, DFT, or semiempirical) and a suitable MM component (e.g., a general-purpose or a more specialized force field). This choice is influenced by the usual considerations of the required accuracy and the available computational resources. There is a need for a hierarchy of QM/MM coupling schemes (e.g., models A–C or even more refined schemes) and for guidelines concerning the placement of the QM/MM boundary (whenever there is no natural choice, as in the solute–solvent case), to achieve optimum performance for a given combination of QM and MM methods. In a semiempirical context, this may include the need to develop special parametrizations for QM/MM interactions and to address subtle problems concerning the link atoms. In view of these issues, hybrid methods will probably not become routine black-box procedures in the near future. It seems more likely that they will be used for a specifically adapted tailor-made modeling of large systems. Given the severe inherent approximations of any QM/MM scheme, semiempirical methods may be useful as QM components in such studies if they include the relevant interactions. Semiempirical QM/MM approaches have the conceptual advantage that they can easily incorporate available experimental data through a specific parametrization (see, e.g., Refs. 222 and 223). More work is needed to establish the potential and the limitations of such approaches.

E.   Parametrizations

It is obvious from the preceding sections that parametrizations with regard to experimental or theoretical reference data are a characteristic feature of semiempirical methods. One may argue that other quantum chemical approaches also involve some kind of parametrization (e.g., the development of basis sets in *ab initio* methods or the derivation of improved exchange-correlation functionals in DFT methods). In these cases, however, the optimization of any parameters is usually a narrowly defined problem that can be solved technically without undue difficulty. By contrast, the parametrization plays a much more important role in semiempirical methods because it determines their accuracy to a large extent. A semiempirical parametrization attempts to reach the limits of accuracy inherent to the underlying semiempirical model (in a given implementation), but it must avoid the danger of overparametrizing the model by striving for too much accuracy in the necessarily limited set of reference data used. Therefore, a successful and balanced parametrization must accept the limits of accuracy imposed by the underlying model, which requires some subjective judgment. The need for such a parametrization complicates methodological developments in semiempirical theory compared with *ab initio* or DFT theory. In the following we comment on recent advances in parametrization techniques and briefly summarize some experience from parametrization work.

The semiempirical parameters are usually optimized in a least-squares sense by minimizing the error function

$$S = \sum_i [(x_i^{\text{calc}} - x_i^{\text{ref}})w_i]^2 , \tag{28}$$

where x_i^{calc} and x_i^{ref} denote the calculated and the reference values for the property of interest, and w_i the associate weighting factor (usually chosen such that S is dimensionless). In general, the minimization of the error function cannot be handled by a manual trial-and-error procedure but requires an automatic optimization algorithm. Traditionally, nonlinear least-squares methods have been used for this purpose [19, 244]. The performance of such methods is greatly improved if the Jacobian matrix is available [i.e., the partial derivatives of the weighted deviations in Eq. (28) with respect to the parameters]. The analytic evaluation of the Jacobian matrix has been described for MNDO-type models [22], but a finite-difference evaluation can easily be implemented for any model [37, 245]. In the latter case, the numerical calculation of the Jacobian matrix is quite time consuming, but the overall performance still benefits considerably from recomputing the Jacobian matrix periodically (e.g. every 10 cycles) [37, 245]. In any event, the nonlinear least-squares algorithms can only find the local minimum in parameter space that is closest to the starting parameters. Therefore, many runs with different starting parameters are normally carried out to scan all reasonable regions of parameter space, hoping that the best parameters are found in this manner. A promising alternative is provided by recent developments in genetic algorithms [246] which have already been applied in semiempirical parametrizations [147]. These algorithms search for the global minimum in parameter space (defined by a predetermined range of allowed parameter values). They do not require derivative information but normally need a very large number of function evaluations until convergence to the global minimum is reached. It would clearly be of great interest to compare the performance of nonlinear least-squares and genetic algorithms in realistic semiempirical parametrizations. In our opinion, genetic algorithms may become the preferred parametrization technique if they can automatically and simultaneously optimize both the parameters and the parametric functions in a semiempirical model.

Even if the global minimum of the error function can technically be found, the optimized parameters will still depend on the choice of reference data in Eq. (28). It is therefore advisable to check the sensitivity of the parameters to changes in the reference data. According to the underlying philosophy of semiempirical methods (see Section II.A), the reference data are normally taken from experiment, and they are selected to be as representative as possible (which may require some subjective judgment). When developing general-purpose semiempirical methods, such representative and reliable experimental reference data are normally available for light elements but are often missing for the heavier elements. This holds especially for thermochemical data which play a central role in the parametrization of MNDO-type methods. Recent advances in *ab initio* theory have alleviated these problems to some extent because they allow the use of certain high-level *ab initio* results as substitutes for experimental reference data. For example, *ab initio* G2 theory [247] has been shown to reproduce the energies of compounds containing first- and second-row elements with a mean absolute error of only 1.2 kcal/mol (125 representative comparisons). This accuracy is sufficient to employ theoretical G2 energies as reference data, as has been done recently in the MNDO/d parametrization [36]. *Ab initio* reference data of similar accuracy are urgently needed for transition metal compounds. Generally speaking, the increasing availability of high-level *ab initio* results will facilitate future semiempirical developments, particularly in areas that have suffered in the past from the lack of reliable experimental reference data.

The last step in a parametrization concerns validation. Obviously, a successful parametrization must perform well in applications beyond those that are covered by the reference data. This has traditionally been confirmed by extensive test calculations and statistical evaluations (see, e.g., Refs. 13 and 19–25). In this context, another point should be emphasized: It is essential to check that the optimized values of the parameters are consistent with the physical meaning of these parameters in the underlying semiempirical model. In the case of atomic parameters, such checks should include the variation of the parameter values across the periodic table. In our opinion, "unphysical" parameter values should not be accepted even if the statistical results for the reference and validation data appear to be satisfactory, because unreasonable results are likely to be found in other applications.

F. Computational Aspects

As discussed previously (see Section II.B), semiempirical calculations can be carried out efficiently using existing programs, and they are much faster than *ab initio* or DFT calculations. Nevertheless, there have been significant computational improvements recently which will further extend the applicability of semiempirical methods. Some of these improvements are summarized in this section.

A conventional semiempirical SCF-MO treatment of a molecule with N basis orbitals contains rate-determining N^3 steps (see Section II.B). In practice, this limits the size of molecules that can reasonably be handled on current hardware to about 1000 nonhydrogen atoms, as exemplified by the reported MNDO geometry optimization of the C_{960} fullerene [74]. Shifting this limit to much larger molecules requires new algorithms that avoid the N^3 steps in the SCF procedure. This has recently been achieved [248, 249] by adopting a divide-and-conquer strategy based on the concept of localized molecular orbitals (LMOs). In essence, 2×2 rotations are applied to annihilate the interactions between occupied and virtual LMOs that are located within a certain cutoff radius (typically 8–10 Å), whereas all other such interactions are considered to be negligible and therefore not treated. Small numerical errors are inevitable in this approach, but they can apparently be controlled (e.g., by a renormalization of the LMOs and a suitable choice of the cutoff radius). The new algorithm is slower than the conventional treatment for small systems, but its computational effort only scales as N^1 to $N^{1.5}$, which results in a much higher computational speed for large systems. Using this algorithm it has been possible to carry out semiempirical SCF-MO calculations on standard workstations for polypeptides with more than 4000 atoms [248, 249].

For variational semiempirical wavefunctions (e.g., closed-shell RHF or UHF), the gradient can be computed at a fraction of the cost for an SCF calculation. Analytic gradients have long been available [250], but a simple finite-difference procedure with a constant density matrix and recalculated two-center integrals is also efficient. For certain nonvariational semiempirical wavefunctions (e.g., open-shell half-electron (HE) RHF [251] or HE-CI), analytic gradients have been introduced more recently [252]. In this case, the coupled-perturbed Hartree–Fock (CPHF) equations need to be solved, which is accomplished in the published implementation [252] by an algorithm that scales as N^4. Recent work in our group has shown [253] that the solution of the CPHF equations for semiempirical HE-RHF wavefunctions can be reformulated to scale as N^3 when making use of the Z-vector method [254]. The implementation of this approach has indeed made the time for gradient evaluation small compared with the SCF time (as in the closed-shell RHF case). This development will greatly facilitate semiempirical HE-RHF studies of open-shell molecules.

Harmonic force fields are commonly calculated from numerical finite differences of gradients in semiempirical programs, and from analytic second derivatives in *ab initio* programs. For both semiempirical and *ab initio* SCF wavefunctions, the analytic second derivatives contain contributions from integral derivatives (direct terms) and from density matrix derivatives (CPHF terms). In the *ab initio* case, the computational cost is usually dominated by the direct terms (which typically require a CPU time of about $10\,T_{SCF}$), so that the analytic evaluation of the second derivatives will normally be more efficient than a numerical evaluation. By contrast, in the semiempirical case, the computational cost is dominated by the CPHF terms since the integral derivatives can be computed with relatively small computational effort. A detailed analysis shows [121] that the analytic calculation of semiempirical harmonic force fields can be competitive to the traditional finite-difference calculation only if the CPHF problem is formulated in the AO basis and is solved iteratively via a suitable Krylov subspace approach. In this case the computational cost for a molecule with M atoms and N basic functions formally scales as MN^3 in both the analytic and numerical evaluations. A careful implementation [121] of analytic second derivatives for MNDO-type semiempirical methods provides typical speedups of four to eight compared with analogous numerical computations and exhibits a reliable convergence over a wide range of molecules. The asymptotic memory and disk storage requirements can be chosen to scale as low as N^2 without significant degradation of performance. It is obvious that these advances will facilitate force constant calculations for large molecules at the semiempirical SCF level.

Modern high-performance computing employs vector machines, moderately parallel systems with shared memory, and massively parallel systems with distributed memory. Most of the computational work in conventional semiempirical SCF codes is spent for matrix operations (e.g., for the N^3 steps of matrix multiplication and diagonalization), which are handled efficiently by modern computer architectures. Much of the remaining work (i.e., mainly the construction of the Fock matrix and the calculation of the integrals) is amenable to vectorization and/or shared-memory parallelization, so that semiempirical SCF codes are very often efficient on such machines. For example, the MNDO SCF geometry optimization of the C_{540} fullerene has been benchmarked [72] to reach a sustained computational speed of 87% (83%) of the hardware peak performance using 1 CPU (8 CPUS) of a Cray Y-MP machine. Concerning massively parallel processors (MPP) with distributed memory, a coarse-grained parallelization can be implemented easily for calculations that consist of many nearly independent tasks (e.g., reaction path or grid calculations and the conventional finite-difference evaluation of harmonic force fields). A general fine-grained parallelization for MPP systems is more difficult due to unfavorable computation-to-communication ratios in important parts of the work and requires a restructuring of the code (e.g., for pseudodiagonalization, Fock matrix construction, and integral evaluation). Work along these lines is in progress [255] and has led to a parallel code with complete data distribution in the SCF routines, which allows calculations on large molecules (e.g., C_{960}) to be carried out on MPP systems with little memory per node (e.g., 64 MB). Generally speaking, however, it should be emphasized that semiempirical computer programs can be well adapted to modern computer architectures. Hence recent advances in computer hardware can be exploited without undue difficulties to extend the applicability of semiempirical methods.

There are other recent computational improvements which are not inherent to semiempirical methods but have been added to semiempirical programs in order to enhance their capabilities to explore potential surfaces. Such examples include the combination of standard molecular dynamics (MD) and semiempirical SCF codes [256, 257]. The MD-MNDO approach has been used to find new fullerene isomers with 20–60 carbon atoms [256] and to study the reactive collisions [257] between C_{60} and C_{60}^+, which may lead to deep inelastic scattering, molecular fusion, and fragmentation (depending on the collision energy) [257–260]. Simulated annealing techniques have been introduced into semiempirical programs to locate stationary points [261]. The corresponding algorithm takes advantage of the specific features of semiempirical methods and mixes simulated annealing and local searches to reduce computational costs [261]. Other algorithmic tools for studying semiempirical potentials surfaces have been extended and made more efficient [262]. This includes the minimization of the energy or the gradient norm by either pseudo-Newton or quadratic procedures, the search for transition states, and the intrinsic reaction coordinate in conjunction with variational transition-state theories [262]. Previous semiempirical work on related topics is also available [263, 264].

IV. OUTLOOK

In recent years there have been significant advances in *ab initio* and DFT methods. Many groups (including our own) employ these techniques to compute molecular properties of chemical interest as accurately as presently possible. What are the perspectives of semiempirical molecular orbital methods under these circumstances? Which role can they play in the near future?

In our opinion, the existing semiempirical methods (see Section II) will continue to serve as electronic structure models and as computational tools. Semiempirical calculations are so much faster than *ab initio* or DFT calculations that they will remain useful for survey studies, for establishing trends, and for scanning a computational problem before proceeding with more expensive and more accurate theoretical studies. Whenever such calculations provide a realistic description, the underlying semiempirical model can be used for explaining and predicting chemical behavior in terms of simple qualitative concepts.

Current developments in semiempirical theory (see Section III) aim at improving the accuracy and applicability of semiempirical methods without compromising their computational efficiency. In our judgment, theoretically guided improvements in the underlying semiempirical models offer the best strategy for developing new general-purpose treatments with better overall performance. In a complementary approach, special parametrizations of established semiempirical models can be found to describe special classes of compounds, reactions, or properties as accurately as possible. This approach holds much promise provided that the necessary parametrizations are performed appropriately. In a similar spirit, semiempirical implementations of continuum solvation models allow a realistic treatment of solvent effects. This can also be achieved by semiempirical QM/MM hybrid methods which may generally be used whenever a large system can be partitioned into an electronically important part and its surroundings.

These and other recent developments (see Section III) promise to open new areas of application for semiempirical methods. For example, new general-purpose methods should be applicable to compounds of heavier elements, including transition metals, and to the photochemistry on the lower excited-state surfaces (see Section III.B). Specific parametrizations may be helpful in cases where the general-purpose parametrizations are not accurate enough (e.g., for the realistic representation of a particular reactive potential surface or for the prediction of properties such as electrostatic potentials and partial atomic charges, see Section III.C). Semiempirical hybrid methods can be used, for example, to treat solvent effects or enzymatic reactions (see Section III.D). Recent computational developments extend the applicability of semiempirical methods to molecules with several thousand atoms and facilitate the exploration of semiempirical potential surfaces through new implementations of analytic derivatives and through techniques such as molecular dynamics or simulated annealing (see Section III.F). Even though the preceding list of examples is far from complete, it clearly indicates that there will be many new types of application for semiempirical methods. This perspective would seem more important than slight gains of accuracy in areas where semiempirical methods have traditionally been applied with success.

In the near future, *ab initio*, DFT, and semiempirical methods are expected to remain the major computational tools of quantum chemistry, with significant improvements in each of these methods. Undoubtedly, the more accurate *ab initio* and DFT approaches will be applied to ever larger molecules, but new applications will also open up for the semiempirical treatments, and the best computational solution of a given chemical problem may indeed involve the synergetic use of several computational tools. Semiempirical methods are expected to be particularly suitable for describing electronic effects in large molecules, either through the use of a general-purpose method that includes all relevant interactions or through a specifically tailored modeling that may require some special parametrization.

ACKNOWLEDGMENTS

The author wishes to thank his co-workers for their contributions, particularly D. Bakowies, A. Gelessus, T. Hansen, M. Kolb, S. Patchkovskii, A. Voityuk, and W. Weber. Financial support of the Deutsche Forschungsgemeinschaft, the Alfried-Krupp-Stiftung, the Fonds der Chemischen Industrie, and the Schweizerischer Nationalfonds is gratefully acknowledged.

REFERENCES

1. R. G. Parr, *The Quantum Theory of Molecular Electronic Structure*, W. A. Benjamin, New York, 1963.

2. L. Salem, *The Molecuar Orbital Theory of Conjugated Systems*, W. A. Benjamin, New York, 1966.

3. M. J. S. Dewar, *The Molecular Orbital Theory of Organic Chemistry*, McGraw-Hill, New York, 1969.

4. J. A. Pople and D. L. Beveridge, *Approximate Molecular Orbital Theory*, Academic Press, New York, 1970.

5. J. N. Murrell and A. J. Harget, *Semiempirical Self-Consistent-Field Molecular Orbital Theory of Molecules*, Wiley, New York, 1972.

6. G. A. Segal, Ed., *Modern Theoretical Chemistry*, Plenum Press, New York, 1977, Vols. 7–8.

7. M. Scholz and H.-J. Köhler, *Quantenchemie*, Heidelberg, Germany, 1981, Vol. 3.

8. T. Clark, *A Handbook of Computational Chemistry*, Wiley, New York, 1985.

9. M. J. S. Dewar, *Science* **187**, 1037 (1975).

10. K. Jug, *Theor. Chim. Acta* **54**, 263 (1980).

11. M. J. S. Dear, *J. Phys. Chem.* **89**, 2145 (1985).

12. W. Thiel, *Tetrahedron* **44**, 7393 (1988).

13. J. J. P. Stewart, *J. Comput.-Aided Mol. Design* **4**, 1 (1990).

14. J. J. P. Stewart, in *Reviews in Computational Chemistry*, K. B. Lipkowitz and D. B. Boyd, Eds., Vol. 1, VCH, New York, 1990, pp. 45–81.

15. M. C. Zerner, in *Reviews in Computational Chemistry*, K. B. Lipkowitz and D. B. Boyd, Eds., Vol. 2, VCH, New York, 1991, pp. 313–365.

16. J. A. Pople, D. P. Santry, and G. A. Segal, *J. Chem. Phys.* **43**, S129 (1965).

17. W. Thiel, *J. Chem. Soc. Faraday Trans. 2* **76**, 302 (1980).

18. T. A. Halgren and W. N. Lipscomb, *J. Chem. Phys.* **58**, 1569 (1973);
    L. Throckmorton and D. S. Marynick, *J. Comput. Chem.* **6**, 652 (1985).

19. M. J. S. Dewar and W. Thiel, *J. Am. Chem. Soc.* **99**, 4899 (1977).

20. W. Thiel, *J. Am. Chem. Soc.* **103**, 1413 (1981).

21. M. J. S. Dewar, E. Zoebisch, E. F. Healy, and J. J. P. Stewart, *J. Am. Chem. Soc.* **107**, 3902 (1985).

22. J. J. P. Stewart, *J. Comput. Chem.* **10**, 209, 221 (1989).

23. M. J. S. Dewar, C. Jie, and J. Yu, *Tetrahedron* **49**, 5003 (1993).

24. A. J. Holder, R. D. Dennington, and C. Jie, *Tetrahedron* **50**, 627 (1994).

25. D. N. Nanda and K. Jug, *Theor. Chim. Acta* **57**, 95 (1980).

26. K. Jug, R. Iffert and J. Schulz, *Intern. J. Quantum Chem.* **32**, 265 (1987).

27. M. C. Zerner and J. Ridley, *Theor. Chim. Acta* **32**, 111 (1973).

28. A. D. Bacon and M. C. Zerner, *Theor. Chim. Acta* **53**, 21 (1979).

29. J. A. Pople and G. A. Segal, *J. Chem. Phys.* **44**, 3289 (1966).

30. J. Del Bene and H. H. Jaffé, *J. Chem. Phys.* **48**, 1807 (1968).

31. J. A. Pople, D. L. Beveridge, and P. A. Dobosh, *J. Chem. Phys.* **47**, 2026 (1967).

32. R. C. Bingham, M. J. S. Dewar and D. H. Lo, *J. Am. Chem. Soc.* **97**, 1285 (1975).

33. W. Thiel and A. A. Voityuk, *Theor. Chim. Acta* **81**, 391 (1992).

34. W. Thiel and A. A. Voityuk, *Intern. J. Quantum Chem.* **44**, 807 (1992).

35. W. Thiel and A. A. Voityuk, *J. Mol. Struct.* **313**, 141 (1994).

36. W. Thiel and A. A. Voityuk, *J. Phys. Chem.*, in press.

37. M. Kolb and W. Thiel, *J. Comput. Chem.* **14**, 37 (1993).

38. W. Weber and W. Thiel, to be published.

39. M. J. S. Dewar and W. Thiel, *Theor. Chim. Acta* **46**, 89 (1977).

40. G. Klopman, *J. Am. Chem. Soc.* **86**, 4550 (1964).

41. K. Ohno, *Theor. Chim. Acta* **2**, 219 (1964).

42. M. G. Hicks and W. Thiel, *J. Comput. Chem.* **7**, 213 (1986).

43. M. J. S. Dewar and W. Thiel, *J. Am. Chem. Soc.* **99**, 4907 (1977).

44. H. L. Hase, G. Lauer, K.-W. Schulte, and A. Schweig, *Theor. Chim. Acta* **48**, 47 (1978).

45. D. Cremer, *J. Comput. Chem.* **3**, 165 (1982).

46. V. Kellö, M. Urban, J. Noga, and G. H. F. Diercksen, *J. Am. Chem. Soc.* **106**, 5864 (1984).

47. W. Thiel, *J. Am. Chem. Soc.* **103**, 1420 (1981).

48. S. Schröder and W. Thiel, *J. Am. Chem. Soc.* **107**, 4422 (1985).

49. S. Schröder and W. Thiel, *J. Am. Chem. Soc.* **108**, 7985 (1986).

50. A. Schweig and W. Thiel, *J. Am. Chem. Soc.* **103**, 1425 (1981).

51. M. Klessinger and T. Pötter, in *Theoretical Models of Chemical Bonding*, Z. B. Maksic, Ed., Vol. 3, Springer-Verlag, Berlin, 1991, pp. 521–544, and references cited therein.

52. M. Klessinger, T. Pötter, and C. van Wüllen, *Theor. Chim. Acta* **80**, 1 (1991).

53. S. Grimme, *Chem. Phys.* **163**, 313 (1992).

54. N. Mataga and K. Nishimoto, *Z. Phys. Chem.* **12**, 335 (1957); **13**, 140 (1957).

55. G. Lauer, K.-W. Schulte, and A. Schweig, *J. Am. Chem. Soc.* **100**, 4925 (1978).

56. R. Schulz, A. Schweig, and W. Zittlau, *J. Mol. Struct.* **121**, 115 (1985).

57. UniChem Chemistry Codes, distributed by Cray Research, Eagan, since 1991.

58. W. Thiel, *Program MNDO93*, Version 4.1, Zürich, 1993.

59. J. Andzelm and E. Wimmer, *J. Chem. Phys.* **96**, 1280 (1992).

60. R. D. Amos, I. L. Alberts, J. S. Andrews, S. M. Colwell, N. C. Handy, D. Jayatilaka, P. J. Knowles, R. Kobayashi, N. Koga, K. E. Laidig, P. E. Maslen, C. W. Murray, J. E. Rice, J. Sanz, E. D. Simandiras, A. J. Stone, and M.-D. Su, CADPAC: *Cambridge Analytical Derivative Package*, Version 5.1, Cambridge, 1993.

61. W. J. Hehre, L. Radom, P. v. R. Schleyer, and J. A. Pople, *Ab Initio Molecular Orbital Theory*, Wiley, New York, 1986.

62. R. Ahlrichs, M. Bär, M. Häser, H. Horn, and C. Kölmel, *Chem. Phys. Lett.* **162**, 165 (1989).

63. D. Bakowies, A. Gelessus, and W. Thiel, *Chem. Phys. Lett.* **197**, 324 (1992).

64. M. Bühl and W. Thiel, *Chem. Phys. Lett.* **233**, 585 (1995).

65. J. J. P. Stewart, P. Csaszar, and P. Pulay, *J. Comput. Chem.* **3**, 227 (1982).

66. W. Krätschmer, L. D. Lamb, K. Fostiropoulos, and D. R. Huffman, *Nature* **347**, 354 (1990).

67. H. W. Kroto, J. R. Heath, S. C. O'Brien, R. F. Curl, and R. E. Smalley, *Nature* **318**, 162 (1985).

68. W. E. Billups and M. A. Ciufolini, Eds., *Buckminsterfullerenes*, VCH, Weinheim, Germany, 1993.

69. A. Hirsch, *The Chemistry of the Fullerenes*, Thieme, Stuttgart, Germany, 1994.

70. J. Cioslowski, *Electronic Structure Calculations on Fullerenes and Their Derivatives*, Oxford University Press, New York, 1995.

71. R. C. Haddon and K. Raghavachari, in Ref. 68, Chap. 7.

72. D. Bakowies and W. Thiel, *J. Am. Chem. Soc.* **113**, 3704 (1991).

73. R. C. Haddon, *Science* **261**, 1545 (1993).

74. D. Bakowies, M. Bühl, and W. Thiel, *J. Am. Chem. Soc.*, **117**, 10113 (1995).

75. G. E. Scuseria, in Ref. 68, Chap. 5.

76. D. Ugarte, *Nature* (*London*) **359**, 707 (1992).

77. D. E. Manolopoulos, *Chem. Phys. Lett.* **192**, 330 (1992); X. Liu, T. G. Schmalz, and D. J. Klein, *Chem. Phys. Lett.* **192**, 331 (1992).

78. K. Raghavachari and C. M. Rohlfing, *J. Phys. Chem.* **96**, 2463 (1992).

79. C. Thilgen, F. Diederich, and R. L. Whetten, in Ref. 68, Chap. 3.

80. K. Raghavachari and C. M. Rohlfing, *Chem. Phys. Lett.* **208**, 436 (1993).

81. D. Bakowies, M. Kolb, W. Thiel, S. Richard, R. Ahlrichs, and M. M. Kappes, *Chem. Phys. Lett.* **200**, 411 (1992).

82. D. A. Dixon, N. Matsuzawa, T. Fukunaga, and F. N. Tebbe, *J. Phys. Chem.* **96**, 6107 (1992).

83. N. Matsuzawa, D. A. Dixon, and T. Fukunaga, *J. Phys. Chem.* **96**, 7594 (1992).

84. N. Matsuzawa, D. A. Dixon, and P. J. Krusic, *J. Phys. Chem.* **96**, 8317 (1992).

85. N. Matsuzawa, T. Fukunaga, and D. A. Dixon, *J. Phys. Chem.* **96**, 10747 (1992).

86. D. Bakowies and W. Thiel, *Chem. Phys. Lett.* **193**, 236 (1992).

87. J. M. Hawkins, A. Meyer, and M. A. Solow, *J. Am. Chem. Soc.* **115**, 7499 (1933).

88. J. M. Hawkins and A. Meyer, *Science* **260**, 1918 (1993).

89. R. L. Murry and G. E. Scuseria, *J. Phys. Chem.* **98**, 4212 (1994).

90. H. Schwarz, *Angew. Chem.* **105**, 1475 (1993).

91. G. von Helden, N.G. Gotts, and M.T. Bowers, *Nature* (*London* **363**, 60 (1993).

92. G. von Helden, M.-T. Hsu, N. G. Gotts, and M. T. Bowers, *J. Phys. Chem.* **97**, 8182 (1993).

93. J. M. Hunter, J. L. Fye, E. J. Roskamp, and M. F. Jarrold, *J. Phys. Chem.* **98**, 1810 (1994).

94. R. I.. Murry, D. L. Strout, G. K. Odom, and G. E. Scuseria, *Nature* (*London*) **366**, 665 (1993).

95. C. Xu and G. E. Scuseria, *Phys. Rev. Lett.* **72**, 669 (1994).

96. R. L. Murry, D. L. Strout, and G. E. Scuseria, in *Fullerenes, Carbon, and Metal–Carbon Clusters*, D. Bohme, Ed., Elsevier, Amsterdam, 1995.

97. A. J. Stone and D. J. Wales, *Chem. Phys. Lett.* **128**, 501 (1986).

98. S. C. O'Brien, J. R. Heath, R. F. Curl, and R. E. Smalley, *J. Chem. Phys.* **88**, 220 (1988).

99. K. F. Freed, in Ref. 6, Vol. 7, pp. 201–253.

100. K. F. Freed, *Acc. Chem. Res.* **16**, 137 (1983).

101. P. Durand and J.-P. Malrieu, *Adv. Chem. Phys.* **67**, 321 (1987).

102. K. F. Freed, *Lecture Notes in Chemistry*, Vol. 52, Springer-Verlag, Berlin, 1989, pp. 1–21.

103. M. G. Sheppard and K. F. Freed, *J. Chem. Phys.* **75**, 4507 (1981).

104. K. F. Freed and H. Sun, *Isr. J. Chem.* **19**, 99 (1980).

105. H. Sun, K. F. Freed, M. F. Herman, and D. L. Yeager, *J. Chem. Phys.* **72**, 4158 (1980).

106. J. J. Oleksik and K. F. Freed, *J. Chem. Phys.* **79**, 1396 (1983).

107. Y. S. Lee and K. F. Freed, *J. Chem. Phys.* **77**, 1984 (1982); **79**, 839 (1983).

108. H. Sun, M. G. Sheppard, K. F. Freed, and M. F. Herman, *Chem. Phys. Lett.* **77**, 555 (1981).

109. H. Sun and K. F. Freed, *J. Chem. Phys.* **80**, 779 (1984).

110. T. Takada and K. F. Freed, *J. Chem. Phys.* **80**, 3253 (1984).

111. J. J. Oleksik, T. Takada and K. F. Freed, *Chem. Phys. Lett.* **113**, 249 (1985).

112. X. C. Wang and K. F. Freed, *J. Chem. Phys.* **86**, 2899 (1987).

113. X. C. Wang and K. F. Freed, *J. Chem. Phys.* **91**, 1151 (1989).

114. Y. S. Lee, K. F. Freed, H. Sun, and D. L. Yeager, *J. Chem. Phys.* **79**, 3862 (1983).

115. C. H. Martin and K. F. Freed, *J. Chem. Phys.* **100**, 7454 (1994).

116. C. H. Martin and K. F. Freed, *J. Chem. Phys.* **101**, 4011 (1994).

117. C. H. Martin and K. F. Freed, *J. Chem. Phys.* **101**, 5929 (1994).

118. C. H. Martin and K. F. Freed, *J. Phys. Chem.* **99**, 2701 (1995).

119. J. E. Stevens, K. F. Freed, M. F. Arendt, and R. L. Graham, *J. Chem. Phys.* **101**, 4832 (1994).

120. J. Schulz, R. Iffert, and K. Jug, *Intern. J. Quantum Chem.* **27**, 461 (1985).

121. S. Patchkovskii and W. Thiel, *J. Comput. Chem.*, in press.

122. *AMPAC 5.0* (1994), Semichem, Shawnee, Kans.

123. W. Thiel and A. A. Voityuk, to be published.

124. *SPARTAN 4.0* (1995), Wavefunction Inc., Irvine, Calif.

125. W. J. Hehre, private communication, 1994.

126. A. J. Holder, private communication, 1994.

127. M. J. Filatov, O. V. Gritsenko, and G. M. Zhidomorov, *Theor. Chim. Acta* **72**, 211 (1987).

128. P. O. Löwdin, *J. Chem. Phys.* **18**, 365 (1950).

129. P. O. Löwdin, *Adv. Quantum Chem.* **5**, 185 (1970).

130. K. R. Roby, *Chem. Phys. Lett.* **11**, 6 (1971); *Chem. Phys. Lett.* **12**, 579 (1972).

131. R. D. Brown and K. R. Roby, *Theory. Chim. Acta* **16**, 175 (1970).

132. G. S. Chandler and F. E. Grader, *Theor. Chim. Acta* **54**, 131 (1980).

133. D. B. Cook, P. C. Hillis, and R. McWeeny, *Mol. Phys.* **13**, 553 (1967).

134. W. Kutzelnigg, *Einführung in die Theoretische Chemie*, Vol. 2, Verlag Chemie, Weinheim, 1978.

135. M. Goeppert-Meyer and A. L. Sklar, *J. Chem. Phys.* **6**, 645 (1938).

136. S. de Bruijn, *Chem. Phys. Lett.* **54**, 399 (1978).

137. S. de Bruijn, in *Current Aspects of Quantum Chemistry*, R. Carbo, Ed., Elsevier, Amsterdam, 1982, pp. 251–272.

138. J. M. L. Martin, J. P. François, and R. Gijbels, *J. Comput. Chem.* **12**, 52 (1991).

139. S. Tseng and C. Yu, *Chem. Phys. Lett.* **231**, 331 (1994); S. Tseng, M. Shen, and C. Yu, *Theor. Chim. Acta* **92**, 269 (1995).

140. K. Y. Burstein and A. N. Isaev, *Theor. Chim. Acta* **64**, 397 (1984).

141. A. Goldblum, *J. Comput. Chem.* **8**, 835 (1987).

142. A. A. Voityuk and A. A. Bliznyuk, *Theor. Chim. Acta* **72**, 223 (1987).

143. M. A. Rios and J. Rodriguez, *J. Comput. Chem.* **13**, 860 (1992).

144. A. Gonzalez-Lafont, T. N. Truong, and D. G. Truhlar, *J. Phys. Chem.* **95**, 4618 (1991).

145. A. A. Viggiano, J. Paschkewitz, R. A. Morris, J. F. Paulson, A. Gonzalez-Lafont, and D. G. Truhlar, *J. Am. Chem. Soc.* **113**, 9404 (1991).

146. Y.-P. Liu, D. Lu, A. Gonzalez-Lafont, D. G. Truhlar, and B. C. Garrett, *J. Am. Chem. Soc.* **115**, 7806 (1993).

147. I. Rossi and D. G. Truhlar, *Chem. Phys. Lett.* **233**, 231 (1995).

148. D. G. Truhlar, R. Steckler and M. S. Gordon, *Chem. Rev.* **87**, 217 (1987).

149. C. Giessner-Prettre and A. Pullman, *Theor. Chim. Acta* **25**, 83 (1972).

150. F. J. Luque, F. Illas, and M. Orozco, *J. Comput. Chem.* **11**, 416 (1990).

151. F. J. Luque and M. Orozco, *Chem. Phys. Lett.* **168**, 269 (1990).

152. C. Alemán, F. J. Luque, and M. Orozco, *J. Comput. Chem.* **14**, 799 (1993).

153. C. Alhambra, F. J. Luque, and M. Orozco, *J. Comput. Chem.*, **15**, 12 (1994).

154. P. L. Cummins and J. E. Gready, *Chem. Phys. Lett.* **174**, 355 (1990).

155. C. A. Reynolds, G. G. Ferenczy, and W. G. Richards, *J. Mol. Struct. (Theochem)* **256**, 249 (1992).

156. G. G. Ferenczy, C. A. Reynolds, and W. G. Richards, *J. Comput. Chem.* **11**, 159 (1990).

157. G. P. Ford and B. Wang, *J. Comput. Chem.* **14**, 1101 (1993).

158. B. Wang and G. P. Ford, *J. Comput. Chem.* **15**, 200 (1994).

159. D. Bakowies, Ph.D. thesis, Universität Zürich, Hartung-Gorre Verlag, Konstanz, Germany, 1994.

160. D. Bakowies and W. Thiel, *J. Comput. Chem.*, in press.

161. G. Rauhut and T. Clark, *J. Comput. Chem.* **14**, 503 (1993).

162. B. Beck, G. Rauhut, and T. Clark, *J. Comput. Chem.* **15**, 1064 (1994).

163. T. P. Straatsma and J. A. McCammon, *Ann. Rev. Phys. Chem.* **43**, 407 (1992).

164. S. M. Bachrach, in *Reviews in Computational Chemistry*, K. B. Lipkowitz and D. B. Boyd, Eds., Vol. 5, VCH, New York, 1993, pp. 171–227.

165. K. B. Wiberg and P. R. Rablen, *J. Comput. Chem.* **14**, 1504 (1993).

166. J. W. Storer, D. J. Giesen, C. J. Cramer, and D. G. Truhlar, *J. Comput.-Aided Mol. Design* **9**, 87 (1995).

167. R. S. Mulliken, *J. Chem. Phys.* **23**, 1833 (1955).

168. A. F. Reed, R. B. Weinstock, and F. Weinhold, *J. Chem. Phys.* **83**, 735 (1985).

169. R. W. F. Bader, *Atoms in Molecules: A Quantum Theory*, Clarendon Press, Oxford, 1990.

170. A. Warshel, *Acc. Chem. Res.* **14**, 284 (1981).

171. S. L. Price and A. J. Stone, *J. Chem. Phys.* **86**, 2859 (1987).

172. D. E. Willims, in *Reviews in Computational Chemistry*, K. B. Lipkowitz and D. B. Boyd, Eds., Vol. 2, VCH, New York, 1991, pp. 219–271.

173. B. H. Besler, K. M. Merz, and P. A. Kollman, *J. Comput. Chem.* **11**, 431 (1990).

174. M. Orozco and F. J. Luque, *J. Comput. Chem.* **11**, 909 (1990).

175. R. T. Sanderson, *J. Am. Chem. Soc.* **74**, 272 (1952); *Science* **121**, 207 (1955).

176. R. T. Sanderson, *Chemical Bonds and Bond Energy*, Academic Press, New York, 1976.

177. R. G. Parr, R. A. Donnelly, M. Levy, and W. E. Palke, *J. Chem. Phys.* **68**, 3801 (1978).

178. P. Politzer and H. Weinstein, *J. Chem. Phys.* **71**, 4218 (1979).

179. J. Gasteiger and M. Marsili, *Tetrahedron* **36**, 3219 (1980).

180. W. J. Mortier, S. K. Gosh, and S. Shankar, *J. Am. Chem. Soc.* **108**, 4315 (1986).

181. A. K. Rappé and W. A. Goddard III, *J. Phys. Chem.* **95**, 3358 (1991).

182. C. I. Bayly, P. Cieplak, W. D. Cornell, and P. A. Kollman, *J. Phys. Chem.* **97**, 10269 (1993).

183. W. D. Cornell, P. Cieplak, C. I. Bayly, and P. A. Kollman, *J. Am. Chem. Soc.* **115**, 9620 (1993).

184. T. R. Stouch and D. E. Williams, *J. Comput. Chem.* **14**, 858 (1993).

185. C. J. Cramer and D. G. Truhlar, in *Reviews in Computational Chemistry*, K. B. Lipkowitz and D. B. Boyd, Eds., Vol. 6, VCH, New York, 1995, pp. 1–72.

186. J. Gao, in *Reviews in Computational Chemistry*, K. B. Lipkowitz and D. B. Boyd, Eds., Vol. 7, VCH, New York, 1995, pp. 119–185.

187. P. A. Bash, M. J. Field, and M. Karplus, *J. Am. Chem. Soc.* **109**, 8092 (1987).

188. J. Gao, *J. Phys. Chem.* **96**, 537, 6432 (1992).

189. J. Gao and J. J. Pavelites, *J. Am. Chem. Soc.* **114**, 1912 (1992).

190. J. Gao and X. Xia, *Science* **258**, 631 (1992).

191. J. Gao, *Intern. J. Quantum Chem. Symp.* **27**, 491 (1993).

192. J. Gao, F. J. Luque, and M. Orozco, *J. Chem. Phys.* **98**, 2975 (1993).

193. J. Gao, *J. Am. Chem. Soc.* **115**, 2930 (1993).

194. J. Gao, *J. Am. Chem. Soc.* **116**, 1563 (1994).

195. V. V. Vasilyev, A. A. Bliznyuk, and A. A. Voityuk, *Intern. J. Quantum Chem.* **44**, 897 (1992).

196. H. Liu, F. Müller-Plathe, and W. F. van Gunsteren, *J. Chem. Phys.* **102**, 1722 (1995).

197. H. Liu and Y. Shi, *J. Comput. Chem.* **15**, 1311 (1994).

198. J. L. Rivail, B. Terryn, and M. F. Ruiz-Lopez, *J. Mol. Struct. (Theochem)* **120**, 387 (1985).

199. M. M. Karelson, A. R. Katritzky, and M. C. Zerner, *Intern. J. Quantum Chem. Symp.* **20**, 521 (1986).

200. M. M. Karelson, A. R. Katritzky, M. Szafran, and M. C. Zerner, *J. Org. Chem.* **54**, 6030 (1989).

201. H. S. Rzepa, M. Y. Yi, M. M. Karelson, and M. C. Zerner, *J. Chem. Soc. Perkin Trans.* 2, 635 (1991).

202. M. M. Karelson and M. C. Zerner, *J. Phys. Chem.* **96**, 6949 (1992).

203. M. Szafran, M. M. Karelson, A. R. Katritzky, J. Koput, and M. C. Zerner, *J. Comput. Chem.* **14**, 371 (1993).

204. J.-L. Rivail, in *New Theoretical Concepts for Understanding Organic Reactions*, J. Bertrán and I. G. Czismadia, Eds., Kluwer, Dordrecht, The Netherlands, 1989, p. 219.

205. D. Rinaldi, J.-L. Rivail, and N. Rguini, *J. Comput. Chem.* **13**, 675 (1992).

206. V. Dillet, D. Rinaldi, J. G. Ángyán, and J.-L. Rivail, *Chem. Phys. Lett.* **202**, 18 (1993).

207. G. P. Ford and B. Wang, *J. Comput. Chem.* **13**, 229 (1992).

208. T. Varnali, V. Aviyente, B. Terryn, and M. F. Ruiz-Lopez, *J. Mol. Struct. (Theochem)* **280**, 169 (1993).

209. M. Negre, M. Orozco, and F. J. Luque, *Chem. Phys. Lett.* **196**, 27 (1992).

210. B. Wang and G. P. Ford, *J. Chem. Phys.* **97**, 4162 (1992).

211. B. Wang and G. P. Ford, *J. Am. Chem. Soc.* **114**, 10563 (1992).

212. T. Fox, N. Rösch, and R. J. Zauhar, *J. Comput. Chem.* **14**, 253 (1993).

213. G. Rauhut, T. Clark, and T. Steinke, *J. Am. Chem. Soc.* **115**, 9174 (1993).

214. A. Klamt and G. Schüürmann, *J. Chem. Soc. Perkin Trans.* 2, 799 (1993).

215. C. J. Cramer and D. G. Truhlar, *J. Am. Chem. Soc.* **113**, 8305, 9901 (E) (1991).

216. C. J. Cramer and D. G. Truhlar, *Science* **256**, 213 (1992).

217. C. J. Cramer and D. G. Truhlar, *J. Comput. Chem.* **13**, 1089 (1992).

218. C. J. Cramer and D. G. Truhlar, *J. Comput.-Aided Mol. Design* **6**, 629 (1992).

219. D. J. Giesen, J. W. Storer, C. J. Cramer, and D. G. Truhlar, *J. Am. Chem. Soc.* **117**, 1057 (1995).

220. D. J. Giesen, C. J. Cramer, and D. G. Truhlar, *J. Phys. Chem.* **99**, 7137 (1995).

221. C. J. Cramer and D. G. Truhlar, private communications, 1995.

222. A. Warshel, *Computer Modeling of Chemical Reactions in Enzymes and Solutions*, Wiley, New York, 1991.

223. J. Aqvist and A. Warshel, *Chem. Rev.* **93**, 2523 (1993).

224. M. J. Field, in *Computer Simulation of Biomolecular Systems: Theoretical and Experimental Applications*, W. F. van Gunsteren, P. K. Weiner and A. J. Wilkinson, Eds., Vol. 2, ESCOM, Leiden, The Netherlands, 1993, p. 82.

225. A. Warshel and M. Levitt, *J. Mol. Biol.* **103**, 227 (1976).

226. U. C. Singh and P. A. Kollman, *J. Comput. Chem.* **7**, 718 (1986).

227. M. J. Field, P. A. Bash, and M. Karplus, *J. Comput. Chem.* **11**, 700 (1990).

228. A. H. de Vries, P. T. van Duijnen, A. H. Juffer, J. C. Rullmann, J. P. Dijkman, H. Merenga and B. T. Thole, *J. Comput. Chem.* **16**, 37 (1995).

229. P. A. Bash, M. J. Field, R. C. Davenport, G. A. Petsko, D. Ringe, and M. Karplus, *Biochemistry* **30**, 5826 (1991).

230. V. V. Vasilyev, *J. Mol. Struct. (Theochem)* **304**, 129 (1994).

231. D. Rinaldi, J.-L. Rivail, and N. Rguini, *J. Comput. Chem.* **13**, 675 (1992).

232. G. G. Ferenczy, J.-L. Rivail, P. R. Surjan, and G. Naray-Szabo, *J. Comput. Chem.* **13**, 830 (1992).

233. V. Thery, D. Rinaldi, J.-L. Rivail, B. Maigret, and G. G. Ferenczy, *J. Comput. Chem.* **15**, 269 (1994).

234. D. Bakowies and W. Thiel, submitted for publication.

235. N. L. Allinger, Y. H. Yuh, and J.-H. Lii, *J. Am. Chem. Soc.* **111**, 8551 (1989).

236. J.-H. Lii and N. L. Allinger, *J. Am. Chem. Soc.* **111**, 8566, 8576 (1989).

237. B. T. Thole, *Chem. Phys.* **59**, 341 (1981).

238. G.-A. Craze and I. Watt, *J. Chem. Soc. Perkin Trans. 2*, 175 (1981).

239. R. Cernik, G.-A. Craze, O. S. Mills, and I. Watt, *J. Chem. Soc. Perkin Trans. 2*, 361 (1982).

240. R. S. Henry, F. G. Riddell, W. Parker, and C. I. F. Watt, *J. Chem. Soc. Perkin Trans. 2*, 1549 (1976).

241. J. E. Eksterowicz and K. N. Houk, *Chem. Rev.* **93**, 2439 (1993).

242. M. J. Sherrod and F. M. Menger, *J. Am. Chem. Soc.* **111**, 2611 (1989).

243. K. P. Eurenius and K. N. Houk, *J. Am. Chem. Soc.* **116**, 9943 (1994).

244. P. K. Weiner, Ph.D. thesis, University of Texas, Austin, 1975.

245. M. Kolb, Ph.D. thesis, Universität Wuppertal, Wuppertal, Germany, 1991.

246. D. E. Goldberg, *Genetic Algorithms in Search, Optimization, and Machine Learning*, Addison-Wesley, Reading, Mass., 1989.

247. L. A. Curtiss, K. Raghavachari, G. W. Trucks and J. A. Pople, *J. Chem. Phys.* **94**, 7221 (1991).

248. J. J. P. Stewart, *Intern. J. Quantum Chem.*, in press.

249. J. J. P. Stewart, *Abstracts of the International Workshop on Electronic Structure Methods for Truly Large Systems*, Braunlage, Germany, 1994.

250. M. J. S. Dewar and Y. Yamaguchi, *Comput. Chem.* **2**, 25 (1978).

251. M. J. S. Dewar, J. A. Hashmall, and C. G. Venier, *J. Am. Chem. Soc.* **90**, 1953 (1968).

252. M. J. S. Dewar and D. A. Liotard, *J. Mol. Struct. (Theochem)* **206**, 123 (1990).

253. S. Patchkovskii and W. Thiel, *Theor. Chim. Acta*, in press.

254. N. C. Handy and H. F. Schaefer, *J. Chem. Phys.* **81**, 5031 (1984).

255. W. Thiel and D. G. Green, in *Methods and Techniques in Computational Chemistry*, METECC-95, E. Clementi and G. Corongiu, Eds., STEF, Cagliari, 1995, pp. 141–168.

256. G. Piccito and R. Pucci, *J. Chem. Phys.* **98**, 502 (1993).

257. X. Long, R. L. Graham, C. Lee, and S. Smithline, *J. Chem. Phys.* **100**, 7223 (1994).

258. E. E. B. Campbell, V. Schyja, R. Ehlich, and I. V. Hertel, *Phys. Rev. Lett.* **70**, 263 (1993).

259. B. L. Zhang, C. Z. Wang, C. T. Chan, and K. M. Ho, *J. Phys. Chem.* **97**, 3134 (1993).

260. G. Seifert and R. Schmidt, *Intern. J. Mod. Phys. B* **6**, 3845 (1992).

261. F. Bockisch, D. Liotard, J.-C. Rayez, and B. Duguay, *Intern. J. Quantum Chem.* **44**, 619 (1992).

262. D. A. Liotard, *Intern. J. Quantum Chem.* **44**, 723 (1992).

263. W. Thiel, *J. Mol. Struct.* **163**, 415 (1988).

264. J. J. P. Stewart, L. P. Davis, and L. W. Burggraf, *J. Comput. Chem.* **8**, 1117 (1987).

AUTHOR INDEX

Numbers in parentheses are reference numbers and indicate that the author's work is referred to although his name is not mentioned in the text. Numbers in *italic* show the pages on which the complete references are listed.

759

SUBJECT INDEX